MW00614150

Topics in Applied Physics
Volume 114

Topics in Applied Physics

Topics in Applied Physics is a well-established series of review books, each of which presents a comprehensive survey of a selected topic within the broad area of applied physics. Edited and written by leading research scientists in the field concerned, each volume contains review contributions covering the various aspects of the topic. Together these provide an overview of the state of the art in the respective field, extending from an introduction to the subject right up to the frontiers of contemporary research.

Topics in Applied Physics is addressed to all scientists at universities and in industry who wish to obtain an overview and to keep abreast of advances in applied physics. The series also provides easy but comprehensive access to the fields for newcomers starting research.

Contributions are specially commissioned. The Managing Editors are open to any suggestions for topics coming from the community of applied physicists no matter what the field and encourage prospective editors to approach them with ideas.

Robert W. Boyd, Svetlana G. Lukishova,
Y. R. Shen (Eds.)

Self-focusing: Past and Present

Fundamentals and Prospects

Robert W. Boyd
University of Rochester
Rochester, NY
boyd@optics.rochester.edu

Svetlana G. Lukishova
University of Rochester
Rochester, NY
sluk@lle.rochester.edu

Y.R. Shen
University of California
Berkeley, CA
shenyr@physics.berkeley.edu

ISSN: 0303-4216
ISBN 978-0-387-32147-9 ISBN 978-0-387-34727-1 (eBook)
DOI 10.1007/978-0-387-34727-1

Library of Congress Control Number: 2008938966

Printed on acid-free paper

springer.com

Preface

Although self-focusing and other self-action phenomena have been studied extensively for nearly 50 years (see, e.g., reviews [1–5] and the books [6–11]), this volume constitutes the first comprehensive treatment devoted entirely to the subject. It combines past and present results on both theory and experiments and includes interpretation of experiments with femtosecond light pulses and their comparison with nanosecond/picosecond self-focusing. The book is intended for scientists and engineers working with lasers and their applications. It consists of overview chapters on self-focusing in different media written by leading scientists from the United States, Russia, France, China, Canada, Israel, Italy, Lithuania, Mexico, and Brazil. Many of our authors participated in a symposium with the same title as this book during the International Quantum Electronics Conference (IQEC 2005) held during July 2005 in Tokyo, Japan.

Modern experiments with femtosecond laser pulses have provided new prospects for this rich field, promising numerous practical applications in terms of experiments that were not possible in the past. One of the current advances in femtosecond self-focusing at laser power levels of several TW is the observation of the generation of the collimated, coherent white light continuum propagating through the atmosphere up to an altitude of more than 10 km and its potential application for remote sensing and lightning control [12–14]. Attosecond pulse generation is another application of femtosecond filamentation [14].

New computer capabilities deepen our knowledge of self-focusing as well, for example, the *nonparaxial* approximation in nonlinear beam propagation showed *nonsingular* behavior at self-focus/self-foci even in the absence of nonlinear absorption and refractive index saturation mechanisms [15–17].

In this Preface, we provide a clarification of terminology, explain the importance of self-focusing studies, outline the book purpose, and briefly describe the main highlights of the chapters on self-focusing in the past and the present.

Terminology in the Classical Case

As described in textbooks on nonlinear optics (e.g., [10, 11]), *self-focusing, self-defocusing, self-trapping*, and *filamentation* are the major self-action effects in which a beam of light modifies its own propagation by means of the nonlinear response of a material medium. The early history of these concepts with relevant references is briefly outlined in [18].

Above a certain *critical power* P_{cr}, an intense light beam with an intensity gradient along its cross-section, propagating in a nonlinear medium, experiences *self-focusing* or *self-defocusing* depending of the sign of the nonlinear susceptibility $\chi^{(3)}$ [19]. For the stationary self-focusing case, a beam with a power above P_{cr} collapses into a focus [5, 7, 20] and/or multiple foci [2, 21] at *the self-focusing distance (length)* [20] within the material. For a pulsed laser, different parts of the pulse with different powers are focused at different distances, so an illusion of a *filament* is observed in the experiment ("moving focus" model [2, 4, 5, 18, 21]). At powers much higher than P_{cr}, the beam breaks up into many components causing *multiple filamentation* (*"small-scale" self-focusing, spatial modulation instability* [20, 22–24]). This process occurs as a consequence of the growth of both amplitude and phase perturbations of the beam wave front by means of forward four-wave mixing amplification of this noise.

In the *self-trapping* process a beam of light propagates with a constant diameter as a consequence of an exact balance between self-focusing and diffraction effects. In 1964, self-trapping (waveguiding) solutions of the wave equation in nonlinear Kerr media were independently found [25, 26]. For the two-dimensional case, the lowest-order solution for a beam with cylindrical symmetry was reported in [25] and is in the form of a bell-shaped curve called the "*Townes profile*." *Higher-order solutions* with spatial profiles containing rings have been reported in [27, 28]. See also books [7, 11] and reviews [5, 17, 29]. Later papers showed that in reality this *spatial soliton* behavior can be stable only in special cases, for example, in a medium with saturation of nonlinearity [5] or in the case of a beam that varies in only one transverse dimension (planar waveguides) where these truly self-trapped filaments were observed experimentally [30–32]. So, in some early papers, the formation of self-focusing filaments in experiments is called self-trapping. For more details on self-trapping and spatial solitons see books [33, 34] and reviews [35–37].

Conical emission [36, 38] manifests itself as the occurrence of single or multiple concentric rings surrounding the pump laser beam. The generated cone can be either at the same frequency as the original beam [39, 40] or spectrally shifted [36, 38]. The physical process leading to conical emission is different under different circumstances [14, 36, 38–43].

Continuum (supercontinuum) generation [14, 38, 41, 44] was first observed in 1970. It is a nonlinear optical process for strong spectral broadening of light and can span from the ultraviolet to the infrared ("white light" spectrum). It has been observed in bulk solids, liquids, gases and fibers including microstructured fibers [44]. An important property of such white light is its spatial coherence. The physical processes behind supercontinuum generation can be very different, for example, self-phase modulation, four-wave mixing, multiphoton ionization with the formation of free electrons, and so on.

Some other terminology used by the authors of this book including its modern development is explained later in this preface.

Importance of Self-Focusing

From a practical point of view, self-focusing effects impose a limit on the power that can be transmitted through an optical medium. Self-focusing also can reduce the threshold for the occurrence of other nonlinear optical processes [4]. Self-focusing often leads to damage in optical materials and is a limiting factor in the design of high-power laser systems [8, 9, 24], but it can be harnessed for the design of optical power limiters and switches.

At a formal level, the equations for self-focusing are equivalent to those describing Bose–Einstein condensates and certain aspects of plasma physics and hydrodynamics. The wave equation governing the self-focusing effect is the prototype of an important class of partial differential equations such as the Ginzburg–Landau equation for type II superconductors and the Schrödinger equation for particles with self-interactions. Important trends in self-focusing theory that have evolved over the past 40 years include *soliton formation* and *wave collapse*. Collapse is a self-compression of a light beam and/or pulse with the formation of a singularity. Analogous to a star undergoing gravitational collapse, nonlinear effects can cause a wave to collapse on itself. Such behavior is intrinsic to plasma physics, hydrodynamics, nonlinear optics, and Bose–Einstein condensates. Analysis of many different types of nonlinear wave equations indicates that a collapsing wave can transform into a universal transverse profile regardless of its initial shape [16, 29, 45, 46].

Remarkable self-focusing effects have been observed recently with femtosecond laser beams propagation in the atmosphere: light filaments ("light strings") in the air with repeated filamentation over distances greater than 10 km [12–14] have been observed. A broad (ultraviolet to mid-infrared) spectrum of radiation is generated from such air filaments, permitting spectroscopy and localized remote sensing of chemical species and aerosols in the lower atmosphere [12–14]. The entire absorption spectrum can be determined by a single pulse from a portable femtosecond laser. Another exciting possibility of the use of these "light strings" containing plasma filaments is to guide lightning away from sensitive sites.

Intent and Outline of This Book

One of the goals of this book is to connect the extensive early literature on self-focusing, filamentation, self-trapping, and collapse with nanosecond and picosecond lasers with more recent studies since approximately the 1990s. These modern studies are aimed at issues such as self-focusing of femtosecond pulses, white light generation [38, 44], and the generation of long filaments in air [12–14, 41, 42, 47, 48].

It should be noted that in explaining femtosecond self-focusing and filamentation, some new terminology and new models were introduced that did not exist in the past. In this connection there has been some debate in the literature [21], so we tried to address this diversity of opinions. We hope that this book will help current and future researchers to understand whether or not this new terminology has analogs in the past or it is connected with purely new phenomena that cannot be deduced from scaling of similar experiments using nanosecond or picosecond pulses.

Because of book-length restrictions, we have omitted or only briefly outlined certain areas. Some of these topics have been discussed in other books, for example, spatial and temporal solitons [33–34]. We did not include thermal self-focusing, spatial self-phase modulation and solitons in liquid crystals, self-focusing in semiconductors, optical power limiting and switching effects, high $\chi^{(3)}$ materials, self-focusing in waveguides and fibers, and so on. The focus of our book is *filamentation* and *self-focusing collapse*.

In addition, we apologize in advance if somebody who made significant contribution to self-focusing is omitted from this book. Initially we planned to reprint the most significant papers on self-focusing, but while selecting such papers we realized that the size of this book would then have become enormous. It was decided to reprint only the first paper on self-focusing by Askar'yan (1962) [49]. To facilitate the understanding of Marburger's notes on early self-focusing theory, we have also reprinted his 1975 review with many citations of the papers of others [5]. Also, we were only able to reference some other excellent reviews of the past [1–4]. Two papers from 1964 that have been widely cited in the literature, but have never been published previously in English (Talanov) [26] or as a full journal publication (only 17 lines in an abstract by Hercher [50]), are included here as well.

The book contains two parts with 11 chapters in the first part, "Self-Focusing in the Past," and 13 chapters in the second part, "Self-Focusing in the Present." Sometimes this division is not rigid: the same chapter may contain the results both from the past and the present.

Next, we describe briefly the content of each of the chapters.

Part I: Self-Focusing in the Past

Part I consists of 11 chapters and contains surveys of experiments on whole-beam self-focusing, filamentation, self-trapping and small-scale self-focusing with nanosecond and picosecond laser pulses in liquids, solids, and atomic vapors. It also treats accompanying processes (e.g., conical emission and white-light supercontinuum generation) and methods of suppression of self-focusing effects. The main theoretical concepts and models from the past are also reviewed. These models are based on analytical approximation and/or numerical modeling of the nonlinear Schrödinger equation (NLSE) [51].

In the first chapter of this book, by Shen, a brief review is provided of early experimental research on self-focusing and filament formation of nanosecond and picosecond light pulses [4, 10]. Similarities and differences are discussed between nanosecond–picosecond self-focusing, and recent experiments with femtosecond pulses with the appearance of long filaments of light in air.

The second chapter, by Marburger, contains the author's comments on early research in self-focusing theory, describing its major advances in light of his two papers and his review of the self-focusing theory before 1975 [5]. A reprint of this review is also provided.

The chapter by Fraiman, Litvak, Talanov, and Vlasov contains a review of self-focusing theory in the paraxial approximation as described in book [7], based on the NLS (parabolic) equation. Various manifestations of self-focusing are described: self-trapped solutions ("homogeneous waves" [7]), instability of plane waves, formation of nonlinear foci in cubic nonlinear media, self-similar collapse solution, suppression of self-focusing in periodic systems, and spectral broadening. Some questions of self-action effects involving femtosecond pulses are also discussed.

Chiao, Gustafson, and Kelley review self-focusing and related phenomena, for example, coupling of waves through the nonlinear index and spatial instability, beam breakup and multiple filament formation, spatial self-phase modulation and estimating the beam self-focusing distance, self-focusing intensity singularity and beam collapse, limitations on blow-up and collapse due to saturation of the nonlinear refractive index and other nonlinearities, self-focusing of pulses and "light bullets," and self-trapping (spatial solitons). They also discuss physical mechanisms that give rise to the refractive index nonlinearity.

Lugovoi and Manenkov address both the theory and experimental confirmation of the multifocus structure model for a stationary case and a moving focus model for laser radiation of nanosecond–picosecond pulse duration [2]. The influence of the beam's spatial profile (Gaussian or super-Gaussian) on the self-focusing threshold and length is discussed. Analysis of the adequacy of these models for self-focusing of femtosecond pulses, discussion of new terminology, and suggested future directions for the study of femtosecond self-focusing in air are provided.

Campillo describes in detail small-scale self-focusing as a result of transverse (spatial) modulational instability. He briefly surveys the essential physics, presents a linearized theory, and summarizes experimental examples illustrating the dependence between optimal spatial frequency, gain coefficient, small-scale self-focusing length, beam intensity, and beam shape including the modern advances in the theory and the experiments with femtosecond lasers. A number of applications are discussed as well as the impact of this field on other scientific disciplines.

Kuznetsov concentrates on a brief review of collapse theory in nonlinear optics described by the NLSE. Emphasis is placed on collapse classification (weak, strong, and black holes regime), self-similarity in collapse, correspondence between solitons and collapses, and the role of group velocity dispersion on the collapse of electromagnetic pulses.

Lukishova, Senatsky, Bykovsky, and Scheulin provide a brief outline of self-focusing as the primary nonlinear optical process limiting the brightness of high-peak-power Nd:glass laser systems. Both small-scale and large-scale self-focusing are considered. The role of Fresnel diffraction at apertures as the source of dangerous spatial scales for small-scale self-focusing is illustrated. Methods for the formation of super-Gaussian laser beams and their self-focusing are presented. Other methods for self-focusing suppression are described in addition to beam apodization.

Zerom and Boyd review self-focusing and other self-action effects in atomic vapors as a medium with a saturable nonlinearity. Experimental results on self-trapping (spatial solitons), conical emission, and filamentation are described. The models for the explanation of conical emission in atomic vapors are presented. Multiple filamentation is discussed in connection with optical pattern formation.

Ni and Alfano consider white-light supercontinuum generation [44] and periodic filamentation in glass controlled by Fresnel diffraction from a circular aperture or a straight edge. Also discussed are self-phase modulation (distortion in phase), self-steepening (distortion of pulse envelope shape), and conical emission in glass as the process of four-wave parametric generation with a nonlinear refractive index change at the surface of the filament. Coherence properties of supercontinuum sources accompanying the filaments generated at different spatial positions are reported.

The last chapter of Part I contains three papers from the past. Askar'yan's paper reprinted from 1962 [49] is the first paper where self-focusing was suggested. Talanov's paper from 1964 [26] is the first English translation of his work on the waveguide (self-trapped or soliton) solution. Hercher's paper is the first publication of his 1964 Optical Society of America Annual Meeting presentation [50] with the first photographs of damage resulting from self-focusing filaments. This paper and the photographs have never been published previously.

Part II: Self-Focusing in the Present

Extensive studies on femtosecond filamentation started after the first report in 1995 of the experimental observation of femtosecond filaments in air at a peak power exceeding a few GW [52]. A 20-m-long filament with an intense core of diameter of approximately 100 microns was observed.

For the explanation of femtosecond filament formation, the following principal models have been suggested: self-channeling, moving focus model, and dynamic spatial replenishment. We do not consider here relativistic and charge displacement self-channeling [53] at incident intensities $\sim 10^{19}$ W/cm^2 when strong ponderomotive forces cause electron cavitation. In the *self-channeling* model [52], self-trapping occurs when self-focusing is counteracted by defocusing owing to medium ionization and diffraction. However, this balance was never identified in numerical simulations showing several cycles of *focusing–defocusing–refocusing*. This process can occur more than once for high-power input pulses. Refocusing events on the trailing edge of the pulse are called "*dynamic spatial replenishment* of light" [54]. In the *moving focus* model [41, 42], which was suggested in the past for nanosecond and picosecond pulses [2, 4, 5], a filament arises from the continuous succession of foci arising from the self-focusing of the various longitudinal slices of a pulse, producing the illusion of long-distance propagation of one self-guided pulse (filament). The same illusion is created in the dynamic spatial replenishment model. The concept of an "*energy reservoir*" was introduced for the background beam that surrounds the filament core and provides energy to it [41–43, 47]. Many practical applications need *regularization* and *control* of the propagation distance of *multiple filaments*. It is achieved by diffraction at apertures, lens arrays, and beam ellipticity [14, 23, 38, 42, 43].

The formation of *spatiotemporal solitons* (STS), sometimes called "*light bullets*," is one of the major goals in the field of nonlinear waves. Although STS in all three space dimensions and time have not been observed experimentally, enormous progress was made in lower space-dimension STS formation in media with cascaded quadratic nonlinearity [55]. Related to STS are *nonlinear X-waves* that are only weakly localized, but manifest a kind of spatiotemporal trapping [55, 56]. X-waves are nondiffractive and nondispersive waves with an X-shaped axial cross-section both in the far-field and the near-field. In the linear case, they can be formed by axicons. In media with normal group-velocity dispersion, the nonlinearity acts as the driving mechanism that reshapes a narrow beam and ultrashort pulse into conical elements with features of X-waves.

Part II consists of 13 chapters. Six chapters contain results on self-focusing and filamentation of *femtosecond* laser pulses and their interpretation by each of several groups actively working in this field. Filamentation both in air and in condensed matter is considered with the emphasis on femtosecond filament formation and on the models explaining its origin and behavior. Four chapters describe the formation of nonlinear X-waves, spatial and temporal dynamics of self-focusing collapse of ultrashort pulses, some aspects of modern self-focusing theory including nonparaxial approximations of the wave equation and its application to the theory of collapse, high-order self-trapped solutions, and spatial modulation instability. Three additional chapters review self-focusing in media with a quadratic nonlinearity, self-trapping in photorefractive crystals, and measurements of nonlinear refraction and its dispersion.

In particular, Couairon and Mysyrowicz review both simulations and experimental results on self-action effects that occur during femtosecond filamentation in gases, liquids, and solids. They outline the evolution of the modeling of femtosecond filamentation in transparent media, and they consider a simple self-trapping model, the moving focus model, and saturation of self-focusing and self-channeling with formation of several focusing–defocusing cycles at intensities sufficient for multiphoton absorption and plasma-induced defocusing. Recently discovered self-action effects are outlined: the generation of single cycle pulses by filamentation in gases, the beam self-cleaning effect, filamentation in an amplifying medium, and the organization of multiple filaments by various methods. They briefly outline the current state of the art of long-distance filamentation in air. Conical emission accompanying a self-guided pulse, continuum generation, and nonlinear X-waves are also discussed.

Zhang, Hao, Xi, Lu, Zhang, Yang, Jin, Wang, and Wei describe experiments on filaments of nearly 160 m in length under propagation of intense femtosecond pulses in air, and they describe in detail their diagnostic techniques. The spatial evolution of the filaments, interactions between filaments, third harmonic generation, lifetime prolongation of the filament by use of a second, sub-ns laser pulse, and laser-guided discharge and laser propulsion are also considered. The dependence of long-distance filamentation on chirp and divergence angle, the role of an energy reservoir, and the comparison of filamentation in focused and unfocused beams are discussed.

Chin, Liu, Kosareva, and Kandidov discuss physical concepts underlying femtosecond laser filamentation as deduced from experiment. These include slice-by-slice self-focusing according to the moving focus model, intensity clamping by plasma-induced defocusing effect, supercontinuum generation and conical emission, the background reservoir concept, multiple refocusing, and multiple filamentation competition. Some important potential applications are also briefly mentioned.

Kandidov, Dormidonov, Kosareva, Chin, and Liu outline the theory of the moving-focus model for the analysis of femtosecond filament formation and the ring formation of a ring pattern in the beam's transverse cross-section near the filament. The authors also provide a comparison of the femtosecond focusing–defocusing–refocusing model with the earlier multifocus model for the stationary case [2]. Their model of filament formation includes the initial stage of filamentation, estimation of the critical power, and the generalization of Marburger's formula for the self-focusing length for femtosecond pulses with elliptical cross-section and for the chirped pulses. The influence of turbulence and of aerosols in the atmosphere on chaotic pulse filamentation is considered as well as the spatial regularization of filaments by the introduction of initial regular perturbation of amplitude and phase. This chapter also contains the first published photograph (1965) of self-focusing filaments [57]. See also [50] and [58, 59] of 1965 for filament photographs.

Matvienko, Bagaev, Zemlyanov, Geints, Kabanov, and Stepanov present an analysis of femtosecond self-focusing in terms of the evolution of the so-called effective parameters (energy transfer coefficient, effective beam radius, effective pulse duration, limiting angular divergence, and effective intensity) describing the regime of formation of a single axial filament. Filamentation behavior of femtosecond laser pulses in the air in the presence of aerosols is also considered.

Gaižauskas, Dubietis, Kudriašov, Sirutkaitis, Couairon, Faccio, and Di Trapani focus on the experimental observations and theoretical investigations of the propagation of intense femtosecond pulses in water and fused silica. They point out that nonlinear losses cause beam distortion which influences self-focusing. Explaining fringe formation during beam propagation with nonlinear losses (see also book [7]), they introduce the concept of a conical (Bessel-like) wave. In the non-waveguiding regime, they find a steady-state solution of the high-intensity central core of the beam containing ~20% of total beam energy. This core takes the energy from the rest of the beam (energy reservoir) surrounding it. Multiple filaments from elliptical beams and nonlinear X-waves are observed experimentally in both water and fused silica.

Conti, Di Trapani, and Trillo overview the main experimental and theoretical results on nonlinear X-waves during filamentation and trapping in normally dispersive media, starting from linear X-waves and X-waves in second-harmonic generation. The role of group velocity dispersion (GVD) during filamentation in Kerr-like media on nonlinear X-wave formation is emphasized.

Gaeta considers recent theoretical and experimental work on spatial and temporal dynamics of self-focusing collapse of ultrashort laser pulses, for example, self-similar evolution, modulation instability versus Townes collapse, and the collapse of super-Gaussian beams. He also discusses pulse splitting in the normal GVD regime, self-focusing in the anomalous GVD regime, optical shock formation and supercontinuum generation, femtosecond filamentation and light strings in air.

Fibich reviews some aspects of modern self-focusing theory, highlighting some pre-1975 results, describes the current understanding of some old concepts, and describes new theoretical challenges. He emphasizes universal and new blow-up profiles of collapsing beams, the blow-up rate, super-Gaussian beam collapse, self-focusing of "low power beams" (~ several or tens of P_{cr}), self-focusing distance, partial beam collapse, the arresting of collapse by nonparaxiality with focusing–defocusing oscillations (multiple foci in [15]), multiple filamentation, and the effect of normal GVD.

Chávez-Cerda, Itube-Castillo, and Hickmann consider the "nonparaxial NLSE" and effects of nonparaxiality on whole-beam self-focusing (see also the chapter by Fibich and [7, 15]). This chapter also shows that nonparaxiality arrests collapse, allowing further beam propagation and providing the opportunity to examine the beam behavior beyond the predicted catastrophic focus. In addition, this chapter provides nonparaxial analogs of stationary (self-trapped) solutions of the NLSE. It is shown that the generation of these modes can be induced by a diffracted beam at a circular aperture, propagating in a self-focusing medium and that, under perturbations, they may break up into hot spots. Modulation-instability patterns are presented for both Gaussian and flat-top beams and are compared with classical experiments of the past (see Campillo's chapter in Part I).

Wise and Moses review self-focusing and self-defocusing of femtosecond pulses in media with cascaded quadratic nonlinearities. Quadratic media appear to be unique in offering a means of impressing a self-defocusing nonlinear phase shift on ultrashort pulses. Self-defocusing nonlinearities are discussed in connection with modelocking in a femtosecond laser, pulse compression, nonlinear polarization rotation and compensation of self-focusing. Saturable self-focusing and space–time solitons ("light bullets") are considered as well.

DelRe and Segev review the current state of the art on both experiments and theory of self-focusing and self-trapping (spatial solitons) in photorefractive media. The study of photorefractive solitons has a significant impact on soliton science. The chapter includes a description of the mechanisms of photorefractive nonlinearity. Some important breakthroughs obtained with solitons in photorefractive crystals are described: soliton interaction and collisions, two-dimensional solitons, vector and composite solitons, and incoherent solitons. Optical induction methods for nonlinear photonic lattices and a series of new solitons are mentioned, for example, 2D lattice ("discrete") solitons, spatial gap solitons, and others.

Van Stryland and Hagan in their chapter on measuring nonlinear refraction and its dispersion describe the beam distortion and Z-scan methods and include a discussion of some possible pitfalls in measuring absolute values of nonlinearities with the Z-scan method. They also describe a method for determining the spectral dependence of the nonlinear change in the refractive index using femtosecond white-light continua as the source for Z-scan. Physical mechanisms leading to nonlinear refraction are also briefly discussed.

Conclusion

We believe that this book will prove useful for academics, researchers, engineers, and students in various disciplines who require a broad introduction to this subject and who would like to learn more about the state-of-the art and upcoming trends in self-focusing, self-trapping, filamentation, and self-focusing collapse. We thank all the contributors who found the time, energy and enthusiasm to write these chapters. We thank Elsevier, Science Direct, American Institute of Physics, and Izvestia Vuzov, Radiophysica for permission to reprint certain papers. We are also grateful to the U.S. Air Force Office for Scientific Research for its support of the IQEC 2005 symposium on this topic, which provided the motivation for the creation of this book.

Rochester, NY, Svetlana G. Lukishova and Robert W. Boyd
Berkeley, CA, Y.R. Shen

References

1. S.A. Akhmanov, R.V. Khokhlov, A.P. Sukhorukov: Self-focusing, self-defocusing and self-modulation of laser beams, *Laser Handbook*, **2**, E3, 1151–1228, F.T. Arecchi, E.O. Schulz-Dubois (Eds.), North-Holland (1972). See also S.A. Akhmanov, A.P. Sukhorukov, R.V. Khokhlov: Self-focusing and diffraction of light in a nonlinear medium, *Sov. Phys. Uspekhi* **10**, 609–636 (1968).
2. V.N. Lugovoi, A.M. Prokhorov: Theory of the propagation of high-power laser radiation in a nonlinear medium, *Sov. Phys. Uspekhi* **16**, 658–679 (1974).
3. O. Svelto: Self-focusing, self-trapping, and self-phase modulation of laser beams: Progress in Optics XII, E. Wolf (ed.), pp. 3–51, North-Holland, Amsterdam (1974).
4. Y.R. Shen: Self-focusing: experimental, *Prog. Quant. Electr.* **4**, 1–34 (1975).
5. J.H. Marburger: Self-focusing: theory, *Prog. Quant. Electr.* **4**, 35–110 (1975). See also Chapter 2 of this book by Marburger.
6. M.S. Sodha, A.K. Ghatak, V.K. Tripathi: *Self-Focusing of Laser Beams in Dielectrics, Plasma, and Semiconductors*, Tata McGraw-Hill, New Delhi (1974).
7. S.N. Vlasov, V.I. Talanov: *Wave Self-Focusing*, Institute of Applied Physics of the Russian Academy of Science, Nizhny Novgorod, 220 pp (1997).
8. D.C. Brown: *High-Peak-Power Lasers*. Chapter 7: Nonlinear effects in high-peak-power Nd: glass laser systems, pp. 188–235, Springer-Verlag, New York (1981).
9. A.A. Mak, L.N. Soms, V.A. Fromzel, V.E. Yashin: *Nd:glass Lasers*. Chapter 6.1: Main physical restrictions of laser power and brightness; Chapter 6.2: Methods of self-focusing suppression, pp. 242–259, Nauka, Moscow (1990).
10. Y.R. Shen: *The Principles of Nonlinear Optics*. Chapter 17: Self-focusing, pp. 303–333, John Wiley, New York (1984).
11. R.W. Boyd, *Nonlinear Optics*. Chapter 7: Processes resulting from the intensity-dependent refractive index, pp. 311–370, Academic Press, San Diego (2003).
12. A.L. Gaeta: Collapsing light really shines, *Science*, **301**, 54–55 (2003).
13. J. Kasparian, M. Rodriguez, G. Méjean, J. Yu, E. Salmon, H. Wille, R. Bourayou, S. Frey, Y.-B. André, A. Mysyrowicz, R. Sauerbrey, J.-P. Wolf, L. Wöste: White-light filaments for atmospheric analysis, *Science*, **301**, 61–64 (2003).

14. A. Couairon et al.: Self-focusing and filamentation of femtosecond pulses in air and condensed matter: Simulations and experiments. *Self-focusing: Past and Present*, R. W. Boyd, S.G. Lukishova, Y.R. Shen (Eds.), (this volume, Chapter 12).
15. M.D. Feit, J.A. Fleck, Jr.: Beam nonparaxiality, filament formation, and beam breakup in the self-focusing of optical beams, *J. Opt. Soc. Am. B* **5**, 633–640 (1988).
16. G. Fibich: Some modern aspects of self-focusing theory. *Self-Focusing: Past and Present*, R.W. Boyd, S.G. Lukishova, Y.R. Shen (Eds.), (this volume, Chapter 17).
17. S. Chávez-Cerda et al.: Diffraction-induced high-order modes of the (2 + 1)-D non-paraxial nonlinear Schrödinger equation. *Self-Focusing: Past and Present*, R.W. Boyd, S.G. Lukishova, Y.R. Shen (Eds.), (this volume, Chapter 22).
18. Y.R. Shen: Self-focusing and filaments of light: Past and present. *Self-Focusing: Past and Present*, R.W. Boyd, S.G. Lukishova, Y.R. Shen (Eds.), (this volume, Chapter 1).
19. E.W. Van Stryland, D.J. Hagan: Measuring nonlinear refraction and its dispersion. *Self-Focusing: Past and Present*, R.W. Boyd, S.G. Lukishova, Y.R. Shen (Eds.), (this volume, Chapter 24).
20. R.Y. Chiao et al.: Self-focusing of optical beams. *Self-Focusing: Past and Present*, R. W. Boyd, S.G. Lukishova, Y.R. Shen (Eds.), (this volume, Chapter 4). For the self-focusing distance see P. L. Kelley: Self-focusing of optical beams, *Phys. Rev. Lett.*, **15**, 1005–1008 (1965).
21. V.N. Lugovoi, A.A. Manenkov: Multi-focus structure and moving nonlinear foci – adequate model of self-focusing of laser beams in nonlinear media. *Self-Focusing: Past and Present*, R.W. Boyd, S.G. Lukishova, Y.R. Shen (Eds.), (this volume, Chapter 5).
22. V.I. Bespalov, V.I. Talanov: Filamentary structure of light beams in nonlinear liquids, *JETP Lett.*, **3**, 307–310 (1966).
23. A.J. Campillo: Small-scale self-focusing. *Self-Focusing: Past and Present*, R.W. Boyd, S.G. Lukishova, Y.R. Shen (Eds.), (this volume, Chapter 6).
24. S.G. Lukishova et al.: Beam shaping and suppression of self-focusing in high-peak-power Nd:glass laser systems. *Self-Focusing: Past and Present*, R.W. Boyd, S.G. Lukishova, Y.R. Shen (Eds.), (this volume, Chapter 8).
25. R.Y. Chiao, E. Garmire, C.H. Townes: Self-trapping of optical beams, *Phys. Rev. Lett.* **13**, 479–482 (1964). Errata, *Phys. Rev. Lett.*, **14** (25), 1056 (1965).
26. V.I. Talanov: Self-focusing of electromagnetic waves in nonlinear media. *Izv. Vuzov, Radiophysica*, **7**, 564–565 (1964). See also Chapter 11.2. *Self-Focusing: Past and Present*, R.W. Boyd, S.G. Lukishova, Y.R. Shen (Eds.), (this volume).
27. H.A. Haus: Higher-order trapped light beam solutions, *Appl. Phys. Lett.*, **8**, (5), 128–129 (1966).
28. Z.K. Yankauskas: Radial field distributions in a self-focusing light beam, *Radiophysics and Quantum Electronics* (Springer, NY), **9**, 261–263 (1966).
29. G.M. Fraiman et al.: Optical self-focusing: Stationary beams and femtosecond pulses. *Self-Focusing: Past and Present*, R.W. Boyd, S.G. Lukishova, Y.R. Shen (Eds.), (this volume, Chapter 3).
30. J.E. Bjorkholm, A. Ashkin: CW self-focusing and self-trapping of light in sodium vapor, *Phys. Rev. Lett.*, **32**, 129–132 (1974).
31. A. Barthelemy, S. Maneuf, C. Froehly: Propagation soliton et auto-confinement de faisceaux laser par non linearité optique de kerr, *Opt. Commun.*, **55**, 201–206 (1985).

32. J.S. Aitchison, Y. Silberberg, A.M. Weiner, D.E. Leaird, M.K. Oliver, J.L. Jackel, E. M. Vogel, P.W.E. Smith: Spatial optical solitons in planar glass waveguides, *J. Opt. Soc. Am. B* **8**, 1290–1297 (1991).
33. S. Trillo, W. Torruellas (Eds): *Spatial Solitons*, 474 pp, Springer-Verlag, New York (2001).
34. Yu.S. Kivshar, G.P. Agrawal: *Optical Solitons: From Fibers to Photonic Crystals*, 544 pp, Elsevier Science, New York (2003).
35. G.I.A. Stegeman, D.N. Christodoulides, M. Segev: Optical spatial solitons: historical perspectives, *IEEE J. Selected Topics in Quant. Electron.*, **6**, 1419–1427 (2000).
36. P. Zerom, R.W. Boyd: Self-focusing, conical emission and other self-action effects in atomic vapors. *Self-Focusing: Past and Present*, R.W. Boyd, S.G. Lukishova, Y.R. Shen (Eds.), (this volume, Chapter 9).
37. E. DelRe, M. Segev: Self-focusing and solitons in photorefractive media. *Self-Focusing: Past and Present*, R.W. Boyd, S.G. Lukishova, Y.R. Shen (Eds.), (this volume, Chapter 23).
38. X. Ni, R.R. Alfano: Periodic filamentation and supercontinuum interference. *Self-Focusing: Past and Present*, R.W. Boyd, S.G. Lukishova, Y.R. Shen (Eds.), (this volume, Chapter 10).
39. E. Garmire, R.Y. Chiao, C.H. Townes: Dynamics and characteristics of the self-trapping of intense light beam, *Phys. Rev. Lett.*, **16**, (9), 347–349 (1966).
40. D. Grishkowsky: Self-trapping of light by potassium vapor, *Phys. Rev. Lett.*, **24**, 866 (1970).
41. S.L. Chin et al.: The Physics of intense femtosecond laser filamentation. *Self-Focusing: Past and Present*, R.W. Boyd, S.G. Lukishova, Y.R. Shen (Eds.), (this volume, Chapter 14).
42. V.P. Kandidov et al.: Self-focusing and filamentation of powerful femtosecond laser pulses. *Self-Focusing: Past and Present*, R.W. Boyd, S.G. Lukishova, Y.R. Shen (Eds.), (this volume, Chapter 15).
43. E. Gaižauskas et al.: On the role of conical waves in self-focusing and filamentation of femtosecond pulses with nonlinear losses. *Self-Focusing: Past and Present*, R.W. Boyd, S.G. Lukishova, Y.R. Shen (Eds.), (this volume, Chapter 19).
44. R.R. Alfano (Ed.): *The Supercontinuum Laser Source: Fundamentals with Updated References*, 2nd ed., 568 pp, Springer, New York (2006).
45. E.A. Kuznetsov: Wave collapse in nonlinear optics. *Self-Focusing: Past and Present*, R.W. Boyd, S.G. Lukishova, Y.R. Shen (Eds.), (this volume, Chapter 7).
46. A. Gaeta: Spatial and temporal dynamics of collapsing ultrashort laser pulse. *Self-Focusing: Past and Present*, R.W. Boyd, S.G. Lukishova, Y.R. Shen (Eds.), (this volume, Chapter 16).
47. J. Zhang et al.: Self-organized propagation of femtosecond laser filamentation in air. *Self-Focusing: Past and Present*, R.W. Boyd, S.G. Lukishova, Y.R. Shen (Eds.), (this volume, Chapter 13).
48. G.G. Matvienko et al.: Effective parameters of high-power laser femtosecond radiation at self-focusing in gas and aerosol media. *Self-Focusing: Past and Present*, R.W. Boyd, S.G. Lukishova, Y.R. Shen (Eds.), (this volume, Chapter 21).
49. G.A. Askar'yan: Effects of the gradient of a strong electromagnetic beam on electrons and atoms, *Sov. Phys. JETP* **15**, 1088–1090 (1962). See also Chapter 11. 1. *Self-Focusing: Past and Present*, R.W. Boyd, S.G. Lukishova, Y.R. Shen (Eds.), (this volume).

50. M. Hercher: Laser-induced damage in transparent media, *J. Opt. Soc. Am.* **54**, 563 (1964). See also Chapter 11.3. *Self-Focusing: Past and Present*, R.W. Boyd, S.G. Lukishova, Y.R. Shen (Eds.), (this volume).
51. C. Sulem, P.L. Sulem: *The Nonlinear Schrödinger Equation*, Springer-Verlag New York (1999).
52. A. Brown, G. Korn, X. Liu, D. Du, J. Squier, G. Moorou: Self-channeling of high-peak-power femtosecond laser pulses in air, *Opt. Lett.*, **20**, 73–75 (1995).
53. A.V. Borovsky, A.L. Galkin, O.B. Shiryaev, T. Auguste: *Laser Physics at Relativistic Intensities*, 218 pp, Springer-Verlag, New York (2003).
54. M. Mlenek, E.M. Wright, J.V. Moloney: Dynamic spatial replenishment of femtosecond pulses propagating in air, *Opt. Lett.*, **23**, 382–384 (1998).
55. F. Wise, J. Moses: Self-focusing and self-defocusing of femtosecond pulses with cascaded quadratic nonlinearities. *Self-Focusing: Past and Present*, R.W. Boyd, S.G. Lukishova, Y.R. Shen (Eds.), (this volume, Chapter 20).
56. C. Conti et al.: X-waves in self-focusing of ultra-short pulses. *Self-Focusing: Past and Present*, R.W. Boyd, S.G. Lukishova, Y.R. Shen (Eds.), (this volume, Chapter 18).
57. N.F. Pilipetskii, A.R. Rustamov: Observation of self-focusing of light in liquids, *JETP Lett.* **2**, 55–56 (1965).
58. P. Lallemand, N. Bloembergen: Self-focusing of laser beams and stimulated Raman gain in liquids, *Phys. Rev. Lett.*, **15**, 1010–1012 (1965).
59. G. Hauchecorne, G. Mayer: Effects de l'anisotropie moléculaire sur la propagation d'une lumière intense, *Comptes Rendus, Acad. Sci. Paris*, **261**, 4014–4017, 15 November (1965).

Contents

Contributors

R.R. Alfano
Institute for Ultrafast Spectroscopy and Lasers,
Physics Department
The City College of City University of New York
USA
e-mail: ralfano@sci.ccny.cuny.edu

G.A. Askar'yan†
P.N. Lebedev Physics Institute
Academy of Sciences, USSR

S.N. Bagaev
Institute of Laser Physics SB RAS
Novosibirsk, Russia

Robert W. Boyd
Institute of Optics, University of Rochester
Rochester, New York, USA
e-mail: boyd@optics.rochester.edu

Nikolai E. Bykovsky
P.N. Lebedev Physical Institute of the Russian Academy of Sciences
Leninsky Prospekt 53, Moscow
119991, Russia
e-mail: nbykovsky@sci.lebedev.ru

Anthony J. Campillo
Naval Research Laboratory
Washington DC 20375, USA
e-mail: campillo@nrl.navy.mil

Sabino Chávez-Cerda
Instituto Nacional de Astrofísica Óptica y Electrónica
Apdo. Postal 51/216, Puebla., Pue.
México 72000
e-mail: sabino@inaoep.mx

R.Y. Chiao
Schools of Natural Sciences and of Engineering
University of California,
Merced California, USA
e-mail: rchiao@ucmerced.edu

See Leang Chin
Centre d'Optique, Photonique et Laser & Département de Physique,
de Génie Physique et d'Optique
Université Laval, Québec
G1V 0A6, Canada
e-mail: slchin@phy.ulaval.ca

Claudio Conti
Research center "Enrico Fermi," Via Panisperna 89/A
00100 Roma, Italy and Research Center SOFT INFM-CNR

University "La Sapienza," P.le Aldo 5,
00185 Roma, Italy
e-mail: claudio.conti@phys.uniroma1.it

A. Couairon
Centre de Physique Théorique
École Polytechnique
CNRS, F-91128,
Palaiseau, France
e-mail: couairon@cpht.polytechnique.fr

E. DelRe
Dipartimento di Ingegneria Elettrica e dell'Informazione
Università dell'Aquila and INFM-CNR CRS SOFT, Italy
e-mail: edelre@ing.univaq.it

A.E. Dormidonov
International Laser Center
Physics Department M.V. Lomonosov
Moscow State University
Moscow
119992 Russia
e-mail: adorm@list.ru

Audrius Dubietis
Department of Quantum Electronics
Vilnius University, Vilnius, Lithuania

Daniele Faccio
Department of Physics and Mathematics
University of Insubria
Como, Italy
e-mail: daniele.faccio@uninsubria.it

Gadi Fibich
School of Mathematical Sciences
Tel Aviv University, Tel Aviv, Israel
e-mail: fibich@tau.ac.il

G.M. Fraiman
Institute of Applied Physics of RAS
Nizhniy Novgorod
e-mail: fraiman@appl.sci-nnov.ru

Alexander L. Gaeta
Cornell University, School of Applied and Engineering Physics
USA
e-mail: a.gaeta@cornell.edu

Eugenijus Gaižauskas
Department of Quantum Electronics
Vilnius University, Vilnius, Lithuania
e-mail: eugenijus.gaizauskas@ff.vu.lt

Yu. E. Geints
Institute of Atmospheric Optics SB RAS
Tomsk, Russia
e-mail: ygeints@iao.ru

T. K. Gustafson
Department of Electrical Engineering and Computer Science
University of California, Berkeley
California, USA
e-mail: tkg@eecs.berkeley.edu

David J. Hagan
University of Central Florida,
CREOL, The College of Optics and Photonics
Orlando, FL 32817 USA
e-mail: hagan@creol.ucf.edu

Zuoqiang Hao
Beijing National Laboratory for Condensed Matter Physics
Institute of Physics, Chinese Academy of Sciences
Beijing 100190, China

Michael Hercher
OPTRA, Inc.,
461, Boston Street,

Topsfield MA 01983-1234
e-mail: michaelhercher@gmail.com

Jandir Miguel Hickmann
Instituto de Física, Universidade Federal de Alagoas
Maceió, Al. Brazil
e-mail: hickmann@optma.org

Zhan Jin
Beijing National Laboratory for Condensed Matter Physics
Institute of Physics, Chinese Academy of Sciences
Beijing 100190, China

Marcelo David Iturbe-Castillo
Instituto Nacional de Astrofísica Óptica y Electrónica
Apdo. Postal 51/216, Puebla., Pue., México 72000
e-mail: diturbe@inaoep.mx

A.M. Kabanov
Institute of Atmospheric Optics SB RAS
Tomsk, Russia

Valerii P. Kandidov
International Laser Center
Department of Physics
Moscow State University, Moscow 119992 Russia
e-mail: kandidov@phys.msu.ru

P.L. Kelley
Department of Electrical and Computer Engineering
Tufts University, Medford
Massachusetts, USA
e-mail: pkelley@tufts.edu

Olga G. Kosareva
International Laser Center
Department of Physics, Moscow State University
Moscow, 119992 Russia
e-mail: kosareva@phys.msu.ru

E.A. Kuznetsov
L.D. Landau Institute for Theoretical Physics
2 Kosygin street, 119334 Moscow
Russia
and
P.N. Lebedev Physical Institute
53 Leninsky ave.,
119991 Moscow, Russia
e-mail: kuznetso@itp.ac.ru

Viačeslav Kudriašov
Department of Quantum Electronics
Vilnius University, Vilnius
Lithuania

A.G. Litvak
Institute of Applied Physics of RAS
Nizhniy Novgorod, Russia
e-mail: litvak@appl.sci-nnov.ru

Weiwei Liu
Institute of Modern Optics
Nankai University
Key Laboratory of Opto-electronic Information Science and Technology
Education Ministry of China
Tianjin 300071, China
e-mail: liuweiwei@nankai.edu.cn

Xin Lu
Beijing National Laboratory for Condensed Matter Physics
Institute of Physics, Chinese Academy of Sciences
Beijing 100190, China

V.N. Lugovoi†
A.M. Prokhorov General Physics Institute of Russian Academy of Sciences
38 Vavilov street, Moscow 119991
Russia

Svetlana G. Lukishova
The Institute of Optics University of Rochester,
Rochester, New York 14627, USA
e-mail: sluk@lle.rochester.edu

A.A. Manenkov
A.M. Prokhorov General Physics Institute of Russian Academy of Sciences
38 Vavilov street, Moscow, 119991 Russia
e-mail: manenkov@kapella.gpi.ru

John H. Marburger
Office of Science and Technology Policy
Executive Office of the President
725 17th Street, Room 5228
Washington, DC 20502, USA

G.G. Matvienko
Institute of Atmospheric Optics SB RAS
Tomsk, Russia

Jeffrey Moses
Department of Applied and Engineering Physics
Cornell University
Ithaca, New York, USA

A. Mysyrowicz
Laboratoire d'Optique Appliquée,
École Nationale Supérieure des Techniques
Avancées–École Polytechnique
CNRS UMR 7639
F-91761 Palaiseau Cedex, France
e-mail: mysyr@ensta.fr

Xiaohui Ni
Institute for Ultrafast Spectroscopy and Lasers
Physics Department, The City College of City University of New York
USA
e-mail: xiaohui@sci.ccny.cuny.edu

Alexander S. Scheulin
S.I. Vavilov State Optical Institute
12 Birghevaya Line, St. Petersburg
199034, Russia
e-mail: angervax@mail.ru

M. Segev
Physics Department, Technion
Haifa, Israel
e-mail: msegev@tx.technion.ac.il

Yury V. Senatsky
P.N. Lebedev Physical Institute of the Russian Academy of Sciences
Leninsky Prospekt 53, Moscow
119991 Russia
e-mail: senatsky@sci.lebedev.ru

Y. Ron Shen
Department of Physics, University of California
Berkeley, California, 94720, USA
e-mail: yrshen@calmail.berkeley.edu

Valdas Sirutkaitis
Department of Quantum Electronics
Vilnius University, Vilnius,
Lithuania

A.N. Stepanov
Institute of applied physics RAS
Nizhnij Novgorod, Russia

Eric W. Van Stryland
University of Central Florida
CREOL, The College of Optics and Photonics
Orlando, FL 32817 USA
e-mail: ewvs@creol.ucf.edu

V.I. Talanov
Institute of Applied Physics of RAS
Nizhniy Novgorod e-mail: v.talanov
@hydro.appl.sci-nnov.ru.

Paolo Di Trapani
Department of Physics and Mathematics
University of Insubria, Como, Italy
Department of Quantum Electronics
Vilnius University
Sauletekio Avenue 9, LT-2040
Vilnius, Lithuania

Stefano Trillo
Department of Engineering
University of Ferrara, Via Saragat 1
44100 Ferrara
e-mail: stefano.trillo@unife.it

S.N. Vlasov
Institute of Applied Physics of RAS
Nizhniy Novgorod
e-mail: vlasov@hydro.appl.sci-nnov.ru

Zhaohua Wang
Beijing National Laboratory for
Condensed Matter Physics
Institute of Physics, Chinese Academy of Sciences
Beijing, 100190, China

Zhiyi Wei
Beijing National Laboratory for
Condensed Matter Physics
Institute of Physics, Chinese Academy of Sciences
Beijing, 100190, China

Frank W. Wise
Department of Applied and Engineering Physics
Cornell University
Ithaca, New York, USA
e-mail: fwise@ccmr.cornell.edu

Tingting Xi
Beijing National Laboratory for
Condensed Matter Physics
Institute of Physics
Chinese Academy of Sciences, Beijing
100190, China

Hui Yang
Beijing National Laboratory for
Condensed Matter Physics
Institute of Physics, Chinese Academy of Sciences
Beijing, 100190, China

Jie Zhang
Beijing National Laboratory for
Condensed Matter Physics
Institute of Physics
Chinese Academy of Sciences
Beijing, 100190, China

Zhe Zhang
Beijing National Laboratory for
Condensed Matter Physics
Institute of Physics
Chinese Academy of Sciences
Beijing, 100190, China

A.A. Zemlyanov
Institute of Atmospheric Optics SB RAS
Tomsk, Russia

Petros Zerom
Institute of Optics, University of Rochester
Rochester, New York, USA

Part I
Self-focusing in the Past

Chapter 1
Self-Focusing and Filaments of Light: Past and Present

Y. Ron Shen

Abstract Early research on self-focusing and filament formation of nanosecond and picosecond light pulses in nonlinear media is briefly reviewed. Emphasis is on physical understanding of characteristic features of the self-focusing phenomenon. Connections with the recent discovery of self-focusing and the appearance of apparently long filaments of light in air created by femtosecond light pulses are made. Similarities and differences between self-focusing of ns-ps pulses and fs pulses are discussed.

1.1 Introduction

In the early development of nonlinear optics, self-focusing and the formation of filaments of nanosecond laser pulses had fascinated many researchers. (See early review articles, [1–4].) The effect often initiates laser damage in transparent media, and is the limiting factor in the design of high-power lasers. It also results in self-structuring and self-spectral-broadening of output pulses. While the main features of the effect are now well understood, some details are still under debate. More recently, observations of intense long filaments created by high-energy femtosecond pulses even in air have aroused another strong wave of interest on the topic [5]. Potential applications of femtosecond filaments include induced lightning, remote triggering and sensing, super-spectral broadening, and attosecond pulse generation. Although the scales of time and distance in self-focusing of ns and fs pulses are drastically different, there are actually many similarities in the two cases, which one would find interesting to compare. We provide here such a comparison with a focus on the underlying physical mechanisms and processes. We begin with a brief review of the early studies on self-focusing and filament formation of nanosecond pulses.

Y.R. Shen (✉)
Department of Physics, University of California, Berkeley, California 94720, USA
e-mail: yrshen@calmail.berkeley.edu

R.W. Boyd et al. (eds.), *Self-focusing: Past and Present*,
Topics in Applied Physics 114, DOI 10.1007/978-0-387-34727-1_1,

1.2 Early History of Self-Focusing and Filaments of Light

Askarian first suggested the possibility of self-focusing and self-trapping of light in 1962 [6], but his paper in Russian was initially not well noticed in the west. Hercher, in early 1964, discovered that the optical damage caused by Q-switched pulses in glass appeared as fine tracks of damage spots [7]. To explain the observation, Chiao et al. proposed the self-trapping model showing that an intense laser beam could be self-trapped in a dielectric waveguide created by the beam itself [8]. Talanov found independently the self-trapping solution of beam propagation in an optical Kerr medium [9]. However, temporal variation of the beam power and stability of self-trapping were not considered in the model. In fact, beam attenuation would readily destroy the condition for establishing a self-trapped filament. Thus, there was difficulty in using self-trapping to explain the optical damage tracks in glass.

Around the same time, stimulated Raman scattering was discovered [10]. In contrast to the well-conceived theory [11], the observed process in solids and liquids appeared to have an anomalous sharp pump threshold, an abnormally large stimulated gain, and an unexpected forward-backward gain asymmetry [12–15]. It was soon realized that self-focusing might be responsible for the anomalies.

Talanov [16] and Kelley [17] showed that if the power of a single-mode laser beam propagating in an optical Kerr medium was larger than a certain threshold value, the beam would self-focus into a very small focal spot at a finite distance inside the medium. The appearance of the focal spot of very high intensity in the medium readily initiated the stimulated Raman process and led to the sharp threshold and high gain of the process. The result was quantitatively verified in experiments on stimulated Raman scattering in liquids [18, 19]. Their analysis, however, was also stationary with the threshold referring to the peak power in the pulse case. The connection between self-focusing and self-trapping in the analysis is that the two become the same when the beam power reaches a critical value at which the self-focused focal spot would appear at infinity.

Shrinking of the beam cross-section upon self-focusing could be directly photographed. In doing so, Garmire et al. observed the appearance of "large-scale filaments" of $\sim$100 μm in diameter in CS_2 upon self-focusing [20]. At sufficiently high input power, there was also a forward conical emission appearing as a ring around each filament. Chiao et al. later reported that the large-scale filament actually contained many "small-scale filaments" of a few μm in diameter [21]. An example is shown in Fig. 1.1(a).

These filaments were believed to be the self-trapped filaments described earlier. Bespalov and Talanov [22] suggested that a beam with power many times more than the threshold power for self-focusing could easily self-focus and break into many filaments. Each weak local maximum on the beam intensity profile resulting from fluctuations or imperfect single mode structure

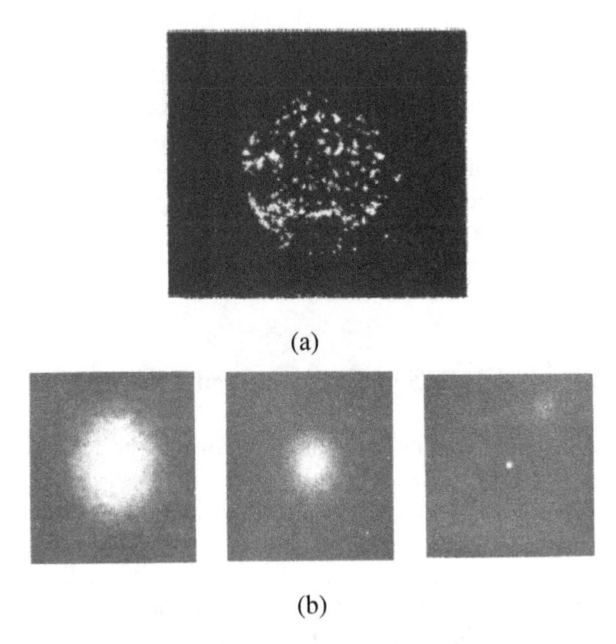

Fig. 1.1 (**a**) Image of small-scale filaments at the exit window of a CS_2 cell created by self-focusing of a multimode laser beam [After S.C. Abbi and H. Mahr, *Phys Rev Lett* **26**, 604 (1971).] (**b**) Image of a self-focused single-mode laser beam at the exit of a toluene cell of different lengths. From left to right: short cell length, beam not yet self-focused (~700 μm); cell length just above the self-focusing threshold, beam self-focused midway to ~50 μm; cell length clearly above self-focusing threshold, beam self-focused to the limiting diameter of 10 μm. (After Ref. [4].)

could self-focus independently and lead to a filament. Indeed, when a good single-mode laser beam with a power no more than two times the threshold power, only a well-centered single filament was observed (Fig. 1.1(b)) [23]. Yet, adopting a stationary theory to explain the filaments was clearly still a problem.

Early experiments on self-focusing and filaments in condensed media were carried out with nanosecond laser pulses. Having the beam power vary with time, self-focusing must be a time-dependent process. An immediate consequence is that the position of the self-focused focal spot must vary in time, as first pointed out by Marburger and Wagner [24]. Lugovoi and Prokhorov suggested that focal spots moving in time could appear as filaments [25]. Loy and Shen proved experimentally that the moving focus model could indeed explain the observed filaments quantitatively [23, 26, 27]. To avoid complications caused by the occurrence of multiple filaments, they used a relatively low-power single-mode Q-switched laser pulse to create a single filament on the beam axis in liquid. Both forward- and backward-moving foci were observed and shown to agree with the predicted trajectory. Zverev et al. also showed that the moving focus model could explain the observed damage tracks created by focused single-mode laser pulses in solids [28]. Using a streak camera,

Korobkin et al. [29] directly observed the backward-moving focus in a CS_2 cell, and Giuliano and Marburger [30] recorded the formation of the damage streak by a moving focus. Thus the basic connection between self-focusing and filament formation was established, and attention turned toward other related areas. What limits the diameter of the focus or filament? How does the transient nonlinear response of the medium affect self-focusing and filaments? What are the characteristics of the output from a filament? How are stimulated Raman and Brillouin scattering influenced by self-focusing?

1.3 Quasi-Steady-State Self-Focusing and Moving Focus

We now describe self-focusing and filament formation in more detail. The physical picture of self-focusing is fairly straightforward [23]. Consider a beam with a bell-shaped cylindrically symmetric beam profile entering an optical Kerr medium with refractive index $n = n_0 + \Delta n$ and $\Delta n = \eta_2 I(\rho)$, where $I(\rho)$ is the beam intensity distribution along the radial coordinate ρ (see Fig. 1.2) and $\eta_2 > 0$. The central part of the beam sees a larger n and therefore moves slower than the edge. Consequently, as the beam propagates into the medium, its wavefront gets increasingly distorted as depicted in Fig. 1.2.

Because the optical rays always propagate perpendicular to the wavefront, the beam appears to focus by itself. A beam with a finite cross-section must diffract. In order to see the self-focusing effect, the self-focusing action from Δn must be strong enough to overcome the diffraction action. (Roughly, for a Gaussian beam of diameter d, this corresponds to $2\Delta n/n_0 > \theta_{dif}^2 \equiv (1.22\lambda/d)^2$, which leads to the criterion on beam power for self-focusing. [16, 17] $(P > P_{cr} \equiv \pi(1.22\lambda)^2/8\eta_2.)$ If the two exactly cancel each other, as is possible for a specific function of $I(\rho)$, then the beam would propagate without any change of its spatial profile. This is the case of self-trapping, also known as a spatial soliton. However, the solution is metastable. A small increase or decrease of the beam intensity would upset the balance for self-trapping and make the beam self-focus or diffract. In this simple picture, self-focusing would continue until $\Delta n = \eta_2 I(\rho)$ no longer holds.

Qualitatively, it is easy to see that the self-focusing distance, z_f, defined as the distance between the position of the self-focused focal spot and the entrance

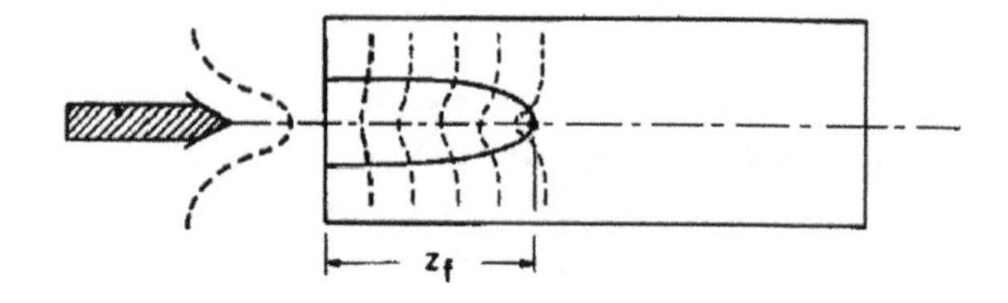

Fig. 1.2 Sketch showing distortion of the wavefront and self-focusing of a laser beam in a nonlinear medium. (After Ref. [4].)

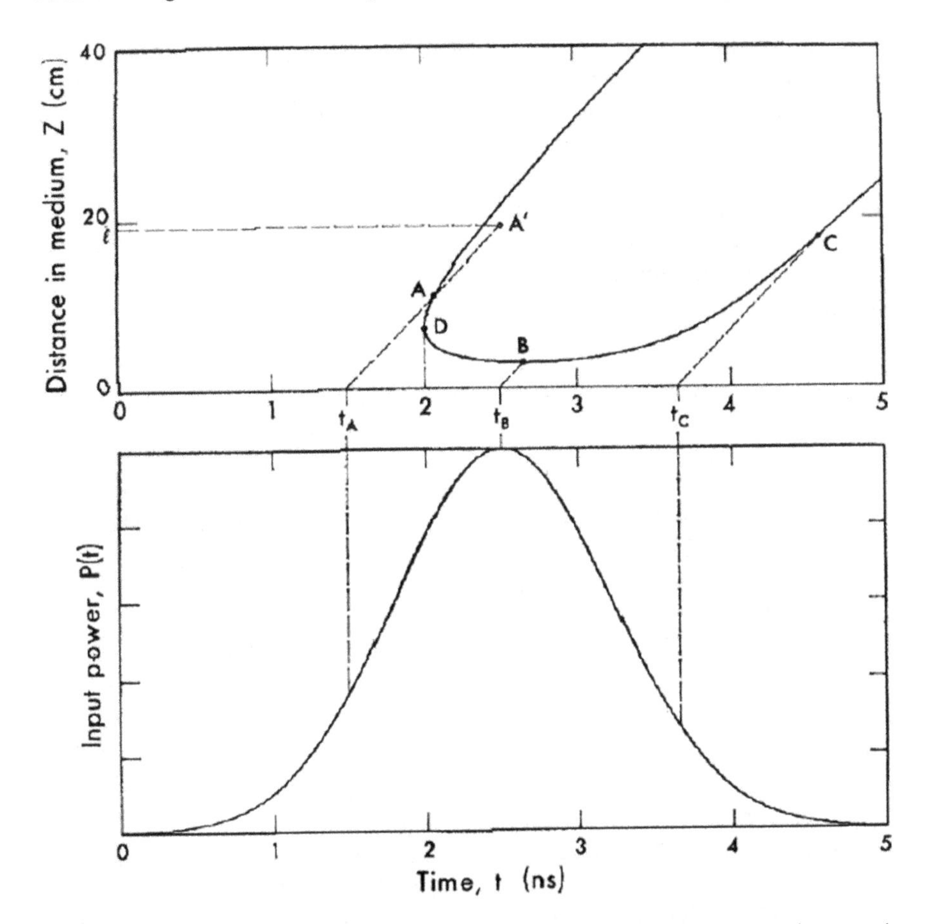

Fig. 1.3 Lower trace describes input power $P(t)$ as a function of time t. Peak power is 42.5 KW and the half-width at 1/e point is 1 ns. The upper trace calculated from Eq. (1.1) describes the position of the focal spot as a function of time. Values of P_0 and K used are 8 KW and 11.6 cm-(KW)$^{1/2}$, respectively, which corresponds roughly to an input beam of 400 μm in diameter propagating in CS_2. The dashed lines with the slope equal to the light velocity indicate how light propagates in the medium along the z-axis at various times. (After Ref. [33].)

face of the medium, must decrease with increase of $I(\rho)$ or P. We can then readily visualize how z_f varies with time (i.e., trajectory of the focal spot) for a pulsed input in the so-called quasi-steady-state self-focusing. This is sketched in Fig. 1.3 for a bell-shaped input pulse [23]. We assume that at time t_A, beam with power $P_A > P_{cr}$ enters the medium, propagates with light velocity, and self-focuses at the end of the medium $z_{fA} = l$. At a later time t_B, beam with $P_B > P_A$, similarly propagating, must self-focus at $z_{fB} < l$. Following such a point-by-point plot, we can obtain z_f versus t in the form of a U curve. As seen in Fig. 1.3, if l is long enough, the trajectory of the moving focus should exhibit two branches. Starting from point D, one branch has the focus moving in the forward direction with an apparent velocity larger than the light velocity, and the other branch has the focus first moving backward, and then forward after it reaches a minimum z_f corresponding to the maximum P. For a nanosecond laser pulse, the focus can traverse a distance of several centimeters in a few nanoseconds. To a slow photographic detection system, it simply appears as a single bright filament.

Quantitatively, for a single-mode Gaussian input beam, the relation between z_f and P was found numerically. For convenience, we take here the asymptotic form [31].

$$z_f = \frac{K}{\sqrt{P} - \sqrt{P_0}} \tag{1.1}$$

with $K = 0.367ka_0^2\sqrt{P_{cr}}$, $P_0 = (0.852)^2 P_{cr}$, and a_0 is the Gaussian beam radius. For $P > 1.25P_{cr}$, Eq. (1.1) is a good approximation to the more correct form. With it, the moving focus trajectory can be determined quantitatively. Experiments on the moving focus in a liquid found that with proper values of K and P_0, Eq. (1.1) described the result well [23, 26].

Establishment of the moving-focus model facilitated understanding of many intriguing phenomena associated with self-focusing, which we briefly summarize below [32, 33].

(A) Stimulated Raman and Brillouin scattering are readily initiated at a focus. Their sharp pump threshold corresponds to the self-focusing threshold at which the focus first appears in the medium.

(B) Forward-stimulated Raman emission is weaker than the backward emission near threshold and becomes stronger at higher input laser power. This asymmetric behavior is a consequence of Raman amplification by a self-focused pump beam that has complex spatial and temporal variations.

(C) The backward-moving focal spot does not turn around and move forward as suggested by the trajectory in Fig. 1.3. It gets terminated because the backward-stimulated Raman and Brillouin emission depletes the lagging part of the incoming laser beam and makes it not strong enough to self-focus.

(D) The moving focus has a limiting diameter of the order of 10 μm depending on the medium. It is most likely because stimulated Raman scattering at the focus depletes the laser energy and stops it from further self-focusing.

(E) The pulse emitting from the focal spot appearing at the exit of the medium is much shorter than the input pulse, and has an appreciably broadened spectrum.

Most of the above features can be qualitatively understood by analyzing the beam paths in Fig. 3 [4]. The last point deserves more explanation below because it is more relevant to self-focusing of femtosecond pulses.

1.4 Effects of Transient Response and Dynamic Self-Focusing

A real focus has finite dimensions. To be more realistic, we should have replaced each focal point on the U-shaped trajectory in Fig. 1.3 by a line with a length representing the longitudinal focal dimension (Fig. 1.4a). This would broaden the trajectory curve, and show explicitly that the focus move across a plane in a finite time. Thus, light emitting from the moving focus as it crosses a plane is a pulse, often much shorter than the input pulse width. For a nanosecond input pulse of sufficiently large power self-focused in a sufficiently long medium, the focus may move toward the end of the medium with a velocity close to the light velocity, and then the emitted pulse from the focus at the exit plane would have a width of only a few ps. We however recall that we have assumed $\Delta n = \eta_2 I(\rho)$, meaning instantaneous response to the beam intensity variation in time and space. In a Kerr liquid, this assumption is certainly not valid on the ps time scale. (For simplicity, we limit our following discussion to media with Δn dominated by the orientational Kerr effect.) Our description of focusing geometry around the forward-moving focal region must be modified to take into account the transient response of Δn, i.e., $\Delta n(t) = \int_{-\infty}^{t} \eta_2(t-t')I(t')dt'$. In Fig. 1.4b, the shaded area around the moving focus trajectory describes the time and space where Δn is significant.

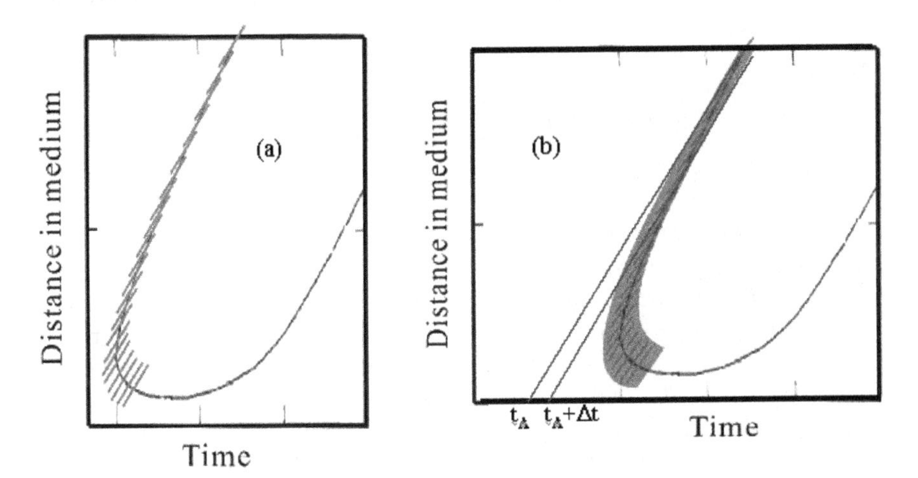

Fig. 1.4 (**a**) A U curve describing the moving focus trajectory with short lines representing the longitudinal focal dimension. (**b**) The shaded area around the U curve indicates that where and when the refractive index change Δn induced by the moving focus is significant. It shows that light in a trailing focus (e.g., the one that originates from the beam entering the medium at $t_A + \Delta t$) experiences a significant transient Δn induced by the leading focus (from the beam entering the medium at t_A), and accumulates a phase increment. The result is a lengthening of the longitudinal focal dimension and a phase increment varying with time. The latter leads to significant spectral broadening of light from the apparent filament

Consider the branch of the forward-moving focus in Figs. 1.3 or 1.4a. The focus formed by the beam entering the medium at t_A appears ahead of that formed by the beam entering at $t_A + \Delta t$ (see Fig. 1.4b). The immediate consequence of the transient Δn induced by a moving focus is a trailblazing effect. It reduces the diffraction of light from the trailing focus and makes its longitudinal focal length effectively longer. Assuming that diffraction is effectively suppressed by residual Δn within its relaxation time T, the longitudinal focal length can be estimated to be $\Delta Z_f = T(n_0/c - 1/v_f)^{-1}$, where $v_f > c/n_0$ is the velocity of the moving focus [26]. We notice that the shorter is the input pulse, the closer v_f is to c/n_0 and the larger ΔZ_f becomes. A large ΔZ_f would make a focus appear like a trapped filament. As an example, one can imagine a 0.5-ns input pulse that has a 0.05-ns section on the leading edge of the pulse to create a forward-moving focus from z_f = 10–20 cm in a medium. For n_0 = 1.5, this yields v_f = (10/9)c/n_0, and a corresponding ΔZ_f= 1.8 cm for T = 10 ps.

In the filament-like focal region, the very intense beam readily induces other nonlinear optical effects. Forward-stimulated Raman scattering along the focus can deplete the self-focused beam and terminate the extension of the focus or the filament [32]. This happens in a sufficiently long medium. Another effect is spectral broadening of the output from the moving focus mentioned earlier. This is due to phase modulation imposed by $\Delta n(\vec{r}, t)$ on the beam in the focal region [27]. As shown in Fig. 1.4b, a beam, entering the medium at t_A, self-focusing at z_A, and exiting the medium at l, accumulates along its path an additional phase of

$$\Delta\Phi \approx \frac{\omega}{c} \int_{z_A}^{l} \Delta n(z, t = t_A + n_0 z/c) dz. \tag{1.2}$$

Beams exiting from the shaded area at l in Fig. 1.4 roughly constitute the output pulse from the focal area. It is obvious from Fig. 1.4 that $\Delta\Phi$ of the beams should vary with time. This phase modulation in time leads to a frequency chirp, $\Delta\omega(t) = -\partial\Delta\Phi(t)/\partial t$, appearing as a spectral broadening mainly on the Stokes side. The maximum broadening is estimated to be $|\Delta\omega|_{max} \sim |\Delta\Phi|_{max}/T$, with $|\Delta\Phi|_{max} \sim (\omega/c)(\Delta n)_{max}(\Delta Z_f)$. Normally, we have $\Delta n_{max} \sim 10^{-3}$ in the focal spot. For $\Delta Z_f \sim$ 2 cm (from ~1 ns input pulse) and T ~ 2 ps (for CS_2), we find $|\Delta\omega|_{max} \sim 10^{13}$/ sec or 100 cm^{-1}. Such a spectral broadening was actually observed in experiment (Fig. 1.5) [34]. Phase modulation on light from a filament can also explain the earlier observation of spectral broadening [38] as depicted in Fig. 1.6.

The above discussion shows that for a picosecond input pulse with sufficient peak power in a Kerr liquid, the longitudinal dimension of the moving focus, ΔZ_f, would easily extend to many tens of cm and spectral broadening to many hundreds of cm^{-1}. If the pulse width is within an order of magnitude of the relaxation time of Δn, the transient response of Δn becomes more significant.

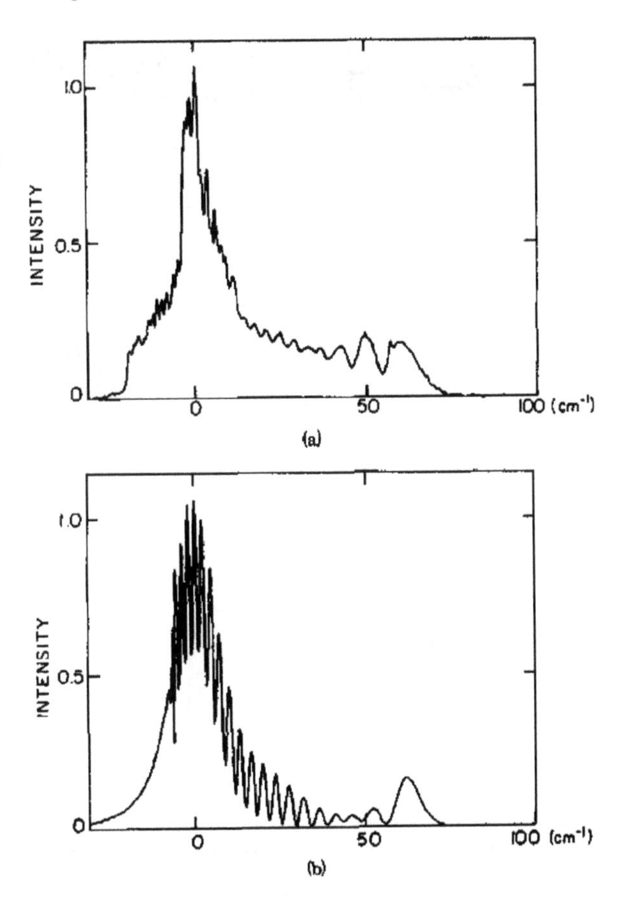

Fig. 1.5 (**a**) Spectrum of light emitted from an apparent filament produced in a 10-cm CS_2 cell by a 1.2-ns ruby laser pulse with a peak power of 27 KW. (**b**) The calculated spectrum using the moving-focus model. The strong central peak comes from phase-unmodulated part of the beam. (After Ref. [34].)

Self-focusing is expected to be more gradual, and the longitudinal focal dimension much longer. We can extend the moving focus description with transient Δn to paint a qualitative picture for this "dynamic self-focusing" process [35, 36].

Figure 1.7 describes how each temporal section of a cylindrically symmetric pulse self-focuses in a medium with a transient Δn that has a response and relaxation time comparable or larger than the laser pulse width [37]. The earliest section, "a-a," hardly experiences any induced Δn, and practically diffracts as if it were in a linear medium, but it induces a small Δn to be seen by the next section "b-b." Accordingly, "b-b" diffracts less than "a-a." The next section, "c-c," then sees a Δn strong enough to self-focus.

With increasing Δn, the subsequent sections, "d-d" to "f-f," self-focus more readily. They soon focus to a limiting diameter set by some other nonlinear process. In this case, diffraction of the self-focused beam must also be gradual, yielding effectively a very long focus. This is especially true for the later sections

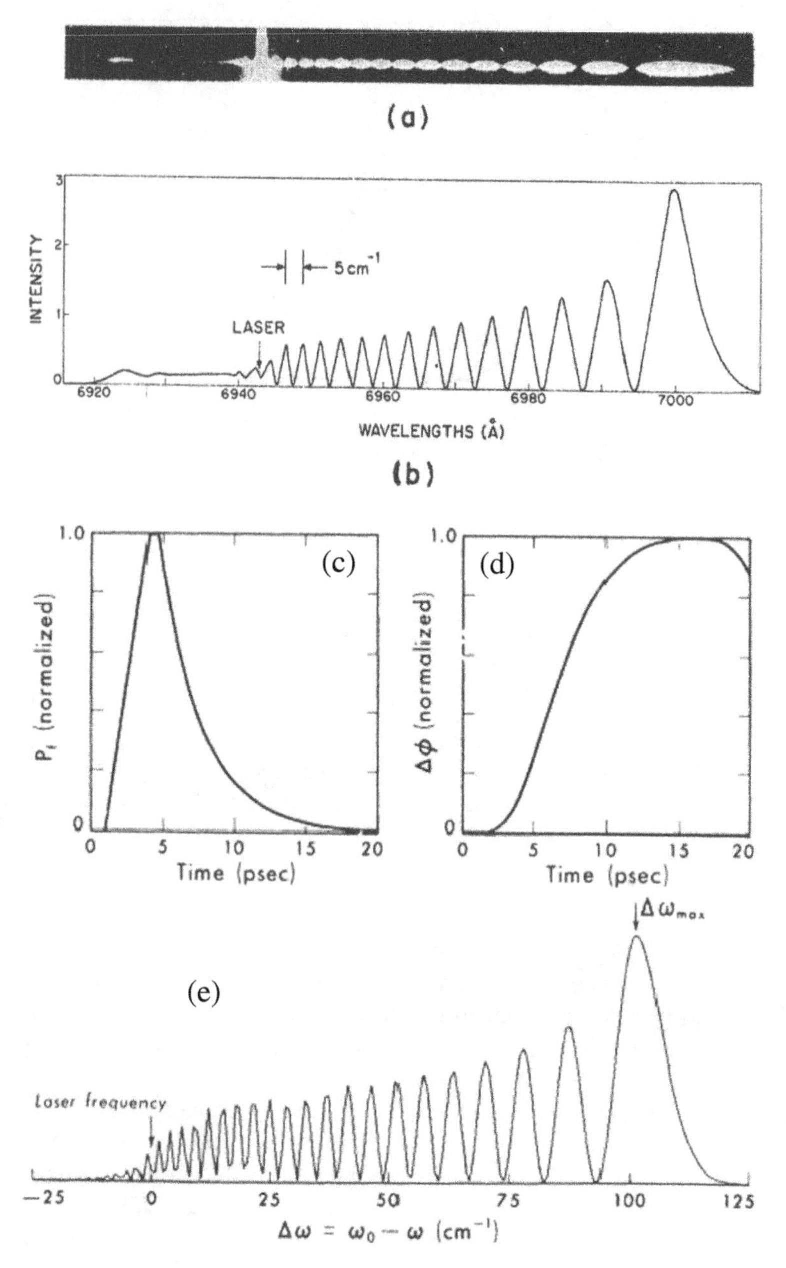

Fig. 1.6 Spectral broadening of light from a filament in a mixture of CS_2 and benzene: (**a**) Experimental spectrum. (**b**) Theoretical fit obtained by assuming a Gaussian pulse of full 1/e width of 5.4 ps in a self-trapped filament. The relaxation time of the medium is assumed to be 9 ps. [After T.K. Gustafson et al, *Phys Rev* **177**, 306 (1969).] Theoretical calculation of the filament pulse obtained from the moving-focus model assumes a 2-ns input pulse propagating in a 22.5-cm CS_2 cell: (**c**) filament pulse; (**d**) phase modulation in time; (**e**) power spectrum of the filament pulse. (After Ref. [27].)

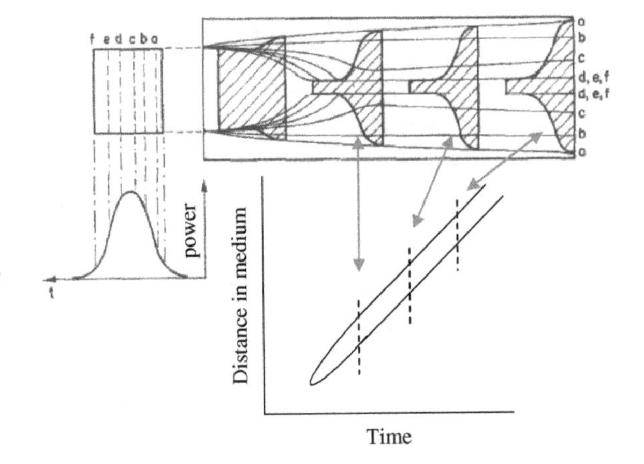

Fig. 1.7 Sketch showing dynamic self-focusing of a picosecond pulse in a Kerr liquid. Different parts (a, b, c, etc.) of the pulse focus and defocus along different ray paths. The pulse first deforms into a horn shape and then propagates with little further change. (After Ref. [39].)

of the pulse as they propagate in the channel of Δn created by the leading foci. We can use the rays propagating in the medium to construct the pulse after certain times or distances of propagation as shown in Fig. 1.6. It is seen that as the trailing part of the pulse shrinks toward a limiting diameter due to self-focusing, the transverse profile of the pulse quickly deforms into a horn shape (of 2 mm long for a 10-ps pulse in a medium with $n_o = 1.5$) and then propagates on over a long distance without significant changes except for an overall expansion following linear diffraction of the front part of the pulse.

This stable form of pulse propagation in a self-focusing medium can be called dynamic trapping. The neck portion of the horn sweeping along the beam axis over a long distance should appear as a bright filament on a time-integrated photograph. Output from the filament is strongly phase-modulated, resulting in significant spectral broadening in the case of picosecond pulses [38]. Note that in the limit of instantaneous response of Δn, the beam construction in Fig. 1.7 would naturally lead to the moving-focus picture described earlier.

Dynamic self-focusing can readily occur with picosecond pulses in ordinary Kerr liquids, but because of the short time scale, is difficult to study quantitatively. However, a liquid crystalline material in the isotropic phase often has a large Δn and a relaxation time T that can be tuned from 1 to 100 ns by varying temperature [37–39]. It is therefore a medium that can be used with ~10-ns laser pulses to study self-focusing from the transient to the quasi-steady-state limit. This was successfully carried out in the experiment by Hanson et al. [37]. With longer T, the focus appearing at the end of the medium became clearly longer. Toward the transient limit, the output pulse indeed was deformed into a horn shape, and the neck of the horn had approximately a constant diameter. The dynamical self-focusing and trapping picture was thus well proven.

1.5 Self-Focusing of Femtosecond Laser Pulses

Having seen the success in understanding self-focusing of ns and ps pulses, one would expect that self-focusing of fs pulses in condensed media is just a further extension of dynamic self-focusing of ps pulses. There are, however, differences in Δn arising from different physical mechanisms. On the fs time scale, electronic contribution to Δn could dominate. Its response is nearly instantaneous ($\sim 10^{-15}$ sec) if only virtual transitions are involved, but could become highly complex if real transitions are involved.

Multiphoton excitation and ionization could further complicate the issue. In any case, a 100-fs pulse is only 30 μm long in space. Self-focusing of the pulse in a medium still must act to modify the spatial profile of the pulse as the pulse propagates along. A high-resolution 3D motion picture would show how the profile of the ultrashort pulse varies in space and time in the medium, and the apparently long intense filaments must result from either moving foci or dynamic trapping of light. Super-spectral broadening of fs pulses from filaments in condensed materials was observed and has been widely used as a fs white-continuum light source [40]. Many of the characteristic features of self-focusing of fs pules in gases and condensed matter are qualitatively similar to self-focusing of ps and ns pulses. In the following, we briefly review the observations on fs self-focusing, discuss their similarities to the ps and ns cases, and comment on their differences.

Self-focusing of femtosecond pulses that leads to filament formation is readily seen not only in condensed matter, but also in gases [5]. The phenomenon in air is most dramatic, as it can produce filament of length from tens of meters to 1 kilometer. It was first discovered in 1995 by Braun et al. using 200-fs Ti:sapphire laser pulses of tens of mJ in energy [41]. Numerical solutions of the wave equation with appropriate Δn appear to be able to describe the phenomenon well, but they do not provide clear physical insight [5]. Here, we focus our discussion on qualitative physical description of self-focusing in air.

As in ns and ps cases, self-focusing of a nearly single-mode fs pulse with relatively low peak power (somewhat higher than the threshold power for self-focusing) leads to the formation of a single filament [42], but with peak power far above the self-focusing threshold, the self-focused beam generally breaks into multiple filaments. This is again due to modulation instability proposed earlier by Bespalov and Talanov [22]. The filaments have limiting diameters because of effective suppression of further self-focusing by the onset of some other nonlinear optical process: multiphoton ionization in the fs case in contrast to stimulated Raman and Brillouin scattering in the ns and ps cases. The multiple filaments can influence one another through their interactions with the surrounding unfocused beam, and remind us of the earlier observation of large-scale filaments in the ns case [20].

The moving-foci picture can again be used to describe self-focusing of a fs pulse [42]. In this case, electronic response of molecules (including plasma generation by multiphoton ionization) is likely the dominating mechanism behind the induced refractive index change Δn in air, although stimulated Raman excitation of vibrational modes may also contribute. Because Δn in air is small, self-focusing is expected to be more gentle. Assuming the Δn response is nearly instantaneous, the critical power, P_{cr}, for self-focusing of a beam of 1 cm in diameter can be estimated to be ~3GW at a wavelength of 800 nm. Then, for a 100-fs, 10-mJ pulse (with a peak power of 100 GW), Eq. (1.1) predicts a moving focal spot traversing a distance from 80 to 600 m in ~1.8 μs when the input pulse has its power vary from 6 to 100 GW in ~100 fs.

Each focus supposedly has a limiting diameter (~100 μm) resulting from defocusing action of the plasma created by multiphoton ionization counteracting the self-focusing action in the focal region [43, 44]. The longitudinal focal dimension can be relatively long, and we simply assume it is 0.1% of the self-focusing distance, z_f, e.g., it is 0.1 m for $z_f = 100$ m.

The above description provides us with some insights. (i) The 100-fs (30-μm) pulse propagates in space like a bullet, but because of self-focusing, its 3D profile varies continuously as it propagates along. (A qualitative picture of how the profile changes can be constructed by the same scheme used in Fig. 1.7.) (ii) Filaments must result from tracks of long moving foci in space. (iii) The effect of transient response of Δn may not be negligible even if the response time of Δn is only ~1 fs (from electronic response), because upon focusing, the local intensity variation of the focused beam in 1 fs can still be significant. The situation is qualitatively similar to that of self-focusing with transient Δn described earlier in Fig. 1.4, but not the dynamic trapping picture in Fig. 1.7 because the response time is much smaller than the pulse width. The leading focus could blaze a channel of Δn to suppress diffraction of light from the lagging focus, effectively extending its longitudinal dimension, ΔZ_f.

As discussed in Section 4, we can use $\Delta Z_f = T(n_0/c - 1/v_f)^{-1}$ and Eq. (1.1) to estimate ΔZ_f of the moving focus of the leading part of the pulse ($v_f > c/n_0$). For a 100-fs pulse and a response time of $T = 1$ fs for Δn, we find $\Delta Z_f \sim 5$ m at a focal distance of $z_f = 100$ m. Again, similar to the ns pulse case discussed earlier, if only light emitted from the focus at z_f was collected, one would obtain a pulse with a pulse width of the order of T. This emitted light pulse would be self-phase-modulated in time because of its traversal through the long Δn channel. The temporal phase modulation further extends spectral broadening of the emitted pulse and yields a white continuum light source, as has been observed [45, 46]. In a recent experiment, it was shown that fs light pulses from filaments had excellent mode quality and compression of the pulses by chirped mirrors led to significant shortening of the pulses [47, 48].

The moving foci model describes how different temporal sections of the input pulse self-focuses into foci at different points in space and time. This explains why a small disk blocking a filament would not be able to stop the reappearance of the filament down stream [49–51], while an aperture allowing a filament to thread through would terminate the filament downstream [51, 52]. Multiple filaments originating from modulation instability can affect one another if they compete with one another in drawing the surrounding beam toward their own foci [53].

Multiple self-focusing–defocusing–refocusing was predicted in the ns case [3], but has never been observed because stimulated light scattering depletes the focused beam energy and terminates the action. It is, however, observable in the fs case [42, 54]. Finally, conical emission associated with a filament has been observed in self-focusing of both ns and fs pulses [20, 55]. It is not yet well investigated and understood in the former case. One possible explanation of the conical emission is that this is Cherenkov radiation from polarization induced in the self-focusing medium by the forward-moving focus that has an apparent velocity faster than light. Another explanation is that the self-focused light passing through a long focus accumulates a spatial phase modulation from experiencing a radially dependent $\Delta n(\rho)$. If the phase difference of the beams at the center and at the edge is larger than 2π, then a ring structure in emission from a filament should appear as a result of constructive interference in the diffracted light [56]. The latter explanation is more plausible for the fs case [57]. The observed ring structure is colored with ring diameter increasing from red to blue, and suggests that spatial phase modulation resulting from plasma formation in the filament is responsible for the observation [57].

We note in passing that the above physical description of self-focusing should also apply to self-focusing of fs pulses in condensed matter, although the self-focusing threshold power, the magnitude and response time of Δn, and the filament formation are quantitatively different.

1.6 Conclusion

There are a great many qualitative similarities between self-focusing of light of different pulse durations. The general picture of self-focusing appears to be applicable to all cases. However, the quantitative details can be very different depending on the peak input power, the magnitude and response times of the induced refractive index change, and other nonlinear optical processes that come in to interfere with self-focusing. The last point is particularly relevant for the femtosecond case because of the very high intensity and long propagation length that have often incurred. It makes self-focusing of femtosecond pulses more complex than the other cases.

Acknowledgments This work was supported by the Director, Office of Energy Research, Office of Basic Energy Sciences, Materials Sciences Division of the U.S. Department of Energy under Contract No. DE-AC03-76SF00098.

References

1. S.A. Akhmanov, R. Khokhlov, A.P. Sukhorukov: Self-focusing, self-defocusing, and self-modulation of laser beams. In: *Laser Handbook*, F.T. Arecchi, E.O. Schulz-Dubois, (Eds.), North-Holland, Amsterdam, 1972, vol. 2, 1152–1228.
2. O. Svelto: Self-focusing, self-trapping, and self-phase modulation of laser beams, *Prog. Optics* **12**, 3–51 (1974).
3. J. H. Marburger: Self-focusing: theory, *Prog. Quant. Electron.* **4**, 35–110 (1975).
4. Y.R. Shen: Self-focusing: experimental, *Prog. Quant. Electron.* **4**, 1–34 (1975).
5. See other chapters on the subject in this volume.
6. G.A. Askar'yan: Effects of the gradient of a strong electromagnetic beam on electrons and atoms, *Soviet. Phys. JETP* **15**, 1088–1090 (1962).
7. M. Hercher: Laser-induced change in transparent media, *J Opt. Soc. Am.* **54**, 563 (1964).
8. R. Chaio, E.M. Garmire, C.H. Townes: Self-trapping of optical beams, *Phys. Rev. Lett.* 13, 479–482 (1964); erratum, ibid **14**, 1056 (1965).
9. V.I. Talanov: Self-focusing of electromagnetic waves in nonlinear media, *Radiophysics* **8**, 254–257 (1964).
10. E.J. Woodbury, W.K. Ng, *Proc. IRE* **50**, 2347 (1962).
11. R.W. Hellwarth: Theory of stimulated Raman scattering, *Phys. Rev.* **130**, 1850–1852 (1963).
12. M. Geller, D.P. Bortfeld, W.R. Sooy: New Woodbury–Raman laser materials, *Appl. Phys. Lett.* **3**, 36–40 (1963).
13. E. Eckhardt, D.P. Bortfeld, M. Geller: Stimulated emission of Stokes and anti-Stokes Raman lines from diamond, calcite, and -sulfur single crystals, *Appl. Phys. Lett.* 3, 137–138 (1963).
14. B.P. Stoicheff: Characteristics of stimulated Raman radiation generated by coherent light, *Phys. Lett.* **7**, 186–188 (1963).
15. R.W. Minck, R.W. Terhune, W.G. Rank: Laser-stimulated Raman effect and resonant four-photon interactions in gases H_2, D_2, and CH_4, *Appl. Phys. Lett.* **3**, 181–184 (1963).
16. V.I. Talanov: Self-focusing of wave beams in nonlinear media, *Radiophysics* **9**, 138–141 (1965).
17. P.L. Kelley: Self-focusing of optical beams, *Phys. Rev. Lett.* **15**, 1005–1008 (1965).
18. Y.R. Shen, Y.J. Shaham: Beam deterioration and stimulated Raman effect, *Phys Rev Lett* **15**, 1008–1010 (1965).
19. P. Lallemand, N. Bloembergen: Self-focusing of laser beams and stimulated Raman gain in liquids, *Phys. Rev. Lett.* **15**, 1010–1012 (1965).
20. E. Garmire, R.Y. Chiao, C.H. Townes: Dynamics and characteristics of the self-trapping of intense light beams, *Phys. Rev. Lett.* **16**, 347–349 (1966).
21. R.Y. Chiao, M.A. Johnson, S. Krinsky et al.: A new class of trapped light filaments, *IEEE J. Quant. Electron.* **QE-2**, 467–469 (1966).
22. V. Bespalov. V.I. Talanov: Filamentary structure of light beams in nonlinear liquids, *JETP Lett.* **3**, 307–309 (1966).
23. M.M.T. Loy, Y.R. Shen: Small-scale filaments in liquids and tracks of moving foci, *Phys. Rev. Lett.* **22**, 994–997 (1969).
24. J.H. Marburger, W.G. Wagner: Self-focusing as a pulse-sharpening mechanism, *IEEE J. Quant. Electron.* **3**, 415–416 (1967).
25. V.N. Lugovoi, A.M. Prokhorov: Possible explanation of small-scale filaments of self-focusing, *JETP Lett.* 7, 117–119 (1968).
26. M.M.T. Loy, Y.R. Shen: Experimental study of small-scale filaments of light in liquids, *Phys. Rev. Lett.* **25**, 1333–1336 (1970).
27. Y.R. Shen, M.M.T. Loy: Theoretical interpretation of small-scale filaments originating from moving focal spots, *Phys. Rev. A* **3**, 2099–2105 (1971).

28. G.M. Zverev, E.K. Maldutis, V.A. Pashkov: Development of self-focusing filaments in solid dielectrics, *JETP Lett.* **9**, 61–63 (1969).
29. V.V. Korobkin, A.M. Prokhorov, R.V. Serov et al.: Self-focusing filaments as a result of motion of focal points, *JETP Lett.* **11**, 94–96 (1970).
30. C.R. Giuliano, J.H. Marburger: Observations of moving self-foci in sapphire, *Phys. Rev. Lett.* 27, 905–908 (1971).
31. J.H. Marburger, E.L. Dawes: Dynamical formation of a small-scale filament, *Phys. Rev. Lett.* **21**, 556–558 (1968).
32. M. Maier, G. Wendl, W. Kaiser: Self-focusing of laser light and interaction with stimulated scattering processes, *Phys. Rev. Lett.* **24**, 352–355 (1970).
33. M.M.T. Loy, Y.R. Shen: Study of self-focusing and small-scale of filaments of light in nonlinear media, *IEEE J. Quant. Electron.* QE-9, 409–422 (1973).
34. G.K.L. Wong, Y.R. Shen: Study of spectral broadening in a filament of light, *Appl. Phys. Lett.* 21, 163–165 (1972).
35. J.A. Fleck, R.L. Carman: *Appl. Phys. Lett.* **20**, 290–293 (1972).
36. F. Shimizu: Numerical calculation of self-focusing and trapping of a short light pulse in Kerr liquids, *IBM J. Res. Develop.* **17**, 286–298 (1973).
37. E.G. Hanson, Y.R. Shen, G.K.L. Wong: Self-focusing: from transient to quasi-state state, *Opt. Commun.* **20**, 45–49 (1977); Experimental study of self-focusing in a liquid crystalline medium, *Appl. Phys.* **14**, 65–77 (1977).
38. R. Cubbedu, R. Polloni, C.A. Sacchi et al: Self-phase modulation and rocking of molecules in trapped filaments of light with picosecond pulses, *Phys. Rev. A* **2**, 1955–1963 (1970).
39. G.K.L. Wong, Y.R. Shen: Transient self-focusing in a nematic liquid crystal in the isotropic phase, *Phys. Rev. Lett.* 32, 527–530 (1974).
40. A. Brodeur. S.L. Chin: Ultrafast white-light continuum generation and self-focusing in transparent condensed media, *J. Opt. Soc. Am. B* **16**, 637–650 (1999).
41. A. Braun et al: Self-channeling of high-peak-power femtosecond laser pulses in air, *Opt. Lett.* **20**, 73–75 (1995).
42. A. Brodeur et al.: Moving focus in the propagation of powerful ultrashort laser pulses in air, *Opt. Lett.* **22**, 304–306 (1997).
43. J. Kasparan, R. Sauerbrey, S.L. Chin: The critical laser intensity of self-guided light filaments in air, *Appl. Phys. B* **71**, 877–879 (2000).
44. A. Becker et al.: Intensity clamping and refocusing of intense femtosecond laser pulses in nitrogen molecular gases, *Appl. Phys. B* **73**, 287–290 (2001).
45. J. Kasparian et al.: Infrared extension of the supercontinuum generated by femtosecond terawatt laser pulses propagating in the atmosphere, *Opt. Lett.* **25**, 1397–1399 (2000).
46. F. Theberge, W. Liu, Q. Luo et al.: Extension of ultrabroadband continuum generated in air up to 230 nm using ultrashort and intense laser pulse, *Appl. Phys. B* **80**, 221–225 (2005).
47. C.P. Hauri et al.: Generation of intense, carrier-envelope phase-locked few-cycle laser pulses through filamentation, *Appl. Phys. B* **79**, 673–677 (2004).
48. A. Couairon et al.: Self-compression of ultrashort laser pulses down to one optical cycle by filamentation, *J. Mod. Optics* **53**, 75–85 (2005).
49. A. Dubietis et al.: Self-reconstruction of light filaments, *Opt. Lett.* **29**, 2893–2894 (2004).
50. S. Skupin, L. Berge, U. Peschel et al.: Interaction of femtosecond light filaments with obscurants in aerosols, *Phys. Rev. Lett.* **93**, 023901 (2004).
51. W. Liu et al.: Experiment and simulations on the energy reservoir effect in femtosecond light filaments, *Opt. Lett.* **30**, 2602–2604 (2005).
52. E. Courvoisier et al.: Ultraintense light filaments transmitted through clouds, *Appl. Phys. Lett.* **83**, 213–225 (2003).
53. S.A. Hosseini et al.: Competition of multiple filaments during the propagation of intense femtosecond laser pulses, *Phys. Rev. A* **71**, 033802 (2004).

54. W. Liu et al.: Multiple refocusing of a femtosecond laser pulse in a dispersive liquid (methanol), *Opt. Commun.* **225**, 193–209 (2003).
55. E.T. Nibbering et al.: Conical emission from self-guided femtosecond pulses in air, *Opt. Lett.* **21**, 62–64 (1996).
56. S.D. Durbin, S. Arakelian, Y.R. Shen: Strong optical diffraction in a nematic liquid crystal with high nonlinearity, *Opt. Lett.* **7**, 145 (1982).
57. O.G. Kosareva et al.: Conical emission from laser plasma interactions in the filamentation of powerful ultrashort laser pulses in air, *Opt. Lett.* **22**, 1332 (1997).

Chapter 2
Notes on Early Self-Focusing Papers

John H. Marburger

Abstract These notes were written in August 30, 2006, about the history of two papers in *Physical Review* by I.W.G. Wagner, H.A. Haus, J.H. Marburger [1]; II. E.L. Dawes, J.H. Marburger [2] and the author's review paper of the year 1975 "Self-focusing: theory" [3], which is reprinted in the present book.

The identification by Chiao et al. [4, I.6*] of self-induced lensing as the likely origin of the then recently observed filamentary damage caused by intense laser propagation in solid materials raised many questions that piqued the interest of both experimenters and theorists. Of all these questions, the one that took longest to resolve was the mechanism responsible for the physical size of the observed small-scale damage tracks. Chiao et al. [4] demonstrated that the stationary self-trapping solutions of the nonlinear wave equation could occur at any scale, so some other mechanism had to be invoked to account for the observed tiny track diameters. Even after Lugovoi and Prokhorov [5] drew attention to the importance of the time dependence of the phenomenon, with consequences analyzed lucidly by Loy and Shen [6] and linked definitively to damage tracks by Giuliano and Marburger [7], the question remained open as to why the tracks have the size they do.

Unraveling this complex behavior was clearly going to require more powerful approximate methods for analyzing the nonlinear wave equation, and it was in this context that paper [1] was produced. Herman Haus, an MIT colleague of the Townes group, had already extended the analysis of Chiao et al. [4] to include stationary self-trapped solutions with radial nodes ([8, I.7]. In 1967, he began a collaboration with William G. Wagner, then in the process of moving from Hughes Research Laboratory in Malibu to the newly forming laser group

* The notation I.6 means reference 6 of paper I, etc.

J.H. Marburger (✉)
Office of Science and Technology Policy, Executive Office of the President,
725 17th Street, Room 5228, Washington, DC 20502, USA

R.W. Boyd et al. (eds.), *Self-focusing: Past and Present*,
Topics in Applied Physics 114, DOI 10.1007/978-0-387-34727-1_2,

at the University of Southern California. Wagner noticed the resemblance of the nonlinear equation for the phase of the electric field in the slowly varying envelope approximation (SVEA) to the Hamilton–Jacobi equation for a particle moving in a potential related to the nonlinear-optical susceptibility, and began preparing a manuscript exploiting this analogy.

I was intrigued by the analytical simplicity of the technique, and Wagner encouraged me to join the collaboration and add to the paper. This was all prior to the moving focus concept, and at the time we suspected the filament scale was related to the saturation of the nonlinear index at high fields, which is explored along with other phenomena in paper I [1]. During this period a number of papers appeared by physicists in the Soviet Union that also developed important approximate techniques for analyzing the nonlinear wave equation. Excellent reviews of this work by Akhmanov et al. appeared in 1968 and 1972 [9]. Many self-induced optical phenomena were first noted and analyzed in the Soviet literature, and contacts between Soviet and American workers accelerated the rate of progress in this field.

Years later the scale of filamentary damage was adequately explained by the rapid onset of electron production in the moving self-focal region where field strengths approached the breakdown threshold. This mechanism, recognized early on by Hellwarth [10] and thoroughly investigated by Yablonovitch and Bloembergen, e.g., [11], nevertheless competes with a host of other nonlinear inelastic processes in the focal region, making the phenomenon of "catastrophic" self-focusing extremely difficult to analyze for comparison with experiment. The reviews by Shen [12] and myself [3] in Volume 4 of *Progress in Quantum Electronics* (1975) summarize the status of the field at the time, and give many references.

The desire to understand the complex physical phenomena accompanying self-focusing was not the only motivation for theoretical studies. The simplest nonlinear medium response leading to self-focusing leads to a wave equation—the nonlinear Schrödinger equation—that appears in many other contexts, including superconductivity and model field theories. It seemed to me that in self-focusing nature has given us the opportunity to explore the consequences of this equation in a readily accessible experimental context. The effect of introducing additional dimensions—the transverse dimensions—on the qualitative properties of the solutions of nonlinear field equations is particularly striking. Plasma physicists had long wrestled with such phenomena in studies of instabilities that plagued magnetic confinement. This was in the days when numerical computation was expensive and limited in power. Probing the qualitative features of solutions of nonlinear partial differential equations required a combination of numerical and analytical techniques. Typically, numerical computation was used to validate and parametrize analytical models. In self-focusing, numerical computation was essential for overcoming the limitations of the "constant-shape" approximations that formed the starting point for analytical work such as in paper I.

Paul Kelley's important paper of 1965 [13, II.1] established the paradigm for this combined analytical/numerical approach. He introduced the concept of a self-focusing length and its importance for understanding the whole complex of nonlinear phenomena that occur near a self-focus. I particularly liked Kelley's use of a simple explicit algorithm for numerical solution of the nonlinear Schrödinger equation because its stability could be related to physically desirable features such as energy conservation. The more stable implicit algorithms permit computation through regions where the discretization grid is too coarse to capture details of the solution, a self-focus, for example.

Kelley's explicit scheme could not be forced through such regions without continual adjustment of grid sizes to follow the details. This enforced a kind of discipline on the computation that made it possible to identify artifacts in the solutions that could be attributed to the numerical algorithm. The USC dissertation of Eddie Lee Dawes, upon which paper II is based, employed this scheme to analyze (among many other things) the progress of transverse beam shapes as they propagated through typical nonlinear media. This permitted us to identify various interesting regimes of the self focusing process, and define critical powers relevant to each regime.

During these investigations we were impressed with the sensitivity of features of the numerical solutions to coding algorithms, and attempted to distinguish between properties of the underlying mathematical structure as opposed to artifacts of the computation. Some such artifacts may simulate real physical phenomena, as when the code loses energy near a self-focus due to insufficient resolution. Many real physical phenomena extract energy from the forward propagating beam in this region, so the "imperfect" code could be physically more realistic than the underlying lossless equation. This appears to be the case with early numerical solutions that exhibited multiple foci along the beam axis (Dyshko et al. [14]), a feature of self-focusing in gases easy to observe with today's high powered lasers, but not well-known or characterized experimentally at the time of the reviews in the early 70's [3, 9, 12].

The combination of cubic nonlinearity and paraxial approximation leads to true singularities at the self-focus. More complex media responses may be invoked to soften this singularity, but it is clear that the paraxial approximation itself breaks down there. Dawes and I, as well as others in our group, notably Dennis White, explored various improvements and modifications to the paraxial approximation, but we were unable to convince ourselves that we understood the correct behavior of the solutions of the full Maxwell equations with a cubic nonlinearity. The question was finally resolved by Joseph Fleck and colleagues who took full advantage of the rapid growth of computing power available at Los Alamos. Under the conditions of their simulation, a true singularity does not form (Feit and Fleck, [15]).

The approximate method of paper I [1] can be derived from a more general variational method employed in a paper that I wrote with Frank Felber on the nonlinear Fabry–Perot interferometer [16]. This paper appeared after my review of self-focusing theory [3], and should be viewed as providing supplementary material on approximate methods.

Erratum for paper I: The axes of Figures 3–5 are labeled incorrectly. The correct labeling is: Fig. 3, $\sqrt{(4/15)}\ \eta_m$ versus $\sqrt{(4/15)}\ \eta_o$; Fig 4, ξ_f/η_o^2 versus $(4/15)\eta_o^2$; Fig. 5, $(4/15)\eta^2$ versus ξ/η_o^2.

References**

1. W.G. Wagner, H.A. Haus, J.H. Marburger: Large scale self-trapping of optical beams in the paraxial ray approximation, *Phys. Rev.* **175**, No. 3, 256–266 (1968).
2. E.L. Dawes, J.H. Marburger: Computer studies in self-focusing, *Phys. Rev.* **179**, No. 3, 862–868 (1969).
3. J.H. Marburger: Self-focusing: theory, *Prog. Quant. Electr.* **4,** 35–110 (1975).
4. R.Y. Chiao, E. Garmire, C.H. Townes: Self-trapping of optical beams, *Phys. Rev. Lett.* **13**, 479–482 (1964).
5. V.N. Lugovoi, A.M. Prokhorov: A possible explanation of the small-scale self-focusing filaments, *JETP Lett.* **7**, 117–119 (1968).
6. M.M. Loy, Y.R. Shen: Small-scale filaments in liquids and tracks of moving foci, *Phys. Rev. Lett.* **22**, 994–997 (1969).
7. C.R. Giuliano, J.H. Marburger: Observation of moving self-foci in sapphire, *Phys. Rev. Lett.* **27**, 905–908 (1971).
8. H.A. Haus: Higher-order trapped light beam solutions, *Appl. Phys. Lett.* **8**, 128–129 (1966).
9. S.A. Akhmanov, R.V. Khokhlov, A.P. Sukhorukov: Self-focusing, self-defocusing, and self-modulation of laser beams. In: *Laser Handbook*, Vol. 2, E3, 1151–1228, F.T. Arecchi, E.O. Schulz-Dubois (Eds.), North-Holland, Amsterdam (1972).
10. R.W. Hellwarth, Role of photo-electrons in optical damage. In: *Damage in Laser Materials*, A.J. Glass, A.H. Guenther (Eds.), National Bureau of Standards Special Publ. 341, **2,** 67–75 (1970).
11. E. Yablonovitch, N. Bloembergen: Avalanche ionization and the limiting diameter of filaments induced by light pulses in transparent media, *Phys. Rev.Lett.* **29**, 907–910 (1972).
12. Y.R. Shen: Self-focusing: experimental, *Prog. Quant. Electr.* **4**, 1–34 (1975).
13. P.L. Kelley: Self-focusing of optical beams, *Phys. Rev. Lett.* **15**, 1005–1008 (1965).
14. A.L. Dyshko, V.N. Lugovoi, A.M. Prokhorov: *JETP Lett.* **6**, 146 (1967).
15. M.D. Feit, J.A. Fleck, Jr.: Beam nonparaxiality, filament formation, and beam breakup in the self-focusing of optical beams, *J. Opt. Soc. Am. B* **5**, 633–640 (1988).
16. J.H. Marburger, F.S. Felber: Theory of a lossless nonlinear Fabry–Perot interferometer, *Phys. Rev. A* **17**, 335 (1978).

** Papers 5–7, 9–12, 14–16 were not cited in I [1] or II [2].

Self-focusing: Theory

J.H. Marburger

Reprinted from Progress in Quantum Electronics, vol. 4, 35–110 (1975) with permission of Elsevier.

Prog. Quant. Electr., Vol. 4, pp. 35-110. Pergamon Press, 1975.

SELF-FOCUSING: THEORY

J. H. MARBURGER
University of Southern California, Los Angeles, California 90007

1. INTRODUCTION

THE THEORY of self-focusing is as interesting as its experimental manifestations because the simplest wave equation describing self-focusing is the prototype of an important class of nonlinear partial differential equations in physics. It closely resembles the Landau Ginsberg equation for the macroscopic wave function of type II superconductors,[1-3] and also Schrödinger's equation for a particle with self-interactions. The self-focusing equation also resembles that describing the gravitational collapse of a star, in a suitable approximation.[4] Its solutions possess a wide variety of structure, including shock waves and other more severe singularities,[5] "soliton" or stationary solutions,[6] and exceptional sensitivity to geometry and initial conditions.

Self-focusing is also the simplest of a class of nonlinear wave phenomena for which plane wave analysis fails to give essential qualitative information about the process. The very fact that the wave amplitude is *not* independent of transverse dimension gives rise to the enhancement of inhomogeneities which is characteristic of the effect. However, the same nonlinearity which causes self-focusing also gives rise to a variety of other phenomena which can be appreciated without recourse to a three-dimensional analysis. We shall discuss these phenomena in the following pages, but in less detail than self-focusing itself.

The emphasis throughout is on the behavior of the optical field, rather than on the nonlinear medium with which it interacts. The variety of mechanisms leading to a reactive nonlinear optical response, which causes self-focusing, all lead to essentially the same qualitative optical phenomena. In fact, it has proven difficult to untangle the various contributions to this response. We shall not deal with particular mechanisms in a very systematic way, but introduce them where necessary for the appreciation of some aspect of self-focusing. A recent review article by O. Svelto gives somewhat more information on microscopic mechanisms.[7]

Our discussion begins with a review of the equations which are normally used to describe self-focusing (Section 2), followed by an outline of the most important features of the nonlinear polarization density (Section 3). The next four Sections (4, 5, 6 and 7) deal with aspects of the "exact" solutions of these equations, both analytical and numerical. Section 8 includes the development of the most important approximate analytical treatments of self-focusing, and their relation to the exact solutions. The last two Sections (9 and 10) include discussions of transient self-focusing and the interplay between self-focusing and other effects, including breakdown and stimulated scattering. These are necessarily incomplete, as detailed studies have not yet been undertaken into all the important aspects of this class of problems. No material on the important phenomena of thermal self-focusing is included. These phenomena, because of their long characteristic time scales, require special techniques for their analysis which would have taken up too much space. Akhmanov *et al.* have reviewed many aspects of thermal self-focusing in ref. 8.

Our presentation is decidedly unhistorical, and only a few remarks regarding the development of the subject will be found in the text. The recent informative review articles by Akhmanov *et al.*[8] and by Svelto[7] may be consulted for more representative views of the self-focusing literature. Here we wish to save the reader some of the pain attendent upon learning self-focusing theory from the literature. We have tried in many instances to include enough detail to make derivations and approximations comprehensible without recourse to the original papers. Indeed some of the material does not appear anywhere else. For the new material we have drawn heavily on unpublished manuscripts and numerical computations prepared by the USC nonlinear optics group during the past 7 years.

Much of the confusion and controversy surrounding self-focusing in the past arose because of inadequate appreciation of the variety of behavior allowed by the theory. Only as this appreciation grew did our language become sufficiently precise to allow unambiguous description of both experimental and theoretical results. It is noteworthy that consecutive advances in self-focusing theory followed a natural mathematical scheme: the earliest paper dealt with self-trapping (one-dimensional) solutions of the basic equation.[9] These were followed by stationary self-focusing studies[10,11] (two dimensions) and then by theories of time dependent but non-transient phenomena[12,13] (two dimensions, time included parametrically). Finally, fully transient analyses (two dimensions plus time) were undertaken.[14,15] The most recent theoretical investigations attempt to include the influence on self-focusing of other phenomena,[16] and detailed applications to systems of technical importance.[17] (References cited here are representative, not exhaustive.)

I am grateful to my colleagues in the USC quantum electronics group, who provided a stimulating context for the preparation of this work. Professor W. G. Wagner, in particular, provided crucial inspiration and advice during my initial exposure to this field. My ex-students, E. L. Dawes and D. R. White, amassed such a wealth of numerical data on self-focusing solutions that years later it still rewards scrutiny.

Research reported here for the first time was sponsored by the Joint Services Electronics Program through the Air Force Office of Scientific Research/AFSC under Contract F 44620-71-C-0067.

2. THE BASIC PROPAGATION EQUATION

The optical electromagnetic field satisfies the vector wave equation implied by Maxwell's equations. In gaussian units, we have

$$\nabla \times \nabla \times \boldsymbol{E} + \frac{1}{c^2}\frac{\partial^2 \boldsymbol{E}}{\partial t^2} = -\frac{4\pi}{c^2}\frac{\partial^2 \boldsymbol{P}}{\partial t^2}. \tag{2.1}$$

The polarization density has a linear and a nonlinear part,

$$\boldsymbol{P} = \boldsymbol{P}^{\mathrm{L}} + \boldsymbol{P}^{\mathrm{NL}} \tag{2.2}$$

where

$$\boldsymbol{P}^{\mathrm{L}}(\boldsymbol{x}, t) = \int_{-\infty}^{t} \hat{\chi}(t - t')\cdot \boldsymbol{E}(\boldsymbol{x}, t')\,\mathrm{d}t'. \tag{2.3}$$

This expression is valid for a homogeneous, optically inactive, stationary medium. Causality demands that $\hat{\chi}$ vanish for negative arguments. The Fourier transform of $\hat{\chi}(t)$ is the usual frequency dependent susceptibility:

$$\boldsymbol{P}^{\mathrm{L}}(\boldsymbol{x}, \omega) = \chi(\omega)\cdot \boldsymbol{E}(\boldsymbol{x}, \omega). \tag{2.4}$$

If the reader introduces $\tau = t - t'$ in (2.3) and uses

$$\boldsymbol{E}(\boldsymbol{x}, t - \tau) = \left[\exp \mathrm{i}\left(\mathrm{i}\frac{\partial}{\partial t}\right)\tau\right]\boldsymbol{E}(\boldsymbol{x}, t),$$

he will find it possible to rewrite (2.3) as

$$\boldsymbol{P}^{\mathrm{L}}(\boldsymbol{x}, t) = \chi\left(\mathrm{i}\frac{\partial}{\partial t}\right)\cdot \boldsymbol{E}(\boldsymbol{x}, t). \tag{2.5}$$

We are interested in nearly monochromatic beams of light, for which the amplitude $\mathscr{E}$ in

$$\boldsymbol{E} = \mathrm{Re}\,\mathscr{E}(\boldsymbol{x}, t)\mathrm{e}^{-\mathrm{i}\omega_0 t}\mathrm{e}^{\mathrm{i}k_0 z} \tag{2.6}$$

varies little in a period $2\pi/\omega_0$. Using similar notation for the polarization density we find, from (2.5),

$$\mathscr{P}^{\mathrm{L}}(\boldsymbol{x}, t) = \chi\left(\omega_0 + \mathrm{i}\frac{\partial}{\partial t}\right)\cdot \mathscr{E}(\boldsymbol{x}, t). \tag{2.7}$$

The series which results from expansion of χ in this equation about ω_0 is supposed to converge rapidly because $\mathscr{E}$ depends weakly upon t. (We assume that ω_0 is far from a resonance of χ.) Nevertheless, the term in $\chi'(\omega_0)$ must be retained in this expansion to recover the correct group velocity with which the envelope $\mathscr{E}$ propagates. The next term in $\chi''(\omega_0)$, leads to pulse spreading in a dispersive medium.

Omitting derivatives of $\mathscr{E}$ with respect to t of third order and higher, we find the following wave equation for the field amplitude:

$$\nabla^2\mathscr{E} - \frac{G}{v_g^2}\frac{\partial^2\mathscr{E}}{\partial t^2} + 2\mathrm{i}k_0\left[\frac{\partial}{\partial z} + \frac{1}{v_g}\frac{\partial}{\partial t}\right]\mathscr{E}$$
$$- \nabla(\nabla\cdot\mathscr{E}) - \mathrm{i}k_0\nabla\mathscr{E}_z - \hat{z}\mathrm{i}k_0(\nabla\cdot\mathscr{E} + \mathrm{i}k_0\mathscr{E}_z)$$
$$= -\frac{4\pi}{c^2}\left(\omega_0 + \mathrm{i}\frac{\partial}{\partial t}\right)^2\mathscr{P}^{\mathrm{NL}}. \tag{2.8}$$

The terms on the second line all arise from the small quantity $\nabla(\nabla\cdot\boldsymbol{E})$. The quantity G is defined in terms of the group velocity v_g by

$$G = 1 - k_0\frac{\partial v_g}{\partial\omega}.$$

We have taken $c^2k_0^2 = [1 + 4\pi\,\chi(\omega_0)]\omega_0^2$, where the quantity in brackets is the squared linear refractive index n at ω_0. Equation (2.8) is not hard to derive using (2.7) and

$$\frac{1}{v_g} = \frac{\mathrm{d}}{\mathrm{d}\omega}\cdot\frac{n\omega}{c}.$$

We have assumed that $\chi(\omega)$ is real here (no linear loss). Imaginary contributions may be included in P^{NL}.

Most studies of self-focusing ignore (a) the terms in the second line of (2.8) and (b) derivatives of the slowly varying amplitudes with respect to z and t of second order and higher. The resulting approximate form is

$$2\mathrm{i}k_0\left(\frac{\partial}{\partial z} + \frac{1}{v_g}\frac{\partial}{\partial t}\right)\mathscr{E} + \nabla_{\mathrm{T}}^2\mathscr{E} = -\frac{4\pi}{c^2}\omega_0^2\mathscr{P}^{\mathrm{NL}} - \frac{8\pi\mathrm{i}\omega_0}{c^2}\frac{\partial\mathscr{P}_{\mathrm{NL}}}{\partial t} \tag{2.9}$$

where $\nabla_{\mathrm{T}}^2 = \partial^2/\partial x^2 + \partial^2/\partial y^2$. The higher derivatives are small if

$$\left|\frac{\partial^2\mathscr{E}}{\partial z^2}\right| \ll k_0\left|\frac{\partial\mathscr{E}}{\partial z}\right|, \qquad \left|\frac{\partial^2\mathscr{E}}{\partial t^2}\right| \ll \omega_0\left|\frac{\partial\mathscr{E}}{\partial t}\right| \tag{2.10}$$

which simply requires the envelope function $\mathscr{E}$ to vary little in a wavelength or an optical period. Some care is required, however, in applying this approximation. It is possible to obtain unphysical results if the derivatives with respect to z and t are not treated with some degree of symmetry. For example, the reader may verify that even in the absence of any dispersion, the solutions of

$$\nabla^2 A + \frac{1}{c^2}\left(\omega_0^2 + 2\mathrm{i}\omega_0\frac{\partial}{\partial t}\right)A = 0$$

imply that a pulse will spread out along its direction of propagation, which is not correct for optical pulses. This equation is obtained from the scalar wave equation by taking $E = A\,\mathrm{e}^{-\mathrm{i}\omega_0 t}$ and discarding second derivatives in t.

To minimize unphysical dispersive effects, one may replace

$$\frac{\partial^2}{\partial z^2} - \frac{1}{c^2}\frac{\partial^2}{\partial t^2}[1 + 4\pi\chi(\mathrm{i}\partial/\partial t)] \tag{2.11}$$

in (2.1) by L_-L_+, where

$$L_\pm = \pm\mathrm{i}\frac{\partial}{\partial z} + k(\mathrm{i}\partial/\partial t) \tag{2.12}$$

are generalized directional derivatives in local frames moving toward increasing (+) or decreasing (−) z values, and k is the square root of the second term of (2.11). Upon introducing slowly varying envelopes, this product contributes the terms

$$(2k_0 + \mathscr{L}_-)\mathscr{L}_+\mathscr{E} \equiv \left[2k_0 - \mathrm{i}\frac{\partial}{\partial z} + \Delta k(\omega_0 + \mathrm{i}\partial/\partial t)\right] \times \left[\mathrm{i}\frac{\partial}{\partial z} + \Delta k(\omega_0 + \mathrm{i}\partial/\partial t)\right]\mathscr{E}$$

where $\Delta k \equiv k(\omega_0 + \mathrm{i}\partial/\partial t) - k_0$. The operators $\mathscr{L}_\pm$ play the same role with respect to $\mathscr{E}$ as $L_\pm$ do with E. Since reflection of signals into the backward direction is expected to be small, higher powers of $\mathscr{L}_-$ can be ignored in any term. This approach discards the higher derivatives in t and z in a consistent way.

Employing this notation in eqn. (2.8) (omitting terms on the second line) one finds

$$(2k_0 + \mathscr{L}_-)\mathscr{L}_+\mathscr{E} + \nabla_T^2\mathscr{E} = -\frac{4\pi}{c^2}(\omega_0 + \mathrm{i}\partial/\partial t)^2\mathscr{P}^{\mathrm{NL}} \tag{2.13}$$

which reduces to

$$2\mathrm{i}k_0\left(\frac{\partial}{\partial z} + \frac{1}{v_g}\frac{\partial}{\partial t}\right)\mathscr{E} + \nabla_T^2\mathscr{E} = -\frac{k_0}{v_g}\frac{\partial v_g}{\partial \omega}\frac{\partial^2\mathscr{E}}{\partial t^2} - \frac{4\pi}{c^2}\left(\omega_0 + \mathrm{i}\frac{\partial}{\partial t}\right)^2\mathscr{P}^{\mathrm{NL}}.$$

To generate correction terms to this lowest approximation, Gustafson *et al.*[18] suggest multiplying (2.13) from the left by $(2k_0 + \mathscr{L}_-)^{-1}$ and expanding in powers of the small operator $\mathscr{L}_-$. This procedure leads to eqn. (1.10a) in their work.

3. NONLINEAR POLARIZATIONS FOR SELF-FOCUSING

A. *Simplest forms*

Self-focusing phenomena arise from the terms in P^{NL} that oscillate near the dominant incident frequency ω_0. The component of P^{NL} in phase with E causes a nonlinear change in refractive index, while the out of phase component leads to nonlinear gain or loss. In the most general case, there is no convenient expression like (2.3) relating P^{NL} to E because of the potentially complicated functional relation between the two. The exact response must be obtained by solving the set of coupled nonlinear equations which describe the dynamics of the polarizable medium. This is actually feasible for the case in which the medium is an assembly of noninteracting two level systems.[19] Unfortunately, it is rarely possible to determine P^{NL} so exactly, and approximate model equations are usually devised for particular situations.

The simplest of all forms of $\mathscr{P}^{\mathrm{NL}}$ which give self-focusing for a single linearly polarized monochromatic beam is

$$\mathscr{P}_j^{\mathrm{NL}} = \eta|\mathscr{E}|^2\mathscr{E}_j, \tag{3.1}$$

where j is a cartesian subscript. This is a special case of the general third order nonlinear polarization

$$\mathscr{P}_j^{\mathrm{NL}}(\omega_4) = D\chi_3^{jklm}(-\omega_4, \omega_1, \omega_2, \omega_3)\mathscr{E}_k(\omega_1)\mathscr{E}_l(\omega_2)\mathscr{E}_m(\omega_3) \tag{3.2}$$

which has been well studied by Maker and Terhune.[20] (For an informative recent discussion, see Owyoung's 1972 dissertation.[21]) Here D is 1, 3 or 6 if, respectively, three, two or none of the incident frequencies ω_1, ω_2, ω_3 are identical. Thus our η in (3.1) is

$$\eta = 3\chi_3^{jjjj}(-\omega, \omega, -\omega, \omega). \tag{3.3}$$

The index change for such a nonlinearity is

$$\delta n = \frac{2\pi\eta}{n}|\mathscr{E}|^2$$

$$= \frac{16\pi^2}{n^2 c}\eta I \tag{3.4}$$

where $I = nc|\mathscr{E}|^2/8\pi$ is the optical intensity in the medium, ignoring corrections for nonlinearity. Table 1 lists a variety of conventions for the coefficient of $|\mathscr{E}|^2$ in (3.4). Throughout this paper, we shall employ n_2 or ϵ_2 defined via

$$\delta n = \tfrac{1}{2}n_2|\mathscr{E}|^2, \quad (n_2 = 4\pi\eta/n) \tag{3.5}$$

$$= \frac{1}{4n}\epsilon_2|\mathscr{E}|^2, \quad (\epsilon_2 = 8\pi\eta). \tag{3.6}$$

Because χ_3 is a fourth-rank tensor, the nonlinear response is sensitive to the polarization of the incident wave. For a wave of arbitrary elliptical polarization incident on an isotropic medium, such as a liquid, the nonlinear polarization amplitude may be written as

$$\mathscr{P}^{\mathrm{NL}} = A\mathscr{E}(\mathscr{E}^* \cdot \mathscr{E}) + \tfrac{1}{2}B\mathscr{E}^*(\mathscr{E} \cdot \mathscr{E}) \tag{3.7}$$

using notation introduced by Maker and Terhune.[20] The state of polarization is more easily determined from the amplitudes and phases of the circular components

$$\mathscr{E}_\pm = \frac{1}{\sqrt{2}}(\mathscr{E}_x \pm i\mathscr{E}_y) \tag{3.8}$$

for a beam polarized in the x, y plane. The corresponding polarization components are

$$\mathscr{P}^{\mathrm{NL}}_\pm = (A|\mathscr{E}_\pm|^2 + (A + B)|\mathscr{E}_\mp|^2)\mathscr{E}_\pm. \tag{3.9}$$

Notice that in this basis the response is always in phase with the field, so that the power in either component is conserved separately.

In media of more general symmetry, it is necessary to use (3.2) and the tensor χ_3^{ijkl} appropriate for the specific symmetry. Tabulations of nonvanishing elements of these tensors for every point symmetry group are given by Maker and Terhune[20] and also by Birss.[22] Owyoung[23] gives expressions for δn for cubic media.

B. *Dispersive effects and transiency*

These simple expressions for $\mathscr{P}^{\mathrm{NL}}$ do not include dispersive effects, which can be important if the envelope function varies rapidly during the response time of the nonlinear medium. A lag in response can arise from either inertial or diffusive retardations. Dispersion in contributions to χ_3 from electronic motion usually comes from inertial effects. Another source of inertial retardation is the lattice or fluid motion associated with electrostriction. An example of diffusive retardation is the redirection of the mean orientation

TABLE 1. Notation for K in $\delta n = K|\mathscr{E}|^2$, where $\mathscr{E}$ is the peak amplitude of a linearly polarized field. Equation numbers are for cited references

Author (reference)	K
This work	$\frac{1}{2}n_2 = \epsilon_2/4n_0 = 2\pi\eta/n_0$
Chiao *et al.*[9]	n_2 in eqn. (1)
Chiao *et al.*[9]	$\epsilon_2/2n_0$ in eqn. (3)
Chiao *et al.*[9]	$\epsilon_2/4n_0$ subsequent eqns.
Talanov[10]	$\epsilon'/8n_0$
Kelley[11]	$n'_2, \epsilon'_2/2n_0$
Akhmanov *et al.*[8]	$n_2, \epsilon_2/2n_0$
Shen[55]	$\frac{1}{2}K_0\lambda_0, \frac{1}{3}K_s\lambda_0$
Hellwarth[34]	$2\pi\eta/n_0$
Dawes and Marburger[47]	$\epsilon_2/4n_0$
C. C. Wang[110]	$6\pi[\chi_3^{1122} + \chi_3^{1212} + \chi_3^{1221}]/n_0 = 6\pi\chi_3^{1111}/n_0 = \pi(2A + B)/n_0$
Wagner *et al.*[46]	$\epsilon'_2/2n_0$ (mks units)

of polarizable anisotropic molecules in a viscous fluid with an applied field. With each such effect is associated a characteristic response time τ during which the polarization "remembers" the previously applied field. In thermally induced effects (diffusive retardation) the response time is very long and the polarization may be a function of the time-integrated intensity passing through the medium.

Analyses of self-focusing have been carried out for both kinds of retardation, although most attention has been given to diffusive response. A model equation universally employed for this case allows the nonlinear index change to decay with characteristic time τ in the absence of any applied field:

$$\tau \frac{\partial \delta n}{\partial t} = \tfrac{1}{2} n_2 |\mathscr{E}|^2 - \delta n. \tag{3.10}$$

The coefficients τ and n_2 can be obtained from a calculation based on a particular microscopic model. For example, eqn. (3.10) can be derived directly from the orientational diffusion equation for anisotropic molecules in a liquid, driven by an applied field. (A detailed derivation is included in ref. 21.) In this model, τ is the rotational relaxation time, which is related to the viscosity of the liquid, η, by

$$\tau \approx \frac{4\pi\eta a^3}{6k_B T}$$

where a is a molecular radius, T the temperature, and k_B is Boltzmann's constant. The precise expression for n_2 in this model depends upon one's choice of local field theory, since the field which exerts torque on an individual molecule is not the macroscopic field. Neither is it the Lorentz local field, as first pointed out by Onsager, since that field includes contributions which turn with the molecule. Nevertheless, a variety of self-focusing calculations have been carried out using the Lorentz local field theory, in varying approximations, for molecular reorientation. In this theory[24]

$$n_2 = \frac{2\pi\rho}{45k_B T}\left(\frac{n_0^2 + 2}{3}\right)^4 \cdot \frac{1}{n_0}[(\alpha_1 - \alpha_2)^2 + \cdots]$$

where ρ is the molecular number density, and $\alpha_1, \alpha_2, \alpha_3$ are the principal values of the molecular polarizability tensor. The dots mean cyclic permutations of the first term.

Equation (3.10) can be integrated if $|\mathscr{E}|^2$ is regarded as a fixed function of time. Then

$$\delta n(t) = \frac{1}{\tau}\int_{-\infty}^{t} \tfrac{1}{2} n_2 |\mathscr{E}|^2 [\exp(t' - t)/\tau]\, dt'. \tag{3.11}$$

If $|\mathscr{E}|^2$ is the parabolic pulse,

$$|\mathscr{E}|^2 = \mathscr{E}_0^2(1 - t^2/T^2)\theta(1 - t^2/T^2) \tag{3.12}$$

(here $\theta(t) = 1$ for $t \geq 0$, $\theta(t) = 0$ for $t < 0$), then for $t^2 \leq T^2$,

$$\delta n(t) = \tfrac{1}{2} n_2 \left\{ |\mathscr{E}(t)|^2 + \frac{2\tau^2 \mathscr{E}_0^2}{T^2}\left[\left(\frac{t}{\tau} - 1\right) + \left(\frac{T}{\tau} + 1\right) e^{-(T+t)/\tau}\right]\right\}, \tag{3.13}$$

and for $t > T$,

$$\delta n(t) = \tfrac{1}{2} n_2 \mathscr{E}_0^2 \cdot \frac{2\tau^2}{T^2} \cdot e^{-t/\tau}\left[e^{T/\tau}\left(\frac{T}{\tau} - 1\right) + e^{-T/\tau}\left(\frac{T}{\tau} + 1\right)\right].$$

It is useful to have some intuition regarding the properties of this solution. The reader may easily verify that the maximum value of this function is less than $\delta n_0 \equiv \frac{1}{2} n_2 \mathscr{E}_0^2$ by the amount

$$1 - \frac{\delta n_{\max}}{\delta n_0} = \left(\frac{t_m}{T}\right)^2, \tag{3.14}$$

where t_m is the time at which $\delta n(t)$ reaches its peak. The value of t_m can be obtained from Fig. 1 by following the prescription given in the figure caption. For long pulsewidths

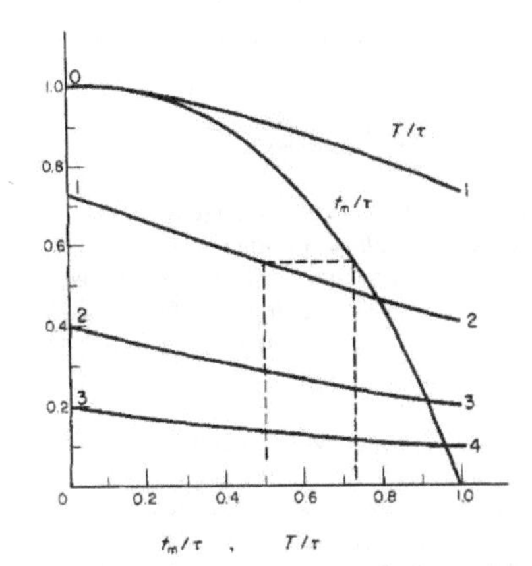

FIG. 1. Graphical solution for time lag t_m between the peak of a parabolic driving pulse and the peak of the nonlinear index change. The curve labeled t_m/τ is a plot of $(1 - t_m/\tau)\exp(t_m/\tau)$ vs t_m/τ. According to eqn. (3.13) this must equal $(1 + T/\tau)\exp(-I/\tau)$, which is the curve labeled T/τ. (This curve is folded to save space. The segments shown cover the ranges $0 \leq T/\tau \leq 1$, $1 \leq T/\tau \leq 2$, etc.) For example, if the ratio of pulse length to relaxation time is $T/\tau = 1.5$, then $t_m/\tau = 0.73$. The maximum pulse delay of $t_m = \tau$ is approached for pulses $T \gtrsim 4\tau$.

$T \gg \tau$, t_m approaches its maximum of τ, and the fractional reduction in response is nearly $(\tau/T)^2$.

An important feature of $\delta n(t)$ for retarded response is its persistence after the field is turned off. Equation (3.13) shows that at the end of the pulse ($t = T$), δn may be appreciable fraction of its maximum. For $T = \tau$, $\delta n(T)/\delta n_{max} = 0.541$. From this residual value, δn decays exponentially with characteristic time τ. Shen and co-workers[25] have emphasized the importance of this lingering index change for self-focusing of the trailing portions of pulses which ordinarily do not possess the intensity to induce their own index change. The importance of such effects, however, decreases as the ratio τ/T decreases. Figure 2 shows δn for $T = \tau$.

A feature of diffusive responses which is not shared by inertial retardation is the appearance of the maximum response during the driving pulse. A system with inertia can attain its maximum polarization after the signal has been turned off. Inertial systems can also exhibit ringing or oscillatory response (often called "coherency" effects in quantum mechanical treatments).

E. L. Kerr has investigated an important inertial self-focusing mechanism in his dissertation on electrostrictively driven self-focusing.[26] In this case the index change obeys

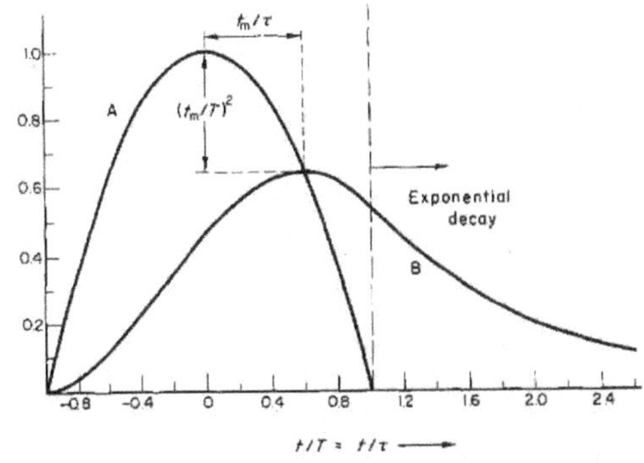

FIG. 2. (A) Parabolic pulse whose duration T equals the response time τ. (B) The corresponding response $\delta n(t)$ normalized to its maximum value $\frac{1}{2}n_2\epsilon_0^2$ in the absence of relaxation.

a wave equation easily derivable from the equation governing the propagation of compression waves in an elastic medium:

$$\left(\nabla^2 - \frac{1}{v_s^2}\frac{\partial^2}{\partial t^2}\right)\delta n = \frac{\rho\alpha^2}{cv_s^2}\nabla^2 I. \tag{3.15}$$

Here v_s is the compression wave velocity, ρ the density, $\alpha = (\partial n/\partial\rho)_s$, and I is the optical intensity in the medium. Clearly a pulsed inhomogeneous signal I will create an index wave which will propagate (in this approximation) with neither dispersion nor attenuation. Kerr develops exact analytical solutions of eqn. (3.15) for $\delta n(t)$ on the symmetry axis of a beam with gaussian transverse cross-section and with a variety of pulse shapes, including parabolic. For the results, which involve integrals of error functions of imaginary argument, we refer the reader to the original papers.[27]

J. Fleck has developed a computer code which solves the slowly varying envelope equation simultaneously with "Bloch" equations describing the driven polarization and population difference in a system of noninteractive resonant two level atoms.[28] The wave propagation equation also includes an off-resonance instantly responding self-focusing term. The coherent response of the two level systems combined with the background nonlinear index gives rise to a variety of interesting inertial transient effects. Fleck and Lane have used this code to simulate the operation of a short pulse Nd glass laser system.[17]

The reader should appreciate that the decision to use a transient or nontransient model for δn cannot be made simply by comparing the input pulse width with the response time of the assumed nonlinear mechanism. One must compare the pulse width *within the nonlinear medium* with the response time. Self-focusing[13] and self-steepening[5] (phase pileup) can cause severe pulse sharpening within the medium even for incident pulses long compared with the transient time τ.

C. *Saturation of the nonlinear polarization*

In the presence of the intense fields generated by the self-focusing process, one expects that higher-order nonlinear contributions to the polarization will become important. We shall call all such higher-order contributions *saturation terms* since they generally limit the magnitude of the response. The origin of saturation is easily perceived in most microscopic models of nonlinear response. In the molecular reorientation model, it arises because once the molecules are aligned with their "easy" axes of polarizability along the field, no increase of field can further increase the effective polarizability. No study has yet been undertaken of self-focusing with a saturable nonlinearity in the transient regime. Here we shall examine several important nontransient saturation mechanisms.

The induced mean dipole moment of uniaxial molecules in a dilute gas may be obtained from the partition function Z if the system is in thermodynamic equilibrium in the presence of the field:

$$\langle m \rangle = k_B T \frac{\partial \ln Z}{\partial \mathscr{E}}.$$

We are assuming that the field is linearly polarized, so the mean moment is aligned along the field, and vector signs are unnecessary. If the energy of the molecule in the field is written

$$u = -\tfrac{1}{2}(\alpha_1 \sin^2\theta + \alpha_2 \cos^2\theta)\cdot\mathscr{E}^2_{\text{rms}},$$

then the partition function may be evaluated exactly in terms of error functions of real or imaginary argument:

$$Z \equiv \int_{-1}^{1} \mathrm{d}(\cos\theta)\exp - u/k_B T.$$

The full expressions may be found in ref. 29. The final result for the mean moment may be written $m(\mathscr{E})$, and appears in the expression for the refractive index as

$$n^2(\mathscr{E}) = 1 + 4\pi\rho m(F)/\mathscr{E}.$$

Here the argument of m has been replaced by the local field F, which according to the Lorentz local field theory is

$$F = \frac{n^2 + 2}{3}\mathscr{E}.$$

Using F as a parameter, we may write

$$\mathscr{E} = F - \tfrac{4}{3}\pi\rho m(F),$$

$$n^2 = \frac{3F + 8\pi\rho m(F)}{3F - 4\pi\rho m(F)}$$

which gives n^2 parametrically as a function of the macroscopic field $\mathscr{E}$, if the function $m(F)$ is known. Figure 3 shows how the nonlinear part of n depends upon the field, both with and without local field corrections. The point is that local field corrections make a very big difference in the saturation properties of the nonlinear index.[30]

Appreciating this sensitivity to local fields, Gustafson and Townes[31] have attempted to refine the theory to include the possibility of mutual steric interference among neighboring molecules as they reorient under a strong field. That is, the molecules, being nonspherical, get in each other's way as they attempt to rotate. For short intense driving pulses, the resulting hindrance of the rotation, and the accompanying local shear stresses, may not have time to relax during the pulse. The result is effective saturation of the nonlinear index at field strengths far below those required to align unhindered molecules. Gustafson and Townes estimate a decrease in the saturated value of δn by a factor of one-tenth for CS_2. If the shear elasticity of the medium on a short time scale were included in this analysis, it would resemble the librational or "rocking" theory advocated by Polloni *et al.*[32] (see also ref. 33). In this theory the hindered molecule executes an angular vibrational (librational) motion about its equilibrium orientation in the presence of the field. The resulting dynamical equation for δn contains second derivatives with respect to time which lead to inertial effects in the transient response. Frenkel gives a good exposition of the elementary theory of molecular rocking.[33] Hellwarth has also attempted to improve on the Lorentz local field theory using a statistical mechanics approach similar to that developed by Kirkwood for the linear dielectric properties of liquids.[34] This theory includes the effect of field-induced particle bunching, which influences the lowest order nonlinear index, but is difficult to extend to higher orders.

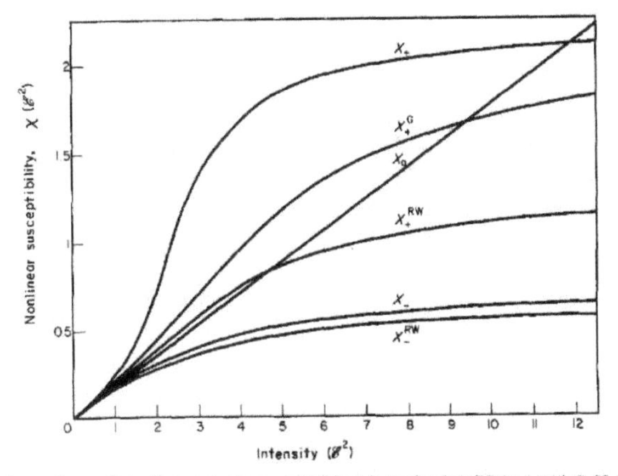

FIG. 3. Dependence of nonlinear response on incident intensity for $CS_2(+)$ and $C_6H_6(-)$, and ignoring saturation effects altogether (0). Superscripts refer to various treatments of local field effects: G (30), RW (41), None (29). The unit of ϵ^2 is $2k_BT/|\alpha_\parallel - \alpha_\perp|$, and that of X is $\rho|\alpha_\parallel - \alpha_\perp|(n^2 + 2)^2/18$. Notice the initial "antisaturation" behavior of the prolate molecule CS_2. (Figure from ref. 29.)

The nonlinear response theory of dense liquids of anisotropic molecules is evidently very difficult and by no means completely developed. But the nonlinear response of gases is not complicated by local field effects, and in some cases can be made large enough by resonant enhancement to be readily detectable. Grischkowsky has observed self-focusing and self-defocusing near electronic resonances in alkali metal vapors.[19,35] The general response theory for two levels in near-resonance with an incident harmonic field is treated in many texts. A useful review by Grischkowsky is also available.[36]

4. SELF-TRAPPING

A. *Scalar self-trapping, no saturation*

Now that we have discussed the origin of the equations for the field and for the nonlinear medium, let us examine the simplest properties of their solutions. First we shall study the solutions of eqn. (2.9) with time independent initial conditions. That is, all time dependence will be assumed to be included in the factor exp $-i\omega_0 t$ of eqn. (2.6). In this section, we discuss solutions whose z dependence is also simply periodic, although not necessarily with wave vector k_0 because of the nonlinear index change.

For the moment, we shall ignore the vector character of $\mathscr{E}$ and write

$$\mathscr{E}(x, y, z) = F(x, y) \exp iqz, \tag{4.1}$$

which gives for eqn. (2.9),

$$-2k_0 qF + \nabla_T^2 F = -\frac{4\pi}{c^2}\omega_0^2 \mathscr{P}^{NL}(F). \tag{4.2}$$

Notice that F can be chosen real if $\mathscr{P}^{NL}$ is real, because the surfaces of constant phase have no curvature. This nonlinear Helmholtz equation has solutions for a limited number of transverse symmetries, if $\mathscr{P}^{NL}$ is real. For example, solutions exist whose constant intensity contours are concentric circles or parallel lines, but there are no solutions for concentric elliptical contour lines.[37] The important circular and slab shaped symmetries were investigated by Chiao *et al.* in their landmark paper on optical self-focusing.[9] At that time these "constant-width", self-trapped solutions were thought to describe the optical field which caused filamentary optical damage in solid materials. Although the actual situation is much more complicated, the intuition that self-focusing is the key to understanding such damage has been fully vindicated.[38]

For slab symmetry, F depends only on a single cartesian coordinate, and satisfies

$$\frac{\partial^2 F}{\partial x^2} = 2k_0 qF - \frac{4\pi\omega_0^2}{c^2}\mathscr{P}^{NL}(F). \tag{4.3}$$

As pointed out by Haus,[39] this resembles Newton's law for the position F of a particle of unit mass moving in the conservative potential

$$U(F) = -k_0 qF^2 + \frac{4\pi\omega_0^2}{c^2}\int^F dF \mathscr{P}^{NL}(F).$$

The coordinate x plays the role of time. If $\mathscr{P}^{NL} = \eta F^3$, then the energy integral leads to a simple solution:

$$x = \int_{F_0}^{F} \frac{dF}{\sqrt{2(E-U)}}$$

$$= (2k_0 q)^{-1/2}\,\text{sech}^{-1}(F/A).$$

Here the total "energy" is zero (because $F = dF/dx = 0$ at $x = \infty$), and $F(0) = A$, where $A^2 = k_0 qc^2/\pi\omega_0^2\eta$. These conditions fix the propagation constant in terms of the axial field:

$$q = \pi\omega_0\eta F^2(0)/nc = \omega_0 n_2 F^2(0)/4c. \tag{4.4}$$

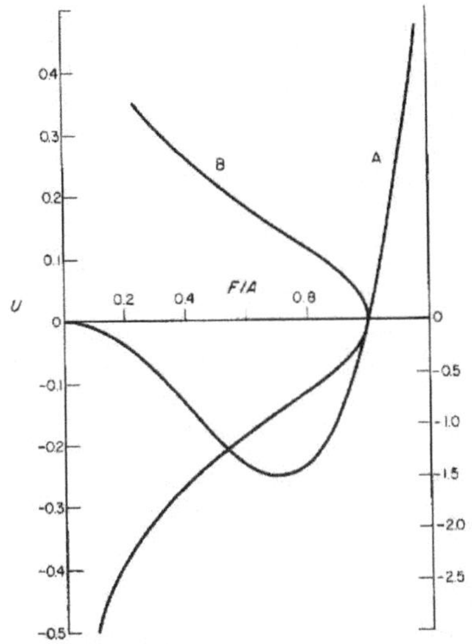

FIG. 4. (A) Potential well for analogous point in mechanical analogue for self-trapping. (B) Corresponding self-trapped profile for linear (slab-shaped) geometry.

Figure 4 shows what is happening here. In the mechanical analogue, the particle is released from F_0 at "t" = 0 and reaches $F = 0$ with zero velocity only after an infinite amount of time. There is only one trajectory that gives $F = 0$ at "t" $= x = \infty$.

Haus actually introduced this mechanical analogue for self-trapped beams of radial symmetry where F depends only on $r = (x^2 + y^2)^{1/2}$, and

$$\frac{\partial^2 F}{\partial r^2} = -\frac{1}{r}\frac{\partial F}{\partial r} + 2k_0 qF - \frac{4\pi\omega_0^2}{c^2}\mathscr{P}^{\mathrm{NL}}(F). \tag{4.5}$$

The first term on the right corresponds to a time-dependent viscous damping proportional to the velocity in the mechanical analogue. This damping slows down the analogous particle so much that there are many starting displacements which lead to zero velocity at $F = 0$ and "t" $= r = \infty$. The "dissipation" is so great that the least starting displacement ($F = 1.56\,A$) which allows the particle to reach the center is off scale in Fig. 4. To each successively larger allowed starting displacement corresponds a solution $F(r)$ with one additional radial null. Figure 5 shows several of these solutions.

An important feature of these higher trapped "modes" is the rather large amount of power contained at large radii. This can be understood from the mechanical analogue,

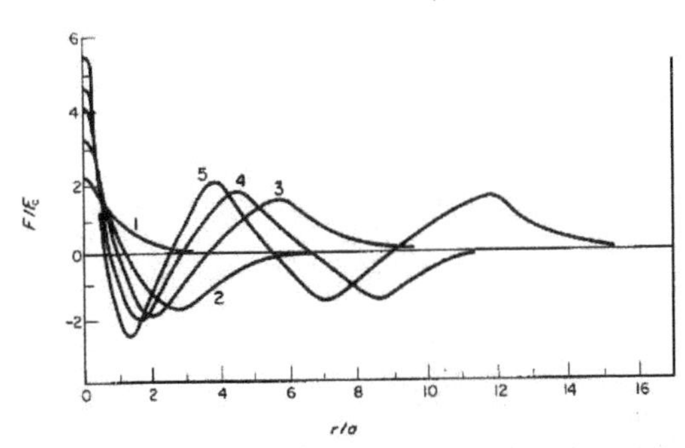

FIG. 5. Self-trapped profiles with axial symmetry. (Redrawn from ref. 39.)

since the viscous term is proportional to $1/r$ and therefore grows smaller and smaller as "time" progresses. The more often the particle oscillates, the less it is damped.

Another important feature of these solutions is that, in the axially symmetric case, the total power contained in a "mode" is nearly independent of the axial field and hence the scale. This is true only for the cubic nonlinearity $\mathscr{P}^{NL} = \eta F^3$, and follows from the invariance of eqn. (4.2) under the transformation $r \to r/\alpha$, $F \to \alpha F$, $q \to \alpha^2 q$. Thus the power, which is proportional to

$$\int F^2 r \, dr$$

is also invariant under this transformation. Or rather it would be if it were not also proportional to the total refractive index, which contains a small nonlinear correction:

$$P = (n_0 + \delta n)\frac{c}{8\pi} \cdot 2\pi \int r \, dr \, F^2(r) \tag{4.6}$$

where $\delta n = cq/\omega_0$, and q is fixed by the on-axis intensity through a relation like eqn. (4.4). For the lowest order axially symmetric mode, Chiao *et al.* found[9]

$$\delta n = n_2 F^2(0)/9.72.$$

The reader should not be surprised that the effective refractive index change is not $n_2 F^2(0)/2$. In fact the nonlinear response changes the phase velocity in two ways. First, it changes the susceptibility directly, and second, it forms a nonlinear wave guide to prevent the field from diffracting. Waves so confined always propagate with phase velocities greater than their unconfined values.

The scale invariance of the mode power (for round beams) implies that each self-trapped mode requires a unique power. Given that power, a mode of any size may be formed. Haus[39] and also Yankauskas[40] have computed several of the lowest "critical" powers. In our notation,

$$P_{cN} = \sigma_N c \lambda_0^2 / 16\pi^2 n_2 \tag{4.7}$$

where λ_0 is the vacuum wavelength and $\sigma_N = 1.86, 12.25, 31.26, 58.57, 94.23$ for the first five modes. Akhmanov *et al.*[8] remark that the critical power increases like $2N^2 - 1$ ($N = 1, 2, \ldots$), for no apparent reason.

B. *The effect of saturation*

Many features of the self-trapping solutions already discussed are also found for saturable nonlinear susceptibilities. There is still only one mode for slab-shaped symmetry, but many for axial symmetry. The most important new feature contributed by saturation is the destruction of the scale invariance of the mode power. Each mode now has a definite size fixed by the saturation behavior of the susceptibility.

What happens can be seen qualitatively from Haus's mechanical analog described above. A typical saturable polarization expression is

$$\mathscr{P}^{NL} = \eta F^3/(1 + F^2/F_s^2) \tag{4.8}$$

where F_s is a saturation field. The corresponding potential function

$$U = (2k_0 q F_s^4/A^2)[(1 - \theta)f^2 - \log(1 + f^2)]$$

has an attractive region only if $\theta < 1$. Here $f = F/F_s$ and $\theta = A^2/2F_s^2$, where A is defined in Section 4A. Thus F_s must be large enough to allow an appreciable index change before a self-trapped mode can form. If $\theta < 1$, then U has a single minimum at

$$F_M = \frac{AF_s}{(2F_s^2 - A^2)^{1/2}} \geq A/\sqrt{2}. \tag{4.9}$$

The slab-shaped trapped mode profile $F(x)$ has an inflexion point at this field strength. The effect of saturation is clearly to make the potential well more shallow and move the $U(F) = 0$ point to higher values of F. Thus the analogous particle, when released at $x = 0$,

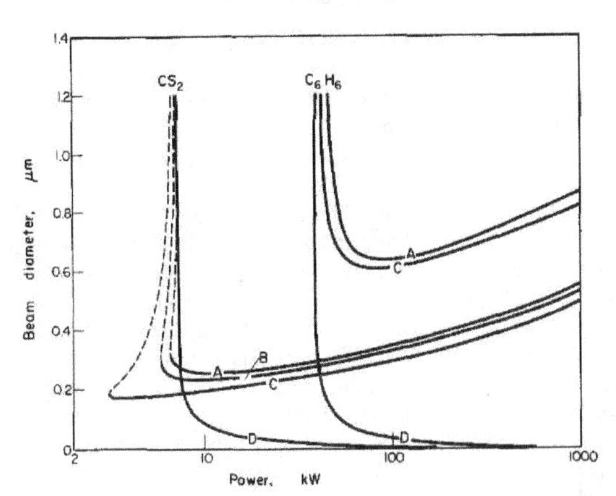

FIG. 6. Relation between beam diameter and power for self-trapped modes in saturable media whose nonlinear susceptibilities are shown in Fig. 3. (A) $X_{\pm}^{\mathrm{kw}}$, (B) X_{+}^{G}, (C) $X_{\pm}$, (D) X_0. (From ref 29.)

will take longer to reach the inflexion point as well as the center at $F = 0$. This means that the mode shape will be broader and flatter, and the on-axis intensity will be greater, than in the comparable nonsaturable case.

These expectations are borne out also for the round modes by numerical solution of the self-trapping equation for a variety of saturable susceptibilities. Most work has been done on the saturable molecular reorientation model discussed in Section 3. Reichert and Wagner[41] show graphs of self-trapped profiles with saturation, and they, as well as Gustafson *et al.*[30] and Marburger *et al.*[29] give the relation between beam diameter and power in the modes for a variety of cases. Figures 6 and 7 show typical results. The saturation of the reorientation mechanism for prolate molecules is interesting because it shows an "antisaturation" effect, the nonlinear response becoming stronger at higher field strengths before finally saturating. This leads to a region of unstable trapped beam solutions containing less than the critical power. Notice that there is a unique stable mode size for every total power above a certain value, but that there is still a critical power in the sense that near this value one can obtain a wide range of mode sizes with very small deviations from the critical power.

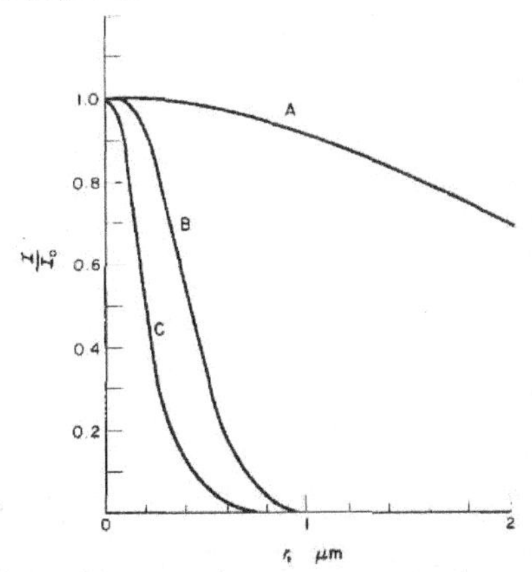

FIG. 7. Radial profiles for self-trapped modes using susceptibility X_{+}^{G} in Fig. 3. Modes A and C have similar powers and lie in the antisaturation region of curve 6B for CS_2. Mode B has about 1000 times more power. Saturation has flattened its central portion compared with curve C. (From ref. 30.)

The motivation for studying saturable self-trapping is to see if the predicted beam sizes agree with typical dimensions of experimental self-focusing phenomena. In nearly all analyses undertaken to date, the saturation fields are so high that the predicted mode sizes are much smaller than typical experimental sizes. Exceptions are the work of Gustafson and Townes on steric interference,[31] and the metal vapor experiments and theory by Grischkowsky *et al.*[19,35] Apart from the fact that the analyses for dense fluids have omitted important competing effects, it is very difficult to know what experimental parameter should be compared with theory. Experiments are rarely done with constant incident powers, and often not even with constant incident mode structure.

Bjorkholm has probably come closest to producing a self-trapped beam in sodium vapor with a nearly resonant CW dye laser source.[42] Although the nonlinear properties of the resonant transition are well understood in this experiment, the role of diffusion of excited atoms out of the beam in determining the CW properties of the nonlinear index has not yet been closely examined. This is a characteristic problem of self-trapping studies: self-trapped beams are not well defined for pulsed sources, but the contributing nonlinear mechanisms are restricted to those with short response times. With CW sources, the nonlinear response includes contributions from many mechanisms with long response times, including thermal and diffusive response.

C. *Self-trapping of vector waves*

Equation (4.2) for the self-trapped mode shape omits the terms arising from $\nabla(\nabla\cdot\boldsymbol{E})$ in the vector wave equation. Indeed, there are no confined wave solutions with purely linear polarization everywhere. It is possible, however, to construct a field configuration for which $\nabla\cdot\boldsymbol{E}$ is identically zero. For example, the field lines $\boldsymbol{E}$ can form concentric circles. Abakarov *et al.*[43] and Pohl[44] have studied this configuration, and Pohl has even managed to produce it experimentally.[45]

The field in this case can be written $\hat{\Phi}\mathscr{E}$, where $\hat{\Phi}$ is a unit vector in the direction of increase of the azimuthal angle:

$$\hat{\Phi} = -\hat{x}\sin\theta + \hat{y}\cos\theta.$$

This gives

$$\nabla^2\mathscr{E} = \left(\nabla^2\mathscr{E} - \frac{\mathscr{E}}{r^2}\right)\hat{\Phi},$$

and our self-trapping wave equation becomes

$$\frac{\partial^2 F}{\partial r^2} = -\frac{1}{r}\frac{\partial F}{\partial r} + \left(2k_0 q + \frac{1}{r^2}\right)F - \frac{4\pi\omega_0^2}{c^2}\mathscr{P}^{\mathrm{NL}}(F), \tag{4.10}$$

where, as before, $\mathscr{E} = F\exp iqz$. The additional term F/r^2 in eqn. (4.10) makes the mechanical potential, in Haus's mechanical analog, time dependent and singular at "t" $= r = 0$. This forces the initial displacement of our zero energy particle to be zero, and therefore the family of self-trapped mode solutions has a null at $r = 0$. Although the initial displacement is fixed, the initial velocity can be varied to generate a set of solutions with off-axis modes. Pohl finds the critical power for the lowest mode (which is scale invariant for cubic nonlinearity) to be about 17 times that of the corresponding scalar wave mode. Figure 8 compares the two modes.

Pohl has also studied a variety of other trapped field configurations with a method which accounts for the correct tensor relation between $\mathscr{P}^{\mathrm{NL}}$ and $\mathscr{E}$. One motivation for studying such modes is that experimental determination of their critical powers allows one to distinguish the relative contributions of different self-focusing mechanisms possessing different symmetries. For example, electrostriction and molecular reorientation lead to different relative values of A and B in eqn. (3.9). The insensitivity of electrostriction to field polarization forces B to vanish in (3.9), whereas B can be shown to be $6A$ for molecular reorientation and redistribution.[21]

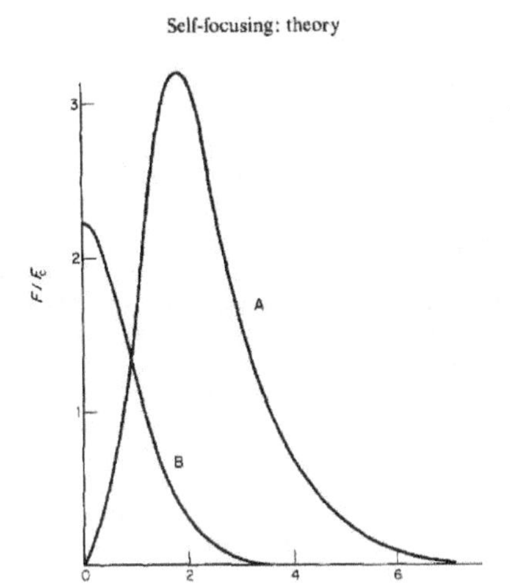

Fig. 8. (A) Lowest order self-trapped mode of eqn. (4.10) with cubic nonlinearity, and (B) lowest order scalar mode of Chiao, Garmire and Townes, for comparison. Scale of field strength same as in Fig. 5. (From ref. 44.)

5. STEADY-STATE SELF-FOCUSING.—I SOME EXACT RESULTS

A. *The ray potential*

In this section we continue to require that the optical field envelope $\mathscr{E}$ be time independent, but allow slow variation along the propagation direction z. Our results should therefore be valid for experiments conducted with CW, or nearly CW, incident laser beams. First we consider scalar $[\nabla(\nabla \cdot E) = 0]$, linearly polarized waves, for which (2.9) implies

$$2ik_0 \frac{\partial \mathscr{E}}{\partial z} + \nabla_T^2 \mathscr{E} = -\frac{4\pi\omega_0^2}{c^2} \mathscr{P}^{NL}. \tag{5.1}$$

Here $\mathscr{E}$ is generally complex, and can be written as

$$\mathscr{E} = A \exp iS$$

where A and S satisfy equations obtained from the real and imaginary parts of (5.1):

$$k_0 \frac{\partial A^2}{\partial z} + \nabla_T \cdot (A^2 \nabla_T S) = -\frac{4\pi\omega_0^2}{c^2} \chi'' A^2, \tag{5.2a}$$

$$2k_0 \frac{\partial S}{\partial z} + (\nabla_T S)^2 - (\nabla_T^2 A)/A - \frac{4\pi\omega_0^2}{c^2} \chi'_{NL} = 0. \tag{5.2b}$$

Here $\mathscr{P}^{NL} = (\chi'_{NL} + i\chi'')\mathscr{E}$, where χ'_{NL} and χ'' are real functions of A. (Recall that linear loss is included in $\mathscr{P}^{NL}$ for convenience.)

Equation (5.2a) is clearly a power conservation equation in the transverse plane. Integrating over this plane, we find

$$\frac{\partial P}{\partial z} = -\frac{4\pi\omega_0}{n_0 c} \iint \chi'' I \, dx \, dy$$

where P is the power and I the intensity obtained by ignoring δn in eqn. (4.6). If $\chi'' = 0$, the total power is conserved along z.

The equation for the phase has the form of a Hamilton–Jacobi equation for the motion of a particle of unit mass moving in two dimensions under the influence of the potential

$$U = -\nabla_T^2 A/2k_0^2 A - 2\pi\chi'_{NL}(A)/n_0^2 \tag{5.3}$$

where $k_0 z$ is regarded as time and $k_0 x, k_0 y$ as spatial coordinates.[46] The corresponding equation for the trajectories is

$$\frac{d^2 \mathbf{r}}{dz^2} = -\nabla_T U \tag{5.4}$$

from which the ray paths can be computed if U, and hence A, is known as a function of x and y. This mechanical analogy should not be confused with that discussed in the previous chapter for self-trapped beams. We are exploiting here the very familiar analogy between optical rays and particle trajectories. Notice that if U = const., eqn. (5.3) becomes the self-trapping equation (4.2).

It is instructive to examine $U(r)$ for a simple case to see how the ray paths will be influenced by the nonlinear response. Suppose that $\chi'' = 0$, $\chi'_{NL} = \eta|\mathscr{E}|^2$, and

$$\mathscr{E} = A_0 \exp -r^2/2a^2$$

corresponding to a gaussian beam whose waist is at $z = 0$, which we assume to be the entrance surface of the nonlinear medium. Then, taking $\rho^2 = r^2/a^2$,

$$k_0^2 a^2 U(\rho) = 1 - \frac{\rho^2}{2} - B\exp(-\rho^2)$$

where

$$B = 2\pi\eta k_0^2 a^2 A_0^2 / n_0^2$$

$$= \frac{4n_2\omega_0^2}{c^3} P.$$

The total power in the beam is $P = n_0 c A_0^2 a^2/8$.

This potential function $U(\rho)$ has a single maximum at $\rho = 0$ for small values of B, leading to the repulsive diffractive force that normally causes finite beams to diverge as they propagate. However, when B exceeds $\frac{1}{2}$, or when the beam power exceeds

$$P_1 = \frac{c^3}{8n_2\omega_0^2} \tag{5.5}$$

there is a dip in $U(\rho)$ at $\rho = 0$, and the near axis rays are focused inward. Figure 9 shows $U(\rho)$ for $P/P_1 = 0, 1$ and 3.72. The third value corresponds to the least critical power for a round scalar self-trapped mode. Numerical solutions of (5.1) with this nonlinearity and initial condition show that for P within a few percent of $P_{c1} = 3.72P_1$, the incident gaussian changes shape during the first diffraction length $z \lesssim k_0 a^2$ until it closely resembles the least self-trapped mode, and then propagates for a long distance without further change of shape or size[47] (see Figs. 10 and 11). Eventually it either diffracts or comes to a self-focus, depending on whether P is less than or greater than a critical power which we call P_2. The value of P_2 seems to be very slightly greater than P_{c1}, but we often set $P_2 = P_{c1}$ for gaussian beams.

All the rays within the maxima of $U(r)$ at

$$r_m = (\log P/P_1)^{1/2} a$$

are initially bent toward the axis, while those at greater radii diffract. The total power in the gaussian beam up to this radius is easily seen to be $P_m = P - P_1$. From this we infer that at the upper critical power P_2 for formation of a self-trapped mode with a gaussian input, about 73% of the power is initially focused toward the axis.

At higher powers, $P > P_2$, a "self-focus" of infinite intensity occurs in the solutions of (5.1) with the nonlinearity and initial condition we have chosen. It is not an easy matter to determine where on the z-axis this singularity appears, because the beam properties at appreciable distances are not simply related to those at $z = 0$. Nor are they accurately represented by the near axis rays, as we shall see later. Apart from a few exact relations, to be discussed next, most quantitative self-focusing theory depends heavily on numerical computation. There are, however, a variety of approximate techniques which give valuable

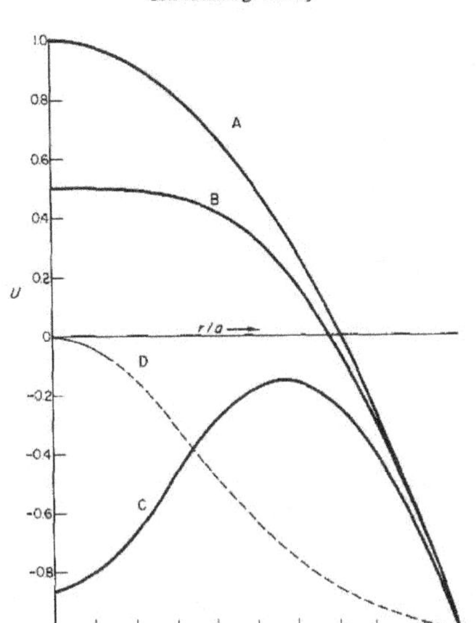

FIG. 9. Ray potentials given by eqn. (5.3) for incident unfocused gaussians of different powers: (A) $P = 0$, (B) $P = P_1$, (C) $P = P_{c1}$, (D) normalized incident beam profile.

insights into the qualitative features of self-focusing, and allow at least rough estimates of their magnitudes. These are discussed in Section 8.

B. *The lens transformation*

The stationary self-focusing equation (5.1), with the nonlinearity $\mathscr{P}^{NL} = \eta|\mathscr{E}|^2\mathscr{E}$, possesses a remarkable invariance property, first pointed out by Talanov.[48] This equation,

$$2ik_0 \frac{\partial \mathscr{E}}{\partial z} + \nabla_T^2 \mathscr{E} = \beta|\mathscr{E}|^2 \mathscr{E} \tag{5.6}$$

is invariant under the "lens transformation"

$$z \to \zeta = \bar{\Lambda}(z)z, \quad \mathbf{r} \to \rho = \bar{\Lambda}(z)\mathbf{r} \tag{5.7}$$

$$\mathscr{E}(\mathbf{r}, z) \to \mathscr{E}'(\rho, \zeta) = \Lambda(\zeta)\mathscr{E}[\Lambda(\zeta)\rho, \Lambda(\zeta)\zeta]\, e^{-ik_0\rho^2\Lambda(\zeta)/2R} \tag{5.8}$$

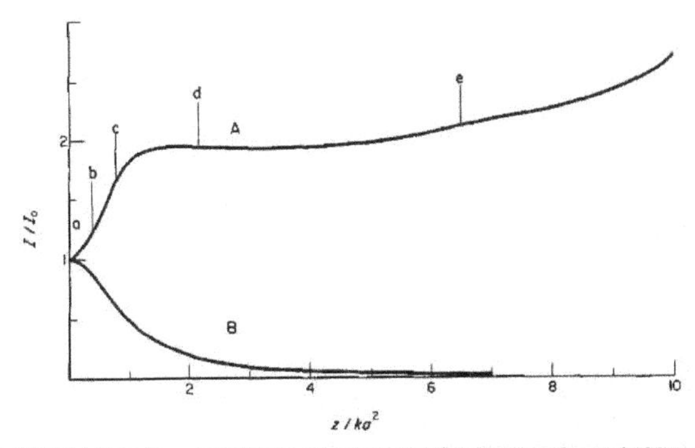

FIG. 10. (A) Axial intensity versus distance along propagation direction for an incident gaussian beam with power $P \approx 1.003P_2$. After initial reshaping in the first diffraction length, the axial intensity is nearly constant to about $6ka^2$. Curve B is the axial intensity without self-focusing. The intensity drops to half its initial value in the first diffraction length.

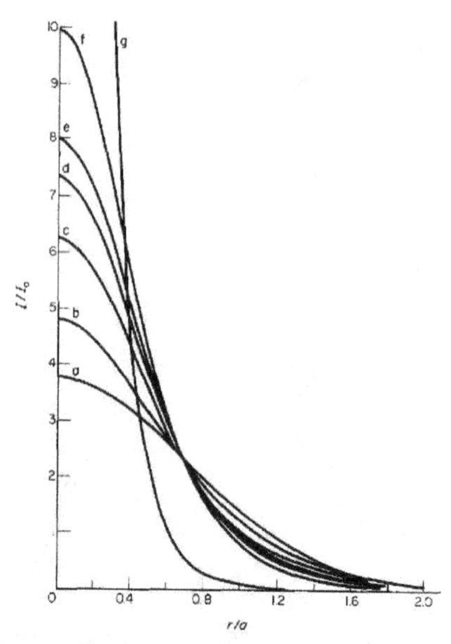

FIG. 11. Radial profiles for indicated points on Fig. 10 corresponding to the following values of z/ka^2: (a) 0, (b) 0.353, (c) 0.713, (d) 2.15, (e) 6.47, (f) 12.23, (g) 18.71. The axial field of (g) exceeds the incident axial field by 25. The self-focal distance is about 19.5 ka^2.

where

$$\bar{\Lambda}(z) = R/(R + z), \quad \Lambda(\zeta) = R/(R - \zeta)$$

or

$$\frac{1}{R} = \frac{1}{\zeta} - \frac{1}{z}. \tag{5.9}$$

Here R is a real constant, $\boldsymbol{r} = (x, y)$, and the choice $\beta = -nn_2\omega_0^2/c^2$ leads to (5.1). The invariance can be verified by substitution.

The significance of this transformation is that, if $\mathscr{E}$ satisfies (5.6) with the initial condition $\mathscr{E}(\boldsymbol{r}, 0) = f(\boldsymbol{r})$, then $\mathscr{E}'$ satisfies the same equation with initial condition

$$\mathscr{E}'(\boldsymbol{r}, 0) = f(\boldsymbol{r}) \exp -ik_0r^2/2R. \tag{5.10}$$

This initial condition could be obtained from the one for $\mathscr{E}$ experimentally by placing a thin lens at $z = 0$ with focal length R, which explains the term "lens transformation". When R is positive, the initial transformed phase front is a "sphere" (a parabola in the slowly varying envelope approximation) of radius R converging on a point on the positive z-axis, so R equals the focal length of a positive lens. Equation (5.9) indicates that ρ, ζ are the coordinates of the object point conjugate to the image point $\boldsymbol{r}, z$ through the lens R (see Fig. 12). The ratio of the intensities at corresponding points is

$$|\mathscr{E}|^2/|\mathscr{E}'|^2 = 1/\Lambda^2(\zeta) = (1 - \zeta/R)^2$$

which is the intensity enhancement predicted by geometrical optics near a focus, where the beam diameter is proportional to $1 - \zeta/R$. The phase factor gives an additional phase curvature of radius $R(1 - \zeta/R)$, which is the geometrical optics radius at z. What is remarkable about (5.8) is that it is "exact" for *diffraction optics* (in the slowly varying envelope approximation) and even for nonlinear diffraction optics with the cubic nonlinearity.

We can use the lens transformation to determine the axial intensity of a beam whose incident intensity profile is that of the lowest order self-trapped mode, but which is focused

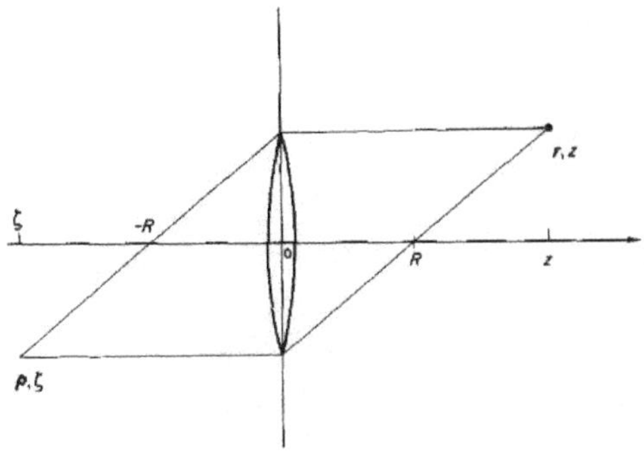

FIG. 12. The lens transformation eqn. (5.7) takes r, z into the conjugate coordinates through a thin lens of focal length R.

into the nonlinear medium through a lens R. The unfocused solution is

$$\mathscr{E}(r,z) = F(r)\exp ipz .$$

Thus the axial intensity is finite at $z = \infty$. The axial intensity for an incident beam with phase curvature, using (5.8), is proportional to

$$|\mathscr{E}'(0,z)|^2 = \frac{R^2}{(R-z)^2}F^2(0) \tag{5.11}$$

which is singular at $R = z$. This is exactly the intensity dependence predicted by geometrical optics in the absence of diffraction. Of course, for the self-trapped modes, the diffractive effects are just cancelled by self-focusing. This cancellation is preserved even when the incident beam is focused, *if* the nonlinearity has the form $\chi_{NL} = \eta|\mathscr{E}|^2$. Otherwise the cancellation is not preserved by focusing.

Equation (5.11) is approximately correct even for a gaussian beam incident with power $P = P_2$ because, as numerical solution shows,

$$\mathscr{E}(r,z) \rightarrow F(r)\exp iqz \tag{5.12}$$

for z large. But near the focal point, $\mathscr{E}'(0,z)$ is proportional to $\mathscr{E}(0,z)$ evaluated at very large distances where eqn. (5.12) is correct. If the incident beam has less than critical power, then its intensity decreases with z, ultimately diminishing like $1/z^2$. This suffices to prevent a singularity in $\mathscr{E}'$ near the geometrical focal point. We conclude that for gaussian incident beams, singular self-foci do not appear anywhere until the beam power exceeds P_2, even if the beam is focused. When a singular focus does appear, it appears first at the geometrical focal point, not at the low intensity beam waist (which occurs prior to the focal point because of diffraction).

What happens beyond the geometrical focal point? Talanov[48] has given the following elegant argument. Equation (5.7) shows that the region $\zeta > R$ in the focused beam is mapped by the lens transformation into $z < 0$ in the unfocused beam. The field in this region can be obtained from that for $z > 0$ if the surface of constant phase through $z = 0$ is plane. In that case, $\mathscr{E}(r,0)$ can be made real. Now $\mathscr{E}(r,-z)$ satisfies the same equation as $\mathscr{E}^*(r,z)$, by inspection of (5.6) for real β. Therefore, since $\mathscr{E}(r,0) = \mathscr{E}^*(r,0)$, we have $\mathscr{E}(r,-z) = \mathscr{E}^*(r,z)$. This implies that if $\mathscr{E}(r,z)$ has no singularity prior to $z = z_f$, then $\mathscr{E}(r,-z)$ has none between $-z_f$ and 0, which maps onto the entire negative ζ-axis plus the positive axis from ζ_f to infinity. That is, for the focused beam, there is no singularity in the entire region beyond ζ_f. For $z_f = \infty$, as for a self-trapped mode, $\zeta_f = R$ and there is no singularity beyond the geometrical focus.

Unfortunately, this argument is not valid because the prescription described above for finding $\mathscr{E}'$ beyond the focus gives the wrong function, even in the case of linear optics. The difficulty is related to the singularity of the lens transformation at $\zeta = R$, and is best appreciated in the context of a simple example: the solution of (5.6) in the variables ρ, ζ, with

$\beta = 0$, and with the initial condition

$$\mathscr{E}''(\rho, 0) = A \exp\{-\rho^2[1 + i(l/R)]/2a^2\}, \tag{5.13}$$

where $l = k_0 a^2$, is well known to be[49]

$$\mathscr{E}'(0, \zeta) = \frac{A \exp - i\psi'(\zeta)}{\{[1 - (\zeta/R)]^2 + (z/l)^2\}^{1/2}}. \tag{5.14}$$

Here we have set $\rho = 0$ for simplicity, and defined

$$\psi'(\zeta) = \tan^{-1}\left[\frac{\zeta/l}{1 - (\zeta/R)}\right]. \tag{5.15}$$

It is important to notice that as ζ goes from $-\infty$ to $+\infty$, $\psi'(\zeta)$ goes from θ to $\theta + \pi$, where $\theta = -\tan^{-1}(R/l)$. The corresponding solution for the unfocused gaussian ($R = \infty$ in eqn. (5.13)) is

$$\mathscr{E}(0, z) = \frac{A \exp - i\psi(z)}{[1 + (z/l)^2]^{1/2}}, \tag{5.16}$$

where $\psi(z) = \tan^{-1}(z/l)$ varies between $-\pi/2$ and $\pi/2$. Applying the lens transformation to (5.16) gives

$$\mathscr{E}''(0, \zeta) = \frac{A \exp - i\psi(A(\zeta)\zeta)}{\{[1 - (\zeta/R)]^2 + (z/l)^2\}^{1/2}}. \tag{5.17}$$

If $\psi(A(\zeta)\zeta)$ is written out, one obtains the form (5.15). But as the argument of ψ goes through infinity, we must specify which branch of the inverse tangent is intended. The functions ψ and ψ' are defined on different ranges, so (5.17) and (5.14) are different for $\zeta > R$ (see Fig. 13). The point is a subtle one, but definitely leads to incorrect results for certain nonlinear problems.

Despite the failure of the lens transformation beyond a geometrical focus, it is very useful for determining prefocal properties from numerical solutions with unfocused initial conditions, or for computing the field near a tight focus from results for a weaker focus.[50]

C. *The moment equations*

The lens transformation and the scale invariance of the self-trapping powers P_{ci} are not the only exact results admitted by the cubic nonlinearity. Vlasov *et al.*[51] have demonstrated the existence of an important constant of the motion for this case, which gives insight into the dynamics of the self-focusing process.

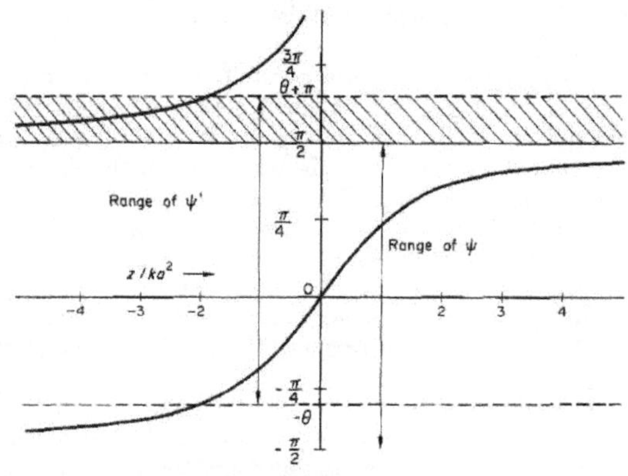

FIG. 13. The function $\psi(z)$ and $\psi'(z)$ for $R = ka^2/2$, $\theta = -\tan^{-1}(R/ka^2)$.

The starting-point for this development is the recognition that the existence of a hierarchy of conservation equations such as

$$\frac{\partial U}{\partial t} = -\nabla\cdot\boldsymbol{P}, \tag{5.18}$$

$$\frac{\partial \boldsymbol{P}}{\partial t} = c^2\nabla\cdot\mathbf{T}, \tag{5.19}$$

$$\frac{\partial}{\partial t}\operatorname{Tr}\mathbf{T} = \nabla\cdot\boldsymbol{Q} \tag{5.20}$$

implies a relation between the conserved quantities and the time derivatives of the moments of U. The reader may verify by direct partial integration that the conserved (time-independent) quantities

$$C_2 = -\int \operatorname{Tr}\mathbf{T}\,\mathrm{d}^3x, \qquad \boldsymbol{C}_1 = \int \boldsymbol{P}\,\mathrm{d}^3x \tag{5.21}$$

satisfy

$$\boldsymbol{C}_1 = \frac{\partial}{\partial t}\int \boldsymbol{x}\,U\,\mathrm{d}^3x \tag{5.22}$$

and

$$C_2 = \frac{1}{2c^2}\frac{\partial^2}{\partial t^2}\int x^2\,U\,\mathrm{d}^3x. \tag{5.23}$$

Equations (5.18)–(5.23) are satisfied by the electromagnetic energy density U, Poynting vector $\boldsymbol{P}\,(=\boldsymbol{Q})$ and stress tensor $\mathbf{T}$ in gaussian units.

Vlasov *et al.* noticed that a similar hierarchy exists for the self-focusing equations. In particular, setting

$$\mathcal{E} = U^{1/2}\exp \mathrm{i}\,S, \tag{5.24}$$

$$\boldsymbol{P} = \frac{1}{k_0}U\nabla_\mathrm{T}S, \tag{5.25}$$

$$\mathbf{T} = \frac{-1}{2k_0^2c^2}[(\nabla\mathcal{E})(\nabla\mathcal{E}^*) + (\nabla\mathcal{E}^*)(\nabla\mathcal{E})] + \frac{1}{4k_0^2c^2}\mathbf{I}\left[\nabla^2|\mathcal{E}|^2 + \frac{8\pi\omega_0^2}{c^2}\left(\chi'_\mathrm{NL}|\mathcal{E}|^2 - \int^{|\mathcal{E}|^2}\chi'_\mathrm{NL}\,\mathrm{d}|\mathcal{E}|^2\right)\right] \tag{5.26}$$

one finds that eqns. (5.18) and (5.19) are satisfied if $\mathcal{E}$ satisfies the self-focusing equation (5.1) and if $t\to z$, $\nabla\to\nabla_\mathrm{T}$, $\chi'' = 0$. Here χ'_NL is a function of $|\mathcal{E}|^2$. The analogue of eqn. (5.20) is satisfied if and only if χ'_NL has the functional form $\eta|\mathcal{E}|^2$, in which case

$$k_0^2c^2\operatorname{Tr}\mathbf{T} = \tfrac{1}{2}(\mathcal{E}^*\nabla^2\mathcal{E} + \mathcal{E}\nabla^2\mathcal{E}^*) + \frac{2\pi\omega_0^2}{c^2}\eta|\mathcal{E}|^4. \tag{5.27}$$

The integral of this quantity, which is a constant of the motion (independent of z), is proportional to

$$H \equiv -c^2\int \operatorname{Tr}\mathbf{T}\,\mathrm{d}^2r = +\frac{1}{k_0^2}\int\left[|\nabla_\mathrm{T}\mathcal{E}|^2 - \frac{2\pi\omega_0^2}{c^2}\eta|\mathcal{E}|^4\right]\mathrm{d}^2r. \tag{5.28}$$

Even if χ'_NL is not proportional to $|\mathcal{E}|^2$, the quantity

$$H' = -\frac{1}{k_0^2}\int\left[|\nabla_\mathrm{T}\mathcal{E}|^2 - \frac{4\pi\omega_0^2}{c^2}\int^{|\mathcal{E}|^2}\chi'_\mathrm{NL}\,\mathrm{d}|\mathcal{E}|^2\right]\mathrm{d}^2r \tag{5.29}$$

is a constant of the motion. In fact H' is proportional to the total field energy derived from the lagrangian density

$$\mathscr{L} = |\nabla_T \mathscr{E}|^2 - ik_0\left(\mathscr{E}^* \frac{\partial \mathscr{E}}{\partial z} - \frac{\partial \mathscr{E}^*}{\partial z}\mathscr{E}\right) - \frac{4\pi\,\omega_0^2}{c^2}\int^{\mathscr{E}\mathscr{E}^*} \chi_{NL}\, d|\mathscr{E}|^2 \tag{5.30}$$

from which the self-focusing equation may be derived via the Euler–Lagrange equations.

The conserved quantity corresponding to $\boldsymbol{C}_1$ in eqn. (5.21) vanishes for a smooth beam profile whose amplitude and phase functions are centrisymmetric. For noncentrisymmetric beams, which experience "self-slewing" in a nonlinear medium due to the induced prism δn, the vector $\boldsymbol{C}_1$ measures the net power flux perpendicular to the initial beam direction. The constancy of $\boldsymbol{C}_1$ implies that the velocity of the centroid of the intensity distribution is constant, no matter how the beam distorts as it diffracts and self-focuses. This result does not necessarily contradict the interesting proposition of Kaplan[52] that such a beam may spiral in to a self-focal point, a conclusion which is based only on an analysis of near-axis rays.

Of greater interest is the moment equation involving C_2, which according to (5.23) implies

$$\frac{\partial^2 a_R^2}{\partial z^2} = 2H/p \tag{5.31}$$

where a_R is the rms radius of the intensity distribution:

$$a_R^2 = \int |\mathscr{E}|^2 r^2\, d^2 r/p, \tag{5.32}$$

$$p = \int |\mathscr{E}|^2\, d^2 r.$$

Equation (5.31) has the solution

$$a_R^2(z) = \frac{H}{p}z^2 + \left.\frac{da_R^2}{dz}\right|_0 z + a_R^2(0), \tag{5.33}$$

in which the coefficient of z vanishes for incident beams with flat phase fronts. If $H < 0$, then a_R^2 can vanish at some value of z, which means that *all* the beam power has collapsed into the point $r = 0$. Beyond this point, no physical solutions exist because $a_R^2 < 0$. This is further evidence that the cubic nonlinearity in the slowly varying envelope approximation without loss or transiency is an unrealistic choice.

Equation (5.28) for H implies a critical power for a given incident beam shape when $H = 0$. We refer to this critical power as P_3 since it is generally higher than either P_1, given by the vanishing of the focusing potential on axis, and also higher than P_2, defined as the least power which gives an axial singularity at infinity:

$$P_3 = \frac{2}{\pi}P_1 \frac{\int |\mathscr{E}|^2\, d^2 r \int |\nabla\mathscr{E}|^2\, d^2 r}{\int |\mathscr{E}|^4\, d^2 r}. \tag{5.34}$$

The integrals are to be evaluated at $z = 0$. When $P = P_3$, then $H = 0$ and the rms radius is independent of z. This is consistent with self-trapping, but by no means requires that the mode shape remain constant in space. Vlasov *et al.*[51] have shown that the function $\mathscr{E}(x, y)$ which minimizes P_3 must satisfy the self-trapping equation, so $P_3 \geq P_{c1}$, where P_{c1} is the least critical power for a self-trapped mode. If $P = P_3 > P_{c1}$, then the rms radius is constant, but the beam cannot be self-trapped, because it has the wrong power. The only allowed shape change is one in which some power is moved toward the axis and some toward the wings in such a way as to keep a_R constant. This is in fact what one expects from the potential $U(r)$ of Fig. 9, and also what is observed in numerical solutions. Numerical solutions also show that self-foci appear at finite distances when $P = P_3$ if $P_3 > P_2$. Table 2 gives values of P_3 for some simple beam shapes.

TABLE 2. Beam parameters and critical powers for several simple beam shapes. CGT denotes the lowest order self-trapped mode first studied by Chiao *et al.* The fundamental radial scale a is defined in eqn. (6·3) and the lower critical power P_1 is defined in (5.5). Other defining equations are indicated in parentheses. A blank space means the value is not available

		CGT	Gaussian	Lorentzian	Parabola
a_R/a	(5.30)		1	∞	$1/\sqrt{2}$
a_T/a	(5.36)		1[a]	$\sqrt{3}$[a]	$1/\sqrt{3}$[a]
a_P/a	(6.2)	1.21	1	$\sqrt{2}$	$\sqrt{2/3}$
σ	(8.31)	2.49	1	3	-1
P_3/P_1	(5.34)	3.72	4	4	20/3
P_2/P_1	(6.4)	3.72	3.77	4.46	

[a] No incident phase curvature.

The distance z_s to the axial singularity predicted by (5.33) for negative H satisfies

$$\frac{P}{P_3} - 1 = \left(\frac{k_0 a_R a_T}{z_s}\right)^2 \tag{5.35}$$

where a_T is another characteristic beam radius defined through

$$a_T^2 = \int |\mathscr{E}|^2 \, d^2 r \bigg/ \int |\nabla \mathscr{E}|^2 \, d^2 r. \tag{5.36}$$

For a focused gaussian beam whose initial field is

$$\mathscr{E} = \mathscr{E}_0 \exp\left\{-\left(1 + i\frac{k a_0^2}{R}\right) r^2/2a_0^2\right\}$$

one finds $a_R = a_0$, and

$$a_T = a_0\left(1 + \frac{k^2 a_0^4}{R^2}\right). \tag{5.37}$$

Thus a_T depends on the incident phase curvature. The exact relation (5.35) predicts a singularity at a distance exceeding by the factor 2.79 that to the first axial singularity found numerically. A deeper investigation of the relation to numerical results will be found in Section 6.

Suydam has extended the moment method for the case of anisotropic nonlinear response, and elliptically polarized incident beams.[53] For this case the tensor $\mathbf{T}$ becomes, in the notation of eqn. (3.2),

$$2k_0^2 c^2 \mathbf{T}_{ij} = -(\partial_i \mathscr{E}_m)(\partial_j \mathscr{E}_m^*) - (\partial_i \mathscr{E}_m^*)(\partial_j \mathscr{E}_m) + \tfrac{1}{2}\delta_{ij}\left[\nabla^2(\mathscr{E}_m^* \mathscr{E}_m) + \frac{24\pi\omega_0^2}{c^2}\chi_3^{stuv}\mathscr{E}_s^*\mathscr{E}_t\mathscr{E}_u^*\mathscr{E}_v\right]. \tag{5.38}$$

D. *Self-similar solutions*

The self-trapped modes discussed in Section 4 have the property that for a fixed critical power, a mode of any size exists. This fact, which is true only for the cubic nonlinearity, is explicitly displayed in the notation

$$|\mathscr{E}(r, z)| = \frac{f(r/a)}{a} \tag{5.39}$$

where a is an arbitrary scale parameter. It is interesting to attempt to generalize this form by allowing a to be a function of z. The resulting form would then represent a *self-similar* solution which retains a constant shape, but changes its scale as z varies. Such solutions do exist, and have been investigated by Glass,[54] Suydam[53] and others.

In a self-similar solution, the rms intensity radius a_R is proportional to the scale radius a (see eqn. (5.32)) so the function $a(z)$ can be written immediately as

$$a^2(z) = \frac{H}{p}z^2 + \frac{da^2}{dz}\bigg|_0 z + a^2(0) \tag{5.40}$$

according to eqn. (5.33). It is convenient to choose the origin of z where $da^2/dz = 0$, which is evidently at the beam waist.

If the reader inserts (5.39) into the self-focusing equation with $\chi'_{NL} = \eta|\mathscr{E}|^2$, and uses (5.40), he will find that self-similar solutions are possible if the phase S of $\mathscr{E}$ has the form

$$S = -\frac{\alpha}{2k_0}\int_0^z \frac{dz}{a^2} - \frac{1}{2}k_0 r^2 \frac{d}{dz}\log a, \tag{5.41}$$

where α is an arbitrary parameter. The resulting equation for f is

$$\frac{d^2 f}{d\rho^2} = -\frac{1}{\rho}\frac{df}{d\rho} + (\alpha + \beta^2\rho^2)f - \frac{4\pi\,\omega_0^2}{c^2}\eta f^3 \tag{5.42}$$

where $\rho = r/a$, and β is another parameter which is related to H,

$$\beta^2 = \frac{H\,k_0^2 a^2(0)}{p} = \frac{a_0^2}{a_T^2}\left(1 - \frac{P}{P_3}\right), \tag{5.43}$$

in the notation of eqn. (5.35). If β vanishes, then the scale is z independent and (5.42) reduces to the self-trapping equation (4.5). Physical solutions occur only for $\beta^2 \geq 0$ as discussed in the previous section. The significance of α is similar to that of q in eqn. (4.5). It is proportional to the axial intensity $f^2(0)$. We are free to choose the scale length a such that

$$f(\rho) = f(0)[1 - \tfrac{1}{2}(r/a)^2 + \ldots]. \tag{5.44}$$

With this choice, eqn. (5.42) can be evaluated at $r = 0$ to give

$$\alpha + 2 = \frac{4\pi\,\omega_0^2\,\eta}{c^2} f^2(0). \tag{5.45}$$

We cannot expect $f(\rho)$ to vanish at $\rho = \infty$ for every choice of α and β. According to the mechanical analogue developed in Section 4, the term in (5.42) containing β acts as a repulsive force which increases with "time" (ρ). For each value of β, there is an infinite sequence of starting displacements $f(0)$ for which $f(\infty) = 0$. Glass[54] has investigated the multivalued function $\alpha(\beta)$, which is sketched in Fig. 14. When the nonlinearity parameter η is small, the solutions of (5.42) approach the usual laguerre–gaussian modes for self-similar propagation in free space. Comparison of (5.40), or

$$a^2(z) = \frac{\beta^2}{k^2 a_0^2}z^2 + a_0^2$$

with the corresponding equation for free gaussian propagation

$$a^2(z) = \frac{1}{k^2 a_0^2}z^2 + a_0^2$$

shows that the effect of the nonlinear medium on a gaussian mode is to increase its Fresnel number ka_0 by the factor β^{-1}. The self-similar solutions are the "modes" of the radiation field in a cavity with spherical mirrors, filled with a self-focusing medium (with $\chi'_{NL} = \eta|\mathscr{E}|^2$).

Suydam[53] has shown that the solutions of (5.42) approach

$$\text{const.} \times K_0(\tfrac{1}{2}\beta\rho^2)$$

for large ρ, where K_0 is a modified Bessel function of the second kind. This information is useful for designing a rapidly converging numerical scheme for determining the $\alpha(\beta)$ diagram.

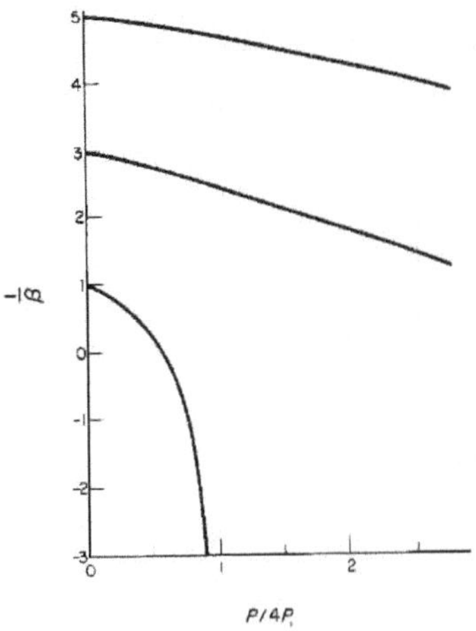

FIG. 14. Relation between self-similar mode parameter β and mode power P determined by Glass.[54] P is uniquely related to α of eqn. (5.45) for each value of β, but the relation has not yet been investigated. When $P \to 0$, the modes are laguerre–gaussians and $1/\beta = 2n + 1$, where n is the mode number.

E. *Thin windows*

Another class of self-focusing problems for which essentially exact analytical results can be obtained is that in which the propagation length is so short that only the optical phase is affected, and not the amplitude. Knowing the phase distortion at the exit of a thin non-linear medium, one may compute the subsequent optical field using linear propagation theory, which is based upon Huygens' principle, or one of its generalizations.

Writing (5.1) as

$$2ik_0 \frac{\partial \mathscr{E}}{\partial z} = -\left(\nabla_T^2 + \frac{4\pi\,\omega_0^2}{c^2}\,\chi_{NL} \right) \mathscr{E} \tag{5.46}$$

we find after a short distance Δz

$$\mathscr{E}(r, \Delta z) \approx \left[\exp \frac{i\Delta z}{2k_0} \left(\nabla_T^2 + \frac{4\pi\,\omega_0^2}{c^2}\,\chi_{NL} \right) \right] \cdot \mathscr{E}(r, 0). \tag{5.47}$$

The first term in the exponential argument is negligible if z is much less than the diffraction length, $z \ll 2k_0 a^2$, where a is a typical transverse dimension of $\mathscr{E}(r, 0)$ ($\mathscr{E}$ is assumed to be a smooth function of r). If we take $\chi_{NL} = \eta|\mathscr{E}|^2$, then the field leaving the thin window is very nearly

$$\mathscr{E}(r, \Delta z) = \exp\left(i\,\delta n \frac{\omega_0}{c}\,\Delta z \right) \mathscr{E}(r, 0) \tag{5.48}$$

where δn is, of course, a function of $\mathscr{E}(r, 0)$ given by (3.4).

Let us examine the linear diffraction properties of this field. To be specific, let us suppose that the beam incident on the window is focused toward the point $z = R$ and has a gaussian transverse intensity profile. We wish to determine the field at the geometrical focal point of the lens (which is very near the beam waist if R does not exceed ka^2). The field at z is given by (5.47):

$$\mathscr{E}(r, z) = \left(\exp \frac{iz}{2k_0}\,\nabla_T^2 \right) \mathscr{E}(r, \Delta z) \tag{5.49}$$

which can be rewritten as

$$\mathscr{E}(r, z) = \frac{-ik_0}{2\pi z} \int d^2 r'\, e^{(ik_0/2z)(r'^2 + r^2)}\, e^{-i(k/z)r \cdot r'}\, \mathscr{E}(r', \Delta z). \tag{5.50}$$

This equation is the slowly varying envelope analogue of the Huygens–Fresnel integral. If this is evaluated at $r = 0$, $z = R$, for the incident field

$$\mathcal{E}(r, 0) = \mathcal{E}_0 \exp(-r^2/2a^2)\exp(-ik_0 r^2/2R),$$

the result is

$$\begin{aligned}\mathcal{E}(0, R) &= \frac{-ik_0 a^2}{2R}\int_0^\infty d(r'^2/a^2)\,\mathcal{E}(r', \Delta z)\\ &= \frac{-ik_0 a^2}{2R}\mathcal{E}_0\int_0^\infty dx\, e^{-x/2}\exp(i\theta e^{-x})\end{aligned}$$

where

$$\theta = n_2\mathcal{E}_0^2\omega_0\Delta z/2c \text{ and } x = r^2/a^2.$$

with the change of variable $y = \theta^{1/2}\exp(-x/2)$, this becomes

$$\begin{aligned}\mathcal{E}(0, R) &= \frac{-ik_0 a^2}{2R}\cdot\frac{\mathcal{E}_0}{\theta^{1/2}}\int_0^{\theta^{1/2}} dy\, e^{iy^2}\\ &= \frac{-ik_0 a^2}{2R}\mathcal{E}_0\left[C(\sqrt{2\theta/\pi}) + iS(\sqrt{2\theta/\pi})\right]/\sqrt{2\theta/\pi}.\end{aligned}$$

Here $C(u)$ and $S(u)$ are the Fresnel cosine and sine integrals, which are the coordinates of the point u on Cornu's spiral. Thus the focal intensity is

$$I(0, R) = \left(\frac{k_0 a^2}{R}\right)^2 I_0\left\{\frac{C^2(\sqrt{2\theta/\pi}) + S^2(\sqrt{2\theta/\pi})}{2\theta/\pi}\right\}. \tag{5.51}$$

The quantity in brackets is the (ratio)^2 of the chord length, from the origin to a point on Cornu's spiral, to the arc length to the same point. It approaches unity for θ small, decreases to one half at $\theta = 0.88\pi$, and approaches $\pi/\theta\sqrt{8}$ for large θ. Typical values for θ are easily obtained from

$$\begin{aligned}\theta &= \frac{1}{2}\frac{P}{P_1}\frac{\Delta z}{k_0 a^2}\\ &\simeq 1.86\frac{P}{P_2}\frac{\Delta z}{k_0 a^2}.\end{aligned}$$

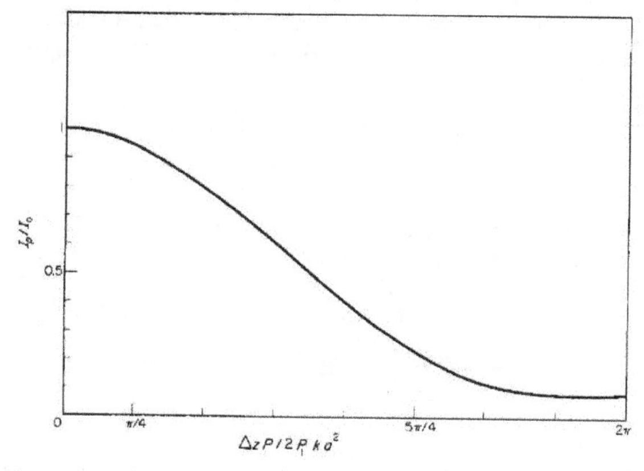

FIG. 15. Degradation of intensity at geometrical focal point of beam focused through thin window with nonlinear refractive index, from eqn. (5.51).

For $\Delta z \approx a \approx 1$ cm, and $\lambda_0 \approx 1\ \mu$m, P must exceed P_2 by about 10^4 before appreciable focal intensity degradation can be expected. Such powers can be achieved for many materials with available laser systems. Figure 15 shows the dependence of focal intensity on beam power.

The analysis described here is also of value for estimating the focal degradation arising from thermal distortion of a thin window exposed to a beam from which it can absorb energy.[56] The local temperature rise, which is proportional to the time integral of the local beam intensity for short times, causes a refractive index change which leads to a phase anomaly. Both the self-focusing phase change and the thermal phase change (ignoring thermal diffusion) have the radial variation of the incident beam intensity. Except under unusual conditions, this radial index variation is not that of a lens with good imaging properties, and the diffraction field beyond the window has strong high-order spherical aberration which leads to off-axis intensity maxima, or rings. We should expect such complicated structure in the nonlinear diffraction field also. This is one reason why conventional near-axis ray optics fails to give an accurate picture of self-focusing phenomena.

6. STEADY-STATE SELF-FOCUSING.—II NUMERICAL SOLUTIONS

A. *Cubic nonlinearity*

Physicists sometimes malign numerical analysis for giving a mass of specific information, but little general knowledge of wider applicability. This need not be the case, and we shall attempt to restrict our attention to numerical results which give insight into the self-focusing process, and which can be used to make quantitative predictions in a variety of situations. Numerical studies are simplified by the scale invariance of the simplest self-focusing equation, which can be written

$$2i\,\frac{\partial \bar{\mathscr{E}}}{\partial z'} + \nabla_{\mathrm{T}}'^{2}\,\bar{\mathscr{E}} + |\bar{\mathscr{E}}|^2\,\bar{\mathscr{E}} = 0 \tag{6.1}$$

if we choose $z' = z/k_0a^2$, $r' = r/a$, and $\bar{\mathscr{E}} = (n_2/n)^{1/2}k_0a\mathscr{E}$, where a is a transverse scale parameter.

Equation (6.1) contains no free parameters, so its solutions are uniquely determined by the properties of the incident beam. These include the intensity and phase functions which, unfortunately, cannot be specified uniquely except by writing out the functional forms. For round beams, however, which have a single maximum and smooth intensity and phase profiles, a few parameters may suffice to give most of the important self-focusing information. The beam scale need not be fixed, because the dimensionless initial field is $\bar{\mathscr{E}}(r/a, 0)$. The most important parameters of the incident intensity profile are its axial curvature (because it governs the diffraction force on near-axis rays), the quantities a_{R} and a_{T} defined in eqns. (5.30) and (5.36), and the "power radius" a_{P}, which we define through

$$\pi a_{\mathrm{P}}^2 = \int_0^\infty \mathscr{E}^2(r/a, 0)\,\mathrm{d}^2r/\mathscr{E}_0^2 \tag{6.2}$$

where $\mathscr{E}_0 = \mathscr{E}(0, 0)$. The axial curvature will be fixed by our scale parameter a, which we define through

$$\frac{1}{a^2} = -\frac{1}{\mathscr{E}}\,\frac{\partial^2\mathscr{E}}{\partial r^2}\bigg|_{r=0} \tag{6.3}$$

Table 2 gives a_{P}, a_{R} and a_{T} for several profiles.

The phase profiles of interest are nearly always spherical with radius of curvature R at $z = 0$. However, some results have been obtained for beams with longitudinal spherical aberration.[57] An approximate analysis of beams with astigmatism will be described in Section 8.

The most important beam parameter is its power. We can define a critical power for any incident beam profile as the least power for which the axial intensity remains finite at z = infinity. For the purposes of numerical computation, z = "infinity" means about ten diffraction lengths and "finite" intensity means comparable to the incident intensity. This is not entirely a satisfactory definition because we would like the critical power to indicate a threshold above which a singular self-focus occurs. For gaussian beams the two conditions appear to be the same, but for other shapes it may be possible for a very intense focus to form at a finite distance at some power, and for the focal intensity to decrease as the focus is formed at farther distances at lower powers. Gaussian beams, at any rate, seem to yield intense foci at arbitrarily great distances.

Numerical solutions of (6.1) show that unfocused gaussians pass over into self-trapped modes at $P = P_2 \approx 1.013P_{c1}$, and that lorentzians appear to do the same at $P \approx 1.20P_{c1}$. Critical powers for other beam shapes have not been obtained, although some numerical evidence indicates that parabolic beams tend to self-focus more readily than gaussians.

Of more importance than the critical power is the position at which singular foci occur, and the properties of the optical field near a focus. The moment equations show that the nonsaturable nonlinear susceptibility $\chi_{NL} = \eta|\mathscr{E}|^2$ leads to truly singular solutions for essentially all incident beam shapes. Such singular behavior prevents a detailed numerical study of the optical field just beyond a self-focal point unless the singularity is softened by weakening the nonlinear response, or unless the numerical method somehow gets around the singularity.

Figure 16 shows axial intensities for a variety of incident powers. In all cases the incident beam is an unfocused gaussian. Experimentally this corresponds to focusing the incident beam and placing the beam waist on the incident plane $z = 0$. Notice that for P near P_2 the axial intensity makes an initial rise, during which the radial profile changes appreciably, then rises more slowly until near the singular point (see Fig. 11). For higher powers, there is no chance of diffraction reshaping and the axial intensity rises directly to its singularity.

The position z_f of the nearest singularity for this case has been determined accurately as a function of power by Goldberg *et al.*[58] and by Dawes and Marburger.[47] The result

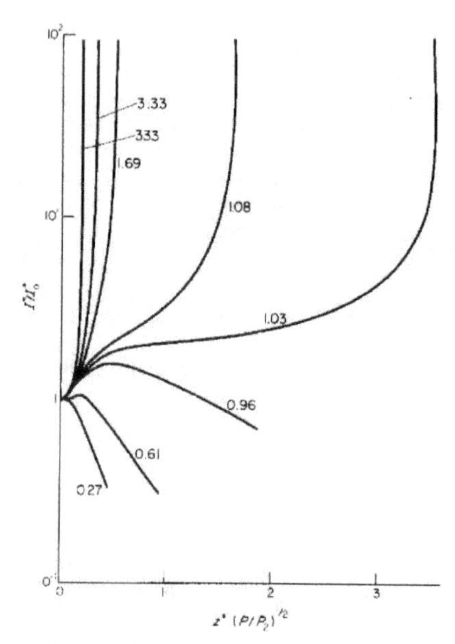

FIG. 16. Axial intensities versus dimensionless distance $z/2ka_0^2$ for the powers P/P_2 indicated next to each curve. For high powers, all the curves lie very close to that for $P/P_2 = 333$ because $z^* \sim (P/P_2)^{-1/2}$. (From ref. 47.)

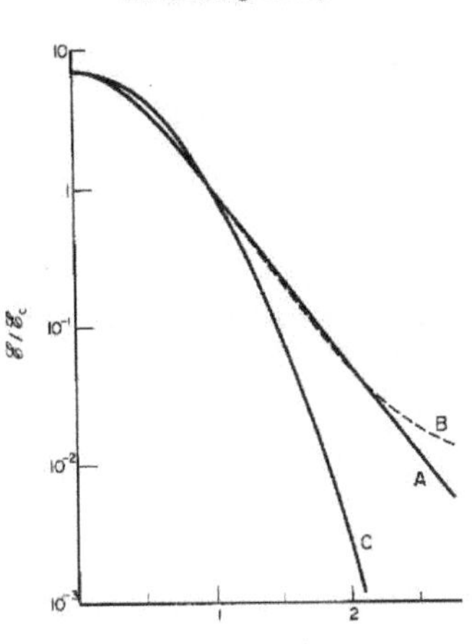

FIG. 17. Comparison of self-trapped mode with solution of self-focusing equation for $P = 1.003P_2$. (A) Self-trapped mode. (B) Self-focusing solution at $z = 1.79ka_0^2$. (C) Gaussian beam with same power as other modes. (From ref. 85.)

is plotted in Fig. 18, which is closely represented by the formula

$$\left[\left(\frac{P}{P_2}\right)^{1/2} - 0.852\right]^2 = 0.0219 + 0.135\left(\frac{k_0 a^2}{z_{\mathrm{f}}}\right)^2, \tag{6.4}$$

which is a hyperbola with asymptote

$$\left(\frac{P}{P_2}\right)^{1/2} = 0.852 + 0.367\frac{k_0 a^2}{z_{\mathrm{f}}}. \tag{6.5}$$

This relation has been repeatedly verified in experiments. C. C. Wang first used a simpler version of it to determine nonlinear indices for several organic liquids.[59]

The functional form of eqn. (6.5) was predicted by Kelley in an important paper in 1965.[11] Kelley argued that for powerful beams the self-focus should be near the focal point of the nonlinear lens induced by the incident intensity profile. Using eqn. (5.48) for the

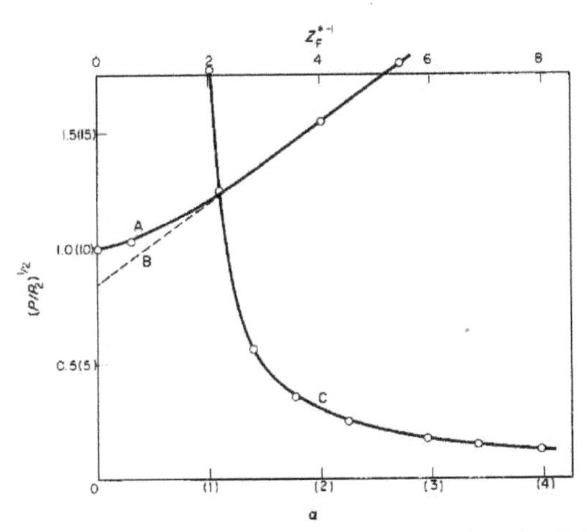

FIG. 18. (A) Dependence of self-focusing length on power. (B) Asymptote of (A). (C) Dependence of exponent α in eqn. (6.8) on power. (From ref. 47.)

phase S of $\mathscr{E}$ after a short distance Δz, we find that near the axis

$$S = \text{constant} - \mathbf{k}_0 r^2/2F$$

where the center of phase curvature is located at $z = F$, and

$$\frac{1}{F} = \frac{n_2}{n}\frac{\mathscr{E}_0^2}{a^2}\Delta z.$$

The distance through which the beam propagates in the nonlinear medium prior to the focus is $z_f = \Delta z \approx F$, so we estimate that

$$\frac{1}{z_f} \approx \sqrt{\frac{n_2 \mathscr{E}_0^2}{na^2}}.$$

This can be written as

$$\left(\frac{P}{P_2}\right)^{1/2} \approx 0.515\frac{k_0 a^2}{z_f}. \tag{6.6}$$

which is satisfyingly close to (6.5). This estimate is in error for large values of z_f because in that case diffraction causes the intensity profile to become distorted, and thus changes the strength of the induced lens.

For the same incident beam, the moment equations predict total beam collapse at z_s, where

$$\left(\frac{P}{P_2}\right)^{1/2} = 1.46\frac{k_0 a^2}{z_s}, \tag{6.7}$$

or $z_s = 3.95\,z_f$. It is not known to what extent the self-focus at z_f is a true mathematical singularity, but painstaking efforts to approach this point numerically indicate that the intensity effectively rises without limit there. This leads us to the following picture of the self-focusing optical field at $P \gg P_2$: The near axis rays converge to a singular focus at z_f, much as predicted by Kelley's simple argument, while the rays farther out in the wings pass by z_f. Between z_f and z_s there may be additional singular points, a singular line, or perfectly regular behavior. But at z_s, *all* the rays, including those which passed through the focus at z_f, converge to a common point, and the subsequent rms radius becomes imaginary. That is, the self-focusing equation (6.1) simply does not possess a physically meaningful solution beyond z_s (Fig. 19).

This picture does not quite agree with numerical work reported by Dyshko *et al.*[60] who discovered a series of axial maxima beyond z_f, with no apparent change in the situation beyond z_s. To understand this disagreement, the reader must appreciate that digital computers cannot solve eqn. (6.1) directly, but rather solve a difference equation whose solutions are defined on a discrete mesh. Hopefully, the discrete solution approaches the con-

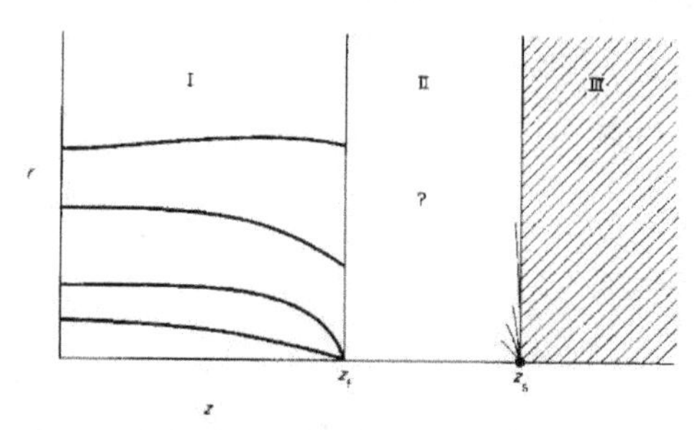

FIG. 19. Schematic representation of rays in I, prefocal region, and II, region beyond first catastrophic focus. The slowly varying envelope equation with cubic nonlinearity and no loss does not possess physical solutions in region III, beyond z_s given by eqn. (6.7). Exact solutions are not available in region II, although it is known that peripheral off-axis rays propagate smoothly into this region.

tinuous solution as the mesh is made finer. In any case, the discretization of the problem imposes a regular small-scale fluctuation on the simulated optical field. The presence of such a fluctuation, even if very small, influences the upper critical power P_3. To simulate the fluctuations induced by discretization, let

$$\mathscr{E}(r,0) = \mathscr{E}_0(1 + \alpha \exp iqr) \exp -r^2/2a^2$$

where $\alpha \ll 1$ and $qa \gg 1$ for a fine mesh. It is possible to compute P_3 exactly for this incident field, but the result becomes simple when qa is large:

$$P_3 \approx 4P_1[1 + (\alpha qa)^2].$$

The corresponding singularity distance is given by (5.35):

$$z_s \approx \sqrt{2}\, k_0 a^2 \left[\frac{P}{4P_1} - (1 + \alpha^2 q^2 a^2)\right]^{-1/2}.$$

If there are N mesh points between $r = 0$ and $r = a$, then $\alpha \lesssim 0.3/N$ and $q \approx 2\pi N/a$, which gives $\alpha qa \lesssim 2$. This can lead to an enhancement of z_s by 10% for $P/P_1 \approx 100$, with greater corrections for smaller powers. Thus the unavoidable presence of fluctuations on the simulated incident beam can push the cutoff singularity z_s beyond the range under investigation.

These unavoidable input ripples may also be responsible for intensity maxima beyond z_f. In an axially symmetric computation, such fluctuations tend to focus into annular rings and then contract toward the axis. As each ring shrinks to zero radius it produces a secondary focus beyond the first, at z_f, which is produced by near-axis rays. This picture is consistent with the published work of Dyshko *et al.*, and also with the experiences of several workers in the United States, including the author, who has constructed a variety of computer codes for studying this problem. Such studies are complicated by the fact that many numerical schemes do not exactly conserve beam power. Usually the mesh must be adjusted to prevent loss of power during the computation. This leads to an awkward situation when a bundle of rays converges to a point, and it is common for nonconservative schemes to lose about one critical power at each consecutive self-focal point. Such artificial "nonlinear absorption" will be discussed in the following section.

In any actual experiment, there will of course be a variety of ripples and fluctuations on the laser beam, and one expects these to self-focus independently of the main beam at high powers. In the most carefully controlled experiments, one still expects quantum noise, and axially symmetric amplitude and phase ripples of the sort commonly found in the solutions of laser cavity mode equations. These can lead to the formation of annular rings and eventual multiple self-foci in nonlinear materials. The actual radii of the rings, and hence the intervals between focal points, are governed as much by the growth rates at various radii as by the incident fluctuation magnitude at that radius. Hence, even if the focal intervals depend on fluctuations for their existence, they may well exhibit reproducibility in both computer solutions and experimental data. At this time, it is not certain whether multiple foci between z_f and z_s actually occur in the exact solutions of (6.1) with incident gaussian beams, and indeed the question is somewhat academic.

If the incident beam is not gaussian, the relation between focal position and power is altered, as well as the critical power P_2, even in the absence of ripples. For $P \gg P_2$, the derivation of (6.6) suggests that we replace a^2 in (6.4) by aa_p. The resulting formula for z_f agrees well with the numerical solutions, at high powers, for parabolic and lorentzian beam shapes. Equation (6.7) for z_s can also be evaluated for other beam shapes, but since it does not refer to the location of a singularity which is easily accessible to numerical computation, it has not been checked. The intensity profile dependence of the relation between P and z_s is very different from that obtained from Kelley's argument for the location of z_f. This lends further support to the view that the cutoff singularity at z_s is not the one which we usually associate with a self-focus.

For $P/P_2 \gtrsim 1.5$, the following analytical formula is a rather good representation of the axial intensity obtained numerically[47] (for gaussian beams)

$$\frac{I(z)}{I(0)} = \frac{1}{[1-(z/z_f)^2]^{\alpha/2}} \tag{6.8}$$

where α is plotted in Fig. 18 and approaches unity for $P \gtrsim 100P_2$. For high powers, an approximate analysis proposed by Kelley leads to this formula. For low powers, $P \lesssim 0.25P_2$, the following formula works well:

$$\frac{I(z)}{I(0)} = \frac{1}{1+[1-(P/P_2)]\,z^2/k_0^2a^4}. \tag{6.9}$$

This expression is similar to one derived using the weak focusing-constant shape approximation. The approximate theories from which these formulas may be obtained are discussed in Section 8.

If the incident phase fronts are not plane, then the self-focal position changes according to the lens formula (5.9). The new self-focus occurs at z_{fR}, where R is the focal length of the incident phase front, and

$$\frac{1}{z_{fR}} - \frac{1}{R} = \frac{1}{z_f}. \tag{6.10}$$

This can be inserted in eqn. (6.4) to obtain the relation between power and focal intensity for focused beams. If $R < 0$ (diverging beam), then the critical power is increased to P_{2R}:

$$\left(\frac{P_{2R}}{P}\right)^{1/2} = 0.852 + [0.0219 + (0.367\,k_0 a^2/R)^2]^{1/2}. \tag{6.11}$$

This is the same as eqn. (6.4) with z_f replaced by R, so Fig. 18 depicts $P_{2R}^{1/2}$ vs $1/R$, as well as $P^{1/2}$ vs $1/z_f$. If $R > 0$ (converging beam) the power required to form a singularity anywhere is still P_2, and the singularity first appears at the geometrical focal point. All these features are in excellent quantitative agreement with the numerical solutions. Figure 20

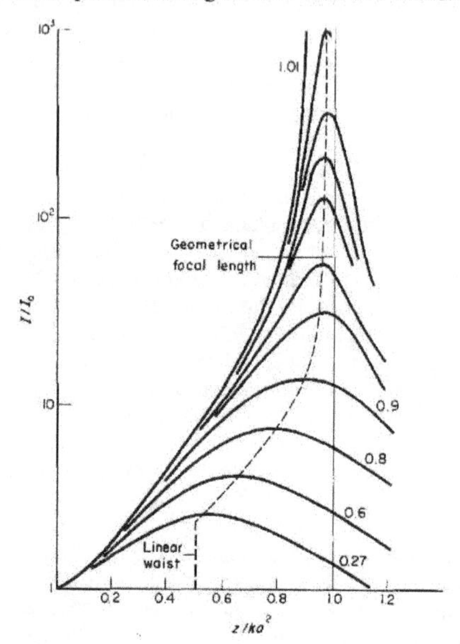

FIG. 20. Axial intensity for beams focused toward the axial point $z = ka^2$. The ratio P/P_2 for the lower power beams is indicated on the right. The point of maximum intensity does not quite reach the geometrical focal point at $P = P_2$, probably because of the relatively coarse numerical scheme employed.

shows the axial intensity in the focal region of a lens with $R = k_0a^2$, and a sequence of increasing powers. Notice that the intensity maximum moves toward the geometrical focal point as the power increases. When $P > P_2$, the self-focal point moves back again.

As pointed out in Section 5, the lens transformation may not be used rigorously to infer the field past the geometrical focal point. Numerical solutions past this point, for $P \approx P_2$, are extremely difficult to obtain and very sensitive to the nature of the numerical computation scheme. However, the physically comprehensible scaling law for the pre-focal field at $P = P_2$ given by eqn. (5.11), and the trend shown in Fig. 20, suggests that the post-focal field should obey the same law. That is, for $P = P_2$ there should be some symmetry about the geometrical focal point. In particular, we do not expect a trapped filament to form beyond the focus.

We have emphasized the strong self-focus which occurs for $P > P_2$, but there is also a weak axial intensity maximum for $P_1 < P < P_2$. This feature is caused by the central dip in the ray potential (5.4) for $P > P_1$, and has been studied analytically by Lugovoi by a method described in Section 8. Referring to Fig. 9, the reader can see that as the power is increased, rays starting from larger radii are focused toward the axis. An elementary annulus of off-axis rays carries more power (for $r \lesssim a$) than a comparable one of smaller radius, so the ratio of focal to input intensity is expected to be greater at higher powers. At the same time, the off-axis rays are focused less sharply toward the axis because of the weakening of the potential near its maximum. Thus we expect the axial intensity maximum to grow and move to farther distances as the power is increased. At $P = P_2$, the maximum becomes indistinct but vestiges of it can be seen at $z \approx 1.5\, k_0a^2$ in Fig. 10.

The lens formula can also be combined with eqns (6.8) or (6.9) to find the dependence of pre-focal intensity on focusing. The result for low powers is

$$I(z)/I(0) = \frac{1}{\left(1 - \frac{z}{R}\right)^2 + \left(1 - \frac{P}{P_2}\right) z^2/k_0^2a^4}. \tag{6.12}$$

This is the axial intensity of a focused gaussian with Fresnel length

$$l = \frac{k_0 a^2}{\left(1 - \frac{P}{P_2}\right)^{1/2}}. \tag{6.13}$$

By focusing strongly at low powers, one can evidently achieve controllable high focal intensities without causing a singular self-focus. Fradin *et al.* have exploited this possibility to measure intrinsic optical breakdown thresholds in solids.[61]

B. *Effects of absorption and saturation*

The simple cubic nonlinearity is certainly unphysical since it possesses solutions with singularities and imaginary rms beam radii. Much of the brief history of self-focusing theory consists of studies of a series of physical effects which alter this simplest nonlinear response, and lead to nonsingular solutions. Linear absorption does not remove the singularities unless the absorption length γ^{-1} is somewhat less than the self-focal length z_f. The "exact" equation (5.2a) gives the power reaching z as

$$P(z) = P(0) \exp - \gamma z$$

where

$$\gamma = \frac{4\pi k_0}{n_0^2} \chi_L''. \tag{6.14}$$

Figure 21 shows how an increasing absorption coefficient first shifts the self-focus to larger distances, and finally removes it altogether.[47] For a fixed power, the critical absorption coefficient γ_{max}, above which a singular self-focus will not form, may be estimated by setting $1/z_f = N\gamma_{max}$ in (6.5), where N should be of order unity. Comparison with numerical solutions gives $N \approx 0.55$ for $P/P_2 \approx 333$ and $N \approx 1.05$ for $P/P_2 \approx 13.4$. Thus for large

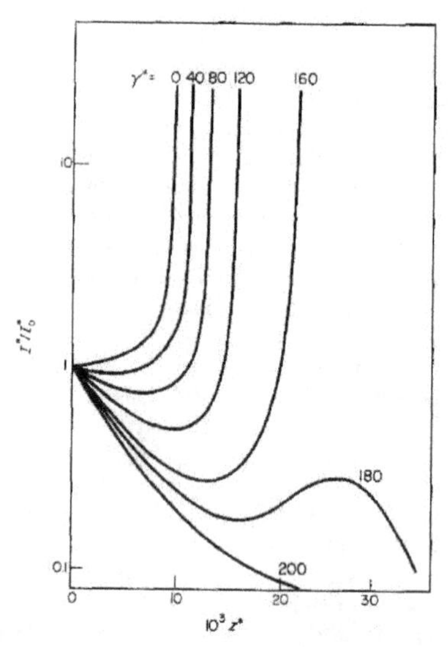

FIG. 21. Axial intensities in a self-focusing medium with absorption. The absorption coefficient $\gamma^* = \gamma/2ka_0^2$ is indicated. $z^* = z/2ka_0^2$. (From ref. 47.)

powers, the absorption must be somewhat stronger to remove the intensity focused toward z_f by the very strong initial nonlinear lens. Nevertheless, the agreement with this simple physical argument is satisfactory. A more detailed approximate theory of self-focusing with absorption will be discussed in Section 8C.

Nonlinear absorption, in which, for example,

$$\chi'' = \gamma_n |\mathscr{E}|^{2\nu}, \tag{6.15}$$

has also been studied by several workers, especially Dyshko *et al.*[62] When $\nu = 1$, catastrophic intensity rises still appear in the numerical solution for the axial intensity for sufficiently large input powers. Apparently nonlinearities greater than $\nu = 1$ are required to limit this intensity rise. This may be associated with the need to break the scaling properties of the general cubic self-focusing equation.

Some schemes for numerical solution of the self-focusing equation without loss actually simulate nonlinear loss if they do not conserve power exactly. Loss of power in such schemes usually occurs in features (spikes, rings) which develop on a scale close to the mesh size, and the loss is significant only if the intensity in such features is very high. Thus the scheme tends to "clip" high intensity peaks, and leave the low intensity features unaffected. This is exactly what nonlinear absorption does. Figure 22 shows the axial intensity obtained with a numerical technique described by Dyshko,[63] and employed without continually decreasing the mesh size to maintain power conservation. The exact spacing of the consecutive foci depends upon the mesh size employed. Presumably, a smaller mesh size simulates a higher value of ν in eqn. (6.15).

Singular self-foci can also be removed by allowing the nonlinear index to saturate. This conserves power and allows the study of the post-focal field for long distances. Figures 23 and 24 show intensity distributions for a nonlinear index change of form

$$\delta n = \frac{\frac{1}{2} n_2 |\mathscr{E}|^2 \mathscr{E}}{1 + |\mathscr{E}|^2/|\mathscr{E}_s|^2} \tag{6.16}$$

where $\mathscr{E}_s$ is a saturation field strength.[47] The period of the post-focal intensity fluctuations, and also the radial scale of the post-focal field can be estimated with rough accuracy by theories developed in Section 8.

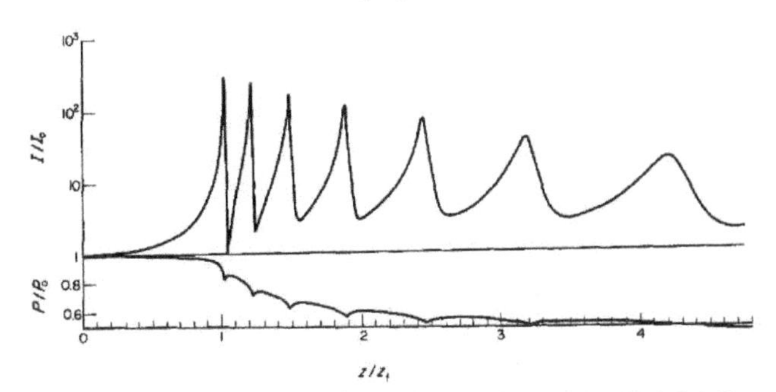

FIG. 22. Axial intensity obtained with code reported in ref. 63 showing multiple foci. The total power is plotted below. The incident power was $7P_2$ and about half the power is lost after seven foci. The intensity was rising again at $z = 5.6z_f$ where the computation was stopped. If the numerical scheme is adjusted to conserve power and retain accuracy, the intensity continues to rise without apparent bound at $z = z_f$.

The post-focal oscillation in Fig. 23 is the closest thing to a transition from a smooth large scale self-focusing beam to a steady-state "self-trapped filament" that the stationary theory is likely to produce. Mechanisms other than saturation which remove the singularity at z_f also remove power from the beam, either by absorption or by stimulated scattering. The saturation mechanism conserves power, but removes the singularity at z_f, introduces radial structure near the self-focal point, and allows the central portion of the beam to trap itself. In most materials, the actual saturation field is so high that dielectric breakdown can be expected to occur before saturation has any effect. Of course the electrons created in the breakdown process are effective in reducing the refractive index, and may themselves be regarded as providing saturation. There are also at least two other cases in which saturation occurs at field strengths less than the breakdown value. The resonant electronic nonlinearity in metal vapors studied by Grishkowsky saturates strongly at accessible field strengths.[19] Less tractable analytically is the steric interference effect treated

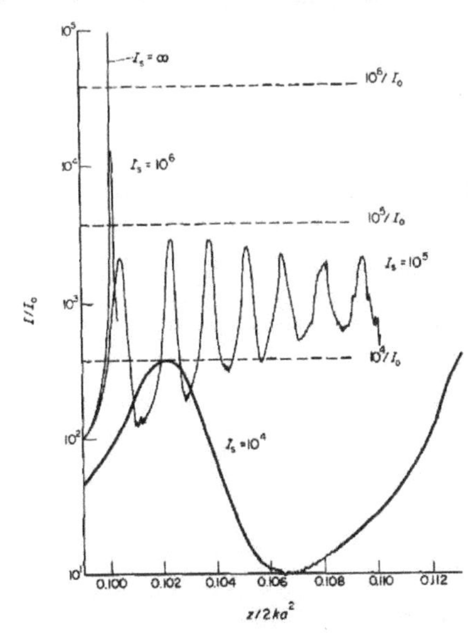

FIG. 23. Axial intensity for self-focusing in a medium with saturable nonlinear index. The incident gaussian beam was the same in all cases with $P = 7P_2$, but the saturation intensity $\mathcal{E}_s^2$ differs as indicated for the three curves. Notice that the z scale does not begin at zero.

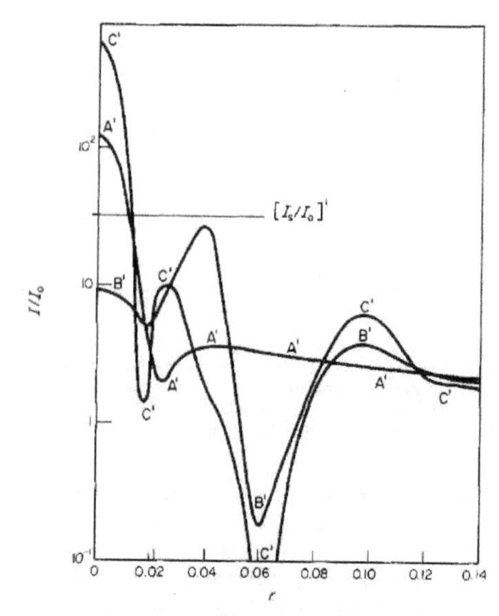

FIG. 24. Radial intensity for the post-focal field in a beam self-focusing in a medium with saturable nonlinear index. The incident gaussian had power $P = 333P_2$, and the saturation field is comparable to the lowest curve in Fig. 23. Curves A' and C' are taken at z-planes near the first and second axial maxima, and B' is near the first axial intensity minimum. The power in the central ring and first maximum is conserved.

by Gustafson and Townes which leads to low saturation fields for the molecular reorientation nonlinearity.[31]

The work of Grishkowsky and co-workers is interesting because the nonlinearity they studied is known with high accuracy to be of the form

$$\chi_{\mathrm{NL}} = -\frac{N\Delta\mathscr{E}_s^2}{[1 + 2(|\mathscr{E}|/\mathscr{E}_s)^2]^{1/2}}, \tag{6.17}$$

where $\Delta = n(\omega - \omega_R)$ may be of either sign depending upon the displacement of the laser frequency ω relative to the electronic resonance ω_R responsible for the nonlinearity. This form is characteristic of the interaction of an oscillating field with a two-level system in the dipole approximation where the envelope function of the applied field changes slowly with respect to the rate $(\omega - \omega_R)$. If the dipole transition operator has matrix element p, the saturation field is $\mathscr{E}_s = |\Delta/p|$. Grishkowsky measured optical intensity distributions at the output of a chamber filled with rubidium vapor, and found excellent agreement with numerical solutions of the self-focusing equation with nonlinearity (6.17) (see Fig. 25). No adjustable parameters need be introduced to improve the agreement.

C. *Self-focusing of vector waves*

The phenomena associated with polarization, and the property of *transversality* are two important features of the vector optical field. The polarization state is altered during passage through an anisotropic medium, while the property of transversality is modified in an inhomogeneous medium. Both effects may be induced by strong fields in a nonlinear medium, which may be homogeneous and isotropic in the linear regime. We shall treat them separately, beginning with polarization effects.

The polarization sensitivity of the nonlinear susceptibility has already been described in Section 3A. Restricting ourselves to isotropic substances, we may use (3.9) in the self-focusing equation to find

$$2ik_0\frac{\partial\mathscr{E}_\pm}{\partial z} + \nabla_\mathrm{T}^2\mathscr{E}_\pm = -\frac{4\pi\,\omega_0^2}{c^2}A\left|\,|\mathscr{E}_\pm|^2 + \left(1 + \frac{B}{A}\right)|\mathscr{E}_\mp|^2\,\right|\mathscr{E}_\pm. \tag{6.18}$$

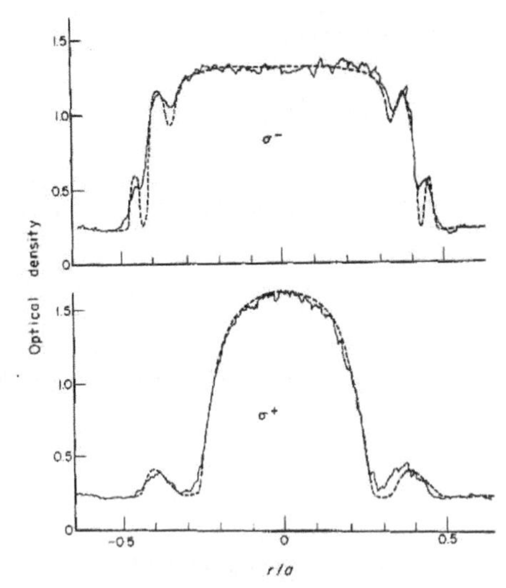

FIG. 25. Grischkowsky's self-defocused profile: theory and experiment. The two curves correspond to right and left circularly polarized near resonant beams incident on rubidium vapor in a magnetic field. For more details, see ref. 35.

Notice that if A/B is positive, the *weaker* circular component sees the *greater* induced lens. This means that the intensities of the two components tend to equalize, so that the polarization state tends toward linearity in the self-focal region. Figure 26 shows the ellipticity $\rho = \tan \chi$ (ratio of minor to major axes) of the polarization ellipse of a self-focusing beam for two different starting values, and $P = 10P_2$. When $\rho_0 = 0.8$, the ellipticity does not seem to evolve to zero, but oscillates about $\rho \approx 0.46$. Extensive investigations of this behavior have not been undertaken.

Maker *et al.*[64] predicted and observed self-induced polarization ellipse rotation, an effect which occurs even in the absence of self-focusing. Equation (6.18) can be integrated easily, if $\mathscr{E}$ does not depend on x and y, to give

$$\mathscr{E}_\pm(z) = \mathscr{E}_\pm(0) \exp \frac{-i2\pi A}{n_0^2} \int_0^z \left| \, |\mathscr{E}_\pm|^2 + \left(1 + \frac{B}{A}\right) |\mathscr{E}_\pm|^2 \right| dz.$$

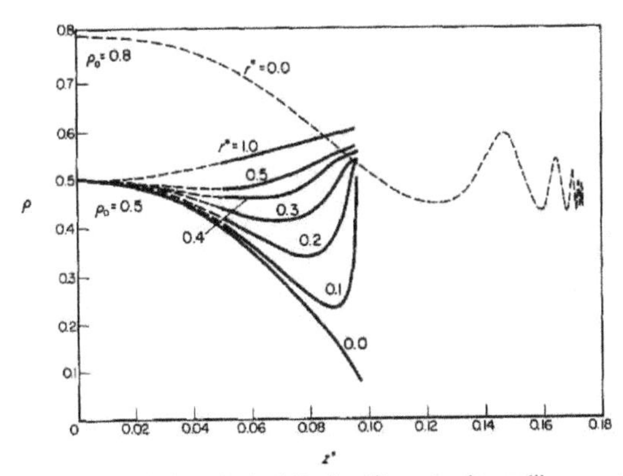

FIG. 26. Ratio ρ of minor to major axis of polarization ellipse, at various radii, versus axial distance along self-focusing beam.

This may be used to infer the rotation angle ψ of the polarization ellipse, which is half the phase difference between $\mathscr{E}_+$ and $\mathscr{E}_-$:

$$\psi - \psi_0 = \frac{\pi k_0 B}{n_0^2} \int_0^z [|\mathscr{E}_+|^2 - |\mathscr{E}_-|^2]\, dz, \tag{6.19}$$

where ψ_0 is the incident orientation. This equation forms the starting-point for analyses of ellipse rotation. Most experiments measuring this effect are carried out at low beam powers where self-focusing is not important, and both $\mathscr{E}_+$ and $\mathscr{E}_-$ propagate as if they were in a linear medium. At high powers, however, self-focusing causes gross departures from this simple situation, and ellipse rotation measurements must be analyzed numerically. Figure 27 shows numerical solutions for a beam at $P = 10P_2$.

The influence of self-focusing on ellipse rotation angle is expected to be small if the self-focusing length z_f greatly exceeds the characteristic ellipse rotation length z_ψ, where

$$\frac{1}{z_\psi} = \frac{2kB}{n^2} [|\mathscr{E}_+|^2 - |\mathscr{E}_-|^2]$$

$$= \frac{2kB}{n^2} |\mathscr{E}|^2 \sin 2\chi. \tag{6.20}$$

Figure 28 shows the relation between the power and the self-focusing length for a variety of ellipticities.

Now let us examine the influence of induced inhomogeneity on a linearly polarized vector wave in a self-focusing medium. The effects of interest vanish when we ignore the difference between $\nabla \times \nabla \times \boldsymbol{E}$ and $-\nabla^2 \boldsymbol{E}$ in the wave equation, or the terms on the second line of eqn. (2.8). Writing

$$\mathscr{E} = \mathscr{E}_T + \mathscr{E}_z \hat{z}, \qquad \mathscr{P}^{NL} = \mathscr{P}_T + \mathscr{P}_z \hat{z}$$

where $\hat{z} \cdot \mathscr{E}_T = \hat{z} \cdot \mathscr{P}_T = 0$, and using

$$\nabla \cdot \boldsymbol{E} = -\frac{4\pi}{n^2} \nabla \cdot \boldsymbol{P}^{NL}, \tag{6.21}$$

we find the slowly varying envelope equations with vector corrections:

$$2ik \frac{\partial \mathscr{E}_T}{\partial z} + \nabla^2 \mathscr{E}_T = -\frac{4\pi \omega^2}{c^2} [\mathscr{P}_T + k^{-2} \nabla_T \Phi], \tag{6.22}$$

$$\mathscr{E}_z = \frac{i}{k} \left| \nabla \cdot \mathscr{E} + \frac{4\pi}{n^2} \Phi \right|. \tag{6.23}$$

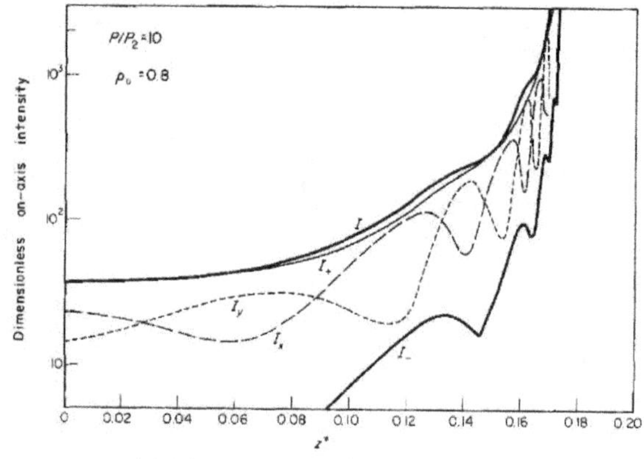

FIG. 27. Axial intensities of linear (I_x, I_y) and circular (I_+, I_-) polarization components of beam with incident ellipticity $\rho_0 = 0.8$ of polarization ellipse.

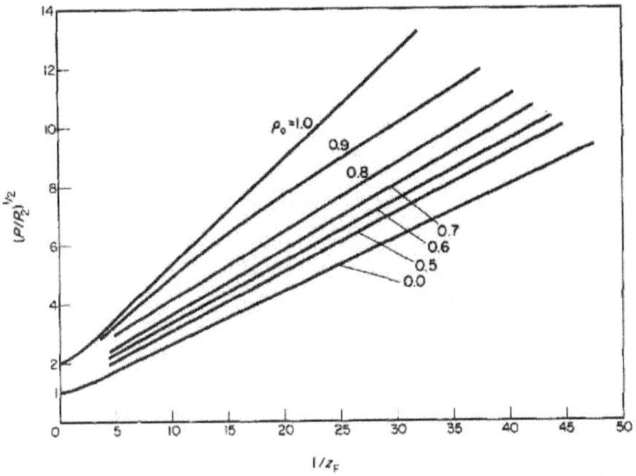

FIG. 28. Relation of self-focusing length to power for gaussian beams of varying elliptical polarization.

$$\phi = e^{-ikz}\nabla\cdot\mathcal{P}\,e^{ikz}. \tag{6.24}$$

Numerical analysis of (6.22) and (6.23) is awkward because the equations no longer preserve cylindrical symmetry and therefore $\mathcal{E}$ is always a function of three variables, x, y, z or r, θ, z. To simplify the analysis, we expand $\mathcal{E}$ and $\mathcal{P}$ in circular functions, e.g.

$$\mathcal{E}(r,\theta,z) = \sum_{\lambda} \mathcal{E}_{\lambda}(r,z)\,e^{-i\lambda\theta}. \tag{6.25}$$

We shall continue to use eqn. (3.7) for $\mathcal{P}^{NL}$. Insertion of (6.25) into (6.22) and (6.23) leads to a set of coupled equations for the coefficients $\mathcal{E}_\lambda(r,z)$ which may be decoupled by ignoring the terms $\lambda > 2$. This amounts to keeping the least orders which give finite corrections to $\mathcal{E}_T$ and $\mathcal{E}_z$. The final equations for the nonvanishing coefficients when the incident beam is linearly polarized along $\hat{x}$ are

$$2ik\frac{\partial\mathcal{E}_{x0}}{\partial z} + \nabla_1^2\mathcal{E}_{x0} = -\frac{4\pi}{n^2}(A+\tfrac{1}{2}B)\Big\{k^2|\mathcal{E}_{x0}|^2\mathcal{E}_{x0} + \left|\frac{\partial\mathcal{E}_{x0}}{\partial r}\right|^2\mathcal{E}_{x0} + \frac{1}{2}\left(\frac{\partial\mathcal{E}_{x0}}{\partial r}\right)^2\mathcal{E}^*_{x0} + \left[\frac{A+B}{2A+B}\right]\mathcal{E}_{x0}\nabla_1^2|\mathcal{E}_{x0}|^2\Big\}, \tag{6.26}$$

$$2k\frac{\partial\mathcal{E}_{y2}}{\partial z} = -\frac{\pi(A+B)}{n^2}\left[\frac{\partial}{\partial r} - \frac{1}{r}\right]\left[\mathcal{E}_{x0}\frac{\partial|\mathcal{E}_{x0}|^2}{\partial r}\right], \tag{6.27}$$

$$2ik\,\mathcal{E}_{z1} = -\frac{\partial\mathcal{E}_{x0}}{\partial r}, \tag{6.28}$$

$$\mathcal{E}_{x2} = -i\mathcal{E}_{y2},\quad \mathcal{E}_{x,-2} = \mathcal{E}_{x2},$$
$$\mathcal{E}_{z,-1} = \mathcal{E}_{z1},\quad \mathcal{E}_{y,-2} = \mathcal{E}_{y2}. \tag{6.29}$$

The total electric field vector is

$$E(x,t) = \mathrm{Re}\,e^{i(kz-\omega t)}[(\mathcal{E}_{x0} + 2\mathcal{E}_{x2}\cos 2\theta)\hat{x} - (2i\mathcal{E}_{y2}\sin 2\theta)\hat{y} + (2\mathcal{E}_{z1}\cos\theta)\hat{z}]. \tag{6.30}$$

Figures 29 and 30 show the radial dependence of the contributions to the electromagnetic energy density of each of the three components $\lambda = 0, 1, 2$, at the points $z \approx 0.45\, z_f$

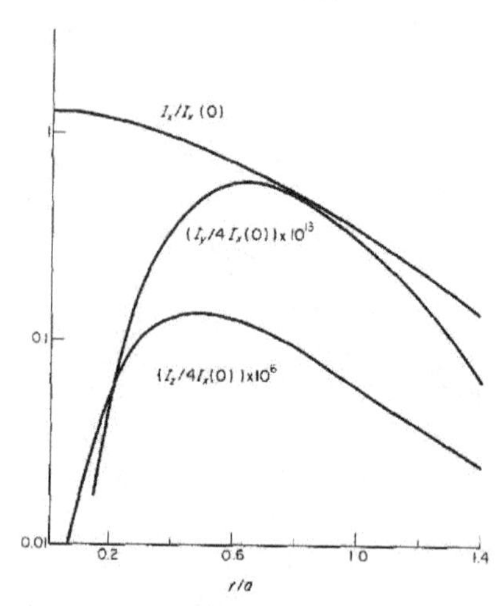

FIG. 29. Radial intensity profiles of polarization components of an initially x-polarized gaussian beam in a self-focusing medium with cubic nonlinearity. Vector corrections were retained in the self-focusing equation. The axial distance is nearly half the distance to the first self-focal point z_f.

and $z \approx 0.99\, z_f$. These curves were obtained by numerical solution of (6.26)–(6.29) with an incident gaussian of power $P = 7P_2$, and $ka = 5 \times 10^3$ ($a \approx 0.8$ mm if $\lambda = 1\ \mu$m). The susceptibility parameters satisfy $B = 6A$, corresponding to the Kerr nonlinearity. Notice that the y and z polarized intensities are greatest near the half width of the dominant x polarized distribution.

Vector corrections to the self-focusing equations have been studied by Abakarov *et al.*[43] using perturbation theory. Kerr has also reported an approximate treatment. The interest in such corrections centers on the possibility that they may remove the catastrophic intensity rise at the initial self-focus, obviating the need for auxiliary limiting

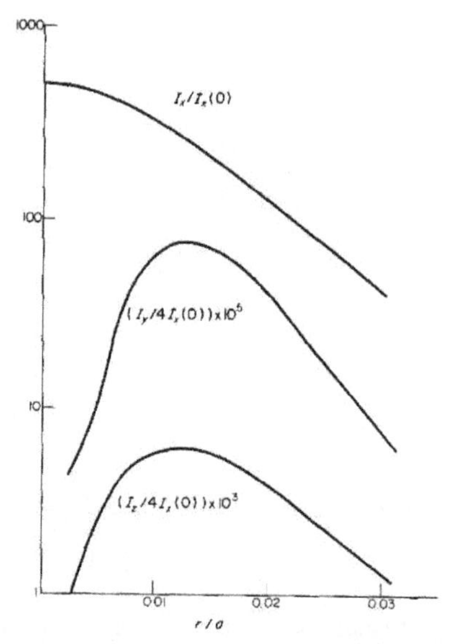

FIG. 30. Same as Fig. 29, but now $z = 0.99z_f$.

mechanisms. D. R. White has performed a series of numerical analyses of the optical field in the behavior of a self-focus, including all vector corrections and also the $\partial^2\mathscr{E}/\partial z^2$ terms usually neglected in the slowly varying envelope approximation. While the critical power for formation of a strong self-focus (P_2) is slightly increased when the additional terms are included, beams with high power still exhibit precipitous axial intensity rises at self-focal points very close to the position predicted by the scalar, slowly varying envelope theory. Despite prodigious computational effort, White never observed a saturation in the intensity rise as such a focal point was approached.

7. TIME DEPENDENT NONTRANSIENT SELF-FOCUSING

A. *Properties of the moving focus*

In the previous sections, the incident field was assumed stationary, with all time dependence in the factor exp $-i\omega t$. But the solutions obtained in this case are frequently also valid when the incident field amplitude varies quasistatically, so transient nonlinear response is unimportant. This important conclusion follows from eqn. (2.9), omitting vector corrections, dispersive pulse distortion, and $(\partial\mathscr{P}^{NL}/\partial t)/\omega_0\mathscr{P}^{NL}$. Introducing the new variable $t' = t - z/v_g$, we find

$$2ik_0\frac{\partial\mathscr{E}'}{\partial z} + \nabla_1^2\mathscr{E}' = \frac{-4\pi\omega_0^2}{c^2}\mathscr{P}'^{NL}, \qquad (7.1)$$

where the primed functions have t' rather than t as independent variable:

$$\mathscr{E}'(x, y, z, t') = \mathscr{E}(x, y, z, t' + z/v). \qquad (7.2)$$

As this has the same form as the stationary self-focusing equation (5.1), its solution may be obtained from the stationary solution f expressed as a functional of the time-dependent initial field. Omitting the vector signs and replacing x, y by r, we have

$$\mathscr{E}'(r, z, t') = f[r, z; \mathscr{E}(r', 0, t')]. \qquad (7.3)$$

This simple and essentially exact result (for quasistatic initial field amplitudes) is the origin of the *moving focus theory*. The self-focal regions, determined by the maxima or singularities of $|f|^2$, move about as the incident field changes during a pulse. As Lugovoi *et al.* first noted,[65] such moving focal regions, when observed with poor time resolution, appear as streaks or filaments, and are probably the cause of most of the filamentary phenomena associated with high-power nonlinear propagation. This judgement is certainly correct if the incident beam is smooth, and not so far above critical power that spatial noise fluctuations grow significantly during propagation. The accompanying review by Shen outlines the relevant experimental investigations of this point.

If the incident field can be written as the product of a mode function and a pulse-shape function,

$$\mathscr{E}(r, 0, t) = P^{1/2}(t)g(r), \qquad (7.4)$$

where $P(t)$ is the incident power, then the positions of particular features in the stationary solution may be plotted as functions of P. The resulting plot may be used as a transfer curve to find the trajectories of these features for arbitrary input pulse shapes. In Fig. 31, for example, we have plotted the incident power required in a self-focusing beam to cause the on-axis field strength at z_B to equal the dielectric breakdown strength $\mathscr{E}_B$ of the medium. This relation is obtained from

$$\mathscr{E}_B = |f[0, z_B; P^{1/2}g]|.$$

The position of the breakdown region (on the axis) at any time during the passage of a pulse is then obtained from the intersection of the pulse profile at time t with the curve P vs z_B, as shown in the figure. There may be several branches of the P vs z_B curve, corresponding to multiple self-foci. The pulse curve may intersect each branch twice, leading to two "breakdown regions".

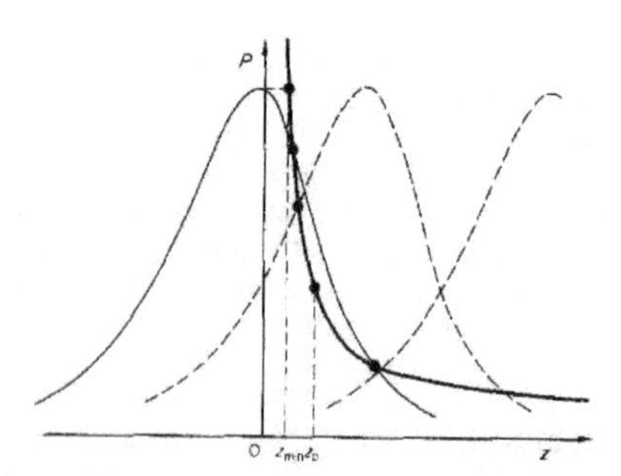

FIG. 31. Graphical determination of trajectory of breakdown region. The intersection of the power pulse curve $P(t - z/v_g)$ with the transfer curve P vs z_B, which gives the axial location of the breakdown intensity for input power P, fixes the location of breakdown regions within the medium at any instant. If the breakdown causes a damage track, it would begin at z_{min}, although the earliest appearance of breakdown occurs at z_0.

The trajectory of the nearest axial singularity at z_f can be obtained for gaussian beams directly from eqn. (6.4), or from its asymptote if z_f is much less than the incident near field length ka^2:

$$z_f(t) \simeq \frac{0.367\, ka^2}{\left[\left[P\left(t - \frac{z_f(t)}{v_g}\right)\Big/P_2\right]^{1/2} - 0.852\right]}. \tag{7.5}$$

This implicit equation may be solved explicitly for $t(z_f)$ if $P(t)$ is the parabolic pulse function discussed in Section 3: $P = P_0\,[1 - (t/T)^2]$, $|t| \leq T$. The result

$$v_g t = z_f \pm v_g T\left[1 - \frac{P_2}{P_0}\left(0.852 + 0.367\,\frac{ka^2}{z_f}\right)^2\right]^{1/2} \tag{7.6}$$

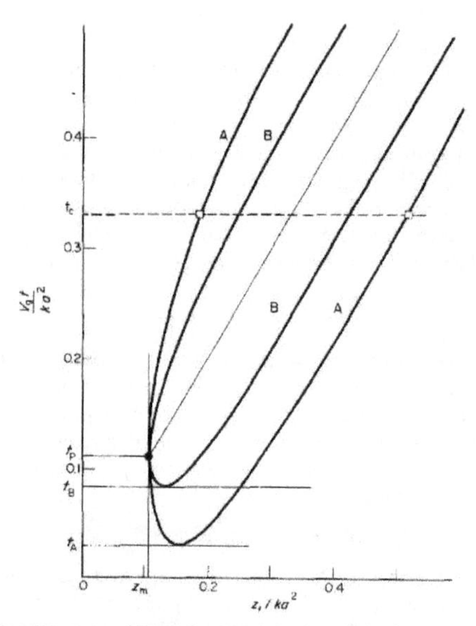

FIG. 32. Self-focal trajectories t vs z_f for two parabolic pulses of duration (A) $T = 0.2$, (B) $T = 0.1$ in time units of ka^2/v_g. z_m is the least focal length, which occurs at t_p. Intense focal points first appear at times t_A and t_B for the two cases.

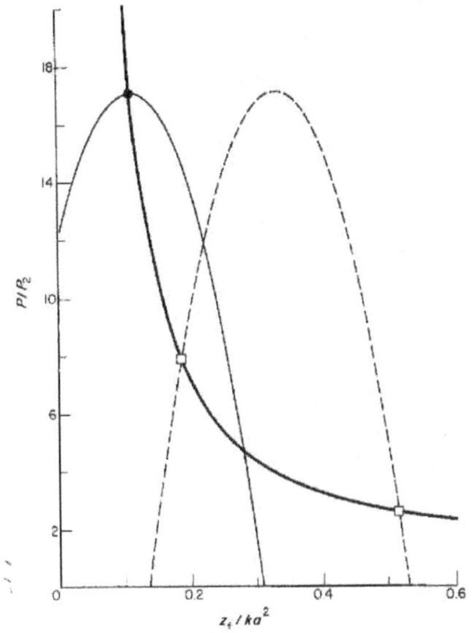

FIG. 33. Transfer curve construction analogous to Fig. 31 for the pulse A shown in Fig. 32. The black dot at z_m and the squares locate corresponding focal positions here and on Fig. 32.

gives the U-shaped curve plotted in Fig. 32. Each horizontal line on this graph intersects the U-curve at z values corresponding to the locations of the self-focal points. Figure 33 shows the corresponding transfer curve construction. It is clear from these figures that the self-focus first appears at (z_0, t_0), and appears to split into two focal regions, one of which moves forward, and one backward to a minimum z_m determined by the peak pulse power, then moves forward again. After a long time, the two focal regions move forward with the group velocity, separated by a constant distance Δ. For unfocused gaussian beams and parabolic pulses, z_m and Δ satisfy

$$\left(\frac{P_0}{P_2}\right)^{1/2} = 0.852 + 0.367\,\frac{ka^2}{z_m} \tag{7.7}$$

$$\Delta = 2v_g T\left|1 - 0.726\,\frac{P_2}{P_0}\right|^{1/2}. \tag{7.8}$$

Analyses similar to these for the moving focus were first reported by Loy and Shen,[66] and Taran and Gustafson[67] in 1969. Other workers have employed them with various degrees of sophistication to interpret damage tracks in solid materials.[68]

The reader will notice that the U-curve of Fig. 32 forms a boundary in the z, t plane outside which no singularities appear. There is another boundary inside this curve formed by the track of the cutoff singularity z_s within which the self-focusing equation with cubic nonlinearity has no physical solution. Between these two curves, the mathematical structure of the solutions of this equation is not known. Physically, however, we know the cutoff singularity is removed by other processes. Furthermore, if breakdown occurs near the self-focal regions, we can expect that subsequent light passing through material visited by these regions will be scattered, thus altering further focal formation. For example, after time t_p in Fig. 33, when the backward-moving focus reverses its motion, this focus must sweep forward through material already exposed to a focal trajectory. Whether a self-focus forms at all in this sensitized material depends upon precisely what happened in the path of the prior focus. Even the forward-moving focus, which always sees fresh material, is subject to self-disruption through stimulated inelastic scattering, or elastic scattering from free electrons generated by the breakdown process.

B. *Local pulse distortion*

The self-focusing field for quasistatic incident pulses is, according to (7.3), a nonlinear function of the incident pulse shape. This causes the local time dependence of the field at a point $z \neq 0$ to differ from the incident time dependence.[13] The intensity pulse at each point r, z can be obtained from the incident pulse shape $P(t)$ if the dependence of the stationary solution upon the initial power P is known:

$$|\mathscr{E}'(r, z, t')|^2 = |f(r, z; P^{1/2}(t')g)|^2. \tag{7.9}$$

(The initial field has form (7.4).) Figure 34 shows how a plot of local intensity at r, z vs incident power can be used as a transfer curve (like an amplifier characteristic) to find the local pulse shape from the incident pulse shape.

Since the IP characteristic always increases sharply near the power P_z forming a self-focus at z, the local pulse is severely distorted there. That is, the local pulse rise time is always much shorter than the incident rise time at points in the medium near a self-focus. In fact the local rise time roughly equals the time in which the incident power climbs to P_z. For a parabolic pulse $P(t) = P_0[1 - (t/T)^2]$, whose peak power greatly exceeds P_z, the local rise time T_z is easily seen to be (from eqn. (6.5))

$$\frac{T_z}{T} = 1 - \left|1 - \frac{P_z}{P_0}\right|^{1/2} \tag{7.10}$$

or

$$\gtrsim \frac{1}{2}\frac{P_2}{P_0}\left|\frac{0.367\,k_0 a^2}{z}\right|^2. \tag{7.11}$$

Thus for $P_0 \gtrsim 50P_2$ and $z = 0.367\,k_0a^2$, the rise time is reduced by 0.01.

This enormous local pulse sharpening causes the quasistatic approximation to be incorrect near a self-focus. That is, even in the absence of material breakdown, saturation or other effects weakening the self-focal singularity, the nonlinear response may be weakened by the inability of the polarization to follow the extremely fast local intensity rise near a self-focus. This point is discussed more thoroughly in Section 9.

Pulse sharpening concomitant with self-focusing has often been observed directly,[66,69] and has also been used to infer the nonlinear susceptibility.[70] One simply reconstructs the $I(z)$ vs P transfer curve by monitoring output axial intensity and input power during a pulse, and then compares the result with the curve obtained from numerical solution of the self-focusing equation with the appropriate experimental incident beam as initial condition. The method requires that the incident beam be accurately known.

C. *Self-phase modulation without self-focusing*

If the incident optical field is nonstationary, then the induced non-linear index, and hence the optical phase, also varies with time.[71,72] This additional time dependence of the phase contributes to the "instantaneous frequency" which we define to be

$$\omega(t) = \omega_0 - \partial\varphi/\partial t \tag{7.12}$$

where φ is the phase of the amplitude function $\mathscr{E}$.

Self-phase modulation is present even when there is no self-focusing, as when the incident field does not depend on x and y. In this case eqn. (7.1) becomes, for the cubic nonlinearity $\mathscr{P}^{NL} = \eta|\mathscr{E}|^2\mathscr{E}$,

$$2ik_0\frac{\partial\mathscr{E}'}{\partial z} = -\frac{4\pi\omega_0^2}{c^2}\eta|\mathscr{E}'|^2\mathscr{E}', \tag{7.13}$$

which has the solution

$$\mathscr{E}'(z, t') = \mathscr{E}_0(t')\exp i(n_2\omega_0|\mathscr{E}_0(t')|^2 z/2c). \tag{7.14}$$

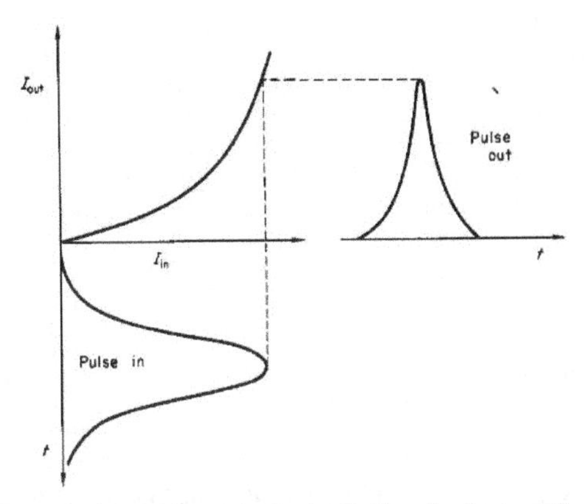

FIG. 34. Characteristic curve for determining local pulse distortion due to self-focusing. A graph of local pulse intensity at a point r, z in the medium versus incident axial intensity may be used to find pulse shape at that point from a given incident pulse shape.

This implies the instantaneous frequency

$$\omega(t) = \omega_0 - \frac{z}{2k_0 a^2} \frac{\partial}{\partial t} \frac{P(t')}{P_1} \tag{7.15}$$

where P_1 is the lower critical power and $P(t)$ the incident power in area πa^2. For a parabolic incident pulse, $P(t) = P_0 [1 - (t/T)^2]\, \theta(1 - (t/T)^2)$, the frequency shift is

$$\omega(t') - \omega_0 = \frac{P_0}{P_1} \frac{zt'}{k_0 a^2 T^2} \theta [1 - (t'/T)^2]. \tag{7.16}$$

Thus we expect an output spectrum with symmetrical Stokes and antiStokes structure extending to about

$$\Delta\omega = \frac{P_0}{P_1} \frac{z}{k_0 a^2} \frac{1}{T} \tag{7.17}$$

on either side of the central frequency. Notice that the frequency spread increases with path length, incident intensity and inverse pulsewidth.

The actual spectrum of the field (7.14) must be obtained by Fourier analysis. Thus the detected intensity in the frequency range ω, $\omega + \Delta\omega$ is

$$I_\omega \Delta\omega = \frac{nc}{4\pi^2} \frac{1}{T_s} \int_\omega^{\omega + \Delta\omega} |\hat{\mathscr{E}}(\omega - \omega_0)|^2 \, d\omega \tag{7.18}$$

where T_s is the time available to determine the spectrum. Here $\hat{\mathscr{E}}(\omega)$ is nearly the Fourier transform of $\mathscr{E}'(z, t')$ at fixed z if T_s is very long:

$$\hat{\mathscr{E}}(\omega - \omega_0) = \int_{-t}^{t} d\tau \, \hat{\mathscr{E}}(z, t') \exp i (\omega - \omega_0) t'. \tag{7.19}$$

This integral may be evaluated numerically, but further analysis is possible if the phase of $\mathscr{E}'$ becomes large compared with π, so that the amplitude $\mathscr{E}_0(t')$ is slowly varying with respect to the exponential.[72] Then the greatest contribution to $\hat{\mathscr{E}}$ comes from times t' such that the phase $(\omega - \omega_0)t' - \varphi(t')$ of the integrand is "stationary". This leads to the condition (7.12). The method of stationary phase then yields[73]

$$\begin{aligned} \hat{\mathscr{E}}(\omega - \omega_0) &\approx \sum_\nu \mathscr{E}_0(t'_\nu) e^{i\varphi_\nu} e^{i(\omega - \omega_0)t'_\nu} \int_{-\infty}^{\infty} \exp \frac{i}{2} \varphi''_\nu (t' - t'_\nu)^2 \, dt' \\ &\approx \sum_\nu \mathscr{E}_0(t'_\nu) e^{i\varphi_\nu} e^{i(\omega - \omega_0)t'_\nu} e^{i\pi/4} [2\pi/\varphi''_\nu]^{1/2} \end{aligned} \tag{7.20}$$

where t'_ν is the νth root of eqn. (7.12) for fixed ω, and $\varphi_\nu, \varphi''_\nu$ are the phase function and its second derivative at t'_ν. Our parabolic pulse function only yields one root of (7.12) for each ω, but realistic pulses always possess at least two roots (or no roots) because $|\partial P/\partial t'|$ rises from zero for large negative t' to a maximum at some value t'_r, and then returns to zero at the pulse peak t'_p. Beyond the pulse peak $|\partial P/\partial t'|$ rises to another maximum at t'_f and vanishes again at large $t' > 0$ (see Fig. 35). The sign of φ'' changes at t'_r and t'_f, which denote the inflexion points on the rise and fall of the pulse shape.

For the pulse of Fig. 35, we can write

$$\hat{\mathscr{E}}(\omega - \omega_0) \approx \mathscr{E}_0(t'_1)\, e^{i(\varphi_1 + (\omega-\omega_0)t'_1)} e^{i\pi/4}(2\pi/|\varphi''_1|)^{1/2} + \mathscr{E}_0(t'_2)\, e^{i(\varphi_1 + (\omega-\omega_0)t'_2)} e^{-i\pi/4}(2\pi/|\varphi''_2|)^{1/2}. \quad (7.21)$$

The corresponding intensity will possess a minimum when the phase difference between the two contributions is an odd multiple of π, or

$$(\varphi_1 - \varphi_2) + (\omega - \omega_0)(t'_1 - t'_2) = (4l - 3)\pi/2 \quad (7.22)$$

where $l = 1, 2, 3, \ldots$, etc. (We have included the additional phase contributed by the sign difference between φ''_1 and φ''_2 which seems to be omitted in the literature.) From this result, one can conclude that the maximum number of minima l_m (on one side of the center frequency ω_0) satisfies

$$\varphi(t'_p) = (4l_m - 3)\pi/2, \quad (7.23)$$

because the points 1 and 2 for maximum phase difference correspond to the peak t'_p and far wing of the pulse, where $\varphi' = 0$ and $\omega = \omega_0$.

Equation (7.23) can be used to infer the maximum nonlinear index change from the spectral properties of the transmitted light, if one is confident that the local pulse shape is as simple as that of Fig. 35. In many substances, however, the transmitted light contains intense inelastically scattered components at well-defined frequency shifts corresponding to Raman, Brillouin and Rayleigh wing processes.[74,75] In these cases, the nonlinear index can be modulated at a beat frequency of two coherent spectral components. The resulting self-phase modulated spectrum is not difficult to compute for two CW incident waves at ω_a, ω_0 (neglecting transverse spatial variations) and possesses an envelope which closely resembles that computed for a single short pulse at ω_0. Taking $\mathscr{E}_0(t)$ in (7.14) equal to

$$\mathscr{E}_0(t) = \mathscr{E}_a e^{-i\Delta t} + \mathscr{E}_b e^{i\Delta t} \quad (7.24)$$

where $\mathscr{E}_a$ and $\mathscr{E}_b$ are real, and $\Delta = \frac{1}{2}(\omega_a - \omega_b)$, $\omega_0 = \frac{1}{2}(\omega_a + \omega_b)$, one finds

$$\mathscr{E}'(z, t) \approx \mathscr{E}_a e^{i\varphi_0} e^{-i\Delta t} \exp\, i(n_2\omega_0 z\mathscr{E}_a\mathscr{E}_b/2c) \cos \Delta t \quad (7.25)$$

where φ_0 is independent of time if $\mathscr{E}_a$ and $\mathscr{E}_b$ are, and we have taken $\mathscr{E}_a \gg \mathscr{E}_b$. This corresponds to a strong signal $\mathscr{E}_a$ at ω_a and a weak one at $\omega_b = \omega_a + 2\Delta$. The resulting spectrum

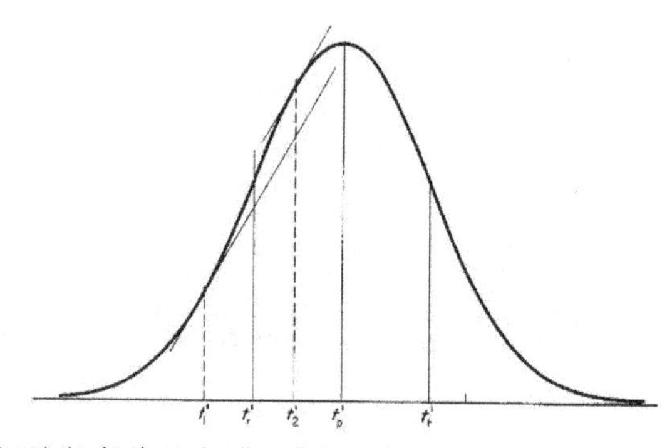

FIG. 35. A typical pulse shape, showing inflexion points at t'_r and t'_f, and two points with equal temporal slope at t'_1 and t'_2 corresponding to two roots of eqn. (7.12).

can be found from the Fourier transform of (7.25), which is

$$\hat{\mathscr{E}}(\omega-\omega_0)=\mathscr{E}_a e^{i\varphi_0}2\pi\{\delta(\omega-\omega_0-\Delta)J_0(\zeta)$$

$$+\sum_{\nu=1}^{\infty} i^\nu J_\nu(\zeta)[\delta(\omega-\omega_0+(\nu-1)\Delta)+\delta(\omega-\omega_0-(1+\nu)\Delta)]\}. \quad (7.26)$$

This yields a line spectrum whose envelope is proportional to $J_\nu^2(\zeta)$, where (compare with (7.15))

$$\zeta = n_2\omega_0\mathscr{E}_a\mathscr{E}_b z/2c$$
$$=\frac{(P_aP_b)^{1/2}}{P_1}\frac{z}{2k_0a^2}.$$

Here P_a and P_b are the powers in the incident beams in area πa^2. Regarded as a function of its order, the Bessel function $J_\nu(\zeta)$ decays exponentially beyond $\nu\approx\zeta$ so the width of the transmitted spectrum is roughly

$$\Delta\omega\approx\frac{(P_aP_b)^{1/2}}{P_1}\frac{z}{k_0a^2}\frac{(\omega_a-\omega_b)}{4} \quad (7.27)$$

which is comparable to (7.17) for a single pulse of width $T=4/(\omega_a-\omega_b)$. Some experimentally observed spectra actually show the substructure with spacing Δ indicated by (7.26). Gustafson *et al.*[18] have pointed out that this substructure could be smeared by other processes, including the time dependence of the envelopes $\mathscr{E}_a$ or $\mathscr{E}_b$, which has been omitted here. We conclude that the self-phase modulated spectrum alone is not a reliable tool for determining properties of the optical field within the medium. Other measurements must be made to narrow the choice of mechanisms for spectral broadening. Figure 36, which is taken from the paper of Gustafson *et al.*,[18] compares the spectra arising from a pulse and from nonlinear mixing of two spectral components.

Assuming that the modulation mechanism is inoperative, Cubeddu *et al.* have suggested that additional interesting properties of the pulse shape $|\mathscr{E}_0(t')|$ can be inferred from the behavior of the self-phase modulated spectrum near its extremes.[73] These extreme frequencies are contributed by the pulse intensity near the maxima of $\partial\varphi/\partial t'$ and t'_r and t'_f (see Fig. 35) where $\varphi''=0$, and eqn. (7.20) fails. Retaining the next term in the expansion of the phase about its inflexion point, we find that (7.20) must be replaced by

$$\hat{\mathscr{E}}(\omega-\omega_0)\approx\mathscr{E}_0(t'_r)e^{i\varphi_r}e^{i(\omega-\omega_0)t'_r}\int_{-\infty}^{\infty}\exp i[\varphi'_r+\omega-\omega_0)t+\tfrac{1}{6}\varphi'''_r t]\,dt$$

$$=\mathscr{E}_0(t'_r)e^{i\varphi_r}e^{i(\omega-\omega_0)t'_r}\frac{2\pi}{(\frac{1}{2}\varphi'''_r)^{1/3}}\,\mathrm{Ai}\left[\frac{(\varphi'_r+\omega-\omega_0)}{(\frac{1}{2}\varphi'''_r)^{1/3}}\right] \quad (7.28)$$

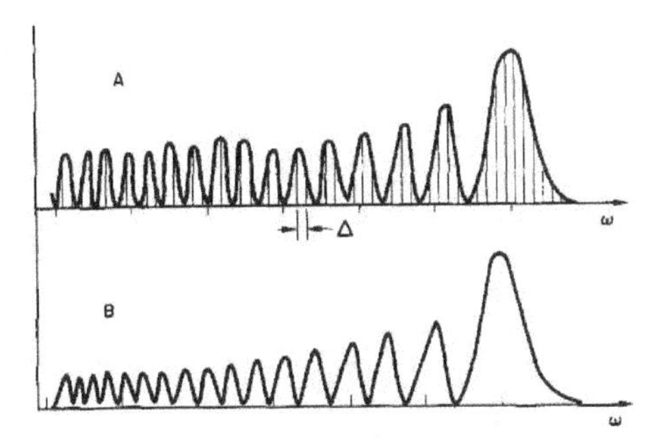

FIG. 36. Self-phase modulated spectra from A: a weakly modulated pulse eqn. (7.24), and B: a gaussian pulse of width comparable to the inverse modulation frequency of A. In A, the spectrum actually consists of discrete lines spaced by Δ [eqn. (7.24)], but the envelope closely resembles spectrum B. (From Gustafson *et al.*[18])

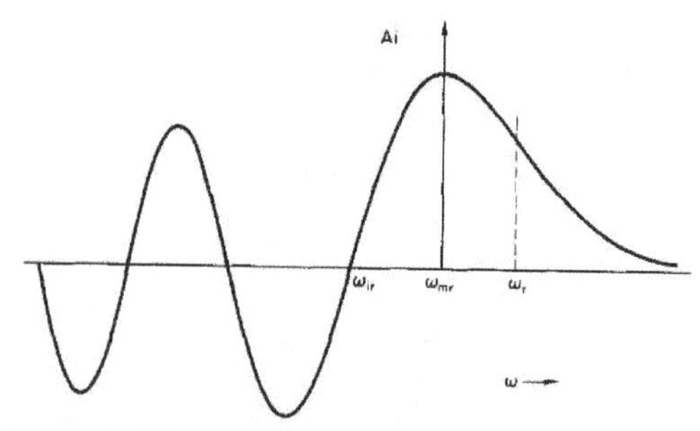

FIG. 37. The Airy function Ai(x) near its turning-point. The self-phase modulated spectrum is proportional to $[\text{Ai}(x)]^2$ here, where $x = (\omega - \omega_0 + \phi_r')/(\frac{1}{2}\phi_r''')^{1/3}$ [*cf.* eqn. (7.28)]. Notice that the maximum occurs prior to $\omega_r = \omega_0 - \phi_r'$.

where Ai(x) is an Airy function, and ω must be near the instantaneous frequency associated with t_r'

$$\omega_r = \omega_0 - \partial\varphi/\partial t'|_{t_r'}. \tag{7.29}$$

Since the Airy function Ai(x) peaks prior to $x = 0$ at $x = -1.02$ (and then decays exponentially) the extreme peak of the spectrum falls somewhat short of ω_r. From (7.28) and Fig. 37 it is clear that φ_r''' may be determined from the frequency shift of the extreme peak at ω_{mr} and the frequency at the first spectral minimum at ω_{1r}:

$$\varphi_r''' = 0.870(\omega_{mr} - \omega_{1r})^3. \tag{7.30}$$

The numerical factor differs somewhat from that reported by Cubeddu *et al.*,[73] presumably because they used an approximate form for (7.28).

Using eqn. (7.28) and the tabulated properties of the Airy function, one may reconstruct the principal features of the phase function, including its first and second derivatives at the inflexion points.

D. *Self-phase modulation with self-focusing*

When self-focusing is present, the optical phase depends upon the transverse variables x, y, and reflects the drastic pulse sharpening associated with the moving focus. Gustafson *et al.*[18] and also Shen and Loy[25] have pointed out that this pulse sharpening will cause much greater spectral broadening through self-phase modulation than one would expect from the incident pulse shape. Thus, for the example following eqn. (7.11) where the rise time is decreased by 0.01, the transmitted Stokes spectral width might be increased by about 100 according to (7.17). Other effects, such as retardation of the nonlinear response and dielectric breakdown, also strongly affect the time variation of the optical phase.

If the incident beam power greatly exceeds the critical power P_{c1}, the beam is unstable with respect to transverse intensity fluctuations, each of which contains power roughly equal to P_{c1}. The temporal development of these small-scale fluctuations has not been studied in detail, but it has necessarily a faster rise time than the incident pulse because the local intensity is enhanced nonlinearly. When the driving pulse power changes slowly, these localized fluctuations can be expected to persist for rather long distances (compared to their effective diffraction lengths) because they contain just enough power to balance diffraction. Assuming that these "filaments" persist for time T_f and therefore have length roughly $z \approx v_g T_f$, we can use eqn. (7.17) to estimate the width of the transmitted spectrum:

$$\begin{aligned}\Delta\omega &\approx P_{c1} v_g / k_0 a_f^2 P_1 \\ &\approx \pi v_g I / k_0 P_1.\end{aligned} \tag{7.31}$$

This suggests that the spectral width of the transmitted light is proportional to the incident intensity, when the self-focusing process is dominated by fluctuations and no other processes disrupt the filament formed from each fluctuation. In many materials, stimulated

Raman, Brillouin or Rayleigh wing scattering or dielectric breakdown may alter the time dependence of the local phase and complicate the situation beyond our ability to derive simple expressions for the spectral width.

If a tube of rays does not converge or diverge very rapidly, its optical phase is approximately

$$\varphi = k_0 z + \frac{\omega_0}{c} \int^z \delta n(z')\,dz'. \tag{7.32}$$

For our simple cubic nonlinearity, this becomes

$$\varphi(z, t') = k_0 z + \frac{n_2 \omega_0}{2c} \int_0^z |\mathscr{E}'(z', t')|^2\,dz' \tag{7.33}$$

in the notation of eqn. (7.2). If all t' dependence is contained in the incident power function $P(t')$, we can replace t' by P in (7.33), as we did in eqn. (7.9). The resulting function $\varphi(z, P)$ can be obtained by numerical solution of the stationary equation, and can be used to construct a transfer curve which gives the local phase as a function of time if the incident pulse shape is known (*cf.* Fig. 34). The instantaneous transmitted frequency is related to the slope of this transfer curve through

$$\omega(t') = \omega_0 - \frac{\partial \varphi}{\partial P} \cdot \frac{\partial P}{\partial t'}. \tag{7.34}$$

Using eqns. (6.5), (6.8) and (7.33), the reader may verify that $\partial\varphi/\partial P$ becomes infinite at the self-focal point $z = z_f$. Therefore, the width of the self-phase modulated spectrum in the presence of self-focusing is determined by the mechanisms which prevent a catastrophic self-focus. These mechanisms are sometimes associated with a threshold field strength, such as the breakdown or saturation field. This suggests that the maximum spectral width may be estimated from (7.34) evaluated at the threshold field. Only the Stokes width can be estimated this way since the antiStokes light is generated by the trailing edge, whose propagation may be affected by the limiting mechanism.

Using eqns. (6.5), (6.8) and (7.33), we find

$$\frac{\partial \varphi}{\partial P} = \frac{0.367}{4} \frac{1}{P_1} \left(\frac{P_2}{P}\right)^{1/2} \left[\sin^{-1}(z/z_f) + \frac{z/z_f}{\left(1 - \frac{z^2}{z_f^2}\right)^{1/2}} \right] \tag{7.35}$$

where $z_f \approx (P_2/P)^{1/2}\, 0.367\, ka^2$. This is consistent with (7.15) in the limit $z \to 0$. If z is near the self-focus, (7.35) becomes approximately

$$\frac{\partial \varphi}{\partial P} \approx \frac{1}{4P_1} \frac{z_f}{k_0 a^2} \frac{I(z)}{I(0)}.$$

The maximum Stokes spectral width is therefore about

$$\Delta\omega = \frac{z_f}{4k_0 a^2} \frac{I_B}{I(0)} \frac{1}{P_1} \left.\frac{\partial P}{\partial t'}\right|_{P_2} \tag{7.36}$$

where $\partial P/\partial t'$ is evaluated at the critical power P_2. This expression differs from (7.17) by the factor $I_B/I(0)$ which can be quite large. Of course, if dielectric breakdown does occur, the subsequent time dependence of the phase must be profoundly affected by the rapidly changing electron density, as first pointed out by Hellwarth,[76] and more recently by Yablonovitch and Bloembergen.[77] This effect will be discussed in Section 10.

8. APPROXIMATE ANALYTICAL TECHNIQUES

A. *Linearized analysis: beam stability*

One of the few general techniques in nonlinear analysis is the derivation of linear equations governing the behavior of small fluctuations about a known steady-state solution of the nonlinear system. The fluctuation amplitudes may oscillate, decay, or grow exponentially with spatial and temporal rates determined by the nonlinearities and the nature of the initial fluctuation. This type of analysis was first performed for the self-focusing equation in 1966 by Bespelov and Talanov,[78] and later by Brueckner and Jorna[79] and others.[80,81] It has gained importance in recent efforts to determine damage thresholds in high-power solid-state laser systems.[82] The problem is that small fluctuations on an otherwise smooth intensity profile may self-focus before the beam as a whole does. Material damage caused by the consequent local intensity enhancement is therefore governed by the self-focusing properties of fluctuations. The method of linearization identifies the scale of fluctuations which have the greatest initial intensity growth rate. For more detailed information, such as the location of the self-focus for a fluctuation of given amplitude, we must employ a method which treats the nonlinear growth more accurately.

For the simplest cubic nonlinearity $\mathscr{P}^{NL} = \eta|\mathscr{E}|^2\mathscr{E}$, the reader may easily verify that eqn. (5.1) possesses the following plane wave solution:

$$\mathscr{E}_s(z) = \mathscr{E}_0 \exp i(n_2\omega_0|\mathscr{E}_0|^2 z/2c), \tag{8.1}$$

with $\mathscr{E}_0$ arbitrary. The linearized equations for the small amplitude and phase fluctuations u and v about this solution are obtained by inserting

$$\mathscr{E}(x, y, z) = [1 + u(x, y, z) + iv(x, y, z)]\mathscr{E}_s(z) \tag{8.2}$$

$$2k_0 \frac{\partial u}{\partial z} + \nabla_T^2 v = 0, \tag{8.3}$$

$$2k_0 \frac{\partial v}{\partial z} - \nabla_T^2 u - \frac{\epsilon_2 \omega_0^2 |\mathscr{E}_0|^2}{c^2} u = 0. \tag{8.4}$$

Taking u and v to be the real parts of

$$u_0 \exp i(qz + \boldsymbol{q}_\perp \cdot \boldsymbol{r}), \qquad v_0 \exp i(qz + \boldsymbol{q}_\perp \cdot \boldsymbol{r}) \tag{8.5}$$

where $\boldsymbol{r} = (x, y)$, $\boldsymbol{q}_\perp = (q_x, q_y)$, the reader will find that q must satisfy

$$4k_0^2 q^2 = q_\perp^2 \left(q_\perp^2 - \frac{2\pi I_0}{P_1} \right) \tag{8.6}$$

if (8.5) is to satisfy (8.3), (8.4). For sufficiently large background intensities $I_0 = nc|\mathscr{E}_0|^2/8\pi$, q becomes purely imaginary and the fluctuation with the corresponding transverse wave vector $\boldsymbol{q}$ grows exponentially. The corresponding e-folding length l of this growth satisfies

$$\frac{2k_0}{q_\perp^2 l} = \left[\frac{2\pi I_0}{q_\perp^2 P_1} - 1 \right]^{1/2} \tag{8.7}$$

which shows the same square root dependence on power as the self-focal length for self-focusing of a whole beam. The background power under one cycle of the periodic fluctuation of (8.5) is $P_0 = I_0 ab$, where

$$q_x = 2\pi/a, \qquad q_y = 2\pi/b.$$

According to eqn. (8.7) a fluctuation will grow only if this power exceeds the critical power

$$P_{cr} = 4\pi\sigma P_1 = 3.38\, P_{c1}\psi, \tag{8.8}$$

where ψ is related to the "aspect ratio" of the fluctuation:

$$\psi = \frac{1}{2}\left(\frac{a}{b} + \frac{b}{a} \right). \tag{8.9}$$

For fixed background intensity, the growth rate is greatest for $a = b$ and

$$a^2 = 8\pi P_1/I_0. \tag{8.10}$$

While eqn. (8.7) predicts a growth rate proportional to $\sqrt{I_0}$ for a given perturbation scale $a = 2\pi/q$, the *optimum* perturbation grows at a rate proportion to I_0. Computing q from (8.10), one finds that (8.7) yields

$$1/l_{opt} = \pi I_0/2k_0 P_1 = P/2k_0 a_0^2 P_1$$

where P is the power in area πa_0^2. Thus self-focusing of a beam with transverse fluctuations is characterized by a length whose dependence on total power differs from that for whole beam self-focusing. Of course, an individual self-focusing region obeys more or less the same equations as before, and the optimum growth rates for two different beam intensities correspond to two different fluctuations. This power dependence of l_{opt} has been verified by Shapiro and co-workers.[83]

We conclude that a smooth beam tends to break up into hot spots, each containing somewhat more than one critical power P_{c1}. The initial intensity rise in a hot spot is exponential, until the linearization approximation breaks down. This happens at roughly

$$z \approx \tfrac{1}{2} l \log(I_0/16 I_f) \tag{8.11}$$

where I_f is the initial fluctuation incremental intensity. At this distance, the hot spot is 50% more intense than the initial background. This failure of the linear approximation is awkward, for it prevents us from determining whether, for a given $q_\perp$, the intensity rise is catastrophic, or simply reaches a modest maximum. Computer solutions reported by Bespalov and Talanov indicate both types of behavior.[78] A more complete picture of the growth of fluctuations in the nonlinear régime does not now exist in the literature. However, the practical importance of this phenomenon suggests that the necessary details will be provided in the near future.

For initial fluctuations so large that diffraction spreading is negligible prior to the self-focus, we may use Kelley's method[11] to find a relation between self-focal length and fluctuation amplitude. A uniform field with a gaussian perturbation

$$\mathscr{E} = \mathscr{E}_0[1 + u\exp(-r^2/2a^2)]$$

induces a nonlinear lens with initial focal length R satisfying (cf. the derivation of eqn. (6.6))

$$1/R = n_2\mathscr{E}_0^2 u(1 + u)\Delta z/na.$$

Taking $z_f = R = \Delta z$, we find

$$\frac{ka^2}{z_f} = \left[u(1 + u)\frac{P_0}{P_1}\right]^{1/2} \tag{8.12}$$

where $P_0 = \pi a^2 I_0$. This is comparable to eqn. (8.7), but shows how the self-focal length is increased if the incident fluctuation amplitude is small ($u < 0.618$).

The linear stability analysis can be generalized in several directions. First, one may employ transverse fluctuation profiles other than sine waves. As Suydam has noted,[81] any eigenfunction of the transverse Laplacian with eigenvalue $-q_\perp^2$ will do. Taking u and v of the form (cf. eqn. (8.5))

$$u_0 J_0(q_\perp r)\exp iqz, \quad v_0 J_0(q_\perp r)\exp iqz, \tag{8.13}$$

one recovers eqn. (8.7) for the growth length. But now the background power within the first hump of the Bessel function J_0 is $P_0 = I_0\pi(j'_{02}/q_\perp)^2$ where j'_{02} is the position of the first minimum of J_0. A fluctuation of form (8.13) will grow if P_0 exceeds

$$P_{cr} = 7.34\, P_1 = 1.97\, P_{c1},$$

which is less than (8.8), indicating that the Bessel function is closer to the shape preferred by a growing instability.

A second generalization allows the background beam to have any polarization state.[78] The reader may verify that eqn. (8.6) is still correct for this case if the lower critical power P_1 for linear polarization is replaced by

$$P'_1 = \frac{P_1(2A + B)}{A + [A^2 + B(2A + B)\cos^2 2\chi]^{1/2}}. \tag{8.14}$$

Here A and B are the susceptibility parameters defined in eqn. (3.7), and $\tan\chi$ is the ratio of the axes (ellipticity) of the incident polarization ellipse. P'_1 is greatest for circular polarization ($|\chi| = \pi/4$) and least for linear polarization ($|\chi| = 0, \pi/2$). There is no polarization dependence for $B = 0$. For $B = 6A$, which is true for the molecular reorientation and redistribution mechanisms, P'_1 becomes

$$P'_1 = 8P_1/[1 + (1 + 48\cos^2 2\chi)^{1/2}]. \quad (8.15)$$

In this case, $P'_1/P_1 = 4$ for circular polarization. From eqn. (8.15) one may infer that, if δ is the fractional intensity admixture of the orthogonal polarization in a nearly circularly polarized beam, then $P'_1/P_1 \approx 4(1 - 48\delta)$. This implies that a 1% departure from circular polarization decreases the critical power by about 50%![46] Thus great care is required in experimental investigations of the polarization dependence of self-focusing.

Other generalizations of the linearized analysis allow the medium to be inhomogeneous,[80] or include additional effects such as gain, loss, or stimulated Brillouin or Raman scattering.[79] Linear optical index inhomogeneities obviously provide a source of phase fluctuations, and scattering centers or absorption inhomogeneities cause amplitude fluctuations. These fluctuations then grow with characteristic rates essentially the same as those derived above.

B. *Constant shape approximation: weak self-focusing*

The most tractable approximate analyses of the self-focusing equation (5.1) eliminate the transverse variables by working with a trial function $\mathscr{E}(x, y, z)$ in which the x, y dependence is explicit. In this section we examine a slight generalization of what Akhmanov *et al.*[8] call the "aberrationless" approximation, in which all z dependence is contained in two scale parameters $a(z)$ and $b(z)$ through

$$\mathscr{E}(x, y, z) \sim \frac{g(x/a(z), y/b(z))}{[a(z)\, b(z)]^{1/2}} \quad (8.16)$$

where g is a fixed function. This shape invariance, or "self-similarity" is not generally a property even of solutions of the linear wave equation, with the important exception of the Laguerre- or Hermite-gaussian modes. With this in mind, let us restrict our discussion initially to gaussian, cylindrically symmetric trial functions

$$A(r, z) = |\mathscr{E}| = \mathscr{E}_0 \frac{a_0}{a(z)} \exp[-r^2/2a^2(z)]. \quad (8.17)$$

The phase function will be specified later. Only the amplitude is required to fix the ray potential U in the ray equation (5.4).

Since the coordinate r of a particular ray is supposed to be proportional to $a(z)$ in our self-similar approximation, the ray equation is actually an "equation of motion" for $a(z)$. If we restrict our attention to rays such that $r(z) \ll a(z)$, that is, to near-axis rays, then the ray equation assumes the following convenient form:

$$\frac{d^2 a}{dz^2} = -\frac{\partial V}{\partial a}. \quad (8.18)$$

Here $\alpha\, \partial V/\partial a$ is simply $\partial U/\partial r$, where U is the potential (5.3) with A given by (8.17), and r is evaluated at $r = \alpha a$, in the limit $\alpha \rightarrow 0$.

$$V = \frac{1}{2k_0^2 a^2} - \frac{2\pi}{n_0^2} \chi'_{NL}(\mathscr{E}_0 a_0/a). \quad (8.19)$$

Wagner *et al.* give a detailed derivation of this result.[46] Its most attractive aspect is that it allows immediate inference of important features of stationary self-focusing from a plot of the potential function $V(a)$.

For the simplest cubic nonlinearity, $\chi'_{NL} = \eta|\mathscr{E}|^2$, the "beam radius potential" V is

$$V(a) = \frac{1}{2k_0^2 a^2}\left(1 - \frac{P}{P_1}\right). \quad (8.20)$$

This is attractive for $P > P_1$, repulsive for $P < P_1$ and implies no force at all for $P = P_1$. Thus when $P > P_1$, the beam is expected to collapse to zero radius. When $P = P_1$, the

Self-focusing: theory 87

beam propagates without change of radius, corresponding to self-trapping. There is no unique self-trapped size, since the force vanishes at all radii when $P = P_1$.

The function $a(z)$ can be obtained from direct integration of the energy formula corresponding to (8.18):

$$\frac{1}{2}\left(\frac{da}{dz}\right)^2 + V = \text{const.} \tag{8.21}$$

If V is given by eqn. (8.20), one finds

$$\frac{a^2}{a_0^2} = \left(1 - \frac{P}{P_1}\right)\frac{z^2}{k_0^2 a_0^4} + \left(1 + \frac{z}{a_0}\left.\frac{da}{dz}\right|_0\right)^2 \tag{8.22}$$

where $a_0 = a(0)$, and the initial "velocity" is simply related to the radius of curvature R of the incident phase front ($R < 0$ for converging beam):

$$\frac{1}{R} = \frac{1}{a_0}\left.\frac{da}{dz}\right|_0. \tag{8.23}$$

When $P > P_1$, and $R = \infty$, eqn. (8.22) predicts a catastrophic self-focus ($a = 0$) at z_f, where

$$\frac{ka_0^2}{z_f} = \left(\frac{P}{P_1} - 1\right)^{1/2}. \tag{8.24}$$

When R is finite, the change in position of the self-focus is given by the lens formula

$$\frac{1}{z_{fR}} = \frac{1}{z_f} + \frac{1}{R} \tag{8.25}$$

which agrees with the exact relation (6.10). The axial intensity ratio $|\mathscr{E}(0, z)|^2/|\mathscr{E}(0, 0)|^2$ is simply $a_0^2/a^2(z)$, which, for $R = \infty$, can be written

$$I(z)/I_0 = \frac{1}{1 + \left(1 - \dfrac{P}{P_1}\right)\dfrac{z^2}{k^2 a_0^4}}. \tag{8.26}$$

How do these results compare with the exact solutions of the full partial differential equation (5.1)? We cannot expect very good agreement because of the severe violation of the self-similarity assumption in the exact solutions, which are initially gaussian (see Fig. 11). Indeed the predicted critical power P_1 is too small by the factor 0.265, and the distance to the self-focal point z_f is too long by the factor 1.4 for $P \gg P_1$ (see eqn. (6.5)). Moreover, the functional dependence of axial intensity upon z given by (8.26) is decidedly incorrect for large powers (despite an explicit statement to the contrary in ref. 8, p. 1166).

In favor of the constant shape approximation, we can say that the lens transformation (8.25) and the functional relation $1/z_f \sim P^{1/2}$ are preserved, as is the scale invariance of the self-trapped solution. The axial intensity function of eqn. (8.26) is not bad for subcritical powers ($P < P_1$). Thus we can trust the qualitative features, but none of the quantitative predictions of the "aberrationless" approximation, unless $P < P_1$.

The solution of the exact moment equation for the rms radius, eqn. (5.31), for the special case of a gaussian incident beam, is identical to eqn. (8.22) only if P_1 is replaced by P_3. (For this case, $P_3 = 4P_1$.) But if the beam preserved its gaussian shape, the rms radius a_R would scale with the $1/e$ radius a, and the two would vanish at the same self-focal distance. The fact that they do not is simply another indication of the shape distortion of an initially gaussian beam.

The phase function in this approximation may be obtained by requiring that surfaces of constant phase be perpendicular to the ray paths which satisfy (8.18). It is sufficient to use the phase function of eqn. (5.41) which is consistent with general self-similar solutions.

Fradin has employed the solution (8.22) to infer the nonlinear index of solid materials from observations of bulk damage induced by focused beams of subcritical power.[61]

Assuming that damage occurs at a fixed value of local intensity I_D, one may focus a beam of fixed power P with a sequence of lenses of decreasing focal length R, until damage is observed. This fixes the maximum intensity in the medium at I_D. The maximum intensity from eqn. (8.22) is

$$I_m \pi a^2/P = 1 + \frac{k_0^2 a_0^4}{\left(1 - \frac{P}{P_1}\right) R^2}$$

which may be solved for P_1 and hence for n_2. If $P \ll P_1$, and if the nonlinear response is instantaneous, then the inferred value of n_2 should be accurate.

Let us examine self-focusing with a saturable nonlinear susceptibility in the constant-shape approximation.[46] Figure 38 shows $V(a)$ for the susceptibility

$$\chi'_{NL} = \frac{\eta|\mathscr{E}|^2}{1 + |\mathscr{E}|^2/\mathscr{E}_s^2} \tag{8.27}$$

where $\mathscr{E}_s$ is the "saturation field strength". The corresponding potential

$$V = \frac{1}{2k_0^2 a^2}\left[1 - \frac{P/P_1}{1 + (a_s/a)^2}\right] \tag{8.28}$$

has a minimum for $P > P_1$ at

$$a_c = \frac{a_s}{\left[\left(\frac{P}{P_1}\right)^{1/2} - 1\right]^{1/2}} \tag{8.29}$$

where $\mathscr{E}_s a_s = |\mathscr{E}(0,0)|a_0$. Thus for any power $P > P_1$ there is a fixed radius a_c at which a beam may propagate without changing size. This self-trapped size is proportional to $P^{1/4}$ for $P \gg P_1$ and therefore varies slowly with self-trapped power. If the incident beam radius differs from a_c, then the beam size oscillates about the minimum of the potential curve (see Fig. 38). The oscillation period, as well as the function $a(z)$, can be evaluated explicitly for this case, but the result involves elliptic integrals. Figure 39 shows how the beam size varies with distance. Notice that the self-focal length (distance to first minimum of beam radius) depends upon the incident radius a_0, increasing to infinity as a_0 approaches its self-trapped value. If a_0 is not too different from a_c, then the distance between consecutive foci is roughly

$$z_s \approx \frac{\pi k a_s^2 (P/P_1)^{1/4}}{[(P/P_1)^{1/2} - 1]^{3/2}}. \tag{8.30}$$

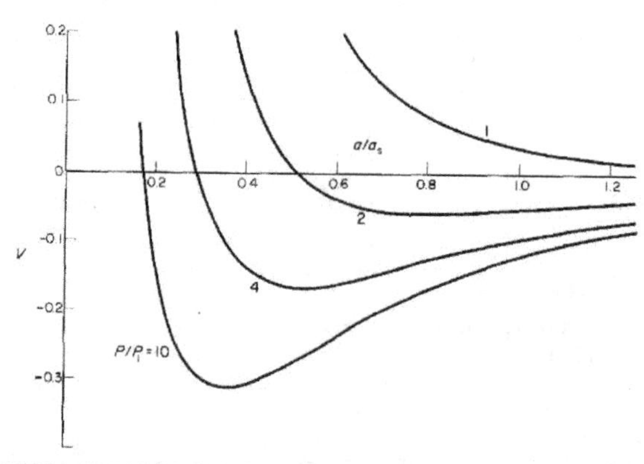

FIG. 38. Effective potential $V(a)$ for beam radius when nonlinearity saturates according to (8.27). The stable self-trapped radius a_c occurs at the minimum of $V(a)$. (From Wagner *et al.*[46])

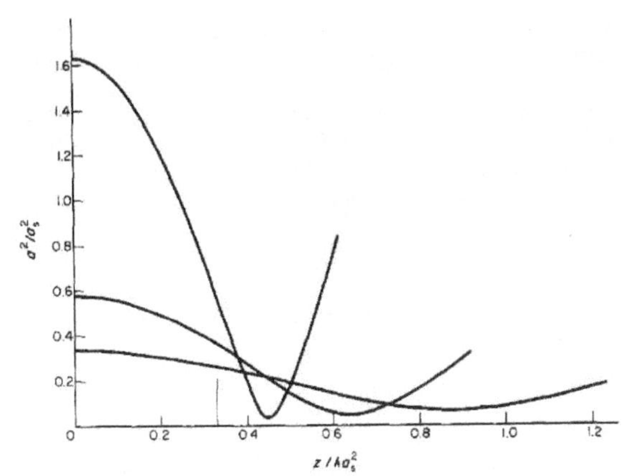

FIG. 39. Oscillation of beam size of a beam self-focusing in the saturable susceptibility whose potential is shown in Fig. 38. The curves are all periodic. (From Wagner *et al.*[46])

More analytical details are provided in the work of Wagner *et al.*[46]

Comparison of these results with numerical solutions of eqn. (5.1) for the same saturable susceptibility reveals enormous discrepancies.[47] Figure 24, for example, shows severe violation of the constant shape assumption, and Fig. 23 shows none of the symmetry about the first intensity maximum that one expects from our simple mechanical analogue. But closer scrutiny suggests that the central portion of the beam in the post-focal region does in fact obey the approximate theory. The power in this central region, which is somewhat greater than P_2, is approximately conserved, according to the numerical solutions, and is distributed sometimes over a central peak plus one ring, and sometimes confined to the central peak. In the latter case, the radius of the central peak oscillates with a period close to eqn. (8.30) about a value close to a_c in eqn. (8.20). During the oscillations, the shape near the axis appears to remain nearly gaussian, which accounts for the success of the approximation.

Figure 40 shows the potential $V(a)$ for the unhindered molecular reorientation model with local field corrections, discussed in Section 3C. The curves shown are for prolate spheroidal molecules, which lead to initial "antisaturation" in the nonlinear response.[84] This behavior admits the possibility of self-trapping at powers below the critical power

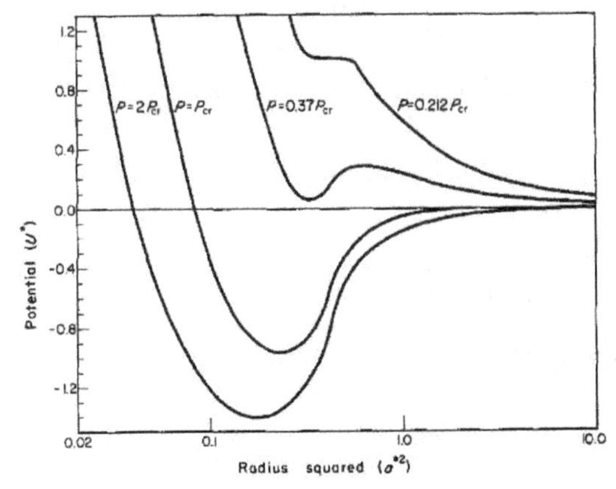

FIG. 40. Effective potential $V(a)$ corresponding to nonlinear susceptibility X_+ of Fig. 3 for reorientation of prolate spheroidal molecules. Notice that the larger of the two equilibrium radii for $P = 0.37P_1$ is mechanically unstable. (From Marburger *et al.*[29])

for self-focusing of a large beam. Numerical solutions of the exact self-trapping equation verify this possibility, but it does not seem to have been definitely established experimentally.

Because high powers of r^2/a^2 in the trial function (8.17) are discarded, we need not restrict our discussion to gaussian intensity profiles. The reader can check that the equation of motion (8.18) for $a(z)$ depends only upon the coefficients of r^2 and r^4 in the expansion of the trial function. For non-gaussian profiles, we define a as in eqn. (6.3) so that

$$\mathscr{E}(r, z) = \mathscr{E}_0\, g(r/a)(a_0/a) \approx \frac{\mathscr{E}_0 a_0}{a}\left[1 - \frac{r^2}{2a^2} + \frac{(\sigma+1)}{16}\frac{r^4}{a^4} - \cdots\right] \tag{8.31}$$

where $\sigma = 1$ for gaussian beams. If the potential $V(a)$ is rederived with this trial function, the result for the simple cubic nonlinearity is

$$V(a) = \frac{\sigma}{2k_0^2 a^2}\left(1 - \frac{P a_0^2}{P_1 \sigma a_P^2}\right)$$

where a_P, the power radius, is defined in eqn. (6.2). Thus the departure from a gaussian profile alters the Fresnel number $k_0 a_0$ of the beam:

$$k_0 a_0 \rightarrow k_0 a_0/\sqrt{\sigma}$$

and also modifies the critical power

$$P_1 \rightarrow (\sigma a_P^2/a_0^2) P_1 \tag{8.32}$$

Table 2 shows that the coefficient in this equation gives the correct trend, but not the correct value of the modification in critical power. Indeed, for parabolic beams, σ is negative. In much better agreement with a variety of computer tests performed by E. L. Dawes[85] is the estimated change in self-focal length for $P \gg P_1$:

$$\frac{k_0 a_0 a_P}{z_f} \approx \left(\frac{P}{P_1}\right)^{1/2}. \tag{8.33}$$

The near-axis constant shape approximation has also been applied to beams of elliptical cross section, for which

$$|\mathscr{E}(x, y, z)| = \mathscr{E}_0 (a_0 b_0/ab)^{1/2} \exp - [(x^2/2a^2) + (y^2/2b^2)],$$

where a and b depend upon z. Vorob'yev[86] has examined the equations of motion for a and b, which are derivable from the two-dimensional potential function (for cubic nonlinearity)

$$2k^2 V(a, b) = \frac{1}{a^2} + \frac{1}{b^2} - \frac{2P/P_1}{ab}. \tag{8.34}$$

Analytical solutions for a and b separately have not been found, but Vorob'yev showed that

$$(a^2 + b^2) = \left[\left(\dot{a}_0^2 + \dot{b}_0^2\right) + \frac{2}{k_0^2 a_0 b_0}\left(\psi - \frac{P}{P_1}\right)\right] z^2 + 2(a_0 \dot{a}_0 + b_0 \dot{b}_0) z + (a_0^2 + b_0^2) \tag{8.35}$$

where ψ is defined in eqn. (8.9) and $\dot{a}_0 = a_0/R_a$ according to eqn. (8.23). The reader may manipulate this formula to obtain the position z_f at which $a^2 + b^2 = 0$ under various conditions. The most important results are that the critical power is enhanced by the factor ψ, and also by astigmatic focusing ($R_a \neq R_b$). Defining z_0 through

$$-\frac{(a_0^2 + b_0^2)}{z_0} \equiv \frac{a_0^2}{R_a} + \frac{b_0^2}{R_b}, \tag{8.36}$$

and

$$\Phi \equiv 1 + \left[\frac{k_0 a_0 b_0}{2\psi}\left(\frac{1}{R_a} - \frac{1}{R_b}\right)\right]^2, \tag{8.37}$$

one finds from (8.35)

$$z_f = \frac{z_0}{1 + \dfrac{z_0}{k_0 a_0 b_0}\left(\dfrac{P}{\psi P_1} - \Phi\right)^{1/2}}. \tag{8.38}$$

Some aspects of this formula have been verified experimentally by Giuliano.[87] Numerical analysis of the self-focusing partial differential equation (5.1) for elliptical beams has just begun, and the reliability of eqn. (8.38) may not be considered established. Shvartsburg has been able to make greater analytical progress using a parabolic rather than a gaussian trial function, and ignoring diffractive effects.[88]

The author feels that this approximate method is interesting only because it is mathematically tractable. It should not be used for quantitative estimates except when the incident beam is accurately gaussian and far below critical power. The method is equivalent to replacing the solution for the self-similar function f determined from (5.42) by a gaussian, which is the solution of (5.42) in a linear medium.

C. *Constant shape approximation: strong self-focusing*

Kelley[11] has proposed an alternative approximation scheme which C. S. Wang[89] has shown is equivalent to assuming the constant shape trial function (for unfocused incident gaussians)

$$\mathscr{E}(r, z) = \mathscr{E}_0 \frac{a_0}{a(z)} \exp[-r^2/2a^2(z)] \exp(\mathrm{i}\, \delta n \omega_0 z/c) \tag{8.39}$$

where, for the simple cubic nonlinearity $\chi'_{NL} = \eta|\mathscr{E}|^2$,

$$\delta n = \frac{2\pi}{n_0} \eta \frac{\mathscr{E}_0^2 a_0^2}{a^2(z)} \exp[-r^2/a^2(z)]. \tag{8.40}$$

This is similar to the weak focusing approximation eqn. (8.17), but requires the phase to be consistent with the induced index change, as in eqn. (5.48). This is *not* consistent with the self-similar phase eqn. (5.41), and ignores some phase corrections associated with diffraction. Nevertheless, for large powers $P \gg P_1$, the self-focal length is expected to be much less than the diffraction length ka_0^2, and the approximation (8.39) should not be too bad.

Inserting (8.39) into the self-focusing equation (5.2) and discarding terms of order r^2/a^2, as before, one finds an equation for $a(z)$ which may be written

$$\frac{\mathrm{d}a^4}{\mathrm{d}z} = -\frac{4P}{k^2 P_1} z. \tag{8.41}$$

The solution is

$$(a/a_0)^4 = 1 - (z/z_f)^2 \tag{8.42}$$

where

$$\frac{ka_0^2}{z_f} = \left(\frac{2P}{P_1}\right)^{1/2} = \frac{1}{0.364}\left(\frac{P}{P_2}\right)^{1/2}. \tag{8.43}$$

This equation agrees closely with the results of numerical computation for $P \gg P_2$, eqn. (6.5). The corresponding axial intensity

$$I(z)/I_0 = \frac{1}{[1 - (z/z_f)^2]^{1/2}} \tag{8.44}$$

also agrees well with the numerical solutions for large powers. The reader should compare these results with eqns. (8.24) and (8.26) which are correct only for $P \ll P_1$.

This method can be generalized in several ways. For example, if $\delta n/n$ is an arbitrary function $G(I)$ of the intensity, then eqn. (8.41) may be replaced by

$$\frac{\mathrm{d}I}{\mathrm{d}(z^2)} = \frac{2\pi}{P} I^3 \frac{\mathrm{d}G}{\mathrm{d}I}. \tag{8.45}$$

This may be integrated explicitly for the simple saturable susceptibility (8.27), a task which we leave to the interested reader.

A more important generalization introduces gain or loss in the trial function through an additional factor $\exp gz$. If g is independent of intensity, then eqn. (8.41) becomes

$$\frac{da^4}{dz} = -\frac{4P}{k^2 P_1} z \exp gz \tag{8.46}$$

where P is the power at $z = 0$. The solution

$$(a/a_0)^4 = 1 - 2[1 - e^{gz}(1 - gz)]/g^2 z_f^2 \tag{8.47}$$

shows explicitly how the self-focal length depends on gain (or loss for $g < 0$). Here z_f is given by (8.43). Kaiser *et al.*[90] who first derived (8.47), reported experimental verification for $g < 0$ in the high power régime. Equation (8.47) also agrees rather well with numerical solutions of the self-focusing equation with loss in the high power regime $P \gg P_2$.[47]

The strong focusing approximation may be extended to focused incident beams simply by applying the lens transformation (Section 5B) to the formulas we have derived above.

D. *Aberration equations*

The "constant-shape" methods discussed previously employ trial functions for the optical field with one scale function $a(z)$ to be determined from a single ordinary differential equation. A natural extension of this approach is to introduce more scale functions, a procedure which yields a set of coupled equations. These equations are, of course, more difficult to analyse, but it is always possible to find the first few terms of the Taylor's series expansion of each scale function in the distance z from the initial plane. This is the procedure suggested by Lugovoi,[91] who employed the trial function

$$\mathscr{E}(r, z) = I^{1/2}(z) \exp[u(r, z) + iv(r, z)] \tag{8.48}$$

where, using a notation slightly different from Lugovoi's,

$$u(r, z) = \sum_{n=1}^{\infty} u_n(z) r^{2n}, \tag{8.49}$$

$$v(r, z) = \sum_{n=0}^{\infty} v_n(z) r^{2n}. \tag{8.50}$$

Inserting this into the self-focusing equation (5.1), and collecting the coefficients of r^{2n}, one finds an infinite coupled set of equations for the *aberration functions* $u_n(z)$, $v_n(z)$. If only the functions I, u_1, v_0 and v_1 are retained, one finds that (8.48) must be identical to the weak focusing, constant shape solution, whose phase is given by the self-similarity formula (5.41).

To improve on this lowest order trial function, Lugovoi retained terms up to $r^6 (n = 3)$ in the amplitude and phase functions, and solved for the axial intensity $I(z)$ through terms in z^4:

$$I(z) \approx I(0) + \frac{z^2}{2} I''(0) + \frac{z^4}{4} I^{(4)}(0).$$

The terms odd in z vanish for a gaussian equiphase beam at $z = 0$. The result, whose derivation is both straightforward and tedious, is simple when $P \approx P_1$:

$$\frac{I(z)}{I(0)} \approx 1 + \left(\frac{P}{P_1} - 1\right)\frac{z^2}{k^2 a_0^4} - \frac{11}{3}\frac{z^4}{k^4 a_0^8} + O\left(\frac{P}{P_1} - 1\right)^2.$$

This has a maximum at $z = 0$ if $P < P_1$, and at

$$z_m^2 = k^2 a^4 \frac{3}{22}\left(\frac{P}{P_1} - 1\right)$$

for $P > P_1$. This is not the catastrophic self-focus predicted by other methods for $P > P_1$, but corresponds to the weak maximum apparent in the numerical solutions for $P_1 < P < P_2$ (see Fig. 16).

This method extends the range of the weak focusing, constant shape approximation slightly, but is not useful for high powers $P \gg P_1$. Notice that the trial function for the strong focusing constant-shape approximation includes all powers of $\mathbf{r}^2$ in the phase function, and that the axial intensity in both the strong and the weak focusing cases involves all powers of z.

9. TRANSIENT AND NONSTATIONARY SELF-FOCUSING

A. *Self-steepening*

The phenomena described in this and subsequent sections have their origin in the time derivatives in eqn. (2.8) that have been omitted until now. The simplest new effects arise by including all the first derivative terms and ignoring transverse variations (no self-focusing). The resulting equation,

$$2ik_0\left(\frac{\partial}{\partial z} + \frac{1}{v_g}\frac{\partial}{\partial t}\right)\mathscr{E} + \frac{8\pi i\omega_0}{c^2}\frac{\partial \mathscr{P}}{\partial t} = -\frac{4\pi}{c^2}\omega_0^2\mathscr{P} \tag{9.1}$$

can be written in a simple form if $\mathscr{P} = \eta|\mathscr{E}|^2\mathscr{E}$. Multiplying by $\mathscr{E}^*$ and adding the result to its complex conjugate, we find

$$\frac{\partial|\mathscr{E}|^2}{\partial z} + \left[\frac{1}{v_g} + \frac{3n_2}{c}|\mathscr{E}|^2\right]\frac{\partial|\mathscr{E}|^2}{\partial t} = 0 \tag{9.2}$$

which has the general solution, in implicit form,

$$|\mathscr{E}(z, t)|^2 = f\left[t - \left(\frac{1}{v_g} + \frac{3n_2}{c}f\right)z\right] \tag{9.3}$$

where $f(t)$ is an arbitrary function. This solution contains the information that different portions of the incident pulse $f(t)$ travel with different velocities depending upon the intensity. This causes the pulse to distort as shown in Fig. 41. Two portions of the pulse initially separated in time by δt may eventually coincide with each other, causing an "optical shock", at distance z_{sh} such that

$$t - \left(\frac{1}{v_g} + \frac{3n_2}{c}f\right)z_{sh} = t + \delta t - \left(\frac{1}{v_g} + \frac{3n_2}{c}(f + f'\,\delta t)\right)z_{sh}$$

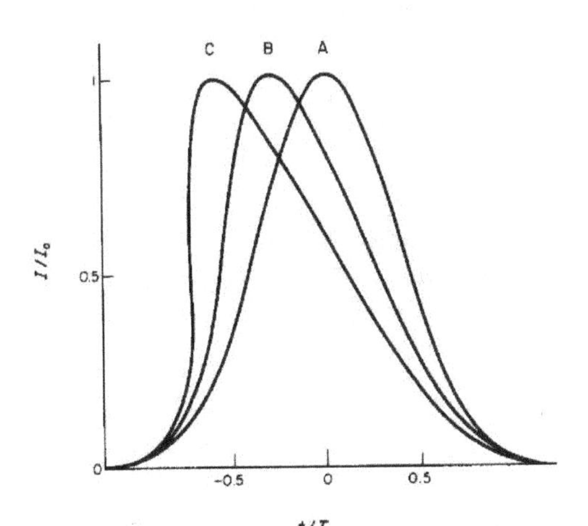

Fig. 41. Pulse distortion due to self-steepening. A: initial pulse shape, B: after propagation through $z = \frac{1}{2}z_{sh}$ given by eqn. (9.4). C: after propagation through $z = z_{sh}$. (From De Martini *et al.*[15])

where $f = f(t)$ and $f' = \partial f(t)/\partial t$. The solution for z_{sh} can be written in terms of the power P in an area πa^2:

$$\frac{k_0 a^2}{z_{sh}} = 3 \frac{\partial P}{\partial t} \Big/ \omega_0 P_1, \tag{9.4}$$

which shows that $z_{sh} \gg k_0 a^2$ unless P exceeds the critical power by the product of the laser frequency and the pulse length. Because z_{sh} is so long, self-steepening is usually masked by other effects. A fuller account of optical self-steepening may be found in ref. 5.

The possibility of electromagnetic shock formation in nonlinear media was recognized long ago, and is an important effect at microwave frequencies where the magnetic response of ferromagnetic materials is strongly nonlinear. References 92 and 93 are valuable treatments of the subject.

B. *Dispersive effects*

If the term in the propagation eqn. (2.13) arising from group velocity dispersion is retained, but that leading to self-steepening omitted, we find

$$2ik_0 \frac{\partial \mathscr{E}'}{\partial z} + \frac{k_0}{v_g^2} \frac{\partial v_g}{\partial \omega} \frac{\partial^2 \mathscr{E}'}{\partial t'^2} = -\frac{4\pi\omega_0^2}{c^2} \mathscr{P}'^{\mathrm{NL}} \tag{9.5}$$

where $t' = t - z/v_g$ as in (7.1). Again we have ignored transverse variations and backward traveling waves. This equation has exactly the same form as that describing the propagation of a slab-shaped beam $\mathscr{E}'(x, z, t')$ in a nondispersive medium:

$$2ik_0 \frac{\partial \mathscr{E}'}{\partial z} + \frac{\partial^2 \mathscr{E}'}{\partial x^2} = -\frac{4\pi\omega_0^2}{c^2} \mathscr{P}'^{\mathrm{NL}}, \tag{9.6}$$

with the important difference that the term involving the second derivative can have either sign. The coefficient of this term, which can also be written

$$\frac{1}{v_g^2} \frac{\partial v_g}{\partial \omega} = -\frac{\partial^2 k}{\partial \omega^2} \equiv -k''$$

is positive only in regions of anomalous dispersion.

The close analogy between (9.5) and (9.6) allows us to exploit our knowledge of the propagation of slab-shaped beams to infer some important properties of dispersive propagation. We consider four cases (cf. Fig. 42):

1. In the absence of nonlinear polarization, a gaussian pulse with no initial phase modulation will suffer dispersive spreading according to the general formula

$$|\mathscr{E}'(z, t')| = \mathscr{E}_0 \frac{\exp\left\{-\frac{t'^2}{2T^2} \Big/ \left[1 + \left(\frac{z}{z_t}\right)^2\right]\right\}}{\left[1 + \left(\frac{z}{z_t}\right)^2\right]^{1/4}} \tag{9.7}$$

where

$$\mathscr{E}'(0, t') = \mathscr{E}_0 \, e^{-t'^2/2T^2}.$$

Here $z_t \equiv -T^2/k''$ plays the role of a "temporal diffraction length" or dispersion length.

2. In the absence of nonlinear polarization, a linear frequency sweep on the incident pulse, corresponding to "temporal focusing" will cause the pulse to *compress* if the relative sign between the sweep rate and $\partial^2 k/\partial \omega^2$ is negative. Thus, if

$$\mathscr{E}'(0, t') = \mathscr{E}_0 \, e^{-t'^2/2T^2} \, e^{+it'^2/2k''S}$$

then

$$|\mathscr{E}'(z, t')| = \mathscr{E}_0 \frac{\exp\left\{-\frac{t'^2}{2T^2} \Big/ \left[\left(1 - \frac{z}{S}\right)^2 + \left(\frac{z}{z_t}\right)^2\right]\right\}}{\left[\left(1 - \frac{z}{S}\right)^2 + \left(\frac{z}{z_t}\right)^2\right]^{1/4}}. \tag{9.8}$$

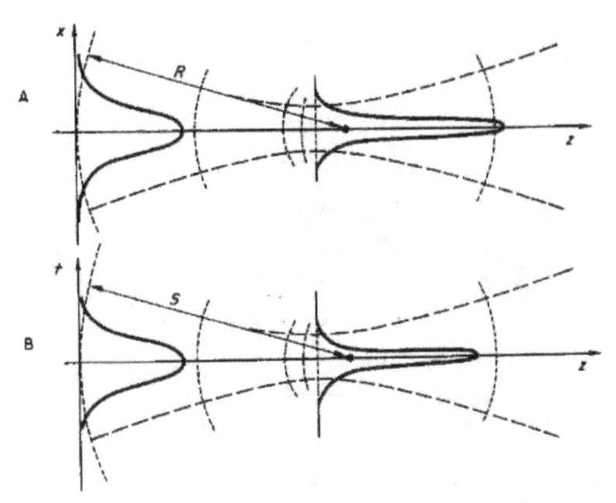

FIG. 42. Analogy between spatial propagation of a slab-shaped beam, and temporal propagation of a pulsed plane wave in a dispersive medium. In A, a gaussian beam is focused toward point R by initial phase curvature, but the beam waist occurs prior to R because of diffraction. In B, a gaussian pulse is focused by initial chirp which introduces temporal phase curvature of radius S.

The length S is analogous to the geometrical focal length of a cylindrical lens which focuses the beam. It is related to the instantaneous frequency by

$$\omega = \omega_0 - t/k''S,$$

or

$$1/S = -k''\mathrm{d}\omega/\mathrm{d}t. \tag{9.9}$$

The pulse achieves its maximum compression just prior to $z = S$, at

$$z_{\mathrm{m}} = \frac{S}{1 + \dfrac{S^2}{z_t^2}} \tag{9.10}$$

where its peak intensity is

$$|\mathscr{E}'(z_{\mathrm{m}}, 0)|^2 = [1 + (T^2\,\mathrm{d}\omega/\mathrm{d}t)^2]^{1/2}|\mathscr{E}'(0, 0)|^2. \tag{9.11}$$

Notice that the intensity enhancement is independent of the dispersion k''. This compression of pulses whose incident phase is swept in time is exploited in chirped radar,[94] and was first observed at optical frequencies by Giordmaine *et al.*[95] It plays an important role in the temporal development of pulses in nonlinear media where self phase modulation produces a frequency sweep during propagation.[96,97]

3. In a medium with cubic nonlinearity and anomalous dispersion, a smooth pulse with no initial phase modulation will self-compress. This is the analogue of self-focusing of a slab-shaped beam. (We are assuming that $\mathscr{P}^{\mathrm{NL}} = \eta|\mathscr{E}|^2\mathscr{E}$, $\eta > 0$.) The self-compression distance is probably most reliably estimated from the strong focusing constant shape approximation developed in Section 8C. Thus near the pulse peak, the field should have the form

$$\mathscr{E}'(z, t') = \mathscr{E}_0\left[\frac{T}{T(z)}\right]^{1/2}\exp[-t'^2/2T^2(z)]\exp(\mathrm{i}\,\delta n\omega_0 z/c), \tag{9.12}$$

where

$$\delta n = 2\pi\eta|\mathscr{E}'|^2/n_0$$

as in eqn. (8.40). Inserting this into the analogue of the power conservation equation (5.2), we find an equation for the local pulse width $T(z)$, whose solution is

$$T^3(z)/T^3 = 1 - \frac{3}{2}\frac{P_0}{P_1}\frac{z^2}{z_t ka^2} \tag{9.13}$$

where P_0 is the peak incident power in area πa^2. The pulsewidth evidently vanishes at the temporal self-focusing length z_{tf} which satisfies

$$\frac{ka^2}{z_{tf}} = \sqrt{\frac{3}{2}\frac{P_0}{P_1}\frac{ka^2}{z_t}}. \tag{9.14}$$

This should be compared with eqn. (8.43) for the spatial self-focusing length z_f of an incident cylindrical gaussian beam. The catastrophic intensity rise due to temporal self-focusing can occur prior to that due to spatial self-focusing if $z_t < \frac{3}{4}ka^2$, or if

$$-k'' = -\partial^2 k/\partial\omega^2 > 4T^2/3ka^2. \tag{9.15}$$

Equation (9.14) must overestimate z_{tf} if $z_{tf} < z_f$ because the pulse-sharpening effect associated with spatial self-focusing assists the compression. Of course, when the pulse length becomes comparable to the relaxation time of the nonlinear response, the compression rate is reduced.

The self-compression of pulses with initial phase modulation cannot be studied with the analogue of the lens transformation for slab-shaped beams. While the linear equation is invariant under this transformation, the nonlinear equation with cubic polarization is not. The reader may enjoy proving, however, that the nonlinear equation with polarization $\mathscr{P}^{NL} = \eta|\mathscr{E}|^4\mathscr{E}$ is invariant under the lens transformation for slab symmetry. (In slab symmetry, the lens transformation is still given by eqns (5.7), (5.8) and (5.9) if the initial factor of $A(\zeta)$ in (5.8) is replaced by $A^{1/2}(\zeta)$.)

4. In a medium with cubic nonlinearity and normal dispersion, a smooth pulse with no initial phase modulation will develop intensity maxima near the inflection points of the pulse shape where the frequency sweep developed by self-phase modulation changes sign. The central portion of the pulse is "temporally defocused", while the leading and trailing edges beyond the inflexion points are temporally focused. The position, but not the strength, of the peaks can be estimated roughly from the temporal self-focusing length associated with the phase of the trial function (9.12) for a gaussian incident pulse [see also eqn. (7.15)]:

$$\begin{aligned}\frac{1}{z_{tf}^2} &\approx \frac{k''}{ka^2}\frac{\partial^2}{\partial t'^2}\frac{P(t')}{P_1}\\ &= \frac{2}{z_t ka^2}\frac{P(t')}{P_1}\left(1 - \frac{2t'^2}{T^2}\right);\end{aligned} \tag{9.16}$$

z_t is negative here so z_{tf} is real only when t'^2 exceeds $\frac{1}{2}T^2$. Although z_{tf} is short for large temporal displacements from the pulse peak, the power is small there. Consequently, we expect to find rather small peaks far in the wings at short propagation lengths (small s) which grow and move closer to the inflexion points on leading and trailing edges as the pulse travels forward. This value for z_{tf} is typically the same as that for pulse compression in an anomalously dispersive medium, except that the peak power in (9.14) should be replaced by the power near the inflexion points of the pulse shape, or about $0.6P_0$. These estimates are very crude, and there are no numerical solutions available yet to refine them. Little attention has been given this case in the literature, although condition (9.15) suggests that it should be observable with wide, or diverging beams and short pulses. The more important transient modification of this case will be discussed in the next section.

From these four basic situations, one can form some intuition for what will happen in more complicated cases. It is important to realize that temporal self-foci (spikes) can form on smooth pulses in normally dispersive nonlinear media as well as in anomalously dispersive substances, if the peak power far exceeds the critical power P_1. The influence of absorption on pulse compression with anomalous dispersion is easily determined in the context of the approximation which led to eqn. (8.47). Dispersion and other temporal effects may also be introduced into the linearized analysis described in Section 8. It is nevertheless difficult to visualize the very strong interaction between spatial and temporal self-focusing. Many authors have assumed that spatial self-focusing forms one or more "filaments", in each of which the temporal development can be treated independently of the spatial behavior. The main justification for this procedure is its mathematical tractability. Although the temporal self-focal length usually exceeds the spatial one for the incident pulse, the two can become comparable in the vicinity of a self-focus of either kind. Some numerical solutions exist which treat both effects simultaneously,[16,98] but they are all simulations of specific cases from which it is difficult to infer general features. These will be discussed in more detail below. More analytical work on the interaction between spatial and temporal self-focusing is certainly possible, but does not seem to have been reported in the literature.

C. *Transient effects without self-focusing*

The general features of the material equations of motion which relate the transient nonlinear response to the applied field have been discussed in Section 3B. Little information exists on the influence of inertial transiency on self-focusing, so we shall confine our treatment to diffusive transiency where the refractive index change δn satisfies equations (3.10):

$$\tau \frac{\partial \delta n}{\partial t} = \tfrac{1}{2} n_2 |\mathscr{E}|^2 - \delta n. \tag{9.17}$$

The most direct consequence of a finite relaxation time τ is the distortion of the "response pulse" $\delta n(t)$ relative to the "driving pulse" $|\mathscr{E}(t)|^2$ as shown in Fig. 2. That figure shows clearly that the rate of change of $\delta n(t)$ is positive during most of the pulse. Therefore the self-phase modulated spectrum, computed from eqn. (7.12), will contain most of its energy in the Stokes (downshifted) sidebands. This influence on the spectrum is the most pronounced observable feature of the diffusive transient response of δn.

Quantitative estimates of the features of the self-modulated spectrum may be obtained in the thin window approximation, where

$$2\mathrm{i}\frac{\partial \mathscr{E}'}{\partial z} = -\frac{2\omega_0}{c}\,\delta n(t')\,\mathscr{E}'. \tag{9.18}$$

We are using the variables x, y, z and $t' = t - z/v_g$ as in eqn. (7.1).

Using eqn. (3.13) for $\delta n(t')$, we may integrate (9.18) for short distances to find the phase of $\mathscr{E}'$, and therefore the instantaneous frequency from (7.12):

$$\omega(t') - \omega_0 = -\frac{\omega_0 z}{c\tau}\left[\tfrac{1}{2} n_2 |\mathscr{E}|^2 - \delta n\right] = \frac{P_0}{P_1}\frac{z\tau}{k_0 a^2 T^2}\left[\frac{t'}{\tau} - 1 + \left(\frac{T}{\tau} + 1\right)\mathrm{e}^{-(T+t')/\tau}\right]. \tag{9.19}$$

The function in square brackets vanishes when $t = t_m$, where t_m can be obtained from Fig. 1 for any value of T/τ, and approaches τ for long pulse lengths. Therefore, the fraction of total energy in the Stokes sidebands is the integral

$$W_{\text{Stokes}}/W_{\text{total}} = \frac{3}{4}\int_{-T}^{t_m}\left(1 - \frac{t^2}{T^2}\right)\mathrm{d}t = \frac{1}{2} + \frac{3}{4}\frac{t_m}{T} - \frac{1}{4}\frac{t_m^3}{T^3} \tag{9.20}$$

This ratio is 90% when $T = \tau$, and is well approximated by $(\frac{1}{2} + 3\tau/4T)$ for $T \gtrsim 4\tau$. The maximum frequency shifts occur when the derivative of (9.19) vanishes (Stokes) or when $t' = T$ (antiStokes). The shifts are decreased from the maximum value given by (7.17) by the factors

$$1 - \frac{\tau}{T}\log\left(1 + \frac{T}{\tau}\right) \qquad \text{(Stokes)}, \tag{9.21}$$

$$1 - \frac{\tau}{T} + \left(1 + \frac{\tau}{T}\right)\mathrm{e}^{-2T/\tau} \qquad \text{(antiStokes)}, \tag{9.22}$$

but the antiStokes radiation is strongly attenuated because it is driven by the weak trailing edge of the pulse [see Fig. 2 and eqn. (9.20)]. To summarize: the transient response narrows the self-phase modulated spectrum, throws most of the energy into the Stokes frequencies, and causes the peak intensity to occur on the Stokes side at the frequency shift [obtained from (9.19) at $t' = 0$]

$$\Delta\omega_\tau = \Delta\omega_0 \frac{\tau}{T}\left[1 - \left(\frac{T}{\tau} + 1\right)\exp - T/\tau\right] \tag{9.23}$$

where $\Delta\omega_0$ is the maximum spectral width when $\tau = 0$, given by eqn. (7.17), and the quantity in brackets is easily evaluated from Fig. 1.

All these features appear in numerical computations of self-phase modulated spectra generated by more complicated pulse shapes. Gustafson *et al.* have examined the spectrum for a gaussian pulse with various relaxation times.[18] Figure 43 from their work shows a typical result. Notice that the oscillations expected from a driving pulse with an inflexion point appear only on the Stokes side (cf. Section 7C). The antiStokes side is generated by the trailing edge of the relaxing index change, whose inflexion point occurs only after the pulse intensity has decreased to small value. Gustafson *et al.* also examined the spectrum arising from two beams of slightly different frequency as in eqn. (7.23), and again found substantial Stokes–antiStokes asymmetry.

If weak linear dispersion is included in eqn. (9.18), we find an equation similar to (9.5):

$$2i\frac{\partial \mathscr{E}'}{\partial z} - k''\frac{\partial^2 \mathscr{E}'}{\partial t'^2} = -\frac{2\omega_0}{c}\delta n(t')\mathscr{E}'. \tag{9.24}$$

This is identical to the formula governing the self-focusing of a slab shaped beam in a medium with a nonlocal nonlinear index, if t' is regarded as a transverse dimension [compare eqns. (9.6) and (9.5)]. In the language of Section 9B, $\delta n(t')$ forms a temporal lens which focuses or defocuses the beam depending upon the signs of $\partial^2\delta n/\partial t'^2$ and k''. The difference between this lens and that for $\tau = 0$ is simply the difference between δn and $\frac{1}{2}n_2|\mathscr{E}|^2$. Thus for anomalously dispersive media (case 3, Section 9B) a smooth pulse with no initial phase modulation will self-compress, but the characteristic self-compression distance z_{tf} will be less than that given by (9.14) because δn has less "temporal curvature" than $\frac{1}{2}n_2|\mathscr{E}|^2$. The effect of finite τ can be estimated for a parabolic pulse by recognizing that in eqn. (9.14) the quantity

$$\frac{P_0}{P_1}\frac{ka^2}{z_1} = -k''\left.\frac{\partial^2\varphi}{\partial t'^2}\right|_{t'=0}$$

can be replaced by the phase curvature with relaxation at the point of maximum δn. The result is

$$\frac{ka^2}{z_{tf}} = \sqrt{\frac{3}{2}\frac{P_m}{P_1}\frac{ka^2}{z_1}\frac{t'_m}{\tau}} \tag{9.25}$$

where P_m is the incident power at time $t' = t'_m$:

$$P_m = P_0(1 - t_m^2/T^2).$$

When $T = \tau$, z_{tf} is increased by the factor 1.6.

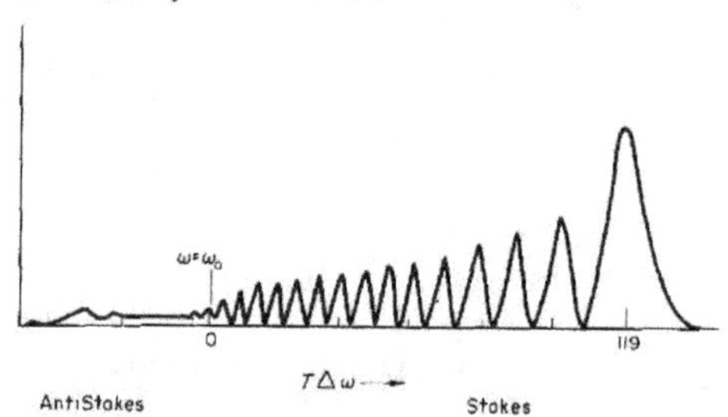

FIG. 43. Self-phase modulated spectrum with relaxation with $\tau/T = 5/3$, $(P_0/P_1)\cdot(z/k_0a^2) = 530$. Equation (9.21) predicts $T\Delta\omega_{max} \approx 115$, whereas the computed value is about 119. The incident pulse is gaussian $\sim\exp(-t^2/T^2)$. Note the strong Stokes–antiStokes asymmetry. This spectrum was calculated by Gustafson *et al.*[18] to fit an experimental spectrum for ruby laser pulses (λ_0 = 0.6943 μm) in CS_2.

The effect of relaxation on the self-compressed spikes near the inflexion points of a pulse in a normally dispersive medium is more dramatic. When the response is instantaneous, these spikes form only in the wings of the pulse beyond the inflexion points where the intensity is low. But when $\delta n(t)$ is transient, its leading edge inflexion point can be close to the pulse peak, where it can cause strong pulse compression.[98] In fact even a parabolic pulse, which shows no compression at all in a normally dispersive medium when $\tau = 0$, may be strongly compressed when τ exceeds the pulse duration, because the function $\delta n(t')$ always has a leading edge inflexion point at

$$\frac{t'_{\text{inf}}}{\tau} = -\frac{T}{\tau} + \log\left(1 + \frac{T}{\tau}\right).$$

When $T = \tau$, this point occurs at $t'_{\text{inf}} = -0.307T$, where the intensity has 90% of its peak value. The focal length $z_{\text{tf}}(t')$ for the formation of temporal spikes at times $t' < t'_{\text{inf}}$ can be estimated in a manner analogous to that which led to eqn. (9.25) [compare with (9.16)]

$$\frac{1}{z_{\text{tf}}^2} \approx -\frac{3}{2}\frac{P_0}{P_1}\frac{1}{z_1 ka^2}\left\{1 - \left(\frac{T}{\tau} + 1\right)\exp[-(T + t')/\tau]\right\}. \tag{9.26}$$

This length is imaginary for $t' > t'_{\text{inf}}$, infinite for $t' = t'_{\text{inf}}$, and minimum when $t' = -T$. Of course, there is no intensity at $t' = -T$, so one finds weak spikes near the leading edge of the pulse at short distances which move closer to the peak and become stronger at longer distances.

Fisher and Bischel have obtained many numerical solutions for pulse shapes in a dispersive medium with diffusive transient nonlinearity.[98] Their results show all the qualitative features implied by the analysis above. Indeed, the numerical work of Shimizu[99] and Fisher and Bischel[98] preceded the analytical work and first suggested the importance of relaxation for self-pulse compression in media with normal dispersion. Figure 44 from ref. 98 shows the evolution of a temporal spike on the leading edge of a short pulse. No corresponding spike forms on the trailing edge, where the inflexion point in $\delta n(t')$ lags behind the driving field, and the "temporal curvature" $\partial^2\delta n/\partial t^2$ is reduced by relaxation.

In concluding this section, we must cite the investigation by DeMartini *et al.* into the effects of relaxation and dispersion upon self-steepening.[5] As one expects, relaxation prevents optical shock formation at finite distances.

D. *Transient effects with self-focusing.*

The pronounced effect of self-focusing upon the local optical pulse shape has already been described in Section 7B. When the local rise time becomes comparable to the relaxation time τ of the nonlinear response, transient effects are important in determining the behavior of the optical field, even if the incident rise time is much greater than τ. Before considering pulsed beams, however, it is instructive to explore the similarity solutions of the transient self-focusing equations.

Shimizu[16] has drawn attention to the important fact that similarity solutions still exist when δn satisfies eqn. (9.17), if only the scale radius satisfies

$$a(z, t') = a_0(z)\exp -t'/\theta, \tag{9.27}$$

where θ is a parameter whose determination is described below, and $a_0(z)$ satisfies

$$a_0^2(z) = \frac{\beta^2}{k^2 a_0^4}z^2 + a_0^2(0) \tag{9.28}$$

when the wave fronts are flat at $z = 0$. The parameter β here plays the same role as that in eqn. (5.42) but is not related to H or to the initial power exactly by eqn. (5.43) because C_2 in (5.41) is not a constant of the motion in the transient case. The amplitude and phase of the self-similar solutions still have the form (5.39) and (5.41), but the amplitude $f(r/a) = a|\mathscr{E}|$ now satisfies the equation

$$\frac{d^2 f}{d\rho^2} = -\frac{1}{\rho}\frac{df}{d\rho} + (\alpha + \beta^2\rho^2)f - \frac{2k_0\omega_0}{c}gf \tag{9.29}$$

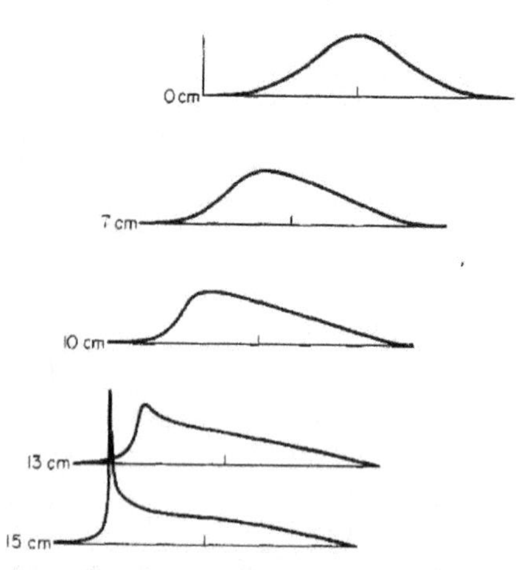

FIG. 44. Evolution of a gaussian pulse in a nonlinear normally dispersive medium with relaxation. The pulse had no initial chirp and the temporal diffraction length is $z_t = -1.67 \times 10^4$ cm. The temporal self-focusing length for this case is $z_{tf} \approx 11$ cm without considering relaxation. Relaxation effects cause a temporal self-focus to form only on the leading edge at a distance greater than z_{tf}. Equation (9.26) is inaccurate here because it is derived for a parabolic pulse. Other parameters are: $T/\tau = 2.5$, $I_0/P_1 = 2.68 \times 10^6\ \text{cm}^{-2}$, $\lambda_0 = 10.6\ \mu\text{m}$. (From Fisher and Bischel.[98])

$$-\frac{\tau}{\theta}\rho\frac{dg}{d\rho} = \tfrac{1}{2}n_2 f^2 - \left(1 + 2\frac{\tau}{\theta}\right)g \tag{9.30}$$

where $\delta n(r, z, t') = g(r/a)/a^2$, and $\rho \equiv r/a$. Expanding f and g near $\rho = 0$ as

$$f = f_0(1 - \tfrac{1}{2}\rho^2 + \ldots),$$

$$g = g_0(1 - \gamma\rho^2 + \ldots)$$

and inserting into (9.28) and (9.29), we find the relation between α and the axial intensity:

$$\alpha + 2 = \frac{(k_0\omega_0/c)n_2 f_0^2}{1 + 2(\tau/\theta)} = (k_0\omega_0/c)n_2 g_0. \tag{9.31}$$

This reduces to (5.45) when $\theta = \infty$. The index curvature is proportional to

$$\gamma = \frac{1 + 2(\tau/\theta)}{1 + 4(\tau/\theta)}. \tag{9.32}$$

The interested reader will enjoy deriving these equations himself, and also proving that self-similar solutions of the same form still exist if δn satisfies an arbitrary linear equation of motion with driving term proportional to $|\mathscr{E}|^2$. Thus even for inertial transiency one finds similarity solutions with exponentially decreasing scale radii. Of course, in this case (9.30) must be modified [g satisfies the same equation as δn, but with $\partial/\partial t'$ replaced by $(2 - \rho\partial/\partial\rho)/\theta$].

As in the nontransient similarity solutions, β governs the phase curvature (if the phase surface is flat at $z = 0$) and α governs the axial index change. The parameter θ is a function of both α and β, and becomes infinite when α and β locate a point on a branch of the curve shown in Fig. 14. These are the steady state similarity solutions. If α and β do not correspond to a steady solution, then $\theta(\alpha, \beta)$ is finite and must be chosen to allow the mode function $f(\rho)$ to vanish at $\rho = \infty$. Shimizu gives a graph of P (related to α) vs θ for $\beta = 0$ (he discusses only the $\beta = 0$ similarity solutions) and remarks that θ is proportional to $1/P$ for large powers. Of course $\theta = \infty$ for $P = P_{c1}$.

The transient similarity solutions represent field configurations with constant power in which the axial intensity increases exponentially with time. The instantaneous axial intensity is weak downstream and strong upstream, and a point of given axial intensity travels

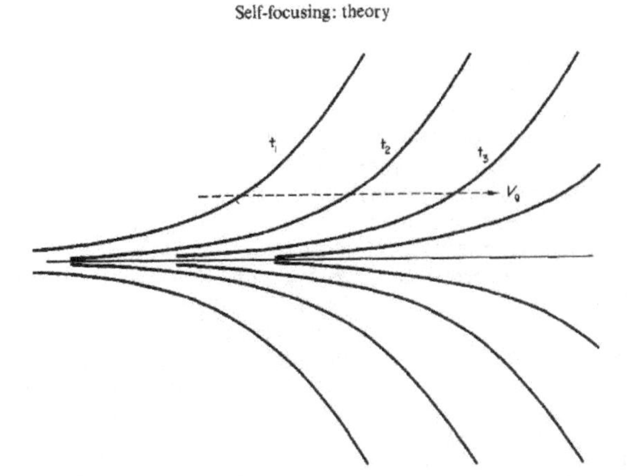

FIG. 45. Scale radius of transient similarity solutions for $\beta = 0$. $a(z, t') = a_0 \exp[(-t + (z/v_g))/\theta]$. The intensity at any point increases exponentially with time, while a point of constant intensity travels to the right with velocity v_g.

to the right with velocity v_g as indicated in Fig. 45. When $\beta = 0$, the axial intensity is finite at finite times. However, if β^2 is negative, which would correspond in the steady case to $P > P_3$, then a^2 vanishes at a singular point z_s determined by eqn. (9.28) or its generalization for initially curved phase fronts. *Thus transient response does not prevent singular self-focusing at finite distances if the index is driven linearly by the field intensity* (cubic nonlinearity in steady state). This contradicts a conclusion of Shimizu who was unaware of the $\beta \neq 0$ solutions. The point is that in the transient régime our notion of critical power must be generalized. The self-focusing process no longer depends so heavily upon the total power, but also on how it is distributed in the transverse plane. All the solutions with $\beta = 0$ must be regarded as generalizations of self-trapped solutions. They have no singularities at finite distances or times, nor do they possess phase curvature. Solutions with $\beta^2 > 0$ are the generalizations of the below-critical-power solutions in the steady state. The self-focusing solutions, with a singularity at finite z_s at *all* times, are obtained for $\beta^2 < 0$.

A transient self-similar mode would be difficult to manufacture in the laboratory. The incident spot size would have to be diminished exponentially with time, and some care would be required to reproduce the transverse mode structure $f(\rho)$. It is not known under what circumstances an arbitrary incident mode profile would evolve toward one of the steady-state solutions, although Shimizu indicates that his numerical solutions resemble them in some régimes.[16] Assuming that such a mode is obtained at some starting time, the upstream axial intensity would increase until some additional mechanism, such as breakdown, spoils the induced lens. This would happen at a critical scale radius a_c, which would propagate through the medium at velocity v_g. Ahead of this critical point, the radius would be comparable to a_c for a distance $\approx v_g\theta$, which Shimizu calls the "filament length". This terminology is plausible for the self-similar solutions with $\beta = 0$, but is not clearly relevant to the confined portion of a pulsed beam.

To what extent do these constant power solutions allow us to understand the more complicated pulsed case? Does a "square" incident pulse with constant power P during pulse length T evolve toward a segment of a similarity solution? We expect that it should qualitatively, because near the leading edge the index change has not yet responded to the incident field, whereas near the trailing edge the rays pass through material in which δn has been evolving for about T seconds. Thus the front of the pulse expands (diffracts) and the rear is self-focused. What we cannot do is predict whether the rear of the pulse contracts to a singularity in a finite distance. Nothing that we know about the similarity solutions suggests that a singular self-focus will not occur. Computer solutions with gaussian incident pulses do show scales much smaller than those observed experimentally, so for all practical purposes we may assume that catastrophic focusing can occur. The available numerical solutions do not seem to show singular self-foci, but the assumed powers are rather low, and the solutions are largely unexplored.

Fleck and Kelley,[15] and later Shimizu,[16] and Fleck and Carman[100] have undertaken the arduous task of numerical analysis of the self-focusing equation coupled with the relaxation eqn. (9.17). Their solutions verify the notion that a propagating pulse develops a horn shape in which the tail self-focuses in the index change created by the broader head. In fact the tail can be confined even if it has negligible power, as stressed by Shen and Loy.[25] Once the index change has been established it remains appreciable for about τ seconds after the field which drove it has passed. Figure 2 for a parabolic pulse shows that even the tail end of the driving pulse sees a large index change. The light in this portion consequently propagates as if it were in a dielectric wave guide formed by the preceding radiation. Such *relaxation trapping* can be detected in the numerical solutions, but it has negligible influence on the observable properties of self-focusing, such as spectral broadening, because of the low trapped power.

A more striking phenomenon, which was not anticipated prior to the numerical investigations, is the oscillation of the beam radius shown in Fig. 46. When the rays are bent inward toward a self-focus, the index change does not respond in time to cancel the strong diffractive force [proportional to a^{-3} according to (8.19)] as the beam radius decreases. Thus the beam reaches a minimum radius and then diverges until the index change catches up. Wong and Shen have observed such oscillations in experiments on slowly relaxing liquid crystals.[101] They also confirmed the dependence of the first minimum radius on incident power predicted by Shimizu's numerical analysis and reproduced here as Fig. 47.

Transient nonlinear response has been of interest primarily because it was thought to offer a natural way of preventing singular self-foci. While it definitely slows the self-focusing process, the existence of transient singular similarity solutions suggests that relaxation does not remove the singular behavior of self-focusing with cubic nonlinear response. Higher-order nonlinear corrections to the index leading to saturation and nonlinear loss, and dielectric breakdown now appear to be the most common mechanisms for determining the optical field in the self-focal region.

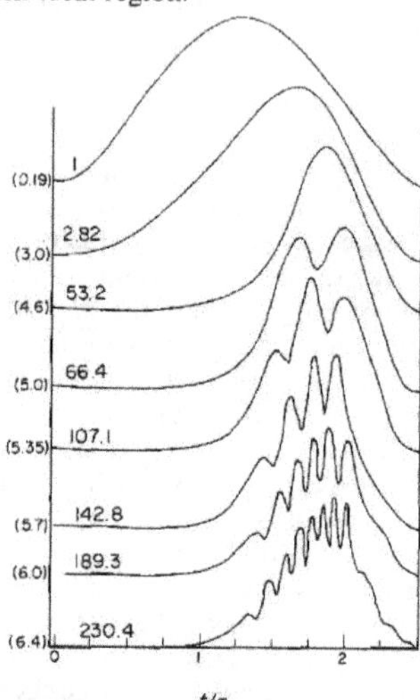

FIG. 46. Axial intensity of a pulsed gaussian beam propagating in a self-focusing medium with relaxation. The incident pulse shape was parabolic with $T/\tau = 1.25$, $P_0 = 16P_1$, $k_0a_0^2 = 15.7$ cm. The expected minimum self-focal length without self-focusing is 3.7 cm according to eqn. (6.5). If n_2 is reduced by eqn. (3.14), then the self-focal length becomes 13.7 cm. The pulse heights for different distances are normalized to the same value, but the values relative to the first pulse shown are indicated on the left. The figure in parentheses is the distance in cm to which the leading edge has penetrated. (Adapted from Fleck and Carman.[100])

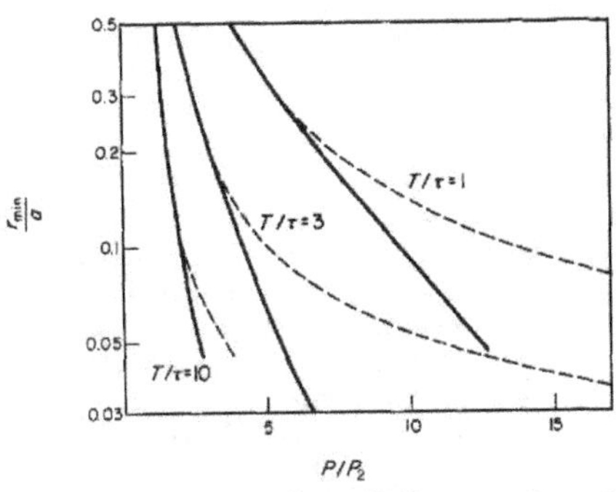

FIG. 47. Dependence of minimum gaussian radius on incident power for a pulsed beam incident on a nonlinear medium with relaxation. The solid lines indicate the absolute minimum radius, whereas the dashed line is the radius at the first intensity maximum. (From Shimizu.[16])

10. INFLUENCE OF OTHER PHENOMENA UPON SELF-FOCUSING

Self-focusing would be of minor interest were it not for its role in enhancing the local optical intensity to extraordinarily high values. In the vicinity of a self-focus, the field strength is large enough to drive a host of nonlinear processes, including multi-photon absorption, stimulated scattering and dielectric breakdown, which might otherwise be absent at the incident intensity. Space does not permit an investigation of the evolution of all these processes once they have been initiated by self-focusing. Indeed, they are just beginning to receive realistic analysis. Our goal here will be simply to indicate qualitatively how they influence the catastrophic intensity rise near a self-focus. General quantitative statements are out of the question except in a few cases.

A. *Dielectric breakdown*

When it occurs, avalanche electron production has the most immediate and most profound influence on beam propagation near a self-focus. Hellwarth[102] and Yablonovitch and Bloembergen[103] have drawn attention to the large index change caused by the creation of even a modest free electron concentration. The frequency dependent dielectric response with free electrons present is

$$\epsilon(\omega) = \epsilon^0 - \frac{\omega_p^2(\omega - i\Gamma)}{\omega(\omega^2 + \Gamma^2)} \tag{10.1}$$

where ϵ^0 is the background dielectric constant without free electrons, and $\omega_p^2 = 4\pi Ne^2/m$ is the squared electron plasma frequency, which is proportional to the electron concentration N. The damping rate Γ of the electronic motion includes contributions for inverse bremsstrahlung, electron–phonon interactions, and other mechanisms that extract energy from the accelerated electrons. The index change corresponding to (10.1),

$$\delta n \approx -\frac{\omega_p^2}{2n_0(\omega^2 + \Gamma^2)}, \tag{10.2}$$

is *negative* and cancels the self-focusing index change when

$$\frac{\omega_p^2}{\omega^2} \approx n_0 n_2 |\mathscr{E}|^2$$

or when the electron density reaches

$$N_c = \frac{2}{r_0}\frac{I}{P_1}. \tag{10.3}$$

Here $r_0 = e^2/mc^2$ is the classical radius of the electron, I is the optical intensity, and P_1 is the lower critical power. For $I\pi a^2 = P_1$ and $a = 1$ cm, $N_c \approx 3 \times 10^{12}/\text{cm}^3$ which is rather small for an electron concentration. For $I\pi a^2 = P$, and $a = 10\,\mu$m, $N_c \approx 3 \times 10^{20}$/cm^2.

Some free electrons may be generated by thermal ionization of shallow traps and other peripheral effects, but at high incident fields large concentrations can be generated by an avalanche created by a few electrons which by chance gain enough energy from the field to initiate the process. This avalanche mechanism is assumed to be the origin of dielectric breakdown, but it is entirely possible for it to occur without causing actual material damage.[104] Damage appears to be caused by the absorption of optical energy associated with the imaginary part of the response (10.1). If the optical intensity exceeds the breakdown intensity only for a very short time, the energy absorbed may be insufficient to cause structural damage. Yablonovitch and Bloembergen have summarized data relating the ionization rate to the optical intensity.[103] Figure 48 shows their estimate of the quantity

$$\eta(|\mathscr{E}|) = \mathrm{d}\log N/\mathrm{d}t, \tag{10.4}$$

from which one can infer the electron density from the intensity:

$$N(t) = N_0 \exp \int^t \eta(|\mathscr{E}|)\mathrm{d}t. \tag{10.5}$$

If the incident pulse shape is known, $|\mathscr{E}(t)|$ at a point prior to the self-focus can be inferred from the transfer curve of Fig. 34. Because $\mathscr{E}$ increases so rapidly near a self-focus, the electron density also rises rapidly until it reaches the value (10.3). After this point, the index change is negative, and no longer follows the applied field, because the electrons once detached diffuse "slowly" away or return slowly to their bound states. Because η is such a steep function of $\mathscr{E}$, the electrons are generated in a very brief time as the field passes through the range $\mathscr{E}_B \approx 10^6 - 10^7$ V/cm, corresponding to the intensity range $I_B \approx 10^9 - 10^{11}$ watts/cm^2 in vacuum (see Fig. 48). This fixes the range of intensities for which (10.3) is satisfied, and allows an estimate of the beam radius a_c at which catastrophic self-focusing is arrested. If P_f is the power associated with the focusing rays, then

$$I_B \pi a_c^2 = P_f.$$

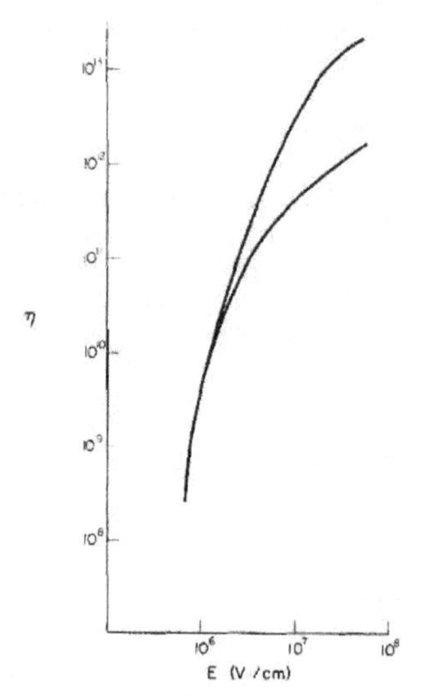

FIG. 48. Ionization rate in s^{-1} versus field strength, from Yablonovitch and Bloembergen.[103] The two branches for high fields may be regarded as limits between which the actual value lies. (The quantity η here corresponds to α in ref. 103.)

The numerical solutions indicate that $P_f \approx P_2$, so the limiting radius is about

$$a_c \approx (P_2/\pi I_B)^{1/2}$$
$$\approx 3.4\lambda/(n_2 I_B)^{1/2} \tag{10.6}$$

where λ is in cm, n_2 in esu, and I_B in W/cm^2. For $\lambda = 1\ \mu$m, and $n_2 = 2 \times 10^{-13}$ (roughly the value for glass), one finds a_c in the range 2–20 μm. This range includes the observed size of filamentary damage phenomena in glass. Yablonovitch and Bloembergen,[103] who developed this argument, point out that self-focal field strengths inferred experimentally by Brewer and Lee[105] are in the breakdown range. If there is no other mechanism such as nonlinear absorption, saturation, or backward scattering to weaken the singular self-focus, avalanche electron production certainly will do it.

The rapid production of free electrons near the singular focus must have a strong influence on the spectral distribution of the radiation passing through the breakdown region. Since the nonlinear index is decreasing, the instantaneous self-phase modulated frequency exceeds the incident frequency. The actual spectrum of a pulse, however, depends sensitively upon the cancellation of the contributions from electron production and the usual nonlinear response in the expression for the instantaneous frequency shift:

$$\Delta\omega = -\frac{\omega_0}{c}\int^z \frac{\partial}{\partial t}\left[\delta n_{NL} - |\delta n_{elect}|\right] dz'. \tag{10.7}$$

Bloembergen[106] has suggested that δn_{NL} can be neglected, while the remaining contribution is of order

$$\Delta\omega \approx 30\,\frac{\omega_0}{c}\,l\cdot\delta n_{max}\,\frac{c}{a_c}, \tag{10.8}$$

where δn_{max} is the index change δn_{NL} associated with the breakdown field strength. a_c is given by (10.6), and l is the absorption length $\approx nc\omega^2/\Gamma\omega_p^2$ of the plasma. The factor 30 is an estimate of the ratio of the electron production rate to a characteristic rate of increase of focal field c/a_c. The resulting frequency shift is of order $\omega_0 c/2\Gamma a_c$ which can be as large as ω_0. These approximations and assumptions[106] are at least dimensionally correct, and give numbers consistent with some experimental data, but are not clearly implied by the theory. It is possible to carry the analytical theory farther to improve these estimates, but numerical studies are ultimately required to determine the relative roles of nonlinear response and electron production in self-phase modulation. Feit and Fleck[107] have reported numerical studies of beam propagation in a gas whose response includes free electron production, but no other nonlinearity.

The potentially large antiStokes frequency shift arising from avalanche electron production contributes to the complexity of the self-focusing process in several ways. First the temporal focusing effect in dispersive media may be enhanced if the "chirp rate" $\partial^2\phi/\partial t^2 \sim 1/T^2$ is large [see eqn. (9.15)]. Second, the various frequency components can mix nonlinearly with each other to produce broad sidebands. Third, the diffraction properties of the frequency shifted light differ from those of the incident light, especially in the small self-focal region. Finally, one should bear in mind that the radial behavior of the intensity and phase functions is complicated and can lead to far field rings of various kinds. Depletion by scattering into other frequencies, localized absorption by the axial plasma, and aberration in the induced lens can all lead to rings in the transmitted radiation. If multiple focal spots occur, the situation is even more complex. Because of the technical importance of dielectric breakdown, however, one can anticipate continued scrutiny of these complexities.

B. *Stimulated scattering*

One of the earliest successes of the self-focusing theory was the explanation of the anomalously low stimulated Raman scattering (SRS) thresholds in certain liquids.[108] The characteristic length in which Raman–Stokes radiation builds exponentially from noise to an intensity comparable to the driving beam in the steady state is[109]

$$l_s \approx \frac{\log(I_L/I_N)}{g_s I_L}. \tag{10.9}$$

Here g_s, the Stokes gain coefficient, is related to the spontaneous scattering cross-section through

$$g_s = \frac{(2\pi)^3 N}{\hbar\omega_L k_s^2}\frac{d^2\sigma_s}{d\Omega d\omega}, \tag{10.10}$$

where N is the density of scatterers, and the subscripts L and s refer to laser and Stokes modes, respectively. In eqn. (10.9) I_N is the noise intensity which initiates the scattering. Self-focusing can enhance I_L and thus decrease l_s dramatically compared to its value without self-focusing. In fact many experiments have used the onset of SRS as an indication that self-focusing has occurred. The minimum length of material required to observe stimulated scattering at the output is taken to be the self-focusing length z_f, and indeed this frequently agrees well with other determinations of z_f.[59,110]

When SRS is enhanced by self-focusing, the driving intensity may vary rapidly compared to the relaxation rate Γ of the Raman oscillation. In this case eqn. (10.9) is not correct, and the characteristic scattering length depends upon the time history of the driving pulse. The scattered intensity I_s grows non-exponentially, although for long propagation distances z it is proportional to [111]

$$I_s(z, t') \approx I_N \exp\left\{-\Gamma t' + 2\left[g_s \Gamma z \int^{t'} I_L \, dt'\right]^{1/2}\right\} \tag{10.11}$$

where $t' = t - z/v_g$, and dispersion of phase and group velocities has been ignored. This result is only accurate for plane waves (no diffraction or focusing) and for large values of the quantity in square brackets. A characteristic transient growth length l_s' implied by eqn. (10.11) is

$$l_s' \approx \frac{\log^2(I_L/I_N)}{4g_s\Gamma \int^{t'} I_L \, dt'}. \tag{10.12}$$

The quantity in the denominator is frequently approximated by $g_s \Gamma t_p I_{L0}$, where I_{L0} is the peak incident intensity, and t_p the incident pulse length.

Stimulated Raman scattering influences the self-focusing process in a variety of ways. First of all, the Stokes gain coefficient is nearly independent of scattering angle in liquids and gases, and the scattered radiation is intense in both forward and backward directions. The backward scattered light can extract energy from the entire forward traveling laser beam, in the medium, and emerge from the entrance face in a pulse with greater peak intensity than the incident beam. An excellent analysis of this effect has been undertaken by Maier *et al.*,[112] who first observed it. Kelley has emphasized that backward scattering can limit the evolution of the laser beam to catastrophic focus by effectively decreasing the incident power.[113] To prevent a singularity, the backward scattering must deplete the incident pulse in the vicinity of the local time t_0' corresponding to the earliest appearance of a self-focus. The depletion must occur *prior* to the actual time $t_0 = t_0' + z_{f0}/v_g$ at which the focus first appears. This is certainly possible because the intensity near z_{f0} increases to arbitrarily high values prior to t_0. The actual limiting focal size depends sensitively upon the precise growth rate of the backward scattered light, which has not been determined when self-focusing is present.

The second way in which SRS alters the evolution of self-focusing is by introducing drastic radial variations in the driving beam. Because the conversion process is exponential, or nearly so, the laser beam is depleted selectively near its intensity maximum. The creation of such a depletion hole in the self-focusing beam leads to strong diffractive forces on rays near the hole. This effect can be studied numerically in the steady state where the computation is not difficult. Rahn and Maier[114] have reported a numerical analysis of the influence of forward steady-state SRS on self-focusing, showing clearly that under some conditions catastrophic self-focusing is inhibited. Good agreement was obtained between calculated and observed spot sizes in organic liquids exposed to moderately powerful 10-ns pulses from a Q-switched ruby laser. Higher powers, shorter pulse lengths and the influence of backward SRS were not examined.

A more detailed analysis of forward SRS is included in Shimizu's work on transient self-focusing.[16] In this non-steady-state treatment, important effects arising from group velocity mismatch and frequency smearing of laser and Stokes pulses are evident. It is nevertheless difficult to infer general quantitative relations among the various intensities, beam sizes, pulse shapes and other parameters from this work. The computations for each case are time consuming, and no systematic parameter study has yet been undertaken.

We conclude that the interaction between self-focusing and stimulated scattering has not yet been completely explored. Indeed, there is very little in the literature even about stimulated scattering with finite beams *without* self-focusing. Depletion and group velocity mismatch effects can lead to a wide variety of phenomena in this context.

Inelastic scattering processes other than SRS differ only in detail in their influence on self-focusing. Stimulated Brillouin scattering (SBS) cross-sections are usually extremely large and lead to large gains for backward scattered light in the steady state. The SBS process has a long transient time, however, and can frequently be ignored for pulses less than 100 ps in duration. Estimates of the growth rates for a variety of scattering processes may be found in a useful paper by Kroll and Kelley.[115]

One process, stimulated Rayleigh wing scattering (SRWS) is closely related to self-focusing, and in fact self-focusing may itself be regarded as an elastic stimulated scattering process. This viewpoint has been advanced by Chiao *et al.* in a paper which will repay the reader's scrutiny.[116] The small signal gain for any stimulated process involving plane waves can be obtained by writing the set of coupled equations for the fields [one equation similar to (2.9) for each relevant noninteracting mode], and then assuming that the incident laser field is fixed and the other fields relatively so small that only terms linear in them need be retained. The resulting set of linear homogeneous equations usually possesses an exponential solution whose characteristic parameters are determined by a secular equation in the usual way. To carry out this procedure for SRWS, one needs a model for the material response. The simplest is provided by eqn. (3.10) which we shall rewrite as

$$\tau \frac{\partial \delta\epsilon}{\partial t} = \epsilon_2 E^2 - \delta\epsilon \tag{10.13}$$

where E is the *total* optical field. If this field includes a driving term E_0 at frequency ω_0 and scattered fields E_1 and E_2 at frequencies $\omega_1 = \omega_0 - \Omega$ and $\omega_2 = \omega_0 + \Omega$, then E^2 has frequencies at 0, $2\omega_0$, $2\omega_1$, $\omega_1 + \omega_0$, Ω, etc. But only frequencies comparable with $1/\tau$ are effective in altering $\delta\epsilon$, so we retain only the terms with frequencies 0, $\pm\Omega$. The resulting solution of (10.13), for plane wave fields whose amplitudes $\mathscr{E}_i$ depend only upon z, is

$$\begin{aligned} \delta\epsilon = {} & \tfrac{1}{2}\epsilon_2 |\mathscr{E}_0|^2 \\ & + \tfrac{1}{2}\epsilon_2 \frac{\mathscr{E}_0 \mathscr{E}_1^*}{1 - i\Omega\tau} e^{iqr - i\Omega t} + \text{c.c.} \\ & + \tfrac{1}{2}\epsilon_2 \frac{\mathscr{E}_0 \mathscr{E}_2^*}{1 + i\Omega\tau} e^{-iqr + i\Omega t} + \text{c.c.} \end{aligned} \tag{10.14}$$

where $q = k_0' - k_1 = k_2 - k_0$. This relation between k_1 and k_2, the wave vectors of the Stokes and antiStokes fields, assures that the spatial periodicity of the scattered fields matches that of the driving polarizations in (10.14). Fields with other wave vectors do not grow to appreciable magnitudes. Using

$$4\pi P_{NL} = \delta\epsilon E,$$

and keeping only the first term on the right of (2.9), the reader will find two coupled equations for $\mathscr{E}_1$ and $\mathscr{E}_2$ from (2.9)

$$2ik_1 \frac{\partial \mathscr{E}_1}{\partial z} - k_{1T}^2 \mathscr{E}_1 = \frac{-\epsilon_2 \omega_1^2}{2c^2(1 + i\Omega\tau)} \{|\mathscr{E}_0|^2 \mathscr{E}_1 + \mathscr{E}_0^2 \mathscr{E}_2^*\},$$

$$-2ik_2 \frac{\partial \mathscr{E}_2^*}{\partial z} - k_{2T}^2 \mathscr{E}_2^* = \frac{-\epsilon_2 \omega_2^2}{2c^2(1 + i\Omega\tau)} \{|\mathscr{E}_0|^2 \mathscr{E}_2^* + \mathscr{E}_0^{*2} \mathscr{E}_1\}. \quad (10.15)$$

Assuming that $\mathscr{E}_1$ and $\mathscr{E}_2^*$ are proportional to exp $i\gamma z$, one finds from the secular equation for γ, that γ has an imaginary part if $k_T^2 = k_{1T}^2 = k_{2T}^2$ is sufficiently small. Ignoring the small quantity Ω^2/ω_0^2, one finds

$$\operatorname{Im} \gamma = \pm \frac{k_T}{2} \left[\frac{\epsilon_2 \mathscr{E}_0^2}{\epsilon_0} \frac{1}{1 + i\Omega\tau} - \frac{k_T^2}{k_0^2} \right]^{1/2}. \quad (10.16)$$

The growth or decay rate of the scattered fields is, of course, twice this value. The properties of the gain equation (10.16) for $\Omega = 0$ have already been examined in connection with (8.6), to which it is identical. That is, the growth of small-scale fluctuations on an otherwise smooth incident beam is caused by elastic ($\Omega = 0$) stimulated Rayleigh wing scattering! The optimum transverse k vector [$q_\perp$ in (8.6)] is independent of frequency shift, but the gain is greatest for zero frequency shift. This identification of the origin of self-focusing with the elementary process of stimulated Rayleigh wing scattering is deeply satisfying. It is analogous to the relation between ordinary refractive index and coherent linear Rayleigh scattering.

REFERENCES

1. K. GROB and M. WAGNER, *Phys Rev. Lett.* **17**, 819 (1966).
2. Y. R. SHEN, M. Y. AU-YANG and M. L. COHEN, *Phys. Rev. Lett.* **19**, 1171 (1967).
3. J. SARFATT, *Phys. Lett.* A **26**, 88 (1967).
4. M. TAKATSUJI, *Phys. Lett.* A **28**, 584 (1969).
5. F. DE MARTINI, C. H. TOWNES, T. K. GUSTAFSON and P. L. KELLEY, *Phys. Rev.* **164**, 312 (1967).
6. A. HASEGAWA and F. TAPPERT, *Appl. Phys. Lett.* **23**, 142, 171 (1973).
7. O. SVELTO in *Progress in Optics*, vol. XI (1974).
8. S. A. AKHMANOV, R. V. KHOKHLOV and A. P. SUKHORUKOV, chap. E3 in the *Laser Handbook*, ed. by F. T. ARECCHI and E. O. SCHULZ-DUBOIS, North Holland, 1972.
9. R. Y. CHIAO, E. GARMIRE and C. H. TOWNES, *Phys. Rev. Lett.* **13**, 479 (1964). Erratum. *ibid.* **14**, 1056 (1965).
10. V. I. TALANOV, *JETP Lett.* **2**, 138 (1965).
11. P. L. KELLEY, *Phys. Rev. Lett.* **15**, 1005 (1965).
12. S. A. AKHMANOV, A. P. SUKHORUKOV and R. V. KHOKHLOV, *Sov. Phys. JETP*, **24**, 198 (1967).
13. J. MARBURGER and W. G. WAGNER, *J. Quant. Electron.* **3**, 415 (1967).
14. S. A. AKHMANOV and A. P. SUKHORUKOV, *JETP Lett.* **5**, 87 (1967).
15. J. A. FLECK and P. L. KELLEY, *Appl. Phys. Lett.* **15**, 313 (1969).
16. F. SHIMIZU, *IBM J. Res. Develop.*, July 1973, p. 286.
17. J. A. FLECK and C. LANE, *Appl. Phys. Lett.* **22**, 467 (1973).
18. T. K. GUSTAFSON, J. P. TARAN, H. A. HAUS, J. R. LIFSITZ and P. L. KELLEY, *Phys. Rev.* **177**, 306 (1969).
19. D. GRISCHKOWSKY, *Phys. Rev. Lett.* **24**, 866 (1970).
20. P. D. MAKER and R. W. TERHUNE, *Phys. Rev.* **137A**, 801 (1965).
21. A. OWYOUNG, *The Origins of the Nonlinear Refractive Indices of Liquids and Glasses*, Dissertation, Calif. Inst. Tech., Pasadena, 1972.
22. R. R. BIRSS, *Proc. Phys. Soc. (London)* **79**, 946 (1962).
23. A. OWYOUNG, *IEEE J. Quant. Electr.* **9**, 1064 (1973).
24. D. H. CLOSE, C. R. GIULIANO, R. W. HELLWARTH, L. D. HESS, F. J. MCCLUNG and W. G. WAGNER, *J. Quant. Electr.* **2**, 553 (1966).
25. Y. R. SHEN and M. M. T. LOY, *Phys. Rev.* **3A**, 2099 (1971).
26. E. L. KERR, *Track Formation in Optical Glass Caused by Electrostrictive Laser Beam Self-Focusing*, Dissertation, New York Univ., New York, 1970.
27. E. L. KERR, *IEEE J. Quant. Electr.* **6**, 616 (1970).
28. J. A. FLECK and R. L. CARMAN, *Appl. Phys. Lett.* **22**, 546 (1973).
29. J. H. MARBURGER, L. HUFF, J. D. REICHERT and W. G. WAGNER, *Phys. Rev.* **184**, 255 (1969).
30. T. K. GUSTAFSON, P. L. KELLEY, R. Y. CHIAO and R. G. BREWER, *Appl. Phys. Lett.* **12**, 165 (1968).
31. T. K. GUSTAFSON and C. H. TOWNES, *IEEE J. Quant. Electr.* **8**, 587 (1972).
32. R. POLLONI, C. A. SACCHI and O. SVELTO, *Phys. Rev. Lett.* **23**, 690 (1969).
33. J. FRENKEL, *Kinetic Theory of Liquids*. Dover Publications, Inc., New York, 1955.
34. R. W. HELLWARTH, *Phys. Rev.* **152**, 156 (1966).
35. D. GRISHKOWSKY and J. A. ARMSTRONG, *Phys. Rev. A* **6**, 1566 (1972).
36. D. GRISHKOWSKY, "Adiabatic following" in Proc. of Conf. on *Physics of Quantum Electronics*, Crystal Mtn., Wash. (sponsored by Univ. of Arizona), June 1973.
37. J. H. MARBURGER, *Optics Commun.* **7**, 57 (1973).

38. See the recent review by A. J. GLASS and A. H. GUENTHER, *Appl. Optics* **12** (April 1973).
39. H. A. HAUS, *Appl. Phys. Lett.* **8**, 128 (1966).
40. Z. K. YANKAUSKAS, *Sov. Radiophysics* **9**, 261 (1966).
41. J. D. REICHERT and W. G. WAGNER, *J. Quant. Electr.* **4**, 221 (1968).
42. J. BJORKHOLM, *Phys. Rev. Lett.*
43. D. I. ABAKAROV, A. A. AKOPYAN and S. I. PEKAR, *Sov. Phys. JETP* **25**, 303 (1967).
44. D. POHL, *Opt. Commun.* **2**, 305 (1970).
45. D. POHL, *Appl. Phys. Lett.* **20**, 266 (1972); *Phys. Rev.* A **5**, 1906 (1972).
46. W. G. WAGNER, H. A. HAUS and J. H. MARBURGER, *Phys. Rev.* **175**, 256 (1968).
47. E. L. DAWES and J. H. MARBURGER, *Phys. Rev.* **179**, 862 (1969).
48. V. I. TALANOV, *JETP Lett.* **11**, 199 (1970).
49. A. E. SIEGMAN, *An Introduction to Lasers and Masers*, McGraw-Hill, 1971.
50. J. A. FLECK (private communication, 1972) has used the lens transformation to assist the numerical computation of fields in the vicinity of a focus.
51. S. N. VLASOV, V. A. PETRISHCHEV and V. I. TALANOV, *IVUZ Radiofiz.* **13**, 908 (1970).
52. A. E. KAPLAN, *JETP Lett.* **9**, 33 (1969).
53. B. R. SUYDAM, Self-Focusing in Passive Media I, Los Alamos Sci. Lab. Report LA-5003-MS (1973).
54. A. J. GLASS (to be published).
55. Y. R. SHEN, *Phys. Lett.* **20**, 378 (1966).
56. M. FLANNERY and J. MARBURGER, Proc. Conf. on *High Power IR Laser Window Materials*, C. S. SAHAGIAN and C. A. PITHA, eds., Air Force Cambridge Res. Labs. Report AFCRL-71-0592, p. 11 (1971).
57. D. R. WHITE, E. L. DAWES and J. H. MARBURGER, *J. Opt. Soc. Am.* **61**, 620 (1971).
58. V. N. GOLDBERG, V. I. TALANOV and R. E. ERM, *IVUZ Radiofiz.* **10**, 674 (1967).
59. C. C. WANG, *Phys. Rev. Lett.* **16**, 344 (1966).
60. A. L. DYSHKO, V. N. LUGOVOI and A. M. PROKHOROV, *JETP Lett.* **6**, 146 (1967).
61. D. W. FRADIN, *IEEE J. Quant. Electr.* **9**, 954 (1973).
62. A. P. DYSHKO, V. N. LUGOVOI and A. M. PROKHOROV, *Sov. Phys. JETP* **34**, 1235 (1972).
63. A. L. DYSHKO, *Zh. Vychisl. Matem. i Matem. Fiz.* **8**, 238 (1968).
64. P. D. MAKER, R. W. TERHUNE and C. M. SAVAGE, *Phys. Rev. Lett.* **12**, 507 (1964).
65. V. N. LUGOVOI and A. M. PROKHOROV, *JETP Lett.* **7**, 117 (1968).
66. M. M. T. LOY and Y. R. SHEN, *Phys. Rev. Lett.* **22**, 994 (1969).
67. J. P. TARAN and T. K. GUSTAFSON, *IEEE J. Quant. Electr.* **5**, 382 (1969).
68. C. R. GIULIANO and J. H. MARBURGER, *Phys. Rev. Lett.* **27**, 905 (1971).
69. G. L. MCALLISTER, J. H. MARBURGER and L. G. DESHAZER, *Phys. Rev. Lett.* **21**, 1648 (1968).
70. G. L. MCALLISTER, Ph.D. Dissertation, University of Southern California, 1970.
71. L. A. OSTROVSKII, *JETP Lett.* **6**, 260 (1967).
72. F. SHIMIZU, *Phys. Rev. Lett.* **19**, 1097 (1967).
73. R. CUBEDDU, R. POLLONI, C. A. SACCHI and O. SVELTO, *Phys. Rev.* A **2**, 1955 (1970).
74. N. BLOEMBERGEN and P. LALLEMAND, *Phys. Rev. Lett.* **16**, 81 (1966).
75. A. C. CHEUNG, D. M. RANK, R. Y. CHIAO and C. H. TOWNES, *Phys. Rev. Lett.* **20**, 786 (1968).
76. R. W. HELLWARTH, "Role of photo-electrons in optical damage", in *Damage in Laser Materials*, NBS Special Publication 341, ed. by A. J. GLASS and A. H. GUENTHER, p. 67 (1970).
77. E. YABLONOVITCH and N. BLOEMBERGEN, *Phys. Rev. Lett.* **29**, 907 (1972).
78. V. I. BESPALOV and V. I. TALANOV, *JETP Lett.* **3**, 307 (1966).
79. K. A. BRUECKNER and S. JORNA, *Phys. Rev. Lett.* **17**, 78 (1966); *Phys. Rev.* **164**, 182 (1967).
80. J. R. JOKIPII and J. MARBURGER, *Appl. Phys. Lett.* **23**, 696 (1973).
81. B. R. SUYDAM, p. 42 in ref. 82.
82. See papers in *Laser Induced Damage in Optical Materials*: 1973, ed. by A. J. GLASS and A. H. GUENTHER, National Bureau of Standards Special Publ. 387 (1973).
83. A. J. CAMPILLO, S. L. SHAPIRO and B. R. SUYDAM, *Appl. Phys. Lett.* **23**, 628 (1973).
84. R. Y. CHIAO, J. DODSON, D. IRWIN and T. GUSTAFSON, *Bull. Am. Phys. Soc.* **12**, 686 (1967).
85. E. L. DAWES, Ph.D. Dissertation, University of Southern California, 1969.
86. V. V. VOROB'YEV, *IVUZ Radiofiz.* **13**, 1905 (1970).
87. C. R. GIULIANO, J. H. MARBURGER and A. YARIV, *Appl. Phys. Lett.* **21**, 58 (1972).
88. A. B. SHVARTSBURG, *IVUZ Radiofiz.* **13**, 1775 (1970).
89. C. S. WANG, *Phys. Rev.* **173**, 908 (1968).
90. W. KAISER, A. LAUBEREAU, M. MAIER and J. A. GIORDMAINE, *Phys. Lett.* **22**, 60 (1966).
91. V. N. LUGOVOI, *Sov. Physics—Doklady* **12**, 866 (1968).
92. R. LANDAUER, *IBM J. Res. Develop.* **4**, 391 (1960). See also S. T. PENG and R. LANDAUER, *ibid.* 299 (1973).
93. I. B. KATAYEV, *Electromagnetic Shock Waves*, Iliffe Books, Ltd., London. 1966.
94. J. R. KLAUDER, A. C. PRICE, S. DARLINGTON and W. J. ALBERSHEIM, *Bell Syst. Tech. J.* **39**, 745 (1960); C. E. COOK, *Proc. IRE* **48**, 310 (1960).
95. J. GIORDMAINE, M. A. DUGUAY and J. W. HANSEN, *J. Quant. Electr.* **4**, 252 (1968).
96. L. A. OSTROVSKY, *Sov. Phys. JETP* **24**, 797 (1967).
97. R. A. FISHER, P. L. KELLEY and T. K. GUSTAFSON, *Appl. Phys. Lett.* **14**, 140 (1969).
98. R. A. FISHER and W. K. BISCHEL, *Appl. Phys. Lett.* **23**, 661 (1973); **24**, 468 (1974).
99. F. SHIMIZU, *IEEE J. Quant. Electr.* **8**, 851 (1972).
100. J. A. FLECK, Jr. and R. L. CARMAN, *Appl. Phys. Lett.* **20**, 290 (1972).
101. G. K. L. WONG and Y. R. SHEN, *Phys. Rev. Lett.* **32**, 527 (1974).
102. R. W. HELLWARTH, "Role of photo-electrons in optical damage" in *Damage in Laser Materials*, ed. by A. J. GLASS and A. H. GUENTHER, NBS Special Publication 341, p. 67 (1970).
103. E. YABLONOVITCH and N. BLOEMBERGEN, *Phys. Rev. Lett.* **29**, 907 (1972).
104. N. BLOEMBERGEN, *IEEE J. Quant. Electr.* **10**, 375 (1974).
105. R. G. BREWER and C. H. LEE, *Phys. Rev. Lett.* **21**, 267 (1968).
106. N. BLOEMBERGEN, *Optics Commun.* **8**, 285 (1973).

107. M. D. FEIT and J. A. FLECK, JR., *Appl. Phys. Lett.* **24**, 169 (1974).
108. P. LALLEMAND and N. BLOEMBERGEN, *Phys. Rev. Lett.* **15**, 1010 (1965).
109. N. BLOEMBERGEN, *Nonlinear Optics*, Benjamin, N.Y., 1965.
110. C. C. WANG, *Phys. Rev.* **152**, 149 (1966).
111. R. L. CARMAN, F. SHIMIZU, C. S. WANG and N. BLOEMBERGEN, *Phys. Rev.* A **2**, 60 (1970).
112. M. MAIER, W. KAISER and J. A. GIORDMAINE, *Phys. Rev.* **177**, 580 (1969).
113. P. L. KELLEY and T. K. GUSTAFSON, *Phys. Rev.* **8A**, 315 (1973).
114. O. RAHN and M. MAIER, *Phys. Rev. Lett.* **29**, 558 (1972).
115. P. L. KELLEY and N. M. KROLL, *Phys. Rev.* A **4**, 763 (1971).
116. R. Y. CHIAO, P. L. KELLEY and E. GARMIRE, *Phys. Rev. Lett.* **17**, 1158 (1966).

Chapter 3
Optical Self-Focusing: Stationary Beams and Femtosecond Pulses

G. M. Fraiman, A.G. Litvak, V.I. Talanov, and S.N. Vlasov

Abstract This chapter provides a systematic account of the theory of self-focusing of waves. The parabolic equation method is applied to solve different self-focusing problems. Various self-focusing manifestations are described, specifically, the existence of homogeneous wave channels, the instability of plane waves, the formation of the regions with abnormally great field amplitudes (nonlinear foci) in cubic nonlinear media, suppression of self-focusing in periodic systems, and the spectral broadening of self-focusing pulses. Some questions of self-action of femtosecond pulses are discussed.

3.1 Introduction

In 1962, the *Journal of Experimental and Theoretical Physics* published a short paper by G.A. Askar'yan called "Effect of the gradient of a strong electromagnetic beam on electrons and atoms" [1]. It was preceded by a series of papers discussing some aspects of the behavior of charged particles in strong inhomogeneous electromagnetic fields [2, 3]. G.A. Askar'yan noted that "the ionizing, heating, and separating effect of an intense-radiation beam on a medium can be so strong that the properties of the medium in the beam and out of it will be different, which will lead to waveguide beam propagation and remove its geometric and diffraction divergences. This interesting phenomenon can be called "*self-focusing* (italicized by G.A.A.) of the beam."

Very soon afterward, G.A. Askar'yan's prediction was confirmed theoretically and experimentally. In 1963, V.I. Talanov obtained a self-consistent solution of the plane-waveguide type, which later was called the spatial soliton [4]. In the following year, the team headed by C.H. Townes found the simplest type of the self-trapped axisymmetric waveguide [5]. Its power, being independent of

G.M. Fraiman (✉)
Institute of Applied Physics of RAS, Nizhniy Novgorod, Russia
e-mail: fraiman@appl.sci-nnov.ru

R.W. Boyd et al. (eds.), *Self-focusing: Past and Present*,
Topics in Applied Physics 114, DOI 10.1007/978-0-387-34727-1_3,

the waveguide dimensions and called "critical," plays the fundamental role in the whole theory of wave beam self-focusing.

In 1965, P.L. Kelly used the quasi-optical equations proposed by V.I. Talanov [6] to demonstrate in numerical experiments that it is possible to compress a light beam infinitely at the power exceeding the critical level (beam collapse) [7]. At first, self-focusing phenomena were observed experimentally by M. Hercher [8] in 1964 and by N.F. Pilipetsky and A.R. Rustamov [9] in 1965. These papers initiated a wide range of theoretical and experimental studies of the self-focusing effect in later years.

The rigorous self-focusing theory was developed mainly for the paraxial approximation of the nonlinear parabolic equation [6, 10–14] with the cubic nonlinearity, which later acquired the name of the nonlinear Schrodinger equation. The main results of this theory are as follows: description of homogenous solutions of this equation (plane and axisymmetric homogenous beams), obtaining the self-similar solutions of this equation, finding a new group of its invariant transforms (self-focusing transform), and proof of the analog of the virial theorem, which provided the opportunity to strongly validate the divergence of the field in nonlinear foci, and to formulate the sufficient criterion of self-focusing of the beams with arbitrary profiles. The main results of the theory of self-focusing stationary beams are presented in a book by S.N. Vlasov and V.I. Talanov [15].

3.2 Nonlinear Parabolic Equation

First, consider time-stationary wave beams with the linear field polarization

$$\bar{E} = \frac{1}{2}[E\exp(-ik_0z + i\omega t) + c.c.],$$

in a medium with dielectric permittivity:

$$\varepsilon_{NL} = \varepsilon_0[1 + \varepsilon' f_{NL}(|E|^2)], \quad \varepsilon' = \text{const}, \quad k_0 = \frac{\omega}{c}(\varepsilon_0)^{1/2}.$$

Let $\Lambda_\perp$ and $\Lambda_\parallel$ be the characteristic scales for variations of the complex field amplitudes across and along the wave beam, respectively. Assuming conditions of

$$k_0\Lambda_\perp >> 1, \quad \Lambda_\parallel \sim k_0\Lambda_\perp^2 >> \Lambda_\perp, \quad \varepsilon' f_{NL} << 1,$$

one can find, from the full electromagnetic wave equation, the equation for a slowly changing amplitude of the transverse component of the field [6]:

$$2ik_0\frac{\partial E}{\partial z} = \Delta_\perp E + k_0^2\varepsilon' f_{NL}(|E|^2)E. \tag{3.1}$$

In a medium with the cubic nonlinearity $f_{NL}(|E|^2) = |E|^2$, Eq. (3.1) can be brought to the dimensionless form by replacing the variables $z_d = k_0 z$, $\vec{r}_{\perp d} = k_0\vec{r}_\perp$, and $\Psi = \sqrt{\varepsilon'}E$ (in what follows, the index d is omitted):

$$2i\frac{\partial \Psi}{\partial z} = \Delta_\perp \Psi + |\Psi|^2\Psi. \tag{3.2}$$

Equation (3.2) is invariant under scale transforms with the coefficient γ: $z \to \frac{z}{\gamma^2}$, $\vec{r}_\perp \to \frac{\vec{r}_\perp}{\gamma}$, and $\Psi \to \gamma\Psi$.

Equation (3.1) is generalized for nonstationary fields in dispersive media [10]:

$$2ik_0\frac{\partial E}{\partial z} = \Delta_\perp E + k_0\frac{dv}{d\omega}E''_{\xi\xi} + k_0^2\varepsilon' f_{NL}(|E|^2)E, \tag{3.3}$$

where $\xi = z - vt$; v is the group velocity. The region with $\frac{dv}{d\omega} > 0$ is called the region with anomalous dispersion of the group velocity, and the region with $\frac{dv}{d\omega} < 0$ is the region with normal dispersion of the group velocity.

In isotropic media with the cubic nonlinearity, the nonlinear polarization is represented in the following form [16]:

$$P^{NL} = A(\vec{E}\vec{E}^*)\vec{E} + \frac{B}{2}(\vec{E}\vec{E})\vec{E}^*, \tag{3.4}$$

where the coefficients A and B are expressed via the components of the fourth-rank tensor of nonlinear susceptibility. For the linearly polarized field, the latter ratio makes it possible to determine the nonlinear dielectric permittivity,

$$\varepsilon^{NL} = \varepsilon_0 + 4\pi\left(A + \frac{B}{2}\right)|E_\omega|^2 = \varepsilon_0(1 + \varepsilon_2'|E_\omega|^2),$$

with the nonlinearity coefficient

$$\varepsilon_2' = \frac{4\pi}{\varepsilon_0}\left(A + \frac{B}{2}\right).$$

The nonlinear optics use the notion of the nonlinear refraction index, which is written as

$$n = n_0 + \tilde{n}_2 < E^2 > . \tag{3.5}$$

Here $n_0 = \sqrt{\varepsilon_0}$ is the linear medium refraction index, $< E^2 > = \frac{|E|^2}{2}$ is the root-mean-square value of the electric-field intensity, and $\tilde{n}_2$ is the nonlinearity coefficient.

For a linearly polarized wave with the intensity $I = \frac{cn_0|E|^2}{8\pi}$, the latter relationship is often rearranged in the form of

$$n = n_0 + n_2 I,$$

where $n_2 = \frac{4\pi\tilde{n}_2}{cn_0}$. Taking into consideration the smallness of the nonlinear correction, the nonlinearity coefficient n_2 can be connected with ε_2':

$$n_2 = \frac{4\pi\varepsilon_2'}{n} = \frac{16\pi^2}{cn_0^2}\left(A + \frac{B}{2}\right).$$

Isotropic media with nonlinear polarization of Eq. (3.4) form are called media with induced anisotropy and are characterized by the parameter of nonlinear anisotropy $\beta = \frac{B}{A}$. For electron nonlinearity having short relaxation, duration ($\tau \sim 10^{-15}\tilde{n}$), $\beta = 1$. For the "Kerr" nonlinearity it is $\beta = 6$ [16], and for the strictional one, $\beta = 0$ [17]. For the relativistic electron nonlinearity in plasma, $\beta = 1$ [18].

One can generalize Eq. (3.2) to the vector field using relationship (3.4). For the functions $\Psi_\pm$ which are the amplitudes of circularly polarized fields $E_\pm = \frac{E_x \pm iE_y}{\sqrt{2}}$, we obtain the equations given in [17]:

$$\Delta_\perp \Psi_\pm - 2i\frac{\partial \Psi_\pm}{\partial z} + [|\Psi_\pm|^2 + (1+\beta)|\Psi_\mp|^2]\Psi_\pm = 0, \tag{3.6}$$

The functions $\Psi_\pm$ are normalized as follows: $\Psi_\pm = \sqrt{\frac{4\pi A}{\varepsilon_0}} E_\pm$.

In the case of linearly polarized counter-propagating waves having the straight amplitude $E_f \exp(-ik_0 z)$ and the counter-propagating amplitude $E_b \exp(ik_0 z)$, the equations for $\Psi_{f,b} = \sqrt{\frac{4\pi}{\varepsilon_0}\left(A + \frac{B}{2}\right)} E_{f,b}$ have the following form [15, 19]:

$$\Delta_\perp \Psi_{f,b} \mp 2i\frac{\partial \Psi_{f,b}}{\partial z} + [|\Psi_{f,b}|^2 + 2|\Psi_{b,f}|^2]\Psi_{f,b} = 0. \tag{3.7}$$

All the equations are traditionally called the nonlinear Schrödinger equations (NLSE). They are the basic equations in the self-focusing theory.

3.3 Homogeneous Wave Beams

Seeking the solution of Eq. (3.2) in the form of the travelling wave beam $\Psi = \bar{\Psi}(r_\perp)\exp(-i\gamma^2 z/2)$ for the function $\bar{\Psi}(r_\perp)$ of the transverse field distribution, we obtain the following equation:

$$\Delta_\perp \bar{\Psi} + [|\bar{\Psi}|^2 - \gamma^2]\bar{\Psi} = 0. \tag{3.8}$$

By making replacements of $\bar{\Psi} = \bar{e}\gamma$, and $r_\perp = \bar{r}\gamma^{-1}$ in Eq. (3.8), we bring it to the form of

$$\Delta_\perp \bar{e} + (|\bar{e}|^2 - 1)\bar{e} = 0. \tag{3.9}$$

The expressions corresponding to self-focusing waveguide channels (single isolated wave beams or spatial solitons) are the solutions of Eqs. (3.2) and (3.9) which decrease at infinity sufficiently fast: $\bar{e} \to 0$ at $\bar{r} \to \infty$. Specifically, for the cubic medium and a plane beam $\left(\Delta_\perp \equiv \frac{\partial^2}{\partial x^2}\right)$ it has the following form [4]:

$$\bar{e} = 2^{1/2}\cosh^{-1}\bar{x}. \tag{3.10}$$

The properties of plane (two-dimensional) and cylindrical (three-dimensional, $\Delta_\perp \equiv \frac{\partial^2}{\partial x^2} + \frac{\partial^2}{\partial y^2}$) beams are significantly different. In the two-dimensional case, Eq. (3.9) has the single localized solution, Eq. (3.10). In the three-dimensional case, the pattern is more complicated. There are beams with the rotating phase $\bar{e} \sim e_{mp}(r)\exp(im\theta)$ $(m, p = 0, \pm 1, \pm 2, \ldots)$ where $r = |\vec{r}_\perp|$ and θ are cylindrical coordinates. Such beams are characterized by two indexes: the radial index p which shows the number of field variations along the r-coordinate, and the azimuth index m which describes field variations during the passage around the beam axis [20, 21].

The power of a stationary beam with the indexes m, p

$$P_{\text{hom},m,p} = \frac{\tilde{n}n_0}{8\pi\varepsilon_2'}\int |\bar{\Psi}_{mp}|^2 dxdy$$

is independent of its width. A homogenous (Townes) beam [5] with the lowest indexes $p = m = 0$ has the power which is called the critical self-focusing power:

$$P_{cr} = 5.850\frac{c\varepsilon_0^{1/2}}{4\pi k_0^2\varepsilon'} = 0.931\frac{\lambda^2}{2\pi n_0 n_2}, \tag{3.11}$$

where λ is the free-space wavelength. The value of the critical power is determined by the nonlinearity mechanism. For the electron nonlinearity, it ranges from several megawatts in glass to several gigawatts in gases under the normal pressure; see, e.g., [22, 23].

In the case of elliptical polarization, the number of plane beams increases, but remains finite. Cylindrical beams with the elliptical polarization [24] are characterized by the matrix of indices $\begin{pmatrix} m & p \\ n & q \end{pmatrix}$, with the columns of the azimuth $\binom{m}{n}$ and radial $\binom{p}{q}$ indices corresponding to the number of oscillations along the azimuth and radial coordinates of the function $\Psi_\pm$ and satisfying Eq. (3.6). The power of such homogenous beams can vary in a certain range of values that depend on the beam type and the anisotropy parameter β. The equations for the circular and linear polarizations can be reduced to the same form. The power of the Townes beam with the linear polarization is lower than the power of the same beam with the circular polarization by $\left(1+\frac{\beta}{2}\right)$ times.

Independence of the power of the homogenous beam on its width is a consequence of the parabolic equation approximation. In the case of a more accurate description taking into account the nonparaxiality and the longitudinal components of the field, the power of the homogenous beam increases with decreasing width [15]. In the framework of the parabolic equation, the power of homogenous channels depend on their width in the media with nonlocal (e.g., thermal [25]) nonlinearities. If the power in the channel grows as its width becomes narrower, formation of stable multisoliton structures is possible [26].

3.4 Filamentation and Modulation Instabilities of Supercritical Beams

A fundamental result in the self-focusing theory is the discovery of the filamentation instability of a plane wave in a nonlinear medium and the existence of an optimal scale for such an instability [27]. The wave of perturbation $\exp(-i\vec{k}_\perp\vec{r}_\perp - ihz)$ in the region of spatial frequencies $0 < k_\perp < 2\varepsilon_2' k_0 E_0^2$ grows with the increment

$$G = ih = \frac{k_\perp}{2k_0}\sqrt{2\varepsilon_2' k_0^2 E_0^2 - k_\perp^2}, \tag{3.12}$$

where E_0 is the amplitude of the initial plane wave. The maximum growth rate

$$G_{\max} = \frac{1}{2}\varepsilon_2' k_0 E_0^2 = \frac{\bar{k}^2}{2k_0} = \frac{2\pi n_2}{\lambda} I, \tag{3.13}$$

is achieved when the spatial perturbation frequency is $k_\perp = \bar{k} = k_0\sqrt{\varepsilon_2'}E_0 = k_0\sqrt{\frac{2n_2 I}{n_0}}$. In this case, with the power of the plane wave being of the order of the critical value, P_{cr} goes into the perturbations with characteristic scale. In the case of an inhomogeneous structure of the initial beam, the instability effect leads, as the beam power grows, to a breakdown of the beam into individual self-focusing filaments. The number of filaments grows in proportion to the ratio of the total beam power P and the critical power, $\frac{P}{P_{cr}}$.

The effect of the transverse self-focusing instability plays a major part in high-power multistage laser systems, because the formation of self-focusing filaments can cause multiple destructions of the medium. The value of the total maximum instability increment growth $B = \int \varepsilon_2' k_0 E_0^2 dz$ is called the break-up integral. Satisfactory operation of a laser system can be ensured if the value of this parameter does not exceed 4–5 units in one nonlinear element.

The solution of the transverse plane-wave nonlinearity problem is easily generalized to the time-modulated waves [10] by using Eq. (3.3). In this case, it is necessary to replace $k_\perp^2$ with

$$k_s^2 = k_0 \frac{dv}{d\omega} \frac{\Omega^2}{v^2}$$

where Ω is the modulation frequency. The necessary condition of the modulation instability is the following inequality:

$$\frac{dv}{d\omega} \varepsilon_2' > 0. \tag{3.14}$$

The modulation instability takes place in self-focusing substances in the regions with abnormal dispersion of the group velocity. The growth rate of the modulation instability is the maximum at the modulation frequency,

$$\Omega = k_0 v \sqrt{\frac{\varepsilon' |E_0|^2}{k_0 \frac{dv}{d\omega}}}. \tag{3.15}$$

The notion of the instability of a linearly polarized plane wave is generalized to the waves with an arbitrary polarization [17]. In the case of an elliptically polarized field, the maximum growth rate is $G = \frac{\bar{k}^2}{2k_0} F(R_0^2, \beta)$, where $\bar{k}^2 = \frac{2k_0^2 n_2}{n_0} I$ is the corresponding value of the transverse wave number of the perturbations for the linearly polarized wave with the same power density, and

$$F(R_0^2, \beta) = \frac{1}{2+\beta} \left\{ 1 + \left[1 + 4\beta(\beta+2) \frac{R_0^2}{(1+R_0^2)^2} \right]^{1/2} \right\}$$

is the correction coefficient depending on the ratio $R_0 = |\frac{\Psi_-}{\Psi_+}| < 1$ of the amplitudes of high-power rotating waves and the induced-anisotropy coefficient β.

For the electron nonlinearity ($\beta = 1$), as the coefficient R_0 decreases from unity (linear polarization) to zero (circular polarization), the value of $G_{\max}$ changes from $\frac{\bar{k}^2}{2k_0}$ to $\frac{1}{3}\frac{\bar{k}^2}{k_0}$. Hence, the waves with the circular polarization are more stable spatially. This is connected to the lower effective nonlinearity parameter. The polarization of maximally growing perturbations is different from that of a high-power wave and closer to the linear polarization as compared with the polarization of the pumping. Thus, an elliptically polarized wave is unstable in terms of its polarization: the wave, which is initially homogenously polarized, breaks down into regions with different polarizations.

The instability of the plane wave can be interpreted as a parametric instability at two-photon pumping The condition of the growth rate maximum coincides with the condition of the pumping wave's being synchronous with the waves amplified in the four-photon parametric amplifier:

$$k_z(\omega + \Omega, \vec{k}_\perp) + \vec{k}_z(\omega - \Omega, k_\perp) = 2k_p(\omega).$$

Taking into account that at small $\frac{\Omega}{\omega}$, $\frac{|\vec{k}_\perp|}{k}$ and $\varepsilon_2'|E|^2$,

$$k_z(\omega \pm \Omega, \vec{k}_\perp) = k(\omega) \pm \frac{\Omega}{v} - \frac{dv}{d\omega}\frac{\Omega^2}{2v^2} - \frac{k_\perp^2}{2k} + k(\omega)\varepsilon_2'|E_0|^2,$$

$$k_p(\omega) = k(\omega) + \frac{k(\omega)\varepsilon_2'|E_0|^2}{2},$$

we obtain the formula for the maximum growth rate, which coincides with Eq. (3.13). From Eq. (3.2) one can determine that the increasing and decreasing perturbation modes have a constant phase shift $\Delta\varphi_\pm$ relative to the high-power wave, which is independent of the longitudinal coordinate and, respectively, equal to

$$\Delta\varphi_\pm = \mp \arctan \frac{\sqrt{\bar{k}^2 - k_\perp^2}}{|k_\perp|}. \tag{3.16}$$

In the presence of the counter-propagating beam the parametric amplification may pass into parametric generation [19, 28]. In a layer with the finite thickness l, this generation has a threshold determined by the value of the integral $B = \varepsilon_2'|E_+|^2 k_0 l$ in the layer. The threshold value $B_{th} = \varepsilon_2'|E_+|^2 k_0 l$ depends on the relative amplitude $|\frac{E_b}{E_f}|$ of the counter-propagating beam. The generation occurs at the threshold value of B_{th} and is manifested by radiation of the waves propagating at a certain angle $\pm\alpha$ to the propagation direction of the pumping waves (E_f and E_b). In the subthreshold regime, when a weak wave is incident on a nonlinear layer with high-power counter-propagating waves at a certain angle $\bar{\alpha}$, four scattered waves are produced, one of which propagates in the direction counter to the incident wave. Such a layer can be used in wave front conjugation systems. The amplitude of the reflected wave can be greater than the amplitude of the incident one.

At small values of $B \ll 1$, the reflection index depends weakly on the angle that corresponds to the ideal conjugation of the wave front. The parametric generation limits those values of the reflection factor, which can be obtained in this case. In the presence of the counter-propagating wave, the transverse instability of a plane wave is possible at any sign of the nonlinearity [19, 29], i.e., also in media with the defocusing nonlinearity.

3.5 Self-Similar Solutions of the Self-Focusing Equation: Lens Transform

Equation (3.2) for the cubic medium permit the existence of self-similar solutions [30, 31] that differ in scale multipliers at different cross-sections. The structures of homogenous beams also belong here. The self-similar solutions of Eq. (3.2) are represented in the form of $\Psi = \Psi_0 \exp(-i\varphi)$ (amplitude Ψ_0 and the phase φ are real-valued functions),

$$\Psi_0 = \frac{1}{a(z)} e(\vec{s}); \quad \varphi = \frac{u(z)s^2}{2} + \varphi_0(z), \tag{3.17}$$

where

$$\begin{aligned} &\vec{s} = \frac{\vec{r}_\perp}{a}, \quad u(z) = C_3 z + C_4, \quad [a(z)]^2 = \frac{(C_3 z + C_4)^2 + C_1}{C_3}, \\ &2\varphi_0(z) = -\frac{C_2}{C_1^{1/2}} \arctan\left\{\frac{C_3 z + C_4}{C_1^{1/2}}\right\} + C_5; \quad C_i(i = 1, \ldots, 5) = \text{const}, \end{aligned} \tag{3.18}$$

and the amplitude structure of the beams is determined by the equation

$$\Delta_{u,v} e + [C_2 - C_1 s^2 + e^2] e = 0, \tag{3.19}$$

which permits localized solutions at $C_1 \geq 0$ and non localized ones at $C_1 < 0$.

Transform (3.17) can be generalized to the case of the arbitrary dependence $a(z)$ [31, 32]. In this case, it appears more convenient to us the variable $\sigma = \frac{1}{a(z)}$, when Eq. (3.2) permits a group of transforms of the coordinates and the field

$$\vec{r}_\perp = \frac{s}{\sigma(\eta)}, \frac{dz}{d\eta} = \frac{1}{\sigma^2(\eta)},$$
$$\psi(\eta, \vec{s}) = \frac{1}{\sigma}\Psi[z(\eta), \vec{r}_\perp(\vec{s}, \eta)] \exp\left[-i\frac{d\ln(\sigma)}{d\eta}\frac{\vec{s}^2}{2}\right], \tag{3.20}$$

under which it is transformed into the following parabolic equation:

$$\Delta_\perp \psi - 2i\frac{\partial\psi}{\partial\eta} + \frac{\vec{s}^2}{\Phi(\eta)}\psi + |\psi|^2\psi = 0, \tag{3.21}$$

which describes the self-effect in a lens-like media with the permittivity $\varepsilon = \varepsilon_0\left(1 + \frac{s^2}{\Phi(\eta)} + \varepsilon_2'|E|^2\right)$. The functions Φ and σ of the variable η are interconnected by the following relationship:

$$\frac{d^2\sigma}{d\eta^2} - \frac{\sigma}{\Phi} = 0. \tag{3.22}$$

These transforms are called the focusing, or lens, transforms. A special case of this transform is the transform for $\sigma = \frac{F-\eta}{F}$ and $z = \frac{F\eta}{F-\eta}$, which describes the focusing of the initial beam by a thin lens with the focal length F at the cross-section $z = 0$ [31]. In the latter case, the transform formulas may be rewritten as follows:

$$\frac{1}{\eta} = \frac{1}{z} + \frac{1}{F}, \; \vec{s} = \frac{\eta}{z}\vec{r}_\perp, \quad \psi(\eta, \vec{s}) = \frac{z}{\eta}\Psi\left[z(\eta), \frac{z}{\eta}\vec{s}\right]\exp\left[-i\left(1 - \frac{z}{\eta}\right)\frac{s^2}{2\eta}\right]. \tag{3.23}$$

In accordance with Eq. (3.23), the focusing of the beam does not change its topology in a nonlinear medium. The existence of the transform group, Eq. (3.19), by virtue of the Noether theorem, demonstrates that Eq. (3.2) has a new integral of motion, which is different from the traditional integrals of the field energy and momentum and is used efficiently in the averaged description of self-focusing beams.

In the simplest case of a homogenous lens-shaped medium $\left(\frac{1}{\Phi(\eta)} = \text{const}\right)$, at $\frac{1}{\Phi(\eta)} < 0$, homogenous beams exist at any power below the critical one. At $\frac{1}{\Phi(\eta)} = 0$, the power of the homogenous beam is equal to the critical value, and at $\frac{1}{\Phi(\eta)} > 0$ there are no homogenous beams. Let $\frac{1}{\Phi(\eta)} = -\mu^2 < 0$, then $z = tg(\mu\eta)/\mu$. Hence, it is seen that at the beam power below the critical value, when the field in the homogenous linear medium is regular, in the lens-shaped medium it is a periodical function with the period $L_p = \frac{\pi}{\mu}$. Specifically, at the distance $L_q = \frac{\pi q}{\mu}$ (where $q = 1, 2, 3, \ldots$), the field has the same structure as the input to the nonlinear medium. At the beam power exceeding the critical value the beam "collapses" at the point

$$\eta_{sf} = \frac{1}{\mu} \arctan \mu z_{sf},$$

where z_{sf} is the point of the nonlinear focus in a homogenous nonlinear medium. In a defocusing medium $\left(\frac{1}{\Phi(\eta)} = \tilde{\mu}^2 > 0\right)$, the self-focusing length increases:

$$\eta_{sf} = \frac{1}{\tilde{\mu}} \arctan h\tilde{\mu} z_{sf}. \tag{3.24}$$

Assuming that $\eta_{sf} = \infty$, we obtain the condition $z_{sf}\left(\frac{P}{P_{cr}}\right) = \frac{1}{\tilde{\mu}}$ which determines the threshold self-focusing power in the defocusing lens-shaped medium.

3.6 Averaged Description of Self-Focusing Beams (Method of Moments). Sufficient Condition of Self-Focusing. Self-Focusing of Supercritical Beams

The moments of intensity [33]

$$r_{mn}(z) = \iint x^m y^n |\Psi|^2 dxdy$$

of the field Ψ which satisfies the linear parabolic equation

$$\frac{\partial \Psi}{\partial z} = \frac{1}{2i} \Delta_\perp \Psi,$$

are the polynomials $P_{m+n}(z)$ of the finite degree $n + m$:

$$r_{mn}(z) = P_{m+n}(z).$$

However, only some of the intensity moments for the solutions of the nonlinear Schrödinger equation, Eq. (3.2), have the polynomial representation of a finite degree [33]. For cylindrical (three-dimensional) beams they are

$$N = \int_{s_\perp} |\Psi|^2 ds_\perp, \tag{3.25}$$

$$\vec{r}_c = \frac{1}{N} \int_{s_\perp} \vec{r}_\perp |\Psi|^2 ds_\perp, \tag{3.26}$$

$$a_{ef}^2 = \frac{1}{N} \int\limits_{s_\perp} (\vec{r}_\perp - \vec{r}_c)^2 |\Psi|^2 ds_\perp. \quad (3.27)$$

They have the sense of the total dimensionless power ($N = \frac{8\pi\varepsilon_2'}{cn_0} P$, where P is power), position of the intensity centre, and the squared effective width of the beam. Using the direct differentiation, one can make certain that the parameters N, $\frac{d\vec{r}_c}{dz}$, and $\frac{d^2 a_{ef}^2}{dz^2}$ are beam invariants, i.e., they do not depend on the coordinate z. This means that the power of the wave beam, Eq. (3.25), is conserved, and the center of the beam intensity in a nonlinear medium always moves along the straight line

$$\vec{r}_c = \vec{r}_c(0) + \vec{b}z$$

determined by the initial profile of the complex beam amplitude $\Psi = \Psi_0 \exp(-i\varphi)$ (here, the amplitude Ψ_0 and the phase φ are real-valued functions)

$$\vec{b} = \frac{1}{N} \iint \Psi_0^2 \nabla_\perp \varphi dxdy|_{z=0};$$

and its effective cross section changes obeying the parabolic law

$$a_{ef}^2(z) = a_{ef}^2(0) + \tilde{B}z + \tilde{A}z^2 \quad (3.28)$$

determined by the same profile

$$\tilde{B} = \frac{2}{N} \iint \Psi_0^2 (\vec{r}_\perp \nabla_\perp \varphi) dxdy|_{z=0},$$

and

$$\tilde{A} = \frac{1}{N} [|\nabla_\perp \Psi_0|^2 + \Psi_0^2 |\nabla_\perp \varphi|^2 - \frac{\Psi_{04}}{2}] dxdy =$$

$$\tilde{A}^L - \frac{1}{N} \iint \left[\frac{\Psi_0^4}{2}\right] dxdy|z = 0. \quad (3.29)$$

Equation (3.29) introduces the parameter

$$\tilde{A} = \frac{1}{N} \iint [|\nabla_\perp \Psi_0|^2 + \Psi_0^2 |\nabla_\perp \varphi|^2] dxdy$$

which characterizes the beam divergence in a linear medium. Because $a_{ef}^2 \geq 0$ by definition, the sufficient self-focusing condition is the negativity of $\tilde{A} < 0$. In this case, a_{ef}^2 goes to zero at a certain point of $z = z^*$ and further becomes negative, which indicates that Eq. (3.2) has no nonsingular solutions then. The singularity occurs starting from some power P^* which can be called "sufficiently critical." When the beam power exceeds P^*, the beam, upon the average, will collapse: $\frac{d^2 a_{ef}^2}{dz^2} < 0$. The sufficient critical power is found from the condition of $\tilde{A} = 0$:

$$\int_{s_\perp} \left(|\nabla\Psi|^2 - \frac{1}{2}|\Psi|^4\right) ds_\perp = 0. \qquad (3.30)$$

Let us determine the sufficient critical power P^* for the collimated ($\varphi \equiv 0$) beam:

$$P^* = \frac{cn}{4\pi k_0^2 \varepsilon'} \frac{\int_{s_\perp} (\nabla_\perp f^2) ds \cdot \int_{s_\perp} f^2 ds}{\int_{s_\perp} f^4 ds}. \qquad (3.31)$$

The sufficient critical power is determined by the medium parameters and the profile $f(r_\perp)$. For the Gaussian beam $f(r_\perp) = \bar{\Psi}_0 \exp\left[-\frac{r_\perp^2}{2a^2}\right]$,

$$P_G^* = \frac{cn}{2k_0^2 \varepsilon'}. \qquad (3.32)$$

The profile of the beam with the minimal critical power $P_{\min}^*$ can be found from Eq. (3.31) by varying it in terms of $f(r_\perp)$. The equation obtained for determination of the profile coincides with Eq. (3.8) for the structure of homogenous beams. The simplest solution of this equation corresponds to the axi-symmetric homogenous beam (Townes beam). Thus, the beam with the lowest sufficient critical power $P_{\min}^* = P_{cr}$ is the beam whose profile coincides with the homogenous-beam profile.

Stationary properties of the functional $P^*[f]$ permit calculating P_{hom} with comparatively high accuracy using the standard variation methods. Even the Gaussian beam yields an almost good approximation for $P_{hom,0,0}$: $P_G^* = 1.07 P_{hom,0,0}$. The obtained value of the sufficient critical power of the Gaussian beam is somewhat higher than the critical power found by numerical calculations. The difference is caused by the fact that the sufficient critical power determines the collapse threshold of the beam as a whole $\left(\frac{d^2 a_{ef}^2}{dz^2} < 0\right)$, while the critical power determines the threshold of the local collapse when the focal point z_{sf} with infinite intensity is formed on the beam axis. In a narrow power interval, $P_{hom,0,0} < P < 1.07 P_{hom,0,0}$, the formation of the focal point is accompanied with an increase in the efficient beam width.

From Eq. (3.28), one can obtain the differential equation for the dependence of the effective beam width a_{ef} on the longitudinal coordinate:

$$\frac{d^2 a_{ef}}{dz^2} = \frac{\tilde{A} a_{ef}^2(0) - \frac{\tilde{B}^2}{4}}{a_{ef}^3}. \tag{3.33}$$

For the Gaussian beam $\Psi = \bar{\Psi}_0 \exp\left[-\frac{r_\perp^2}{2a^2}\right]$, Eq. (3.33) is simpler:

$$\frac{d^2 a_{ef}}{dz^2} = \frac{1 - \frac{P}{P_G^*}}{a_{ef}^3}.$$

The same equation is obtained also in the aberration-free self-focusing theory [6] assuming that a_{ef} is the beam width.

From Eq. (3.28), for $P > P^*$, it is possible to determine the self-focusing length z^* as the distance to the point where $a_{ef} = 0$. For the collimated Gaussian beam,

$$z^* = \frac{a^2}{\sqrt{\frac{P}{P_G^*} - 1}}.$$

When we correlate this expression with the value of the self-focusing length determined from numerical calculations as the distance to the focal point with infinite intensity on the axis [34]

$$z_{sf} = \frac{0.366 a^2}{\{[(P/P_{cr})^{1/2} - 0.825]^2 - 0.03\}^{1/2}}, \tag{3.34}$$

we see that $z^* > z_{sf}$ at any value of $\frac{P}{P_{cr}}$. Correspondingly, the points z_{sf} and $z*$ can be interpreted as the points of the local and absolute collapses, respectively.

Note that for the value $I(z) = P a_{ef}^2$ which is analogous to the moment of inertia for the system of bodies with the space-distributed density $\rho = |\Psi|^2$, an analogy of the virial theorem takes place:

$$\frac{d^2(a_{ef}^2)}{dz^2} = 2H,$$

where $H = \int\limits_{s_\perp} (|\nabla\Psi|^2 - \frac{1}{2}|\Psi|^4) ds_\perp$ is the Hamiltonian of Eq. (3.2), being the motion integral $H = \text{const}$. Specifically, for the wave beam with the structure of the homogenous (Townes) beam, the condition $H < 0$ is fulfilled for $P > P_{cr}$. The condition $H < 0$ is only a sufficient criterion for self-focusing: in the beams with arbitrary profiles, the singularity can also appear at $H > 0$.

3.7 Self-Focusing of Wave Beams in Periodic Systems

Self-focusing in periodic systems, which in the simplest case consist of layers of linear and nonlinear media with thicknesses L and l, respectively, has a number of distinctive features. In such media, development of a self-focusing instability can be decreased. According to Eq. (3.16), the increasing and decreasing perturbation modes in a nonlinear medium have a constant phase shift $\Delta\varphi_{\pm}$ relative to the high-power wave, which does not depend on the longitudinal coordinate. In a linear medium, when the wave propagates to the distance z, the phase between the perturbation waves and the high-power wave changes by

$$\delta\varphi = \sqrt{k^2 - k_{\perp}^2}\,z.$$

Because of this, the wave having a favorable phase shift that is amplified in one of the nonlinear-medium layers can dephase away from the pumping, having passed through the linear-medium section, and can start decreasing in the next layer of the nonlinear medium. At certain ratios of the thicknesses l and L, the self-focusing instability is suppressed for some spatial scales [35]. For small values of $G_{\max}l$ on one nonlinear element, $G_{\max}l \ll 1$, the growth rate maximum per unit of nonlinear-medium length in the simplest periodical system cannot be decreased, but the amplification band for the maximally growing perturbations can be changed: it can be made $\sim \frac{L}{l}$ times narrower, and the scale of the perturbations increasing at the maximum rate increases by the same degree.

The self-focusing instability can be suppressed completely in periodical systems with relays transferring images. The relays are formed by a pair of lenses with the focal distances $F = \frac{l}{2}$, which are placed between nonlinear layers at the distance l from each other. The conditions for instability suppression are the equality of the lengths of the relay and the nonlinear medium, $l = L$, and the restriction of the maximum increment $G_{\max}l < \frac{\pi}{2}$ at one nonlinear layer [36, 37].

In accordance with the spatio-temporal properties of Eq. (3.3), the systems with relays are analogous to fiber optic waveguides in which the dispersion on their linear parts is compensated completely, within a certain frequency range, by the dispersion on the nonlinear parts (systems with variable dispersion).

Periodic systems formed by layers of linear and nonlinear dielectrics can be used for channeling of high-power light beams. The power of quasi-homogenous wave beams propagating in the simplest of such systems exceeds the critical value by $\sim \frac{L}{l}$ times [38].

3.8 Field Structure in the Vicinity of the Nonlinear Focus; Self-Similar Collapse

Strictly speaking, it is inappropriate to describe the field in the vicinity of the nonlinear focus within the NLSE framework: this equation is inapplicable in this region. However, specific estimates show that this extrapolation is quite acceptable, because the transform to the asymptotics can occur while NLSE is still applicable. Besides, studying special NLSE solutions is of fundamental interest for understanding the whole solution structure: singular fields play here approximately the same role as solitons in plane problems: in the long run, the total initial wave field breaks into them.

Precise self-similar solutions of Eq. (3.2) are known for the cylindrical beams having a singularity: a nonlinear focus at $z = z_{sf}$ [30] . They are described by the equalities in Eqs. (3.17)–(3.19) for a certain choice of constant C_i. There is a solution for $\tilde{N}_1 = 0, \tilde{N}_2 < 0$. It is localized, has the structure of a compressing Townes beam, and decreases exponentially, $e \sim \exp[-\sqrt{-C_2}s]$, at $s \to \infty$. Its power is equal to the critical value, $P = P_{cr}$. As it approaches the nonlinear focus, the field grows: $|\Psi(0,z)| = \frac{d_1}{z_{sf}-z}$, where d_1 is a constant independent of the z coordinate.

There is a self-similar solution of Eq. (3.19) for $\tilde{N}_1 < 0$. It oscillates along the s coordinate and decreases, $e \sim \frac{\sin(\int \sqrt{-C_1}sds)}{s}$, at $s \to \infty$. The power of the beam described by this solution is infinite, $P \sim \ln s$. All of it is concentrated in the nonlinear focus. The field on the axis $r_\perp = 0$ grows:

$$|\Psi(0,z)| \approx \frac{1}{\sqrt{(z_{sf} - z)}}. \tag{3.35}$$

Even in early numerical experiments [34] it was noted that in the beams with limited power, the power close to the critical one is localized in the vicinity of nonlinear foci. Later it was shown that the modulus of the field near the nonlinear focus grows obeying the law of Eq. (3.35), approximately [39]. The difference is caused by the difference in the beam structure beyond the external boundary of the light pipe with its power equal to the critical one.

Whereas at the boundary of this light pipe the field $\Psi(r_\perp, z)$ is finite and differs from zero, $|\Psi(0,z)| \sim \sqrt{-\frac{\ln(z_{sf}-z)}{z_{sf}-z}}$ when the focal point is approached [39]. In this case, the compressing field within the light pipe can be "matched" with the field beyond this area [40]. Whereas the field $|\Psi(r_\perp, z)|$ grows infinitely at the boundary of this light pipe for $z \to z_{sf}$, in the vicinity of the nonlinear focus [41–43]:

$$|\Psi(0,z)| \sim \sqrt{\frac{\ln[-\ln(z_{sf}-z)]}{z_{sf}-z}}$$

The structure of the field in the nonlinear focus can be determined by describing the field more precisely as compared with the description yielded by the parabolic equation. The correction is connected with both the account of the changes in the electric-field polarization, and the more accurate account of the diffraction. Taking these corrections into account leads to disappearance of the singularity in the solution: the field at the nonlinear focus compresses to a wavelength, and in this case appearance of the longitudinal component of the electric field plays a more significant role [15, 44–45].

3.9 Nonstationary Self-Focusing; Distributed Collapse

Formation of nonlinear focal points with the critical power coming to each focus is a specific feature of the cubic nonlinearity and time-stationary beams [46–48]. Figure 3.1 shows the pattern of the beams for the case of two consecutive focal points, $z_{sf}^{(1)}$ and $z_{sf}^{(2)}$.

In principle, the scenario of nonstationary self-focusing of a spatially bounded pulse in a nondispersive medium with inertia-free nonlinearity $\Delta\varepsilon \approx \varepsilon_2'|E|^2$ can be obtained on the basis of solving the stationary problem, because in this case time is only a parameter in the relationship P/P_{cr}. Here, the focus positions change depending on time, and they can draw high-intensity tracks in space [49]. The motion of the foci during nonstationary self-focusing can be both direct and reversed and happen with different velocities, including super-light ones. In dispersive media, nonstationary self-focusing should be studied on the basis of the three-dimensional self-focusing equations, Eq. (3.3). In the regions with anomalous dispersion of the group velocity $v_\omega'>0$ in self-focusing media, the phenomenon of the distributed three-dimensional collapse is possible. Unlike the self-focusing of axisymmetric beams, in the three-dimensional case ($D = 3$) the collapse passes two stages: pre-focal and focal. At the first stage, a field structure is formed, whose energy is absorbed at the second stage [50].

To describe this process, we use Eq. (3.3), in which we make the replacements

$$z_d = k_0 t, \quad \xi = \sqrt{\frac{1}{k_0}\frac{dv}{d\omega}}z, \quad \vec{r}_{\perp d} = k_0\vec{r}_\perp, \quad \Psi = \sqrt{\varepsilon_2'}E,$$

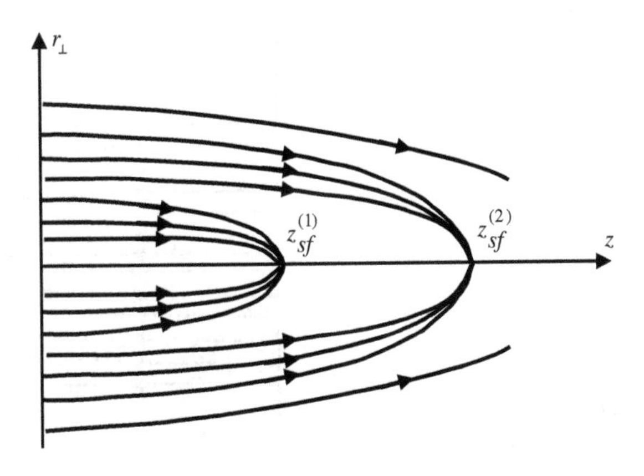

Figure 3.1 Beam structure of self-focusing for axiallv symmetric beams

and reduce it to the form

$$\Delta\Psi - 2i\frac{\partial\Psi}{\partial t} + |\Psi|^2\Psi = 0, \tag{3.36}$$

Here Δ is the Laplacian in the space (x,y,z). The field structure at the prefocal stage is described by solving Eq. (3.36) with the field singularity at the point t_{sf} [51]

$$\Psi \approx \frac{1}{\sqrt{t_{sf} - t}}.$$

In this case, the energy near the point t_{sf}, which is absorbed or scattered there, tends to zero. The pattern of spatio-temporal beams $t$in the case of the three-dimensional collapse is shown in Fig. 3.2. At $r \neq 0$, the beams stay aligned toward the axis even after the singularity has appeared. This means that the energy flow stays directed towards the axis. This field structure with the energy flowing towards the axis, in a spherically symmetric case, is described by the equation

$$-2i\frac{\partial\Psi}{\partial t} + \frac{2}{r}\frac{\partial\Psi}{\partial r} + \frac{\partial^2\Psi}{\partial r^2} + |\Psi|^2\Psi = 0. \tag{3.37}$$

For the time-stationary processes $\Psi = \Psi_0(r)\exp(i\Omega t)$, this equation is reduced to the form

$$2\Omega\Psi_0 + \frac{2}{r}\frac{\partial\Psi_0}{\partial r} + \frac{\partial^2\Psi_0}{\partial r^2} + |\Psi_0|^2\Psi_0 = 0.$$

The latter equation has solutions with the singularity $|\Psi_0| = \frac{1}{r\sqrt{-2\ln|\tilde{N}|r}}$ for $r \to 0$ and with the constant radial energy flow $S = 4\pi|C|$ (where C is the real

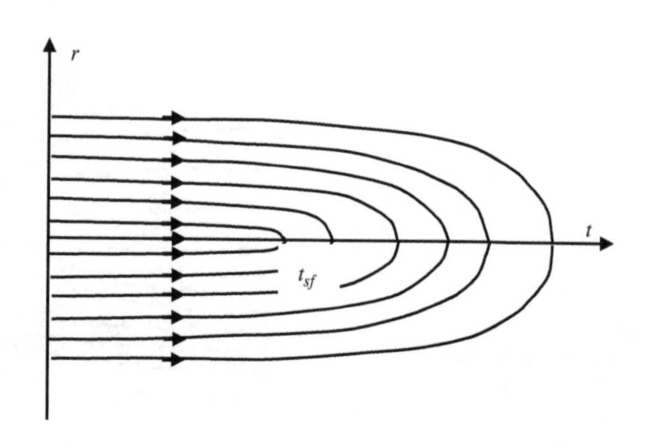

Fig. 3.2 Beam structure of distributed collapse

constant) [40] towards the center. In the case of the three-dimensional self-focusing in the cubic medium described by Eq. (3.37), the focal points turn into finite-extent focal filaments, which can be called a distributed collapse.

In the media with normal dispersion of the group velocity, the diffraction operator becomes hyperbolic:

$$\left(\Delta_\perp - \frac{\partial^2}{\partial z^2}\right)\Psi - 2i\frac{\partial \Psi}{\partial t} + |\Psi|^2\Psi = 0, \tag{3.38}$$

and the self-action dynamics change significantly [52]. Depending on the type of initial conditions, either the longitudinal-splitting regime or the distributed collapse of tubular beams is realized [53]. However, currently there is no analytical proof of the collapse regime.

3.10 Spectral Broadening

One of the most important features of nonstationary self-focusing is the increase in the phase self-modulation of the pulse and, correspondingly, the broadening of its spectrum due to the transverse compression of the beam. The expression for the frequency deviation of the pulse with the time dependence of the power $P(t)$, the initial width a_0, and the finite width a in the vicinity of the nonlinear focus [54],

$$\Omega = -\frac{dP}{dt}\frac{1}{(PP_{cr})^{1/2}}\left(\frac{a_0}{a}\right)^2, \tag{3.39}$$

shows that in a self-focusing beam, a faster variation of the dielectric permittivity caused by the transverse channel compression increases the deviation by $\frac{1}{2}\left(\frac{a_0}{a}\right)^2$ times as compared with the deviation in the homogenous channel with the radius a_0. In the medium with the relaxing nonlinearity (time of relaxation is τ), the total frequency deviation can be up to the values of the order of $1/\tau$ [55]. Even the first experiments on self-focusing of laser pulses with the Q-switching in Kerr-type liquids revealed the spectral broadening of the order of hundreds of Ångstroms. This broadening is explained by the light phase self-modulation in thin self-focusing filaments due to the nonstationary character of the dielectric permittivity.

In 1970, a new type of broadening was discovered, which was called the abnormally large, super broadening, or continuum generation [56, 57]. In the experiments [57], the broadening of the spectrum of a partially synchronized pulse radiated by a Nd laser ($\lambda = 1.06\,\mu$m) was studied after self-focusing in glass rods and formation of the nonlinear focus in them. The spectrum width of the incident laser radiation was 5–10Å. In this case, no filament-type destructions were observed in the glass. Figures 3.3a_1, 3a_2, and 3.3b show patterns of the spatio-temporal spectral distributions of the field at the rod end recorded by means of various light-sensitive films (see [57]).

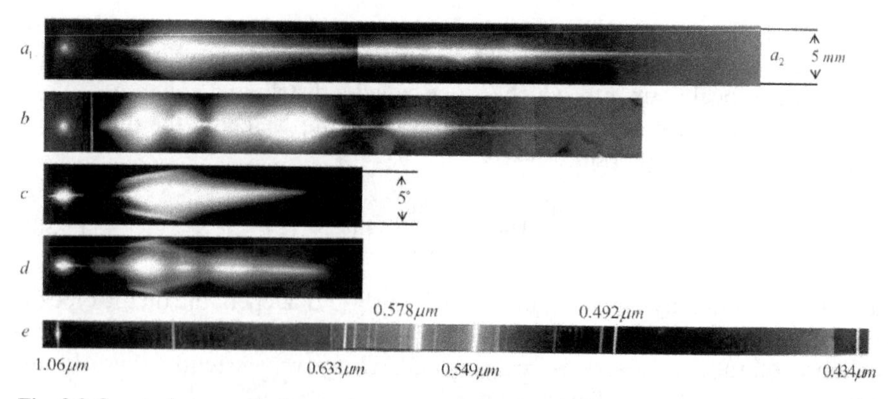

Fig. 3.3 Spectral scan radiation at the output of the glass rod
a_1, a_2 - spectrograms of the image of the end face of the rod LK;
a_1, a_2 - the different types of films;
b - spectrograms of the image of the end face of the rod LGS-228;
c - spectrograms of the far field after self-focusing, the rod LK;
d - spectrograms of the far field after self-focusing, the rod LGS-228;
e - spectrograms of mercury lamp with add lines $\lambda = 0.63$ μm and $\lambda = 1.06$ μm.

For the sake of comparison, Fig. 3.3*e* shows the spectrum of a hydrogen lamp. The bright axial line (here, a_1 and a_2 are LK glass, and *d* is LGS-228) is the spectralized image of the nonlinear-focus structure at the end of the rod. For the LK glass, two types of films, a_1 and a_2, were used for different ranges of scattered-wave lengths. The spectrum of the scattered radiation encompasses a wide range that covers the whole visible area and reaches the frequencies that significantly exceed the second harmonic of the laser radiation, unlike [56].

Figure 3.3*c* and *d* show also the spectral patterns of the far radiation field after self-focusing for the same glass samples. At the long-wave end of the spectrum, there are two intense beams radiated from the vicinity of the point $\lambda = 1.06$ μm. Both the self-focusing and the spectrum transformation in the glass is caused by the small-inertia mechanism of the electron nonlinearity with $\tau \approx 10^{-15} s$. The observed spectrum broadening can be achieved, in principle, by using the phase self-modulation and modulation instability on combination frequencies [57, 58] in a compressing channel. The anti-Stokes shift of the frequency of axial radiation can be explained by the change in the nonlinearity sign in super-strong fields arising in the focusing filaments. The off-axis anti-Stokes radiation is explained by the mechanism of the four-photon interaction of the pumping with the axial stokes radiation for the Cherenkov radiation type.

3.11 Self-Action of Femtosecond Pulses

The factor that acquires basic importance for the femtosecond pulses is accounting for the dependence of the wave-packet group velocity on its intensity [59–61]. Evidently, the simplest model for description of such effects looks as follows:

$$\frac{\partial}{\partial \bar{\tau}}\left(\frac{\partial \bar{E}}{\partial \bar{z}} + 3\bar{E}^2\frac{\partial \bar{E}}{\partial \bar{\tau}} - \bar{b}\frac{\partial^3 \bar{E}}{\partial \bar{\tau}^3} - \bar{c}\frac{\partial^2 \bar{E}}{\partial \bar{\tau}^2}\right) + \bar{a}\bar{E} = \Delta_\perp \bar{E}. \tag{3.40}$$

This equation describes the reflection-free propagation of wide-band radiation $\bar{E}(\bar{z}, \bar{\tau} = t - \bar{z}, \vec{r}_\perp)$ along the z-axis with the group velocity depending on the amplitude. The parameters $\bar{a}, \bar{b}$, and $\bar{c}$ determine the low- and high-frequency dispersions of the group velocity and damping, respectively. For the case of long quasi-monchromatic pulses, Eq. (3.40) after averaging over the carrier frequency is transformed to the traditional Schrödinger equation (3.3). However, with shortening of the pulse duration, the effect of the nonlinear dispersion of the group velocity becomes essential.

For the linear dispersion relation $k_0 = k_z$ parameter, one can find $\partial^2 k_0/\partial\omega^2 = 6\bar{b}\omega - 2\bar{a}/\omega^3$. Hence, it is seen that for $\bar{a} = 0$ and $\bar{b} > 0$, Eq. (3.40) describes the dynamic of ultra-short pulse self-action in a medium with the normal group velocity dispersion ($\partial^2 k_0/\partial\omega^2 > 0$). Another limiting case ($\bar{b} = 0$, $\bar{a} > 0$) corresponds to the medium with the abnormal dispersion ($\partial^2 k_0/\partial\omega^2 < 0$). In the case of $\bar{c} = 0$ (in the absence of damping), Eq. (3.40) can be reformulated in terms of the variation principle for the potential Φ of the field $\bar{E} = \Phi_\tau$ with the density of the Lagrange function,

$$L_l = \frac{1}{2}\Phi_{\bar{z}}\Phi_{\bar{\tau}} + \frac{1}{4}\Phi_{\bar{\tau}}^4 + \frac{\bar{b}}{2}\Phi_{\bar{\tau}\bar{\tau}}^2 - \frac{(\nabla_\perp\Phi)^2}{2} - \bar{a}\frac{\Phi^2}{2}, \tag{3.41}$$

which makes it possible to determine the main motion integrals of the problem:

$$N = \int \Phi_{\bar{\tau}}^2 d\bar{\tau} d\vec{r}_\perp, \tag{3.42}$$

and

$$H = \int [(\nabla_\perp\Phi)^2 - \bar{b}\Phi_{\bar{\tau}\bar{\tau}}^2 - \frac{1}{2}\Phi_{\bar{\tau}}^4 + \bar{a}\Phi^2] d\bar{\tau} d\vec{r}_\perp. \tag{3.43}$$

Using the method of moments, it can be shown that for three-dimensional wave packets there is a relationship analogous to Eq. (3.33):

$$\frac{d^2 a_{ef}^2}{d\bar{z}^2} = \frac{8H + 8\int [\bar{b}\Phi_{\bar{\tau}\bar{\tau}}^2 - \bar{a}\Phi^2] d\bar{\tau} d\vec{r}_\perp}{N}, \tag{3.44}$$

which describes the variation of the effective transverse size of the wave field $a_{ef}^2 = \frac{\int r_\perp^2 \Phi_{\bar{\tau}}^2 d\bar{\tau} d\vec{r}_\perp}{N}$ in the process of its evolution. It is seen that in the absence of the linear dispersion ($\bar{a} = \bar{b} = 0$) there is a collapse at $H < 0$.

In the one-dimensional variant, this problem is integrable and has been studied in detail (see, e.g. [60]). The numerical analysis [62] of the complete problem shows that in the case of ultra-short pulses (several half-waves long), the dynamics of self-focusing is modified significantly. Specifically, in a dispersion-free medium ($\bar{a} = \bar{b} = 0$), the collapse is caused primarily by breaking each of the half-waves.

Taking elastic friction ($\tilde{n} \neq 0$) into account yields that there is no field increase at the axis due to focusing correction, and the in-flowing energy dissipates at the shock fronts. Upon the whole, this dynamics correspond to a great pulse spectrum widening (pulse self-compression). In the case of circular polarization and in the media with dispersion, no breaking is observed in each of the half-waves, but a shock wave of the envelope is formed. Then, the field at the front of the shock wave increases infinitely, and the spectrum becomes significantly wider [59].

3.12 Conclusions

The above theory is closed within the quasi-optical approximation. The methods developed currently to study nonparaxial beams and wideband wave packets [11, 15, 45, 63–67] are based on the use of decomposition of wave equations and pseudo-differential operators. The resulting approximate equations have received different names: the generalized nonlinear Schrödinger equation, evolution equations for wave packets, equations for the pulses in the spectral region, or the one-directional wave-packet equation. The developed approaches permit describing short pulses with a small amount of periods, up to video light pulses, as well as the wide-angle diffraction.

References

1. G.A. Askarian: Effect of the gradient of a strong electromagnetic beam on electrons and atoms, *Sov. Phys. JETP* **15**, 1088–1090 (1962). See also current book.
2. A.V. Gaponov, M.A. Miller: Potential wells for charged particles in high-frequency electromagnetic field, *Sov. Phys. JETP* **7**, 168–169 (1958).
3. H.A.H. Boot, S.A. Self, R.B.R, Shersby-Harvie: Containment of a fully-ionized plasma by radio-frequiency fields, *J. Electron. Controls* **4**, 434–453 (1958).
4. V.I. Talanov: On self-focusing of electromagnetic waves in nonlinear media, *Izv. VUZ. Radiophysica* 7, 564–565 (1964). See also current book.
5. R.Y. Chiao, E. Garmire, C.H. Townes: Self-trapping of optical beams, *Phys. Rev. Lett.* **13**, 479–482 (1964).
6. V.I. Talanov: On self-focusing of the wave beams in nonlinear media, *JETP Lett.* **2**, 138–139 (1965).
7. P.L. Kelley: Self-focusing of optical beams, *Phys. Rev. Lett.* **15**, 1005–1008 (1965).
8. M. Hercher: Laser-induced damage in transparent media, *J. Opt. Soc. Am.* **54**, 563 (1964). See also current book.
9. N.F. Pilipetsky, A.R. Rustamov: Observation of self-focusing of light in liquids, *JETP Lett.* **2**, 55–56 (1965).

10. A.G. Litvak, V.I. Talanov: A parabolic equation for calculating the fields in dis-persive nonlinear media, *Radiophys. Quant. Electron.* **10**, 296–302 (1967).
11. V.I. Talanov, S.N. Vlasov: The parabolic equation in the theory of wave propa-gation (on the 50th anniversary of its publication), *Radiophys. Quant. Electron.* **38**, 1–12 (1995).
12. J.H. Marburger: Self-focusing theory, *Prog. Quant. Electr.* **4**, 35–110 (1975).
13. S.A. Akhmanov, A.P. Sukhorukov, R.V. Khokhlov: Self-focusing and diffraction of light in a nonlinear medium, *Sov. Phys. Uspekhi.* **10**, 609–636 (1968).
14. S.A. Akhmanov, R.V. Khokhlov, A.P. Sukhorukov: Self-focusing, self-defocusing, and self-modulation of laser beams. In: *Laser Handbook*, vol. 2, pp. 1151–1228, F.T. Arecchi, E.O. Schulz-Dubois (Eds.), North-Holland, Amsterdam (1972).
15. V.I. Talanov, S.N. Vlasov: *Wave Self-Focusing*. IAP RAS, N. Novgorod (1997).
16. R.W. Hellwarth: Third-order optical susceptibilities of liquids and solids, *Prog. Quant. Electr.* **5**, 1–68 (1978).
17. V.I. Bespalov, A.G. Litvak, V.I. Talanov: Self-action of electromagnetic waves in cubic isotropic media. In: *Nonlinear Optics*, R.V. Khokhlov, G.V. Krivoschekov (Eds.), Nauka, Novosidirsk, pp. 428–462 (1968).
18. A.G. Litvak: Finite-amplitude wave beams in magnetoactive plasma, *Sov. Phys. JETP*, **30,** 344 (1970).
19. A.G. Litvak, G.M. Fraiman: Interaction of opposite electromagnetic wave beams in a transparent nonlinear medium, *Radiophy. Quant. Electron.* **15**, 1024–1029 (1972).
20. Z.K. Yankauskas: Radial field distributions in a self-focusing light beam, *Radiophys. Quant. Electron.* **9**, 261–263 (1966).
21. H.A. Haus: High-order trapped light beam solutions, *Appl. Phys. Lett.* **8**, 128–129 (1966).
22. N.G. Bondarenko, I.V. Eremina, A.I. Makarov: Measurement of coefficient of nonlinearity of an optical and laser glass, *Sov. J. Quant. Electron* **8**, 482–484 (1978).
23. V.S. Averbach, A.A. Betin, V.A. Gaponov et al.: Effects of the stimulated scattering and self-action in gases and effect them on propagation of optical radiation, *Radiophys. Quant. Electron.* **21**, 755–775 (1978).
24. S.N. Vlasov, V.A. Gaponov, I.V. Eremina et al.: Self-focusing of beams with elliptic polarization, *Radiophys. Quant. Electron.* **21**, 358–362 (1978).
25. A.G. Litvak, G.M. Fraiman, A.D. Junakovskii: Thermal self-action of wave beams in plasma with not local nonlinearity, *Phys. Plasma*, **1**, 60–71 (1975).
26. G.A. Markov, V.A. Mironov, A.M. Sergeev et al.: Multi-beam self-channeling of plasma waves, *Sov. Phys. JETP* **53**, 1183 (1981).
27. V.I. Bespalov, V.I. Talanov: About filamentation of beams of light in a nonlin-ear liquid. *JETP Lett.* **3**, 307–310 (1966).
28. V.I. Talanov, S.N. Vlasov: About some features of scattering of signal wave on opposite pumping beams at degenerate four-photon interaction. In: *The Conjugate Wave Front Generation of Optical Radiation in Nonlinear Media.* V.I. Bespalov (Ed.), IAP RAS, Gorky, pp. 85–91 (1979).
29. H. Pu, N.P. Bigelow: Collective excitations, metastability, and nonlinear response of a trapped two-species Bose–Einstein condensate, *Phys. Rev. Lett.* **80**, 1134–1137 (1998).
30. V.I. Talanov: Self-similar wave beams in nonlinear dielectric, *Radiophys. Quant. Electron.* **9**, 260–261 (1966).
31. V.I. Talanov: About self-focusing of light in cubic media, *JETP Lett.* **11**, 199–201 (1970).
32. S.N. Vlasov, S.N. Gurbatov: The self-action theory of intense light beams in smoothly imhomogeneouse media, *Radiophys. Quant. Electron.* **19**, 811–816 (1976).
33. S.N. Vlasov, V.A. Petrischev, V.I. Talanov: Average description of wave beams in linear and nonlinear media, *Radiophys. Quant. Electron.* **14**, 1062–1070 (1971).
34. V.N. Goldberg, V.I. Talanov, R.E. Erm: Self-focusing of axiall-symmetrical wave beams, *Radiophys. Quant. Electron.* **10**, 368–375 (1967).
35. S.N. Vlasov: Instability of an intensive plane wave in periodically nonlinear media, *Sov. J. Quant. Electron.* **6**, 245–246 (1976).

36. S.N. Vlasov: Stabilization of instability of a plane wave in a periodic system, *Technical Physics Letters (Pis'ma v Zhurnal Tekhnicheskoi Fisiki)* **4**, 795–800 (1978).
37. S.N. Vlasov, V.E. Yashin: Self-focusing suppression in neodymium–glass laser systems by means of relay, *Sov. J. Quant. Electron.* **11**, 313–318 (1981).
38. S.N. Vlasov, V.A. Petrischev, V.I. Talanov: Nonlinear quasi-optical system, *Radiophys. Quant. Electron.* **15**, 886–894 (1972).
39. L.V. Piskunova, V.I. Talanov, S.N. Vlasov: Field structure near a singularity produced on self-focusing in a cubic medium, *Sov. Phys. JETP* **48**, 808–812 (1978).
40. D. Wood: The self-focusing singularity in nonlinear Schrodinger equation, *Stud-ies in Appl. Math.* **71**, 103–115 (1984).
41. G.M. Fraiman: Asymptotic stability of manifold of self-similar solutions on self-focusing, *Sov. Phys. JETP* **61**, 228 (1985).
42. M.J. Landman, G.C. Papanicolaou, C. Sulem et al.: Rate of blowup for solution of the nonlinear Schrodinger equation at critical dimension, *Phys. Rev. A* **38**, 3837–3843 (1988).
43. B.J. Mesurier, G.C. Papanicolaou, C. Sulem et al.: Local structure on the self-focusing singularity of the cubic Schrodinger equation, *Physica D Nonlinear Phenomena* **32**, 210–226 (1988).
44. S.N. Vlasov: The structure of the circularly polarized wave beam field in the vicinity of a nonlinear focus in a cubically nonlinear medium, *Sov. J. Quant. Electron.* **17**, 1191–1193 (1987).
45. M.D. Feit, J.A. Fleck, Jr.: Beam nonparaxiality, filament formation, and beam break-up in self-focusing of optical beams, *J. Opt. Soc. Am. B* **5**, 633–640 (1988)
46. V.N. Lugovoi, A.M. Prokhorov: Theory of the propagation of high-power laser radiation in a nonlinear medium, *Sov. Phys. Uspekhi.* **16**, 658–679 (1974)
47. A.L. Dyshko, V.N. Lugovoi, A.M. Prokhorov: Self-focusing of intense light beams, *JETP Lett.* **6**, 146–148 (1968)
48. A.L. Dyshko, V.N. Lugovoi, A.M. Prokhorov: Multifocus structure of a light beam in a nonlinear medium, *Sov. Phys. JETP* **34**, 1235–1241 (1972)
49. V.N. Lugovoi, A.M. Prokhorov: A possible explanation of the small-scale self-focusing filaments, *JETP Lett.* **7**, 117–119 (1968)
50. L.V. Piskunova, V.I. Talanov, S.N. Vlasov: Three-dimensional wave collapse in a nonlinear Schrodinger equation model, *Sov. Phys. JETP* **68**, 1125–1128 (1989).
51. E.A. Kuznetsov, V.E. Zakharov: Quasi-classical theory of three-dimensional wave collapse, *Sov. Phys. JETP* **64**, 773 (1986).
52. A.G. Litvak, T.A. Petrova, A.M. Sergeev et al.: About multiple splitting of wave structures in nonlinear media, *JETP Lett.* **44**, 12–15 (1986).
53. A.G. Litvak, V.A. Mironov, N.A. Zharova: The structural features of wave collapse in medium with normal group velocity dispersion, *Sov. Phys. JETP* **96**, 643–652 (2003).
54. V.A. Petrischev, V.I. Talanov: About nonstationary self-focusing of light, *Sov. J. Quant. Electron.* **1**, 587–592, (1972).
55. V.I. Talanov: Some questions of self-focusing theory, *Sov.t Phys. Uspekhi* **15**, 521–522 (1972).
56. R.R. Alphano, S.L. Shapiro: Observation of self-phase modulation and small-scale filaments in crystal and glasses, *Phys. Rev. Lett.* **24**, 592–594 (1970).
57. N.G. Bondarenko, I.V. Eremina, V.I. Talanov: Broadening of spectrum in self-focusing of light beams in crystals, *JETP Lett.* **12**, 85–87 (1970); Errata **12**, 267.
58. V.I. Talanov, S.N. Vlasov: Modulational instability at sum and difference frequencies in media with a high-order nonlinearity, *JETP Lett.* **60**, 639–642 (1994).
59. A.G. Litvak, V.A. Mironov, S. A. Skobelev: Dynamics of self-action of super-short electromagnetic pulses, *JETP Lett.* **82**, 119–123 (2005).
60. D.J. Kaup, A.C. Newell: An exact solution for a derivative nonlinear Schrödinger equation, *J. Math. Phys.* **19**, 798 (1978).
61. H. Stendel: The hierarchy of multi-solitons of the derivative nonlinear Schrödinger equation, *J. Phys. A Math. Gen.* **36**, 1931 (2003).

62. A.G. Litvak, V.A. Mironov, N.A. Zharova: Selfaction of shock wave envelopes in nonlinear dispersive medium, *Sov. Phys. JETP* **103**, 15–22 (2006).
63. S.A. Kozlov, Yu.A. Shpolyanskiy, A.O. Oukrainski et al: Spectral evolution of propagation extremely short pulses, *Phys. Vibration* **7**, 19–27 (1999).
64. Th. Brabec, F. Krausz: Nonlinear optical pulse propagation in the single-cycle regime. *Phys. Rev. Lett.* **78**, 3282–3285 (1997).
65. M. Kolesic, J.V. Moloney: Nonlinear optical pulse propagation simulation: from Maxwell's to unidirectional equations, *Phys. Rev.* **E-70**, 036604 (2004).
66. A. Ferrando, M. Zacares, P.F. Cordoba et al.: Forward–backward equations for nonlinear propagation in axially invariant optical systems, *Phys. Rev.* **E-71**, 016601 (2005).
67. S.N. Vlasov, E.V. Koposova, V.I. Talanov: Use of decomposition of the wave equations and pseudo-differential operators for the description of nonparaxial beams and broadband wave packets. *Radiophys. Quant. Electron.*, **49**, 258–301 (2006).

Chapter 4
Self-Focusing of Optical Beams

R.Y. Chiao, T.K. Gustafson, and P.L. Kelley

Abstract The nonlinear index of refraction is responsible for self-action effects in optical beam propagation, including self-focusing. We review self-focusing and related phenomena and discuss physical mechanisms that give rise to the refractive index nonlinearity.

4.1 Introduction

Optical self-action effects occur when an electromagnetic field induces a refractive index change in the medium through which the field propagates. The change in index then exhibits a back-action on the field so as to influence its propagation characteristics. The principal effects are shown in the following table.

	Spatial	Temporal
Instabilities	Light-by-Light Scattering	Modulational Instability
	Spatial self-phase modulation	Temporal self-phase modulation – self-chirping
Envelope Effects	Self-focusing – whole-beam and multimode	Self-compression Self-decompression – self-dispersion Self-steepening
	Self-trapping – spatial solitons	Temporal solitons
Combined	Light Bullets	

The consequences of these nonlinear effects can be significant. In beam propagation, self-focusing and self-trapping lead to lowering of thresholds for

R.Y. Chiao (✉)
Schools of Natural Sciences and of Engineering, University of California, Merced, California, USA
e-mail: rchiao@ucmerced.edu

R.W. Boyd et al. (eds.), *Self-focusing: Past and Present*,
Topics in Applied Physics 114, DOI 10.1007/978-0-387-34727-1_4,

other nonlinear processes, such as stimulated Raman and Brillouin scattering, self-phase modulation, and optical damage. Nonlinear index effects impact the design of very high energy laser systems such as those required for laser fusion and the implementation of long-distance fiber optical communication systems.

4.2 Early History

Two years after self-focusing and self-trapping in plasmas was proposed in 1962 [1], the theory of self-trapping was developed [2,3]. In reference [2], a numerical solution was given for a two-transverse dimension beam; the critical power was calculated; mechanisms for the nonlinearity were discussed; and the effect was used to explain anomalous Raman gain and optical damage. In both 1964 references the one-transverse-dimension hyperbolic secant soliton solution was given. The theory of self-focusing was developed in 1965 [4,5]. Reference [5] estimates the self-focusing distance, derives the nonlinear Schrödinger equation (NLSE), and uses this equation to obtain numerical results on self-focusing. That same year the first observation of self-focusing was made [6] and it was experimentally related to anomalous stimulated Raman gain [7,8]. In 1966, the theory of nonlinear instability was given [9,10] and the effect observed experimentally [11]. In this same year, multi-filament structure in multimode beams was observed, and the influence on spectral broadening was considered [12]; this was followed several years later by experiments in which regular filament structure was seen on carefully prepared beams [13].

Since this initial work a number of important contributions have been made. As of 2006, there were about 1000 papers with self-focusing in the title and more than 500 papers with self-trapping or spatial solitons in the title. The early work on self-focusing has been reviewed [14–17] and considerable attention has been given to mathematical methods for dealing with blow-up of solutions to the NLSE and related equations in physics (see [18] and [19] for a discussion of these problems). Modification of self-focusing theory due to the finite duration of pulses has also been considered [20–25]. For a recent source of references on self-focusing and self-trapping, see [26]. A unified approach to self-action effects is given in [27].

In our review, we take an approach that is opposite to much of the historical sequence. We will first discuss instabilities, then beam self-focusing, and finally self-trapping.

4.3 Nonlinear Polarization and the Nonlinear Refractive Index

Self-action effects arise from the third-order nonlinear polarization. Slow molecular motions can contribute to the nonlinear response because the nonlinear polarization includes terms for which the slowly varying part of the square of the field amplitude drives the material system. This can lead to very strong nonlinearities.

In most cases, the refractive index increases with increasing light intensity. A very simple example occurs for ensembles of anisotropic molecules. Because the molecules have lowest energy when they are aligned in the direction(s) of highest polarizability, they experience a torque that increases the fractional alignment and the index of refraction. Increasing density of an initially uniform fluid in the region of the beam (electrostriction) or moving molecules of higher polarizability than a surrounding fluid into a beam also lowers the energy and results in a positive nonlinear refractive index.

Neglecting dispersion, the polarization can be written,

$$P = \varepsilon_0 \chi E \tag{4.1}$$

where E is the electric field, and χ is the electric susceptibility including non-linear terms. For simplicity, we assume the various frequency and wavevector components of the field are all polarized in the same direction. We write the susceptibility

$$\chi = \chi^{(1)} + \chi^{(3)} \langle E^2 \rangle, \tag{4.2}$$

where $\langle E^2 \rangle$ is the average of E^2 over a few optical cycles. We have assumed there are no contributions to the polarization in even powers of the field and have neglected any contributions by odd powers beyond the third-order term. We can replace $\chi^{(1)}$ and $\chi^{(3)}$ by the linear and nonlinear refractive index, n_0 and n_2, where

$$n = n_0 + 2n_2 \langle E^2 \rangle, \tag{4.3}$$

The linear refractive index is given by

$$n_0 = \sqrt{1 + \chi^{(1)}} \tag{4.4}$$

while the nonlinear refractive index is given by

$$n_2 = \frac{\chi^{(3)}}{4n_0}, \tag{4.5}$$

assuming the nonlinear term is much smaller than the linear term. We can then write the polarization as

$$P = P^{\mathrm{L}} + P^{\mathrm{NL}} \tag{4.6}$$

where

$$P^{\mathrm{L}} = \varepsilon_0 (n_0^2 - 1) E \tag{4.7}$$

and

$$P^{\mathrm{NL}} = 4\varepsilon_0 n_0 n_2 \langle E^2 \rangle E. \tag{4.8}$$

Equation (4.8) is the induced polarization to third power in the electric field. We have neglected polarization terms that occur at the sum of three optical frequencies and have kept terms that are close to at least one of the driving frequencies. This neglect of "third harmonic" terms can be justified by assuming they are small or not phase matched. However, we do not assume contributions due to material resonances at the sum of two driving frequencies are negligible; these terms do contribute to n_2.

4.4 The Nonlinear Schrödinger Equation

To facilitate our understanding of the self-focusing problem we use an approximation to the scalar nonlinear wave equation which has the familiar form of a Schrödinger equation [5]. We assume a single frequency and use the slowly varying envelope approximation. We first write the field in terms of its positive and negative frequency components

$$E = \tfrac{1}{2}(E^{+} + E^{-}), \tag{4.9}$$

where $E^{-} = E^{+*}$. The positive frequency component is given by

$$E^{+}(\mathbf{r}, t) = \mathcal{E}(\mathbf{r})e^{i(kz-\omega t)} \tag{4.10}$$

where $k = \omega n_0(\omega)/c$ and we have assumed the wave is traveling in the $+z$ direction. In the slowly varying amplitude approximation, it is assumed that $|\partial^2\mathcal{E}/\partial z^2| \ll k|\partial\mathcal{E}/\partial z|$. This leads to the following approximation to the wave equation

$$i\frac{\partial\mathcal{E}}{\partial z} + \frac{1}{2k}\nabla_{\perp}^{2}\mathcal{E} + k\frac{n_2}{n_0}|\mathcal{E}|^2\mathcal{E} = 0. \tag{4.11}$$

4.5 Four-Wave Mixing, Weak-Wave Retardation, Instability

In this section, we discuss the coupling of waves through the nonlinear index. This will lead to a discussion of spatial instabilities in monochromatic plane waves. From a Fourier component point of view, the nonlinear interaction involves the coupling of three field components to produce a nonlinear polarization that drives a fourth field component. The basic process can be described as light-by-light scattering or four-wave mixing.

We separate the field into weak and strong parts. This is convenient as it allows us to linearize the problem in the weak field. It is appropriate when a strong optical beam from a laser initially enters the nonlinear medium and the weak field is assumed to be much smaller than the strong field. In fact, the initial weak field can be the zero point field so that the weak wave is initiated by spontaneous emission. We can write

$$\mathcal{E} = \mathcal{E}_s + \mathcal{E}_w, \tag{4.12}$$

where s and w stand for strong and weak, respectively. Neglecting terms in the weak field beyond the linear term we have for the strong field nonlinear polarization.

$$\mathcal{P}_s^{\mathrm{NL}} = 2\varepsilon_0 n_0 n_2 |\mathcal{E}_s|^2 \mathcal{E}_s \tag{4.13}$$

and for the weak field nonlinear polarization

$$\mathcal{P}_w^{\mathrm{NL}} = 2\varepsilon_0 n_0 n_2 \left(2|\mathcal{E}_s|^2 \mathcal{E}_w + \mathcal{E}_s^2 \mathcal{E}_w^*\right) \tag{4.14}$$

This result has two interesting aspects. First, the weak wave experiences twice the nonlinear polarization and index of refraction change [the first term in Eq. (4.14)] as the strong field induces on itself [Eq. (4.13)]. The weak wave retardation can compensate for the breaking of phase matching by diffraction [10]. Second, the positive and negative parts of the weak field are coupled to each other [the second term in Eq. (4.14)]. Because of the cross-coupling, phase conjugation occurs and coupled frequency and propagation vector sidebands grow on the strong field.

To calculate the instability gain, we assume a strong plane-wave together with two weak components traveling very nearly parallel to the strong wave as shown in Fig. 4.1. The weak-wave retardation effect is also shown in the figure.

From the NLSE [Eq. (4.11)] the exponential growth and decay constants for the weak-wave power are found to be [9]

$$g = \pm k_\perp \left(\frac{4n_2}{n_0} |\mathcal{E}_s|^2 - \frac{k_\perp^2}{k^2} \right)^{1/2} \tag{4.15}$$

where $k_\perp$ is the component of the scattered wave that is perpendicular to the strong wave. Because $k_\perp \ll k$, for all the cases of interest, $k_\perp$ can be replaced by $k\theta$ where θ is the angle between the strong wave and the weak waves. A plot of g vs. θ is given in Fig. 4.2. Note that the maximum value of g is $g_{\max} = 2kn_2|\mathcal{E}_s|^2/n_0$

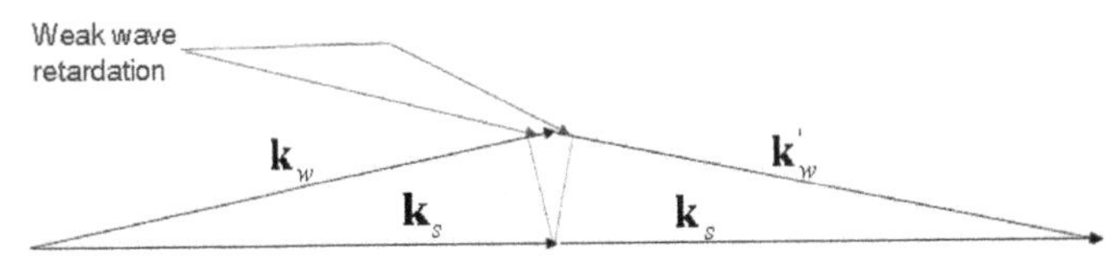

Fig. 4.1 Light-by-light scattering. Strong forward wave ($\mathbf{k}_s$) interacting with two weak waves ($\mathbf{k}_w$, $\mathbf{k}'_w$) whose magnitudes are increased by weak-wave retardation

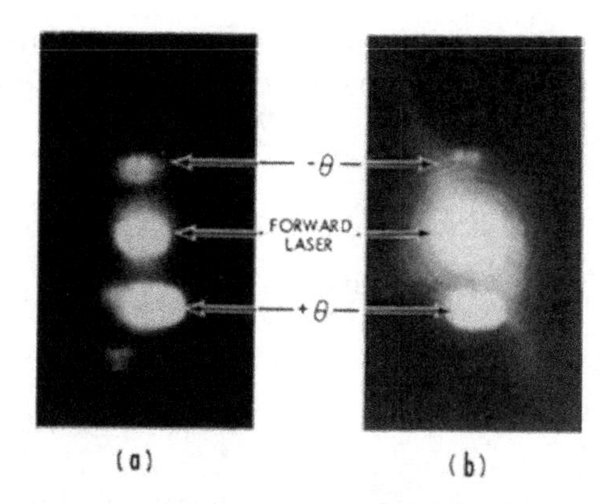

Fig. 4.2 (**a**) Typical data for a 3-mm cell length where no self-focusing occurs. Due to overexposure, the actual intensity ratios are not faithfully reproduced. (**b**) Simultaneous 1.5-cm spacer Fabry–Perot interferometer analysis of all three beams from the 3-mm cell

and the angle at which this occurs is $\theta_{\text{opt}} = 2\sqrt{2}n_2|\mathcal{E}_s|^2/n_0$. Because of the cross coupling, the normal Fourier components of the scattered waves grow (or decay) in pairs, one at θ and the other at $-\theta$. As we shall see from the discussion of self-focusing given below, the instability gain is related to the beam self-focusing distance.

An example of light-by-light scattering is shown in Fig. 4.2. A strong forward beam is sent into a short cell containing nitrobenzene, a medium with large n_2. To obtain the strong beam, a Q-switched ruby laser with 240 MW power was focused to an area of 20 mm^2. This is the central peak in the figure. When a weak beam was sent in at an angle $+\theta$ to the strong beam, a second weak beam at $-\theta$ appeared.

4.6 Spatial Self-Phase Modulation and Estimating the Beam Self-Focusing Distance

In order to understand the nonlinear propagation of spatially finite beams, we first consider the effect on a beam of the nonlinear polarization alone. If diffraction is neglected in the wave equation, Eq. (4.11), the solution for the nonlinear phase build-up from an input boundary at $z = 0$ is

$$\Phi_{\text{NL}}(\mathbf{r}, z) = kz\frac{n_2}{n_0}|\mathcal{E}(\mathbf{r}, 0)|^2 \tag{4.16}$$

where $\mathbf{r}$ is the coordinate vector transverse to z. We can define the nonlinear distance by setting the phase deviation across the beam $\Delta\Phi_{\text{NL}}$ equal to 1, so that

$$z_{\text{NL}} = \frac{n_0}{kn_2|\mathcal{E}(0,0)|^2} = \frac{2}{g_{\max}} \tag{4.17}$$

From the wave equation we see that there is also a characteristic distance for diffraction. We take this to be

$$z_{\mathrm{DIF}} = k{r_0}^2 \tag{4.18}$$

where r_0 is a distance characteristic of the beam radius. Note that z_{DIF} is the Rayleigh range or the Fresnel length of the beam.

The transverse component of the wave-vector at a transverse point $\mathbf{r}$ in a beam is

$$\mathbf{k}_\perp(\mathbf{r}, z) = \nabla_\perp \Phi_{\mathrm{NL}}(\mathbf{r}, z) = kz\frac{n_2}{n_0}\nabla_\perp|\mathcal{E}(\mathbf{r}, 0)|^2 \tag{4.19}$$

The variation of $\mathbf{k}_\perp$ with $\mathbf{r}$ can be viewed as a spatial chirp. A plot of $k_\perp$ is shown in Fig. 4.3.

For a beam that is symmetric with either circular or slab symmetry, the angle the wave-vector makes with the z-axis for small angles is

$$\theta(r, z) = z\frac{n_2}{n_0}\frac{\partial|\mathcal{E}(r, 0)|^2}{\partial r} \tag{4.20}$$

a result that is independent of wavelength. Assuming the beam is most intense at the center, θ is negative and the beam will focus when we include the diffraction term. An estimate of the focusing distance can be found by setting $\theta(r, z) = r/z$, the angle that a ray starting at r will reach the beam axis at z. When $z_{\mathrm{DIF}} \gg z_{\mathrm{NL}}$ this gives a self-focusing distance [5]

$$z_{\mathrm{SF}} = \left.\sqrt{\frac{-n_0 r}{n_2\partial|\mathcal{E}|^2/\partial r}}\right|_{r\to 0.} \tag{4.21}$$

Assuming a Gaussian beam with r_0 equal to the intensity $1/e$ halfwidth and using Eqs. (4.17) and (4.18), we find

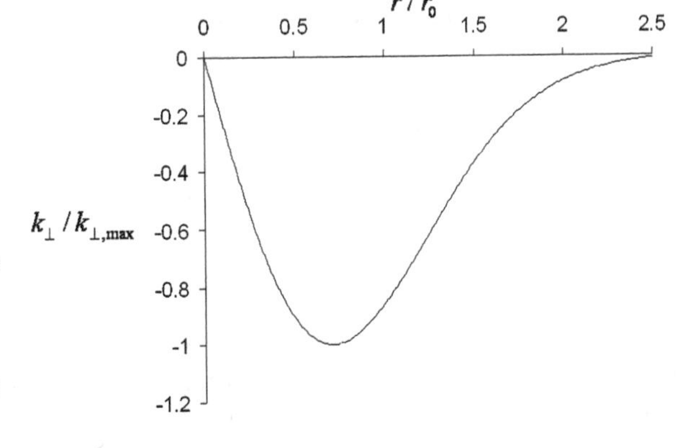

Fig. 4.3 $k_\perp$ versus r for a Gaussian beam. r_0 is the $1/e$ intensity half-width and $k_{\perp,\max} = kzn_2|\mathcal{E}(0,0)|^2 \sqrt{2}e^{\frac{1}{2}}/n_0 r_0$. Near $r = 0$, the spatial chirp is linear as it is for a conventional lens

$$z_{SF} = \sqrt{\frac{z_{NL} z_{DIF}}{2}}. \tag{4.22}$$

A similar result for the self-focusing distance can be found by simply inserting the angle corresponding to the Fourier inverse of the beam radius into Eq. (4.5) for the instability gain.

4.7 Self-Focusing Intensity Singularity and Beam Collapse

If a beam is has circular symmetry, it will nonlinearly focus without limit to an intensity singularity. On the other hand, slab beams that are confined in only one transverse dimension will not focus to a singularity. This difference between one- and two- dimensional focusing is easy to understand by simple scaling arguments. From the NLSE, Eq. (4.11), we see that in both cases the diffraction term scales as the inverse of the beam width squared. In the two-dimensional transverse confinement case, the nonlinear term also scales as the reciprocal of the beamwidth squared because this term is proportional to intensity and power is conserved. Because the scaling of the two terms is the same, the dominant term will remain dominant and in this situation the beam will come to a catastrophic focus when the nonlinear term dominates.

For confinement in one transverse dimension, the nonlinear term scales as the inverse of the beamwidth because the power per unit distance in the unconfined direction is conserved. Because the nonlinear term is only proportional to the inverse first power of the beamwidth, the nonlinear term will grow more slowly with decreasing radius than the diffraction term. The diffraction term comes into balance with the focusing term and there is self-focusing without an intensity singularity.

We will examine the two-transverse-dimensions case in further detail. Although there are a number of ways to analyze the problem, we will numerically solve the NLSE for cylindrical symmetry using a finite-difference method. In [5], the finite difference solution involved directly calculating the transverse derivatives. Here we will use the split-step method where the transverse diffraction term in the NLSE is calculated in Fourier space and the nonlinear term in the NLSE is calculated in coordinate space. To efficiently solve the cylindrically symmetric problem we use a version of the fast Hankel transform given in [28]. The method can be used to come close to the initial singularity in a very computationally efficient fashion. With the split-step method we can also avoid the paraxial approximation by using the nonlinear Helmholtz equation to directly calculate the contribution of the transverse diffraction term to the axial phase factor in Fourier space.

The result of a typical calculation for an initially Gaussian beam is shown as the outer curve in Fig. 4.4. The estimated self-focusing distance, as given by Eq. (4.21), is found to be smaller than the numerically calculated distance in

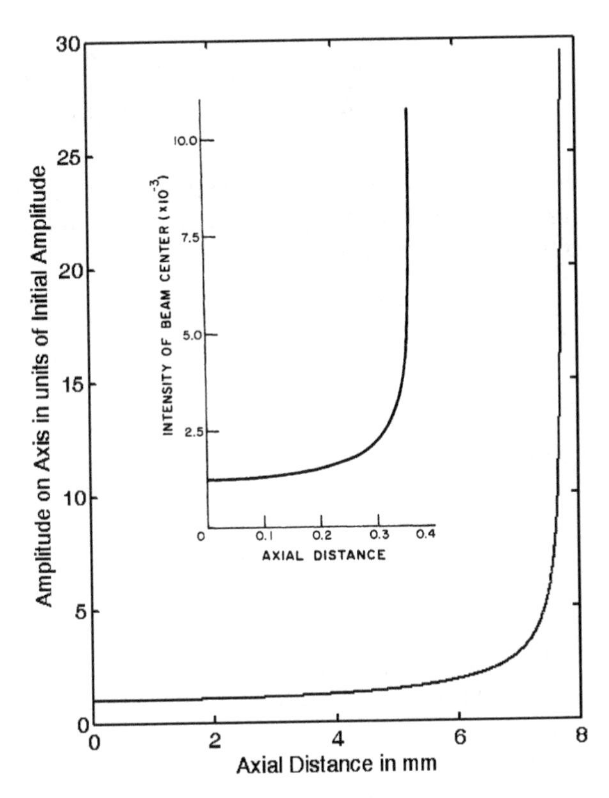

Fig. 4.4 Numerical calculations of the approach to the self-focus. For the outer curve, the self-focusing distance is found to be ≈ 7.8 mm. The initially Gaussian beam at $\lambda = 1\ \mu$m with $r_0 = 70.7$ mm has $z_{\mathrm{DIF}} = 31.4$ mm, $z_{\mathrm{NL}} = 3.14$ mm, and from Eq. (21) $z_{\mathrm{SF}} = 7.02$ mm. The inset shows an early numerical calculation of self-focusing [5]

part because the estimated distance does not take into account the effect of diffraction in lengthening the self-focusing distance. Comparing the present split-step Hankel with the early calculation [5], there is a least order of magnitude decrease in computation time and about a factor of one hundred increase in intensity near the focus. This improvement can be attributed to both the efficiency of the present algorithm and the increase in computer speed.

In addition to the intensity singularity, it is possible to show that beams undergoing self-focusing collapse in the NLSE approximation. Collapse occurs when the entire beam shrinks to a point. Reference [29] obtained a second-order differential equation for the RMS beam radius which in the two-transverse-dimension case can be written in terms of the ratio of two invariants. Because of the Schrödinger character of Eq. (4.11), the two invariants correspond to the expectation value of the Hamiltonian and the expectation value of the unnormalized probability density. In the cylindrically symmetric case, the equation for the average beam radius is:

$$\frac{d\langle r^2\rangle}{dz^2} = 2\frac{\int\left(\frac{1}{k^2}|\nabla_\perp\mathcal{E}|^2 - \frac{2n_2}{n_0}|\mathcal{E}|^4\right)rdr\Big|_{z=0}}{\int|\mathcal{E}|^2 rdr\Big|_{z=0}}, \tag{4.23}$$

where the numerator is the Hamiltonian term and the denominator is the unnormalized probability density. On carrying out the integrations for an initially Gaussian beam we find,

$$\langle r(z)^2\rangle = r_0^2 + \left(\frac{1}{k^2 r_0^2} - \frac{n_2|\mathcal{E}(0,0)|^2}{2n_0}\right)z^2. \tag{4.24}$$

When $n_2|\mathcal{E}(0,0)|^2/2n_0 > 1/k^2 r_0^2$ the beam will collapse. From this equation, we can obtain the collapse distance

$$z_{\mathrm{COL}} = \sqrt{\frac{2z_{\mathrm{NL}}z_{\mathrm{DIF}}}{1 - \dfrac{2z_{\mathrm{NL}}}{z_{\mathrm{DIF}}}}}. \tag{4.25}$$

When $z_{\mathrm{DIF}} \gg z_{\mathrm{NL}}$, we find that $z_{\mathrm{COL}} = \sqrt{2z_{\mathrm{NL}}z_{\mathrm{DIF}}}$, a factor of two greater than the self-focusing distance given in Eq. (4.22). For the case of Fig. 4.4, where $z_{\mathrm{DIF}} = 31.4$ and $z_{\mathrm{NL}} = 3.14$, we find from Eq. (4.25) that $z_{\mathrm{COL}} = 15.7$ which is also about a factor of two larger than the distance where blow-up is observed in the numerical calculation.

4.8 Limitations on Blow-Up and Collapse

It should be apparent that much of the analysis given here is of mathematical significance rather than representing actual physics because the NLSE involves several limitations, including the use of the scalar wave equation, the slowly varying envelope assumption, the absence of saturation of the nonlinear index, and the neglect of other nonlinearities. These nonlinearities include stimulated light scattering, optical damage, and breakdown. That these other nonlinear effects become important is evidenced by the orders of magnitude increase in intensity due to self-focusing as shown, for example, in Fig. 4.4. In addition, beams with small-scale structure exhibit a complex breakup process as discussed in the next section.

The stabilization of self-focusing resulting in self-trapping or periodic focusing requires consideration of processes not included in the simple expression [Eq. (4.3)] for the nonlinear dielectric response. For liquids, observed focal spots can vary from a few microns in size to a few tens of microns depending upon the liquid. The stabilizing process either limits the nonlinear increase in the index of refraction or depletes the forward beam in a strongly nonlinear fashion. The former includes saturation of the nonlinear index [30–32] and the latter stimulated light scattering [33] or a negative contribution to the index due to the production of electrons on ionization of the liquid in the intense light field [34].

In air, in particular, the critical power is approximately 2 GW for a diffraction limited Gaussian beam [35]. When the intensity approaches 10^{13} to 10^{14} W/cm^2, multiphoton ionization occurs. The resultant intensity-dependent reduction of the refractive index arising from the consequent underdense plasma can stabilize the beam. Thin intense filaments over tens (and up to hundreds) of meters has been observed. A theoretical description has been given by generalizing Eq. (4.11) to include both ionization and a Kerr effect which has both a nondispersive and a dispersive contribution [36].

An understanding and control of the nonlinear optical interaction resulting in self-focussing and self-trapping as well as the associated phenomena in the table shown on the first page of this chapter has resulted in the ability to nonlinearly shape optical beams in a controlled manner.

4.9 Beam Breakup and Multiple Filament Formation

Small-scale beam structure can play an important role in self-focusing [12,13]. Figure 4.5 shows small-scale filament formation during self-focusing of a beam. This effect is particularly significant when the beams have structure at angles near θ_{opt}. For most beams where self-focusing is significant, $\theta_{\text{opt}} \gg \theta_{\text{av}} = 1/kr_0$ so that $g_{\max} z_{\text{SF}} \gg 1$. It then becomes a question of whether a beam is

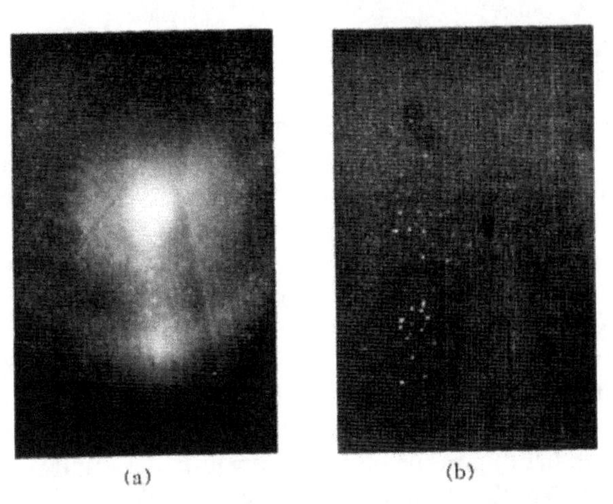

Fig. 4.5 (**a**) Image of a laser beam emerging from a 50-cm cell of CS_2 and exhibiting large- and small-scale trapping. Magnification is 30×. The bright central portion is the large-scale trapped beam; the many small bright filaments demonstrate the small-scale trapping. The broad disk and ring of light are the untrapped beam diffracting from the initial pinhole. (**b**) Raman Stokes radiation under conditions similar to (a). Magnification 50×. From [12]

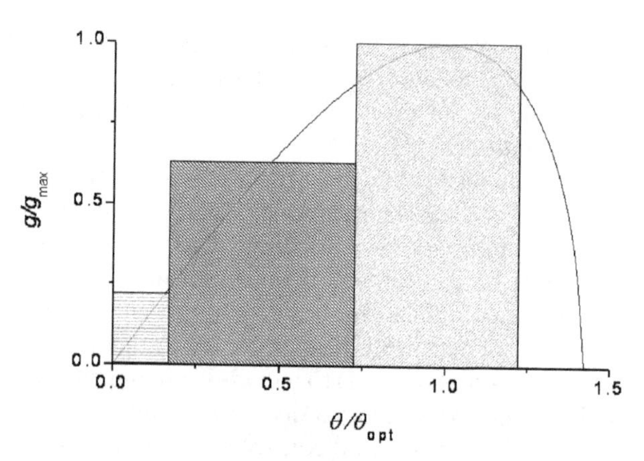

Fig. 4.6 Weak-wave gain, g, versus angle, θ, between the weak and strong waves as calculated from Eq. (4.15). The contributions of the three shaded regions are described in the text

sufficiently smooth initially that whole-beam focusing can dominate beam breakup into multiple small-scale filaments. To further understand this, we return to the instability analysis discussed earlier. Figure 4.6 gives a rough schematic of the different contributions to the self-focusing of a typical beam: the small box on the left includes the angular region corresponding to the transverse Fourier components for a smooth beam; the middle box encompasses the region where beam fine structure contributes; and the right box corresponds to Fourier components that must grow from noise including zero point oscillations. The height of each box corresponds to the typical gain in each of the three regions. Increasing gain with angle is offset by decreasing initial Fourier amplitude.

4.10 Self-Focusing of Pulses: Light Bullets

When an optical pulse transverses the nonlinear medium, the focusing distance becomes time-dependent as well as radially dependent. The most obvious consequence of this is that the most intense portions are focused at a shorter distance than the weaker leading and trailing edges. References [20] and [21] first considered the motion of the focal regions for temporal pulses. Reference [22] included retardation and theoretically deduced the trajectories of the moving focal regions for a Gaussian temporal amplitude profile with a 1/e halfwidth of t_p at the entrance boundary. Due to the retardation, the distance at which the focal region first appears exceeds z_{SF}, the focussing distance for the peak of the pulse. This emanates from a portion of the pulse shifted to the leading edge by an amount dependent upon the initial pulse width. This shift is $t_p(ct_p/2n_0z_{\mathrm{SF}})$ for a pulse whose length t_p/n_0 is much less than z_{SF}. For such a pulse the initial focussing distance exceeds z_{SF} by a distance equal to $[(c/n_0t_p)^2/4z_{\mathrm{SF}}]$. The focal region subsequently expands forward, and also expands backwards to the minimum focussing distance, z_{SF}, at time $z_{\mathrm{SF}}n_0/c$ corresponding to the peak. Subsequently the focal region continues to expand as it propagates through the medium. The details of this time dependent behavior have been investigated both experimentally [23,24] and theoretically [25].

When anomalous group velocity dispersion is included in the NLSE then a balance can occur between the dispersion and the phase buildup due to the nonlinearity and a three-dimensional trapped pulse—or "light bullet"—can propagate [37]. There has been an extensive investigation of such temporally and spatially trapped light.

4.11 Self-Trapping, Spatial Solitons

If a beam has circular symmetry it will focus to an intense singularity. When $z_{\mathrm{DIF}} \approx z_{\mathrm{NL}}$, a solution to the wave equation for a beam can be found whose transverse intensity profile does not vary in z. Because both of these parameters (z_{DIF} and z_{NL}) are proportional to the radius squared, the self-trapped solution in two transverse dimensions occurs at a critical power independent of the beam radius. Using the values of z_{NL} and z_{DIF} given in Eqs. (4.17) and (4.18) the critical power can be estimated by setting these two characteristic parameters equal obtaining

$$P_{\mathrm{CR}} = \frac{\varepsilon_0 c \lambda^2}{8\pi n_2}. \tag{4.26}$$

This can be compared with numerical results [2].

In case of confinement in one transverse direction, the following constraint applies

$$\frac{dP}{dy} r_0 \approx \frac{\varepsilon_0 c \lambda^2}{(2\pi)^2 n_2} \tag{4.27}$$

where dP/dy is the power per unit distance in the transverse direction perpendicular to the confinement direction. In this case, there is no critical power per unit distance. Whatever value of dP/dy one chooses there is an r_0 such that Eq. (4.27) is satisfied. The lowest order hyperbolic secant solution has been known since 1964 [2,3]. and inverse scattering theory has been used to find higher-order analytic solutions [38].

References

1. G.A. Askar'yan: Effects of the gradient of a strong electromagnetic beam on electrons and atoms, *JETP* **15**, 1088–1090 (1962).
2. R. Chaio, E.M. Garmire, C.H. Townes: Self-trapping of optical beams, *Phys. Rev. Lett.* **13**, 479–482 (1964); erratum, ibid, **14**, 1056 (1965).

3. V.I. Talanov: Self-focusing of electromagnetic waves in nonlinear media, *Radiophys.* **8**, 254–257 (1964).
4. V.I. Talanov: Self-focusing of wave beams in nonlinear media, *Radiophys.* **9**, 138–141 (1965).
5. P.L. Kelley: Self-focusing of optical beams, *Phys. Rev. Lett.* **15**, 1005–1008 (1965).
6. N.F. Pilipetskii, A.R. Rustamov: Observation of self-focusing of light in liquids, *JETP Lett.* **2**, 55–56 (1965).
7. Y.R. Shen, Y.J. Shaham: Beam deterioration and stimulated Raman effect, *Phys. Rev. Lett.* **15**, 1008–1010 (1965).
8. P. Lallemand, N. Bloembergen: Self-focusing of laser beams and stimulated Raman gain in liquids, *Phys. Rev. Lett.* **15**, 1010–1012 (1965).
9. V. Bespalov, V.I. Talanov: Filamentary structure of light beams in nonlinear liquids, *JETP Lett.* **3**, 307–309 (1966).
10. R.Y. Chiao, P.L. Kelley, E.M. Garmire: Stimulated four-photon interaction and its influence on stimulated Rayleigh-wing scattering, *Phys. Rev. Lett.* **17**, 1158–1161 (1966).
11. R.L. Carman, R.Y. Chiao, P.L. Kelley: Observation of degenerate stimulated four-photon interaction and four-wave parametric amplification, *Phys. Rev. Lett.* **17**, 1281–1283 (1966).
12. R.Y. Chiao, M.A. Johnson, S. Krinsky et al.: A new class of trapped light filaments, *IEEE J. Quantum Electron.* **QE-2**, 467–469 (1966).
13. A.J. Campillo, S.L. Shapiro, B.R. Suydam: Periodic breakup of optical beams due to self-focusing, *Appl. Phys. Lett.* **23**, 628–630 (1973).
14. S.A. Akhmanov, R. Khokhlov, A.P. Sukhorukov: Self-focusing, self-defocusing and self-modulation of laser beams. In: *Laser Handbook*, F.T. Arecchi, E.O. Schulz-Dubois, (Eds.), North-Holland, Amsterdam (1972), vol. 2, pages 1152–1228.
15. O. Svelto: Self-focusing, self-trapping and self-phase modulation of laser beams. In: *Progress in Optics*, Vol. XII, E. Wolf (Ed), North Holland, Amsterdam (1974), pages 3–51.
16. Y.R. Shen: Self-focusing: experimental, *Prog. Quant. Electron.* **4**, 1–34 (1975).
17. J.H. Marburger: Self-focusing: theory, *Prog. Quant. Electron.* **4**, 35–110 (1975).
18. J.J. Rasmussen, K. Rypdal: Blow-up in nonlinear Schrödinger equations: I A general review, *Physica Scripta* **33**, 481–497 (1986).
19. K. Rypdal, J.J. Rasmussen: Blow-up in nonlinear Schrödinger equations: II Similarity structure of the blow-up singularity, *Physica Scripta* **33**, 498–504 (1986).
20. A.L. Dyshko, V.N. Lugovoi, A.M. Prokhorov: Self-focusing of intense light beams, *JETP Lett.* **6**, 146–148 (1967).
21. V.N. Lugovoi, A.M. Prokhorov: A possible explanation of the small scale self-focusing light filaments, *JETP Lett.* **7**, 117–119 (1968).
22. J.-P.E. Taran, T.K. Gustafson: Comments on the self-focusing of short light pulses, *IEEE J. Quant. Electron.* **5**, 381–382 (1968).
23. M.M.T. Loy, Y.R. Shen: Small-scale filaments in liquids and tracks of moving foci, *Phys. Rev. Lett.* **22**, 994–997 (1969).
24. M.M.T. Loy, Y.R. Shen: Experimental study of small-scale filaments of light in liquids, *Phys. Rev. Lett.* **25**, 1333–1336 (1970).
25. Y.R. Shen, M.M.T. Loy: Theoretical interpretation of small-scale filaments originating from moving focal spots, *Phys. Rev. A* **3**, 2099–2105 (1971).
26. Y.S. Kivshar, G.P. Agrawal: *Optical Solitons*, Academic Press, Boston (2003).
27. P.L. Kelley: Nonlinear index of refraction and self-action effects in optical propagation, *IEEE Sel. Top. Quant. Electron.* **6**, 1259–1264 (2000).
28. M. Guizar-Sicairos, J.C. Gutierrez-Vega: Computation of quasi-discrete Hankel transforms of integer order for propagating optical wave fields, *J. Opt. Soc. Am.* **A21**, 53–58 (2004).

29. S.N. Vlasov, V.A. Petrishchev, V.I. Talanov: Averaged description of wave beams in linear and nonlinear media (the method of moments), *Quantum Electron. Radiophys.* **14**, 1062–1070 (1971).
30. T.K. Gustafson, P.L. Kelley, R. Chiao et al.: Self-trapping of optical beams, *Appl. Phys. Lett.* **12**, 165–168 (1968).
31. J. Marburger, L. Huff, J. Reichert et al.: Stationary self-trapping of optical beams in dense media with Lorentz local-field corrections, *Phys. Rev.* **184**, 255–259 (1969).
32. T.K. Gustafson, C.H. Townes: Influence of steric effects and compressibility on nonlinear response to laser pulses and the diameters of self-trapped filaments, *Phys. Rev. A* **6**, 1659–1664 (1972).
33. P.L. Kelley, T.K. Gustafson: Backward stimulated light scattering and the limiting diameters of self-focused light beams, *Phys. Rev. A* **8**, 315–318 (1973).
34. E. Yablonovitch, N. Bloembergen: Avalanche ionization and the limiting diameter of filaments induced by light pulses in transparent media, *Phys. Rev. Lett.* **29**, 907–910 (1972).
35. J. Kasparian, R. Sauerbrey, S.L. Chin: The critical laser intensity of self-guided light filaments in air, *Appl. Phys. B* **71**, 877–879 (2000).
36. A. Couairon, L. Berge: Light filaments in air for ultraviolet and infrared wavelengths, *Phys. Rev. Lett.* **71**, 35003 1–4 (2002).
37. Y. Silberberg: Collapse of optical pulses, *Optics Lett.* **15**, 1282–1284 (1990).
38. V.E. Zakharov, A.B. Shabat: Exact theory of two-dimensional self-focusing and one-dimensional self-modulation of waves in nonlinear media, *JETP* **34**, 62–69 (1972).

Chapter 5
Multi-Focus Structure and Moving Nonlinear Foci: Adequate Models of Self-Focusing of Laser Beams in Nonlinear Media

V.N. Lugovoi[†] and A.A. Manenkov

Abstract This chapter presents a review of the theory of the multi-focus structure (MFS) and moving nonlinear foci (MNLF) models of self-focusing. It also reviews some experimental results on self-focusing of nanosecond-duration-pulses that support these models. Some experimental and simulation results of studies of femtosecond self-focusing in air are discussed. It is concluded that the main features of MFS-MNLF models can be applied to beam filamentation in different nonlinear media and over a wide range of pulse durations, including ultra-short pulses. Future directions for the study of femtosecond self-focusing in air are suggested.

5.1 Introduction

Great interest in self-focusing phenomena was shown in the 1960s and 1970s of the last century. A large number of papers were published during the last four decades on the elucidation of mechanisms of self-focusing of laser beams propagating in nonlinear optical media. Several models of self-focusing were proposed and investigated, both theoretically and experimentally, and were widely discussed in the literature (see, in particular, the review papers [1–6]). Among these models, the multi-focus structure (MFS) and moving nonlinear foci (MNLF) models proposed by V.N. Lugovoi and A.M. Prokhorov [7, 8] have been theoretically the most well developed and grounded, and have been confirmed in many experiments, especially for laser pulses in the nanosecond duration range. In particular, we refer to experiments [9–11] where self-focusing in various nonlinear optical media (liquids and solids) were studied using different detection methods. In Sections 5.2, 5.3, and 5.4 we review the theory of the MFS–MNLF

A.A. Manenkov (✉)
A.M. Prokhorov General Physics Institute of Russian Academy of Sciences,
38 Vavilov Street, Moscow, 119991, Russia
e-mail: manenkov@kapella.gpi.ru

R.W. Boyd et al. (eds.), *Self-focusing: Past and Present*,
Topics in Applied Physics 114, DOI 10.1007/978-0-387-34727-1_5,

models and describe some early experimental studies of self-focusing of nanosecond pulses that were performed to verify the models.

A new wave of interest in self-focusing phenomena was created recently by reports [12–14] of the observation of *super-long light and plasma filaments* in air due to self-focusing of femtosecond laser pulses. Based on experimental results and computer simulations, several models have been proposed [12–24] for the explanation of the observed filamentation. Mechanisms for the formation of such filaments were under discussion in the literature and at many conferences during the last decade, and interest in this problem still continues. In Section 5 we conclude that MFS–MNLF models can be applied to the explanation of these experiments as well. Future directions of development of the self-focusing in air are also outlined.

5.2 Review of the Theory of MFS–MNLF Models

The MFS concept has been formulated in [7] based on the results of a numerical solution of the nonlinear wave equation (in parabolic approximation) for a Gaussian light beam propagating in a Kerr-type nonlinear medium:

$$\frac{\partial^2 E}{\partial r^2} + \frac{1}{r}\frac{\partial E}{\partial r} + 2ik\frac{\partial E}{\partial z} + n_2 k^2 |E|^2 E = 0 \tag{5.1}$$

Here E is the electric field strength of the light beam, r and z are radial and longitudinal coordinates, k is the wave number, n_2 is the coefficient of the nonlinear refractive index according to the relation

$$n = n_0 + n_2 |E^2| \tag{5.2}$$

It has been shown that for an incident beam power P exceeding the critical power $P_{\text{cr}}^{(1)} = cn_0 N_1{}^2/8n_2k^2$, where c is the speed of light in vacuum and N_1 is a numerical coefficient equal approximately to 2, a *self-guided (or self-trapped)* propagation mode, supposed in some previous works (see review papers [1, 2, 6] and references therein), does not take place in such the media , but instead a *multi-focus structure* is formed.

According to the MFS model, the light beam, propagating in the nonlinear medium, splits into ring zones in the transverse direction, and each such zone is focused at different distances along the propagation z-axis. Only a part of the initial light beam power, comparable with the critical power $P_{\text{cr}}^{(1)}$, flows into the first nonlinear focus. This portion of the beam power is partly absorbed in the nonlinear focal region and partly diffracts after passing through the focus at relatively large angles to the light beam axis.

A similar process takes place for other ring zones (slices) of the beam. Thus, formation of the subsequent nonlinear foci seems to be recurrent. The formation of a multi-focus structure is shown schematically in Fig. 5.1 for a stationary, i.e., time-independent power beam.

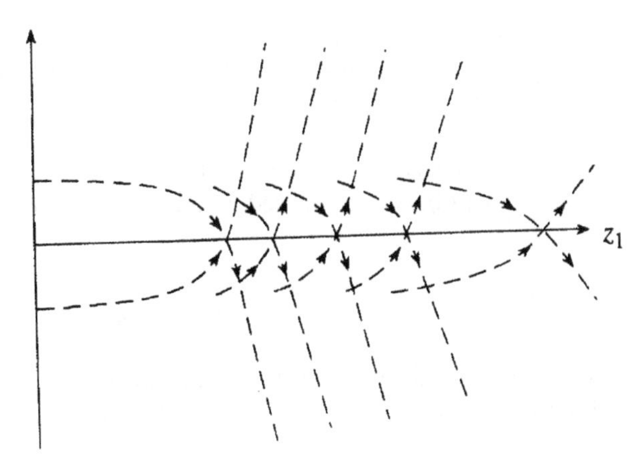

Fig. 5.1 Schematic longitudinal cross-section of the laser beam propagating in the Kerr-type nonlinear medium, illustrating the formation of the multi-focus structure on the beam z-axis for $P > P_{cr}$. Here $z_1 = z/l_{sf}$, where l_{sf} is the self-focusing length

The dimensions, both transverse and longitudinal, of the nonlinear foci and their relative locations can depend on additional physical processes in the nonlinear media (nonlinear absorption, ionization, etc.) which can change the form of the refractive index nonlinearity and limit the growth of the light energy density in the nonlinear focal regions. However, theoretical investigations of many of these effects have shown (see review paper [2]) that the multi-focus structure of a self-focused beam turns out to be universal, i.e., it can be observed in various physical conditions. For illustration, Fig. 5.2 shows the on-axis and

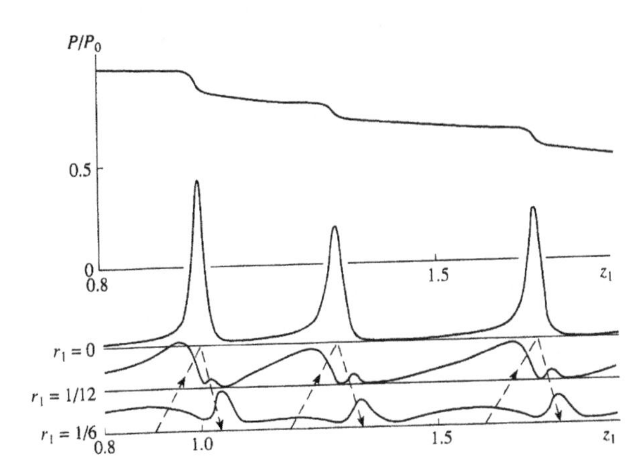

Fig. 5.2 Computation results for on-axis ($r_1 = 0$) and near-axis ($r_1 = 1/12$ and $r_1 = 1/6$) distributions of the beam intensity along the propagation z-direction in a Kerr-type nonlinear medium with three-photon absorption (on bottom) and for the evolution of the beam power P along this direction (on top). Here $r_1 = r/\bar{a}_0$, where $\bar{a}_0$ is the radius of the incident Gaussian beam at the 1/e intensity level, and $z_1 = z/l_{sf}$ where l_{sf} is the self-focusing length. The three-photon absorption parameter μ_4 and the incident beam power P_0 were taken in this computational example equal to $\mu_4 = 0.05$ and $P_0 = 9\ P_{cr}$. The parameter μ_4 is connected with the three-photon absorption coefficient m_4 by the relationship $\mu_4 = m_4 E_0^2 N/n_2$ where n_2 is the Kerr coefficient of the nonlinear refractive index, E_0 is the electric field strength of the incident beam, and $N = 2(P_0/P_{cr})^{1/2}$

near-axis distributions of the light intensity along the longitudinal coordinate $z_1 = z/l_{sf}$ for three-photon absorption in a medium with some definite value of the parameter $\Delta = (P/P^{(0)})^{1/2} - 1$, where $P^{(0)} = cn_0/8n_2k^2$ and l_{sf} is a self-focusing length.

The location $\xi_{\Phi m}$ of the nonlinear foci on the z-axis as a function of the beam parameters can be approximated by the following analytical expression [2]:

$$\xi_{\Phi m} = \frac{\chi_m}{N_m} \frac{k\bar{a}_0^2}{\sqrt{P_0/P_{cr}^{(m)}} - 1} \tag{5.3}$$

where m denotes the order of the nonlinear foci, $P_{cr}^{(m)}$ is the critical power for m-focus, χ_m and N_m are numerical parameters, and $\bar{a}_0$ is the radius of the

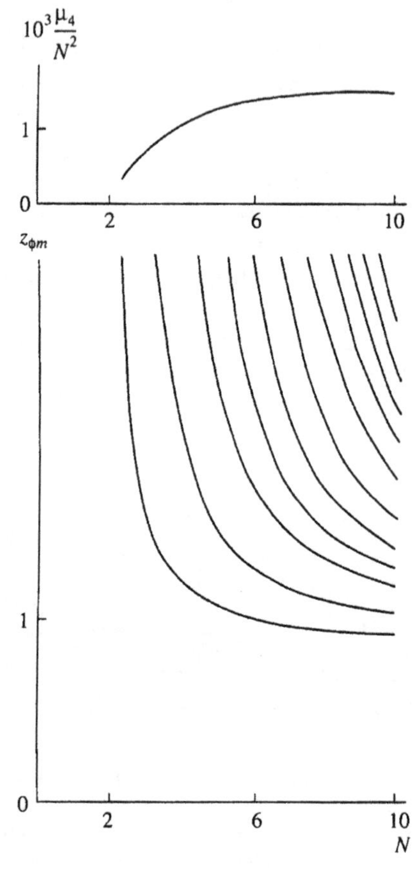

Fig. 5.3 The location on the z-axis of the nonlinear foci in a multi-focus structure as a function of the incident power P_0 (parameter $N = 2\,(P_0/P_{cr})^{1/2}$) for a Gaussian beam propagating in the Kerr-type nonlinear medium with three-photon absorption characterized by the parameter μ_4 defined in the previous figure caption. Here $z_{\Phi m} = \xi_{\Phi m}/l_x, l_x = \bar{a}_0/\sqrt{n_2 E_0^2}$. [See expression (5.3) in the text for details.]

incident beam. Numerical calculations [2] have shown that for rather weak three-photon absorption $P_{cr}{}^{m} \cong mP_{cr}{}^{(1)}$.

Figure 5.3 shows the location on the z-axis of the nonlinear foci of the multi-focus structure as a function of the power excess parameter $N = \Delta + 1$ for the same (three-photon absorption) mechanism limiting the light energy density in the nonlinear focal regions. It is obvious that for pulsed beams with a smooth variation of the power with time, the location of the nonlinear foci should change during the pulse in accordance with relationship (5.3) in which $\xi_{\Phi m}$ and P_0 are now time-dependent functions, that is, $\xi_{\Phi m} = \xi_{\Phi m}(t)$ and $P_0 = P_0(t)$. From this qualitative consideration it follows that the moving nonlinear foci model can be realized also for nonstationary light beams. The total number of foci in this MNLF structure at time t is determined by the condition $P_0(t) > P_{cr}{}^{(m)}$.

Figure 5.4 illustrates the formation of a moving multi-focus structure in a nonlinear medium with the fast-response three-photon absorption. The location of the nonlinear foci (solid curves) and the light pulse shape $N(t,z)$ (dashed curve) are plotted in this figure in $z/k\bar{a}_0^2$, N coordinates. The pulse shape function $N(t)$ is given by [2]:

$$N(t,z) = \frac{1}{E_{cr}}\left|E_0\left(t - \frac{k\bar{a}_0^2}{\nu}\frac{z}{k\bar{a}_0^2}\right)\right| \tag{5.4}$$

where E_{cr} is the critical field strength corresponding to the critical power for self-focusing $P_{cr} = cn_0/2n_2k^2$ and v is the speed of light in the medium. The values of $z/k\bar{a}_0^2$ corresponding to the crossing of the solid curves and dashed curves determine the locations (in the corresponding scale) of the nonlinear foci on the beam axis at time t.

For further characteristics of the moving foci model, we refer to the review paper [2] where some important features of the model (in particular, the structure and speed of the moving foci for ultra-short laser pulses) are considered in detail.

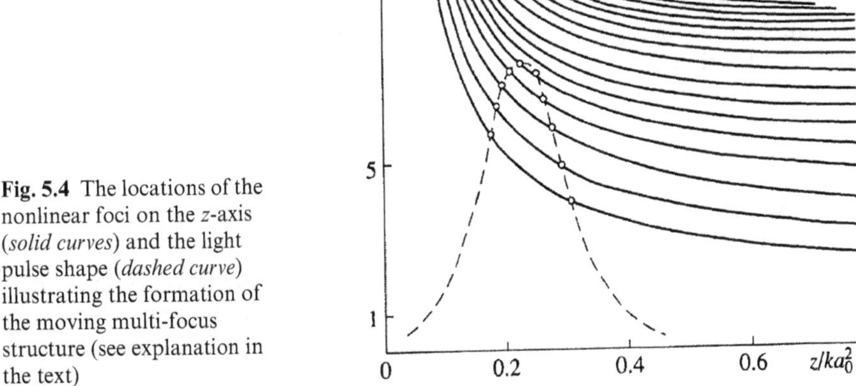

Fig. 5.4 The locations of the nonlinear foci on the z-axis (*solid curves*) and the light pulse shape (*dashed curve*) illustrating the formation of the moving multi-focus structure (see explanation in the text)

5.3 Self-Focusing of Super-Gaussian Laser Beams

The multi-focus structure and moving nonlinear foci models reviewed in Section 2 above deal with the self-focusing of Gaussian beams. There is considerable interest in how the spatial profile of the incident beam affects the details of these models. This question was investigated by Danileiko et al. [25] for Gaussian and super-Gaussian spatial profiles of the incident laser beams:

$$|E_0|^2 = \exp\{-\alpha(r/a_0)^\beta\}, \tag{5.5}$$

where the parameter β takes on the values 2, 4, 6, 8, and 10, and the coefficient α is chosen from the condition of normalizing the total energy integral $P(r) = ½$ at $r \to \infty$ (for Gaussian beams $\alpha = 1$, and $\beta = 2$). Based on a numerical solution of the parabolic wave equation (5.1), the influence of the intensity spatial profile on the self-focusing threshold and on the self-focusing length was investigated. The dependence of the self-focusing threshold on the divergence of the incident beam was also investigated.

The main results of this study are summarized as follows.

- The qualitative character of self-focusing is unchanged for all intensity spatial profiles investigated, that is, a multi-focus structure of the intensity distribution occurs along the propagation z-axis for $P_0 > P_{cr}$.
- The self-focusing threshold P_{th} varies slightly with the spatial profile parameter of the incident beam: the coefficient S in the expression $P_{th} = S\lambda^2 c/8\pi^2 n_2$ equals 0.924, 0.988, and 1.105 for Gaussian ($\beta = 2$) and super-Gaussian ($\beta = 4$ and $\beta = 8$) beams, respectively (here λ is the wave length of the incident beam).
- The power trapped into the first nonlinear focus does not depend on the spatial profile of the incident beam and remains equal to $P_{cr} = 0.932\ \lambda^2 c/8\pi^2 n_2$.
- The spatial intensity profile of the incident beam significantly affects the distance between the nonlinear foci in the multi-focus structure formed along the propagation z-axis for $P_0 > P_{cr}$; it becomes shorter for the super-Gaussian profiles than that for the Gaussian profile.
- The initial divergence of the incident beams does not appreciably increase the self-focusing threshold.

5.4 Experimental Verification of MFS-MNLF Models

The multi-focus structure and moving nonlinear foci models have been reliably verified in experiments [9–11] where self-focusing of nanosecond laser pulses in condensed nonlinear media (liquids and solids) were studied using different approaches and detection methods.

Loy and Shen [9] investigated the self-focusing of a single-mode ruby laser beam of 8 ns pulse duration in toluene and in CS_2 by observing the spatial and temporal evolution of the beam inside and at the end of the nonlinear liquid cell. The evolution of the beam in this study was analyzed for a single (first) non-linear focus for various incident beam powers including values as high as 18 P_{cr}, where P_{cr} for toluene was taken to be 30 kW. From this analysis the authors concluded that the observed features are definitely consistent with the moving nonlinear focus model (this conclusion is discussed in detail by Shen in this book).

Korobkin et al. [10] studied self-focusing of a single-mode ruby laser beam of 15 ns pulse duration in nitrobenzene and CS_2, applying a high-speed time-resolved detection technique to observe the temporal evolution of the beam intensity along the propagation direction. At high incident beam powers exceeding the critical power, the authors observed a moving multi-focus structure of the self-focused beam in liquids.

In both works [9, 10] just described above, bell-shaped temporal pulses of Q-switched lasers were used in the experiments. They thus correspond to non-stationary conditions of self-focusing, and the results obtained were interpreted in terms of the moving nonlinear foci model. It is important to recall here that many researches interpreted light filaments, observed under similar conditions, in terms of an alternative model, the *nonlinear wave guide model* (see the review paper [2] for references and discussion of such the interpretation).

In this context, the approach proposed and realized by Lipatov et al. [11] for investigating the dependence of self-focusing characteristics on the temporal shape of the incident beam laser pulse seems to be a very direct and single meaning method to distinguish between the two interpretations mentioned above and hence to elucidate the preferred model of self-focusing.

Three temporal pulse shapes, rectangular (flat-top), bell-like, and saw-tooth-like, were used in these experiments [11] to investigate self-focusing of a ruby laser beam in the nonlinear optical glass TF-105. The rectangular and saw-tooth-like laser pulses were obtained from a single-mode Q-switched laser pulse having initially a bell-like shape, by passing it through an electro-optic modulator controlled with electric voltage pulses of the corresponding temporal shape. The laser beam was focused inside the glass sample under study. The shapes of the incident and transmitted pulses were recorded through use an oscilloscope. Also, light scattered from the interaction region within the sample was photographed for various incident beam powers P_i .

Laser-induced damage was observed in the sample for $P_i > P_d$, where P_d is the damage threshold power. The damage morphology observed for various temporal pulse shapes was quite different: it was a point-like structure in the case of the rectangular (flat-top) shape and a filamentary structure in the cases of the saw-tooth-like and bell-like pulses. The number of point-like damage spots, their relative locations along the beam axis, and their dimensions depended on the excess ratio P_i/P_d. In particular, the number of the damage points (spots) increased from one to three when P_i/P_d varied from 1 to 6

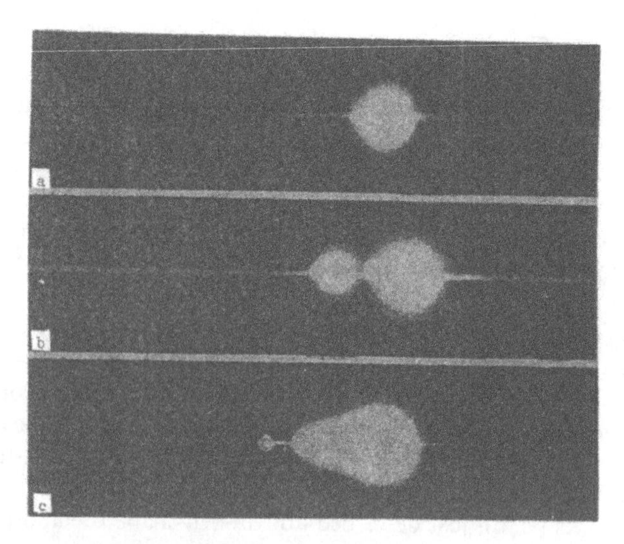

Fig. 5.5 Pattern of scattering of ruby laser radiation (rectangular pulse) in TF-105 glass for various incident powers P_i : (**a**) $P_i \approx P_d$, (**b**) $P_i = 3P_d$, and (**c**) $P_d \approx 6P_d$, where P_d is the laser-induced damage threshold. The distance between scattering center in part (c) is $\approx$5 mm. Bright scattering halos are seen around the damage points, as well as a weak trace of the ordinary scattering from the glass. The laser beam propagates from left to right

(see Fig. 5.5). Also, the lengths of the filaments, observed in the cases of the saw-tooth like and bell-shaped pulses, increased with increasing P_i.

These experimental results have been interpreted [11] on the basis of the MFS-MNLF models. According to this interpretation, the laser-induced damage morphology observed in the case of the rectangular flat-top laser pulses, is due to the steady-state self-focusing of the laser beam in glass, whereas the filamentary damage observed in the case of the saw-tooth-like and bell-shaped pulses is the consequence of the movement of the nonlinear focal points. It was thus concluded [11] that the experimental results on the observation of the laser-induced damage morphology with various temporal shapes of the laser pulses qualitatively well-confirm the MFS-MNLF models of self-focusing

5.5 Comments on Self-focusing of Femtosecond Laser Pulses in Air and Possible Future Directions of This Field

Since the first observation of self-focusing of femtosecond laser pulses in air [12], many papers have been published on this subject (the observation of self-focusing effects in air were reported earlier for nanosecond and picosecond laser pulses [26, 27]). Based on experimental and simulation results, several models have been formulated [12–24] to explain the observed features of femtosecond self-focusing.

In this section we make some comments regarding the results of references [12–24] and suggest some possible future directions for research in this field.

Experimental results published to date on the self-focusing of femtosecond laser pulses in air, are not informative enough for a complete understanding of the mechanisms and processes involved in laser beam filamentation and the related laser-induced plasma formation In particular, the observations of the filaments were temporally and spatially integrated. Also, measurements of the light energy distribution along the propagation direction using the diaphragms with apertures (0.5–1 mm) much larger than the observed filament diameter ($\sim$80 μ) seem to be too rough for resolving the possible fine structure of the filament.

Theoretical studies are based on the numerical solution of the nonlinear wave equation (NLWE) with the Kerr-type term for the nonlinear refractive index of air, the terms attributed for the multi-photon absorption, and laser-induced plasma formation. There are, however, some important differences in formulations of different authors related to the account of the group velocity dispersion and to the description of the two-component Kerr-nonlinearity model. In particular, we note an incomplete description of the group velocity dispersion, and the inadequate interpretation of the delayed response term in the Kerr-type nonlinear refractive index. Indeed, the refractive index dispersion of the plasma component of the medium (neutral + laser-ionized air) was not taken into account, whereas its contribution becomes comparable with that of the neutral air term at plasma densities (electron concentrations) of the order $\approx 2\times10^{16}$ cm^{-3} estimated from the experiments. In the two-component Kerr-nonlinearity model the delayed-response term was assigned to rotational stimulated Raman scattering (SRS). We note, however, that this term does not relate to SRS but, in fact, it is only a phenomenological model with physically uncertain relaxation times, τ_1 and τ_2.

The terminology used to interpret femtosecond self-focusing processes sometimes differs from the terminology in the earlier literature, although the underlying physical concepts are similar. For example, "*self-guided light strings*" seems to describe a similar picture as nonlinear moving foci; the "*nonlinear robust mode*" seems to refer to a portion of the beam flowing through a nonlinear focus; and the "*dynamic spatial replenishment*" seems to correspond to the process of sequential formation of the nonlinear foci on the beam axis in the *multi-focus structure* model described in Section 5.2 above.

We believe that the observed characteristics of femtosecond self-focusing can be explained on the basis of the multi-focus structure: moving nonlinear foci (MFS-MNLF) models that are well-developed and well-grounded for "long" (ns-ps) pulse self-focusing (see Sections 5.2–5.4). But further studies are required for a more complete understanding of the mechanisms of femtosecond self-focusing in air.

Taking into account above comments, the following experimental and simulation works would be desirable:

- The spatial distribution along the propagation direction of the plasma density generated by the moving nonlinear foci should be measured in the *single*-pulse propagation regime (observation of the filaments in a *multi*-pulse regime makes the interpretation of the effect somewhat ambiguous).
- A numerical solution corresponding to the experimental conditions of references [12, 14] should be obtained for the fast-response Kerr-nonlinearity (one-component) model. Also, the plasma density distribution along the propagation direction generated by the moving foci should be obtained in this simulation.
- In the two-component Kerr nonlinearity model, the fitting parameters including the multi-photon absorption coefficient (multi-photon ionization rate) needs to be chosen to be consistent with experimentally observed value of the moving focus diameter (filament size) in the stabilization range (if such a range is, indeed, realized). The plasma density distribution along the propagation direction has to be obtained in this simulation, similarly to that of the one-component model indicated in the previous item.

A comparison of such theoretical and experimental results would make it possible to draw more reliable conclusions on the mechanisms and the adequate model for the femtosecond self-focusing in air.

Acknowledgments Regrettably, V.N. Lugovoi, who made fundamental contributions to the understanding of the phenomenon of self-focusing, passed away in 2005. His name is included as a co-author of the present work because he contributed to the early drafts of the chapter and because much of the present chapter is based on the publications of Lugovoi and Manenkov.

References

1. S.A. Akhmanov, A.P. Sukhorukov, R.V. Khokhlov: Self-focusing and diffraction of light in a nonlinear medium. *Uspekhi Fiz. Nauk*, **93**, 19–70 (1967) [*Sov. Phys. Uspekhi*, 1968, **10**(9), 609–636].
2. V.N. Lugovoi, A.M. Prokhorov: Theory of the propagation of high-power laser radiation in a nonlinear medium . *Uspekhi Fiz Nauk*, **111,** (2), 203–247 (1973) [*Sov. Phys. Uspekhi*, 1974, **16**(5), 658–679].
3. Y.R. Shen, Self-focusing: experimental, *Progr. Quant. Electr.*, **4**, 1–34 (1975).
4. J.H. Marburger: Self-focusing: theory. *Progr. Quant. Electr.*, **4**, 35–110 (1975)
5. Y.R. Shen, *The Principles of Nonlinear Optics*, Chapter 17: Self-focusing, pp. 303–333, John Wiley & Sons, New York, 1984.
6. R.W. Boyd, *Nonlinear Optics*, Chapter 6: Processes resulting from the intensity-dependent refractive index, pp. 241–286, Academic Press, San Diego (1992).
7. A.L. Dyshko, V.N. Lugovoi, A.M. Prokhorov: Self-focusing of intense light beams. *Pis'ma Zhurnal Ekper Teor Fiz* **6**, 655 (1967) [*Sov Phys JETP Lett* **6**, 146 (1967)].
8. V.N. Lugovoi, A.M. Prokhorov: Possible explanation of small-scale filaments of self-focusing. *Pis'ma Zhurn Eksper Teor Fiz* **7**, 153 (1968) [*Sov Phys JETP Lett* **7**, 117 (1968)].
9. M.M.T. Loy, Y.R. Shen: Small-scale filaments in liquids and tracks of moving foci. *Phys. Rev. Lett.*, **22**, 994 (1969).

10. V.V. Korobkin, A.M. Prokhorov, R.V. Serov et al.: Schelev: Self-focusing filaments as a result of moving focal points . *Pis'ma Zhurnal Eksper. Teor. Fiz.*, **11**, 153 (1970) [*Sov Phys JETP Lett* **11**, 94 (1970)].
11. N.I. Lipatov, A.A. Manenkov, A.M. Prokhorov: Standing pattern of self-focusing points of laser radiation in glass. *Pis'ma Zhurnal Eksper Teop Fiz* **11**, 444 (1970) [*Sov Phys JETP Lett* **11**, 300 (1970)].
12. A. Braun, G. Korn, N. Lin et al.: Self-channeling of high-power femtosecond laser pulses in air. *Opt. Lett.*, **20**, 73, (1995).
13. E.T.J. Nibbering, P.F. Curley, G. Grillon et al.: Conical emission from self-guided femtosecond pulses in air. *Opt. Lett.*, **21**, 62, (1996).
14. A. Brodeur, C.Y. Chien, F.A. Ilkov, S.L. Chin, O.G. Kosareva, V.P. Kandidov: Moving focus in the propagation of ultrashort laser pulses in air. *Opt. Lett.*, **22**, 304 (1997).
15. O.G. Kosareva, V.P. Kandidov, A. Brodeur, S.L. Chin: From filamentation in condensed media to filamentation in gases. *J. Nonlinear Optical Phys. Mater.*, **6**, 485, (1997).
16. M. Mlejnek, E.M. Write, J.V. Moloney : Dynamic spatial replenishment of femtosecond pulses propagating in air. *Opt. Lett.*, **23**, 382 (1998).
17. M. Mlejnek, M. Kolesik, J.V. Moloney, E.M. Wright: Optically turbulent femtosecond light guide in air. *Phys. Rev. Lett.*, **83**, 2938 (1999).
18. J.V. Moloney, M. Kolesik, M. Mlejnek, E.M. Wright: Femtosecond self-guided atmospheric light strings. *Chaos*, **10**, 559 (2000).
19. V.P. Kandidov, O.G. Kosareva, E.I. Mozhaev, M.H. Tamarov: Femtosecond nonlinear optics of atmosphere. *Opt. Atmospher. Okean*, **13**, 429 (2000) [*Atmos. Oceanic Opt.*, **13**, 394 (2000)]
20. I.S. Golubtsov, V.P. Kandidov, O.G. Kosareva, et al.: Conical emission of powerful femtosecond laser pulse in atmosphere. *Opt. Atmos. Okean*, **14**, 335 (2001). [*Atmos. Oceanic Opt.*, **14**, 303 (2001)].
21. S.L. Chin, A. Talebpour, J. Yang, S. Petit, V.P. Kandidov, O.G. Kosareva, M.P. Tamarov: Filamentation of femtosecond laser pulses in turbulent air. *Appl. Phys. B*, **74**, 67 (2002).
22. K.Yu. Andrianov, V.P. Kandidov, O.G. Kosareva, S.L. Chin, A. Talebpour, S. Petit, W. Liu, A. Iwasaki, M.C. Nadeau: Influence of beam quality on filamentation of high-power femtosecond laser pulse in air. *Izvestia Akad. Nauk, Ser. Fiz.*, **66**, 1091 (2002). [*Bull. Rus. Acad. Sci., Phys.*, **66**, (8), 1192 (2002)].
23. S.L. Chin, S. Petit, W. Liu, A. Iwasaki, M.-C. Nadeau, V.P. Kandidov, O.G. Kosareva, K.Yu. Andrianov: Interference of transverse rings in multi-filamentation of powerful femtosecond laser pulses in air. *Opt. Commun.*, **210**, 329–341 (2002).
24. V.P. Kandidov, O.G. Kosareva, A.A. Koltun: Nonliner-optical transformation of powerful femtosecond laser pulses in air. *Quant. Electron.*, **33**, 69 (2003).
25. Yu.K. Danileiko, T.P. Lebedeva, A.A. Manenkov, A.M. Prokhorov: Self-focusing of laser beams at various spatial profiles of incident radiation. *Zhurnal Ekper. Teor. Fiz.*, **80**, (2), 487–495 (1981) [*Sov. Phys. JETP*, **53**(2), 247–252, 1981)].
26. V.V. Korobkin, A.J. Alcock, Self-focusing effects associated with laser-induced air breakdown, *Phys. Rev. Lett.*, **21**, 1433–1436 (1968).
27. N.G. Basov, P.G. Kryukov, Yu.V. Senatskii, S.V. Chekalin, Production of powerful ultra-short light pulses in a neodimium glass laser, *Zhurnal Eksper. Teor. Fiz.*, **57**, 1175–1183 (1969) [*Sov. Phys. JETP*, **30**, 641–645 (1970)].

Chapter 6
Small-Scale Self-focusing

Anthony J. Campillo

Abstract At high enough power, a beam propagating in a positive n_2 material will spontaneously break up into multiple filaments as a result of a transverse modulational instability. This chapter briefly surveys the essential physics, presents a linearized theory and summarizes experimental examples illustrating the dependence between optimal spatial frequency, gain coefficient, self-focusing length, beam intensity, and beam shape. A number of applications are discussed as well as the impact of this field on other scientific disciplines.

6.1 Introduction

An intriguing optical phenomenon, with parallels in other areas of physics (hydrodynamics, plasma waves, etc.), is the spontaneous multifilament breakup of a high-power beam propagating through a positive n_2 medium (see Fig. 6.1). This form of self-focusing differs from total beam collapse [1–7] in that it does not rely on the shape of the beam to induce a nonlinear lens and typically occurs at powers greater than 50 P_{cr}, where P_{cr} is the critical power. It is now generally accepted that such small-scale self-focusing [8] is described in terms of an instability model proposed by V.I. Bespalov and V.I. Talanov [9] and later extended by others [10–17]. Alternatively, the growth of the instability can be visualized in terms of stimulated four-wave mixing [18].

The instability model predicts that even a uniform plane wave is unstable to amplitude and phase perturbations, and consequently will break up into multiple beamlets. In effect, a beam amplitude spatial ripple develops and grows exponentially with distance until it becomes comparable in strength to that of the initial plane-wave field, whereupon it quickly collapses to form multiple filaments.

A.J. Campillo (✉)
Naval Research Laboratory, Washington, DC 20375, USA
e-mail: campillo@nrl.navy.mil

R.W. Boyd et al. (eds.), *Self-focusing: Past and Present*,
Topics in Applied Physics 114, DOI 10.1007/978-0-387-34727-1_6,

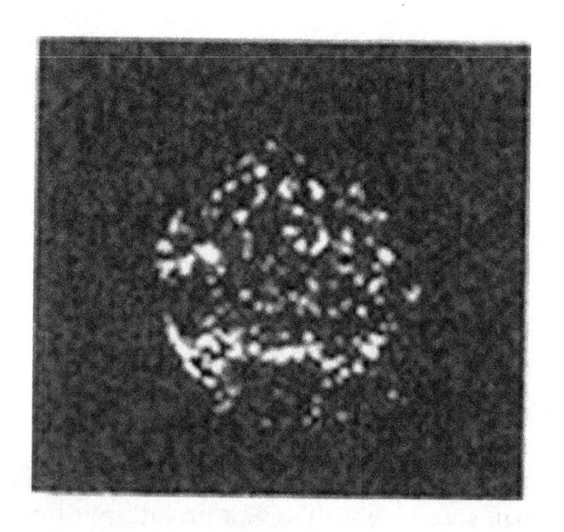

Fig. 6.1 High-power beam breakup and multi-filament formation at the exit of a CS_2 cell. After Abbi and Mahr [21]

In a smooth beam, the breakup arising from such ripples should form a transverse periodic array of filaments. However, early breakup patterns invariably displayed a random character (as shown in Fig. 6.1) due to the poor beam quality and multimode behavior of solid-state lasers prior to the 1970s. This led to early confusion as to the operative mechanism of small-scale self-focusing; that is, whether a Bespalov–Talanov modulational instability dominated or whether observations could be explained using "whole beam" self-focusing theory applied to transverse hot spots and temporal spikes [19].

Eventually, Chilingarian [20] and later Abbi and Mahr [21] demonstrated that the measured self-focus length, z_f, varied with the inverse of the intensity expected of an instability mechanism and differing from whole beam self-focusing which has a characteristic inverse $I^{1/2}$ dependency [1]. Also, Campillo et al. [22, 23] directly observed the transverse periodic breakup of the beams, i.e., the telltale signature of a spatial instability mechanism. They also verified that the thresholds, intensity dependences of self-focusing length, and optimum spatial frequencies of the perturbations all agreed with the Bespalov–Talanov theory. Bliss [24], using amplitude-modulated beams, measured the growth of the instability in glass samples. They showed that the measured gain coefficient as a function of the **k** vector agreed well with a full linearized analysis.

6.2 Modulational Instability

A seminal work related to small-scale self-focusing is the theoretical paper on light-by-light scattering by Chaio et al. [18]. The growth of a spatial instability is a special case of this more general 4-wave mixing process. Briefly, the geometry outlined in Fig. 6.2 is considered. Two beams, a weak beam (I_1 in Fig. 6.2b) at θ to the z-axis and a strong beam (I_o) propagating along z, mix and generate a second weak beam at $-\theta$ (I_2 in Fig. 6.2b). When θ satisfies a phase matching

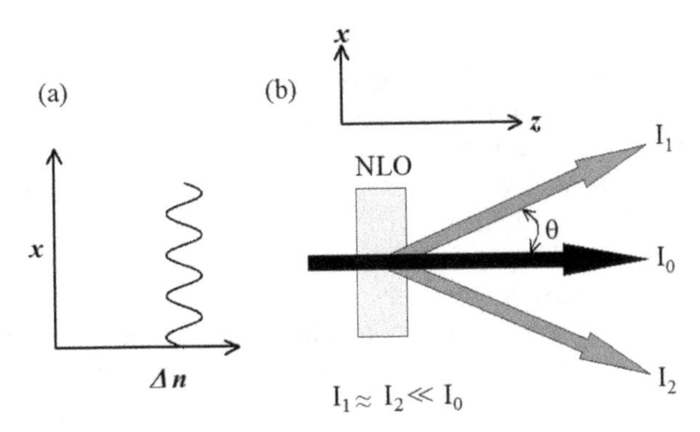

Fig. 6.2 (**a**) Noise superimposed on the laser amplitude induces a transverse spatial ripple on the index of refraction; (**b**) the resultant phase grating couples light from the intense forward beam to the two weak side beams at angle θ. When ripples are in the appropriate spatial frequency range (see text), the side beams grow at the expense of the forward strong beam

condition ($\theta \sim \theta_{\text{opt}}$) and the strong beam's intensity exceeds a threshold dependent on n_2, both weak beams subsequently grow exponentially at the expense of the strong beam. The resulting interference pattern caused by mixing of all three beams appears as an amplitude/phase ripple (or instability) on the strong beam, which grows with z (see Fig. 6.3). Carman et al. [25] verified many features of the theory by performing the two-beam mixing experiment to generate the third beam as in Fig. 6.2b.

In the case of small-scale self-focusing, the two weak side beams arise, in effect, from noise. Any light beam, no matter how perfect, will have at least

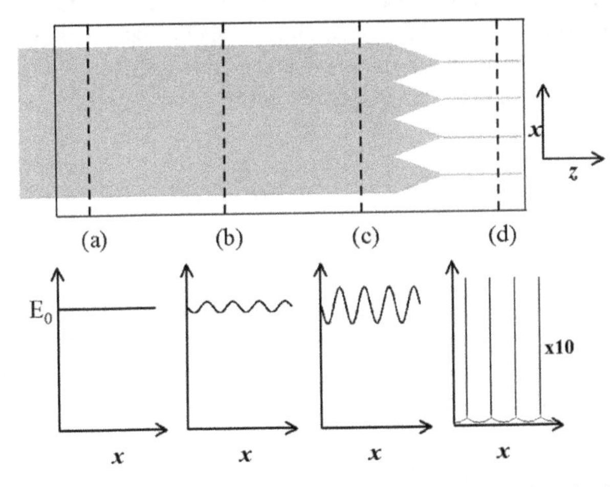

Fig. 6.3 Evolution of a one-dimensional transverse spatial intensity ripple as the beam propagates in the z direction. The beam intensity near the front (**a**) of the nonlinear medium shows the incident uniform-intensity beam. Not visible is a small amount of superimposed noise. As the beam propagates, a ripple of optimum spatial frequency begins to grow at the expense of the background field. Ultimately, catastrophic self-focusing occurs, at right (**d**)

some degree of transverse spatial amplitude and phase variation due to refractive index inhomogenieties, absorbing centers (dust, dirt) as well as diffraction fringes caused by truncation of the beam as it passes through optical elements [26]. Such perturbations can be Fourier decomposed into a sum of sinusoidal ripples of varying spatial wavelengths, amplitudes and phases. Alternatively, each perturbation may be decomposed into sets of three beams at various θ as in Fig. 6.2, θ being dependent on the spatial frequency of any one ripple. While the noise has two transverse dimensions, Figs. 6.2 and 6.3 illustrate the physics occurring in one transverse dimension. The conditions for the growth of the instability are quantified in the next section.

6.2.1 Linearized Theory

We closely follow Siegman's linearized analysis [27], which nicely summarizes previous theoretical treatments of the breakup of beams in nonlinear media [9–18] as well as in high gain media such as laser diodes [28–32] (discussed in Section 6.4). We consider an optical medium with a positive nonlinear refractive index n_2 and a saturable gain/loss. Using Siegman's notation, the complex propagation constant $\tilde{k}$ is approximated:

$$\tilde{k} = [k_0 - k_2 I] - j[g_o - g_2 I] \tag{6.1}$$

where $I = I(x, y, z)$ is the intensity of light in the medium, g_o is the low-intensity gain, $g_2 I$ the saturable gain term, $k_o = 2\pi n_o/\lambda$ where n_o is the real part of the index of refraction, and $k_2 = 2\pi n_2/\lambda$. Here, n_2 is defined through the relation $n = n_o + n_2 I$. The nonlinear paraxial wave equation for the complex wave amplitude $\tilde{u}(x, y, z)$of a beam propagating in the z direction may be expressed as

$$\left[\nabla^2_{xy} - 2jk_0\frac{\partial}{\partial z} + 2jk_0 g + 2k_0(k_2 - jg_2)|\tilde{u}|^2\right]\tilde{u}(x, y, z) = 0 \tag{6.2}$$

The Fig. 6.2b geometry is approximated by plane waves and limiting analysis to one transverse direction, x. The intense beam propagates in the z direction and has an amplitude u_o at $z = 0$. The two weak side beams travel at small angles $\pm\,\theta$ and have amplitudes $\tilde{c}_1(z)$ and $\tilde{c}_2(z)$. The complex beam amplitude can then be expressed as

$$\tilde{u}(x, z) = u_o e^{g(z) - j\phi(z)} \times [1 + \tilde{c}_1(z)e^{j\kappa x} + \tilde{c}_2(z)e^{-j\kappa x}] \tag{6.3}$$

where $\kappa = k_o \sin\theta$ is a transverse k vector, $g(z)$ is a measure of the gain as the beam propagates, and $\phi\ (z)$ is the phase shift in the beam due to the optical Kerr effect.

To simplify the analysis, we express the transverse perturbations in both cosine and sine form, with complex amplitudes of $\tilde{c}(z)$ and $\tilde{s}(z)$, respectively:

$$\tilde{u}(x,z) = u_o\, e^{g(z)-j\phi(z)} \times [1 + \tilde{c}(z)\cos\kappa x + \tilde{s}(z)\sin\kappa x] \qquad (6.4)$$

Because the sine and cosine terms differ only by a lateral shift, we need only consider one of them (the cosine term). Substituting Eq. (6.4) into the nonlinear Eq. (6.2) and separating like terms yields the following three equations

$$\frac{dg(z)}{dz} = g_o - g_2 u_o^2 \exp[2g(z)] \qquad (6.5)$$

$$\frac{d\phi(z)}{dz} = k_2 u_o^2 \exp[2g(z)] \qquad (6.6)$$

$$\frac{d\tilde{c}(z)}{dz} = j\frac{\kappa^2}{2k_o}\tilde{c}(z) - j(k_2 - jg_2)u_o^2 e^{2g(z)}[\tilde{c}(z) + \tilde{c}*(z)] \qquad (6.7)$$

Equation (6.5) describes the amplitude growth taking into account the saturated gain. Equation (6.6) simply describes the induced phase shift produced in the Kerr medium by the intense wave. Equation (6.7), which describes the first-order growth of the cosine coefficient, is more important for this analysis. If we separate $\tilde{c}$ (z) into a real, $c_r\,(z)$ and imaginary part, $c_i\,(z)$, and write the expression for the total intensity

$$I(x,z) = |\tilde{u}(x,z)|^2 \cong u_o^2 e^{2g(z)}[1 + 2c_r(z)\cos\kappa x] \qquad (6.8)$$

it becomes clear that the real part, $c_r\,(z)$, describes the periodic amplitude ripple on the beam, while the imaginary part, $c_i\,(z)$, represents a periodic phase ripple on the beam wavefront. Separating the growth equation into real and imaginary parts yields:

$$\frac{dc_r}{dz} = -\left(\frac{\kappa^2}{2k_o}\right)c_i - 2g_2 u_o^2 c_r \qquad (6.9)$$

and

$$\frac{dc_i}{dz} = \left(\frac{\kappa^2}{2k_o} - 2k_2 u_o^2\right) \qquad (6.10)$$

Equations (6.9) and (6.10) indicate that in the absence of a nonlinearity ($k_2 = g_2 = 0$), the coefficients $c_r\,(z)$ and $c_i\,(z)$ both oscillate with distance as a result of the well-known Talbot phase shift factor, $\kappa^2/2\,k_o$; i.e., the ripple pattern oscillates periodically between pure amplitude and pure phase. Keeping $g_2 = 0$ but giving k_2 a positive value, yields solutions to Eqs. (6.9) and (6.10) where both $c_r(z)$ and $c_i(z)$ grow exponentially with distance with a growth coefficient, γ, given by

$$\gamma^2 = \left(\frac{\kappa^2}{2k_o}\right)\left[2k_2u_o^2 - \left(\frac{\kappa^2}{2k_o}\right)\right] \tag{6.11}$$

The growth rate γ has a maximum value of

$$\gamma_{\max} = k_2u_o^2 \tag{6.12}$$

at

$$\kappa_{\max} = \sqrt{2k_o\,k_2\,u_o^2} \tag{6.13}$$

The above equations show that the Talbot effect plays an important role in the exponential buildup of the ripples by periodically converting phase ripples into amplitude ripples, and vice versa [27]. The self-focusing length, z_f, can be determined by using Eq. (6.12) to determine the z needed for a ripple of $\tilde{c}$ (0) to grow to a value comparable with u_o .

$$z_f \cong \frac{1}{\gamma_{\max}}\log_e\left[\frac{u_o(0)}{\tilde{c}(0)}\right] \cong \frac{1}{k_2u_o^2}\log_e\left[\frac{u_o(0)}{\tilde{c}(0)}\right] \tag{6.14}$$

As can be seen, the self-focusing length varies inversely with the intensity, which differs markedly from the whole-beam case [1]. Also, the self-focusing distance is relatively insensitive to the initial size of the ripples. Therefore, even small-amplitude noise components of $\kappa_{\max}$ will quickly overtake larger but slower-growing ripples and so dominate the final self-focusing pattern.

6.2.2 Experimental Confirmation of the Modulation Instability

Several early experimental studies [20–25] verified much of Bespalov and Talanov's theory. Campillo et al. [22] were the first to observe the periodicity typical of ripple instabilities. In their experiments, they gave the instability a head start by impressing a modulation in one spatial dimension.

Figure 6.4 shows the results of breakup of a beam with a radial ripple. This was obtained by using a circular aperture to spatially truncate a smooth Gaussian beam at the 1/e point and passing the beam through a CS_2 cell with a Fresnel number of 7. The breakup pattern of the beam was photographed in Raman shifted light generated in the focal spots. Beams with Fresnel number 7 have a sharp central peak surrounded by three high intensity rings. As threshold is exceeded for this pattern, the center peak and the intense outer ring self-focus

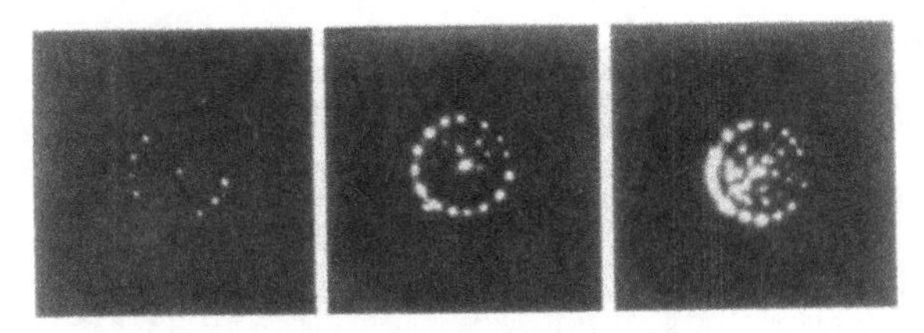

Fig. 6.4 Spatial periodicity of focal spots is evident in photograph showing beam breakup of an apertured beam at Fresnel number 7. Left to right shows progressively higher beam intensities. From Campillo et al. [22]

before the rest of the annular ring structure. As the power is increased further, these additional rings also break up.

The dominant and striking feature of these photographs is the regularity of the spacing of the focal spots circling the rings. Clearly, an azimuthal instability of optimum spatial frequency developed from noise, causing this periodic breakup. By measuring the powers in each ring, it was determined that each focal spot contained 25 kW or about four critical powers. To test that the effects observed were not due to the circular apertures, other diffraction patterns with noncircular symmetry were tested for self-focusing, e.g., straight edge and slit diffraction patterns. In all cases, regular periodic breakup of the beam was observed.

In another set of experiments by Campillo et al. [23], amplitude modulated a smooth Gaussian beam in one dimension by passing it through a glass wedge of small angle. This produces an intensity modulation of 16% of the form sin $(\kappa_x x)$. When the modulated beam passes through a CS_2 cell, the κ_y component grows from noise and freely adjusts itself so that the total κ vector is of optimum size for the available beam intensity as set by Eq. (6.13). Figure 6.5 shows the subsequent self-focusing pattern formed at the end of a CS_2 cell is a uniform periodic array. Each spot evolved from an area roughly 200 × 200 μm containing 3–4 critical powers. The κ vector of the instability mode may be determined from photos such as Fig. 6.5 simply by measuring the spacing between the focal

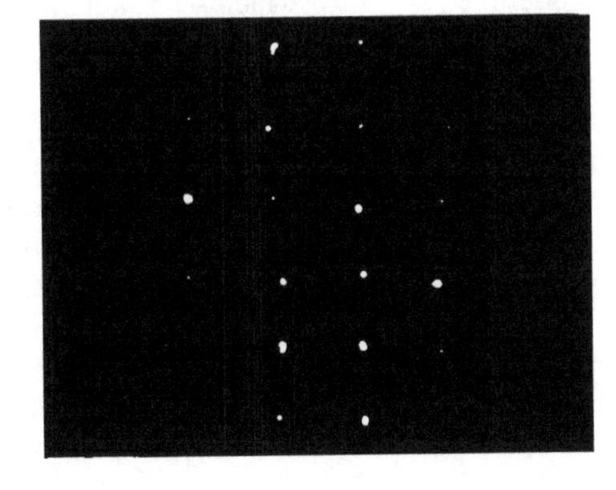

Fig. 6.5 Self-focusing pattern at the exit face of a CS_2 cell. This pattern originates because of the rapid growth of a spatial perturbation. From Campillo et al. [23]

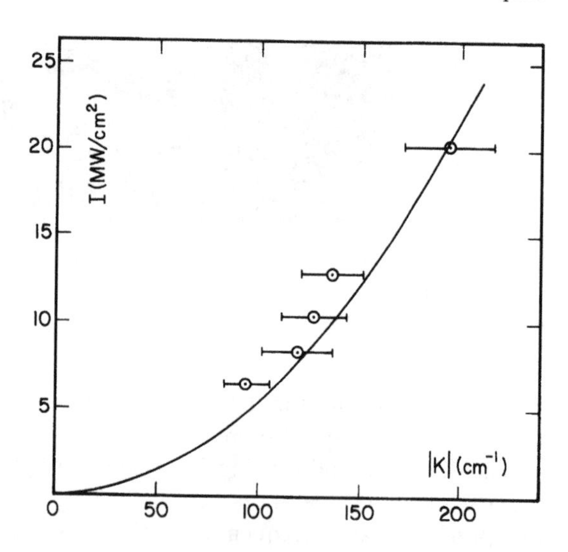

Fig. 6.6 Threshold intensity for self-focusing is plotted as a function of spatial wave vector, κ. Solid line is calculated from Eq. (6.13). The threshold intensity is dependent on κ^2, in agreement with instability theory. From Campillo et al. [23]

spots in the x and y directions. This was repeated for a series of experiments using different length CS_2 cells and varying the intensity of the beam to achieve self-focusing. The absolute magnitude of the κ vector which develops at the end of each CS_2 cell is plotted as a function of threshold intensity in Fig. 6.6. Error bars exist mainly because the intensity near the peak of the beam is not uniform but Gaussian, which leads to a κ-vector dependence with position (2–10% correction), and because of a nonlinear lens effect due to Gaussian profile (3% correction). Figure 6.6 shows that κ varies with the square root of the intensity as predicted by Eq. (6.13). The experimental data differs in absolute magnitude by about 18% from that calculated using the accepted value of n_2 for CS_2. Even so, the agreement is satisfying as the simple theory of Section 6.2 is only meant to apply to beams of uniform intensity. According to linear theory all modes with the same value of κ will grow at the same rate, and those with $\kappa_{\max}$ will grow the fastest. However, Suydam [11] has pointed out that after the perturbation becomes large, it starts to interact with itself. During the nonlinear phase its growth becomes more rapid than exponential, and the most compact modes (i.e., those for which $\kappa_x = \kappa_y$) grow fastest.

Figure 6.7 shows the threshold power for self-focusing plotted as a function of cell length. The self-focusing distance is seen to vary inversely with intensity in agreement with Eq. (6.14).

Bliss et al. [24] measured growth rates of instabilities of various spatial frequencies in glass. Before the beam reaches a sample of unpumped ED-2 laser glass, a shear plate impresses a one-dimensional sinusoidal intensity modulation; in effect a second weak beam at small angle is generated and propagates with the strong beam. Figure 6.8 shows the growth of the instability as the 5 GW/cm² beam propagates through 24 cm of glass. A comparison of measured gain coefficients to the simple linearized theory dashed line obtained from Eq. (6.11) is shown in Fig. 6.9. To compare the measured growth of the

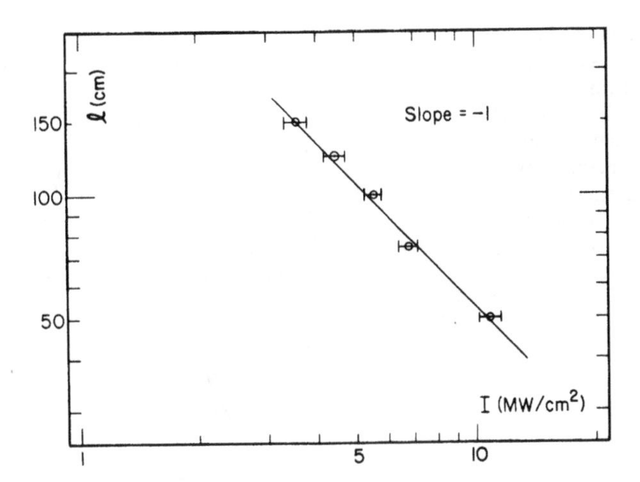

Fig. 6.7 Self-focusing length plotted as a function of average intensity. Slope of −1 in log plot is consistent with theory prediction of self-focusing length being inversely dependent on intensity. From Campillo et al. [23]

sinusoidal intensity modulation and the growth predicted by Eq. (6.11), both the input intensity and gain coefficient are time-averaged over the measured temporal profile of the input pulse. The solid curve is computed using a full linearized analysis [33] in which the gain is not approximated as exponential and in which proper note is taken of the initial relationship between the modulations of amplitude and phase generated by the shear plate. Agreement was good with this extended theoretical treatment.

During the 1970s, there was considerable interest in small-scale beam breakup because of its detrimental impact on the development of high-energy

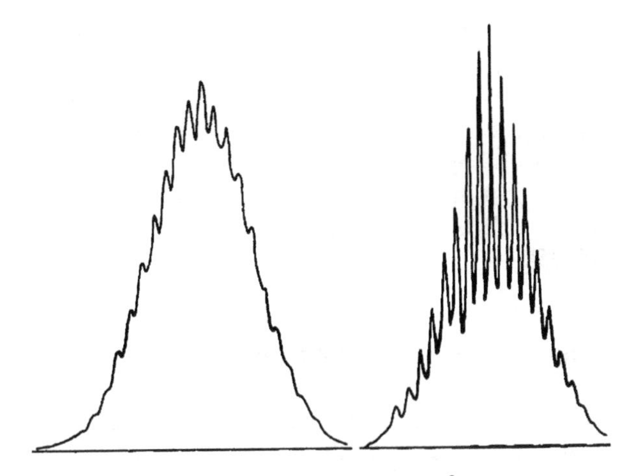

Fig. 6.8 Beam profiles showing fringe growth on 5 GW/cm^2 beam in glass. Left profile is at input face; right profile is output after propagating 24 cm in glass. Spacing between fringes is 2 mm. After Bliss et al. [24]

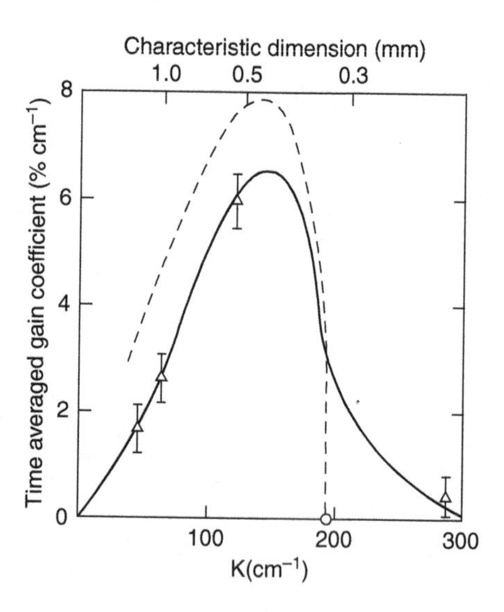

Fig. 6.9 Experimental and theoretical values of the time-averaged gain coefficient for the growth of intensity modulations on a pulse with a peak intensity of 5 GW/cm^2. Δ, experiment. Dashed line, exponential approximation to linearized theory. Solid line, full linearized theory. From Bliss et al. [24]

glass lasers for inertial confinement fusion [15]. The difficulty of determining the magnitude of the initial perturbation and the complications of dealing with multiple amplifying stages, air gaps, and lenses, led researchers to introduce a parameter called the breakup integral, B [15, 34, 35]. Here it is assumed that a ripple grows as $\exp(B)$, where B is given by

$$B = \frac{2\pi n_2}{\lambda} \int_0^L I \, dz \tag{6.15}$$

B represents the logarithmic gain of the fastest growing mode in a nonlinear medium of length L and was found to be a useful criterion for designing laser amplifiers against beam instability. Because of the differences in noise sources, beam divergence, and other important properties, upper limits on the breakup integral must be regarded as a somewhat system-dependent parameter. However, even perturbations of 10^{-5} will grow to a size comparable to I with a B of 11. In practice, efforts were made to keep B well below such values, e.g., typically 2–4 [15, 35].

6.3 Beam Shape, Polarization, and Pulse Duration Effects

Although a preponderance of experimental evidence exists to support the modulational instability model of multiple filament formation, the Bespalov–Talanov theory gives an incomplete picture because of its assumption of plane waves. It further assumes the existence of long-duration pulses and a Kerr-type nonlinearity as well as a lack of dependence on polarization. If realistic beams (Gaussian, super-Gaussian), ultrashort (femtosecond) pulses and/or saturated nonlinearities are encountered, the linearized theory fails to provide adequate agreement with experiment. Even after the initial instability pattern is established, further evolution occurs due to competition between nonlinear effects, diffraction, and saturation leading to pattern changes and even filament coalescence. Group-velocity dispersion becomes important for ultrashort pulses, causing them to temporally broaden [36]. As peak intensity is reduced as a result of the temporal broadening, multiple filamentation is delayed or even averted.

A number of works have extended the theory for commonly encountered beam shapes [37–50]. It is beyond the scope of this chapter to review the varying theoretical nuances. However, we will attempt to point out where new physics or differing behavior becomes relevant. Several studies [44, 46] have shown that relatively clean Gaussian beams tend toward whole-beam collapse rather than form multiple filaments for intensities as high as 100 P_{cr}. The Gaussian profile readily transforms to a Townes profile, which is very robust [51].

Fibich et al. [46] have shown even that with 10% random noise, the transition from whole-beam collapse to multiple filaments occurs at about 40 P_{cr}. Super-Gaussian beams, characterized by a flat top, on the other hand, readily form multiple filaments at intensities as low as 10 P_{cr} [44, 49]. These beams tend to collapse into a ring-shaped profile [52–55]. Ring patterns are inherently azimuthally unstable [22, 37–40] and readily form multiple evenly spaced filaments around the rings (as shown in Fig. 6.4).

Grow et al. [49] have found that the number of such filaments around the ring is proportional to the square root of the input power. In general, any cylindrically symmetric beam other than a Gaussian will form rings. As previously discussed, circular diffraction effects due to aperturing by optical elements also play a role in creating rings. The collapse of optical vortices has recently been studied and finds that the number of filaments is a function solely of the power and topological charge [56].

Departure from cylindrical symmetry also has a significant effect. Grow and Gaeta [47] found that increasing the ellipticity of the initial beam decreases the power required for multiple filamentation. They found that for high ellipticities, bands are formed in the short direction before filamentation occurs, and these are independent of noise and polarization. However, the beam breakup along the long direction is consistent with the instability model.

Modeling by Fibich and Ilan [42, 43] suggests that vectorial effects (i.e., coupling to the axial component of the electric field and the grad-div term) should cause multiple filamentation independent of and at a lower threshold than noise under some conditions (e.g., beam size on the order of a wavelength). This would result in a polarization dependency for the orientation of the filament patterns. This has not yet been experimentally confirmed. Fibich et al. [45] have also shown that astigmatism in the beam will suppress noise-induced filament formation. By controlling the degree of astigmatism by use of a tilted lens setup they were able to control the number and pattern of filaments.

Siegman has shown [27] that strongly tapered beams, either converging or diverging, resist small-scale breakup. It has been found that intensity ripples do not display exponential small-signal growth under any circumstances, while phase ripples show only weak and limited growth. This has been attributed analytically to the loss of the Talbot phase shift term, which in a collimated beam provides a mechanism to periodically convert phase ripples into amplitude ripples, and vice versa.

6.4 Filamentation in Lasers

Solving Eqs. (6.9) and (6.10) for the case of a saturable gain or loss medium (yet keeping $k_2 = 0$), yields the following solution for the growth of a spatial instability:

$$\gamma = -g_2 u_o^2 \pm \sqrt{(g_2 u_o^2)^2 - (\kappa^2/2k_o)^2} \tag{6.16}$$

This illustrates that such beam breakup is not unique to traditional NLO materials, but is readily observed in high-gain laser amplifiers as well. Paxton and Dente [28] extended the simple linear analysis presented in Section 6.2.1 to analyze single quantum-well semiconductor amplifiers with saturable gain including details such as carrier diffusion and linewidth enhancement factor. Their results were very similar to the Kerr NLO case. They found the perturbations grow if the spatial frequency is below a critical value that depends on the intensity. For spatial frequencies above a critical value, the perturbations die out. The critical spatial frequency decreases as the intensity increases above a certain value, decreasing the tendency to filament. Subsequent experiments verified many aspects of the theory [29]. Figure 6.10 shows the buildup of filaments in a GaAlAs amplifier as a function of amplifier current [30].

Filament formation in broad area semiconductor amplifiers becomes a serious problem whenever feedback is present as in the case of a double-pass amplifier [31, 32]. The presence of a counterpropagating beam causes behavior

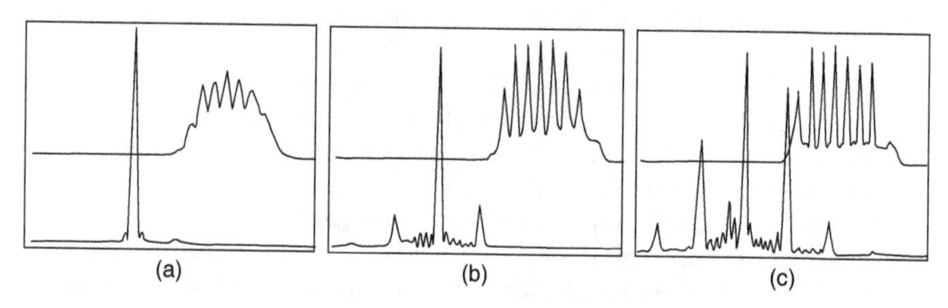

Fig. 6.10 Near field (*upper traces*) and far field (*lower traces*) distributions of laser diode amplifier output at (**a**) 1.0 amp, (**b**) 3.0 amp, (**c**) 4 amp. After Goldberg et al. [30]

distinctively different from the single pass and leads to filament formation at very low intensities. Theory [32] shows that the instability gain becomes infinite after the amplifier gain reaches a threshold value. The counterpropagating beam induces a longitudinally periodic oscillation whose period is the Talbot distance, and for this reason these authors call this a nonlinear Talbot effect.

6.5 Applications/Impact

Over the years a few high-profile applications have provided the motivation to study small-scale self-focusing in depth. Historically, high-energy and high-power laser development was a major driver—initially for inertial confinement fusion (ICF) and then later for military applications (laser weaponry, countermeasures). This led to methods to alleviate the onset of multiple filamentation (see Lukishova et al.'s chapter in this book), including major design changes of the lasers themselves [35, 38], development of apodizers to smooth and shape the beams [52–61] and random phase plates [62] and plasma filters [63] to both prevent multiple filamentation and to allow more energy to be placed onto the target [64]. As an example, multiple filamentation places an upper limit on the laser's intensity. Consequently, to achieve higher energies, designers have had to expand the gain medium's cross-section. In glass, an otherwise smooth beam with an energy density of 10^{10} W/cm^2 will self-focus in about 40 cm. Thus, in a one-kilojoule, 200-ps system, the last amplifier must have a total aperture of at least 500 cm^2. It is largely avoidance of multiple filament formation that has necessitated the huge scale of the ICF laser systems [65].

More recently, femtosecond laser propagation in the atmosphere has strongly influenced the field [66–69]. As these beams propagate they self-focus and channel through the atmosphere, emitting white light. By chirping the pulse, the breakdown distance can be adjusted to be kilometers away and so the self-focus region provides a remote light source to probe the composition of the air. Furthermore, much of the emitted/scattered light is backscattered, making such a probe a useful LIDAR for environmental sensing [66]. Researchers, by better understanding multiple filament formation, have devised clever ways to control and extend the useful range of such standoff detection systems [45, 50, 66]. Research is also in progress using ultrafast self-focusing in air as a means to control lightning [70].

Small-scale self-focusing research has impacted other scientific fields as well as benefited in turn from research in other areas. The characteristic instability is ubiquitous in many areas of physics: plasma physics, hydrodynamics, and Bose–Einstein condensates as well as nonlinear optics. Much of the mathematical development shares similarities. In terms of optics, small-scale self-focusing is historically the first experimental example of pattern formation and competition in nonlinear optics, now an active area [71]. Indeed, a laser beam propagating through a Kerr medium is the simplest system one could devise that spontaneously forms a transverse pattern. By simple changes in geometry (symmetry, double pass, use of cavity, propagation through fibers, etc.) or NLO sample (photorefractives, $\chi^{(2)}$ materials, saturable nonlinearities, etc.) many other patterns form. For example, as a beam self-focuses in atomic sodium vapor, a characteristic honeycomb pattern is formed [72]. The connection between filament formation and optical spatial solitons [73–75] is close as well. Self-focusing represents a case in which nonlinear lensing overwhelms diffractive effects. Solitons represent a case where nonlinear effects precisely balance diffraction, resulting in a stable propagating profile. Often the difference between the two cases is due to beam geometry or NLO characteristics of the medium.

References

1. P.L. Kelley: Self-focusing of optical beams, *Phys. Rev. Lett.* **15**, 1005–1008 (1965).
2. V.I. Talanov: Self-focusing of wave beams in nonlinear media, *JETP Lett.* **2**, 138–141 (1965).
3. M. Hercher: Laser-induced damage in transparent media, *J. Opt. Soc. Am.* **54**, 563–570 (1964).
4. P. Lallemand, N. Bloembergen: Self-focusing of laser beams and stimulated Raman gain in liquids, *Phys. Rev. Lett.* **15**, 1010–1012 (1965).
5. E. Garmire, R.Y. Chiao, C.H. Townes: Dynamics and characteristics of the self-trapping of intense light beams, *Phys. Rev. Lett.* **16**, 347–349 (1966).
6. Y.R. Shen: Self-focusing: experimental, *Prog. Quant. Electr.* **4**, 1–34 (1975).
7. J.H. Marburger: Self-focusing: theory, *Prog. Quant. Electr.* **4**, 35–110 (1975).
8. R.Y. Chiao, M.A. Johnson, S. Krinsky et al.: A new class of trapped light filaments, *IEEE J. Quant. Elec.* **2**, 467–469 (1966).
9. V.I. Bespalov, V.I. Talanov, Filamentary structure of light beams in nonlinear liquids. *JETP Lett.* **3**, 307–310 (1966).
10. K.A. Bruekner, S. Jorna: Linear instability theory of laser propagation in fluids, *Phys. Rev. Lett.* **17**, 78–81 (1966).
11. B.R. Suydam: Self-focusing of very powerful laser beams. in *Laser-Induced Damage in Optical Materials*, A.J. Glass, A.H. Guenther (Eds.), National Bureau of Standards Special Publ. **387**, 42–48 (1973).
12. B.R. Suydam: Self-focusing of very powerful laser beams II, *IEEE J. Quant. Elec.* **10**, 837–843 (1974).
13. B.R. Suydam: Effect of refractive-index nonlinearity on the optical quality of high-power laser beams, *IEEE J. Quant. Elec.* **11**, 225–230 (1975).
14. R. Jokipii, J. Marburger: Homogeniety requirements for minimizing self-focusing damage by strong electromagnetic waves, *Appl. Phys. Lett.* **23**, 696–698 (1973).
15. J.A. Fleck, J.R. Morris, E.S. Bliss: Small-scale self-focusing effects in a high-power glass laser amplifier, *IEEE J. Quant. Elec.* **14**, 353–363 (1978).
16. S.C. Abbi, N.C. Kothari: Theory of filament formation in self-focusing media, *Phys. Rev. Lett.* **43**, 1929–1931 (1979).
17. S.C. Abbi, N.C. Kothari: Growth of Gaussian instabilities in Gaussian laser beams, *J. Appl. Phys.* **5**, 1385–1387 (1980).
18. R.Y. Chiao, P.L. Kelley, E. Garmire: Stimulated four-photon interaction and its influence on stimulated Rayleigh-wing scattering, *Phys. Rev. Lett.* **17**, 1158–1160 (1966).

19. S.C. Abbi, H. Mahr: Correlation of filaments in nitrobenzene with laser spikes, *Phys. Rev. Lett.* **26**, 604–606 (1971).
20. Yu.S. Chilingarian: Self-focusing of inhomogeneous laser beams and its effect on stimulated scattering, *Sov Phys. JETP* **28**, 832–835 (1969).
21. S.C. Abbi, H. Mahr: Optical filament formation in nitrobenzene from laser intensity inhomogenieties, *Appl. Phys. Lett.* **19**, 415–417 (1971).
22. A.J. Campillo, S.L. Shapiro, B.R. Suydam: Periodic breakup of optical beams due to self-focusing, *Appl. Phys. Lett.* **23**, 628–630 (1973).
23. A.J. Campillo, S.L. Shapiro, B.R. Suydam: Relationship of self-focusing to spatial instability modes, *Appl. Phys. Lett.* **24**, 178–180 (1974).
24. E.S. Bliss, D.R. Speck, J.F. Holzrichter et al.: Propagation of a high-intensity laser pulse with small-scale intensity modulation; *Appl. Phys. Lett.* **25**, 448–450 (1974).
25. R.L. Carman, R.Y. Chiao, P. L. Kelley: Observation of degenerate stimulated four-photon interaction and four-wave parametric amplification, *Phys. Rev. Lett.* **17**, 1281–1283 (1966).
26. A.J. Campillo, J.E. Pearson, S.L. Shapiro et al..: Fresnel diffraction effects in the design of high-power laser systems, *Appl. Phys. Lett.* **23**, 85–87 (1973).
27. A.E. Siegman: Small-scale self-focusing effects in tapered optical beams, Memo for File, www.stanford.edu/~siegman/self_focusing_memo.pdf, 1–13 (2002).
28. A.H. Paxton, G.C. Dente: Filament formation in semiconductor laser gain regions, *J. Appl. Phys.* **70**, 2921–2925 (1991).
29. M. Tamburrini, L. Goldberg, D. Mehuys: Periodic filaments in reflective broad area semiconductor optical amplifiers, *Appl. Phys. Lett.* **60**, 1292–1294 (1992).
30. L. Goldberg, M.R. Surette, D. Mehuys: Filament formation in a tapered GaAlAs optical amplifier, *Appl. Phys. Lett.* **62**, 2304–2306 (1993).
31. R.L. Lang, D. Mehuys, A. Hardy et al.: Spatial evolution of filaments in broad area diode laser amplifiers, *Appl. Phys. Lett.* **62** 1209–1211 (1993).
32. R.L. Lang, D. Mehuys, D.F. Welch et al.: Spontaneous filamentation in broad-area diode laser amplifiers, *IEEE J. Quant. Elec.* **30**, 685–694 (1994).
33. S.A. Akhmanov, R.V. Khokhlov, A.P. Sukhorukov: Self-focusing, self-defocusing and self-modulation of laser beams. In: *Laser Handbook*, F.T. Arecchi, E.O. Schulz-Dubois (Eds.), North-Holland, **2**, E3, 1151–1228 (1972).
34. D.T. Attwood, E.S. Bliss, E.L. Pierce et al.: Laser frequency doubling in the presence of small-scale beam breakup, *IEEE J. Quant. Elec.* **12**, 203–204 (1976).
35. J. Bunkenberg, J. Boles, D.C. Brown et al.: The Omega high power phosphate–glass system; design and performance, *IEEE J. Quant. Elec.* **17**, 1620–1628 (1981).
36. K.D. Moll, A.L. Gaeta: Role of dispersion in multiple-collapse dynamics, *Opt. Lett.* **29**, 995–997 (2004).
37. K. Konno, H. Suzuki: Self-focusing of laser beam in nonlinear media, *Phys. Scripta* **20**, 382–386 (1979).
38. M.D. Feit, J.A. Fleck: Beam nonparaxiality, filament formation, and beam breakup in the self-focusing of optical beams, *J. Opt. Soc. Am. B* **5**, 633–640 (1988).
39. J.M. Soto-Crespo, D.R. Heatley, E.M. Wright et al.: Stability of the higher-bound states in a saturable self-focusing medium, *Phys. Rev. A* **44**, 636–644 (1991).
40. J.M. Soto-Crespo, E.M. Wright, N.N. Akhmediev: Recurrence and azimuthal-symmetry breaking of a cylindrical Gaussian beam in a saturable self-focusing medium, *Phys. Rev. A* **45**, 3168–3175 (1992).
41. P.L. Kelley: The nonlinear index of refraction and self-action effects in optical propagation, *IEEE J. Select Topics Quant. Elec.* **6**, 1259–1264 (2000).
42. G. Fibich, B. Ilan: Deterministic vectorial effects lead to multiple filamentation, *Opt. Lett.* **26**, 840–842 (2001).
43. G. Fibich, B. Ilan: Vectorial effects in self-focusing and in multiple filamentation, *Physica D*, **157**, 112–146 (2001).

44. L. Berge, C. Gouedard, J. Schjodt-Eriksen et al.: Filamentation patterns in Kerr meadia vs. beam shape robustness, nonlinear saturation, and polarization states, *Physica D* **176**, 181–211 (2003).
45. G. Fibich, S. Eisenmann, B. Ilan et al.: Control of multiple filamentation in air, *Opt. Lett.* **29**, 1772–1774 (2004).
46. G. Fibich, S. Eisenmann, B. Ilan et al.: Self-focusing distance of very high power laser pulses. *Opt. Express* **13**, 5897–5903 (2005).
47. T.D. Grow, A.L. Gaeta: Dependence of multiple filamentation on beam ellipticity, *Opt. Express* **13**, 4594–4599 (2005).
48. J. Garnier: Statistical analysis of noise-induced multiple filamentation, *Phys. Rev. E* **73**, 046611–1–046611-11 (2006).
49. T.D. Grow, A.A. Ishaaya, L.T. Vuong et al.: Collapse dynamics of super-Gaussian beams, *Opt. Express*, **14**, 5468–5475 (2006).
50. G. Fibich, Y. Sivan, Y. Erlich et al.: Control of the collapse distance in atmospheric propagation, *Opt. Express* **14**, 4946–4957 (2006).
51. K.D. Moll, A.L. Gaeta, G. Fibich: Self-similar optical wave collapse: Observation of the Townes profile, *Phys. Rev. Lett.* **90**, 203902-1–203902-4 (2003).
52. V.R. Costich, B.C. Johnson: Apertures to shape-high power laser beams, *Laser Focus* **10**(9), 43–46 (1974).
53. W.W. Simmons, W.F. Hagen, J.T. Hunt et al.: Performance improvements through image relaying, Laser Program Annual Report, 1976, UCRL-50021-76, Part 2-1.4, pp 2-19–2-28, LLNL (1977).
54. D.C. Brown: High-peak-power lasers, Springer, New York (1981).
55. V.I. Kryzhanovskii, B.M. Sedov, V.A. Serebryakov et al.: Formation of the spiral structure of radiation in solid-state laser systems by apodizing and hard apertures, *Sov. J. Quant. Electron.* **13**, 194–198 (1983).
56. L.T. Vuong, T.D. Grow, A. Ishaaya et al.: Collapse of optical vortices, *Phys. Rev. Lett.* **96**, 133901-1–133901-4 (2006).
57. S.G. Lukishova, I.K. Krayuk, P.P. Pashinin et al.: Apodization of light beams as a method of brightness enhancement in neodymium glass laser installations. In: *Formation and Control of Optical Wave Fronts*, Proc. of the General Physics Institute of the USSR Academy of Science, P.P. Pashinin, Ed., **7**, 92–147, Nauka Publ., Moscow (1987).
58. E.W.S. Hee: Fabrication of apodized apertures for laser beam attenuation, *Opt. Laser Technol.* **7**(2), 75–79 (1975).
59. A.J. Campillo, B. Carpenter, B.E. Newman et al.: Soft apertures for reducing damage in high-power laser systems, *Opt. Comm.* **10**, 313–315 (1974).
60. A.J. Campillo, S.L. Shapiro: Toward control of self-focusing, *Laser Focus* **10(**6), 62–65 (1974).
61. J. Auerbach, V. Karpenko: Serrated-aperture apodizers for high-energy laser systems, *Appl. Opt.* **33**, 3179–3183 (1994).
62. Y. Kato, K. Mima, S. Aringa et al.: Random phasing of high-power lasers for uniform target acceleration and plasma-instability suppression, *Phys. Rev. Lett.* **53**, 1057–1060 (1984).
63. N.K. Moncur: Plasma spatial filter, *Appl. Opt.* **16**, 1449–1451 (1977).
64. A.J. Campillo, R.A. Fisher, R.C. Hyer et al.: Streak camera investigation of the self-focusing onset in glass, *Appl. Phys. Lett.* **25**, 408–410 (1974).
65. G.H. Miller, E.I. Moses, C.R. Wuest: The national ignition facility: Enabling fusion ignition for the 21st century, *Nucl. Fusion* **44**, S228–S238 (2004).
66. J. Kasparian, M. Rodriguez, G. Mejean et al.: White-light filaments for atmospheric analysis, *Science* **301**, 61–64 (4 July 2003).
67. A.L. Gaeta: Catastrophic collapse of ultrashort pulses, *Phys. Rev. Lett.* **84**, 3582–3584 (2000).

68. A. Couairon, L. Birge: Light filaments in air for ultraviolet and infrared wavelengths, *Phys. Rev. Lett.* **88**, 135003-1–135003-4 (2002).
69. J. Schwarz, P. Rambo, J.-C. Diels et al.: Ultraviolet filamentation in air, *Oct. Comm.* **180**, 383–390 (2000).
70. M. Rodriguez, R. Sauerbrey, H. Wille et al.: Triggering and guiding megavolt discharges by use of laser-induced ionized filaments, *Opt. Lett.* **27**, 772–775 (2002).
71. F.T. Arecchi, S. Boccaletti, P.L. Ramazza: Pattern formation and competition in nonlinear optics, *Phys. Rep.* **318**, 1–83 (1999).
72. R.S. Bennink, V. Wong, A.M. Marino et al.: Honeycomb pattern formation by laser-beam filamentation in atomic sodium vapor, *Phys. Rev. Lett.* 88, 113901-1–113901-4 (2002).
73. M. Remoissenet: *Waves Called Solitons*. Springer-Verlag, Berlin (1996).
74. G.I.A. Stegeman, D.N. Christodoulides, M. Segev: Optical spatial solitons: Historical perspectives, *IEEE J. Select. Topics Quant. Elec.* **6**, 1419–1427 (2000).
75. Y.S. Kivshar, D.E. Pelinovsky: Self-focusing and transverse instabilities of solitary waves, *Phys. Rep.* **331**, 117–195 (2000).

Chapter 7
Wave Collapse in Nonlinear Optics

E.A. Kuznetsov

Abstract In this chapter, we give a brief review of collapses in nonlinear optics with and emphasis on their classification (weak, strong, and black holes), correspondence between solitons and collapses and effects of the group velocity dispersion as well.

7.1 Introduction

Wave collapse is the process of singularity formation in a finite time for smooth initial conditions. Historically the notation "collapse" in physics was first intended in general relativity for the description of catastrophic compression of astrophysical objects. In the 1970s–1980s of the last century the word "collapse" came to be applied to wave systems. The first time this term was used was in 1972 by V.E. Zakharov in his classical paper [1] about the prediction of the collapse of Langmure waves in isotropic plasma. Later this notation became to be used widely also in nonlinear optics not only for description of stationary self-focusing but also for the nonstationary self-compression of light pulses. See [2–5]. What kind of singularities appear as the result of the collapse development depends on a physical model. For instance, for self-focusing of light the point foci are formed where the light intensity becomes very large. For water waves such singularities are surface wedges. For these singularities the second derivative of surface profile is infinite.

Collapses play very significant role in various fields of physics, not only in nonlinear optics but also in plasma physics, hydrodynamics, physics of atmosphere and ocean, and in solid-state physics as well. As a finite-time process, collapse in many physical situations represents one of the most effective mechanisms of the wave energy dissipation into heat. For instance, collapse

E.A. Kuznetsov (✉)
L.D. Landau Institute for Theoretical Physics, 2 Kosygin str., 119334 Moscow, Russia
and P.N. Lebedev Physical Institute, 53 Leninsky ave., 119991 Moscow, Russia
e-mail: kuznetso@itp.ac.ru

R.W. Boyd et al. (eds.), *Self-focusing: Past and Present*,
Topics in Applied Physics 114, DOI 10.1007/978-0-387-34727-1_7,

of plasma waves defines efficiency of various collective methods of fusion plasma heating resulting mainly in increasing of electron component energy. In nonlinear optics for moderate intensities this process is stopped by multi-photon absorption and for higher amplitudes by ionization of atoms. Therefore collapse in nonlinear optics can be used to study light-matter interaction in a variety of intensities including, in particular, processes of matter ionization by strong electromagnetic fields and the creation of relativistic particles as well. Another possibility in optical systems is to use collapse to shorten a pulse duration, for instance, to get ultra short pulses.

In this chapter the main attention will be paid to collapse in nonlinear optics described by the nonlinear Schrodinger equation (NLSE) and its generalizations which can be applied to study self-focusing of light, both stationary and nonstationary. This model in dimensionless variables is written in the form:

$$i\psi_t + \Delta\psi + |\psi|^2\psi = 0. \tag{7.1}$$

In the nonlinear optics content ψ has the meaning of the wave envelope of the electric field with definite polarization (e.g., linear), time t means a distance along the wave propagation. Here and everywhere below the subscript t denotes a partial derivative over time. The second term in (7.1) describes both diffraction and the group velocity dispersion (only in the case of anomalous dispersion). For the normal dispersion the operator Δ should be changed by the hyperbolic operator $\Delta_\perp - \partial_z^2$. The nonlinear term $|\psi|^2\psi$ in (7.1) corresponds to account of the Kerr addition to the refractive index.

It is necessary to say that the nonlinear Schrodinger equation, which is a generic equation modeling a variety of wave phenomena, plays a central role in the wave collapse theory. Especially it became evident since prediction of self-focusing of light [6] and development of the corresponding theory [7–10]. The influence of this model in the wave collapse theory is not settled by self-focusing of light only, it has a big variety of applications. Usually the NLSE and its modifications appear as a result of the reduction of equations of motion for some nonlinear media to a solution in the form of quasi-monochromatic waves, namely it assumes the application of some average over rapid in space and in time oscillations. Therefore the NLSE can be considered very often as an equation for the wave packet envelope. We refer here to the whole monograph [11] devoted to this equation where except physical background it is possible to find many mathematical details.

Equation (7.1) is often called the Gross–Pitaevskii equation [12]. In particular, this equation fairly accurately describes long-wavelength oscillations of the condensate of a weakly imperfect Bose gas with negative scattering length. At present, Eq. (7.1) is the basic model for studying the nonlinear dynamics of Bose condensates (see, e.g., [13,14]). In this case (7.1) represents the Schrodinger equation and ψ is the wave function. Respectively, the equation describes the motion of a quantum-mechanical particle in a self-consistent potential with attraction: $U = -|\psi|^2$. Just the attraction is a cause of the singularity formation.

From the quantum-mechanical viewpoint, the collapse within the framework of the nonlinear Schrodinger equation can be interpreted as the fall off a particle to an attracting center in a self-consistent potential [15].

However, collapse in the NLSE is possible not for all spatial dimensions, but only for $D \geq 2$. For the one-dimensional (1D) case, as it was shown by Zakharov and Shabat in 1971 [16], this equation can be integrated exactly by means of the inverse scattering transform. This theory demonstrates that solitons as localized stationary objects play a very essential role in dynamical behavior of the 1D system. They turn out to be structurally stable entities, i.e., stable not only with respect to small perturbations but also against the finite ones. In particular, solitons scatter on each other elastically without change their forms. This very attractive property was put in the base of applications of optical solitons to fiber communications [17]. Now this idea is being realized already in practice (see, e.g., [18]).

At higher dimensions, however, solitons play another role.

7.2 Solitons Versus Collapses

As is well known, Eq. (7.1) belongs to the Hamiltonian type, and can be rewritten as follows,

$$i\psi_t = \frac{\delta H}{\delta \psi^*} \tag{7.2}$$

where the Hamiltonain

$$H = \int |\nabla \psi|^2 d\mathbf{r} - \frac{1}{2}\int |\psi|^4 d\mathbf{r} \equiv X - Y. \tag{7.3}$$

Besides H, this equation has two other simple integrals of motion: number of quasi-particles (coinciding, up to some constant multiplier, with the wave packet energy),

$$N = \int |\psi|^2 d\mathbf{r}, \tag{7.4}$$

and momentum,

$$\mathbf{P} = \frac{i}{2}\int (\psi \nabla \psi^* - \psi^* \nabla \psi) d\mathbf{r}. \tag{7.5}$$

The conservation of N is connected with gauge symmetry: $\psi \to \psi e^{i\alpha}$, and $\mathbf{P}$ due to the translation symmetry. These two symmetries provide invariance of (7.1) with respect to the Gallilean transformation. In particular, the simplest solution of the equation (7.1),

$$\psi_s = \psi_0(\mathbf{r})e^{i\lambda^2 t} \tag{7.6}$$

corresponds to the standing soliton. The moving soliton solution hence can be easily constructed by applying to (7.6) the Gallilean transform.

One can easily establish also that soliton solution (7.6) represents a stationary point of the Hamiltonian H for fixed number of particles (compare with [19]):

$$\delta(H + \lambda^2 N) = 0 \tag{7.7}$$

where λ^2 is a Lagrange multiplier. This variational problem is equivalent to the stationary NLSE:

$$-\lambda^2\psi + \Delta\psi + |\psi|^2\psi = 0. \tag{7.8}$$

Hence one can get the dependence of N on the soliton solutions (7.6):

$$N_s = \lambda^{2-D} N_0, \; N_0 = \int |f(\xi)|^2 \mathbf{d}\xi, \tag{7.9}$$

where f obeys the equation:

$$-f + \Delta f + |f|^2 f = 0.$$

The dependence (7.9) turns out to be the key one for the linear stability criterion. It says: If

$$\partial N_s / \partial \lambda^2 < 0, \tag{7.10}$$

then such solitons are unstable. In the opposite case solitons will be stable. This is the Vakhitov–Kolokolov (VK) criterion [20]; for more details see the review [19]. It has simple physical meaning. In Eq. (7.8) the quantity $-\lambda^2$ has the meaning of the energy ε of a soliton as a bound state. For a positive derivative in (7.10), i.e., when the addition of a single particle decreases the energy ε, the soliton is stable. In the opposite case, where a level is expelled as N increases, the soliton becomes unstable [21]. This criterion gives stability for one-dimensional solitons and instability at $D = 3$. In two dimensions this criterion shows that solitons are neutrally stable, i.e. there is no exponential instability but, as shown in [22], the power-type growth of perturbations is possible. See also the review [19].

All these facts can be understood if one considers the scaling transformation remaining the number of waves, $N = \text{const}$:

$$\psi(\mathbf{r}) \to a^{-D/2}\psi_s\left(\frac{\mathbf{r}}{a}\right).$$

Under this transform H becomes the function of the scaling parameter a:

$$H(a) = \frac{X_s}{a^2} - \frac{Y_s}{a^D}. \tag{7.11}$$

Hence one can see that for $D = 1$ the function has a minimum at $a = 1$ corresponding to one-dimensional soliton At $D = 2$ $H(a) \equiv 0$. This straight line (on $H - a$-plane) shows that solitons can be treated as separatrices between collapsing and noncollapsing solutions. In the three-dimensional geometry the function $H(a)$ attains its maximum on the three-dimensional soliton that indicates its instability. Notice also that the Hamiltonian becomes unbounded as $a \to 0$. It is necessary to underline that unboundedness of H represents one of the main criteria for wave collapses [15] (see also [23]). In such a case collapse can be considered as the nonlinear stage of soliton instability.

To clarify the latter we apply the variational approach, taking a trial function for the NLSE (7.1) in the form

$$\psi(\mathbf{r}, t) = a^{-3/2}\psi_s\left(\frac{\mathbf{r}}{a}\right)\exp(i\lambda^2 t + i\mu r^2),$$

where $a = a(t)$ and $\mu = \mu(t)$ are assumed to be unknown functions of time. After substitution of this anzats into the action

$$S = \frac{i}{2}\int(\psi_t\psi^* - c.c.)dt d\mathbf{r} - \int H dt$$

and integration over spatial variables we arrive at the Newton equation for a,

$$C\ddot{a} = -\frac{\partial H}{\partial a}, \tag{7.12}$$

where $C = \int \xi^2|\psi_0(\xi)|^2$ plays a role of a particle mass and the function (7.11) has a meaning of the potential energy. Behavior of $a(t)$ depends on the total energy,

$$E = C\frac{\dot{a}^2}{2} + H(a)$$

and the dimension D. At $D = 1$ soliton realizes the minimal value of the potential energy $H(a)$ and it is a reason why 1D soliton is stable. At $D = 3$ if a "particle" stands at the maximal point of $H(a)$ initially then depending on the its motion direction (toward or upwards the center $a = 0$) the system will collapse ($\psi \to \infty$) or expand ($\psi \to 0$). For collapsing regime (falling at the center) $a(t)$ behaves near singularity like

$$a(t) \sim (t_0 - t)^{2/5} \tag{7.13}$$

where t_0 is the collapse time. As shown in [15], this asymptotic behavior for $a(t)$ near singular time coincides with that following from the exact semi-classical collapsing solution which asymptotically (as $t \to t_0$) tends to the compact distribution:

$$|\psi| \to \lambda\sqrt{1-\xi^2} \qquad \text{for} \qquad \xi = r/a(t) \leq 1$$

with $\lambda \sim (t_0 - t)^{-3/5}$.

Hence we can make a few conclusions. First, the influence of nonlinearity grows with increase spatial dimension D. As a sequence, stable solitons are intrinsic for low-dimensional systems while for higher dimensions instead of solitons we have to expect blow-up events. Secondly, one of the main criteria of collapse is unboundedness of the Hamiltonian. In this case, the collapse can be interpreted as the fall-off a particle to an attracting center in a self-consistent potential [15].

Thus, the Hamiltonian unboundedness can be considered as one of the main criteria of the wave collapse existence, and the collapse in such systems can be represented as a process of falling down of some "particle" in a self-consistent unbounded potential. However, the picture is more complicated than considered above. From the very beginning we have a spatially distributed system with an infinite number of degrees of freedom and therefore, rigorously speaking, it is hardly feasible to describe such a system by its reduction to a system of ODEs like (7.12). The NLSE is the wave system and therefore, first of all, here we deal with waves. Waves may propagate, may radiate, and so on. To illustrate the importance of this point, let us try to understand the influence of wave radiation on the wave collapse.

Let Ω be an arbitrary region with a negative Hamiltonian $H_\Omega < 0$. Then using the mean value theorem for the integral Y,

$$\int_\Omega |\psi|^4 d\mathbf{r} \leq \max_{x\in\Omega} |\psi|^2 \int_\Omega |\psi|^2 d\mathbf{r},$$

we have

$$\max_{x\in\Omega} |\psi|^2 \geq \frac{|H_\Omega|}{N_\Omega}. \tag{7.14}$$

This estimate shows that radiation of waves promotes collapse: far from the region Ω radiative waves can be considered almost linear; nonlinear effects are small for them. These waves carry out the positive portion of the Hamiltonian, making H_Ω more negative with simultaneous vanishing of the number of waves N_Ω that results in growth of the r.h.s. of (7.14) [1,2,4]. This is why we can say that wave radiation promotes collapse which plays the role of friction in the nonlinear wave dynamics. Simultaneously radiation turns out to accelerate compression of the collapsing area with the self-similarity,

$$r \sim (t_0 - t)^{1/2}, \tag{7.15}$$

which is different from that given by the semiclassical answer (7.13).

7.3 Collapse

7.3.1 Virial Theorem

The exact criterion for singularity formation within the NLSE can be obtained from the virial theorem. In classical mechanics the virial theorem can be easily obtained if one first calculates the second time derivative from the moment of inertia and then averages the obtained result. It gives the relation between mean kinetic and potential energies of particles if the interaction between particles is of the power type.

In 1971, Vlasov et al. [24] found that this theorem can be applied also to the two-dimensional nonlinear Schrodinger equation. The relation obtained is written for the mean square size $\langle r^2 \rangle = N^{-1} \int r^2 |\psi|^2 d\mathbf{r}$ of the distribution as follows:

$$\frac{d^2}{dt^2} \int r^2 |\psi|^2 d\mathbf{r} = 8H. \tag{7.16}$$

This equality is verified by the direct calculation. In this relation $N\langle r^2 \rangle$ has the meaning of the inertia moment.

Since H is a conserved quantity, Eq. (7.16) can be integrated two times to yield

$$\int r^2 |\psi|^2 d\mathbf{r} = 4Ht^2 + C_1 t + C_2, \tag{7.17}$$

where $C_{1,2}$ are the additional integrals of motion. The existence of these integrals is explained by two Noether symmetries: the lens transform (this fact was established by V.I. Talanov [8]) and the scaling transformation [22,25].

Hence one can easily see that the mean square size $\langle r^2 \rangle$ of any field distribution with negative Hamiltonian

$$H < 0, \tag{7.18}$$

independently on $C_{1,2}$ vanishes in a finite time, which, with allowance for the conservation of N, means the formation of a singularity of the field ψ [24]. This is the famous [24] (VPT) criterion; nowadays it is a cornerstone in the theory of wave collapses. This was the first rigorous result for nonlinear wave systems with dispersion, which showed the possibility of the formation of a wave-field singularity in finite time, despite the presence of the linear dispersion of waves, the effect impeding the formation of point singularities (focii) in the linear optics.

7.3.2 *Strong Collapse*

Notice that $H = 0$ corresponds to the soliton for the number of waves $N = N_s$. Moreover, if $N < N_s$ then $H \geq X(1 - N/N_s) > 0$ and collapse is impossible [26]. All waves are spread due to dispersion (diffraction) vanishing as $t \to \infty$. Thus, we can say that solitons in this case represent separatrices between collapsing and noncollapsing submanifolds.

From Eq. (7.17) for the collapsing regime ($H < 0$) we can see that the characteristic size a of the collapsing area behaves like

$$a \sim (t_0 - t)^{1/2}$$

in the correspondence with the self-similar law (7.15).

However, the exact analysis [27] performed by Fraiman shows that

$$a^2(t) \sim \frac{(t_0 - t)}{\log|\log(t_0 - t)|},$$

and asymptotically the spatial profile of ψ tends to that for the soliton one (see also [5, 28–31]). Exponentially small radiation is accompanied by collapse in this case. The power (up to some multiplier, coinciding with N) captured into the singularity occurs *finite*, equal to the power of the 2D soliton. This is why such collapse is called a *strong collapse* [15] and, respectively, the 2D equation (7.1) is a critical nonlinear Schrodinger equation.

7.3.3 *Weak Collapse*

In the three-dimensional case ($D = 3$), Eq. (7.16) is replaced by

$$\frac{d^2}{dt^2}\int r^2|\psi|^2 d\mathbf{r} = 4(2H - Y), \tag{7.19}$$

and equality (7.17), by the inequality

$$N\langle r^2\rangle < 4Ht^2 + C_1 t + C_2, \tag{7.20}$$

where $C_{1,2}$ are the integration constants determined by the initial conditions. The last relationship yields the same sufficient criterion of the collapse as for $D = 2$: $H<0$ [1].

However, the criterion $H<0$ is not sharp. As was shown in [32] and [33], this criterion can be improved. The sharper criterion of collapse is given by the conditions:

$$H<H_s \text{ and } X<X_s,$$

so that the inequality following from Eq. (7.19) is of the form,

$$\frac{d^2}{dt^2}\int r^2|\psi|^2 d\mathbf{r} = 8(H - H_s), \tag{7.21}$$

and, respectively,

$$N\langle r^2\rangle < 4(H - H_s)t^2 + C_1 t + C_2.$$

Here H_s is the value of H on the ground soliton solution (without nodes).

Now we consider the self-similar (corresponding to (7.15)) solution of Eq. (7.1), determined by the substitution,

$$\psi = \frac{1}{(t_0 - t)^{1/2+i\alpha}}\chi\left(\frac{r}{(t_0 - t)^{1/2}}\right). \tag{7.22}$$

Here we will deal only with spherically symmetric solutions, for which the function χ satisfies the equation

$$i[(1/2 + i\alpha)\chi + \frac{1}{2}\xi\chi_\xi] + \frac{1}{2}\chi_{\xi\xi} + \frac{1}{\xi}\chi_\xi + |\chi|^2\chi = 0. \tag{7.23}$$

We are interested only in a regular solution of (7.23) that decreases as $\xi \to \infty$ and satisfies the equation

$$(1/2 + i\alpha)\chi + \frac{1}{2}\xi\chi_\xi = 0. \tag{7.24}$$

Consequently χ has an asymptote

$$\chi \to \frac{C}{\xi^{1+2i\alpha}}, \tag{7.25}$$

where C is a certain constant that can be assumed, without loss of generality, to be positive and real.

The requirement that the solution be regular eliminates the ambiguity in the choice of the constants α and C. We have in fact a nonlinear eigenvalue problem for Eq. (7.23). The numerical solution of this problem yields

$$\alpha = 0.545, C = 1.01. \tag{7.26}$$

We proceed to interpret the self-similar solution. We note first that for any fixed point of physical space with coordinate r, the corresponding self-similar coordinate ξ tends to infinity as $t \to t_0$. The self-similar solution goes over then into its asymptote (7.25), which takes in the physical variables r the singularity,

$$\psi \to \frac{C}{r^{1+2i\alpha}}. \tag{7.27}$$

which is independent on t.

The self-similar solution is realized in physical space in a certain region with coordinate $r < r_0$ where r_0 is the region size. The integrable singularity (7.27) "grows out" at the center of this region as $r \to 0$. At first glance, the self-similar substitution (7.22) leads to nonconservation of the integral

$$N = \int |\psi|^2 d\mathbf{r} \tag{7.28}$$

and corresponds by the same token to a value $N = \infty$ for the integral. This is indeed the case if the self-similar solution is considered in all of space. In any finite region $r < r_0$, however, the value of the integral N remains constant. Indeed, after substituting (7.22) in (7.28) we have for the region $r < r_0$

$$N = (t_0 - t)^{1/2} 4\pi \int_0^{\xi_*} \xi^2 |\chi(\xi)|^2 d\xi, \xi_* = r_0 (t_0 - t)^{-1/2}. \tag{7.29}$$

When account is taken of the asymptote (7.25), the integral in (7.29) diverges on the upper limit and tends to a finite value as $t \to t_0$. Let now r_0 be large enough. The value of the integral N in the region $r < r_0$ should be close to its value in this region at the instant of the collapse $t = t_0$. In other words, the following equation should hold:

$$\int_0^{\infty} \{|\chi|^2 - C^2/\xi^2\} \xi^2 d\xi = 0. \tag{7.30}$$

This equation was verified with high accuracy for the computer-generated $\chi(\xi)$ and C.

The solution constructed here corresponds to *weak collapse*; formally speaking, zero energy enters the singularity at $r = 0$ [15]. This means in fact that if the characteristic amplitude value at which energy absorption in the collapse sets in and Eq. (7.1) no longer holds is of the order of ψ_0, then the amount of energy absorbed in one collapse act is

$$\Delta N \sim \psi_0^2 r_0^3 \sim 1/\psi_0.$$

Here $r_0 \sim 1/\psi_0$ is the characteristic size of the absorption region.

To conclude this part, we note that the weak regime of wave collapse was first demonstrated in numerical experiments [34] and later in [35].

7.3.4 *Black Hole Regime*

As we see above weak collapse regime results in formation of the time-independent singularity (7.27) that corresponds to the attraction potential

$$U = -\frac{C^2}{r^2}.$$

The constant C^2 here is larger than 1. As known from quantum mechanics (see [36]) in this case a particle can fall into the potential center. In its final stage this falling becomes more quasi-classical. This potential can, thus, serve as an attractor for the waves surrounding the singularity, where these waves can be captured as in a funnel. Because originally in the weak collapse regime the energy captured into the singularity formally is equal to zero, such post-collapse falling can be considered as almost stationary. This process is characterized by a finite flux of particles into the singular region. As was shown by Vlasov et al. (1989) and [38] (see also [39]) near the singular point the density $|\psi|^2$ in this regime has the form:

$$|\psi|^2 = \frac{1}{2r^2|\log r|}.$$

The constant flux is provided by waves captured into the singularity that results in existence of long-time burning points supported by wave energy from outside. This regime is also called the *black hole regime*.

7.4 Role of Dispersion in Collapse

If before we treated collapse for electromagnetic waves with anomaluos dispersion now we consider the case of normal dispersion. Formally this case corresponds to the NLSE (7.1) in which the Laplacian operator is changed by the hyperbolic operator $\Delta_\perp - \partial^2/\partial z^2$, where $\Delta_\perp = \partial_{xx} + \partial_{yy}$. Therefore the equation

$$i\frac{\partial\psi}{\partial t} + \Delta_\perp\psi - \psi_{zz} + |\psi|^2\psi = 0 \tag{7.31}$$

is often called the hyperbolic nonlinear Schrodinger equation. The different sign in the second-order operator appears according to the sign of the derivative of the group velocity relative to the frequency, $\partial V_{gr}/\partial\omega$. For normal dispersion this derivative is negative.

The hyperbolic operator standing in (7.31) changes dramatically the character of nonlinear interaction. If for anomalous dispersion all "directions" are equivalent, for the normal dispersion we get different behavior along the z direction: instead of attraction for (7.1) we have repulsion while in the transverse plane (corresponding to $\Delta_\perp$) attraction remains. Respectively, the quasi-particles are attracted to each other in the transverse direction and tend to shrink the beam across the z-axis. The masses of quasi-particles in the longitudinal direction are negative, so the nonlinearity tends to increase the packet size along the z-axis. In this respect, the main question is whether there is enough time for the singularity formation due to the transverse shrinking, despite the longitudinal expansion of the quasi-particles.

Concerning the hyperbolic nonlinear Schrodinger equation (7.31) in the three-dimensional case, we show how an analysis of inequalities of the virial type yields the conclusion that the collapse of a wave packet as a whole is impossible at the stage when a pulse compresses in all three directions. We should note that, strictly speaking, the proof outlined below does not exclude the possibility of singularity formation for initial conditions different from those considered in this section.

As Eq. (7.1) the equation (7.31) belongs to the Hamiltonian ones with

$$H = \int |\nabla_\perp \psi|^2 d\mathbf{r} - \int |\psi_z|^2 d\mathbf{r} - \frac{1}{2}|\psi|^4 d\mathbf{r} \equiv I_\perp - I_z - Y. \qquad (7.32)$$

Different signs of first and second terms in H correspond to different characters of nonlinear interaction in transverse and longitudinal directions.

Consider variations in the wave-packet average square sizes $\langle r_\perp^2 \rangle$ and $\langle z^2 \rangle$ along and across the z-axis, respectively. Calculations similar to (7.16) give

$$N \frac{d^2}{dt^2} \langle r_\perp^2 \rangle = 4\left[2 \int |\nabla_\perp \psi|^2 d\mathbf{r} - \int |\psi|^4 d\mathbf{r} \right], \qquad (7.33)$$

$$N \frac{d^2}{dt^2} \langle z^2 \rangle = 8 \int |\psi_z|^2 d\mathbf{r} + 2 \int |\psi|^4 d\mathbf{r}. \qquad (7.34)$$

The quantities $\langle r_\perp^2 \rangle$ and $I_\perp$ from (7.34), as well as $\langle z^2 \rangle$ and I_z from Eq. (7.34), obey the uncertainty relations

$$I_\perp \langle r_\perp^2 \rangle \geq N, I_z \langle z^2 \rangle \geq N/4. \qquad (7.35)$$

With the help of these relations and using the definition (7.32) of H, one can estimate the right-hand sides of Eqs. (7.33) and (7.34):

$$N\frac{d^2}{dt^2}\langle r_\perp^2\rangle = 8H + 8I_z \geq -4H + 2\frac{N}{\langle z^2\rangle}, \tag{7.36}$$

$$N\frac{d^2}{dt^2}\langle z^2\rangle = -4H + 4I_z + 4I_\perp > -4H + 4\frac{N}{\langle r_\perp^2\rangle} \tag{7.37}$$

Consider now the regime of shrinking in all directions when

$$\frac{d}{dt}\langle r_\perp^2\rangle < 0, \frac{d}{dt}\langle z^2\rangle < 0,$$

which is most favorable from the collapse viewpoint, and show that a collapse, treated as a decrease in the average square transverse and/or longitudinal sizes to zero ($\langle r_\perp^2\rangle \to 0$, $\langle z^2\rangle \to 0$), is impossible in this case.

At first, we prove that the average longitudinal square size z^2 of the wave packet cannot vanish if $\frac{d}{dt}\langle z^2\rangle < 0$. Consider Eq. (7.34), from which, with account of Eq. (7.35), the closed inequality for $\langle z^2\rangle$ can be obtained:

$$N\frac{d^2}{dt^2}\langle z^2\rangle \geq 8I_z \geq 2\frac{N}{\langle z^2\rangle}. \tag{7.38}$$

This relationship can be integrated once over time if $\frac{d}{dt}\langle z^2\rangle < 0$:

$$\xi(t) = \frac{1}{2}\left(\frac{d\langle z^2\rangle}{dt}\right)^2 - 2\log\langle z^2\rangle \leq \xi(0), \tag{7.39}$$

where $\mathcal{E}(0)$ is the initial value of $\mathcal{E}(t)$. If $\langle z^2\rangle \to 0$, then the left-hand side of this inequality increases to infinity due to the logarithmic term, that contradicts the inequality (7.39). Thus, compression $\langle z^2\rangle \to 0$ is impossible.

Let us now show that the collapse to zero in the transverse direction is also impossible. For this aim we multiply the inequality (7.36) by $\frac{d}{dt}\langle z^2\rangle < 0$, and the inequality (7.37) by $\frac{d}{dt}\langle r_\perp^2\rangle < 0$, sum the results, and then integrate the obtained expression over time from zero to t. As a result, we get

$$E(t) = N\frac{d\langle r_\perp^2\rangle}{dt}\frac{d\langle z^2\rangle}{dt} - 8H\langle z^2\rangle + 4H\langle r_\perp^2\rangle - 2N\log\langle z^2\rangle - 4N\log\langle r_\perp^2\rangle \leq E(0), \tag{7.40}$$

where $E(0)$ is the value of $E(t)$ at the initial instant of time. It follows immediately from this inequality that collapse at the shrinking stage is impossible because the first term on the right-hand side of Eq. (7.40) is positive by definition, the terms proportional to H are finite, and the logarithmic term turns out to be infinitely large for $\langle r_\perp^2\rangle \to 0$. This does not conform to the fact that the function $E(t)$ is bounded from above by the initial value $E(0)$. Hence, the collapse of a three-dimensional wave-field distribution as a whole is impossible at the most "dangerous" stage of shrinking in all directions [40] (see also [41]). Does this mean that collapse in such a system is entirely impossible? Strictly speaking, no, because, firstly, criteria similar to the Vlasov-Petrishchev-Talanov criterion are sufficient and, secondly, the above analysis shows that collapse, if possible, should be searched for regimes corresponding to the outflow of the gas of quasi-particles in the longitudinal direction, which stipulates the increase in the longitudinal size of the wave packet. However, despite significant interest in this problem this question remains open at present, in spite of the recent numerical simulations suggest that normal dispersion can arrest collapse (see, e.g., the papers [42–45] concerning this issue).

7.5 Conclusion

Thus, in this short review on the example of the nonlinear Schrodinger equation we have demonstrated that:

- The role of nonlinearity increases with the spatial dimension growth. It is why stable solitons can be observed in low-dimensional systems, and collapses are intrinsic to higher dimensions.
- The Hamiltonian unboundedness can be considered as one of the main criteria of the wave collapse existence, and the collapse in such systems can be represented as a process of falling down of some "particle" in a self-consistent unbounded potential.
- Solitons in collapsing systems represent separatrices between collapsing and spreading distributions. Solitons in this case define the collapse threshold.
- Classification of wave collapses is given that includes weak and strong collapses and the black hole regime.
- Collapse of electromagnetic pulses in nonlinear dielectrics is very sensitive to linear dispersion. Collapse of pulses in media with normal dispersion as the process of compression in all direction is impossible.

Acknowledgments This work was supported by the Russian Foundation for Basic Research (grant No. 06-01-00665) and by the Council for the State Support of the Leading Scientific Schools of Russia. The author wishes to thank du Laboratoire de Physique et Chimie de l'Environnement du CNRS (Orleans), where this work was finalized, for its kind hospitality during our visit in the fall of 2006.

References

1. V.E. Zakharov, Collapse of Langmuir waves, *Zh. Eksp. Teor. Fiz.* **62**, 1745–1751 (1972)[*Sov. Phys. JETP* **35**, 908–914 (1972)].
2. V.E. Zakharov, Collapse and self-focusing of Langmuir waves, in: *Handbook of Plasma Physics, Vol. 2, Basic Plasma Physics*, eds. A.A. Galeev, R.N. Sudan, Elsevier, North-Holland, (1984), pp. 81–121.

3. J.J. Rasmussen, K. Rypdal, Blow-up in nonlinear Schrodinger equations: I. A general review, *Phys. Scr.* **33**, 481–504 (1986).
4. E.A. Kuznetsov, Wave collapse in plasmas and fluids, Chaos, **6**, 381–390 (1996).
5. L. Berge, Wave collapse in physics: Principles and applications to light and plasma waves, *Phys. Rep.* **303**, 259 (1998).
6. G.A. Askaryan, Effect of the gradient of a strong electromagnetic beam on electrons and atoms, *Zh. Eksp. Teor. Fiz.* **42**, 1567 (1961) (*Sov. Phys. JETP* **15**, 1088–1090 (1962)].
7. V. I. Talanov, On self-focusing of the wave beams in nonlinear media, *ZhETF Pis'ma* **2**, 218–222 (1965) [*JETP. Lett.* **2**, 138–142 (1965)].
8. V. I. Talanov, About selffocusing of light in cubic media, *ZhETF Pis'ma* **11**, 303–305 (1970) [*JETP Lett.* **11**, 199–201 (1970)].
9. R.Y. Chiao, F. Garmire, C.H. Townes, Self-trapping of optical beams, *Phys. Rev. Lett.* **13**, 479–482 (1964).
10. E.L. Dawes, J.H. Marburger, Computer studies in self-focusing, *Phys. Rev.* **179** 862–868 (1969).
11. C.Sulem, P.L. Sulem, *The Nonlinear Schrodinger Equation*, Springer-Verlag, New York (1999).
12. E.P. Gross, Structure of quantized vortex, *Nuovo Cim.*, **20** 454–461 (1961); L.P. Pitaevskii, Vortex lines in an imperfect Bose gas. *Sov. Phys. JETP*, **13**, 451–454 (1961).
13. V. Flambaum, E. Kuznetsov, Nonlinear dynamics of ultra cold gas, in: Proc. NATO ARW "Singularities in Fluids, Optics and Plasmas," Eds. E. Caflisch and G. Papanicolaou, Kluwer (1993), pp. 197–203.
14. L. Berge, J.J. Rasmussen, Collapsing dynamics of attractive Bose–Einstein condensates, *Phys. Lett.* **A 304**, 136–142 (2002).
15. V.E. Zakharov, E.A. Kuznetsov, Quasi-classical theory of three-dimensional wave collapse, *Sov. Phys. JETP* **64**, 773–780 (1986).
16. V.E. Zakharov, A.B. Shabat, Exact theory of two-dimensional self-focusing and one-dimensional self-modulation of waves in nonlinear media, *Zh. Eksp. Teor. Fiz.* **61**, 118–134 (1971) [*Sov. Phys. JETP* **34**, 62–69 (1972)].
17. A. Hasegawa, F. Tappert, Transmission of stationary nonlineat optical pulses in dispersive dielectric fibers. II. Normal dispersion, *Appl. Phys. Lett.*, **23** 142–144 (1973).
18. Yu.S. Kivshar, G.P. Agrawal, Optical Solitons: From Fibers to Photonic Crystals, Academic Press, Amsterdam (2003).
19. E. A. Kuznetsov, A.M. Rubenchik, V.E. Zakharov, Soliton stability in plasmas and fluids, *Phys. Rep.* **142**, 103–165 (1986).
20. N.G. Vakhitov, A.A. Kolokolov, Stationary solutions of the wave equation in media with nonlinearity saturation, *Izv. Vyssh. Uchebn. Zaved., Radiofiz.* **16**, 1020 (1973) [*Radiophys. Quantum Electron.* 16, 783 (1975)].
21. E.A. Kuznetsov, Hard soliton excitation: Stability investigation, *Zh. Eksp. Teor. Fiz.* **116**, 299–317 (1999) [JETP, **89**, 163–172 (1999)].
22. E.A. Kuznetsov, S.K. Turitsyn, Talanov transformations for selffocusing problems and instability of waveguides. *Phys Lett A* **112**, 273–275 (1985).
23. E.A. Kuznetsov, S.L. Musher, Effect of the collapse of sound waves on the structure of the collisionless shocks in a magnetized plasma, *Sov. Phys. JETP* **64**, 947–955(1986).
24. S.N. Vlasov, V.A. Petritshev, V.I. Talanov, Averaged description of wave beams in linear and nonlinear media (the method of moments), *Radiophys. Quant. Electr.* **14**, 1062–1070 (1971).
25. S.N. Vlasov, V.I. Talanov, private communication (1984).
26. M.I. Weinstein, Nonlinear Schrodinger equations and sharp interpolation estimates, *Comm. Math. Phys.* **87** 567–576 (1983).
27. G.M. Fraiman, Asymptotic stability of manifold of self-similar solutions in self-focusing, *Zh. Eksp. Teor. Fiz.* **88**, 390 (1985) [*Sov. Phys. JETP* **61**, 228 (1985)].

28. V.E. Zakharov, V.F. Shvets, Nature of wave collapse in the critical case, *JETP Lett.* **47**, 275–278 (1988).
29. M.J. Landman, G.C. Papanicolaou, C. Sulem et al., Rate of blowup for solutions of the nonlinear Schrodinger equation at critical dimension, *Phys. Rev.* A **38**, 3837–3843 (1988).
30. A.M. Smirnov, G.M. Fraiman , The interaction representation in the self-focusing theory, *Physica D* **52**, 2–15 (1992).
31. L. Berge, D. Pesme, Time-dependent solutions of wave collapse, *Phys. Lett. A* **166**, 116–122 (1992).
32. S.K. Turitsyn, Nonstable solitons and sharp criteria for wave collapse, *Phys. Rev. E* **47**, R1316 (1993).
33. E. A. Kuznetsov, J.J. Rasmussen , K. Rypdal et al., Sharper criteria of the wave collapse, *Physica D*, **87**, 273–284 (1995).
34. O.B. Budneva, V.E. Zakharov, V.S. Synakh, Certain models for wave collapse, *Sov. J. Plasma Phys.* **1**, 335–338 (1975).
35. V.E. Zakharov , E.A. Kuznetsov, S.L. Musher, Quasiclassical regime of a three-dimensional wave collapse, *JETP Lett.*, **41**, 154–156 (1984).
36. L.D. Landau, E.M. Lifshitz, *Quantum Mechanics*, 3rd edi., Pergamon, New York (1977).
37. S.N. Vlasov, L.V. Piskunova, V.I. Talanov, Talk on the IV International Workshop on Nonlinear and Turbulent Processes in Physics, Kiev, October 9–22, 1989).
38. V.E. Zakharov, N.E. Kosmatov, V.F. Shvets, Ultra-strong wave collapse, *JETP Lett.* **49**, 492–495 (1989).
39. N.E. Kosmatov, V.F. Shvets, V.E. Zakharov, Computer simulation of wave collapses in the nonlinear Schrodinger equation, *Physica D* **52**, 16–35 (1991).
40. L. Berge, E.A. Kuznetsov, J.J. Rasmussen, Defocusing regimes of nonlinear waves with negative dispersion, *Phys Rev E*, **53**, R1340–1343 (1996).
41. E.A. Kuznetsov, Integral criteria of wave collapses, Izv. VUZ. Radiofiz., vol. *XLVI*, 342–359 (2003) [*Radiophys Quant Electron.*, **46**, 307–322 (2003)].
42. L. Berge, K. Germaschewski, R. Grauer et al., Hyperbolic shock waves of the optical self-focusing with normal group-velocity dispersion, *Phys. Rev. Lett.* **89**, 153902 (2002).
43. N.A. Zharova, A.G. Litvak, V.A. Mironov, On the collapse of wave packets in a medium with normal group velocity dispersion, *JETP Lett.*, **75**, 539–542 (2002).
44. N.A. Zharova, A.G. Litvak, V.A. Mironov, The structural features of wave collapse in medium with normal group velocity dispersion, *JETP*, **96**, 643 (2003).
45 G. Fibich, W. Ren, X.-P. Wang, Numerical simulations of self-focusing of ultrafast laser pulses, *Phys. Rev. E* **67**, 056603 (9 pp) (2003).

Chapter 8
Beam Shaping and Suppression of Self-focusing in High-Peak-Power Nd:Glass Laser Systems

Svetlana G. Lukishova, Yury V. Senatsky, Nikolai E. Bykovsky, and Alexander S. Scheulin

Abstract Laser-beam shaping and suppression of beam self-focusing in high-peak-power Nd:glass laser systems are discussed. The role of Fresnel diffraction at apertures at the laser facility as the source of dangerous spatial scales for small-scale self-focusing is illustrated. Methods of the formation of super-Gaussian laser beams for suppressing diffraction ripples on the beam profile and their self-focusing are presented. A brief outline of self-focusing as the primary nonlinear optical process limiting the brightness of Nd:glass laser installations is given. Methods for self-focusing suppression are presented.

8.1 Introduction

In 1964, Basov and Krokhin published the concept of using lasers for thermonuclear fusion [1]. After the first fusion neutrons from laser irradiation of targets containing deuterium had been detected in 1968 by Basov et al. [2] and were confirmed by Floux et al. in 1970 [3], several laser-fusion centers with high-peak-power Nd:glass laser systems (NGLS) have been established in different countries (USSR, USA, France, UK, Japan, and China). These laser systems operated with light pulses of 10^{-8} to 10^{-11} s durations with output powers for a typical laser-amplifier channel from several gigawatts to the terawatt range [4–9]. Modern multi-channel NGLS for fusion applications, e.g., upgraded OMEGA Laser at the University of Rochester's Laboratory for Laser Energetics (LLE), has 30 kJ at the third harmonic in 60 beamlines [10]; the National Ignition Facility at Lawrence Livermore National Laboratory (LLNL) will have at the same wavelength (351 nm) $\sim$1.8 MJ, 500 TW in 192 beamlines [11].

Laser-beam self-focusing is the primary nonlinear optical process that affects the performance of powerful NGLS and limits the output power [4–7].

S.G. Lukishova (✉)
The Institute of Optics, University of Rochester, Rochester, NY 14627, USA
e-mail: sluk@lle.rochester.edu

R.W. Boyd et al. (eds.), *Self-focusing: Past and Present*,
Topics in Applied Physics 114, DOI 10.1007/978-0-387-34727-1_8,

Suppression of beam self-focusing may be regarded as one of the main problems in the design of high-peak-power NGLS. On the other hand, NGLS, with the length of nonlinear medium (glass) of up to several meters, served as excellent test-beds to study laser beam self-focusing. The first observations of a collimated laser beam self-focusing directly in the laser media itself were made beginning in the 1960s. Self-focusing was found to be a process that restricts the laser power and causes the filamentary damage of the laser medium [12, 13].

It is significant that power levels of laser beams for high-temperature plasma heating and laser fusion exceed the critical power P_{cr} for the whole-beam self-focusing "at infinity" by a factor of more than 10^3–10^5 ($P_{cr} \approx 3.5 \times 10^6$ W for laser glass at the wavelength $\lambda = 1.06$ μm [14]). As a rule, long before the distance where a whole beam is self-focused in the laser medium, the beam decays into multiple filaments. Bespalov and Talanov were the first to consider this beam instability [15]. Minor transverse amplitude and phase perturbations of a light beam of a certain size are unstable and can grow in a nonlinear medium, giving rise to the light-beam collapse. This effect is called small-scale self-focusing and occurs in high-power NGLS at powers exceeding approximately 1 GW/cm^2 [4, 5]. Perturbations of a different nature may take place in a real laser facility. Small obstacles of an accidental nature (such as inclusions in the neodymium glass or in other optical elements of the laser, dust particles, etc.) can, in this way, seriously perturb the intensity distribution of the laser beam, producing amplitude and/or phase perturbations on the laser beam profile, which grow in the nonlinear medium, giving a characteristic "mosaic" structure at the powerful NGLS beam cross section (Fig. 8.1) [16]).

Fresnel diffraction on different apertures in a laser installation [17], including the hard-edges of Nd:glass active elements, plays a prominent role in the origin of intensity perturbations [17–21]. Analysis reveals [17, 19, 20] that the spatial scale of nonuniformities attributable to Fresnel diffraction on the laser-rod edges is within an order of magnitude of the spatial scale most easily focused due to small-scale self-focusing (10–500 μm). An isolated diffraction ring may, in this case, be unstable with respect to the azimuthal angle and will decay into spatial cells [20, 22, 23].

Both axial and concentric-ring damage of laser rods with thin filaments extended into the rod bulk [18–20] were observed at intensities substantially below the glass damage threshold (Fig. 8.2). In addition to multiple filamentary damages in the bulk of active elements as well as on the surfaces, small-scale self-focusing also produces an energy redistribution in the laser-beam cross section and scatters up to 80% of the energy into large angles [13, 24, 25], so that the beam cross section may be surrounded by a halo of scattered radiation at the output of amplifiers.

To suppress hard-edge Fresnel diffraction ripples, but at the same time use an amplifier cross section more efficiently, a super-Gaussian intensity profile

$$I(r)/I_0 \sim \exp\left[-(r/r_0)^N\right] \tag{8.1}$$

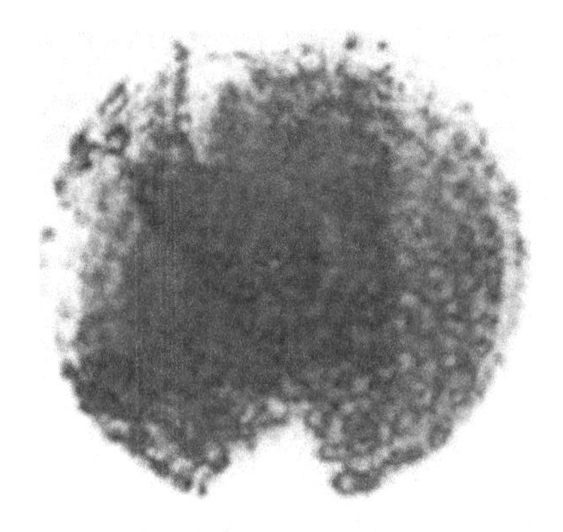
Fig. 8.1 A mosaic structure of a beam cross-section at the output of a high-peak-power Nd:glass laser amplifier system. From Ref. [16]

is one of the solutions. Here I_0 is intensity in the beam center, r is the distance from the beam center along the beam radius, r_0 is the distance at which $I = I_0/e$, and N is the numeric coefficient, $N > 2$. In this profile with a flat top, intensity diminishes monotonically from the beam center to the edges. The use of these profiles in NGLS was initially suggested at LLNL, and devices producing profiles with smoothed edges were called apodized [26], or soft apertures, in LLNL terminology [16, 27].

This chapter is structured as follows: Fresnel diffraction on hard-edge and soft apertures in linear media will be discussed in Section 8.2, as well as methods of preparation of soft apertures. The whole-beam self-focusing in NGLS including super-Gaussian beam collapse will be briefly outlined in Section 8.3. The small-scale self-focusing in NGLS will be discussed in Section 8.4. Section 8.5 is devoted to the methods of suppression of self-focusing in NGLS. Section 8.6 is a summary of the chapter.

It should be noted that this chapter is not a comprehensive review of self-focusing effects in NGLS; e.g., spectral broadening of the amplified pulse [25]

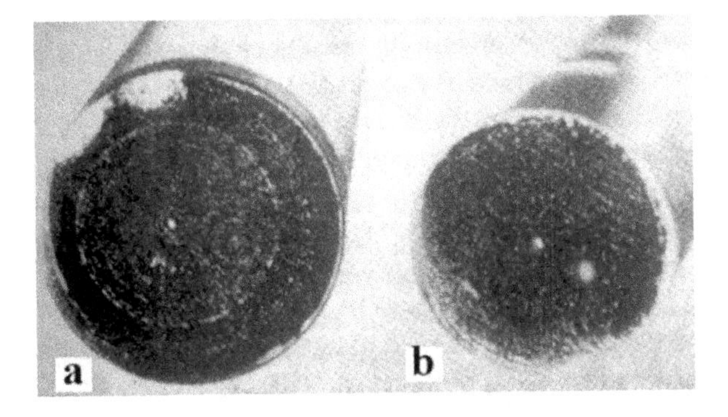

Fig. 8.2 Damage at the exit ends of Nd:glass rods: (**a**) 45 mm in diameter and 680 mm long and (**b**) 30 mm in diameter and 680 mm long, which were used in the 1- to 2-ns pulse amplifier. From Ref. [20]

and altering the temporal-pulse profile [13, 14, 28] will not be considered here. Nor will we discuss self-focusing in unfocused beams in air [12, 29] (this has already been observed in the 1960s at the output of NGLS [12]) or other nonlinear effects in collimated beams in air on the path of ~100 m (e.g., stimulated rotational Raman scattering) in these systems; see, e.g., Ref. [30]. In addition, the restricted length of this chapter does not permit us to include more details in Sections 8.2–8.5 as well. More detailed reviews of some topics covered here can be found in books by Brown [4], Mak et al. [5], an issue of selected reprints (edited by Soures) [6], and a review by Lukishova et al. [16].

8.2 Fresnel Diffraction on Apertures in Linear Media

8.2.1 Fresnel Diffraction on Hard-Edge Apertures

Interest in Fresnel diffraction of laser beams arose from a connection to nonlinear effects in high-peak-power NGLS for high-temperature heating of plasma and in particular, in connection with the studies of the role of beam small-scale self-focusing in these systems [17–20].

We assume that electric field $E\,(x, y, z, t)$ is a wave of the type

$$\frac{1}{2}[U(x,y,z)\exp(-i\omega t)+\text{c.c.}]=\frac{1}{2}[E(x,y,z)\exp(-i\omega t+ikz)+\text{c.c.}], \tag{8.2}$$

where $E\,(x, y, z)$ is the slowly varying complex-field amplitude, $k = \omega/c$, and c.c. denotes the complex conjugate.

The problem of electromagnetic-radiation diffraction by a certain aperture of a monochromatic wave reduces to a strictly determined mathematical problem: finding the solutions of the time-independent Helmholtz equation

$$\nabla^2 U + k^2 U = 0 \tag{8.3}$$

with the appropriate boundary conditions [31].

Ordinarily in laser physics the Fresnel diffraction for type (8.2) waves is described in a parabolic approximation by the equation [32]

$$2ik\frac{\partial E}{\partial z}+\frac{\partial^2 E}{\partial x^2}+\frac{\partial^2 E}{\partial y^2}=0. \tag{8.4}$$

The solution of the parabolic Eq. (8.4) for a plane wave takes the form of the Huygens integral in Fresnel approximation [20, 32]

$$E(x,y,z)=\frac{k}{2\pi iz}\iint dx'dy'E(x',y',0)\exp\left\{\frac{ik}{2z}\left[(x-x')^2+(y-y')^2\right]\right\}, \tag{8.5}$$

where

$$|x - x'|/z \ll 1, |y - y'|/z \ll 1 \tag{8.6}$$

is the small angle approximation; x' and y' are the present coordinates in the aperture plane.

We recall that the diffraction phenomena represent Fresnel diffraction when

$$\lambda z/a^2 \sim 1, \tag{8.7}$$

where a is the aperture radius.

For an axially symmetric case the Huygens integral is written [33] as

$$E(r,z) = \frac{k}{2\pi iz} \int_0^a \left\{ r' E(r',0) \exp\left[\frac{ikr'^2}{2z}\right] \int_{-\pi}^{\pi} \exp\left[-(ikr'r\cos\theta)\frac{1}{z}\right] d\theta \right\} dr'. \tag{8.8}$$

Using the integral representation in the form of zeroth-order Bessel functions we obtain from (8.8)

$$E(r,z) = \frac{k}{iz} \exp\left(\frac{ikr^2}{2z}\right) \int_0^a r' E(r',0) J_0\left(\frac{kr'r}{z}\right) \exp\left(\frac{ikr'^2}{2z}\right) dr'. \tag{8.9}$$

For the case of incidence of a Gaussian beam on a circular aperture

$$E(r) = E_0 \exp\left[-(r/r_0)^2\right], \tag{8.10}$$

$$E(r,z) = \frac{kE_0}{iz} \exp\left(\frac{ikr^2}{2z}\right) \int_0^a r' J_0\left(\frac{kr'r}{z}\right) \exp\left[\left(\frac{ik}{2z} - \frac{1}{r_0^2}\right) r'^2\right] dr'. \tag{8.11}$$

An exact expression of the integral (8.11) [20, 31] exists for the case of incidence on the circular aperture of a plane wave with a constant field E_0 along the cross section:

$$\begin{aligned} E(r,z) = E_0 \Bigg\{ 1 - \exp\left[\frac{ik}{2z}(a^2 + r^2)\right] \\ \times \left[\sum_{s=0}^{\infty} (-1)^s \left(\frac{r}{a}\right)^{2s} J_{2s}\left(\frac{kra}{z}\right) + i \sum_{s=0}^{\infty} (-1)^s \left(\frac{r}{a}\right)^{2s+1} J_{2s+1}\left(\frac{kra}{z}\right) \right] \Bigg\}. \end{aligned} \tag{8.12}$$

The wave amplitude along the axis of a beam limited by the circular aperture will be

$$E(z) = E_0\left[1 - \exp(ika^2/2z)\right]. \tag{8.13}$$

Numerical integration of (8.11) was carried out in Ref. [33]. Some studies on Fresnel diffraction in laser systems provide a direct numerical solution of Eq. (8.4) for different Fresnel numbers [17, 19, 34–36]

$$F = a^2/\lambda z. \tag{8.14}$$

Some LLNL simulation codes and approaches of solving Eq. (8.4) were outlined in Ref. [7].

For a spherical wave the Fresnel diffraction problem can be reduced to a Fresnel diffraction problem in a plane wave. The initial parabolic equation for a spherical wave can, by substitution of the variables

$$\tilde{z} = \frac{zf}{f+z} \equiv \frac{z}{M(z)}, (\tilde{x}, \tilde{y}) \equiv \tilde{\mathbf{r}}_\perp = \frac{\mathbf{r}_\perp f}{z+f} \equiv \frac{\mathbf{r}_\perp}{M(z)} \tag{8.15}$$

be reduced to a form identical to that of Eq. (8.4). Here

$$z = \frac{\tilde{z}f}{f-\tilde{z}}, (x, y) \equiv \mathbf{r}_\perp = \frac{\tilde{\mathbf{r}}_\perp f}{f-\tilde{z}}. \tag{8.15a}$$

In relations (8.15) and (8.15a) $f > 0$ for a positive lens and $f < 0$ for a negative lens.

Invariance of Eq. (8.4) to the transformation of coordinates (8.15) means that the diffraction of a wave with spherical divergence in the range $0 < z < \infty$ is equivalent to the diffraction of a corresponding plane wave in the limited range $0 < z < f$ [20].

Therefore we will focus solely on the plane-wave case. Interestingly, the number of intensity maxima in the beam cross section along the aperture diameter is equal to the Fresnel number. For Fresnel diffraction, the number of maxima grows at close proximity to the aperture, while it diminishes with increasing distance z. The central maximum stands out sharply in the case of odd Fresnel numbers, while the relative levels of the remaining maxima differ little. In the case of even Fresnel numbers there are an even number of maxima and the center intensity is equal to zero. If a beam with a Gaussian intensity distribution along the cross section diffracts on the aperture, the fundamental mechanisms of Fresnel diffraction remain the same as in the case of incidence of a beam with a uniform cross-sectional intensity distribution.

It is important to note that the level of the diffraction peak at the beam center I', for the case of cutting the beam by a circular aperture at a level χI_0 (where I_0 is the center intensity of the beam incident on the aperture) is [20]

$$I' = (1 + \sqrt{\chi})^2 I_0 = (1 + 2\sqrt{\chi} + \chi) I_0. \tag{8.16}$$

For $\chi = 1$ we have $I' = 4\,I_0$; for $\chi = e^{-1}$, $I' = 2.5\,I_0$; for $\chi = 0.1$, $I' = 1.73\,I_0$. For $\chi \ll 1$ the center intensity grows by $2\sqrt{\chi}I_0$. For example, cutting the beam to 1% of I_0 will cause center intensity growth by 20% of I_0 [20].

The effect of the shape of the hard-edge aperture on the diffraction peaks in the Fresnel diffraction region was examined in Ref. [20]. The intensity in the diffraction peak is greatest when a circular aperture is used. Sharp peaks can be substantially suppressed by replacing the curvilinear aperture edge with a rectilinear edge.

Experimental results of hard-edge Fresnel diffraction on a circular aperture in the optical range and their comparison with the theory have been reported in Ref. [17].

8.2.2 Fresnel Diffraction by Soft Apertures and Propagation of Super-Gaussian Beams

It is generally well-known that in order for the diffraction pattern at the aperture output with the radius a to be smoothed out at a certain distance z', it is necessary for the changes in the Fresnel number F induced by edge irregularities or blurring to be of the order of 1, e.g.,

$$\Delta F = 2a\Delta a/\lambda z' = (2\Delta a/a)F \sim 1. \tag{8.17}$$

It is then possible to find the necessary dimensions of the edge irregularities or the level of blurring of the soft aperture $\Delta a \sim a/2F = \lambda z'/2a$.

Apodization by means of edge irregularities is very limited in nature and can be used solely for Fresnel numbers determined by formula (8.17). Diffraction by each separate irregularity will appear for short distances (and large Fresnel numbers), while for long distances the dimensions Δa will already be too small to avoid diffraction ripples. In spite of such restrictions, the so-called serrated-tooth metal apertures are very popular in high-power laser design because of their simplicity, see, e.g., Refs. [37, 38]. Using photolithography, the shape of the serrations can be made for the production of a super-Gaussian spatial profile of determined N. Our experiments showed that a serrated aperture with $a = 1$ cm and $\Delta a = 1$ mm operated effectively at $F \sim 5$.

Apodized apertures with gradually variable transmission profiles and with the appearance of Fresnel diffraction ripples at much larger distances than conventional hard-edge apertures will be outlined in Section 8.2.3. Hadley [35] made a comparative analysis of Fresnel diffraction ripples from the apodized apertures with different degrees of apodization. Optimal profile for beam propagation in NGLS was discussed in Ref. [39]. This profile satisfies both the criteria of propagation without ripples on the length of the laser system and maximum uniformity to obtain higher total gain from the amplifiers. The so-called "fill-factor" F_{fill} of the beam profile was introduced. We are using here a definition of Ref. [40]:

$$F_{\mathrm{fill}} = \frac{\int_0^{r'} 2\pi r I(r) dr}{I_0 \pi r'^2}. \tag{8.18}$$

Here radius r' is usually selected so that $I(r') = 10^{-3}$ to $10^{-2}\ I_0$. The optimum transmission profile of a soft aperture is then selected accounting for two facts: (1) a good fill factor and (2) diffraction peaks below a particular value (e.g., 2%). Calculations made at LLNL provided the optimal values of N for super-Gaussian beams: $5 \leq N_{\mathrm{opt}} \leq 10$.

Propagation of super-Gaussian beams with $N = 10$ at different distances from a hard-edge circular aperture with a 2-cm diameter at $\lambda = 1.06$ μm and with a $10^{-3}\ I_0$ level of cutting the beam by the hard-edges was considered in Ref. [39]. The evolution of the beam profile with different N for the case of a circular hard-edge aperture for various cutting levels χ (0, 10^{-2}, 10^{-3}) was modeled in Ref. [36] (for Fresnel numbers of $F = 5$ and 20, see Fig. 8.3).

Figure 8.4 shows the experimental results of the comparison of beam propagation at the output of induced-absorption soft aperture in Refs. [16, 41, 42], made of doped fluorite (calcium fluoride) crystal with a super-Gaussian transmission profile with $N = 4.62$ and $r_0 = 2.2$ mm (left set of images), and a hard-edge aperture (right set of images). No Fresnel diffraction ripples were observed with this soft aperture even at a distance of 4 m from the aperture. At the same time it is easy to see these ripples already at 20 cm at the output of the hard-edge aperture.

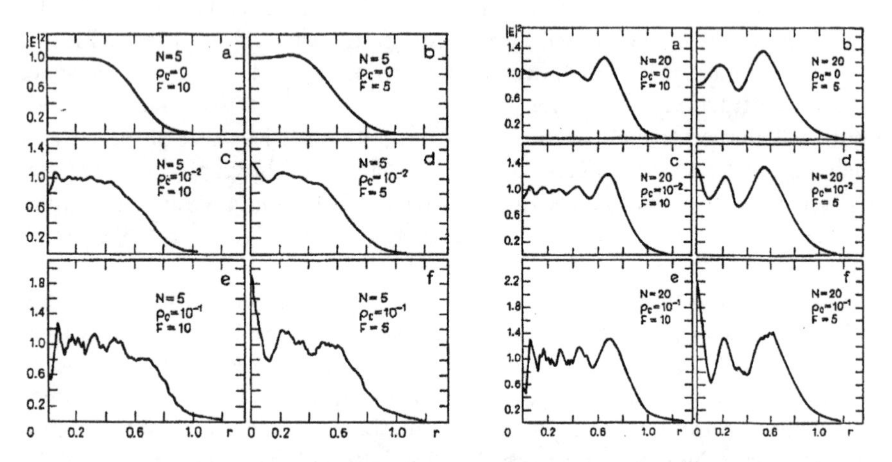

Fig. 8.3 Spatial intensity distribution for super-Gaussian beams with $N = 5$ (*left*) and $N = 20$ (*right*) with different levels $\chi = \rho_c$ of aperturing at $z = 0$, at the distances corresponding to Fresnel numbers $F = 5$ and 10. From Dubik and Sarzyński [36]

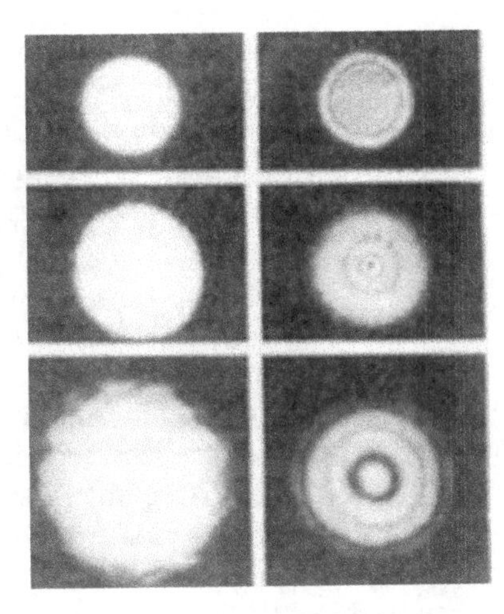

Fig. 8.4 Photographs of beam cross-sections at the output of a color-center CaF_2:Pr soft aperture [42] (*left*) with a super-Gaussian transmission profile of $N = 4.62$ and $r_0 = 2.2$ mm, and a hard-edge aperture with a 5-mm diameter (*right*) at different distances from the apertures: (*top*) at the soft-aperture output plane and at 20 cm from a hard-edge aperture; (*center*) at 1.5 m; (*bottom*) at 4 m from both apertures ($\lambda = 0.633$ μm)

Figure 8.5 shows the beam profiles at 60 cm from the output of 45-mm-diam induced-absorption doped fluorite crystal apodizers [43] with different transmission at the edges ($\chi = 10^{-2}$, 10^{-1}, 0.5). Ripples appear as a result of diffraction of the uniform input beam on the crystal edges in a low-absorption case.

8.2.3 Preparation of Super-Gaussian Beams

In addition to serrated apertures considered in Section 8.2.2, there are a large number of devices used to achieve continuous apodization (gradually variable transmission from the center to edges): (1) photographic plates or films with a controllable blackening level [44–46]; (2) monolayer dielectric coatings of

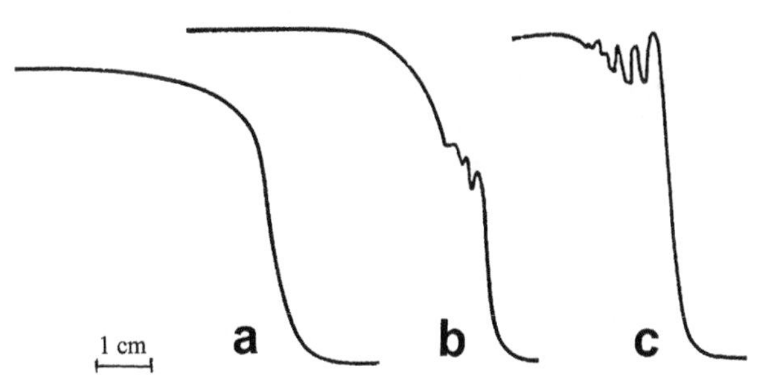

Fig. 8.5 Beam profiles at 60 cm from the output of a 45-mm-diam color-center CaF_2:Pr apodizer with induced absorption at the edges [43] and with different transmission at the edges [$\chi = 10^{-2}$ (a), 10^{-1} (b), 0.5 (c)]

variable thickness [27]; (3) multilayer dielectric coatings with variable thickness (graded reflectance mirrors) [27, 47–51]; (4) deposited metallic layers of variable thickness [26, 52, 53] or pixellated metal apodizers fabricated by photolithography [54]; (5) liquids doped with dyes or other dispersed absorbers in cuvettes of variable thickness [27, 55–60]; (6) solid absorbers of variable thickness (specifically a high-temperature sintered design combining an optical absorbing glass filter and a transparent glass substrate [61, 62]; (7) a Pockels cell in an inhomogeneous electric field [63, 64]; (8) a Faraday rotator in an inhomogeneous magnetic field [27, 64]; (9) a design employing the optical activity of a quartz crystal [65]; (10) a design employing crystal birefringence [66]; (11) a design employing the total internal reflection on a lens surface [67]; (12) a design employing two aspherical lenses [68, 69]; (13) a design based on the diffusion of light-absorbing ions in glass [70]; (14) cross-sectional profiling of the population inversion in the active elements of the amplifier [71]; (15) introduction of phase inhomogeneity at the aperture edges [21, 72, 73]; (16) fabrication of a diffusely scattering layer in the aperture, in particular using a sand-blasting technique or laser treatment of a glass plate [60, 74–79]; (17) induced absorption on the edges of glasses and crystals by the ionizing radiation [16, 41–43, 80–83]—see Figs. 8.4 and 8.5; (18) photooxidation of divalent rare-earth ions in fluorite crystal or bleaching color centers in glasses and crystals under optical irradiation [84, 85]; (19) additive coloration at the edges of doped fluoride crystals as a result of annealing in a reducing atmosphere [86–88]; (20) photochemical reactions of color centers and colloids in additively colored doped fluoride crystals under ultraviolet-light irradiation [89–91]; (21) frustrated total internal reflection [87, 92–94]; (22) cholesteric liquid crystals [95–97]; (23) dye bleaching in doped polymers [98]; (24) microlenses [99]; and (25) spatial light modulators. Detailed reviews of these techniques are carried out in Refs. [16, 27]. See also the book by Mills et al. [100] with selected papers on apodization.

8.3 Whole-Beam Self-focusing

A standard NGLS of the 1970s and 1980s for plasma-heating experiments had a pulse duration of $\tau \sim 10^{-11}$ to 10^{-8} s [4–6, 101–103] and consisted of a master oscillator and a preamplifier with an output intensity $I_{\mathrm{out}} \leq 1$ GW/cm^2 (beam diameter ~5–10 mm), multiple amplifier stages with a large active medium volume (30–60 mm in diameter for rod active elements and 40 × 240 mm rectangle for slabs, or 460–740 mm in diameter in the case of disk amplifiers), and I out $\leq \sim$10 GW/cm^2 [4].

Such intensities give rise to an intensity-dependent change of the local refractive index. The existence of a nonlinear index n_2 in glass is responsible for effects of whole-beam self-focusing, small-scale self-focusing, focal zooming, self-phase modulation, and some other nonlinear effects routinely observed in large laser systems [4, 5]. The intensity I is related to the time average $\langle E^2 \rangle$ in electrostatic units by

$$I = \frac{cn_0}{4\pi}\langle E^2 \rangle = \frac{cn_0}{8\pi}|E|^2, \tag{8.19}$$

where n_0 is the refractive index of the medium, $E\,(x, y, z)$ is the slowly varying complex amplitude of the quasimonochromatic field. At intensities of $I \sim 1$ GW/cm^2, E can reach $\sim 10^6$ V/cm. For isotropic glass and centrosymmetric crystals the refractive index n is written as

$$n = n_0 + n_2\langle E^2 \rangle = n_0 + (n_2/2)|E|^2 = n_0 + \gamma I. \tag{8.20}$$

In different papers n_2 is introduced differently by a factor of 2 or n_0; see, e.g., Table 1 of the Marburger review [104], or n_2 is denoted as γ in relation (8.20); see, e.g., Boyd's book [105]. We have employed the terminology used in Ref. [104] and in experimental studies devoted to measurements of n_2 in laser glass [4, 5, 14, 106, 107]. Because the value of n_2 is usually given in electrostatic units (esu), γ is given by [4]

$$\gamma = \frac{4\pi n_2}{n_0 c}10^7 = \frac{(4.19 \times 10^{-3})n_2}{n_0}\left(\frac{\text{cm}^2}{\text{W}}\right). \tag{8.21}$$

The following dimensional relation is easily obtained from (8.20):

$$\Delta n = 4.19 \times 10^6 \frac{n_2}{n_0}(\text{esu})I(\text{GW/cm}^2). \tag{8.22}$$

The experimental values of n_2 for standard silicate and phosphate laser glasses are $\sim$(1–1.5) $\times\ 10^{-13}$ esu [4, 5]; here n_0 is $\sim$1.5–1.6 [4, 5]. At intensities $I \sim$ 1–10 GW/cm^2, $\Delta n \sim 5 \times 10^{-7}$ to 5×10^{-6} for glasses. However, since the total glass length of the optical path of the amplifiers may reach several meters, such an insignificant variation of the refractive index of the medium may substantially distort both the amplitude and phase profiles of a light beam.

The most significant formation mechanism of n_2 in laser glasses for pulse durations less than nanoseconds ($\sim$70–85% [106, 107]) is the electron Kerr effect associated with the deformation of the electron clouds around the nucleus (the characteristic nonlinear polarization time is $\sim 10^{-15}$–10^{-16} s). Another formational mechanism of n_2 at such pulse durations in glass—nuclear polarizability—has a buildup time of 10^{-12} s and its contribution to n_2 is substantially lower [106, 107].

Before treating whole-beam self-focusing in laser amplifiers, we refer the general whole-beam self-focusing results. In 1962, Askar'yan [108] first suggested the possibility of self-focusing due to the gradient of a strong electromagnetic field. Hercher [109] and Pilipetsky and Rustamov [110] first showed photographs of self-focusing filaments. Comprehensive reviews of self-focusing in different media were offered by researchers significantly contributing to the understanding of self-focusing: experiments—by Shen [111, 112], Svelto [113], Akhmanov et al. [114, 115]; theory—by Marburger [104], Lugovoi and Prokhorov [116], Vlasov and Talanov [117]. See also the books by Sodha et al. [118] and Boyd [105].

Talanov [119] (see also this book) and Chiao et al. [120] first calculated radial distribution of the electric field in a self-trapped electromagnetic field. For cubic media it has a soliton-type behavior, and for beam with cylindrical symmetry the so-called "Townes profile" has the form of a bell-shaped curve.

The critical power P_{cr} at which self-focusing of the beam as a whole is initiated (self-trapping) was introduced by Chiao et al. [120]. It is determined in the following manner [5]:

$$P_{cr} \approx c/(2n_2k_{vac}^2) = c\lambda^2/(8\pi^2 n_2), \tag{8.23}$$

where $k_{vac} = \omega/c$ is the wave vector in the absence of the medium. For laser glasses and wavelength in vacuum $\lambda = 1.06$ µm, $P_{cr} \approx 3.5$ MW [14].

The self-focusing length (distance) L_{self} was introduced by Kelley [121] and for a laser beam of radius a and divergence b is [122]

$$L_{self} = \frac{ka^2}{\sqrt{(P/P_{cr}) - 1} - kab}. \tag{8.24}$$

As power $P \to P_{cr}$ for the case of diffraction divergence only from formula (8.24), we have $L_{self} \to \infty$. In real media, whole-beam (large-scale) self-focusing may be observed at a threshold power $P_{th} \approx$ (10–100) P_{cr} [123] over a finite optical path length. Danileiko et al. [123] provide the values of P_{th} for different lengths of the nonlinear medium over which self-focusing is manifested for different N of super-Gaussian profiles. P_{cr} is independent of N.

For a stationary case, beam propagation in the amplifier medium can be described in quasioptical approximation [117]. The slowly varying complex amplitude E (x, y, z) of an electrical field propagating on the z-axis of the wave will satisfy the parabolic (nonlinear Schrödinger) equation

$$\begin{gathered}\Delta_\perp E + 2ik\left(\frac{\partial E}{\partial z}\right) + \frac{k^2 n_2}{n_0}|E|^2 E + igkE = 0, \\ \Delta_\perp = \frac{\partial^2}{\partial x^2} + \frac{\partial^2}{\partial y^2}.\end{gathered} \tag{8.25}$$

Here $k = (\omega/c)\, n_0$ is the wave number in the medium and g is the amplifier gain per unit length.

For a stationary case where $g = 0$ for P much higher than P_{cr}, numerical calculations of self-focusing by Luguvoi and Prokhorov [116], Dyshko and colleagues [124, 125], Danileiko et al. [123], Lebedeva [126], and Amosov et al.

[127] for the case of Gaussian [123–126] and super-Gaussian [123, 126, 127] beams, have revealed that an axially symmetric beam in a nonlinear medium acquires a multifocus structure (in the direction of beam propagation). Each focus contains power equal to P_{cr}. The number of foci is equal to P/P_{cr}.

Feit and Fleck obtained a similar multifocus structure for nonparaxial approximation [128] without invoking nonlinear absorption or saturation mechanisms to avoid a singularity.

In a nonstationary case, when a laser pulse is changing in time, filaments are observed instead of stationary focal points [109, 110, 129–134] as a result of moving foci. The moving-focus model for the explanation of filaments was introduced by Lugovoi and Prokhorov [131]. Marburger independently suggested the rapid change in the focusing length during the laser pulse [135]. Loy and Shen have shown experimental evidence of the moving-focus model [133]. Multiple foci have been observed experimentally in Lipatov et al. [129] and Korobkin et al. [130].

It should be emphasized that, different from the Gaussian beam profile which collapses in a nonlinear medium into a self-similar Townes profile [136, 137] in each nonlinear focus, the super-Gaussian profiles can collapse into a ring structure [123, 138–140]. Already in a near-field its behavior is different from a Gaussian profile: it may evolve into a spatial distribution with fringes [27, 62].

Figure 8.6 [27] shows the near-field evolution of the super-Gaussian profile in a nonlinear medium of high-peak power NGLS into a ring-shape structure

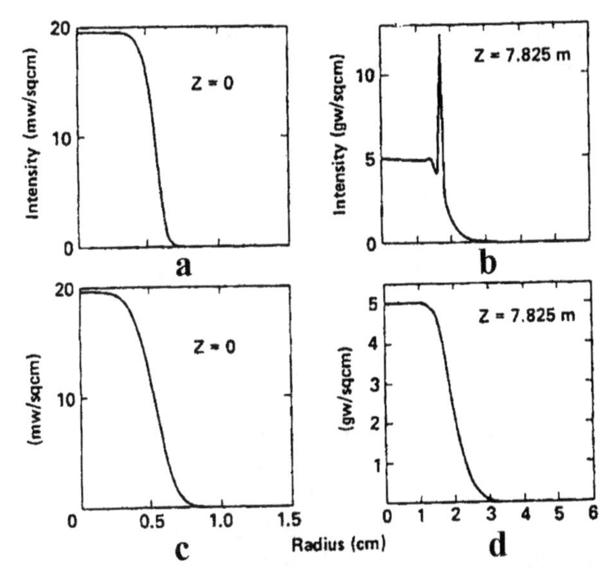

Fig. 8.6 Nonlinear propagation of spatially shaped laser beams. Input beam profile (**a**) develops self-focusing spike (**b**) after propagation about 8 m. Appropriate adjustment of input shape (**c**) eliminates the effect (**d**). From Costich and Johnson [27]

("horns" [4]) which is very dangerous because small-scale self-focusing can occur [4, 62]. (See also Sections 8.5.1 and 8.5.2.). Observation of a ring-shape profile and its filamentation for self-focusing of femtosecond pulses as a result of self-focusing of a super-Gaussian beam was reported recently in Ref. [139] for Fresnel numbers ~ 1. Recent numerical simulations of Fibich et al. [140] showed that for $P \gg P_{cr}$ a noiseless super-Gaussian profiles collapse in a nonlinear medium with self-similar ring profile, rather than with the Townes profile. A super-Gaussian beam with power a few times the critical power will collapse toward a Townes profile similar to a Gaussian profile [139, 140]. With the addition of noise, the ring breaks apart into filaments [139, 140]. It should be noted, that in the presence of noise, the super-Gaussian beams undergo multiple filamentation at much lower powers ($P \sim 10\ P_{cr}$) [139] than the Gaussian beams ($P \geq 100\ P_{cr}$) [23, 139].

Talanov [141] has noted the equivalence of self-focusing of plane and spherical waves in media with Kerr nonlinearity. An important relation is

$$1/L_{self}^{spher} = 1/f + 1/L_{self}, \tag{8.26}$$

where L_{self}^{spher} is the large-scale self-focusing length in the spherical wave case.

References [4, 5, 7, 24, 25, 27, 142, 143] have described large-scale self-focusing in high-power neodymium glass laser installations, although the main self-focusing effects in such installations, whose power level exceed P_{cr} by a factor of 10^3 to 10^4, are ordinarily manifested as small-scale self-focusing at distances substantially shorter than the collapse point of the entire beam. The length of development of the small-scale instability [15] l_{self} is proportional to $ka^2(P/P_{cr})^{-1}$. Recalling relation (8.24), we obtain

$$l_{self}/L_{self} \sim \sqrt{P_{cr}/P}. \tag{8.27}$$

In high-power NGLS with imaging optics, L_{self} is outside the laser system, but a whole-beam self-focusing introduces far-field aberrations [142–145]. After passing through a nonlinear medium and being brought to focus by means of a lens, beam focal position will be changed, depending on the magnitude of the nonlinearity and the intensity. Because the intensity is time dependent, at various times during the pulse duration, rays will focus at different locations, and thus the best focus will "zoom," giving rise to the phenomenon of "focal zooming" in spatial filters and on the target. Such intensity-dependent, focal-beam diameter changes were recorded using the streak camera (see Refs. [145, 146]). It is obvious that the beam shape has a profound influence on the magnitude of zooming as well as on the intensity distribution. Reference [4] provides the results of numerical modeling of focal zoom for a super-Gaussian beam with $N = 5$, which shows complex intensity and phase variations in the region of the focus of the lens. The so-called "modified quadratic profile" [4] is better for focusing by the lens, although it has a low fill factor.

8.4 Small-Scale Self-focusing Effects in High-Peak-Power Nd:Glass Laser Systems

Under actual experimental conditions of high-peak-power neodymium laser systems, small-scale self-focusing is often manifested as a halo of scattered radiation with a divergence exceeding that of the main beam [13, 24, 25]. A detailed analysis of such a halo was carried out in Ref. [24]. The amplified beam with a 100-ps pulse duration and a 45-mm output diameter remained homogeneous through intensities of 3–5 GW/cm^2 at which minor ripples appear in the cross-sectional intensity distribution; upon further amplification, the ripples developed into a mosaic structure. Such a structure is shown in Fig. 8.7 (see also Fig. 8.1). The beam brightness in Ref. [24] diminishes rapidly with increasing intensity above 10 GW/cm^2, so any further increase of gain was not advantageous (Fig. 8.8). These phenomena have not been accompanied by damage of the active medium.

The onset of damage of the medium at such intensities depends in this case on pulse duration. Damage to the exit face of a circular 45-mm-diam neodymium glass rod used in a high-power laser amplifier is shown in Fig. 8.2. Both axial and ring damage are caused by small-scale self-focusing of diffraction perturbations [19–21]. Small-scale self-focusing of the interference fringes from the optical elements of the laser system [20, 21] may also produce such damage, but only in the form of stripes. Thin small rings (Fig. 8.2) are the result of self-focusing induced by perturbations from minor inclusions in the neodymium glass (small bubbles, stones, face dust, etc.). Filamentary damage normally

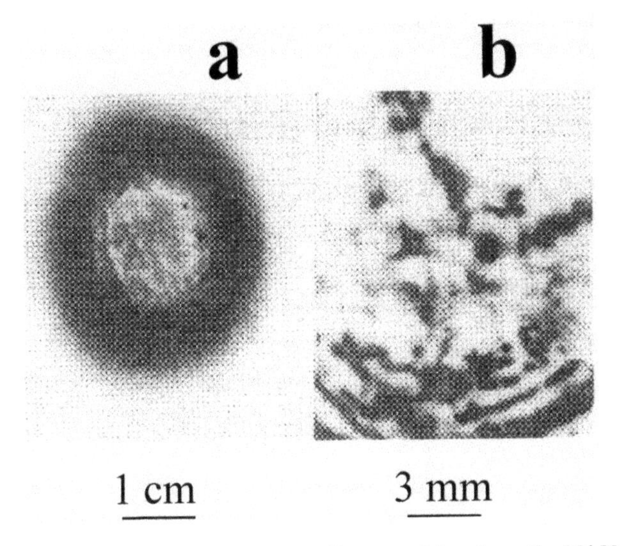

Fig. 8.7 Images of the beam cross-section at a distance of 1 m from the Nd:YAG amplifier output: (**a**) pulse energy 3.62 J, pulse duration 100 ps, output intensity ~10 GW/cm^2; (**b**) magnified central part of the spot in Fig. 8.7(a). From Zherikhin et al. [24]

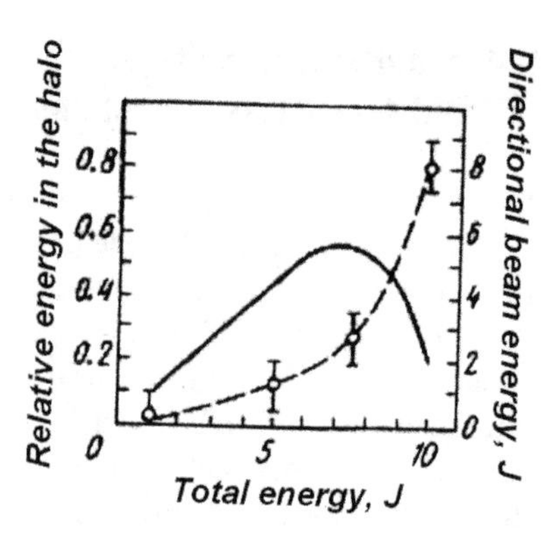

Fig. 8.8 Dependence of the relative energy in the halo (*dotted curve*) and of the energy in the directional beam (*solid curve*) on the total energy in the pulse. From Zherikhin et al. [24]

extends from the surface damage deep into the rod bulk in nanosecond and picosecond laser amplifier systems.

The instability of a light wave in a nonlinear medium was first considered by Bespalov and Talanov [15]. It causes minor transverse amplitude and phase perturbations of a light beam of a certain scale in a nonlinear medium to grow exponentially with the increment Γ (see also Refs. [4, 5, 19–25, 101, 105, 111–118, 147–161], and Campillo's chapter in this book).

A parabolic equation for the stationary case has been examined with $g = 0$ for a slow complex amplitude $E\,(x, y, z)$:

$$\Delta_{\perp}E + 2ik\frac{\partial E}{\partial z} + k^2\frac{n_2}{n_0}|E^2|E = 0. \tag{8.28}$$

We will closely follow the Bol'shov linearized analysis [153]. For a plane wave an exact solution of this equation exists:

$$E(r_{\perp}, z) \equiv E(z) = E_0 \exp\left(\frac{i\omega n_2}{2c}|E_0|^2 z\right). \tag{8.29}$$

We will search the solution of Eq. (8.28) as

$$E(r_{\perp}, z) \equiv E_0 \exp\left(\frac{i\omega n_2}{2c}|E_0|^2 z\right)[1 + u(r_{\perp}, z)], \tag{8.30}$$

where u is the minor perturbation $u \ll 1$.

For a perturbation of the type $u = (E_1 + iE_2)\cos(kr_{\perp} + \varphi)$, linearization of Eq. (8.28) results in the following equation set:

$$\begin{bmatrix} E_1(z) \\ E_2(z) \end{bmatrix} = \begin{pmatrix} \cosh \Gamma z & \dfrac{|\kappa_{\perp}|^2 \sinh \Gamma z}{2k\Gamma} \\ \dfrac{2k\Gamma}{|\kappa_{\perp}|^2}\sinh \Gamma z & \cosh \Gamma z \end{pmatrix} \begin{bmatrix} E_1(0) \\ E_2(0) \end{bmatrix}. \tag{8.31}$$

Here $E_1(0)$ and $E_2(0)$ are perturbations of the real $E_1(z)$ and imaginary $E_2(z)$ part of electrical field $E(z)$ at the entrance to the nonlinear medium.

The instability increment is equal to

$$\Gamma = \frac{|\kappa_\perp|}{2}\sqrt{\frac{2n_2}{n_0}|E_0|^2 - \frac{|\kappa_\perp|^2}{k^2}}. \tag{8.32}$$

Perturbation growth takes place only for a transverse perturbation wave number

$$0 < \kappa_\perp < \kappa_{\perp cr}, \tag{8.33}$$

where

$$\kappa_{\perp cr} = k\sqrt{\frac{2n_2}{n_0}|E_0|^2}. \tag{8.34}$$

In addition to the transverse perturbation wave number $\kappa_\perp$ it is possible to introduce a perturbation propagation angle with respect to a pumping beam:

$$\theta_\perp = \kappa_\perp / k. \tag{8.35}$$

Maximum growth is attained with the instability increment

$$\Gamma_{\max} = \frac{k\,n_2}{2\,n_0}|E_0|^2 = \frac{8\pi^2 n_2 I}{\lambda c n_0} \tag{8.36}$$

at the transverse perturbation wave number

$$\kappa_{\perp\max} = k\sqrt{\frac{n_2|E_0|^2}{n_0}} = \kappa_{\perp cr}/\sqrt{2}. \tag{8.37}$$

The range of unstable vectors is within the circle

$$|\kappa_\perp| \le \kappa_{\perp\max}\sqrt{2}. \tag{8.38}$$

The wavelength of the transverse intensity perturbation corresponding to $\kappa_{\perp\max}$ is

$$\Lambda_{\max} = \frac{\pi}{\kappa_{\perp\max}} = \lambda\sqrt{\frac{c}{32\pi n_2 I}}, \tag{8.39}$$

where λ is the wavelength of laser radiation in a vacuum. Calculating power carried by each of the filaments shows that each filament carries power of the order of P_{cr}.

The small-scale self-focusing length is

$$l_{self} \sim 1/\Gamma_{max} = \frac{\lambda c n_0}{8\pi^2 n_2 I}. \quad (8.40)$$

For $\lambda = 1.06$ μm we obtain

$$\Lambda_M = 0.58\sqrt{n_2 I}, l_{self} = 0.4 n_0/(n_2 I), \Gamma_{max} = 2.48 n_2 I/n_0. \quad (8.41)$$

In relations (8.41) n_2 is in units of 10^{-13} esu, I is in GW/cm^2, Γ_{max} is in m^{-1}, l_{self} is in meters, and Λ_M is in millimeters.

We will perform the estimates for $I = 10$ GW/cm^2, $n_0 = 1.5$, and $n_2 = 1.5 \times 10^{-13}$ esu. From relations (8.41) we obtain $\Gamma_{max} = 24.8$ m^{-1}, $\Lambda_M = 150$ μm, and $l_{self} = 4$ cm.

Analyzing beam instability in actual amplifier conditions requires accounting for changes in beam intensity along the optical path as well as variations in the properties of the optical medium (the glass amplifiers, air gaps, etc.).

In actual high-power laser systems the small-scale self-focusing is ordinarily not characterized by the quantity of $\Gamma_{max}\, l$ but rather the break-up integral [4, 5]

$$B = \frac{8\pi^2}{c\lambda}\int_0^l \frac{n_2}{n_0} I(z)dz. \quad (8.42)$$

Here the level of the spatial frequency with the greatest increment grows as

$$I = I_0 e^{2B}. \quad (8.43)$$

The value for $\lambda \sim 1$ μm is often encountered in the literature using the same units as in relations (8.41)

$$B = 2.62\int_0^l \frac{n_2}{n_0} I(z)dz. \quad (8.44)$$

Experiment and numerical modeling suggest that in conditions typical for a high-power neodymium-glass-laser installation ($I \sim 1$–10 GW/cm^2) spatial irregularities on a scale of 10–500 μm represent the most dangerous irregularities from the viewpoint of the development of self-focusing [20, 21]. Experimental analyses have revealed that values of $B \leq 2$–3 [4, 5] are permissible in laser amplifier beams ($B = 0.5$–4 in Ref. [161]).

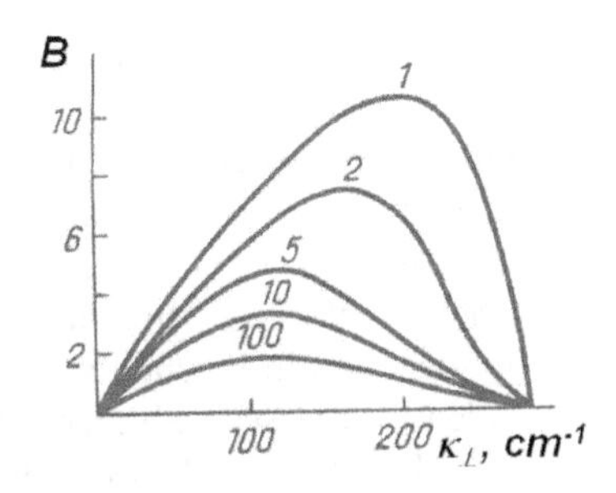

Fig. 8.9 Break-up integral B as a function of inverse ripple size $\kappa_\perp$ for 30-cm ED-2 glass ($n_2/n_0 = 0.9 \times 10^{-13}$esu) at the output intensity 20 GW/cm^2 under different gain G. From Mak et al. [5]. See also Ref. [4]

It is, however, important to remember that the equal B -factor for two different configurations of a nonlinear medium does not infer that their nonlinear losses are equal and, although their characterization by the integral B is quite popular in the literature, it remains an estimate.

One of the ways to minimize small-scale self-focusing effects is maximizing amplifier gain $G = \exp(gz)$ [4, 5]. Numerical modeling showed that amplifiers that have G greater than 6.434 have $\kappa_{\perp\max}$ (or $\theta_{\perp\max}$) independent of the gain and $\kappa_{\perp\max}/\kappa_{\text{cr}} \approx 0.394$ in this case [4, 5]. For amplifiers with gain less than that value, $\kappa_{\perp\max}$ increases to the passive value at $G = 1$ given by (8.37). Figure 8.9 shows these results where the B-integral is shown as a function of $\kappa_\perp$ for different values of G for a 30-cm-long rod of ED-2 glass with an output intensity of 20 GW/cm^2 [4, 5].

Mak et al. [5] suggested the use of long pulses to avoid self-focusing effects in NGLS with the subsequent pulse compression at the output using stimulated scattering. The current state-of-the-art of self-focusing suppression, which almost all of the highest peak-power lasers in the world are using, is increasing the pulse duration and, after amplification, it is compressing using chirped-pulse amplification (CPA) first introduced by Strickland and Mourou [162]. Some examples of NGLS using pulse stretching and compressing gratings are the Vulcan Petawatt Upgrade (Rutherford Appleton Laboratory) [163], the Gekko Petawatt Laser (Institute of Laser Engineering at Osaka University) [164], the OMEGA EP Laser (University of Rochester) [165], and the now dismantled petawatt line on the former Nova Laser (LLNL) [166].

8.5 Methods of Suppression of Self-focusing

In addition to CPA and maximizing the amplifier gain mentioned in Section 8.4, there are several other methods for suppressing self-focusing effects in the amplifiers of high-peak-power NGLS. Common techniques include: (1) using spatial filters and relay imaging devices; (2) eliminating the effect of Fresnel diffraction on the uniformity of the light beams by means of apodization; (3) using diverging beams; (4) insertion of air-spaces into nonlinear medium; (5) using circular polarization; and (6) limiting the degree of coherence of the laser radiation. There are also several other methods of suppressing self-focusing. These include: diminishing n_2 (for fluoroberyllium glass $n_2 \approx 0.4 \times 10^{-13}$ esu [167]); improving the uniformity of the laser glass; using self-defocusing media with $n_2 < 0$ to compensate self-focusing in the optical elements of the installation [168, 169]; using nonlinear dissipative interaction of a high-power laser beam with an optical medium in which self-focusing is not the primary energy-dissipation mechanism but rather some other nonlinear process plays this role (such as second-harmonic conversion [170, 171]); and using a plasma spatial filter [172]. Phase-conjugation methods also exist (see, e.g., Refs. [173–176]) to achieve self-compensation of distortions attributable to amplifier nonuniformities [176].

Mak et al. [5, 177] used a small-scale phase aberrator with a phase inhomogeneity prepared from an etched plate. The divergence introduced by such an aberrator θ_{aber} should be greater than $\theta_{\perp max}$. The small-scale self-focusing threshold with such an aberrator was four times greater than in the case of uniform beams [177].

Mailotte et al. [178] proposed to suppress small-scale self-focusing using structured beams. If two beams intersect in a nonlinear medium at an angle $\theta_{inc} \geq \theta_{max}$, small-scale self-focusing ripples can be suppressed. Propagation will be dominated by diffraction off the nonlinear index grating with period d written within the material. In this case, the small-scale self-focusing threshold will be raised by a factor of

$$\left(\frac{\theta_{inc}}{\lambda/d}\right)^2 . \tag{8.45}$$

Population trapping can eliminate self-focusing as well. Jain et al. [179] experimentally demonstrated the elimination of the self-focusing filaments for a copropagating pair of intense laser beams whose frequencies differ by a Raman resonance. In this case the atoms are forced into a population-trapped state and their contribution to n_2 is eliminated.

We will consider the common methods for suppressing self-focusing in high-peak-power NGLS in more detail.

8.5.1 Application of Spatial Filters and Relay Imaging Optics

Spatial filters represent a necessary element in modern high-power NGLS even with CPA (see, e.g., Refs. [4, 5, 143, 157, 161, 180, 181]). The spatial filter is fabricated as a Kepler telescopic system consisting of two positive lenses with a pinhole placed at the common focus. The pinhole rejects high spatial frequencies in the focal plane of the lens. As a rule, the pinhole is fabricated from ruby or sapphire, Teflon, or metal. The spatial filter is normally constructed as an evacuated chamber below 10^{-5} to 10^{-6} Torr pressure in order to eliminate breakdown of the medium [4]. Calculation of an optimum pinhole diameter is difficult. Usually, the best procedure is to experimentally determine the diameter for each system. Typical values of the pinhole diameter are in the range of 0.5–1.5 mm [4].

The spatial filter also serves as a relay imaging system by imaging the entrance aperture through the entire amplifier chain [4, 5]. When several relay imaging systems are used through the amplifier chain, the effective length of the amplifier system for diffraction and self-focusing ripple growth can be reduced to zero. This method leads to the achievement of a much more constant beam profile and a larger fill factor than that obtainable without spatial filters.

Figure 8.10 (lower part [180]) shows diminishing diffraction and small-scale self-focusing ripples in the amplifier system using relay imaging. The beam-profile fill-factor is preserved. Figure 8.10 (upper part) shows beam profiles in the same system without relay imaging. In this case an intensity-dependent index of refraction leads to steepening of gradients of an initially smooth beam profile, leading to the development of "horns" [4] (rings) on the beam.

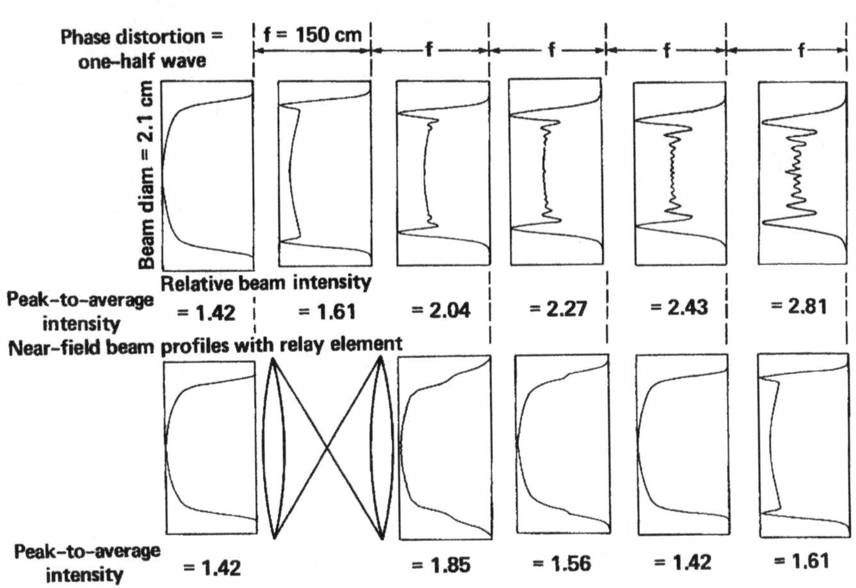

Fig. 8.10 The combination of diffraction and nonlinear phase distortions produces large intensity gradients. With further propagation in a nonlinear medium, self-focusing enhances these gradients, reducing the effective filling factor and, ultimately, focusable power. The use of a two-lens combination reduces diffraction effects as shown in the lower figures, thereby minimizing the growth of intensity gradients due to nonlinear phase distortions. The effective filling factor is maintained for component placement through a post-relay distance of $2f$. From Ref. [180]

The following relation must be satisfied in this case for each relay imaging system [5]:

$$\frac{L_1 f_2}{f_1} + \frac{L_2 f_1}{f_2} - f_1 - f_2 = 0, \tag{8.46}$$

where f_1 and f_2 are the focal lengths of the lenses, L_1 is the distance from the aperture or its image plane to the first lens, and L_2 is the distance from the second lens to the image plane. If Eq. (8.46) is satisfied for each relay imaging system, the beam will be imaged, or relayed, through the entire amplifier system; e.g., an apodized or hard aperture will be imaged through the entire system to near or at the location of the target.

8.5.2 Elimination of Fresnel Diffraction Effects and Small-Scale Self-focusing Using Apodizing Devices

The simplest method of eliminating diffraction effects on light beam propagation is to underfill the cross section of the amplifier active element with the beam. To further increase the output intensity it is necessary to fill the amplifier aperture completely. In this case Fresnel diffraction rings appear in the beam's cross section. To avoid it, apodized or soft apertures are used. Figure 8.6 from Ref. [27] shows beam profile transformation for two super-Gaussian profiles with different fill factors.

Analytical expression was derived in Ref. [182] of the dependence of the small-scale self-focusing length in NGLS on the super-Gaussian beam profile, system geometry (air gap length), and incident intensity.

In Ref. [62], the experimental investigations were carried out of two schemes for shaping the spatial structure of beams with a 0.3-ns pulse duration and a 30-mm diameter. One of them contained a spatial filter with relay imaging and an apodized aperture. The second scheme contained a hard-edge aperture.

An apodized aperture with a high-damage threshold was fabricated by sintering a light-absorbing filter and transparent glass. The diameter of the apodized aperture at 10^{-2} I_{max} was 10.5 mm with a transmission function $\sim \exp\left[-6.91(r/r_1)^{10}\right]$, where $I(r_1) = 10^{-3} I_0$. The hard aperture of an equivalent size had the form of a circular hole in a metal disk. The apertures were mounted at 47 and 255 cm from the input of a 63-cm-long amplifier. The corresponding Fresnel numbers were 61 and 11. Figure 8.11 shows near-field photographic images at $\lambda = 1.064$ μm of the output of apodized and hard-edge apertures at an output fluence of 0.4, 1.2, and 1.6 J/cm^2 at a pulse duration of 0.3 ns in the case of a 255-cm aperture distance from the amplifier input.

At high power, the beam cross section acquires a mosaic structure resulting from the growth of small-scale intensity perturbations from Fresnel diffraction along the edges of the hard aperture. The spatial scale of the mosaic structure

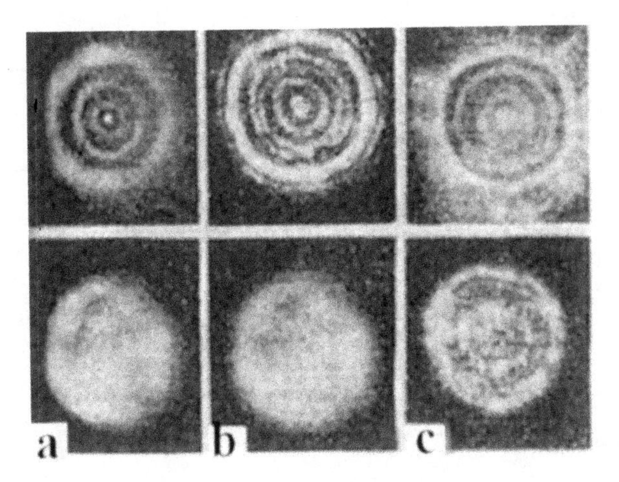

Fig. 8.11 Photographs of the beam cross-sections in the near-field zone for an apodized beam (lower series of photographs) and at the output of a hard-edge aperture (upper series) at the output of the amplifier in the case of output-energy densities (**a**) 0.4 J/cm^2; (**b**) 1.2 J/cm^2, and (**c**) 1.6 J/cm^2. The distance from the amplifier was 255 cm. Pulse duration was 0.3 ns. The sintered glass apodizer [61, 62] had a super-Gaussian transmission profile with $N = 10$ and $r = 5.25$ mm at the level of 10^{-2} $I_{\max}$. From Kryzhanovskii et al. [62]

depends on the spatial beam profile at the entrance to the nonlinear medium. At the fluence ~1 J/cm^2, large-angle laser-radiation scattering occurs for the hard-edge aperture. Such scattering is not observed for an apodized beam, although the extent of its decay into the mosaic structure also increases with increasing intensity. In the case of a 255-cm-aperture distance using the apodizer, the value of transmitted energy in the angle close to the diffraction limit (~0.3 mrad) increased *up to three times* in comparison with the hard-edge aperture. No damage was identified in this experiment at any intensity levels.

Based on the experiments described above, it is suggested in Ref. [62] that hard apertures can be used over shorter paths than can apodized apertures; several relay-imaging systems are required for long paths and hard apertures. Apodized apertures allow greater flexibility in designing multi-element amplifier systems. The best solution for enhancement of laser-beam uniformity is using apodized apertures with relay imaging [4, 62].

Super-Gaussian-transmission soft apertures have been investigated in the 1.06-μm Delfin laser installation of the P.N. Lebedev Physical Institute [91]. Smaller-diameter color-center/colloid apodizer in a SrF_2:Na crystal [89–91] provided $N=20$ with a r_0 = 5.3-mm transmission profile. A larger-diameter induced-absorption CaF_2:Pr apodizer [42, 81] with crystal diameter of 45 mm produced a beam profile with $N = 10$ and $r_0 = 12.5$ mm.

Figure 8.12 shows a schematic of the Delfin laser preamplifier system with 30-J, 3-ns output and transmission profiles of soft apertures. The figure illustrates the results of using these soft apertures in different parts of the Delfin laser. Beam cross sections in Fig. 8.12(a) are compared at a distance of 8 m from

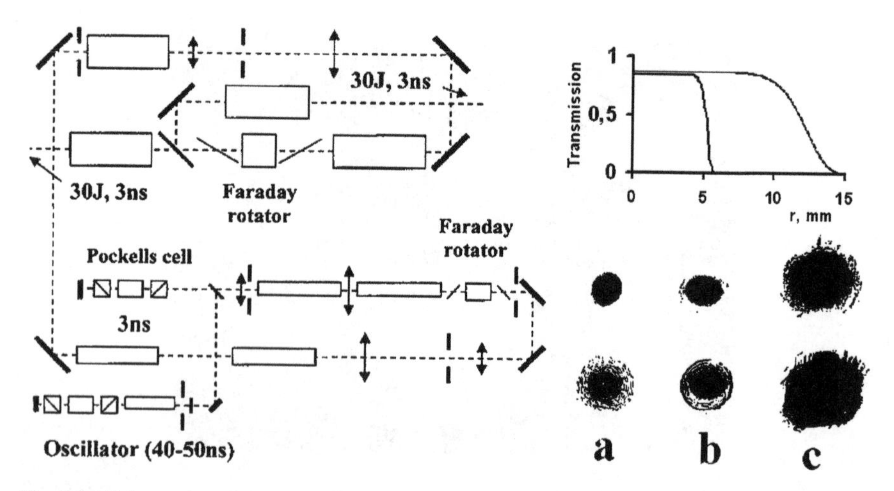

Fig. 8.12 Schematics of a preamplifier system of the Delfin laser; soft apertures' transmission profiles and beam cross sections at different distances from soft apertures (*top spots*) and hard-edge apertures (*bottom spots*). From Ref. [91]. See explanations in text

the smaller diameter soft aperture (top spot) and a similar-diameter hard-edge aperture (bottom spot) after beam passage through two amplifiers with 30 mm × 630 mm rods. Apertures were placed after the Faraday rotator. Figure 8.12(b) shows beam cross sections at 16 m from the output of the same soft aperture (top spot) after beam passage through four amplifiers: two with 20 mm × 630 mm rods and two 30 mm × 630 mm rods. In this case the soft aperture was placed between the Pockells cell and the first spatial filter.

For comparison, the bottom spot in Fig. 8.12(b) shows beam cross sections in the same system, but with the soft aperture removed. Figure 8.12(c) shows beam cross sections at 16 m from the larger-diameter soft aperture (top image), and in the same place without any additional apertures (bottom image). After the soft aperture, the beam passed through three amplifiers with rod dimensions of 45 mm × 630 mm. The larger-diameter soft aperture was placed near the input of the first 45 mm × 630 mm amplifier. We observed significant improvement in beam homogeneity with the soft apertures of Fig. 8.12.

Figure 8.13 shows no ripples in the beam cross section at 2 m from the induced-absorption apodized aperture made of CaF_2:Pr crystals with a 30-mm-beam diameter [42, 81], which is different from a 30-mm-diam, hard-edge aperture at an intensity of ~0.5 GW/cm^2 ($\lambda = 1.06$ μm).

We emphasize that apodized apertures do not hinder the formation of small-scale inhomogeneities resulting from laser-radiation scattering by random inclusions in the glass nor by dust on the face surfaces of the optical elements. Spatial filters, which today represent the primary optical elements in laser systems for small-scale self-focusing suppression, are employed to eliminate such laser-beam inhomogeneities.

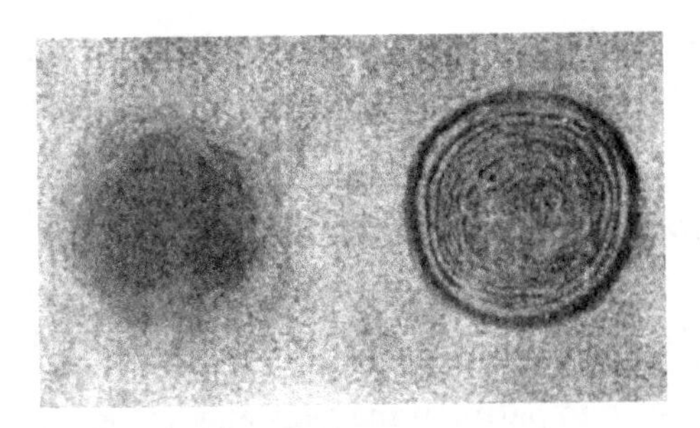

Fig. 8.13 The burn patterns on a photosensitive paper of the beam's cross section at 2 m from the output of CaF_2: Pr-induced absorption apodizer with a super-Gaussian transmission profile with $N = 5$ and $r_0 = 11$ mm (left) and a hard-edge aperture of 25 mm in diameter. Incident intensity ~0.5 GW/cm^2

8.5.3 Application of Divergent Beams

The application of divergent beams in laser systems is analyzed in Refs. [183] and [184]. The equivalence of plane and spherical beams [104, 141] will cause the small-scale self-focusing length $l_{\text{self}}^{\text{sphere}}$ and the increment Γ_{sphere} in the spherical-beam case to be expressed by the same formulas as l_{self} and Γ in the plane-beam case, which use $\tilde{E}^2$ instead of E^2. Because the intensity ratio at conjugate points is

$$\tilde{E}^2/E^2 = (f+z)^2 f^{-2} \equiv (f-\tilde{z})^{-2} f^2 \equiv M^2, \tag{8.47}$$

while the modulus of the wave number of the transverse perturbation in the case of a spherical wave is $|\kappa_{\text{sphere}}| = 2\pi/|\tilde{r}_\perp| = |\kappa| M(z)$, it follows

$$l_{\text{self}}^{\text{sphere}} = [M(z)]^{-2} l_{\text{self}}, \tag{8.48}$$

$$\Gamma^{\text{sphere}} = [M(z)]^2 \Gamma. \tag{8.49}$$

In the case of a divergent beam ($f < 0$), $l_{\text{self}}^{\text{sphere}} > l_{\text{self}}$, while $\Gamma_{\text{sphere}} < \Gamma$, is due to the diminishing beam intensity. The easiest focused transverse scale in the spherical-beam case is determined by the distance between the same beams as in a plane-wave beam, although it varies as $[M(z)]^{-1}$ with increasing z.

The most effective experimental technique is to employ a divergent beam for large-scale self-focusing suppression [24]. In the case of small-scale self-focusing the filamentary damage threshold grew when amplifying beams with a 10^{-2} rad divergence compared to the case when a beam with a 10^{-4} rad divergence is amplified [25].

Siegman considered small-scale self-focusing suppression using tapered (strongly converging or diverging) beams [159] using a model involving a combination of Newton's rings and nonlinearly induced Fresnel zone plates.

No exponential growth is predicted because of the lack of any Talbot effects in strongly tapered optical beams [159].

8.5.4 Partitioning of the Active Medium

The idea to split a nonlinear optical medium into segments separated by air gaps in order to suppress self-focusing was already discussed in the 1970s [21, 185]. For the NGLS it is possible to partition the glass amplifying medium into disks of 2–4 cm thickness [4, 5]. Small-scale self-focusing in a piecewise-continuous medium with air gaps was investigated experimentally in Refs. [21, 101]. It was demonstrated that nonlinear losses attributable to self-focusing are lower in disk elements than in rods for an identical total glass thickness due to the defocusing action of the air gaps (due to diffraction) on small-scale perturbations.

Recently an experiment with the propagation of femtosecond light pulses with powers of 2–6 P_{cr} through the layered glass medium had demonstrated the suppression of the whole-beam self-focusing [186]. A number of theoretical studies devoted to the analysis of self-focusing in a piecewise-continuous medium exist [5, 185–189]. It is, however, important to remember that a set of spatial frequencies having substantial gains through the system exists for disks with equidistant air gaps. The total small-scale self-focusing loss level will diminish for a nonequidistant configuration of the disks [189]. This is because the nonequidistant configuration will smooth out the frequency response of the spatial filter formed by the equidistant air gaps [189]. Changing the thickness and related positions of disk or short rod fragments along the powerful laser beamline can bring an up to 10 times decrease of power losses due to self-focusing [189]. This offers a flexible instrument for configuration optimization of the optical elements in a powerful laser channel.

There is another aspect to a physical picture of the nonlinear interaction of the powerful laser beam propagated in a medium with sliced optical elements. A phenomenon of formation of "hot images" (and related laser damage) in the sequence of disks was observed experimentally [21, 190]. This effect was explained as a nonlinear formation of holographic images of obscurations in laser beams inside the downstream disks [190, 191].

8.5.5 Using Circular Polarization

Isotropic nonlinear media, e.g., glass, become anisotropic under high-power laser radiation, and the axis of elliptical polarization of a beam rotates by an angle Ψ that is proportional to the beam intensity. This means that n_2 depends on a polarization state of high-power laser beams. In addition, if some intensity perturbations present in the beam, Ψ will depend on a beam-cross-sectional position. A depolarization of radiation may be observed in high-power neodymium laser systems [5].

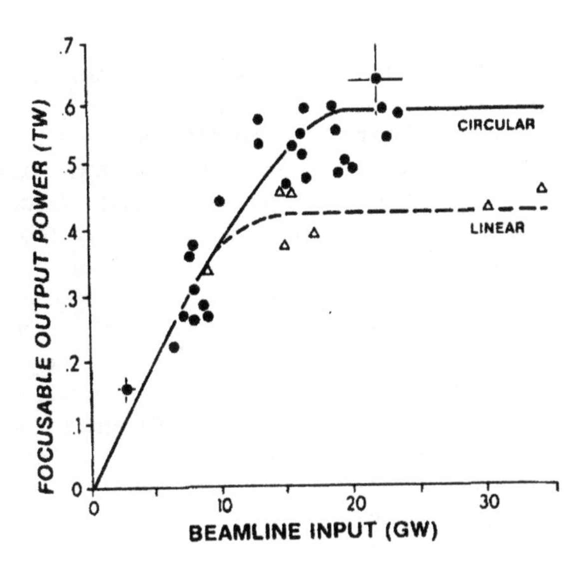

Fig. 8.14 Focusable output power as a function of the beamline input power for the GDL laser system of the University of Rochester [8] using linear and circularly polarized pulses of a 50-ps duration. From Seka et al. [193]

Circular polarization results in a lower n_2 than the linear polarization case. In Ref. [5] (see also [192]) the break-up integral value was calculated for the elliptically polarized beam. When the value of the ellipticity coefficient [192] increases from 0 (linear polarization) to 1 (circular polarization), the value of B diminishes by 1.5 times [5].

Figure 8.14 from book by Brown [4] and Ref. [193] shows both theoretical calculations using the *RAINBOW* code and experimental results obtained on the GDL system (University of Rochester) [8] of focusable output power at the terawatt level as a function of beamline input power for linear and circular polarization and a pulse duration of 50 ps. This GDL beamline consisted of seven rod amplifiers. Similar results were described in book by Mak et al. [5], where twice-higher than average brightness for circular polarization was reported, in comparison with linear polarization [192]. See also Ref. [194] and recent theoretical paper of Fibich et al. [195].

8.5.6 Coherence Limiting of Laser Radiation

An interest in coherence limiting of powerful NGLS radiation arose in connection with the requirements of the most homogeneous laser-intensity distribution on the target for achieving a large target compression ratio in inertial confinement fusion experiments. Combining spatial and temporal incoherence of laser radiation permits a reduction of the irradiation nonuniformity at the focal plane of the laser and on the target [196]. The requirement of decreasing the radiation coherence has prominent consequences for the optical scheme and properties of the powerful laser itself and in regard to suppression of self-focusing in the laser medium. The possibility of reducing the rate of growth of small-scale perturbations in the laser medium has been shown theoretically in Ref. [197] by diminishing the degree of coherence of the laser radiation. According to these calculations, the break-up integral is approximately proportional to the degree of the coherence [197].

Intensive research on beam smoothing on the target provided several different techniques and a better understanding of the role of radiation coherence on self-focusing suppression in the laser facility. The beam-smoothing methods use laser temporal incoherence itself, or induced spatial incoherence, or both. For instance, at incident intensities $\sim$1 GW/cm^2 the focused energy was increased by 50% when temporal coherence was reduced, by increasing a spectral width from 7 nm to 10.6 nm [198] because of small-scale self-focusing suppression. A method of induced spatial incoherence with echelons using temporal incoherence [199, 200] was provided, e.g., 30% exceeding focused energy level on the target in comparison with a coherent case over a broad energy range [201]. Beam smoothing by a multimode optical fiber was accomplished at the Phebus Nd:Glass Laser (Centre d'Études de Limeil-Valenton) [202]. Another active method for temporal and spatial incoherence in a laser system is smoothing by spectral dispersion (SSD) [203]. This method of producing spectrally dispersed broad-bandwidth light uses phase-modulators, gratings, and distributed phase plates.

Recently, the concept of developing partially coherent lasers has received a new experimental embodiment. The Nd:glass laser with a controllable function of mutual coherence of radiation was built in the P.N. Lebedev Physical Institute [204]. The design of the Q-switched multimode master oscillator allows one to control the degree of the spatial and temporal coherence and to change the divergence of the output radiation within the range 2×10^{-2} to 3×10^{-3} rad [204]. The amplifier cascades use long glass rods (680-mm length) and no spatial filters in the scheme. The absence of spatial filters between cascades, as well as the absence of complex optical devices for beam-wavefront corrections with a small number of active elements, allows for a simplified laser scheme decreasing the total number of optical elements [204–206].

The output-radiation parameters are: beam diameter, 45 mm; pulse energy, 150 J; pulse duration, 2.5 ns. The degree of spatial coherence is 0.05–0.015, the degree of temporal coherence is 3×10^{-3} to 10^{-4}. The average laser intensity on target is up to 5×10^{14} W/cm^2 [204–206]. The fluence of 10 J/cm^2 attained at the laser output is related to a substantial suppression of the beam small-scale self-focusing in the optical medium of this laser with coherence-limited radiation [204–206]. The risk of dangerous ripples of intensity at the laser-beam spatial profile due to Fresnel diffraction of the radiation on apertures or small obstacles is decreased greatly in the low-coherence laser. Apertures with both hard and soft profiles may be used for beam shaping in such a type of laser. The formation of radiation-intensity distribution with different profiles in the amplified channel, at the laser output, and in the focal plane at the target, was demonstrated in the low-coherence powerful Nd:glass laser using a relay-imaging technique [204].

It should be noted, however, that some of the induced-incoherence techniques involve intensity space–time modulations in the amplifiers [207]. For instance, an anomalous, unexpected intensity-saturation effect in the amplification of intense incoherent pulses at 2.5 GW/cm^2, ~1.5-ns pulse duration was observed on the Phebus laser [202, 208]. With an optical fiber for coherence reduction, the output power was limited to 4 TW, which is about half of the laser's capability, by a saturation-like effect in the large amplifier section. These temporally and spatially incoherent laser pulses showed a strong limitation in output power probably because of the near-field instantaneous nonuniform patterns, which seed self-focusing effects in the glass amplifiers [208, 209].

On the other hand, the SSD method apparently does not increase nonlinear effects. It was shown in this case, that amplitude perturbations due to phase-modulated propagation of such beams do not appreciably suffer self-focusing when injected into a properly relay-imaged nonlinear medium. Angular dispersion and frequency dispersion in the medium prevent nonlinear effects from contributing at the same location on all wavefronts along the direction of propagation [210].

Most research on self-focusing is based on the assumption of *coherent* radiation interaction with matter. The specific character of self-focusing, laser damage, stimulated scatterings, and other aspects of the nonlinear interaction of the intensive radiation of low coherence with laser glass and other optical media must be the subject for new research.

8.6 Summary

The problems of beam shaping and self-focusing in high-peak-power Nd:glass laser systems are considered. The authors discuss self-focusing as the primary nonlinear optical process limiting the brightness of such laser installations. Fresnel diffraction effects, whole-beam, and small-scale self-focusing, as well as methods of suppression of self-focusing, have been outlined. The super-Gaussian beam as an optimal beam profile with a relatively high fill factor and the lower self-focusing ripples are considered. Methods of preparation of super-Gausian beams are listed. Coherence limiting of laser radiation as a method to suppress self-focusing and to control the laser beam profile and the intensity distribution on target is briefly described.

Acknowledgments S.G.L. thanks P.P. Pashinin and A.M. Prokhorov for initiating some of the described research, I.K. Krasyuk for technical assistance in some experiments, Yu.K. Danilejko and T. Kessler for discussions, M.V. Pyatakhin for his suggestions (see also his book, Ref. [211]), and J. Kelly for providing some papers. The authors thank colleagues from the N.G. Basov Department of Quantum Radiophysics of the P.N. Lebedev

Physical Institute for support in some experiments, numerical modeling, and helpful discussions.

Some of the presented research of S.G.L. has been supported in part by the International Science (Soros) Foundation (Long-Term Grant Nos MF2000 and MF2300), the Government of Russian Federation (Grant No MF2300), and the Russian Foundation for Basic Research (Grant No 96-02-18762). S.G.L. acknowledges travel grant support to the IQEC 2005 symposium, "Self-Focusing: Past and Present," provided by the U.S. Air Force Office for Scientific Research.

References

1. N.G. Basov, O.N. Krokhin: Conditions for heating up of a plasma by the radiation from an optical generator: *Sov. Phys. JETP* **19**, 123–125 (1964).
2. N. Basov, P. Kriukov, S. Zakharov et al.: Experiments on the observation of neutron emission at a focus of high-power laser radiation on a lithium deuteride surface, *IEEE J. Quant. Electr.* **QE-4**, 864–867 (1968).
3. F. Floux, D. Cognard, L-G. Denoeud et al.: Nuclear fusion reactions in solid-deuterium laser-produced plasma, *Phys. Rev A* **1**, 821–824 (1970).
4. D.C. Brown: *High-Peak-Power Lasers*, Springer, Berlin, Heidelberg, NY (1981).
5. A.A. Mak, L.N. Soms, V.A. Fromzel et al.: *Nd:glass Lasers*, Nauka, Moscow, 288 pp. (1990).
6. J.M. Soures (Ed.): Selected *Papers on High-Power Lasers, SPIE Milestone Series of Selected Reprints* MS **43**, 711 pp. (1991).
7. W.W. Simmons, J.T. Hunt, W.E. Warren: Light propagation through large laser systems, *IEEE J. Quant. Electr.* **QE-17**, 1727–1744 (1981).
8. W. Seka, J.M. Soures, S.D. Jacobs et al.: GDL: a high-power 0.35-μm laser irradiation facility, *IEEE J. Quant. Electr.* **QE-17**, 1689–1693 (1981).
9. J. Bunkenberg, J. Boles, D.C. Brown et al.: The Omega high-power phosphate–glass system: design and performance, *IEEE J. Quant. Elec.* **17**, 1620–1628 (1981).
10. T.R. Boehly, R.S. Craxton, T.H. Hinterman et al.: The upgrade to the OMEGA laser system, *Rev. Sci. Instrum.* **66**, 508–510 (1995).
11. G.H. Miller, E.I. Moses, C.R. Wuest: The National Ignition Facility: enabling fusion ignition for the 21st century, *Nucl. Fusion* **44**, S228–S238 (2004).
12. N.G. Basov, P.G. Kryukov, Yu.V. Senatskii et al.: Production of powerful ultrashort light pulses in a neodymium glass laser, *Sov. Phys. JETP* **30**, 641–645 (1970).
13. N.G. Basov, I. Kertes, P.G. Kryukov et al.: Nonlinear losses in generators and amplifiers of ultrashort light pulses, *Sov. Phys. JETP* **33**, 289–293 (1971).
14. N.G. Bondarenko, I.V. Eremina, A.I. Makarov: Measurement of the coefficient of electronic nonlinearity in optical and laser glass, *Sov. J. Quantum Electron.* **8**, 482–484 (1978).
15. V.I. Bespalov, V.I. Talanov: Filamentary structure of light beams in nonlinear liquids, *JETP Lett.* **3**, 307–310 (1966).
16. S.G. Lukishova, I.K. Krasyuk, P.P. Pashinin et al.: Apodization of light beams as a method of brightness enhancement in neodymium glass laser installations. In: *Formation and Control of Optical Wave Fronts*, Proceed. of the General Physics Institute of the USSR Academy of Science, P.P. Pashinin (Ed.), vol. 7, 92–147, Nauka Publ., Moscow (1987).
17. A.J. Campillo, J.E. Pearson, S.L. Shapiro et al.: Fresnel diffraction effects in the design of high-power laser systems, *Appl. Phys. Lett* **23**, 85–87 (1973).
18. N.G. Basov, A.R. Zaritskii, S.D. Zakharov et al.: Generation of high-power light pulses at wavelengths 1.06 and 0.53 μm and their application in plasma heating. II. Neodymium–glass laser with a second harmonic converter, *Sov. J. Quant. Electron.* **2**, 533–535 (1973).

19. J.A. Fleck, Jr., C. Layne: A study of self-focusing damage in a high-power Nd:glass rod amplifier, *Appl. Phys. Lett.* **22**, 467–469 (1973).
20. N.B. Baranova, N.E. Bykovskii, B.Ya. Zel'dovich et al.: Diffraction and self-focusing during amplification of high-power light pulses. I. Development of diffraction and self-focusing in an amplifier, *Sov. J. Quant. Electron.* **4**, 1354–1361 (1975).
21. N.B. Baranova, N.E. Bykovskii, B.Ya. Zel'dovich et al.: Diffraction and self-focusing during amplification of high-power light pulses. II. Suppression of harmful influence of diffraction and self-focusing on a laser beam, *Sov. J. Quant. Electron.* **4**, 1362–1366 (1975).
22. A.J. Campillo, S.L. Shapiro, B.R. Suydam: Periodic breakup of optical beams due to self-focusing, *Appl. Phys. Lett.* **23**, 628–630 (1973).
23. A.J. Campillo, S.L. Shapiro, B.R. Suydam: Relationship of self-focusing to spatial instability modes, *Appl. Phys. Lett.* **24**, 178–180 (1974).
24. A.N. Zherikhin, Yu.A. Matveets, and S.V. Chekalin: Self-focusing limitation of brightness in amplification of ultrashort pulses in neodymium glass and yttrium aluminum garnet, *Sov. J. Quant. Electron.* **6**, 858–860 (1976).
25. P.G. Kryukov, Yu.A. Matveets, Yu.V. Senatskii et al.: Mechanisms of radiation energy and power limitation in the amplification of ultrashort pulses in neodymium glass lasers, *Sov. J. Quant. Electron.* **3**, 161–162 (1973).
26. P. Jacquinot, B. Roizen-Dossier: Apodization. In: *Progress in Optics*, E. Wolf (Ed.), vol. 3, 29–186, North–Holland: Amsterdam (1964).
27. V.R. Costich, B.C. Johnson: Apertures to shape high-power laser beams, *Laser Focus* **10**, (9), 43–46 (1974).
28. A.N. Zherikhin, P.G. Kryukov, Yu.A. Matveets et al.: Origin of the temporal structure of ultrashort laser pulses, *Sov. J. Quant. Electron.* **4**, 525–526 (1974).
29. D.M. Pennington, M.A. Henesian, R.W. Hellwarth: Nonlinear index of air at 1.053 μm, *Phys. Rev. A* **39**, 3003–3009 (1989).
30. M.A. Henesian, C.D. Swift, J.R. Murray: Stimulated rotational Raman scattering in nitrogen in long air paths, *Opt. Lett.* **10**, 565–567 (1985).
31. M. Born and E. Wolf, *Principles of Optics*, Cambridge University Press, Cambridge, UK, 952 pp. (2005).
32. A.E. Siegman: *Lasers*, University Science Books, Mill Valley, CA, 1283 pp. (1986).
33. J.R. Campbell, L.G. De Shazer: Near fields of truncated Gaussian apertures, *J. Opt. Soc. Amer.* **59**, 1427–1429 (1969).
34. J. Trenholme: A user-oriented axially symmetric diffraction code, *LLNL: Semiannual Report*, January–June, UCRL-50021-73-1, 46–47 (1973).
35. G.R. Hadley: Diffraction by apodized apertures, *IEEE J. Quant. Electron.* **QE-10**, 603–608 (1974).
36. A. Dubik, A. Sarzyński: A study of the problem of propagation and focusing of laser radiation in an apertured super-Gaussian form, *J. Techn. Phys. Polish Acad. Sci.* **25** (3–4), 441–445 (1984).
37. J.M. Auerbach, V.P. Karpenko: Serrated-aperture apodizers for high-energy laser systems, *Appl. Opt.* **33**, 3179–3183 (1994).
38. B.M. Van Wonterghem, J.R. Murray, J.H. Campbell et al.: Performance of a prototype for a large-aperture multipass Nd:glass laser for inertial confinement fusion, *Appl. Opt.* **36**, 4932–4853 (1997).
39. J. Weaver: Initial beam shaping for a fusion laser, *LLNL, Semiannual Report*, January–June, UCRL–50021–73–1, 48–52 (1973).
40. V.N. Alekseev, A.D. Starikov, V.N. Chernov: Optimization of the spatial profile of a high-power optical beam in the amplifier channel of a neodymium–glass laser system, *Sov. J. Quant. Electron.* **9** 1398–1402 (1979).
41. I.K. Krasyuk, S.G. Lukishova, D.M. Margolin et al.: Induced absorption soft apertures, *Sov. Tech. Phys. Lett.* **2**, 577–581, Leningrad (1976).

42. S.G. Lukishova: Apodized apertures for visible and near-infrared band powerful lasers, SPIE Milestone Series, *Selected Papers on Apodization: Coherent Optical Systems*, J.P. Mills and B.J. Thomson, (Eds.), MS **119**, 334–341 (1996).
43. S.G. Lukishova, N.R. Minhuey Mendez, V.V. Ter-Mikirtychev, T.V. Tulajkova: Improving the beam quality of solid-state laser systems using both outside and inside cavity devices with variable optical characteristics along the cross section, SPIE Milestone Series, *Selected Papers on Apodization: Coherent Optical Systems*, J.P. Mills and B.J. Thomson, (Eds.), MS **119**, 362–374 (1996).
44. A.J. Campillo, B. Carpenter, B.E. Newman et al.: Soft apertures for reducing damage in high-power laser amplifier systems, *Opt. Commun.* **10** (4), 313–315 (1974).
45. A.J. Campillo, B.E. Newman, S.L. Shapiro et al.: Method and apparatus for reducing diffraction–induced damage in high-power laser amplifier systems, *U.S. Patent* No. 3,935,545, January 27, 1976.
46. E.W.S. Hee: Fabrication of apodized apertures for laser beam attenuation, *Opt. Laser Technol.* **7** (2), 75–79 (1975).
47. Five-gigawatt coating reduces self-focusing in laser glass, *Laser Focus* **9** (9), 46 (1973).
48. G. Emiliani, A. Piegari, S. de Silvestri et al.: Optical coatings with variable reflectance for laser mirrors, *Appl. Opt.* **28**, 2832–2837 (1989).
49. G. Duplain, P.G. Verly, J.A. Dobrowolski et al.: Graded-reflectance mirrors for beam quality control in laser resonators, *Appl. Opt.* **32**, 1145–1153 (1993).
50. S.G. Lukishova, S.A. Chetkin, N.V. Mettus et al.: Techniques for fabrication of multilayer dielectric graded–reflectivity mirrors and their use for enhancement of the brightness of the radiation from a multimode Nd^{3+}:YAG laser with a stable cavity, *Quant. Electron.* **26**, 1014–1017 (1996).
51. *PureAppl Opt*, Special Issue, **3** (4), 417–599, Papers presented at the workshop on laser resonators with graded reflectance mirrors, 8–9 September 1993 (1994).
52. S.B. Arifzhanov, R.A. Ganeev, A.A. Gulamov et al.: Formation of a beam of high optical quality in a multistage neodymium laser, *Sov. J. Quant. Electron.* **11**, 745–749 (1981).
53. P. Giacomo, B. Roizen-Dossier, S. Roizen: Préparation, par evaporation sous vide, d'apodiseurs circulaires, *J. Phys. (France)* **25** (1–2), 285–290 (1964).
54. C. Dorrer, J.D. Zuegel: Design and analysis of beam apodizers using error diffusion, Optical Society of America, Technical Digest Series, *Conf. on Lasers and Electro Optics CLEO 2006*, paper JWD1 (2006).
55. E.S. Bliss, D.R. Speck: Apodized aperture assembly for high-power lasers, *U.S. Patent* No. 3,867,017, February 18, 1975.
56. R.L. Nolen, Jr., L.D. Siebert: High-power laser apodizer, *U.S. Patent* No. 4,017,164, April 12, 1977.
57. S. Jorna, L.D. Siebert, K.A. Bruekner: Apodizer aperture for lasers, *U.S. Patent* No. 3,990,786, November 9, 1976.
58. L.M. Vinogradsky, V.A. Kargin, S.K. Sobolev et al.: Soft diaphragms for apodization of powerful laser beams, *Proceed. SPIE* **3889**, 849–860 (2000).
59. Yu.V. Senatsky, N.E. Bykovsky, L.M. Vinogradsky et al.: Apodizers for single-mode lasing, *Bull. Russ. Acad. Sci. Phys.* **66**, 1008–1012 (2002).
60. L.M. Vinogradsky, S.K. Sobolev, I.G. Zubarev et al.: Development of the nonlinear optical element for light beam apodization and large-aperture laser amplifier decoupling, *Proc. SPIE* **3683**, 186–193 (1998).
61. A.D. Tsvetkov, N.I. Potapova, O.S. Shchavelev et al.: Apodized glass aperture with a super-Gaussian transmission function, *Zh. Prikl. Spektrosk. (Sov. J. Appl. Spectrosc.)* **45**, 1022–1025, Minsk (1986).
62. V.I. Kryzhanovskĭ, B.M. Sedov, V.A. Serebryakov et al.: Formation of the spatial structure of radiation in solid-state laser systems by apodizing and hard apertures, *Sov. J. Quant. Electron.* **13**, 194–198 (1983).

63. W. Simmons, B.C. Johnson, A. Glass: Design and potential uses of nonuniform Pockels cells, *Laser Program, Annual Report*: 1974, 138–142, UCRL-50021-74, LLNL (1975).
64. W.W. Simmons, G.W. Leppelmier, B.C. Johnson: Optical beam shaping devices using polarization effects, *Appl. Opt.* **13**, 1629–1632 (1974).
65. S.B. Papernyĭ, V.A. Serebryakov, V.E. Yashin: Formation of a smooth transverse distribution of intensity in a light beam by a phase-rotating plate, *Sov. J. Quant. Electron.* **8**, 1165–1166 (1978).
66. G. Giuliani, Y.K. Park, R.L. Byer: Radial birefringent element and its application to laser resonator design, *Opt. Lett.* **5**, 491–493 (1980).
67. G. Dube: Total internal reflection apodizers, *Opt. Commun.* **12**, 344–347 (1974).
68. J.A. Hoffnagle, C.M. Jefferson: Beam shaping with a plano–aspheric lens pair, *Opt. Eng.* **42**, 3090–3099 (2003).
69. J.A. Hoffnagle, C.M. Jefferson: Refractive optical system that converts a laser beam to a collimated flat-top beam, *U.S. Patent* No. 6,295,168, September 25, 2001.
70. Y. Asahara, T. Izumitani: Process of producing soft aperture filter, *U.S. Patent*, No. 4,108,621, August 22, 1978.
71. M.E. Brodov, F.F. Kamenets, V.V. Korobkin et al.: Controlled-inversion-profile amplifier as soft aperture, *Sov. J. Quant. Electron.* **9**, 224–225 (1979).
72. B.J. Feldman, S.J. Gitomer: Annular lens soft aperture for high-power laser systems, *Appl. Opt.* **15**, 1379–1380 (1976).
73. N.I. Potapova, A.D. Tsvetkov: Apodization of laser radiation by phase apertures, *Sov. J. Quant. Electron.* **22**, 419–422 (1992).
74. M.A. Summers, W.F. Hagen, R.D. Boyd: Scattering apodizer for laser beams, *U.S. Patent* No. 4,537,475 (1985).
75. N. Rizvi, D. Rodkiss, C. Panson: Apodizer development, *Rutherford Appleton Lab., Annual Rep.*, The Central Laser Facility CLF-87-041, 113–114 (1987).
76. M.V. Pyatakhin, Yu.V. Senatsky: Formation of the intensity distribution in laser beams due to diffraction on structures of small size optical inhomogeneities, *J. Russian Laser Res.* **23** (4), 332–346 (2002).
77. Yu.V. Senatsky: Laser beam apodization by light scattering, Optical Society of America, Technical Digest Series, *Conf. on Lasers and Electro Optics CLEO 2001*, Paper CTuM8, p. 160, May 6–11, Baltimore (2001).
78. Yu. Senatsky: Soft diaphragm for lasers, *Russian Patent* No. 2,163,386, March 19, 1999.
79. I. Zubarev, M. Pyatakhin, Yu. Senatsky: Method of soft diaphragm formation, *Russian Patent*, No. 2,140,695, April 23, 1998.
80. V.N. Belyaev, N.E. Bykovskĭ, Yu V Senatskĭ et al.: Formation, by penetrating radiation, of absorbing layers in the optical medium of a neodymium laser, *Sov. J. Quant. Electron.* **6**, 1246–1247 (1976).
81. B.G. Gorshkov, V.K. Ivanchenko, V.K. Karpovich et al.: Apodizing induced-absorption apertures with a large optical beam diameter and their application in high-power 1.06 μm laser systems, *Sov. J. Quant. Electron.* **15**, 959–962 (1985).
82. V.K. Ivanchenko, S.G. Lukishova, D.M. Margolin et al.: The method of fabrication of apodized apertures, *The Soviet Invention Certificate*, No. 1,098,409 (1984).
83. I.K. Krasyuk, S.G. Lukishova, B.M. Terentiev et al.: The method of fabrication of apodized apertures, *The Soviet Invention Certificate*, No. 1,019,583 (1985).
84. S.G. Lukishova, N.R. Minhuey Mendez, T.V. Tulaikova: Investigation of a soft aperture formed by photooxidation of a rare-earth impurity in fluorite and used as an intracavity component in a YAG : Er^{3+} laser, *Quant. Electron.* **24**, 117–119 (1994).
85. S.G. Lukishova, A.Z. Obidin, S.Kh. Vartapetov et al.: Photochemical changes of rare-earth valent state in gamma-irradiated CaF_2:Pr crystals by the eximer laser radiation: investigation and application, *Proceed. SPIE* **1503**, 338–345 (1991).
86. S.G. Lukishova: Some problems of spatial and temporal profile formation of laser radiation, *Ph.D Thesis*, Moscow Inst. of Phys. & Technology (1977).

87. S.G. Lukishova, P.P. Pashinin, S.Kh. Batygov et al.: High-power laser beam shaping using apodized apertures, *Laser Part. Beams* **8** (1–2), 349–360 (1990).
88. V.A. Arkhangelskaya, S.Kh. Batygov, S.G. Lukishova et al.: The method of making of amplitude filters, *The Soviet Invention Certificate*, No. 1,647,044A1 (1990). See also A.S. Shcheulin, T.S. Semenova, L.F. Koryakina et al.: Additive coloration of crystals of calcium and cadmium fluorides, *Opt. and Spectrosc.* **103**, 660–664 (2007).
89. A.E. Poletimov, A.S. Shcheulin, I.L. Yanovskaya: Apodizing apertures for visible and IR lasers, *Sov. J. Quant. Electron.* **22**, 927–930 (1993).
90. V.A. Arkhangelskaya, S.G. Lukishova, A.E. Poletimov et al.: Apodized aperture for high–peak power near infrared and visible lasers without phase shift at the edges, Optical Society of America, Technical Digest Series, *Conf. on Lasers and Electro Optics CLEO 92*, paper CThQ6 (1992).
91. S.G. Lukishova, N.E. Bykovsky, A.E. Poletimov, A.S. Scheulin: Apodization by color centres apertures on the Delfin laser, Optical Society of America, Technical Digest Series, *Conf. on Lasers and Electro Optics CLEO 94*, **8**, 135–136 (1994).
92. J.C. Diels: Apodized aperture using frustrated total reflection, *Appl. Opt.* **14**, 2810–2811 (1975).
93. I.K. Krasyuk, S.G. Lukishova, P.P. Pashinin et al.: Formation of the radial distribution of intensity in a laser beam by "soft" apertures, *Sov J. Quant. Electron.* **6**, 725–727 (1976).
94. S.G. Lukishova, S.A. Kovtonuk, A.A. Ermakov et al.: Dielectric film deposition with cross-section variable thickness for amplitude filters on the basis of frustrated total internal reflection, SPIE Milestone Series, *Selected Papers on Apodization: Coherent Optical Systems*, J.P. Mills and B.J. Thomson, (Eds.), MS **119**, 447–458 (1996).
95. J.-C. Lee, S.D. Jacobs, K.J. Skerrett: Laser beam apodizer utilizing gradient–index optical effects in cholesteric liquid crystals, *Opt. Eng.* **30**, 330–336 (1991).
96. S.D. Jacobs, L.A. Cerqua: Optical apparatus using liquid crystals for shaping the spatial intensity of optical beams having designated wavelengths, *U.S. Patent*, No. 4,679,911, July 14, 1987.
97. J.-C. Lee, S. Jacobs: Gradient index liquid crystal devices and method of fabrication there of, *U.S. Patent*, No. 5,061,046, October 29, 1991.
98. A. Penzkofer, W. Frohlich: Apodizing of intense laser beams with saturable dyes, *Opt. Commun.* **28**, 197–201 (1979).
99. D.R. Neil, J.D. Mansell: Apodized micro-lenses for Hartmann wavefront sensing and method for fabricating desired profiles, *U.S. Patent* No. 6,864,043 (2003).
100. J.P. Mills, B.J. Thomson, (Eds.): *Selected Papers on Apodization: Coherent Optical Systems*, SPIE Milestone Series of Selected Reprints MS **119** (1996).
101. N.B. Baranova, N.E. Bykovsky, Yu.V. Senatsky et al.: Nonlinear processes in the optical medium of a high-power neodymium laser, *J. Sov. Laser Res.* **1**, 53–88 (1980).
102. N.G. Basov, V.S. Zuev, P.G. Kryukov et al.: Generation and amplification of high-intensity light pulses in neodymium glass, *Sov. Phys. JETP* **27**, 410–414 (1968).
103. J.L. Emmet, W.F. Krupke, J.B. Trenholme: Future development of high-power solid-state laser systems, *Sov. J. Quant. Electron.* **13**, 1–23 (1983).
104. J.H. Marburger: Self–focusing: theory, *Prog. Quant. Electr.* **4**, 35–110 (1975). See also this volume, Chapter 2 for this reprint
105. R.W. Boyd: *Nonlinear Optics*. Chapter 7: Processes resulting from the intensity-dependent refractive index, pp. 311–370, Academic Press, San Diego (2003).
106. R. Hellwarth, J. Cherlow, T.-T. Yang: Origin and frequency dependence of nonlinear optical susceptibilities of glasses, *Phys. Rev.* **B 11**, 964–967 (1975).
107. R. Hellwarth: Third-order optical susceptibilities of liquids and solids, *Prog. Quant. Electron.* **5**, 1–68 (1977).
108. G.A. Askar'yan: Effects of the gradient of a strong electromagnetic beam on electrons and atoms, *Sov. Phys. JETP* **15**, 1088–1090 (1962). See also this volume, Chapter 11.1 for this reprint.

109. M. Hercher: Laser-induced damage in transparent media, *J. Opt. Soc. Am.* **54**, 563 (1964). See also this volume, Chapter 11.3 for this reprint.
110. N.F. Pilipetskii, A.R. Rustamov: Observation of self–focusing of light in liquids, *JETP Lett.* **2**, 55–56 (1965). See also Chapter 15, Figure 15.1 of this volume.
111. Y.R. Shen: Self-focusing: experimental, *Prog. Quant. Electr.* **4**, 1–34 (1975).
112. Y.R. Shen: *The Principles of Nonlinear Optics*. Chapter 17: Self-focusing, pp. 303–333 , John Wiley & Sons, New York (1984).
113. O. Svelto: Self-focusing, self-trapping, and self-phase modulation of laser beams. In: *Progress in Optics XII*, E. Wolf (Ed.), 3–51, North–Holland, Amsterdam (1974).
114. S.A. Akhmanov, A.P. Sukhorukov, R.V. Khokhlov: Self-focusing and diffraction of light in a nonlinear medium, *Sov. Phys. Uspekhi* **10**, 609–636 (1968).
115. S.A. Akhmanov, R.V. Khokhlov, A.P. Sukhorukov: Self-focusing, self-defocusing and self-modulation of laser beams. In: *Laser Handbook*, vol. 2, E3, 1151–1228, F.T. Arecchi, E.O. Schulz-Dubois (Eds.), North-Holland, Amsterdam (1972).
116. V.N. Lugovoi, A.M. Prokhorov: Theory of the propagation of high-power laser radiation in a nonlinear medium, *Sov. Phys. Uspekhi* **16**, 658–679 (1974).
117. S.N. Vlasov, V.I. Talanov: *Wave Self-focusing*, Institute of Applied Physics of the Russian Academy of Science, Nizhny Novgorod, 220 pp. (1997).
118. M.S. Sodha, A.K. Ghatak, V.K. Tripathi: *Self-focusing of Laser Beams in Dielectrics, Plasma, and Semiconductors*, Tata McGraw-Hill, New Delhi (1974).
119. V.I. Talanov: Self-focusing of electromagnetic waves in nonlinear media. *Izv Vuzov Radiophysica*, **7**, 564 –565 (1964). See also this volume, Chapter 11.2 for the first English translation of this paper.
120. R.Y. Chiao, E. Garmire, C.H. Townes: Self-trapping of optical beams, *Phys Rev Lett* **13**, 479–482 (1964).
121. P.L. Kelley: Self-focusing of optical beams, *Phys. Rev. Lett.* **15**, 1005–1008 (1965).
122. W.G. Wagner, H.A. Haus, J.H. Marburger: Large-scale self-trapping of optical beams in the paraxial approximation, *Phys. Rev.* **175**, 256–266 (1968).
123. Yu.K. Danilejko, T.P. Lebedeva, A.A. Manenkov et al.: Self-focusing of laser beams with various spatial profiles of incident radiation, *Sov. Phys. JETP* **53**, 247–252 (1981).
124. A.L. Dyshko, V.N. Lugovoi, A.M. Prokhorov: Self-focusing of intense light beams, *JETP Lett.* **6**, 146–148 (1967).
125. A.L. Dyshko, V.N. Lugovoi, A.M. Prokhorov: Multifocus structure of a light beam in a nonlinear medium, *Sov. Phys. JETP* **34**, 1235–1241 (1972).
126. T.P. Lebedeva: The effect of the amplitude and phase profile of a laser beam on self-action processes, *Ph.D. Thesis*, General Physics Institute, Moscow (1984).
127. A.A. Amosov, N.S. Bakhvalov, Ya.M. Zhileikin et al.: Self-focusing of wave beams with a plateau-shaped intensity distribution, *JETP Lett.* **30**, 108–111 (1979).
128. M.D. Feit, J.A. Fleck, Jr.: Beam nonparaxiality, filament formation, and beam breakup in the self-focusing of optical beams, *J. Opt. Soc. Am. B* **5**, 633–640 (1988).
129. N.I. Lipatov, A.A. Manenkov, A.M. Prokhorov: Standing pattern of self-focusing points of laser radiation in glass, *JETP Lett.* **11**, 300–302 (1970).
130. V.V. Korobkin, A.M. Prokhorov, R.V. Serov et al.: Self-focusing filaments as a result of the motion of focal points, *JETP Lett.* **11**, 94–96 (1970).
131. V.N. Lugovoi, A.M. Prokhorov: A possible explanation of the small-scale self-focusing filaments, *JETP Lett.* **7**, 117–119 (1968).
132. V.V. Korobkin, A.J. Alcock: Self-focusing effects associated with laser-induced air breakdown, *Phys. Rev. Lett.* **21**, 1433–1436 (1968).
133. M.M. Loy, Y.R. Shen: Small-scale filaments in liquids and tracks of moving foci, *Phys. Rev. Lett.* **22**, 994–997 (1969).
134. G.M. Zverev, E.K. Maldutis, V.A. Pashkov: Development of self-focusing filaments in solid dielectrics, *JETP Lett.* **9**, 61–63 (1969).

135. J.H. Marburger: Self-focusing as a pulse sharpening mechanism, *IEEE J. Quantum Electron.* **QE–3**, 415–416 (1967).
136. G.M. Fraiman: Asymptotic stability of manifold of self-similar solutions in self-focusing, *Sov Phys JETP* **61**, 228–233 (1985).
137. K.D. Moll, A.L. Gaeta and G. Fibich: Self–similar optical wave collapse: observation of the Townes profile, *Phys. Rev. Lett.* **90**, 203902 (2003).
138. L. Bergé, C. Gouédard, J. Schjodt-Eriksen et al.: Filamentation patterns in Kerr media vs. beam shape robustness, nonlinear saturation and polarization states, *Physica D* **176**, 181–211 (2003).
139. T.D. Grow, A.A. Ishaaya, L.T. Vuong et al.: Collapse dynamic of super-Gaussian beams, *Opt. Express* **14**, 5468–5475 (2006).
140. G. Fibich, N. Gavich, X.P. Wang: New singular solutions of the nonlinear Schrödinger equation, *Physica D* **211**, 193–220 (2005).
141. V.I. Talanov: Focusing of light in cubic media, *JETP Lett.* **11**, 199–201 (1970).
142. E.S. Bliss, J.T. Hunt, P.A. Renard et al.: Effect of nonlinear propagation on laser focusing properties, *IEEE J. Quant. Electron.* **12**, 402–406 (1976).
143. J.T. Hunt, J.A. Glaze, W.W. Simmons et al.: Suppression of self-focusing through low-pass spatial filtering and relay imaging, *Appl. Opt.* **17**, 2053–2057 (1978).
144. J.T. Hunt, P.A. Renard, R.G. Nelson: Focusing properties of an aberrated laser beam, *Appl. Opt.* **15**, 1458–1464 (1978).
145. E.S. Bliss, J.T. Hunt, P.A. Renard et al.: Whole-beam self-focusing, focal zoom, *Laser Program, Annual Report*, 1975, UCRL-50021-75, 225–227, LLNL (1976).
146. A.J. Campillo, R.A. Fisher, R.C. Hyer et al.: Streak camera investigation of the self-focusing onset in glass, *Appl. Phys. Lett.* **25**, 408–410 (1974).
147. A.J. Glass: A rational definition of index nonlinearity in self-focusing media. In: *Laser-Induced Damage in Optical Materials*, A.J. Glass, A.H. Guenther (Eds.), National Bureau of Standards Special Publ. **387**, 36–41 (1973).
148. J. Marburger, R. Jokipii, A.J. Glass et al.: Homogeneity requirements for minimizing self-focusing damage. In: *Laser-Induced Damage in Optical Materials*, A.J. Glass, A.H. Guenther (Eds.), National Bureau of Standards Special Publ. **387**, 49–56 (1973).
149. B.R. Suydam: Self-focusing of very powerful laser beams. In: *Laser-Induced Damage in Optical Materials*, A.J. Glass, A.H. Guenther (Eds.), National Bureau of Standards Special Publ. **387**, 42–48 (1973).
150. E.S. Bliss, D.R. Speck, J.F. Holzrichter et al.: Propagation of a high-intensity laser pulse with small-scale intensity modulation, *Appl. Phys. Lett.* **25**, 448–450 (1974).
151. J.B. Trenholme: Review of small-signal theory, *Laser Program, Annual Report*, 1974, UCRL-50021-74, LLNL, 179–191 (1975).
152. K.A. Bruekner, S. Jorna: Linear instability theory of laser propagation in fluids, *Phys. Rev. Lett.* **17**, 78–81 (1966).
153. L.A. Bol'shov, L.M. Degtryaryev, A.N. Dykhne et al.: Numerical investigation of small-scale self-focusing of light pulses in neodymium glass amplifiers, *Preprint*, the Keldysh Institute of Applied Mathematics, No. 109, 31 pp, Moscow (1979).
154. B.R. Suydam: Self-focusing of very powerful laser beams II, *IEEE J. Quant. Elec.* **10**, 837– 843 (1974).
155. B.R. Suydam: Effect of refractive-index nonlinearity on the optical quality of high-power laser beams, *IEEE J. Quant. Elec.* **11**, 225–230 (1975).
156. R. Jokipii, J. Marburger: Homogeniety requirements for minimizing self-focusing damage by strong electromagnetic waves, *Appl. Phys. Lett.* **23**, 696–698 (1973).
157. S.C. Abbi, N.C. Kothari: Theory of filament formation in self-focusing media, *Phys. Rev. Lett.* **43**, 1929–1931 (1979).
158. S.C. Abbi, N.C. Kothari: Growth of Gaussian instabilities in Gaussian laser beams, *J. Appl. Phys.* **5**, 1385–1387 (1980).

159. A.E. Siegman: Small-scale self-focusing effects in tapered optical beams, *Memo for File*, www.stanford.edu/~siegman/self_focusing_memo.pdf , 1–13 (2002).
160. D.T. Attwood, E.S. Bliss, E.L. Pierce et al.: Laser frequency doubling in the presence of small-scale beam breakup, *IEEE J. Quant. Elec.* **12**, 203–204 (1976).
161. V.N. Alekseev, A.D. Starikov, A.V. Charukhchev et al.: Enhancement of the brightness of radiation from a high-power phosphate–glass Nd^{3+} laser by spatial filtering of the beam in an amplifying channel, *Sov. J. Quant. Electron.* **9**, 981–984 (1979).
162. D. Strickland, G. Mourou: Compression of amplified chirped optical pulses, *Opt. Commun.* **56**, 219–221 (1985).
163. C.N. Danson, P.A. Brummitt, R.J. Clarke et al.: Vulcan petawatt: an ultra-high-intensity interaction facility, *Nucl Fusion* **44**, S239–S249 (2004).
164. Y. Kitagawa, H. Fujita, R. Kodama et al.: Prepulse-free petawatt laser for a fast ignitor, *IEEE J. Quantum Electron.* **40**, 281–293 (2004).
165. L.J. Waxer, D.N. Maywar, J.H. Kelly et al.: High-energy petawatt capability for the Omega laser, *Opt. Photonics News* **16**, 30–36 (2005).
166. M.D. Perry, D. Pennington, B.C. Stuart et al.: Petawatt laser pulses, *Opt. Lett.* **24**, 160–162 (1999).
167. M.J. Weber, C.B. Layne, R.A. Saroyan et al.: Low-index fluoride glasses for high-power Nd lasers, *Opt. Commun.* **18**, 171–172 (1975).
168. R.H. Lehmberg, J. Reintjes, R.C. Eckardt: Two-photon resonantly enhanced self-defocusing in cross-section vapor at 1.06 μm, *Appl. Phys. Lett.* **25**, 374–376 (1974).
169. O.A. Konoplev, D.D. Meyerhofer: Cancelation of B -integral accumulation for CPA lasers, *IEEE J. Selected Topics Quant. Electron.* **4**, 459–469 (1998).
170. N.E. Bykovskii, Yu.V. Senatskii: Enhancement of the spatial homogeneity of the intensity distribution in high-power laser beams, *Preprint of the P.N. Lebedev Physical Institute*, No. 15, 9 pp., Moscow (1977).
171. J.P. Caumes, L. Videau, C. Rouyer: Direct measurement of wave-front distortion induced during second-harmonic generation: application to breakup-integral compensation, *Opt Lett* **29**, 899–901 (2004).
172. C.E. Max, W.C. Mead, I.J. Thomson: Mechanics of the plasma spatial filter for high-power lasers, *Appl. Phys. Lett.* **29**, 783–785 (1976).
173. B.Ya. Zel'dovich, N.F. Pilipetsky, V.V. Shkunov: *Principles of Phase Conjugation*, 250 pp, Springer-Verlag, New York (1985).
174. V.I. Bespalov, G.A. Pasmanik: *Nonlinear Optics and Adaptive Systems*, Nauka, Moscow, 136 pp (1986).
175. R.A. Fisher, (Ed.): *Optical Phase Conjugation*, Academic Press, New York, 636 pp. (1983).
176. O.Yu. Nosach, V.I. Popovichev, V.V. Ragul'skii: Cancellation of phase distortions in an amplifying medium with a "Brillouin mirror," *JETP Lett.* **16**, 435–438 (1972).
177. A.A. Mak, V.A. Serebryakov, V.E. Yashin: Suppression of self-focusing in spatially incoherent light beams, in J.M. Soures, (Ed.), *Selected Papers on High–Power Lasers*, SPIE Milestone Series of Selected Reprints MS **43**, 460–461 (1991).
178. H. Maillotte, J. Monneret, A. Barthelemy et al.: Laser beam self-splitting into solitons by optical Kerr nonlinearity, *Opt. Commun.* **109**, 265–271 (1994).
179. M. Jain, A.J. Merriam, A. Kasapi et al.: Elimination of optical self-focusing by population trapping, *Phys. Rev. Lett.* **75**, 4385–4388 (1995).
180. W.W. Simmons, W.F. Hagen, J.T. Hunt et al.: Performance improvements through image relaying, *Laser Program Annual Report*, 1976, UCRL-50021-76, Part 2-1.4, pp. 2-19–2-28, LLNL (1977).
181. W.W. Simmons, J.E. Murray, F. Rainer et al: Design, theory, and performance of a high-intensity spatial filter, *Laser Program, Annual Report*, 1974, UCRL-50021-74, 169–174, LLNL (1975).
182. N.N. Rozanov, V.A. Smirnov: Small-scale self-focusing of confined beams, *Sov. J. Quant. Electron.* **8**, 1429–1435 (1978).

183. Yu.V. Senatskii: Active elements for high-power neodymium lasers, *Sov. J. Quant. Electron.* **1**, 521–523 (1972).
184. M.P. Vanyukov, V.I. Kryzhanovskii, V.A. Serebryakov et al.: Laser systems for the generation of picosecond high-irradiance light pulses, *Sov. J. Quant. Electron.* **1**, 483–488 (1972).
185. J.B. Trenholme: Theory of irregularity growth on laser beams, *Laser Program, Annual Report*, 1975, UCRL-50021-75, 237–242, LLNL (1976).
186. M. Centurion, M.A. Porter, P.G. Kevrekidis, D. Psaltis: Nonlinearity management in optics: experiment, theory, and simulation, *Phys. Rev. Lett.* **97**, 033903 (2006).
187. S.M. Babichenko, N.E. Bykovsky, Yu.V. Senatsky: Laser beam self-focusing in nonlinear medium with local inhomogeneities, *Preprint* N 14, 17 pp., P.N. Lebedev Physical Institute, Moscow (1981).
188. S.M. Babichenko, N.E. Bykovskiĭ, Yu.V. Senatskiĭ: Feasibility of reducing nonlinear losses in the case of small-scale self-focusing in a piecewise-continuous medium, *Sov. J. Quant. Electron.* **12**, 105–107 (1982).
189. N.E. Bykovsky, V.V. Ivanov, Yu.V. Senatsky: Intensity profiles of local perturbations in a laser beam propagating in a nonlinear medium, *Proceedings of the P.N. Lebedev Physical Institute* **149**, 150–161 (1985), Nauka Publ., Moscow.
190. C.C. Widmayer, D. Milam, S.P. deSzoeke: Nonlinear formation of holographic images of obscurations in laser beams, *Appl. Opt.* **36**, 9342–9347 (1997).
191. J.H. Hunt, K.R. Manes, P.A. Renard: Hot images from obscurations, *Appl. Opt.* **32**, 5973–5982 (1993).
192. S.N. Vlasov, V.I. Kryzhanovskiĭ, V.E. Yashin: Use of circularly polarized optical beams to suppress self-focusing instability in a nonlinear cubic medium with repeaters, *Sov. J. Quant. Electron.* **12**, 7–10 (1982).
193. W. Seka, J. Soures, O. Lewis et al.: High-power phosphate–glass laser system: design and performance characteristics, *Appl. Opt.* **19**, 409–419 (1980).
194. D. Auric, A. Labadens: On the use of circulary polarized beam to reduce the self-focusing effect in a glass rod amplifier, *Opt. Commun.* **21**, 241–242 (1977).
195. G. Fibich, B. Ilan: Self-focusing of circularly polarized beams, *Phys. Rev. E* **67**, 036622 (2003).
196. H.T. Powell, T.J. Kessler, (Eds.): *Laser Coherence Control: Technology and Applications, Proc. SPIE* **1870**, 200 pp (1993).
197. I.V. Alexandrova, N.G. Basov, A.E. Danilov et al.: The effects of small-scale perturbations on the brightness of laser radiation in laser fusion experiments, *Laser Part. Beams* **1**, pt. 3, 241–250 (1983).
198. A.E. Danilov, V.V. Orlov, S.M. Savchenko et al.: Investigation of the influence of the spectral composition of radiation on amplification in a neodymium glass, *Sov. J. Quant. Electron.* **15**, 139–140 (1985).
199. R.H. Lehmberg, S.P. Obenschain: Use of induced spatial incoherence for uniform illumination of laser fusion targets, *Opt. Commun.* **46**, 27–31 (1983).
200. R.H. Lehmberg, A.J. Schmidt, S.E. Bodner: Theory of induced spatial incoherence, *J. Appl. Phys.* **62**, 2680–2701 (1987).
201. A.E. Danilov, V.V. Orlov, S.M. Savchenko et al.: Investigation of the effect of the spatial coherence of laser radiation on the brightness properties of high-power neodymium glass lasers, *Preprint* No. 136, 13 pp, P.N. Lebedev Physical Institute, Moscow (1985).
202. D. Veron, G. Thiell, C. Gouedard: Optical smoothing of the high power PHEBUS Nd–glass laser using the multimode optical fiber technique, *Opt. Commun.* **97**, 259–271 (1993).
203. S. Skupsky, R.W. Short, T. Kessler et al.: Improved laser beam uniformity using the angular dispersion of frequency-modulated light, *J. Appl. Phys.* **66**, 3456–3462 (1989).

204. S.I. Fedotov, L.P. Feoktistov, M.V. Osipov et al.: Laser for ICF with a controllable function of mutual coherence of radiation, *J. Sov. Laser Res.* **25**, 79–92 (2004).
205. A.N. Starodub, S.I. Fedotov, Yu.V. Korobkin et al.: Nonlinear conversion of laser radiation with controllable coherence into second harmonic, *Book of Abstracts of the 29th ECLIM*, Madrid, June 11–16, 274 (2006).
206. A.N. Starodub, S.I. Fedotov, Yu.V. Korobkin et al.: Coherence of laser radiation and laser–matter interaction, *Book of Abstracts of the 29th ECLIM*, Madrid, June 11–16, 200 (2006).
207. J. Garnier, L. Videau, C. Gouédard et al.: Propagation and amplification of incoherent pulses in dispersive and nonlinear media, *J. Opt. Soc. Am. B* **15**, 2773–2781 (1998).
208. P. Donnat, C. Gouedard, D. Veron et al.: Induced spatial incoherence and nonlinear effects in Nd:glass amplifiers, *Opt. Lett.* **17**, 331–333 (1992).
209. H.T. Powell: Broadband development for Nova, *LLE/LLNL Workshop Laser Science and ICF Target Science Collaborative Research*, 16–17 June 1992, Rochester, NY.
210. P.W. McKenty, J.H. Kelly, R.W. Short et al.: Self-focusing of broad bandwidth laser light, *LLE/LLNL Workshop Laser Science and ICF Target Science Collaborative Research*, 16–17 June 1992, Rochester, NY.
211. M.V. Pyatakhin, A.F. Suchkov: *Spatiotemporal Characteristics of Laser Emission*, Nova Science Publishers, New York, 203 pp (1994).

Chapter 9
Self-focusing, Conical Emission, and Other Self-action Effects in Atomic Vapors

Petros Zerom and Robert W. Boyd

Abstract A broad overview of self-action effects in atomic vapors, such as self-focusing, self-trapping and pattern-formation, is presented. Different theoretical models that describe conical emission in atomic media are discussed, together with supporting experimental results.

9.1 Introduction

Atomic vapors have played a key role in the development of the understanding of self-action effects in nonlinear optics. The intrigue of using atomic vapors lies both in the very large nonlinear response that can be obtained from atomic vapors and from the fact that the optical properties of atomic systems can be calculated accurately using fundamental physical laws. In this chapter, we present a broad overview of the field of self-action effects in atomic vapors. We begin with a brief discussion of the standard processes of self-trapping and self-focusing in atomic media. We then proceed to a treatment of pattern formation including conical emission in atomic media.

We begin with a very brief summary of the theory of the nonlinear response in atomic vapors as it applies to self-focusing experiments. Most experiments on self-focusing are conducted with laser light detuned sufficiently far into the wings of an atomic absorption line that the optical response can be treated with the neglect of Doppler broadening. This conclusion holds because of the well-known fact that Gaussian lineshapes decrease far more rapidly in their wings than do Lorentzian lineshapes. In this limit, and for a radiatively broadened medium, we can represent the third-order nonlinear susceptibility in Gaussian units in the simple form [1]

$$\chi^{(3)} = \frac{2}{3}\frac{N\mu^4}{\hbar^3\Delta^3}, \tag{9.1}$$

R.W. Boyd (✉)
Institute of Optics, University of Rochester, Rochester, NY, USA
e-mail: boyd@optics.rochester.edu

R.W. Boyd et al. (eds.), *Self-focusing: Past and Present*,
Topics in Applied Physics 114, DOI 10.1007/978-0-387-34727-1_9,

where N is the atomic number density, μ is the electric dipole transition moment, and $\Delta = \omega - \omega_0$ is the detuning of the optical field from the atomic resonance frequency ω_0. Note that $\chi^{(3)}$ is negative on the low-frequency side of resonance and is positive on the high-frequency side of resonance. Typical values of the parameters in this expression might be $N = 2 \times 10^{13}$ cm^{-3}, $\Delta = 3\pi \times 3$ GHz, and $\mu = 5 \times 10^{-18}$ esu, leading to a value of the nonlinear response of $\chi^{(3)} = 1.4 \times 10^{-6}$ esu. Because the nonlinear coefficient n_2 in cm^2/W is related to $\chi^{(3)}$ in esu by $n_2 = 0.04\chi^{(3)}$ (for $n = 1$), we find for this example that $n_2 = 5 \times 10^{-8}$ cm^2/W. This is a relatively large value of n_2. For example, the critical power for self-focusing, which is given in general by

$$P_{\text{crit}} = \frac{\lambda^2}{8n_2}, \tag{9.2}$$

takes on the value $P_{\text{crit}} = 10$ mW for this value of n_2. In the same limits as those of (9.2), we can write a simple expression for the linear absorption coefficient as

$$\alpha = \frac{8\pi^2 N \mu^2}{\hbar \lambda \Delta^2 T_2}, \tag{9.3}$$

where λ is the vacuum wavelength of the radiation and T_2 is the dipole dephasing time. For the typical values $\lambda = 0.6$ μm and $T_2 = 32$ ns and for N, μ and Δ given as above, we find that $\alpha = 0.1$ cm^{-1}.

When more precise predictions of the size of the nonlinear response are required, it is necessary to include the possibility of saturation of the optical response and the influence of Doppler broadening. First, we note that in the absence of Doppler broadening the refractive index and absorption coefficient, including saturation effects, can be represented as

$$n = 1 - N \frac{3\lambda^3}{8\pi^2} \frac{T_2}{2T_1} \frac{\Delta T_2}{1 + \Delta^2 T_2^2 + I/I_s}, \tag{9.4}$$

$$\alpha = N \frac{3\lambda^2}{2\pi} \frac{T_2}{2T_1} \frac{1}{1 + \Delta^2 T_2^2 + I/I_s}. \tag{9.5}$$

Here λ is the vacuum wavelength of the radiation, T_2 is the dipole dephasing time, T_1 is the upper state lifetime, I is the laser intensity, and I_s is the saturation intensity.

We now introduce Doppler broadening by replacing Δ with $\Delta - kv$ and averaging the expression over a Maxwellian distribution of velocities. We find that [2,3]

$$n = 1 - N \frac{3\sqrt{\ln 2}\lambda^3}{4\pi^{3/2} T_2 \Delta\omega_D} \frac{T_2}{2T_1} \text{Im}\left[w(\xi + i\eta/2)\right] \tag{9.6}$$

$$\alpha = N \frac{3\eta\lambda^2}{4\pi^{1/2}(1 + I/I_s)} \frac{T_2}{2T_1} \mathrm{Re}\,[w(\xi + i\eta/2)], \tag{9.7}$$

where $\Delta\omega_D = (2\pi/\lambda)\sqrt{8 \ln 2 k_B T/m}$ is the Doppler linewidth (FWHM), k_B is the Boltzmann constant, T is the temperature, and m is the mass of one atom; $\xi = 2\sqrt{\ln 2}\Delta/\Delta\omega_D$, $\eta/2 = (2\sqrt{\ln 2}/T_2\Delta\omega_D)\sqrt{1 + I/I_s}$ and $w(z) = e^{-z^2}\mathrm{erfc}(-iz)$ is the complex error function.

9.2 Self-focusing and Self-trapping in Atomic Vapors

The term self-focusing usually is used to refer to the tendency of an initially collimated or nearly collimated beam of light to come to a focus within a nonlinear material with n_2 positive as a result of nonlinear effects. The term self-trapping usually is used to refer to the tendency of a beam of light to propagate over large distances (much greater than the Rayleigh range) as a result of a balance between diffraction and self-focusing effects. However, when reading the literature, great care should be exercised in determining the authors' intended meaning, as not all authors follow the terminology used here. In fact, problems of this sort were especially prevalent in the early literature before the nature of self-focusing and self-trapping were well understood.

The first observation of self-focusing of light in an atomic vapor appears to be that of Grischkowsky [4]. He observed self-focussing of a light beam tuned $12\,\mathrm{cm}^{-1}$ to the high-frequency side of the atomic potassium resonance line. The frequency of the light beam was established by Raman shifting the output of a ruby laser through use of nitrobenzene. The experimental results were shown to be in good qualitative agreement with the predictions of Javan et al. [5] based on a steady-state model of the atomic response, although detailed comparison was made based on an adiabatic following model that takes account of the pulsed nature of the laser excitation. The adiabatic following model can reliably be applied for laser pulses of duration considerably shorter than the population relaxation time T_1 and dipole moment relaxation time T_2 of the atomic medium. According to this model, the nature of the nonlinear response is ascribed to the tendency of the atomic dipole moment pseudovector to follow adiabatically on the Bloch sphere the time evolution of the electric field amplitude of the applied laser field. The adiabatic following model was developed further by Grischkowsky et al. [6], who used it to interpret their experimental investigation of defocusing of light in atomic rubidium vapor. Self focusing of light describable by the steady-state model was observed later by Bjorkholm et al. [7]. They studied self-focusing, self-trapping and self-defocusing of cw laser light in atomic potassium vapor.

Stable self-trapping in bulk media is only possible in the presence of some form of saturation of the nonlinear change in the refractive index. Bjorkholm et al. were the first to observe self-trapping (spatial solitons) because of the saturation of nonlinearity in atomic potassium vapor.

A related effect is that of self-steepening. This effect was reported experimentally by Grischkowsky et al. [8]. Self-steepening can occur in an atomic vapor by either of two processes or by a combination of the two. These processes are (1) the (instantaneous) intensity dependence of the group velocity of light and (2) self-phase modulation, which broadens the spectrum of the light combined with group-velocity dispersion. Of these two processes it is believed that the second is more important for atomic systems. In their experimental studies of this effect, the authors launched pulses of approximately 10 ns duration into an atomic rubidium vapor. The pulses were detuned approximately $0.2\,\text{cm}^{-1}$ to the low-frequency side of the rubidium resonance line, and strong reshaping effects such as self-steepening and possible shock formation were observed.

More complicated sorts of interactions can also lead to self-action effects. For example, Tam et al. have observed the long-range interaction between cw self-focused laser beams in an atomic vapor as a consequence of the diffusion of polarized atoms [9].

9.3 Conical Emission

Conical or cone emission (CE) has a long and controversial history since its first observation by Grischkowsky in a potassium vapor [4]. CE is one of many examples of transverse nonlinear effects and is observed normally when a strong laser beam, tuned to the blue side of a resonant atomic transition, propagates through a dense medium of two-level atoms. The generated cone is spectrally shifted (in most cases red-shifted) from the line center of the transition, or it can be at the same frequency as the original beam and manifests itself as a single and multiple concentric rings (in the far-field) around the pump laser beam.

CE has been observed using both continuous wave and pulsed laser beams, with pulse duration mostly in the nanosecond (ns) regime, in various atomic vapors. Although CE has been observed in glasses [10] and other media, we will focus mainly on studies of conical emission in atomic vapors. Such experimental observations in atomic vapors include: sodium [11–13], cesium [14], strontium [15–17], calcium [18], potassium [4], barium [14,19,20] using pulsed laser beams, and in sodium [21] using a continuous wave (cw) dye laser.

The physical process leading to conical emission can be different under different circumstances, and several different models have been proposed to explain conical emission. These models include: a hypothetical parametric four-wave mixing (FWM) process [19], coupling of the field to the transient response of the atomic media [22], four-wave mixing enhanced by the ac-Stark effect in self-trapped filaments of light [23], CE due to four-photon parametric scattering of self-focused exciting radiation [24], Cerenkov emission due to laser-induced moving polarization [11,25], an anomalous CE due to a type of parametric down-conversion process [26], free-induction-decay emission following three-photon scattering [27], conical emission due to spatial-temporal breakup of the laser pulse into solitary waves [28], a theory that includes Doppler-broadened Raman-gain amplification of Doppler-broadened resonance fluorescence, four-wave mixing, propagational coupling, self-trapping of the pump beam, and diffraction and pump-induced refraction of new frequencies [21], CE as a result of scattering of intrinsic modes from a filament formed due to self-focusing of a strong light beam in a resonant media [29], CE due to the low-frequency Rabi sideband of an atomic resonance [30], CE due to cooperative, Cerenkov-like, spontaneous emission of collisionally perturbed, optically dressed atoms [31], generation of transient Rabi sidebands in pulse propagation as a possible source of CE [32], CE due to Raman resonant four-wave mixing including self- and induced-phase modulations [33], vacuum fluctuations as a possible source of conical emission [34], and conical emission as a result of spatial self-phase modulation under near-resonant condition [14,35].

In this section, we consider some of the representative models used to explain the physics behind the generation of both pulsed and cw conical emission using a single-input pump laser and single-photon resonant atomic transitions (Sect. 9.3.1), multi-photon allowed transitions (Sect. 9.3.2), and CE generation with multiple pump beams (Sect. 9.3.3).

9.3.1 Conical Emission with a Single Pump Beam and One-Photon Resonant Transition

The most commonly quoted model is that due to Harter et al., which is based on resonantly enhanced nondegenerate four-wave mixing and refraction (FWMR) at the surface of a self-trapped filament [12]. This waveguide model [12,13,23,36,37] can be described as follows. We assume that an intense laser beam of amplitude E_1 and frequency ω_1 tuned to the high-frequency (self-focusing) side of the atomic resonance propagates through an atomic vapor. This wave experiences self-action effects and forms a self-trapped filament as shown in Fig. 9.1(a). In Fig. 9.1(a), $\Delta = \omega_1 - \omega_{ba}$ is the pump detuning from the

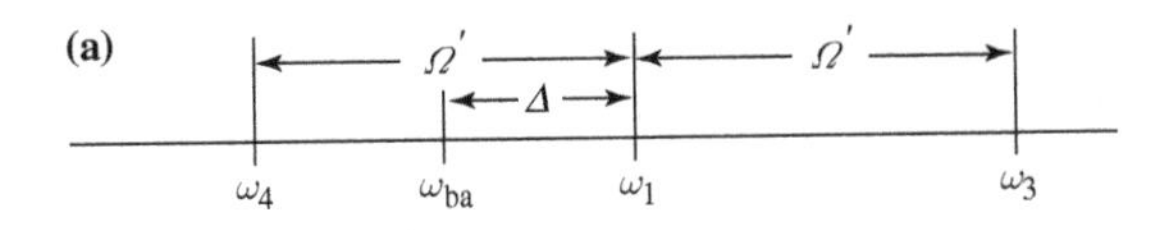

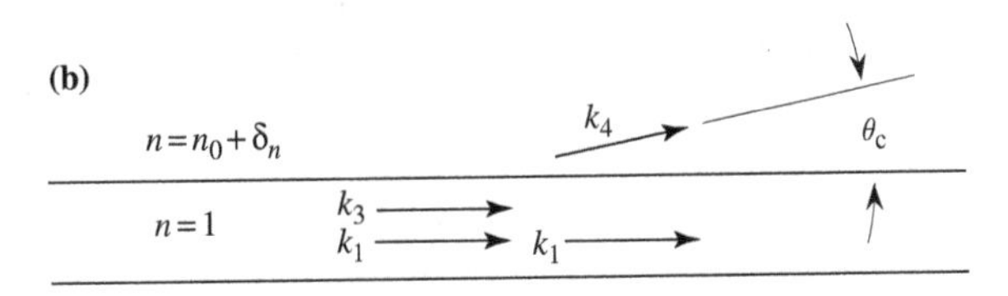

Fig. 9.1 (**a**) Frequencies present in Rabi-sideband generation. (**b**) The four-wave mixing, waveguide model of conical emission

unperturbed atomic-resonance frequency ω_{ba} of a two-level system and $\Omega' = (\Omega^2 + \Delta^2)^{1/2}$ is the generalized Rabi frequency, where $\Omega = \mu_{ba} E_1/\hbar$ is the Rabi frequency and μ_{ba} denotes the transition dipole matrix element connecting the unperturbed atomic levels. We assume that because of saturation effects the refractive index inside of the filament is nearly equal to 1 and outside of the filament is equal to $1 - \delta n_1$, where δn_1 is a positive quantity. Within the filament, the propagation constant of this wave will be equal to $k_1 = (\omega_1/c)(1 - \delta n_1)$; the modification of the propagation constant given by the second factor is a well-known consequence of guided-wave propagation. Within this filament, light is generated at frequency ω_3 by the gain of the three-photon effect [38, and references therein] and at frequency ω_4 by a phase-matched four-wave mixing process. These frequencies are thus related by $2\omega_1 = \omega_3 + \omega_4$. By the properties of the three-photon effect (see Fig. 9.1(b)), ω_3 will be emitted at a frequency larger than ω_1, and ω_4 will thus be emitted at a frequency lower than both ω_1 and the atomic resonance frequency. Light at frequency ω_4 is thus anti-guided and is ejected from the filament. The emission angle θ_c can be estimated by applying Snell's law at the surface of the filament. In particular, one requires that the longitudinal components of the wavevectors inside and outside the filament be equal. The value outside is given by $k_4(\text{out}) = (\omega_4/c)(1 + \delta n_4)(1 - \theta_c^2)$, where the refractive index at frequency ω_4 is represented as $1 + \delta n_4$. The value inside is found from the phase matching condition as $k_4(\text{in}) = 2k_1 - \omega_3/c$. Note that we do not apply a guided-wave correction to the propagation constant of the wave at frequency ω_3 because it is not a guided wave. By equating these two expressions we find that $\theta_c^2 = \delta n_4 + 2\delta n_1$. Finally, we note that under many circumstances, the pump wave and cone emission are found to be symmetrically detuned from the atomic resonance frequency, that is $\omega_1 - \omega_{ba} = \omega_{ba} - \omega_4$. Under these conditions, δn_1 and δn_3 are equal, so that the cone emission angle can be expressed as

$$\theta_c = (6\delta n_1)^{1/2}.$$

Most experimental studies of conical emission involve the propagation of a strong laser radiation tuned close to the two-level atomic resonance of a dense atomic vapor. The far field spectral and angular distribution of the generated radiation in the form of a cone is studied as a function of different experimental parameters such as atomic number density, detuning from the two-level system or power density of the irradiating pump laser. Harter et al. used a single-mode rhodamine-6G dye laser, pumped by a frequency-doubled Nd:YAG laser, which produces 2–7 ns pulses [13]. Typical experimental values used to study the spectrum of the transmitted radiation are: laser power density of $\sim 1 \times 10^8\, W/cm^2$, a sodium number density of $1 \times 10^{14} cm^{-3}$, a laser detuning of 2.4 Å to the short-wavelength side of the $3^2S_{1/2} - 3^2P_{3/2}$ (D_2) resonance line and an argon buffer gas of 1 Torr pressure. The resulting spectrum of the

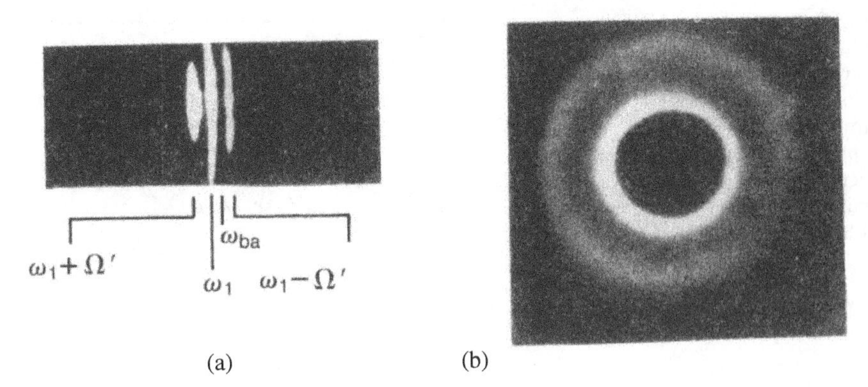

Fig. 9.2 (**a**) Spectrum of the transmitted radiation. (**b**) Observed cone with the central pump blocked [13]

transmitted light is shown in Fig. 9.2(a) and at a relatively higher number density, the laser pump beam self-focuses and results in a cone as shown in Fig. 9.2(b).

Another common feature in conical emission experiments is the cone angle dependence on the number density of the medium (N) and the detuning from the two-level atomic resonance (Δ). The exact angle of the conical emission, which is determined by phase matching and refraction at the filament boundary, depends on the internal angle θ_0 and the change in refractive index across the filament $\delta n(\omega_4)$ through $\theta = [\theta_0 + 2\delta n(\omega_4)]^{1/2}$ [13, and references therein]. The internal angle depends on the dispersion characterstics of the light within the self-trapped filaments and is mostly experimentally inaccessible [17]. A typical dependence of the cone half-angle on the two parameters N and Δ is shown in Fig. 9.3.

Another approach that has attracted favorable attention is the Cerenkov-type model [25]. According to this model a near resonant pulsed laser beam (of frequency ω_L) propagating through a medium induces a polarization moving with velocity $c/n(\omega_L)$. This induced polarization produces coherent

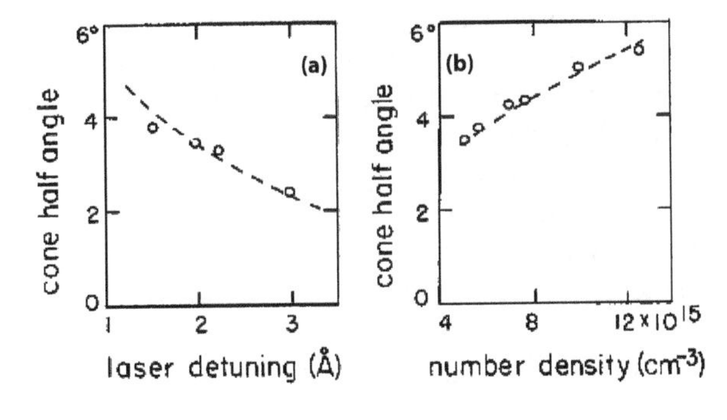

Fig. 9.3 Cone half-angle as a function of (**a**) laser detuning (for a number density of $5.0 \times 10^{15} cm^{-3}$) and (**b**) atomic number density (for a laser detuning of 2 Å) [12]

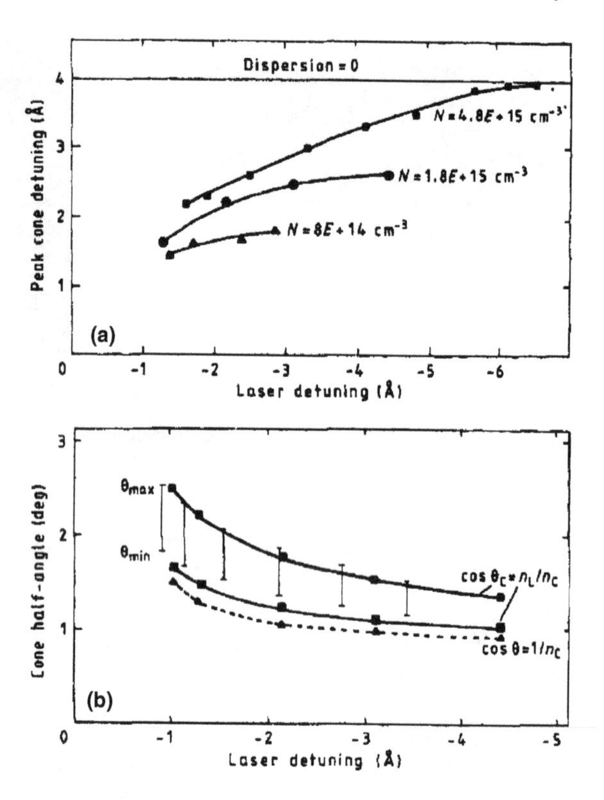

Fig. 9.4 (**a**) Cone peak intensity wavelength as a function of laser detuning from the D_2 resonance transition. (**b**) Cone half-angle as a function of laser detuning for sodium number density of $1.8 \times 10^{15} cm^{-3}$ [25]

radiation (of frequency $\omega_L - \Omega'$) propagating with velocity $c/n(\omega_L - \Omega')$. This velocity difference is responsible for CE of the $\omega_L - \Omega'$ light, with a cone angle given by $\cos\theta_C = n(\omega_L)/n(\omega_L - \Omega')$. Figure 9.4 shows the cone peak intensity wavelength and the cone half-angle as functions of laser detuning. The vertical bars in Fig. 9.4(b) represent the values of θ_c between the measured maximum and minimum of the cone half-angle and the dotted lines values of θ_c according to the FWMR model with $\theta_0 = 0$.

The most complete description of the process of conical emission, both theoretically and experimentally is due to Valley et al. [21] and Paul et al. [17,39]. Valley et al. have studied the formation of cw conical emission in sodium vapor. Their theoretical model incorporates the propagation and quasi-trapping of a strong cw pump beam $E_1(\omega_1)$ and the ensuing generation of new frequencies through Raman gain amplification of the resonance fluorescence $E_3(\omega_3)$ and propagational four-wave mixing between the strong pump and $E_3(\omega_3)$ generating $E_4(\omega_4)$, with $\omega_3 > \omega_1$ and $\omega_4 = 2\omega_1 - \omega_3 < \omega_1$. Optical-Stark-shifted absorption causes division of the field envelope at frequency ω_4, i.e. A_4, into A_{4L} and A_{4R}, which is accompanied by the division of A_3 into A_{3L} and A_{3R} by four-wave mixing coupling. The propagation of the A_{4L} through the spatially dependent index of refraction prepared by the strong

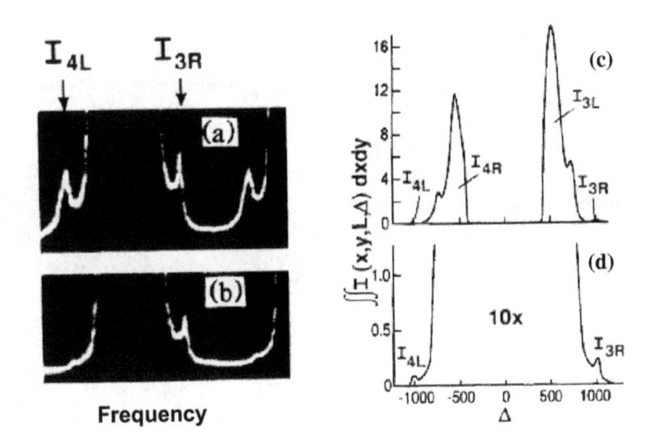

Fig. 9.5 **(a) (b)** Spectra of the generated new frequencies (I_{4L} and I_{3R}) in a continuous wave conical emission process in sodium vapor with the cone unblocked and blocked, respectively. **(c) (d)** Computed spectrum of the new frequencies [21]

pump results in the formation of a cone. No such cone formation was observed, either experimentally or computationally, for the $A_3(\omega_3)$ field components.

Figure 9.5 shows the experimentally observed spectra and the computed spectrum of the generated frequencies (at ω_3 and ω_4). For the experimental result shown in Fig. 9.5(a) and (b), the pump power at the entrance of the cell containing the sodium vapor was about 470 mW and the pump is detuned by $+3.4\,\mathrm{GHz}$ from the $^2S_{1/2} - ^2P_{3/2}$ sodium D_2 transition line. Some of the parameters used for the computational result shown in Fig. 9.5(c) and (d) are Rabi frequency (at the entrance of the cell) of 680 and pump detuning from atomic resonance of $+430$ (both frequencies in units of $(2\pi T_1)^{-1}$). Blocking the generated cone results in the disappearance of the I_{4L} component from the spectra with no effect on the I_{3R} frequency components as shown in Fig. 9.5(a) and (b). They concluded that the cone is formed by pump-induced radially dependent refraction of new frequencies close to the Stark-shifted resonance.

A detailed study of the formation of conical emission from a *single*, self-trapped filament of light in strontium (Sr) vapor, both theoretically and experimentally, was conducted by Paul and coworkers [17,39]. They used the $5s^2\ ^1S_0 - 5s5p\ ^1P_1$ transition line of Sr and studied cone generation as a function of pulsed-laser energy (E), Sr number density (n_{Sr}), ratio of the dipole ($\tilde{\gamma}$) to the population (Γ) decay rates ($g = \tilde{\gamma}/\Gamma$), and the laser input beam diameter (d_{in}). They found that CE occurs when the blue-detuned input pulsed-laser is predominantly self-trapped.

Figure 9.6 shows contour plots of the total energy radiated into an angle θ at a frequency ν, $E(\theta, \nu)$. The numbers on the top right part of the graph are the detuning of the laser (Δ), d_{in}, g, the measured energy in the cone (E_c) and the energy in the laser (E) respectively. The region between the two solid curves is integrated to find the energy in the cone. The dashed curve is a fit of $k|2\delta n(\nu_r)|^{1/2}$ to the shape of the cone for relative detunings between about -2.0 and -1.0, with the fit parameter $k \approx 2.0$ and where $\delta n(\nu_r)$ is the change in the index-of-refraction between saturated and unsaturated vapor at the cone

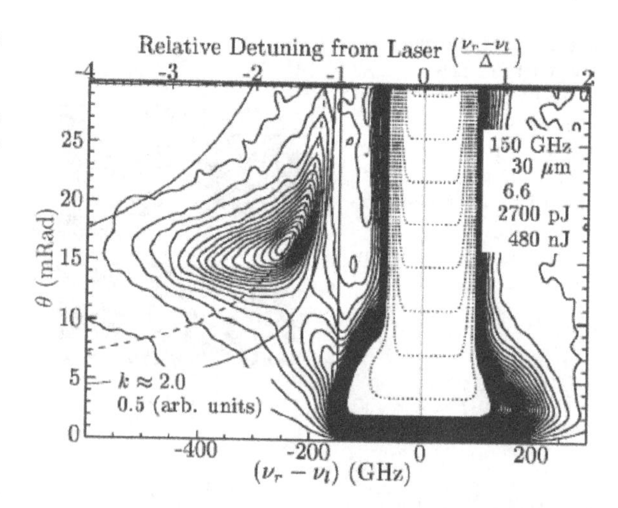

Fig. 9.6 Contour plot of the emission into angle θ at frequency ν. The bottom axis is the detuning from the laser while the top axis is the relative detuning from the laser. The *atomic line* is the *vertical* line at a relative detuning of -1 [17]

frequency ν_r. Although there is a reasonable agreement of the cone angle dependence on $k\sqrt{n/\Delta}$, there is a large angular spread at a given frequency ν and also at relative detuning past -2 there is a region where θ is constant over a range of ν_c values [17].

Figure 9.7 shows cone half angle in mrad as a function of a dimensionless laser beam radius ($\rho_{1/2}$) and a dimensionless scaled power (Φ/Φ_K), where Φ_K is a dimensionless Kerr power [17]. The gray region is where the beam is predominantly self-trapped. The phase velocity (v_1) of the self-trapped laser beam varies according to $v_1/c = \{1 + \beta[n(\nu_1) - 1]\}^{-1}$, where $\beta = 1(0)$ at small (large) values of Φ/Φ_K and $n(\nu_1)$ refers to the unsaturated vapor. Unlike most of the above described models, where the cone angle depends on the phase velocity of the laser beam and of the cone light, the cone angle does not decrease as Φ/Φ_K increases for a constant frequency (horizontal line in Fig. 9.7).

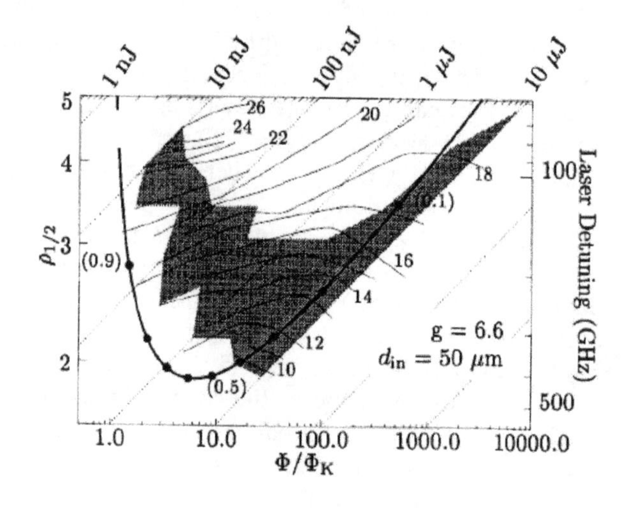

Fig. 9.7 Cone half-angle in mrad as measured by Paul et al. [17]. The black dots along the stationary filament curve represent the positions where β takes on the values 0.1, 0.2, ..., 0.9. The *diagonal lines* represent the input energy of the pulsed laser

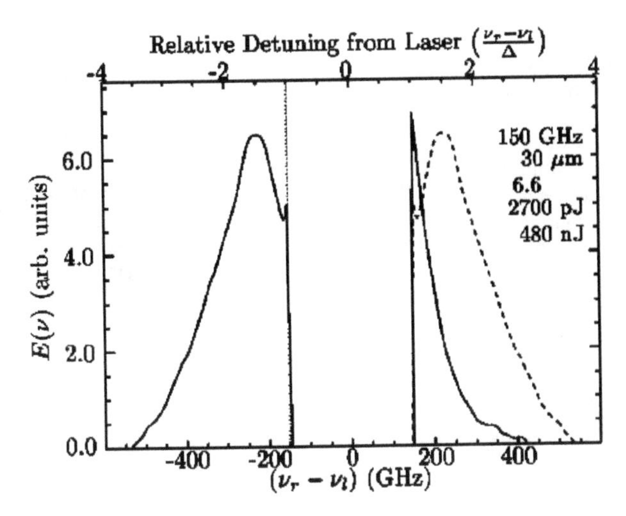

Fig. 9.8 Total energy integrated over the subtended cone angle of 30 mrad. The data between the atomic line and twice the detuning of the laser have been removed [17]

Paul et al. have also compared their experimental result to the FWM theory, which predicts as much energy scattered into the blue as into the red Rabi sidebands, and found that the blue sideband drops off in intensity much faster than the red sideband [17]. In Fig. 9.8, the dotted line is a reflection of the red-detuned sideband about the laser frequency, and thus a simple FWM theory is not adequate to explain the spectrum observed in CE experiments. The experimental parameters on the top right are defined the same way as in Fig. 9.6.

We now present some of the commonly observed features of conical emission and describe some of the strength and some of the drawbacks of the FWMR model. These common features include the position of the peak of the CE spectrum from the atomic resonance, the dependence of the cone angle on the laser detuning, and the atomic number density [16]. (i) The laser detuning from the atomic resonance is approximately equal to the magnitude of the detuning of the conical component from the same atomic resonance. (ii) The angle of the most intense part of the cone (θ_c) depends on the number density (N) and the laser detuning from the atomic resonance (Δ) through $\theta_c \propto \sqrt{N/\Delta}$ (Paul et al. and Chalupczak et al. have also found a different cone angle dependence [17, 30]). (iii) Conical emission occurs when the laser forms self-trapped filaments (Valley et al.'s experiment on cw CE shows that their pump was quasi-trapped [7,21]) (iv) CE occurs when the laser is blue-detuned from the atomic transition. Some of the drawbacks of the FWMR model include [16,40–42] (a) lack of the complementary blue-shifted component of the red-detuned cone, which is expected from a four-wave mixing process (see Fig. 9.9); (b) the dependence of the cone angle ($\theta_c(\omega)$) over the broad spectral width does not have the predicted $2|\delta n(\omega)|^{1/2}$ form, where ω is the emitted frequency [16]; (c) observation of conical emission when the pump laser is red-detuned from

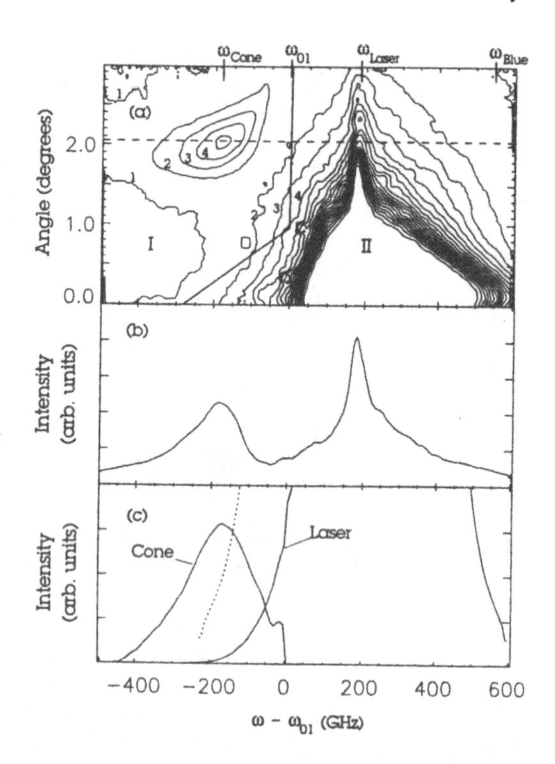

Fig. 9.9 (**a**) Intensity contour plot of emission from a strontium cell with number density 7×10^{14} cm^{-3}, laser intensity 0.25 MW/cm^2 and detuning of +185 GHz near the 461-nm atomic resonance (ω_{01}) transition. The value ω_{cone} is the peak of the cone emission and ω_{blue} is the detuning from the laser for which a FWM complement to ω_{cone} should occur. (**b**) Intensity profile through the dotted line in (**a**) showing the spectrum at an angle of 2.05°. (**c**) Angular integration of data over regions I and II in (**a**). The dotted line is the blue side of the laser spectrum reflected around ω_{laser} [16]

an atomic resonance for which filament formation and refraction at a boundary cannot occur [16].

9.3.1.1 Multiple Filamentation in Conical Emission

A related phenomenon that is observed in CE experiments is laser beam breakup into many self-trapped filaments [4,13,15,17,19,41,43–46]. A comprehensive theoretical model, which employs the Maxwell–Bloch equations for a two-level atomic system, for the propagation of intense laser light in a fully saturable medium, was described by Dowell and others [39,47]. The propagation eigenfunctions for this model are the stationary filament solutions that propagate unchanged over very large distances. According to this model, the characterstic propagation distance over which self-focused beam modifications occur is inversely proportional to η, and the stationary filament half-power radius $R_s \propto (4\pi\eta/\lambda)^{-1/2}$. The self-focusing parameter is defined as $\eta \equiv (3\lambda^2\Gamma/8\pi)(N_{Sr}/\Delta)$, where λ is the laser wavelength and Γ is the natural linewidth. Dowell et al. used an Nd:YAG pumped single-mode, pulsed dye laser tuned to the blue of the 460.7 nm strontium (Sr) resonance transition and studied self-focusing and filament formation as a function of vapor density (N_{Sr}), laser detuning (Δ), and laser input power [46].

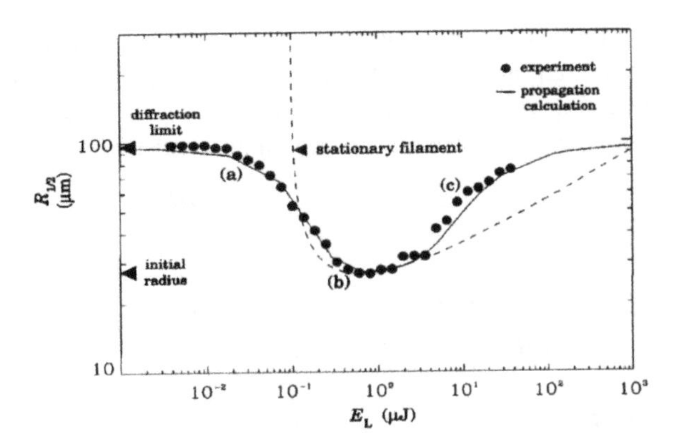

Fig. 9.10 Half-power radius versus laser energy. Here $N_{Sr} = 1.1 \times 10^{13}$ cm^{-3} and $\Delta = 50$ GHz. **(a)** $\Omega_0/\Delta = 0.5$ (weak saturation), **(b)** $\Omega_0/\Delta = 2$ (moderate saturation), and **(c)** $\Omega_0/\Delta = 10$ (strong saturation), where Ω_0 is the on-axis Rabi frequency at the cell entrance [46]

Figure 9.10 shows the half-power radius, i.e., the radius at which half of the power lies within $r = R_{1/2}$, versus the laser energy E_L for the self-focusing parameter $\eta = 1.8$ cm^{-1}. The top and bottom arrows correspond to the beam half-power radius at the strontium cell entrance and exit, the latter in the absence of self-focusing. The solid curve corresponds to the theoretical predictions for steady-state Gaussian beam propagation and the dotted line represents the stationary filament solution.

Paul et al. have also studied the propagation behavior of an input beam with nearly Gaussian spatial profile through strontium vapor [17]. Figure 9.11(a) shows the $\Phi/\Phi_K, \rho_{1/2}$ space for an input beam diameter $d_{\text{in}} = 50$ μm, and strontium vapor density of $N_{Sr} = 1.0 \times 10^{14}$ cm^{-3}. The white area between the gray regions is where the beam is predominantly self-trapped. A plot of

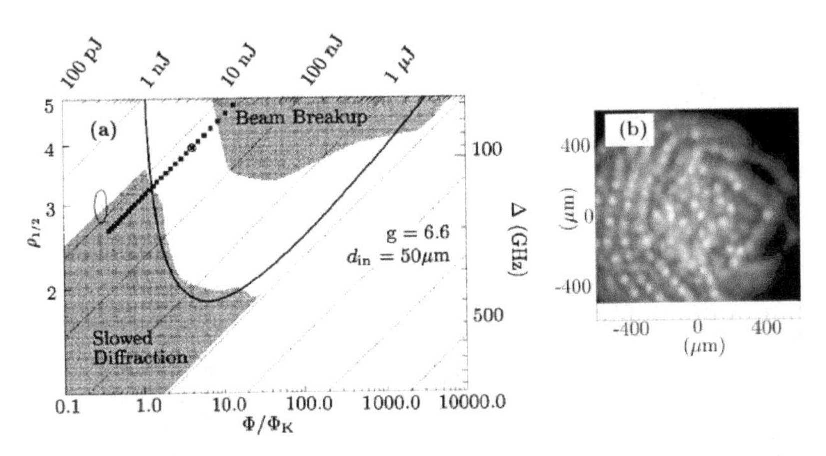

Fig. 9.11 **(a)** Stationary filament curves with self-focusing data (see text for experimental parameters). The diagonal lines represent input pulse laser energy. **(b)** Images of the filaments in the exit plane of the Sr cell. Here $N_{Sr} = 1.4 \times 10^{14}$ cm^{-3}, $g = 1.68$, $\Delta = 90$ GHz, $E = 3.2$ μJ, and $d_{\text{in}} = 860$ μm [17]

the output beam diameter d_{out} versus the laser detuning Δ reveals that there are four regions of interest: free diffraction (at large Δ), slowed diffraction, self-trapping and beam breakup. At smaller Δ, self-focusing overpowers diffraction and the beam breaks up into many self-trapped filaments [see Fig. 9.11(b)] [17,47].

9.3.2 *Conical Emission with a Single Pump Beam and Multi-Photon Resonant Transitions*

The process of conical emission has also been studied for multi-photon allowed resonant transitions in atomic vapors, the origin of which has been attributed to a phase-matched four-wave mixing process [48–52]. Krasinski et al. used the $3S - 3D$ two-photon allowed resonant transition of atomic sodium [48]. An excimer-laser pumped dye-laser, with typical pulse energies of 200 μJ in a 1.5 ns pulse and high spectral purity, was focused into a heat-pipe vapor cell containing sodium atoms with number densities ranging from 1 to 15×10^{15} cm^{-3}. The spectrally resolved spectrographs of the generated conical emission is shown in Fig. 9.12. Note the on-axis emission of radiation in the forward direction near the 3p→3 s (and their complementary components near the 3d→3p) transitions, which occurs for a perfectly phase-matched four-wave mixing process and refractive index exactly equal to unity. They have also found an excellent agreement between the experimentally obtained dependence of the cone angle on the emission wavelength of the generated fields near the 3d→3$p_{1/2}$ and 3$p_{1/2}$ →3 s transitions with the predictions of a model based on the phase-matched FWM process.

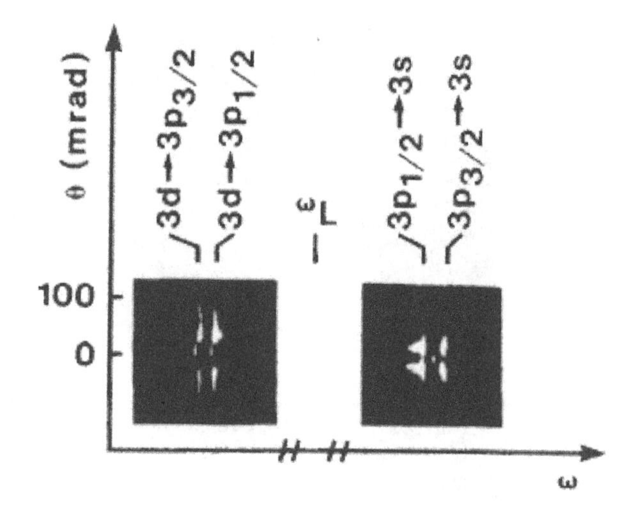

Fig. 9.12 Angular versus spectral distribution of the conical emission at an atomic number density of 1.8×10^{16} cm^{-3} [48]

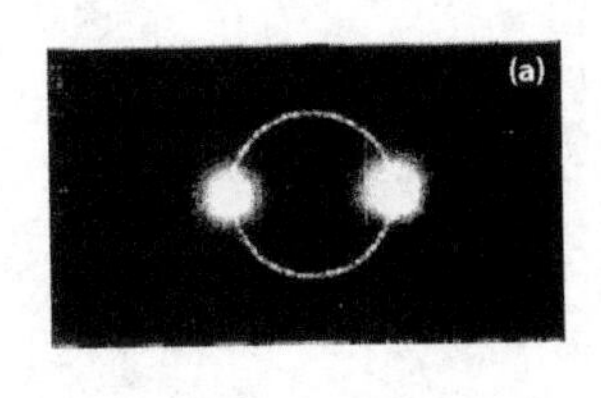

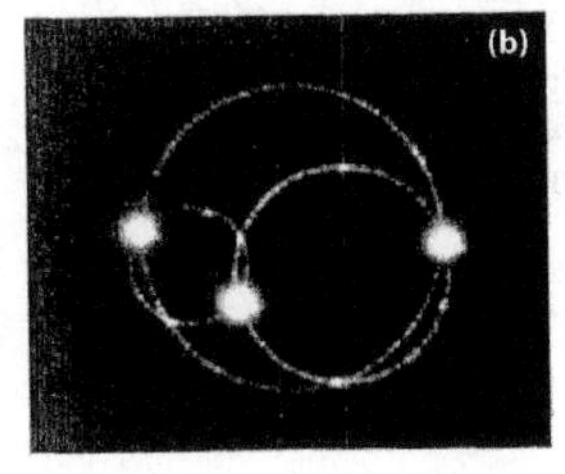

Fig. 9.13 Two- and three-beam excited conical emission in sodium vapor [56]

9.3.3 Conical Emission Generation Using Two (or More) Pump Beams

Conical emission has been observed with two (or more) pump beams interacting in a nonlinear medium [53–58]. One of the main features that distinguishes CE generated from multiple pump beams from what has been discussed so far in Sect. 9.3.1 and 9.3.2 is its spectral characterstics. The spectrum of CE from co- (or counter-) propagating pump beams is degenerate in wavelength with the pump laser (for the case of degenerate pump beams). Another feature is the cone angle which is determined solely by the geometry of the interacting pump beams.

Kauranen et al. report on the generation of CE using two degenerate pump beams as they pass through sodium vapor [56]. For the two-beam excited CE shown in Fig. 9.13(a), the laser was detuned by 40 GHz to the red side of the D_2 line and the pulse energy of each beam was 50 μJ, with a 5° crossing angle between them. The wavelength and the polarization of the generated cone was the same as the pump beams, and blocking one of the pump beams resulted in the vanishing of the cone. More generally, the cone was generated for pump beams red- or blue-detuned from either the D_1 or the D_2 resonance lines of sodium. Figure 9.13(b) shows the intersection of three different degenerate pump beams of equal intensity, all propagating in the forward direction, but each pair with a distinct crossing angle. As can be seen, these result in three cones of different diameters. A theoretical analysis shows that any two diametrically opposite parts of the cone are coupled through a FWM process. This effect can be seen as bright spots in Fig. 9.13(b).

9.4 Self-focusing and Pattern Formation

In addition to the self-focusing effects described above, a number of other sorts of self-action effects leading to pattern formation have been reported by various workers [59–72, and references therein]. Some of these effects are described briefly here.

A striking example of pattern formation is the spontaneous generation of a honeycomb pattern illustrated in Fig. 9.14 [59]. For this experiment, an input

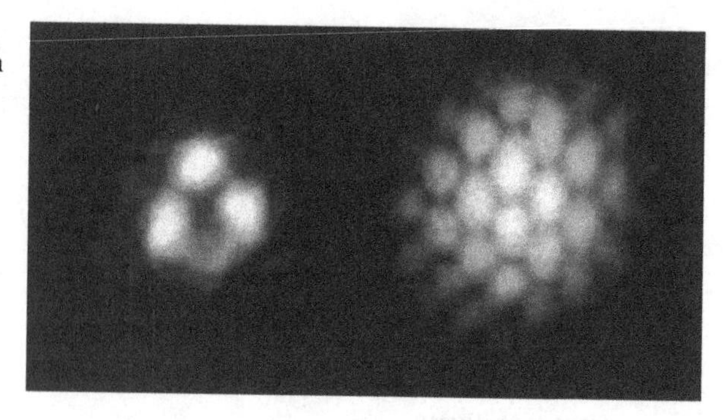

Fig. 9.14 Honeycomb pattern formation in sodium vapor without any feedback [59]

laser beam of 160 μm diameter carrying 150 mW of power was used. The laser was tuned 2 GHz to the high-frequency side of the sodium D_2 resonance. A cell of 7 cm length contained sodium vapor with an atomic number density of 8×10^{12} cm^{-3}. Under carefully controlled experimental conditions, the incident laser beam tends to break up into three components of nearly equal intensity arranged in an equilateral triangle, as seen on the left side of the figure. This image shows the intensity distribution at the output window of the sodium cell. The image on the right of the figure shows the intensity distribution in the far field of the cell. The honeycomb pattern is the Fourier tansform of the near-field optical distribution. In the published paper, these authors develop a mathematical model that describes many of the features of this sort of pattern formation.

Another topic of recent interest has been the propagation of laser beams carrying orbital angular momentum [73, and references therein]. Such light fields are of conceptual interest because they carry angular momentum of two different sorts (spin angular momentum, related to circular polarization, and orbital angular momentum, associated with the azimuthal dependence of the phase of the optical field), and of practical interest because the orbital angular momentum degree of freedom can be used to encode information on a light beam for applications in classical and quantum information processing. Beams carrying orbital angular momentum are necessarily ring shaped, and there has been considerable interest in determining the stability characteristics of such beams upon propagation. Indeed, the character of many self-focusing processes is profoundly different for ring-shaped beams rather than the usual Gaussian-shaped beams.

Theoretical and numerical work of Firth et al. [74] demonstrated that the primary instability mechanism of these beams is an azimuthal modulational instability that leads to the fragmentation of the beam into individual beams that propagate as fundamental solitons. In particular, these authors showed that a beam carrying orbital angular momentum of $m\hbar\omega$, that is, a beam with an azimuthal field dependence of $e^{im\phi}$, will tend to break up into $2m$ filaments that drift away tangentially from the original ring.

An experimental investigation of this effect has been reported by Bigelow et al. [60]. In this experiment, laser beams in the form of Laguerre–Gauss beams with m values ranging from 1 to 3 were prepared by sending a laser beam through a computer-generated hologram. These beams were then sent through a sodium vapor cell and the beam profile of the transmitted beam was recorded. Some of the results of this investigation for $m = 2$ are shown in Fig. 9.15. For this measurement, the laser was tuned 46.7 GHz to the high-frequency side of the sodium D_2 resonance with a pulse energy of 234 nJ. These observations were also found to be in good agreement with a numerical model based on the saturable optical response of a collection of two-level atomic systems. These results confirmed the predictions of Firth et al. [74]. Related experimental results have been reported by Tikhonenko et al. [75].

An extremely sensitive all-optical switch based on the properties of spontaneous pattern formation was recently reported by Dawes et al. [61]. The basis of this switch is shown in Fig. 9.16. Two counterpropagating laser beams interact in an atomic rubidium vapor. Above a certain threshold intensity, a spatial instability can develop leading to the emission of radiation in a structured pattern surrounding the transmitted laser beams [76].

Dawes et al. realized that they could control the orientation of this pattern by applying an additional, very weak laser beam to the interaction region. In the situation shown schematically in the figure, the orientation of the emission pattern can be rotated by 60° by applying this additional laser beam. In the laboratory demonstration of this effect, the control beam carried an intensity approximately 6000 times weaker than that of the instability pattern itself. The

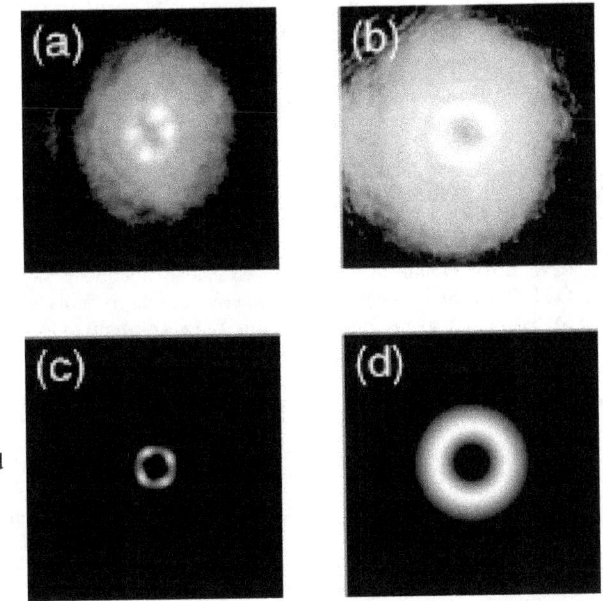

Fig. 9.15 (**a**) $m = 2$ vortex beam after propagating through a resonant sodium cell. The beam has broken into four components. (**b**) When the laser is detuned from resonance, the beam does not break up. (**c**) and (**d**) Computer simulations of the results shown in (**a**) and (**b**), respectively [60]

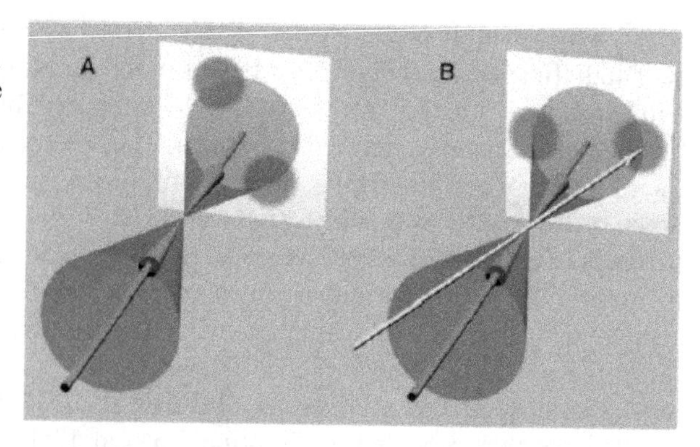

Fig. 9.16 The off and on state of an all-optical switch in rubidium vapor. The state of polarization of the control beam (in B) is linear and orthogonal to the counter-propagating pump beams [61]

ability to control a strong beam with a weak beam lies at the heart of practical applications of all-optical switching. In this particular measurement, it was found that a control beam containing as few as 3000 photons was adequate for switching a much stronger beam. Thus, this work constitutes an important step toward the goal of constructing switches that can work with control fields containing a single photon.

9.5 Concluding Remarks

In summary, as a consequence of their large nonlinear response, atomic vapors have played a key role in the development of a conceptual understanding of self-action effects. Some of the processes observed in atomic vapors include: self-focusing, self-defocusing, self-trapping, self-steepening, all-optical switching, cone emission and other pattern formations (with and without feedback), optical instability, and chaos and optical bistability.

References

1. R.W. Boyd: *Nonlinear Optics*, 2nd ed., Academic Press, San Diego (2003); see Section 6.3.
2. D.H. Close: Strong-field saturation effects in laser media, *Phys. Rev.* **153**, 360–371 (1967).
3. M. Saffman: Private communication. We are grateful to M. Saffman for pointing out some small errors in the formulas in [60]. The form of the equations used in the present document are those of Dr. Saffman.
4. D. Grischkowsky: Self-focusing of light by potassium vapor, *Phys. Rev.* Lett. **24**, 866 (1970).
5. A. Javan, P.L. Kelley: Possibility of self-focusing due to intensity-dependent anomalous dispersion, *IEEE J. Quant. Electron.* **QE-2**, 470 (1966).
6. D. Grischkowsky, J.A. Armstrong: Self-defocusing of light by adiabatic following in rubidium vapor, *Phys. Rev. A* **6**, 1566 (1972).
7. J.E. Bjorkholm, A. Ashkin: cw self-focusing and self-trapping of light in sodium vapor, *Phys. Rev. Lett.* **32**, 129 (1974).

8. D. Grischkowsky, E. Courtens, W. Armstrong: Observation of self-steepening of optical pulses with possible shock formation, *Phys. Rev. Lett.* **31**, 422–425 (1973).
9. A.C. Tam, W. Happer: Long-range interactions between cw self-focused laser-beams in an atomic vapor, *Phys. Rev. Lett.* **38**, 278–282 (1977).
10. R.R. Alfano, S.L. Shapiro: Emission in the region 4000 to 7000 Å via four-photon coupling in glass, *Phys. Rev. Lett.* **24**, 584–587 (1970).
11. I. Golub, R. Shuker, G. Erez: On the optical characteristics of the conical emission, *Opt. Commun.* **57**, 143–145 (1986).
12. D.J. Harter, P. Narum, M.G. Raymer et al.: Four-wave parametric amplification of rabi sidebands in sodium, *Phys. Rev. Lett.* **46**, 1192–1195 (1981).
13. D.J. Harter, R. W. Boyd: Four-wave mixing resonantly enhanced by ac-stark-split levels in self-trapped filaments of light, *Phys. Rev. A* **29**, 739–748 (1984).
14. M.L. Ter-Mikaelian, G.A. Torossian, G.G. Grigoryan: Conical emission in the quasi-resonant media as a result of self-phase modulation, *Opt. Commun.* **119**, 56–60 (1995).
15. G. Brechignac, P. Cahuzac, A. Debarre: Anomalous off-axis emissions on the resonance strontium line, illuminated by a quasi-resonant pulsed laser-light, *Opt. Commun.* **35**, 87–91 (1980).
16. R.C. Hart, Y. Li, A. Gallagher et al.: Failures of the four-wave mixing model for cone emission, *Opt. Commun.* **111**, 331–337 (1994).
17. B.D. Paul, M.L. Dowell, A. Gallagher et al.: Observation of conical emission from a single self-trapped atom, *Phys. Rev. A* **59**, 4784–4796 (1999).
18. M. Fernández Guasti, J.L. Hernández Pozos, E. Haro Poniatowski et al.: Anomalous conical emission in calcium vapor, *Opt. Commun.* **108**, 367–376 (1994).
19. C.H. Skinner, P.D. Kleiber: Observation of anomalous conical emission from laser-excited barium vapor, *Phys. Rev. A* **21**, 151–156 (1980).
20. W. Chalupczak, W. Gawlik, J. Zachorowski: Degenerate parametric emission in dense barium vapor, *Opt. Commun.* **111**, 613–622 (1994).
21. J.F. Valley, G. Khitrova, H.M. Gibbs et al.: cw conical emission: first comparison and agreement between theory and experiment, *Phys. Rev. Lett.* **64**, 2362–2365 (1990).
22. M. Leberrerousseau, E. Ressayre, A. Tallet: Self-induced generation of an off-axis frequency shifted radiation from atoms, *Opt. Commun.* **36**, 31–34 (1981).
23. D.J. Harter, R.W. Boyd: Conical emission due to four-wave mixing enhanced by the ac-stark effect in self-trapped filaments of light, *Opt. Lett.* **7**, 491–493 (1982).
24. A.I. Plekhanov, S.G. Rautian, V.P. Safonov et al.: Frequency-angular diffusion of intense quasiresonant radiation, *JETP Lett.* **36**, 284 (1982).
25. I. Golub, G. Erez, R. Shuker: Cherenkov emission due to laser-induced moving polarization in sodium, *J. Phys. B* **19**, L115–L120 (1986).
26. I. Golub, R. Shuker, G. Erez: Anomalous blue-shifted emission near the $D1$ transition from laser-excited sodium vapor, *J. Phys. B* **20**, L63–L68 (1987).
27. Y. Shevy, M. Rosenbluh: Multiple conical emissions from a strongly driven atomic system, *J. Opt. Soc. Am.* B **5**, 116–122 (1988).
28. M.E. Crenshaw, C.D. Cantrell: Conical emission as a result of pulse breakup into solitary waves, *Phys. Rev. A* **39**, 126–148 (1989).
29. A.A. Afanas'ev, B.A. Samson, R. Yakite: Conical emission at light self-focusing in resonant media, *Laser Physics* **1**, 399 (1991).
30. W. Chalupczak, W. Gawlik, J. Zachorowski: Conical emission in barium vapor, *Opt. Commun.* **99**, 49–54 (1993).
31. W. Chalupczak, W. Gawlik, J. Zachorowski: Conical emission as cooperative fluorescence, *Phys. Rev. A* **49**, R2227–R2230 (1994).
32. J. Guo, J. Cooper, A. Gallagher: Generation of transient Rabi sidebands in pulse propagation: A possible source of cone emission, *Phys. Rev. A* **52**, R3440–R3443 (1995).
33. A. Dreischuh, V. Kamenov, S. Dinev et al.: Spectral and spatial evolution of a conical emission in Na vapor, *J. Opt. Soc. Am. B* **15**, 34 (1997).

34. L. You, J. Mostowski, J. Cooper et al.: Cone emission from laser-pumped two-level atoms, *Phys. Rev. A* **44**, R6998–R7001 (1991).
35. W.K. Lee, Y.C. Noh, J.H. Jeon J. Lee et al.: Conical emission as a result of self-phase modulation in samarium vapor under the near-resonant condition, *J. Opt. Soc. Am. B* **18**, 101–105 (2001).
36. R.W. Boyd, M.G. Raymer, P. Narum et al.: Four-wave parametric interactions in a strongly driven two-level system, *Phys. Rev. A* **24**, 411–423 (1981).
37. P. Narum, R.W. Boyd: (unpublished).
38. M.T. Gruneisen, K.R. Macdonald, R.W. Boyd: Induced gain and modified absorption of a weak probe beam in a strongly driven sodium vapor, *J. Opt. Soc. Am. B-Opt. Phys.* **5**, 123–129 (1988).
39. B.D. Paul, J. Cooper, A. Gallagher et al.: Theory of optical near-resonant cone emission in atomic vapor, *Phys. Rev. A* **66**, 063816 (2002).
40. W. Gawlik, R. Shuker, A. Gallagher: Temporal character of pulsed-laser cone emission, *Phys. Rev. A* **64** (2001).
41. L.A. Chauchard, Y.H. Meyer: On the origin of the so-called conical emission in laser-pulse propagation in atomic vapor, *Opt. Commun.* **52**, 141–144 (1984).
42. D. Sarkisyan, B.D. Paul, S.T. Cundiff et al.: Conical emission by 2-ps excitation of potassium vapor, *J. Opt. Soc. Am. B* **18**, 218–224 (2001).
43. Y.H. Meyer: Multiple conical emissions from near resonant laser propagation in dense sodium vapor, *Opt. Commun.* **34**, 439–444 (1980).
44. A.I. Plekhanov, S.G. Rautian, V.P. Safonov et al.: The nature of frequency-angular diffusion of powerful quasiresonant radiation, *JETP Lett.* **61**, 249–254 (1985).
45. Y. Shevy, M. Rosenbluh, S. Hochman et al.: Polarization dependence of resonance-enhanced 3-photon scattering, *Opt. Lett.* **13**, 1005–1007 (1988).
46. M.L. Dowell, R.C. Hart, A. Gallagher et al.: Self-focused light propagation in a fully saturable medium: Experiment, *Phys. Rev. A* **53**, 1775–1781 (1996).
47. M.L. Dowell, B.D. Paul, A. Gallagher et al.: Self-focused light propagation in a fully saturable medium: *Theory, Phys. Rev. A* **52**, 3244–3253 (1995).
48. J. Krasinski, D.J. Gauthier, M.S. Malcuit et al.: Two-photon conical emission, *Opt. Commun.* **54**, 241–245 (1985).
49. V. Vaicaitis, A. Piskarskas: Tunable four-photon picosecond optical parametric oscillator, *Opt. Commun.* **117**, 137–141 (1995).
50. T. Efthimiopoulos, M.E. Movsessian, M. Katharakis et al.: Study of the $5P_{3/2} - 4S_{1/2}$ emission in K under two-photon $4S_{1/2} - 6S_{1/2}$ excitation, *J. Phys. B* **29**, 5619–5627 (1996).
51. V. Vaicaitis, S. Paulikas: Resonantly enhanced parametric four-wave mixing in sodium vapour, *Opt. Commun.* **247**, 187–193 (2005).
52. V. Vaicaitis, S. Paulikas: Resonantly enhanced parametric four-wave mixing in sodium vapour, *Opt. Commun.* **247**, 187–193 (2005).
53. G. Grynberg, E. Lebihan, P. Verkerk et al.: Observation of instabilities due to mirrorless four-wave mixing oscillation in sodium, *Opt. Commun.* **67**, 363–366 (1988).
54. J. Pender, L. Hesselink: Conical emissions and phase conjugation in atomic sodium vapor, *IEEE J. Quantum Electron.* **25**, 395–402 (1989).
55. J. Pender, L. Hesselink: Degenerate conical emissions in atomic-sodium vapor, *J. Opt. Soc. Am.* B **7**, 1361–1373 (1990).
56. M. Kauranen, J.J. Maki, A.L. Gaeta et al.: Two-beam-excited conical emission, *Opt. Lett.* **16**, 943–945 (1991).
57. A.A. Afanasev, B.A. Samson: Multiconical emission at light counterpropagation in a resonant medium, *Phys. Rev. A* **53**, 591–597 (1996).
58. M. Fernández Guasti, J.L. Hernández Pozos, E. Haro Poniatowski, L.A. Julio Sánchez: Anomalous conical emission: Two-beam experiments, *Phys. Rev. A* **49**, 613–615 (1994).

59. R.S. Bennink, V. Wong, A.M. Marino et al.: Honeycomb pattern formation by laser-beam filamentation in atomic sodium vapor, *Phys. Rev. Lett.* **88**, 113901 (2002).
60. M.S. Bigelow, P. Zerom, R.W. Boyd: Breakup of ring beams carrying orbital angular momentum in sodium vapor, *Phys. Rev. Lett.* **92**, 083902 (2004).
61. A.M.C. Dawes, L. Illing, S.M. Clark et al.: All-optical switching in rubidium vapor, *Science* **308**, 672–674 (2005).
62. F. Huneus, B. Schapers, T. Ackemann et al.: Optical target and spiral patterns in a single-mirror feedback scheme, *Appl. Phys. B-lasers Optics* **76**, 191–197 (2003).
63. A. Aumann, T. Ackemann, E.G. Westhoff et al.: Transition to spatiotemporally irregular states in a single-mirror feedback system, *Int. J. Bifurcation Chaos* **11**, 2789–2807 (2001).
64. T. Ackemann, A. Aumann, E.G. Westhoff et al.: Polarization degrees of freedom in optical pattern forming systems: alkali metal vapour in a single-mirror arrangement, *J. Optics B-quantum Semiclassical Opt.* **3**, S124–S132 (2001).
65. T. Ackemann, T. Lange: Optical pattern formation in alkali metal vapors: Mechanisms, phenomena and use, *Appl. Phys. B-lasers Optics* **72**, 21–34 (2001).
66. W. Lange, T. Ackemann, A. Aumann et al.: Atomic vapors: a versatile tool in studies of optical pattern formation, *Chaos Solitons & Fractals* **10**, 617–626 (1999).
67. Z.H. Musslimani, L.M. Pismen: Resonant optical patterns in sodium vapor in a magnetic field, *Phys. Rev. A* **59**, 1571–1576 (1999).
68. A. Aumann, E. Buthe, Y.A. Logvin et al.: Polarized patterns in sodium vapor with single mirror feedback, *Phys. Rev. A* **56**, R1709–R1712 (1997).
69. A.J. Scroggie, W.J. Firth: Pattern formation in an alkali–metal vapor with a feedback mirror, *Phys. Rev. A* **53**, 2752–2764 (1996).
70. T. Ackemann, W. Lange: Nonhexagonal and nearly hexagonal patterns in sodium vapor generated by single-mirror feedback, *Phys. Rev. A* **50**, R4468–R4471 (1994).
71. G. Grynberg, A. Maitre, A. Petrossian: Flowerlike patterns generated by a laser-beam transmitted through a rubidium cell with single feedback mirror, *Phys. Rev. Lett.* **72**, 2379–2382 (1994).
72. N.B. Abraham, W.J. Firth: Overview of transverse effects in nonlinear-optical systems, *J. Opt. Soc. Am. B-optical Phys.* **7**, 951–962 (1990).
73. L. Allen, S.M. Barnett, M.J. Padgett: *Optical Angular Momentum*, Taylor and Francis, New York (2003).
74. W.J. Firth, D.V. Skryabin: Optical solitons carrying orbital angular momentum, *Phys. Rev. Lett.* **79**, 2450 (1997).
75. V. Tikhonenko, J. Christou, B. Luther-Davies: Three-dimensional bright spatial soliton collision and fusion in a saturable nonlinear-medium, *Phys. Rev. Lett.* **76**, 2698–2701 (1996).
76. D.J. Gauthier, M.S. Malcuit, A.L. Gaeta et al.: Polarization bistability of counterpropagating laser-beams, *Phys. Rev. Lett.* **64**, 1721–1724 (1990).

Chapter 10
Periodic Filamentation and Supercontinuum Interference

Xiaohui Ni and R.R. Alfano

Abstract In this chapter, supercontinuum (SC) generation and filamentation in BK7 glass were controlled by Fresnel diffraction from a circular aperture or a straight edge. We demonstrated the salient coherent property of multiple SC sources by the periodic filamentation.

10.1 Introduction

The filamentation process involves a balance between self-focusing and defocusing [1, 2]. High-power laser beams break up into an unpredictable distribution of small-scale filaments due to, for example, beam irregularities and refractive index inhomogeneities. Scientists in the early 1970 s were concerned about filamentation problems in the design of high-power laser systems [3]. Today, many applications rely on the efficient generation and control of filaments; for example, remote sensing of chemical molecules in air [4–6], and fabriacation of three-dimensional photonic structures such as optical data storages, waveguides, grating, and couplers inside a wide variety of transparent materials [7].

Motivated by those applications, researchers have aimed to control filamentation overcome the irregularities. The propagation distance at which filamentation occurs has been controlled by adding negative chirp to the pulse [8]. Méchain et al. achieved a degree of control by introducing field gradients and phase changes on the input beam in air [9]. Schroeder et al. performed similar investigations for controlling filaments in water [10, 11]. Cook et al. produced a coherent array of SC filaments by diffractive microlenses [12]. Other methods for filament control have utilized the effects of beam astigmatism and ellipticity [13, 14].

The self-focusing and filamentation of ultrashort pulse in various media can give rise to one incredible and startling nonlinear optical phenomenon: the supercontinuum (SC), which is characterized by a dramatic "white light"

R.R. Alfano (✉)
Physics Department, Institute for Ultrafast Spectroscopy and Lasers, The City College of City University of New York, New York, NY, USA
e-mail: ralfano@sci.ccny.cuny.edu

R.W. Boyd et al. (eds.), *Self-focusing: Past and Present*,
Topics in Applied Physics 114, DOI 10.1007/978-0-387-34727-1_10,

spectrum that can span from the ultraviolet to the infrared [15–18]. The SC is an unusual “white light” source because of its distinction with a conventional white light source. The SC has a high degree of spatial coherence and is usually produced in short burst. There are many applications for the SC sources that depend on coherence, such as optical coherence tomography [19], optical frequency metrology [20], communications [21], and femtosecond and attosecond pulse generation [22, 23]. The coherent nature of the SC generation process is important for ensuring the spectral mode structure of the frequency comb associated with laser pulses to be transferred coherently to the SC. It is shown that the SC generation from both a single filament and multiple filaments exhibit a high degree of spatial coherence [24, 25]. The SC mutual coherence was quantified by means of a time-delay pulsed method with an interferometer [26–28].

This chapter describes the periodic filamentation controlled by diffraction, supercontinuum generation, conical emission, and supercontinuum interference with periodic filaments. The SC will be the ultimate light source.

10.2 Self-phase Modulation and Conical Emission

An optical pulse traveling through a medium with an intensity-dependent refractive index [$\delta n(t) \propto I$] becomes distorted in phase (self-phase modulation) [29–33] and in envelope shape (self-steepening). The phase change and the produced additional frequency components are given by:

$$\Delta\varphi = (\omega_c z/c)\delta n(t) \quad \text{and} \quad \delta\omega(t) = -\partial(\Delta\varphi(t))/\partial t \tag{10.1}$$

The self-phase modulation (SPM) process for an incident Gaussian pulse is shown in Fig.10.1. The frequency shift within the pulse shape is proportional to the derivative of the pulse envelope, which corresponds to the generation of new

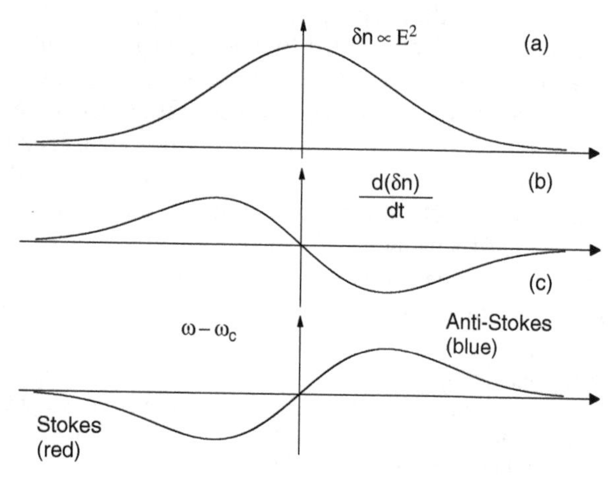

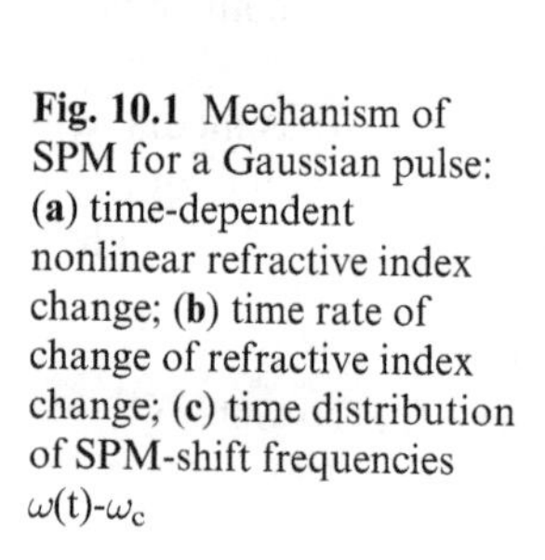

Fig. 10.1 Mechanism of SPM for a Gaussian pulse: (**a**) time-dependent nonlinear refractive index change; (**b**) time rate of change of refractive index change; (**c**) time distribution of SPM-shift frequencies $\omega(t)$-ω_c

frequencies resulting in wider spectra. Figure 10.1(a) shows the time-dependent nonlinear refractive index change; Fig. 10.1(b) shows the time rate of change of refractive index change; Fig. 10.1(c) shows the frequency distribution. The leading edge, the pulse peak, and the trailing edge are red-shifted, nonshift, and blue-shifted, respectively. Extreme spectral broadening was observed by Alfano and Shapiro [16] in crystals and glasses under picosecond pulse excitation, which contributed to an electronic mechanism (e.g., electronic cloud distortion, and molecular libration).

Self-steepening of pulse arises from the intensity dependence of the group velocity. It tends to compact and sharpen up the most intense part of the pulse to the tail, which generates the asymmetric blue-shifted broadening of the SPM [34, 35].

Self-focusing and filamentation increase the local intensity, which in turn assists the intensity dependent SPM process. The high-power picosecond or femtosecond lasers combined with filamentation can produce a large enough change in the refractive index and result in a significant spectral broadening, i.e., a supercontinuum.

Figure 10.2 shows a typical angular emission structure selected from a white light filament. The center portion arises from self-phase modulation. The outer ring structure corresponds to part of the conical emission. The salient feature of the angular distribution of the anti-Stokes emission shown in Fig. 10.2 is that the angle of the ring increases as the emission wavelength decreases, a trend that is opposite to that expected from normal beam diffraction.

The observed ring structure can arise from the anti-Stokes emission from the surface of a small-scale filament, which is classified as a class-II emission. Historically two types of anti-Stokes emission have been reported. Class-I emission was first observed in calcite [36] and obeys the volume phase-matching equation:

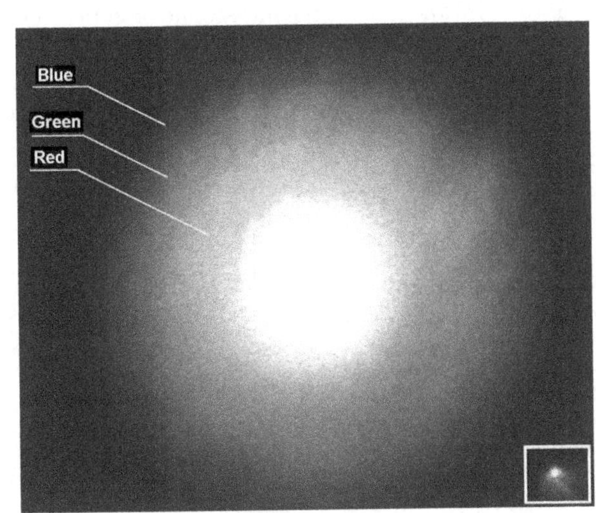

Fig. 10.2 Supercontinuum generation and conical emission in BK7 glass. The insert is the photograph of filament formed in the BK7 glass. A filament in the central region is shown in this photograph [44]

$$\vec{K_L} + \vec{K_L} = \vec{K_A} + \vec{K_S}, \tag{10.2}$$

where K_A, K_S, and K_L are the anti-Stokes, Stokes, and laser wave vector, respectively. Class-II emission fails to obey the volume phase-matching relation and originates in filaments. Detailed explanations of anti-Stokes radiation from filaments have been given by Shimoda [37, 38] and Sacchi et al. [39]. Following Shimoda [37] the anti-Stokes angle φ_A is calculated from

$$\varphi_A^2 = 2\Delta K_A/K_A - a^2/2l^2, \tag{10.3}$$

where $\Delta K_A = K_A + K_S - 2K_L$ is the collinear phase change, a is the filament radius, and l is the filament length.

An important correction to the emission angle is produced by the nonlinear change of refractive index induced by the short, intense laser pulses. This correction tends to both anti-Stokes and Stokes emission angles described as class-III. Chiao et al. [40] have shown that the refractive index increase for weak waves is twice that of the strong wave. Thus instead of Eq. (10.3) for the radiation angle for the filaments we write a new ΔK_A, $\Delta K_A'$, which takes into account the nonlinear increase in the refractive index, i.e., $\Delta K_A' = (K_A + 2\Delta K) + (K_S + 2\Delta K) - 2(K_L + \Delta K)$ so that

$$\varphi_A^2 = 2\Delta K_A/K_A + 4\Delta K/K_A - a^2/2l^2 \tag{10.4}$$

where $\Delta K = \Delta n_L \omega_L/c$ and $\Delta n_L = \varepsilon_2|E_L|^2/4n_L$ are the induced changes in the refractive index, ε_2 is the electronic Kerr coefficient, $|E_L|^2$ is the laser intensity, n_L is the refractive index for the laser light in the medium, and c is the speed of light in vacuum. This model was used by Alfano and Shapiro to fit the conical emission from picosecond pulse in solids (borosilicate BK-7 glasses) as shown Fig. 10.3 [15] and late from femtosecond laser in an ethylene glycol medium [41]. In Fig. 10.3, the circles were experimental points; curve I (dash-dot) was calculated for class-I emission; curve II (dash) was calculated for class-II emission; curve III (solid line) was calculated for small-scale filaments plus a

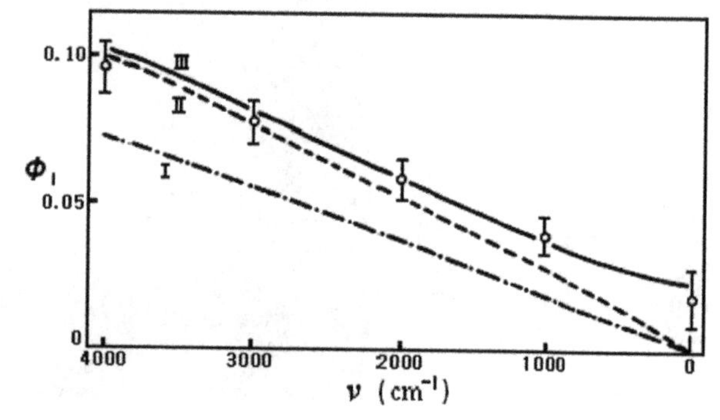

Fig. 10.3 Anti-Stokes angle outside material: circles, experimental points; dash-dot, curve I calculated for class-I emission; dash, curve II calculated for class-II emission; solid line, curve III calculated for small scale filaments plus a non-linear refractive index change [15]

nonlinear refractive index change. The excellent fitting to data indicates that the conical emission is from the process of four-photon parametric generation with a nonlinear refractive index change at the surface of filament.

10.3 Periodic Filamentation and Supercontinuum Generation from Diffraction

Breakup of a high-power laser in a nonlinear medium can arise from modulational instability [42], which leads to multiple filamentation with randomized spatial distribution. Many practical applications call to control the propagation distance at which filamentation occurs and overcome the randomization of spatial distribution.

Campillo et al. first observed periodic filamentation in CS_2 by introducing diffraction components into the beam path of nanosecond pulses [43]. Figure 10.4 shows the self-focusing patterns that they photographed at the exit face of the CS_2 cell. The ring pattern in Fig. 10.4(a) was induced by an aperture in the pathway of a Q-switched ruby laser beam and the breakup along the straight lines in Fig. 10.4(b) came from a straight-edge diffraction. The periodicity was explained by an extension of the instability theory [42, 43].

Schroeder et al. used a two-photon fluorescence technique to visualize the controlled formation of filamentation by launching femtosecond pulses through water [11]. Figure 10.5 shows the well-defined 1-D array of filaments that they obtained by superimposing a slit in the beam pathway.

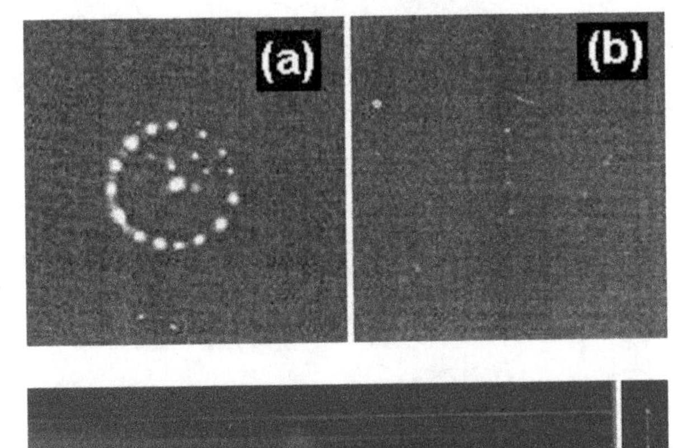

Fig. 10.4 Photographs showing spatial periodicity of focal spots: (**a**) focal spot pattern from aperture; (**b**) focal spot pattern from straight edge [Reprinted with permission from ref. 43, Copyright 1973, American Institute of Physics]

Fig. 10.5 Creation of a space-fixed one-dimensional filament array in water behind a slit [11]

Advancement of femtosecond laser technology made possible the broadband SC generation accompanying the filamentation process. Here, we demonstrated the formation of periodic filamentation in BK7 glass using diffraction components. The creation of the periodic filaments is used to investigate the coherent property of the supercontnuum generation.

Diffraction and interference are two important interconnecting effects taking place in optics and optical phenomena to control the phase of waves. Both processes can rearrange the spatial distribution of the incident energy and make it possible to alter nonlinear processes at a sample site such as SC generation at a lower incident energy and even help with the generation of even shorter ultrashort pulses. Fresnel diffraction describes the near-field diffraction pattern and more intensity can occur spatially at well-defined distances for a light beam from a small aperture than for a full-beam nonaperture case [44, 45].

When a collimated beam propagation through an aperture, the most important parameter needed for classifying near-field diffraction pattern is the Fresnel number N. The number of zones seen by the observation point in the aperture is given by well-known Fresnel numbers:

$$N = R^2/\lambda L, \tag{10.5}$$

where R is the radius of the aperture, λ is incident wavelength, L is the on-axis distance between the observation point and aperture. The sum of the optical field disturbance on-axis from odd-numbered Fresnel zones at P is:

$$E \approx |E_1|/2 + |E_N|/2, \tag{10.6}$$

where E_N is the contribution from the electric field intensity of the Nth Fresnel zone. An even-numbered Fresnel zones leads to field at P given by:

$$E \approx |E_1|/2 - |E_N|/2, \tag{10.7}$$

The near-axis intensity I at this point P corresponding to one Fresnel zone is

$$I \approx 4I_0, \tag{10.8}$$

where I_0 is the unobstructed intensity. A large intensity enhancement happens at the observation point P corresponding to odd Fresnel zones. The first zone gives largest intensity; see Eq. (10.8).

The experimental arrangement to observe the effect of Fresnel diffraction on SC generation and interference is shown in Fig. 10.6. A Ti:sapphire amplifier system produces 0.35 mJ, 160 fs (FWHM) at 800-nm pulses with a repetition rate of 1 KHz. The 1/e^2 diameter of the beam is 1 cm. Collimated beam propagates through an aperture of diameter 300 μm. A BK7 glass with the thickness of 6 mm is placed at a distance L from the aperture. A lens is used to collect the

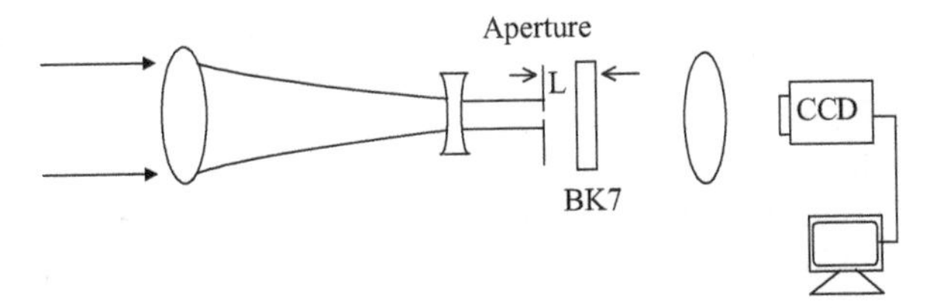

Fig. 10.6 Schematic diagram of the SC interference setup

SPM and conical emission from the BK7. The intensity of the incident beam is set below the threshold of self-focusing and SC generation in the BK7 without the aperture. The collecting lens and the BK7 are moved together toward the aperture along its axis at distances corresponding to $N = 1, 2, 3\ldots$ zones.

The supercontinuum generation first appears when $N=1$. A typical snapshot of the observed emission pattern is shown in Fig. 10.7. The insert shows the filaments pattern at the exit surface of the glass. SPM contributes to the white light spot in the central region [16]. The nondegenerate four-photon stimulated emission [15] results in the colorful cone (red to violet). The blue portion near the central white light comes from the Rayleigh scattering in BK7. According to Eq. (10.8) for $N=1$, the intensity will be four times higher than the unobstructed light intensity at the generation spot. The appearance of SC is consistent with the calculation of this radiation distribution.

When the collecting lens and the BK7 moved further toward the aperture, the SC generation will disappear and reappear again at the distance corresponding to $N=3$, as expected from the radiation distribution along the axis. The observed SC pattern is shown in Fig. 10.2, consisting of a colorful conical emission with white light in the central region.

Figures 10.2 and 10.7 demonstrate that SC is formed at distances from the aperture that coincide with local intensity maxima ($N=1,3$) in the near-field diffraction pattern. The difference between the emission patterns displayed in

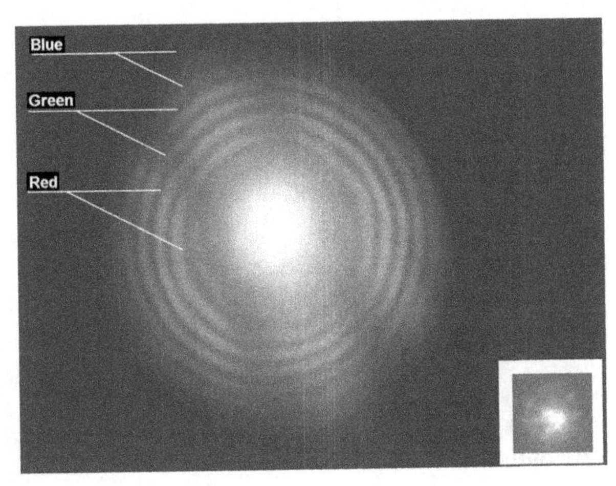

Fig. 10.7 Supercontinuum interference in BK7 glass resulting from energy enhancement of ultrashort beam propagation through a 300-μm-diam aperture at a distance corresponding to Fresnel number $N = 1$. The insert is the photograph of filaments formed in the BK7 glass. A central filament and some small-scale filaments around the central one are shown in this photograph

Figs. 10.2 and 10.7 is the interference ring structure appearing inside the conical emission of Fig. 10.7. This interference ring structure arises from the supercontinuum generation inside the BK7 glass along the axis and along the first intense outer annular ring of diffraction pattern. As the intensity is enhanced and exceeds the threshold in the field distribution at the distance corresponding to Fresnel number $N = 1$, the central portion of the beam self-focuses to form a bright filament, and the first intense diffracted annular ring at the sample site P also self focuses but breaks up to small-scale filaments.

Both the brighter filament in the central region and small-scale filaments in the annular ring can result in SC generation. Due to the intensity difference in the diffraction field of the central region and the outer first annular ring, the central filament is developed earlier than the small-scale filaments from the annular ring in the propagation. A small adjustment of the collecting lens position showed a change in the contrast of the central filament and small-scale filaments around it, giving an estimation of 0.9 mm separation distance between the two set of filaments created. The SC generation at the two different positions along the beam propagation direction contributed to the interference ring observed on the CCD.

Calculations based on the interference ring spacing have confirmed the 0.9 mm separation between the filament sources in the glass. This kind of interference pattern can be also observed near the focus of a spherical lens for ultrafast pulses of well-prepared intensity. The field at the focus has the similar pattern as the Fresnel diffraction. In Fig. 10.2 ($N=3$), the intensity in the outer ring of the diffraction field was not enough to form filaments. The SC generation is present in the central region, and no interference rings occur in the conical emission.

The inserts in Figs. 10.2 and 10.7 show the filaments pattern photographed in the BK7 glass. Both photographs have a clear filament pattern generated by self-focusing in the central region. In Fig. 10.7, a periodic break-up of small-scale filaments is observed along the outer ring. The imposed diffraction aperture overcomes the randomization of filamentation, which allows the exploitation of the coherent properties of multiple filamentation and SC sources. This periodicity was explained by Campillo et al. with an extension of the instability theory [43]. They presented a clearer periodic filamentation pattern as shown in Fig. 10.4 due to the absence of significant background conical emission for a longer pulse.

The periodic filamentation and SC interference can also occur from a diffraction straight edge. Figure 10.8 shows SC interference in BK7 glass when an ultrashort intense laser propagates through a straight edge. In this case, the incident laser pulses were adjusted to be not strong enough to generate SC without the edge. The first Fresnel diffraction maximum from the edge is well-known to be ~1.4 times higher than the incident intensity [46]. This increase in intensity creates filaments and SC with a lower power incident laser pulse. Fringes that are perpendicular to the edge are observed in the SC emission. This interference phenomenon can be understood from the intensity distribution of the straight-edge diffraction in the BK7 glass. Self-focusing can

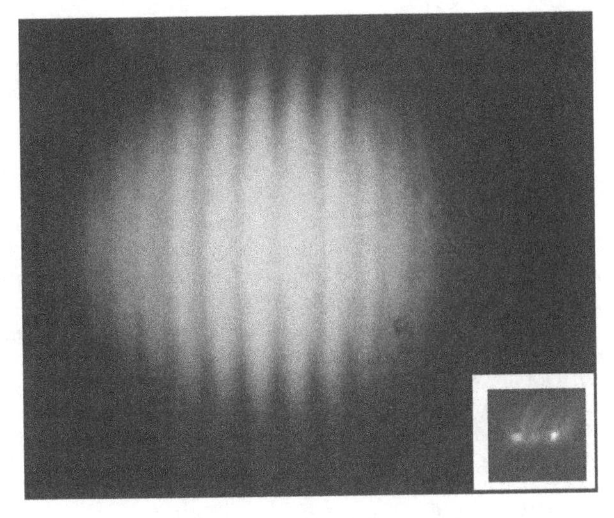

Fig. 10.8 Supercontinuum interference in BK7 glass resulting from energy enhancement of ultrashort beam propagation through a straight edge. The insert is the photograph of filaments in the BK7 glass. Two filament parallel to the edge are formed

be formed along the first maximum fringe in the diffracted field and break up into small-scale filaments along the edge direction. The spatial distribution of filaments inside the BK7 glass is photographed and presented in the insert of Fig. 10.8. Two distinctive filaments are formed parallel to the edge. These two filaments produce interference fringe pattern perpendicular to the edge.

A similar fringe pattern as shown in Fig. 10.8 has also been demonstrated by Bellini et al. [47] using a double-source interference experiment, and by Cook et al. [48] using a focusing cylindrical lens to produce a horizontal spatial array of SC filaments. The divergence of the pattern and the number of fringes have been calculated by Cook et al. to agree with the predicted values assuming the filaments are a pair of spatially coherent sources of SC analogous to Young's slit experiment [48].

The observed ring and fringe interference patterns in Figs. 10.7 and 10.8 demonstrate the highly coherent nature of SC and the phase stability of multiple filaments.

10.4 Conclusion

In summary, we illustrated the formation of periodic filamentation by the superimpose of diffraction components in the light pathway. The onset of filamentation and SC generation is also controlled by this method. We showed the coherent property of SC sources accompanying the filaments generated at different spatial positions.

Acknowledgments This research is supported in part by DOD Center for Nanoscale Photonic Emitters and Sensors and NASA URC Center for Optical Sensing and Imaging (COSI) at CCNY (NASA Grant No.: NCC-1-03009).

References

1. A. Couairon, L. Bergé: Light filaments in air for ultraviolet and infrared wavelengths. *Phys. Rev. Lett.* **88**, 135003 (2002).
2. V.P. Kandidov, O.G. Kosareva, I.S. Golubtsov et al.: Self–transformation of a powerful femtosecond laser pulse into a white–light laser pulse in bulk optical media (or supercontinuum generation). *Appl. Phys. B* **77**, 149–165 (2003).
3. A.J. Campillo, J.E. Pearson, S.L. Shapiro et al.: Fresnel diffraction effects in the design of high-power laser systems. *Appl. Phys. Lett.* **23**, 85–87 (1973).
4. J. Kasparian, M. Rodriguez, G. Méjean et al.: White-light filaments for atmospheric analysis. *Science* **301**, 61–64 (2003).
5. P. Rairoux, H. Schillinger, S. Niedermeier et al.: Remote sensing of the atmosphere using ultrashort laser pulses. *Appl. Phys. B* **71**, 573–580 (2000).
6. Q. Luo, W. Liu, S.L. Chin: Lasing action in air-induced by ultrafast laser filamentation. *Appl. Phys. B* **76**, 337–340 (2003).
7. W. Watanabe, T. Asano, K. Yamada et al.: Wavelength division with three-dimensional couplers fabricated by filamentation of femtosecond laser pulses. *Opt. Lett.* **28**, 2491–2493 (2003).
8. H. Wille, M. Rodriguez, J. Kasparian et al.: Teramobile: A mobile femtosecond–terawatt laser and detection system. *Eur. Phys. J. AP* **20**, 183–190 (2002).
9. G. Méchain, A. Couairon, M. Franco et al.: Organizing multiple femtosecond filaments in air. *Phys. Rev. Lett.* **93**, 035003 (2004).
10. H. Schroeder, S.L. Chin: Visualization of the evolution of multiple filaments in methanol. *Opt. Commun.* **234**, 399–406 (2004).
11. H. Schroeder, J. Liu, S.L. Chin: From random to controlled small-scale filamentation in water. *Opt. Express* **12**, 4768–4774 (2004).
12. K. Cook, R. McGeorge, A.K. Kar et al.: Coherent array of white-light continuum filaments produced by diffractive microlenses. *Appl. Phys. Lett.* **86**, 021105 (2005).
13. G. Fibich, S. Eisenmann, B. Ilan et al.: Control of multiple filamentation in air. *Opt. Lett.* **29**, 1772–1774 (2004).
14. A. Dubietis, G. Tamosăuskas, G. Fibich et al.: Multiple filamentation induced by input-beam ellipticity. *Opt. Lett.* **29**, 1126–1128 (2004).
15. R.R. Alfano, S.L. Shapiro: Emission in the region 4000 to 7000 Å via four-photon coupling in glass. *Phys. Rev. Lett.* **24**, 584–587 (1970).
16. R.R. Alfano, S.L. Shapiro: Observation of self-phase modulation and small-scale filaments in crystals and glasses. *Phys. Rev. Lett.* **24**, 592–594 (1970).
17. R.R. Alfano, S.L. Shapiro: Direct distortion of electronic clouds of rare-gas atoms in intense electric fields. *Phys. Rev. Lett.* **24**, 1217–1220 (1970).
18. R.R. Alfano: *The Supercontinuum Laser Source*. Springer, New York (2005).
19. I. Hartl, X.D. Li, C. Chudoba et al.: Ultrahigh-resolution optical coherence tomography using continuum generation in air–silica microstructure optical fiber. *Opt. Lett.* **26**, 608–610 (2001).
20. S.A. Diddams, D.J. Jones, J. Ye et al.: Direct link between microwave and optical frequencies with a 300-THz femtosecond laser comb. *Phys. Rev. Lett.* **84**, 5102–5105 (2000).
21. H. Takara: Multiple optical carrier generation from a supercontinuum source. *Opt. Photon. News* **13**, 48–51 (2002).
22. A. Baltuška, T. Fuji, T. Kobayashi: Visible pulse compression to 4 fs by optical parametric amplification and programmable dispersion control. *Opt. Lett.* **27**, 306–308 (2002).
23. A. Baltuška, M.Uiberacker, E. Goulielmakis et al.: Phase-controlled amplification of few-cycle laser pulses. *IEEE JSTQE* **9**, 972–989 (2003).
24. S.L. Chin, S. Petit, F. Borne et al.: The white light supercontinuum is indeed an ultrafast white light laser. *Jpn. J. Appl. Phys.* **38**, L126–L128 (1999).

25. W. Watanabe, K. Itoh: Spatial coherence of supercontinuum emitted from multiple filaments. *Jpn. J. Appl. Phys.* **40**, 592–595 (2001).
26. X. Gu, M. Kimmel, A.P. Shreenath et al.: Experimental studies of the coherent of microstructure–fiber supercontinuum. *Opt. Express* **11**, 2697–2703 (2003).
27. F. Lu, W.H. Knox: Generation of a broadband continuum with spectral coherence in tapered single-mode optical fibers. *Opt. Express* **12**, 347–353 (2004).
28. I. Zeylikovich, V. Kartazaev, R.R. Alfano: Spectral, temporal, and coherence properties of supercontinuum generation in microstructure fiber. *J. Opt. Soc. Am. B* **22**, 1453–1460 (2005).
29. N. Bloembergen, P. Lallemand: Complex intensity-dependent index of refraction, frequency broadening of stimulated Raman lines, and stimulated Rayleigh scattering. *Phys. Rev. Lett.* **16**, 81–84 (1966).
30. R.G. Brewer: Frequency shifts in self-focused light. *Phys. Rev. Lett.* **19**, 8–10 (1967).
31. A.C. Cheung, D.M. Rank, R.Y. Chiao et al.: Phase modulation of Q-switched laser beams in small-scale filaments. *Phys. Rev. Lett.* **20**, 786–789 (1968).
32. W.J. Jones, B.P. Stoicheff: Inverse Raman spectra: induced absorption at optical frequencies. *Phys. Rev. Lett.* **13**, 657–659 (1964).
33. F. Shimizu: Frequency broadening in liquids by a short light pulse. *Phys. Rev. Lett.* **19**, 1097–1100 (1967).
34. A.L. Gaeta: Catastrophic collapse of ultrashort pulses. *Phys. Rev. Lett.* **84**, 3582–3585 (2000).
35. N. Aközbek, M. Scalora, C.M. Bowden et al.: White-light continuum generation and filamentation during the propagation of ultra-short laser pulses in air. *Opt. Commun.* **191**, 353–362 (2001).
36. R. Chiao, B.P. Stoicheff: Angular dependence of maser-stimulated Raman radiation in calcite. *Phys. Rev. Lett.* **12**, 290–293 (1964).
37. K. Shimoda: Angular distribution of stimulated Raman radiation. *Jpn. J. Appl. Phys.* **5**, 86–92 (1966).
38. K. Shimoda: Gain, frequency shift, and angular distribution of stimulated Raman radiations under multimode excitation. *Jpn. J. Appl. Phys.* **5**, 615–623 (1966).
39. C.A. Sacchi, C.H. Townes, J.R. Lifsitz: Anti-Stokes generation in trapped filaments of light. *Phys, Rev.* **174**, 439–447 (1968).
40. R.Y. Chiao, P.L. Kelley, E. Garmire: Stimulated four-photon interaction and its influence on stimulated Rayleigh-wing scattering. *Phys. Rev. Lett.* **17**, 1158–1161 (1966).
41. Q. Xing, K.M. Yoo, R.R. Alfano: Conical emission by four-photon parametric generation by using femtosecond laser pulses. *Appl. Opt.* **32**, 2087–289 (1993).
42. V.I. Bespalov, V.I. Talanov: Filamentary structure of beams in nonlinear liquids. *JETP Lett.* **3**, 307–310 (1966).
43. A.J. Campillo, S.L. Shapiro, B.R. Suydam: Periodic breakup of optical beams due to self-focusing. *Appl. Phys. Lett.* **23**, 628–630 (1973).
44. X. Ni, C. Wang, X. Liang et al.: Fresnel diffraction supercontinuum generation. *IEEE JSTQE* **10**, 1229–1232 (2004).
45. K. Cook, A.K. Kar, R.A. Lamb: White-light filaments induced by diffraction effects. *Opt. Express* **13**, 2025–2031 (2005).
46. M. Born, E. Wolf: *Principles of Optics*. 6th ed., Pergamon Press, Oxford (1980).
47. M. Bellini, T.W. Hänsch: Phase-locked white-light continuum pulses: toward a universal optical frequency–comb synthesizer. *Opt. Lett.* **25**, 1049–1051 (2000).
48. K. Cook, A.K. Kar, R.A. Lamb: White-light supercontinuum interference of self-focused filaments in water. *Appl. Phys. Lett.* **83**, 3861–3863 (2003).

Chapter 11
Reprints of Papers from the Past

Chapter 11.1
Effects of the Gradient of a Strong Electromagnetic Beam on Electrons and Atoms

G.A. Askar'yan†

Reprinted from Sov. Phys. JETP 15, 1088–1090 (1962) with permission of the American Institute of Physics

First paper on self-focusing and self-trapping

SOVIET PHYSICS JETP VOLUME 15, NUMBER 6 DECEMBER, 1962

EFFECTS OF THE GRADIENT OF A STRONG ELECTROMAGNETIC BEAM ON ELECTRONS AND ATOMS

G. A. ASKAR'YAN

P. N. Lebedev Physics Institute, Academy of Sciences, U.S.S.R.

Submitted to JETP editor December 22, 1961

J. Exptl. Theoret. Phys. (U.S.S.R.) **42**, 1567-1570 (June, 1962)

It is shown that the transverse inhomogeneity of a strong electromagnetic beam can exert a strong effect on the electrons and atoms of a medium. Thus, if the frequency exceeds the natural frequency of the electron oscillations (in a plasma or in atoms), then the electrons or atoms will be forced out of the beam field. At subresonance frequencies, the particles will be pulled in, the force being especially large at resonance. It is noted that this effect can create either a rarefaction or a compression in the beam and at the focus of the radiation, maintain a pressure gradient near an opening from an evacuated vessel to the atmosphere, and create a channel for the passage of charged particles in the medium.

It is shown that the strong thermal ionizing and separating effects of the ray on the medium can be used to set up waveguide propagation conditions and to eliminate divergence of the beam (self-focusing). It is noted that hollow beams can give rise to directional flow and ejection of the plasma along the beam axis for plasma transport and creation of plasma current conductors. The possibilities of accelerating and heating plasma electrons by a modulated beam are indicated.

IN addition to the "light" pressure (radiation reaction), the electromagnetic field can exert a force on the particles through the mean square electric field.[1,2] Usually, in inhomogeneous high frequency (hf) fields, this force is many times ($\lambda\!\!\!^{-2}/r_0L$) larger than the force of the light pressure (here $\lambda\!\!\!^{-}$ is the wavelength, L is the dimension of the field inhomogeneity, and r_0 is the classical radius of the charged particle).

In contrast with the problems considered[1-3] in the acceleration and retardation of charged particles by the nodes of hf fields in waveguide fields, we consider the behavior of charged and neutral particles in the region of a boundary inhomogeneity of the field of an intense, highly directional or focused beam.

1. FORCING OF ELECTRONS OUT OF THE BEAM. HEATING OF THE PLASMA ELECTRONS

We assume that a beam is propagated in a plasma with a frequency which greatly exceeds the plasma frequency. We examine the forcing of the plasma out of the beam field.

The amplitude of the wave field of a beam emanating from a circular aperture of radius a is*

$$\mathbf{E} = \frac{a^2\omega^2}{2\pi c^2 R_0}[[\mathbf{P}_0\mathbf{n}]\,\mathbf{n}]\;\frac{J_1(\omega c^{-1}a\sin\theta)}{\omega c^{-1}a\sin\theta},$$

where $\mathbf{P}_0$ is the amplitude vector of the dipole moment with unit radiation surface, R is the distance to the point of reception, J_1 is the Bessel function.

Making use of the well-known expression for the averaged force[1,2]

$$\mathbf{f} = -(e^2/2m\omega^2)\,\nabla(E^2)_{\text{av}},$$

we get an expression for the average potential that expels the electrons in a direction perpendicular to the beam:

$$U(\rho) \approx \frac{e\omega^2}{m}\,\frac{P_0^2a^4}{8\pi^2c^4R_0^2}\,\frac{J_1^2(\omega a\rho/cR)}{(\omega a\rho/cR)^2} = \frac{2\pi e^2}{mc\omega^2}w,$$

where $\rho \approx \theta R$ is the distance to the axis of the beam and w is the energy flux density. The size of the region of strong inhomogeneity of the field is $\Delta\rho \sim \lambda R/a$ (use of a set of antennas makes it possible to increase the directivity of the beam).

Strong ejection from the beam leads to the appearance of a space charge which brings about

*$[[\mathbf{P}_0\mathbf{n}]\,\mathbf{n}] = (\mathbf{P}_0\times\mathbf{n})\times\mathbf{n}$.

escape of the ions. Thus the beam will expel the plasma from the volume of its field, which can decrease the diffractive spreading of the beam.

Modulation or motion of such a beam leads to entrainment and acceleration of parts of the plasma by the fringe field gradients; this can be used for volume heating of the plasma. The intensity of the oscillation buildup and of volume heating of the plasma should increase sharply as the modulation frequency approaches the plasma frequency.

2. DRAWING OUT AND EJECTION OF THE PLASMA IN A HOLLOW BEAM

In a number of cases it is necessary to create concentrated plasma jets in a vacuum, capable of carrying large currents for energy transfer. If there is an external magnetic field (for example, the earth's magnetic field), then, upon initiating a stream along the lines of force, it is possible to decrease the transverse spreading of the jet. In the absence of a magnetic field (for example, in outer space) hollow radio beams can be used for directed spreading of the plasma. A typical example of the field of such a hollow beam can be obtained by using the field of radiation from an annular aperture or a pencil of rays.

In a hollow beam, the averaged potential $U = -(e^2/2m\omega^2)E^2$ has a well near the axis. This ensures retention of the plasma on the beam axis.

We now consider the ejection of the plasma along such a beam. Upon rotation or rocking of such a beam, the plasma is accelerated or guided along the ray under the action of the centrifugal force. The equation for the acceleration of the plasma along the beam is $\ddot{R} = R\Omega^2$, where Ω is the angular velocity of rotation or rocking of the beam. If we assume $\Omega^2 = \text{const}$ (Ω can change sign in the case of rocking of the beam), then, by multiplying both sides of the equation by $\dot{R}$ and integrating, we get

$$\dot{x} = |\Omega|\sqrt{x^2 - 1}, \qquad x + \sqrt{x^2 - 1} = \exp\{|\Omega| t\},$$

where $x = R(t)/R(0)$. The ejection of the plasma along the beam can improve the plasma-conducting properties of the hollow beam.

3. FORCING OF ATOMS OUT OF THE BEAM. CREATION OF RAREFACTIONS IN THE MEDIUM

We now consider the formation of a condensation or a rarefaction in a gaseous medium by means of a sharply-directional or focused beam.

The total mean force acting on the electrons of an atom from the external hf field (compare with [3]) is

$$\mathbf{F}_{av} = \frac{e^2}{m}\left\{\sum_k{}' \frac{\omega_{0k}^2 - \omega^2}{(\omega_{0k}^2 - \omega^2)^2 + \gamma_k^2}\right\} \nabla (E^2)_{av},$$

where ω_{0k} and γ_k are the resonance frequency and the damping of the k-th electron. In view of the layered structure of the electronic shells of the atoms, and the significant differences of ω_{0k} in the different electron shells, one can easily arrange it so that $\omega > \omega_{0k}$ for the outside electrons, and $\omega < \omega_{0k}$ for the inside.

If the radiation frequency ω exceeds the natural coupling frequency of the external atomic electrons $\omega_0 \sim 10^{15}$, then a mean force will act on these electrons and force them out of the stronger field into the weaker one, while the force exerted on the other electrons of the atom can be neglected because of their great coupling (i.e., low polarizability). It should be noted that the condition $\omega > \omega_0$ is necessary chiefly to change the sign of the force (see [3]); therefore, small excesses of ω over ω_0 are also possible, i.e., a resonant increase in the force is possible.

The fact that a certain ionization occurs in such a beam increases the effect of removal of matter from the beam, since the free electrons are knocked out of the strong field region, and the ions will be removed by the excess of space charge.

We now estimate the gradient of the density of the neutral molecules, neglecting the ionization effect. Using the expression obtained for the force acting on the atoms, we get for the density of the atoms on the axis of the beam or at the center of the focus (from the Boltzmann formula)

$$\frac{n(0)}{n(\infty)} \approx \exp\left\{-\frac{U}{kT}\right\} = \exp\left\{\frac{\nu e^2 E_0^2}{m(\omega_0^2 - \omega^2)kT}\right\},$$

where ν is the number of electrons in the outer shells of the atoms. It is easy to see that the medium becomes rarefied at the center when $\omega > \omega_0$, and condensed when $\omega < \omega_0$. For example, far off resonance and at $E_0 \sim 10^7$ V/cm (concentrated beam from a laser), for $\omega_0 \sim 10^{15}\,\text{sec}^{-1}$, $kT \sim 0.03$ eV and $\nu \sim 1$, we get an exponent ~ 10.

The formula obtained for the density decrease can also be obtained from the general formula for confining forces, (e.g., see [4]), by taking the time average and setting $\rho(\partial\epsilon/\partial\rho) \approx \epsilon - 1$, which is valid for a gaseous medium.

Generally speaking, there is interest in a detailed study of the effect of strong pulsating confining stresses in dense media in the strong fields of a focused beam from a laser, since such an effect will lead to microdestruction, to the emission of hypersonic and shock waves, and to local heating of the medium. The effects considered in gases are more general because the concentrated effect of the radiation will transform all substances into the gaseous state. Moreover, the gaseous medium is more frequently encountered in technical problems (the transmission of concentrated radiant energy over large distances).

For the first experiments on the ejection of atoms from the hf field regions one could use a laser beam focused in the gas. It is possible to create radiation plugs with large pressure gradients in front of small apertures.

The small dimensions of the concentration zone of the beam (in gases at normal pressures, the zone dimensions can be commensurate with the length of the atomic mean free path) lead to a large temperature gradient and to a strong heat removal. The closeness of the cold gas regions, from which the atoms move into the concentration zone of the beam, justifies the application of the resultant formulas with the value of the mean temperature of the medium, even at elevated temperatures in the focal regions. However, it is not possible to use these expressions in the case of strong absorption or for too high field amplitudes, when saturation effects become significant. For small fields, noticeable separation effects can be obtained at low gas temperatures.

The creation of radiation channels with decreased density of the medium produces in them, for example, charged-particle currents with small energy dissipation.

It is interesting to note that the ionizing thermal and separating actions of intense radiation in the medium can be so strong that a gradient is produced in the properties of the medium in the beam and outside of it, which results in the waveguide propagation of the beam and removes the geometric and diffraction divergences—this interesting phenomenon can be called the self-focusing of the electromagnetic beam.

I express my gratitude to M. S. Rabinovich for support and for discussion of the results.

[1] H. A. H. Boot and R. B. R. S. Harvie, Nature **180**, 1187 (1957), J. Electr. Control **4**, 434 (1958).

[2] A. V. Gaponov and M. A. Miller, JETP **34**, 242, 751 (1958), Soviet Phys. JETP **7**, 168, 515 (1958).

[3] G. A. Askar'yan, Atomnaya énergiya **4**, 71 (1958).

[4] L. D. Landau and E. M. Lifshitz, Elektrodinamika sploshnykh sred (Electrodynamics of Continuous Media) Gostekhizdat, 1957, p. 307.

Translated by R. T. Beyer
262

Chapter 11.2
On Self-focusing of Electromagnetic Waves in Nonlinear Media

V.I. Talanov

First time publication of English translation from the Russian journal Izvestia Vuzov, Radiophysica, 7, 564–565 (1964) with a permission of the journal.

Izvestia Vuzov, Radiofizika (Radiophysics and Quantum Electronics)
volume 7, number 3, pages 564-565, 1964 (in Russian)

On self-focusing of electromagnetic waves in nonlinear media

V.I. Talanov

(Received 25 February 1963)

As is well known [1], the effect of a high-power, high-frequency field on a plasma causes redistribution of the densities of electrons and ions depending on the field amplitude. In its turn, this effect can lead to self-focusing of electromagnetic waves [2]. Such behavior is illustrated below in the example of an isothermal equilibrium hydrogen plasma[1] in a monochromatic field with frequency ω. In the steady state, the dielectric permeability ε of the plasma (with no account for collisions and spatial dispersion) depends on the field intensity E as follows [3]:

$$\varepsilon = \varepsilon_0[1 - q^2 \exp(-|\mathrm{E}|^2)], \tag{1}$$

where $q^2 = \frac{\omega_{p_0}^2}{\omega^2}$, $\omega_{p_0}^2 = \frac{e^2 N_0}{m_e \varepsilon_0}$, $\mathrm{E} = \frac{E}{E_p}$, N_0 is the electron density in the absence of the field, $E_p^2 = 8\kappa T \omega^2 m_e/e^2$, κ is the Boltzmann constant, T is temperature and m_e and e are the mass and charge of the electron, respectively.

Consider a traveling two-dimensional TE-mode wave of the form

$$\mathrm{E}_y = \mathrm{E}(x) \exp[i(\omega t - hz)] \tag{2}$$

in a medium with dielectric permeability (1). The amplitude $\mathrm{E}(x)$of such a wave satisfies the equation

$$\mathrm{E}'' + k_0^2[\frac{\varepsilon(\mathrm{E}^2)}{\varepsilon_0} - \gamma^2]\mathrm{E} = 0. \tag{3}$$

Under the condition that the field decreases at $x \to \infty$ (condition of field localization), the first integral of the above equation can be written down in the form

[1] In the case of multi-charge ions, the numerical factor $(Z+1)/2Z$, where Z is the ion charge, will appear in the expression for E_p in (1).

$$(\mathrm{E}')^2 = k_0^2\{\mathrm{E}^2[\gamma^2 - 1] + q^2[1 - \exp(-\mathrm{E}^2)]\}, \tag{4}$$

where $k_0 = \omega\sqrt{\mu_0\varepsilon_0}$, $\gamma = \frac{h}{k_0}$. Integrating Eq. (4) once more, we obtain

$$k_0 x = \pm \int_{\mathrm{E}}^{\mathrm{E}_m} \frac{d\mathrm{E}}{\sqrt{q^2[1 - \exp(-\mathrm{E}^2)] - [1 - \gamma^2]\mathrm{E}^2}}. \tag{5}$$

Here E_m is the maximum value of the field amplitude (assuming that it is reached at $x = 0$) which is determined from Eq. (4) at $\mathrm{E}' = 0$:

$$[1 - \gamma^2]\mathrm{E}_m^2 = q^2[1 - \exp(-\mathrm{E}_m^2)]. \tag{6}$$

Equation (6) can be treated as the characteristic expression that links the propagation constant γ and the field amplitude E_m at the maximum of the wave. It has the nontrivial solution $\mathrm{E}_m \neq 0$ only under the condition

$$\gamma^2 > 1 - q^2 = 1 - \frac{\omega_{p0}^2}{\omega^2}, \tag{7}$$

which means that wave (2) is slow in comparison with a plane uniform wave in non-perturbed plasma. For $q^2 > 1$, when plane uniform waves cannot propagate in the plasma, wave (2) will be propagating, if the field at its maximum exceeds some critical value determined by Eq. (6) for $\gamma = 0$.

In the practically significant case of rather weak fields ($\mathrm{E}_m^2 << 1$), the dependence $\mathrm{E}(x)$ can be obtained from Eq. (5) in the explicit form:

$$\frac{\mathrm{E}}{\mathrm{E}_m} = \{\cosh[k_0\sqrt{\gamma^2 - (1 - q^2)}x]\}^{-1}. \tag{8}$$

The field structure $\mathrm{E}(x)$corresponds to the formation of a waveguide in the plasma, which is maintained by the action of the field itself. It can be shown that the waveguide is formed at any power of the wave[2]. The power level determines the effective width of the waveguide (along x) and the structure of the field in it.

The consideration of non-uniform two-dimensional TM-modes in a medium with dielectric permeability (1) is almost analogous. In this case, there are also solutions of the traveling-wave type, $\mathrm{E}_x = \mathrm{E}_x(x)\exp[-ihz)]$ and $\mathrm{E}_z = \mathrm{E}_z(x)\exp[-ihz)]$, which vanish at $x \to \pm\infty$ with the field maximum in the finite region of space.

The generalization for three-dimensional fields encounters greater difficulties of integrating nonlinear partial equations with nonseparating variables. It should be

[2] We consider here only equilibrium self-focusing waveguide channels, without touching upon the issues of their establishment and stability. The latter require a combined solution of the equations of the electromagnetic field and hydrodynamic equations for the medium.

taken into account, however, that for cylindrical wave beams, separation of variables can be performed if one consider the traveling waves of the form

$$\{\vec{\mathbf{E}}, \vec{\mathbf{H}}\} = \{\vec{\mathbf{E}}_0, \vec{\mathbf{H}}_0\} \exp[-ihz - im\phi],$$

(where m is an integer), which cause axisymmetric deformations of the medium only. Here, the symmetric $(m = 0)TE_0$ mode is the analog of the two-dimensional TE mode which has been already considered. In the cases such that the dielectric medium is almost completely displaced out of the region occupied by the high-power, high-frequency field, the problem can be reduced to finding the position of the equilibrium of the waveguide channel wall with the account for the pressure of the surrounding medium.

The formation of the self-focusing waveguide under the effect of a high-frequency field is connected, in the cases under consideration, with the increase of the dielectric permeability ε, as the field amplitude grows. One can easily see that such a dependence takes place also in an arbitrary isotropic medium, if it is assumed that the field effect leads to an isothermal deformation that changes the density ρ of such a medium and that the changes of the density ρ and the pressure p are proportional to each other.

For the dependence $\varepsilon(E)$ of such a medium, we will have the ratio similar to Eq. (1):

$$\varepsilon = \varepsilon^{(0)} + \rho_0 (\frac{\partial \varepsilon}{\partial \rho})_{T,\rho_0} [\exp(|E|^2 / \tilde{E}^2) - 1], \tag{9}$$

where $\tilde{E}^2 = 4/(\frac{\partial \varepsilon}{\partial \rho})_{T_0,\rho_0}(\frac{\partial p}{\partial \rho})_{T_0,\rho_0}$, and ρ_0 and $\varepsilon^{(0)}$ are equilibrium parameters of the medium in the absence of the field. It follows from Eq. (9) that the required condition of wave self-focusing $\partial\varepsilon/\partial(|E|^2) > 0$ is fulfilled for either sign of $(\frac{\partial \varepsilon}{\partial \rho})_{T_0,\rho_0}$.

It is also of interest to study the possibility of formation of self-focusing waveguides in anisotropic (in terms of both electrical and mechanical parameters) media, including gyrotropic ones.

Specifically, it relates to plasma in a magnetic field.

The influence of weak attenuation in a nonlinear medium on the conditions of formation and the structure of self-focusing waveguides can be accounted for in two extreme cases using the conventional energy method[3]. At a small value of the transferred power, when it can be assumed that the loss coefficient does not depend on the field amplitude, the attenuation leads to gradual spreading of the waveguide. On the contrary, when the power of the wave flow that displaces the medium (i.e., the plasma) from the region occupied by the field is great, the losses are concentrated mainly at the boundaries of the waveguide. In this case, the waveguide becomes narrower as the transferred power decreases. In the case of negative absorption in

[3] It should be noted that the presence of losses can lead to a non-equilibrium distribution of the temperature in the region occupied by the field.

the medium, the pattern of wave propagation in the self-focusing will correspond to the time-reversed pattern of the propagation in an absorbing medium. The latter can take place, e.g., when a focused optical beam propagates in a medium with non-equilibrium difference of level populations at the signal frequency.

References

1. A.V. Gaponov, M.A. Miller, ZhETF, **34**, 242 (1958).
2. G.A. Askar'yan, ZhETF, **42**, 1567 (1962).
3. H.A.H. Boot, S.A. Self, R.B.R. Shersby- Harvie, J. Electron. and Control, **4**, 434 (1958).

Chapter 11.3
Laser-induced Damage in Transparent Media

M. Hercher

This paper presents the first laboratory observation of self-focusing. It is published here for the first time in its entirety. Previously only the abstract had been published in J. Opt. Soc. Am. 54, 563 (1964).

Laser Induced Damage in Transparent Media

Michael Hercher
Institute of Optics, University of Rochester
Rochester, NY

This paper was presented at the Spring Meeting of the Optical Society of America in Washington, D.C., 1964

One of the major characteristics of the solid state laser which distinguishes it from the more elegant gas laser is its ability to generate almost unbelievably high peak powers. Today you can go out and buy a Q spoiled ruby laser with a peak power output in excess of 500 MW (and this figure can, of course, be further increased by the use of optical amplifiers). One naturally wonders whether there is an upper limit to the peak power that can be realized with such a laser and, if so, what mechanism is responsible for this limitation.

At the moment there is no doubt that the immediate obstacle to unlimited peak power is material failure that is, failure of the laser medium, the resonator mirrors or of associated optical components. This chapter reports an early (1964) investigation of the nature of the failure of transparent optical components in the presence of peak power densities on the order of thousands of megawatts per square centimeter.

Figure 1 shows the Q switching arrangement used to obtain relatively short pulses of radiation having peak powers on the order of a megawatt—corresponding to an output power density, or irradiance, of about 2.5 MW/cm^2. This power level turns out to be relatively convenient because it is high enough to be useful for our purposes without being so high as to damage any of the laser components. One can easily identify this power level by the fact that it is just possible to obtain a spark in air (Terhune spark) using an uncoated f/16 lens.

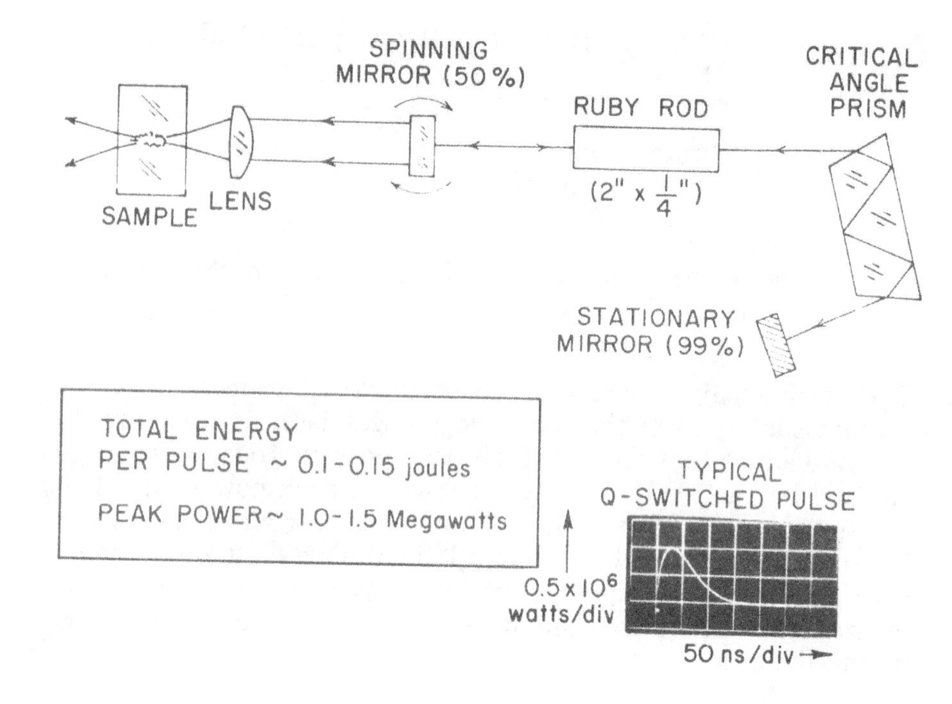

Fig. 1. Q-switched ruby laser

In order to produce damage in transparent media the laser radiation was focused within the material using a simple lens. The power density within the medium is then a function of position relative to the focal point of the lens. As was pointed out in the last paper,[1] there exist well-defined field gradients in the vicinity of the focus of a lens. While it appears that in many cases these electric field gradients are closely related to the damage mechanism, we shall see that it is possible in a number of cases to obtain significant damage even in the absence of such highly localized gradients.

Figure 2 shows a typical example of the damage that was produced when the Q-switched laser emission was focused in glass using a 35-mm lens. This figure, in which the incident light traveled

[1] "Mechanism for Energy Transfer between a Focused Laser Beam and a Transparent Medium Involving Electromagnetic Field Gradients," Marc S. Bruma (Paper WF15 presented at the Spring Meeting of the Optical Society of America, Washington, DC, 1964)

from left to right, illustrates a number of the characteristic features of such damage: first, there is a central region in which the glass appears to be pulverized. Surrounding this is an area where the glass is fractured and where one can see interference colors between adjacent surfaces. Surrounding this fractured region is an area where the glass was seen to be noticeably strained when viewed between crossed polarizers. The final striking feature of the damage is the narrow track emerging from the region of gross damage and pointing in the same direction as the incident light. This track will be shown to be a fundamental characteristic of this type of damage.

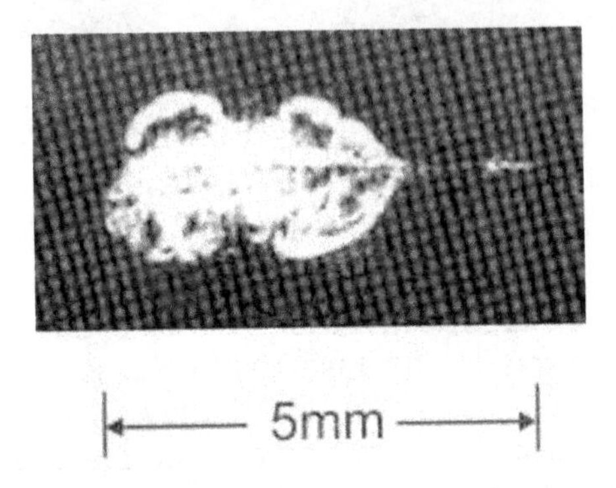

Fig. 2. Typical damage in glass produced by focused radiation from a Q-switched ruby laser

Figure 3 shows damage in the same material when the f-stop. of the lens is decreased from f/6 to f/20. The same general features are present but the length of the damage track is considerably longer now. As the f-stop. is further decreased (i.e., as the focal length of the lens is increased for a fixed beam diameter) the length of the track becomes considerably greater—reaching values in excess of 5cm for f/numbers on the order of f/256.

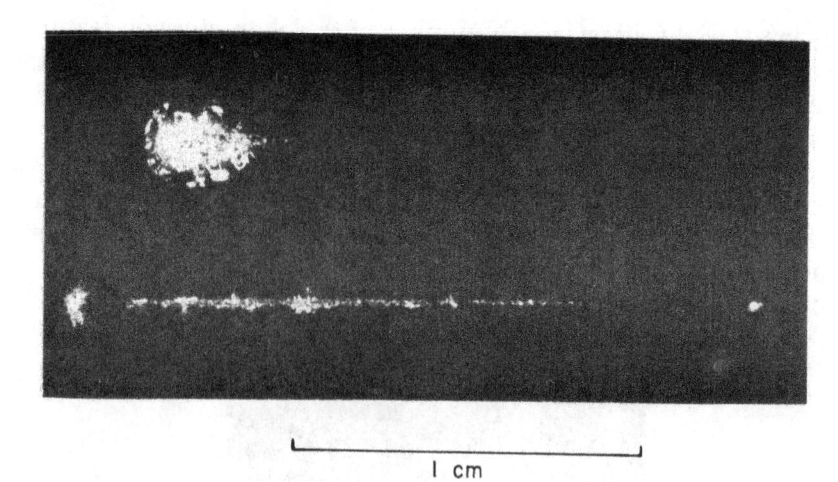

Fig. 3. Effect of changing the f/no. of the focused beam on the damage produced in glass (top, f/6; bottom, f/20).

A striking feature of the mechanism by which glass is damaged by intense laser radiation is the fact that light is emitted from the damaged region at the same time the damage occurs. A small flash of light also appears at the point where the laser radiation strikes the back surface of the sample. Figure 4 shows some of the characteristics of the light emitted by a damaged sample. The major features of this emission are the continuous nature of its spectrum (which approximates a BB at around 5000K) and the raggedness of its time development. Evidently this emission lasts somewhat longer than the laser pulse. In addition to the continuum there is a significant amount of scattered laser radiation at 6943A. This indicates that the damage is well underway before the laser pulse has been completely transmitted.

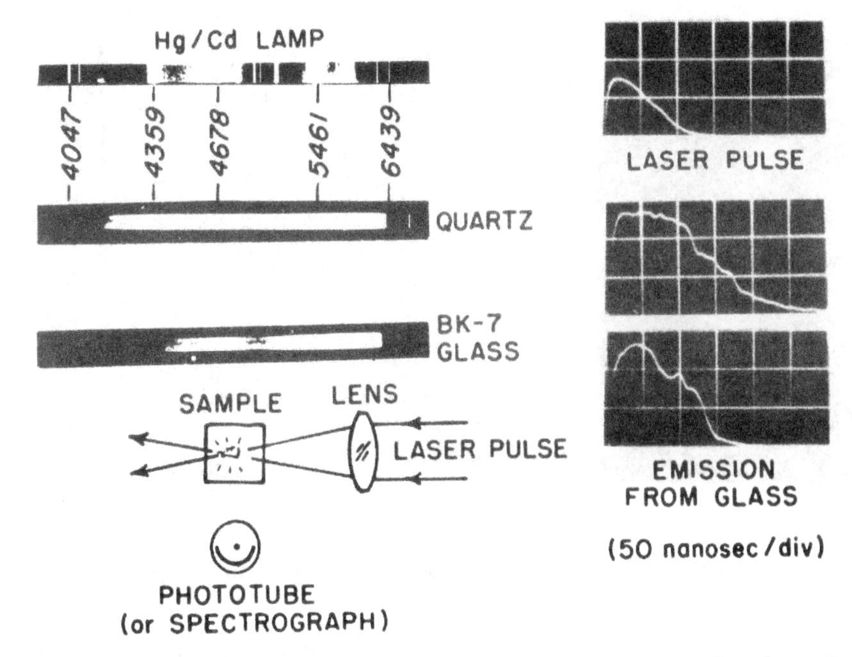

Fig. 4. Radiation from damaged glass (spectra and pulse shape)

Figure 5 illustrates a method for estimating the number of W/cm² at which damage in glass begins to occur. This slide shows the damage in a glass sample for increasing laser output powers. In each case the geometrical focus lies on a vertical line just to the right of the left edge of the photograph, but as you can see the damage begins further and further to the right of the focus as the incident power is increased. The fact that the damaged region shifts towards the lens as the power is increased suggests that there is a threshold power density at which damage begins to occur. By knowing the total power and the cross sectional area of the focused beam at the point where damage first occurs it is possible to obtain a reasonably accurate figure for this threshold irradiance. For optical crown glass this figure turned out to be between 1000 and 2000 MW/cm². It should be emphasized that this measurement involved damage at a point considerably removed from the focal point of the lens.

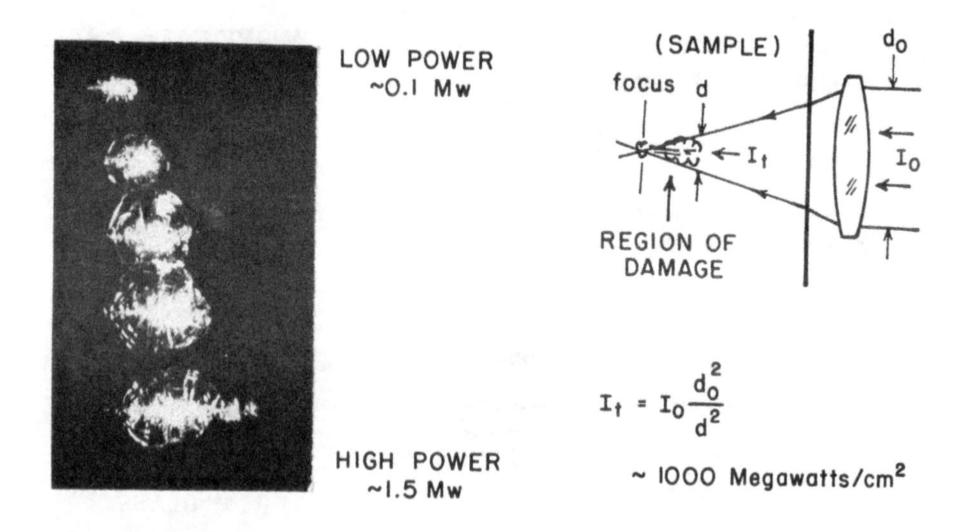

Fig. 5. Q-switched laser damage in glass as a function of peak power

One of the most surprising of our findings was that the nature and extent of the damage caused by focused laser radiation is largely independent of the physical characteristics of the transparent media. Figure 6 shows damage incurred under identical circumstances in three different glass types: (DF 3), a dense flint, (BK 7), an interferometric quality crown, and Pyrex®. In each case the light is incident from the left and a 35-mm lens with an effective f/number of f/8 was used. The top photographs were obtained by observing the actual emission of light from the sample through a blue filter. The interaction between the laser pulse and the transparent medium is probably most accurately represented by these top photographs and, as you can see, there is little difference between them. Note that the fine tracks emanating from the central region of damage emit light: in fact, all of the light is emitted by what is essentially a narrow line pointing in the direction of the incident light.

The lower photographs were obtained by reflected light when the damaged regions were examined with a low-power microscope. The prominent damage in this mode of observation consist s of a region surrounding the central damage track in which the glass is pulverized and fractured. This damage is almost certainly the result of the thermal expansion of the material in the regions from which light was emitted. This viewpoint is supported by the fact that there is considerably less fracturing in the Pyrex than in the other two samples.

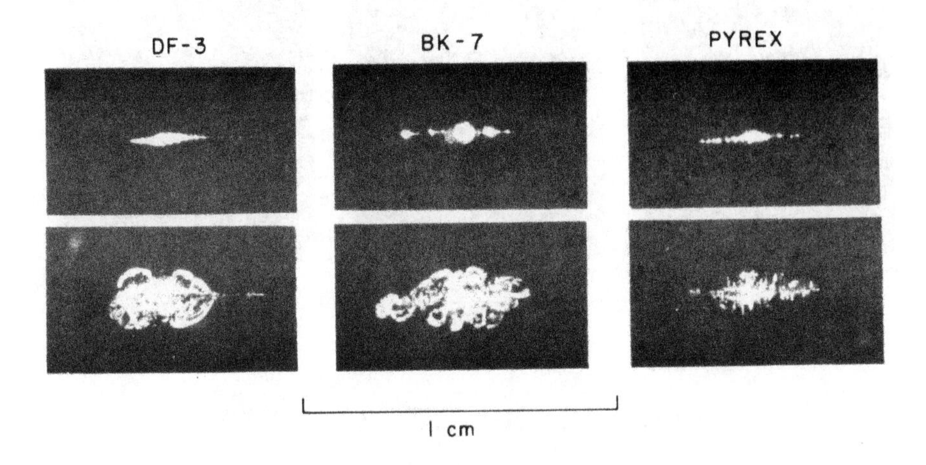

DAMAGE IN VARIOUS GLASSES

TOP : EMISSION BY DAMAGED GLASS (BLUE FILTER)
BOTTOM : MACROPHOTOGRAPH OF SAME DAMAGE

Fig. 6. Laser-induced damage in different types of glass

Figure 7 shows the corresponding results, on the same scale, for samples of fused quartz, crystal quartz and Lucite®. In the case of the two quartz samples the upper damage tracks are essentially the same as those obtained in glass. The thermal fracturing in fused quartz was, however, less than in any of the other samples tested. The damage in Lucite was rather different in nature and more extensive than that in the glass and quartz samples.

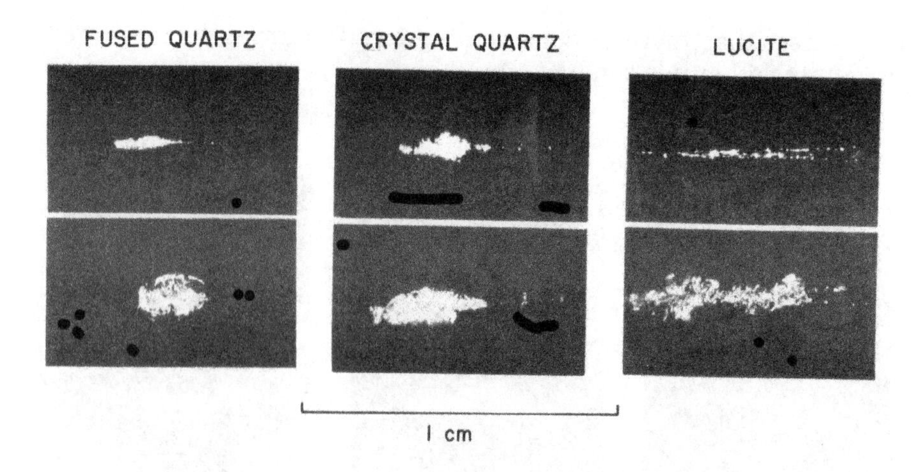

DAMAGE IN QUARTZ & LUCITE

TOP : EMISSION
BOTTOM : MACROPHOTOGRAPH

Fig. 7. Laser-induced damage in quartz and Lucite

One sees the same sort of damage when the laser light is focused within samples of sapphire or ruby. The nature of the damage in ruby is particularly surprising in view of the fact that ruby has a resonant absorption at precisely the wavelength of the laser emission. By varying the angle between the polarization vector of the light and the optical axis of the ruby sample it was possible to obtain different values for this absorption coefficient: in everycase the damage to the ruby was essentially the same. This rather strikingly suggests that the nature and magnitude of extremely high power radiation damage within semi transparent and transparent materials does not involve optical absorption as we usually think of it. Further investigation showed that it was generally not possible to increase the damage in a transparent material by increasing its absorption at the laser wavelength – at least in glasses and crystals.

For example, it was quite difficult to obtain any damage at all in a sample of blue filter glass. This was probably because the light was uniformly absorbed before it could do much damage. As was indicated earlier, the central damage tracks become longer as the f/no of the focusing system is made numerically larger. Figure 3 showed two adjacent damaged regions in a sample of optical crown glass. As the focal length is further increased the tracks become still longer, as shown in Figure 8. Although it is not apparent in this slide, a central damage track runs practically the entire length of the damage in each of the three cases, linking the regions of more obvious damage.

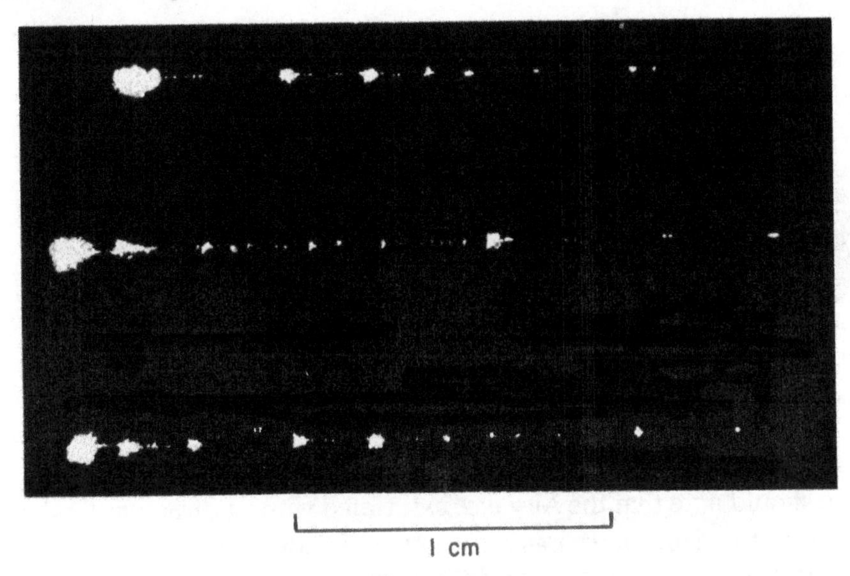

Fig. 8. Damage tracks in glass produced by successive Q-switched laser pulses at f/32

Figure 9 shows another damage track in somewhat more detail. As you can see the damage track is quite large in diameter near the beginning of the damage. To all appearances the track is hollow in this vicinity. As the track continues to the right it becomes smaller, and for about the last third of its length it cannot be resolved. Examination under higher magnification shows that the track becomes extremely narrow towards the end. In general, we have observed that the tracks also become increasingly narrow as the focal length of the lens is increased. The diameter of the track is apparently a decreasing function of the power density or irradiance. Figure 10 shows a section of a very narrow damage track obtained using a 40cm lens. At the bottom is a photograph of some blood cells (approximately 7 μm in diameter) at the same magnification for comparison purposes. Although the track is not very well resolved, it is clearly no more than a few wavelengths in diameter, and probably less. The track itself appears to be a hollow capillary with occasional breaks. The microphotograph in the center is of another portion of the same track where the damage is somewhat different in nature. Most of the track, however, looks more like the top photograph.

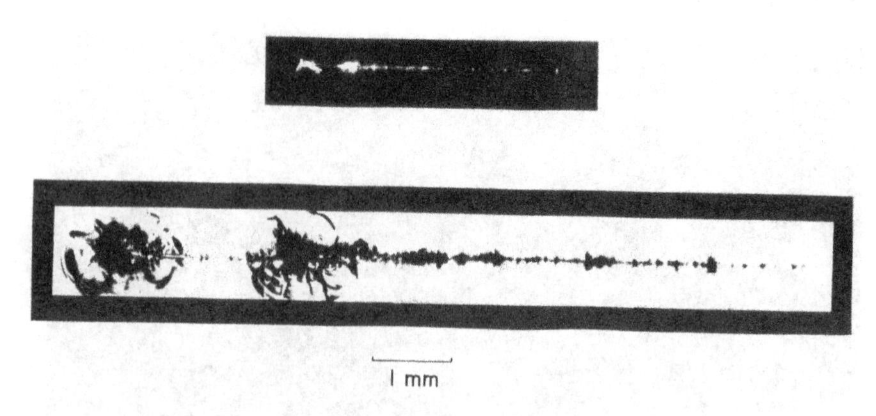

Fig. 9. Damage track in glass for an f/20 beam

One should note that the Airy disc associated with a diffraction-limited beam of this f/number. has a diameter of about 100 μm or at least 100 times greater than that of the damage track. The length of the track is of the same order of magnitude as the diffraction-limited depth of focus. For all intents and purposes the beam is practically collimated in the vicinity of the focus and has a diameter approximately equal to the f/number in microns. Thus, by varying the f/number, it is also possible to vary the power density in the vicinity of the focus. We found that it was no longer possible to observe damage in glass with f/numbers in excess of about f1300. This would correspond to an axial power density of approximately 1500 MW/cm^2 which is in good agreement with the threshold power density cited earlier.

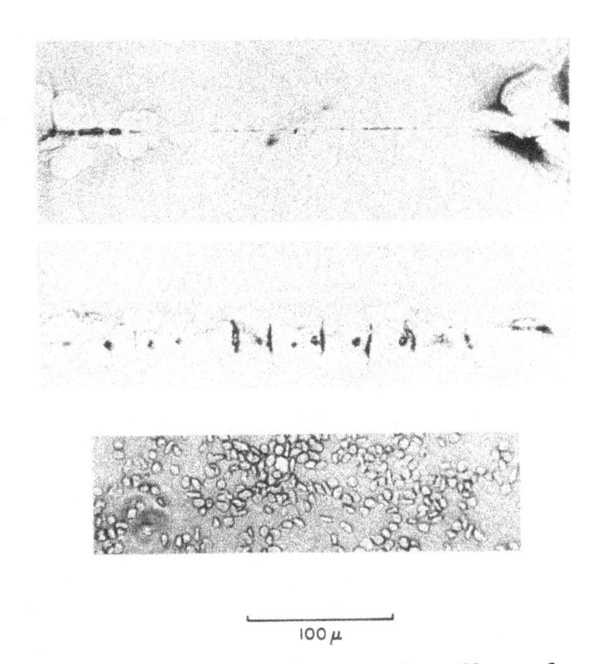

Fig. 10. Top and center: microphotographs of laser damage tracks produced using a 40-cm lens; Bottom: red blood cells (~7 μm diameter) at the same magnification for compare-son.

Figure 11 gives some expressions from which the peak irradiance electric field amplitude and radiation pressure may be calculated in terms of the incident power, effective f/number and refractive index of the transparent medium. These expressions were used in arriving at the data in the table at the bottom of the slide, where an incident power of 1 MW was assumed. Although we have included figures for radiation pressure, it is doubtful that this phenomenon plays a major role in the absence of a reflecting or absorbing surface. The tabulated data covers the range of f/numbers and peak irradiances involved in the majority of our experiments.

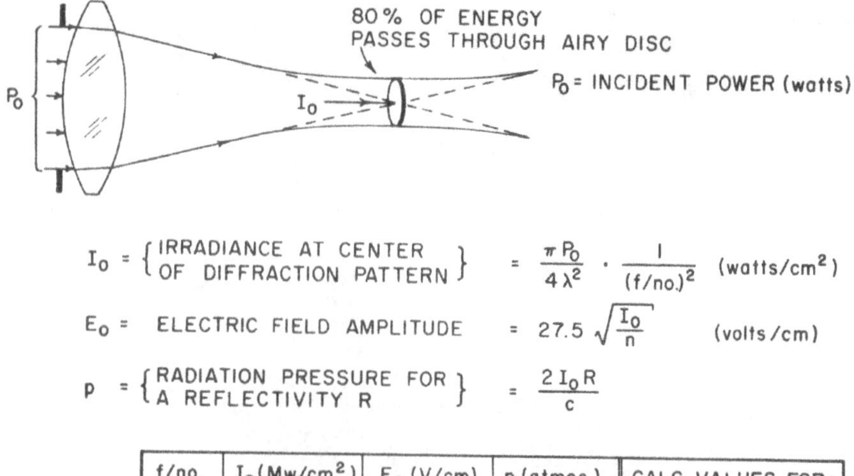

f/no.	I_0 (Mw/cm²)	E_0 (V/cm)	p (atmos.)	CALC. VALUES FOR:
f/2	2.0×10^7	$\sim 10^8$	6000	R ~ 50%
f/8	2.5×10^6	3.5×10^7	800	P_0 ~ 1 Megawatt
f/32	1.5×10^5	10^7	50	n ~ 1.5
f/256	2.5×10^3	1.2×10^6	0.8	λ = 6943 Å

Fig. 11. Calculation of focused beam parameters

An interesting aspect of this study was the unexpected appearance of the long damage tracks in transparent material. Quite frankly, we have so far been unsuccessful in satisfactorily explaining the mechanism behind their formation. We do, however, know a number of their characteristics. For example it is quite clear that energy or momentum is transmitted down the damage track in the direction of the incident light. This is evidenced by the fact that whenever the damage track intersected an interface going from glass to air, it would invariably eject a sizable fragment of glass at the point of intersection.

Figure 12 shows an instance In which a long damage track in glass has undergone an internal reflection here, too, a fragment was ejected from the surface at the point of reflection, In many cases the exchange of energy at the surface was sufficient to prevent the propagation of a reflected track. The glass samples were often fractured at the surface when focused laser light was incident from within the sample: in no case, however, has any similar damage been observed at an air to glass interface.

This is, of course, the same sort of behavior one observes when a high-velocity projectile strikes a thick plate of glass, and suggests that these damage tracks might be at least partially explained in terms of acoustics. To see if the tracks propagated at the speed of sound (rather than at the speed of light) we optically contacted two glasses of different properties and attempted to obtain a damage track extending from one into the other. Had we been able to do this, we could have immediately told whether or not the track was propagated at the speed of sound by simply observing its angle of refraction. However, every time a track was produced in the first medium it would stop at the interface, where it would damage the two surfaces and break the optical contact over a small area.

At the moment the most promising possibility Is that the tracks are caused by an expanding cylindrical shock wave having a singularity along its axis, where the damage track occurs. (This shock wave would not actually be cylindrical but would be conical with a cone angle approximately equal to the ratio of the speed of sound to the speed of light within the medium.) Further experiments are currently underway and we hope to have this problem solved shortly.

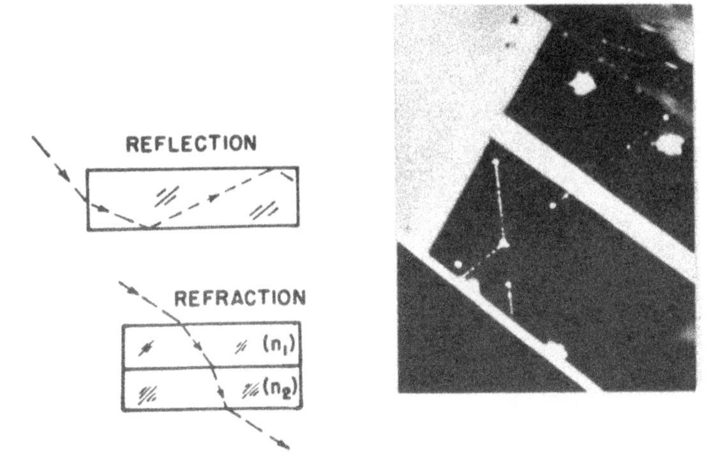

Fig. 12. Total internal reflection of a damage track. Attempts to observe a refracted damage track were unsuccessful (see text).

Our findings can be briefly summarized:

1. All of the commonly used transparent media we have examined exhibit essentially the same type of damage when irradiated above a certain threshold irradiance.

2. The threshold irradiance above which damage occurs is approximately 1000–2000 MW/cm^2 for glasses, quartz, and sapphire.

3. The nature and extent of the damage does not appear to depend in any significant manner on the value of the optical absorption coefficient.

This work was supported in part by the U.S. Army Research Office (Durham).

Part II
Self-focusing in the Present

Chapter 12
Self-focusing and Filamentation of Femtosecond Pulses in Air and Condensed Matter: Simulations and Experiments

A. Couairon and A. Mysyrowicz

Abstract We review the evolution of the modeling of femtosecond filamentation in transparent media from the monochromatic self-trapping model to the development of numerical models including all relevant physical effects. We discuss four self-action effects which occur during filamentation: i) the generation of single cycle pulses by filamentation in gases, ii) the beam self-cleaning effect associated with filamentation, iii) filamentation in an amplifying medium and iv) organization of multiple filaments by various methods.

12.1 Introduction

The subject of *femtosecond* filamentation was developed in 1995 when researchers at Michigan University showed that a 200-fs IR laser pulse with a peak power exceeding a few GW is able to propagate nonlinearly over several tens of meters by forming a structure with an intense core of diameter of $\sim 100\,\mu\text{m}$ [1]. It has been shown since that the core has an intensity (of nearly 10^{14}W/cm^2), sufficient to ionize air molecules, leaving in its trail a plasma channel with a density of about $10^{16}\ \text{cm}^{-3}$. The core contains only a fraction of the energy of the total beam and is surrounded by an energy reservoir, the role of which is to refill the hot core undergoing the main energy losses. This structure is called a filament because it propagates over several tens of Rayleigh lengths in an apparently self-guided way, i.e., by maintaining an intensity at the verge of the ionization threshold with weak energy losses [2,3]. This laser– matter interaction over a long propagation distance constitutes an interesting property in itself, which opens the way to several applications involving the transport of high intensities over long paths [4]. The phenomenon of filamentation is generally accompanied by a characteristic

A. Couairon (✉)
Centre de Physique Théorique, École Polytechnique, CNRS, F-91128, Palaiseau, France
e-mail: couairon@cpht.polytechnique.fr

R.W. Boyd et al. (eds.), *Self-focusing: Past and Present*,
Topics in Applied Physics 114, DOI 10.1007/978-0-387-34727-1_12,

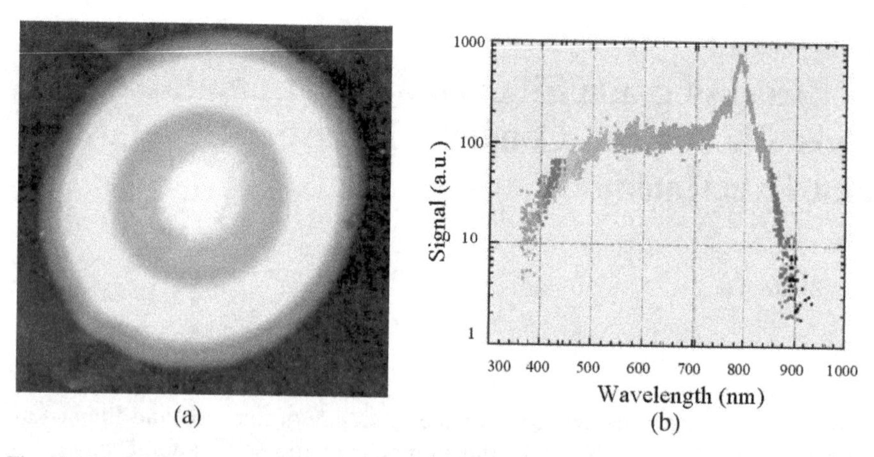

Fig. 12.1(a) (**a**) Conical emission accompanying a self-guided pulse. A central white core (the filament) is surrounded by Newton's rings having a divergence of the order of the mrad [3]. (**b**) Spectrum of a femtosecond laser beam ($\lambda_0 = 800$ nm; $\Delta t = 70$ fs, $P = 3$ TW) after propagation over a distance of 10 m in air. The fast fall between 800 and 900 nm corresponds to the fall of the detectivity of the measuring apparatus (see [6])

effect represented in Fig. 12.1(a). A transverse section of the beam at a propagation distance of 50 m reveals an intense white central part that corresponds to the filament core surrounded by a conical emission of light with colored rings appearing in an order opposite to what is expected from diffraction (bluer rings on the outside) [3,5]. To date, the detailed explanation of this phenomenon remains a topic of debate.

The considerable broadening of the pulse spectrum in the filament core, due to self-phase modulation over a long distance, is of great importance for several applications: Fig. 12.1(b) shows measurements of the spectrum of an 800 nm, 70 fs, 3 TW laser pulse after propagation and filamentation over 10 m in air. It is seen that the spectrum initially centered at 800 nm with a spectral bandwidth of 10 nm extends over the entire visible range after propagation over 10 m. On the infrared side, the spectrum extends up to 4 μm although the intensity is lower than in the visible domain. The super continuum covers the absorption band of many components of the atmosphere. This property associated with those of long propagation distances [7] and of filamentation at low pressures [8,9] made it possible to consider LIDAR applications to detect the presence of pollutants in the atmosphere [10]. Figure 12.2(a) shows the image of the white continuum created by vertical filamentation detected by the telescope located in Tautenburg, Germany. The beam traversed a thin cloud layer located at an altitude of 5 km which gives a zone of Mie scattering visible in Fig. 12.2(a). A part of the spectrally and time-resolved backscattered white light is shown in Fig. 12.2(b), together with the calculations from a hytran code [10]. Most of the lines can be assigned to water vapor and to oxygen. From these data, important information for meteorology such as the humidity and temperature of the atmosphere

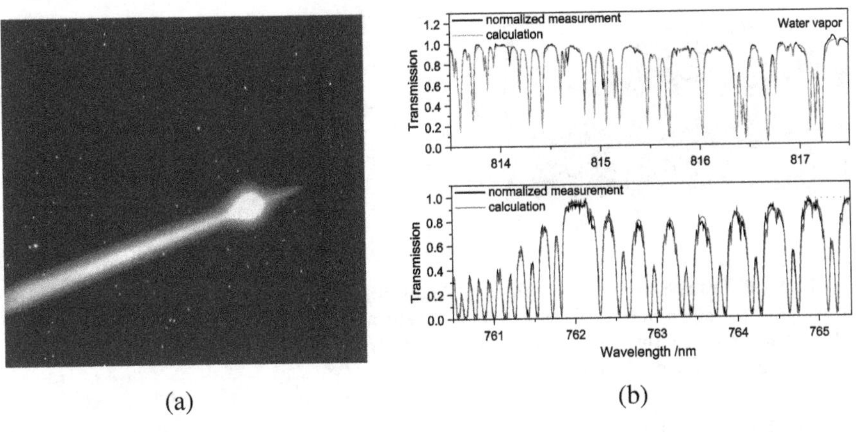

Fig. 12.2 (**a**) Vertical propagation in the sky of the teramobile laser beam (peak power of 5 TW at 800 nm; see [21]): backward scattering of the beam is recorded using the telescope with a 2-m aperture, located in Tautenburg, Germany. One distinguishes a zone of Mie scattering from a vapor cloud located at an altitude of 9 km. (**b**) High-resolution atmospheric absorption spectrum from an altitude of 4.5 km measured in a LIDAR configuration

can be obtained at a distance reaching several kilometers. One also notices the presence of unidentified lines, showing the potential of white continuum LIDAR for broadband detection of pollutants.

Filaments and their signatures were observed for different laser wavelengths. Several teams showed that it is possible to form filaments at laser wavelengths of 1.06 μm [11], 527 nm [12], or in the UV domain (400 nm) [13] and 248 nm [14–17]. Femtosecond filamentation does not take place only in gases but also in transparent liquids [18] or solids [19,20]. Figure 12.3 shows the formation of a filament in fused silica [19]. The beam was focused on the entrance face of the sample. If it propagated linearly according to the laws of Gaussian optics, it would have diffracted as indicated by the dotted lines. However, for an energy of 2 μJ, the filament spans the entire length of the 2-cm-long fused silica sample. The main differences from filamentation in gases are the following: (1) The energy necessary is on the order of μJ per pulse of 100 fs duration. (2) The diameter of the filament core is a few microns to a few tens of microns. (3) The plasma corresponds to electron–hole pairs, instead of electrons–ions in air. The density of electrons–hole pairs generated is 10^{18} to 10^{19} cm^{-3}, depending whether the beam is loosely or tightly focused [22,23]. These features are well reproduced by numerical simulations discussed later, as shown in Fig. 12.3(b). The possibility of forming in-depth filaments over distances exceeding the Rayleigh length in a transparent material opens the way to the generation of optical components directly written in the bulk of optical materials and could find medical applications, e.g., in ophtalmology.

In the present chapter, we will briefly review the evolution of the modeling of femtosecond filamentation in transparent media from the simplest self-trapping

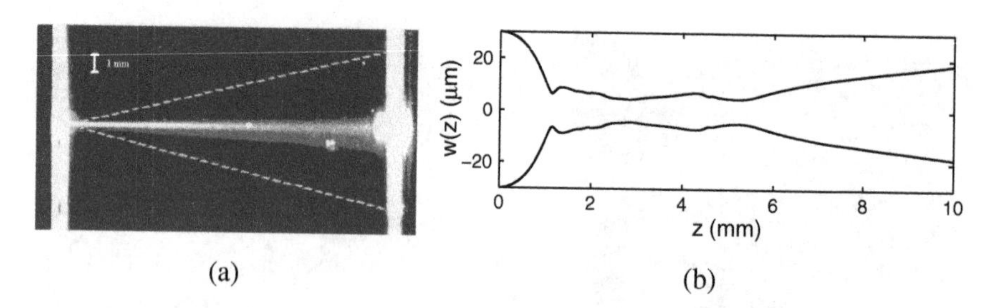

Fig. 12.3 (**a**) Transverse photograph of a 2-cm-long filament generated in fused silica by focusing a 2 μJ, 800 nm, 160 fs laser pulse on the entrance face of the sample. (**b**) Full width at half maximum of the fluence distribution obtained from numerical simulations in the conditions of the experiment. From [19]

model to the development of numerical models including all relevant physical effects. We will then discuss recently discovered self-action effects that occur during filamentation: (1) the generation of single-cycle pulses by filamentation in gases; (2) the self-cleaning effect associated with filamentation; (3) the filamentation in an amplifying medium; and (4) the organization of multiple filaments by various methods.

12.2 Models

12.2.1 Self-trapping

Self-focusing due to the optical Kerr effect competes with the natural diffraction of the beam. The refractive index of air n in the presence of an intense electromagnetic field depends not only on its frequency, but also on the space- and time-dependent intensity $I(r,t)$ of the laser according to the law: $n = n_0 + n_2 I(r,t)$. The quantity n_2 denotes the coefficient of the nonlinear Kerr index related to the third-order susceptibility $\chi^{(3)}$ by $\chi^{(3)} = 4\epsilon_0 c n_2 n_0^2/3$. The coefficient n_2 is usually positive. In the presence of intense radiation, this leads to an increase of the refractive index. Standard beams have their peak intensity on axis, at the center of the beam. The refractive index acts as a lens by making convergent an initially collimated beam. In addition, this effect is cumulative: in the absence of saturating effects, the enhancement of curvature of the wavefront can lead to a catastrophic collapse of the beam on itself at a finite propagation distance. This effect is represented in a diagrammatic way in Fig. 12.4(a). The significant parameter for self-focusing and collapse is the power of the beam (not its intensity). Self-focusing dominates diffraction when the power exceeds a critical value (3 GW in air at 800 nm). Self-focusing balances diffraction only for a peculiar beam profile known as the Townes mode which contains a power exactly equal to P_{cr} [24]. Unfortunately, this mode is unstable in the sense that an excess power will lead to collapse at a finite distance. The position of the collapse is closer to the laser when the power of the beam is large.

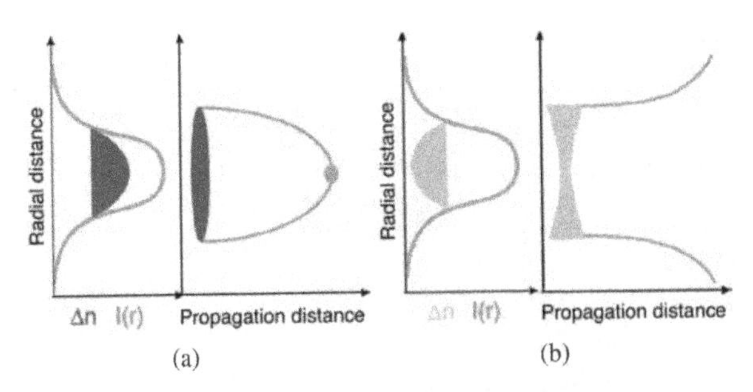

Fig. 12.4 (**a**) Self-focusing of a beam by the optical Kerr effect. The refractive index of the medium depends on the intensity of the laser and acts as a lens by making convergent an initially collimated beam. Focusing prevails over diffraction when the power of the beam exceeds a critical power P_{cr} ($P_{cr} = 3.2$ GW for air at the wavelength of 800 nm) and leads, in the absence of other nonlinear effects, to the collapse of the beam on itself. (**b**) Defocusing of the beam by the presence of a plasma. The ionization of the medium initially takes place in the center of the beam, where the intensity is most significant. The creation of an under-dense plasma decreases the local index of the medium, which causes beam defocusing

12.2.2 Moving Focus

The extension of these concepts to a pulsed beam led to the moving focus model. In the moving focus model [25,26], the laser pulse is stacked into time slices that are considered to be independent from each other. This picture relies on the assumption that the influence of collapsed time slices on the subsequent propagation can be neglected. This approach is valid when the physical effects coupling the various time slices together, such as, e.g., GVD, remains weak. In air, the typical distance for GVD for a $t_p = 100$ fs infrared pulse is $z = 250$ m, much larger than the typical self-focusing and collapse distances. The approximation of independent time slices seems therefore justified at first sight. Each time slice contains a specific power. All the central slices of a pulse with peak power above critical are self-focused at distances that become larger as the corresponding power is closer to critical value. The slices with power below P_{cr} diffract. In this picture, the perception of a filament is constituted by the collection of nonlinear foci corresponding to the different temporal slices of the pulse comprised between the time slices which correspond to a power P_{cr} on the ascending (resp. descending) slope of the pulse envelope. These foci do not appear simultaneously and can be viewed as a moving focus. Figure 12.5 shows the locations of the focus corresponding to the peak power of the pulse and those corresponding to a smaller power in the leading or in the trailing part of the pulse, as predicted by the Marburger formula for the position of beam collapse as a function of its power [27]. Applying the same formulation for all time slices with power above critical and assuming the existence of any saturation mechanism that prevents the collapse, the evolution of the pulsed beam into a filament [bold line in Fig. 12.5(b)] is predicted from this simple model

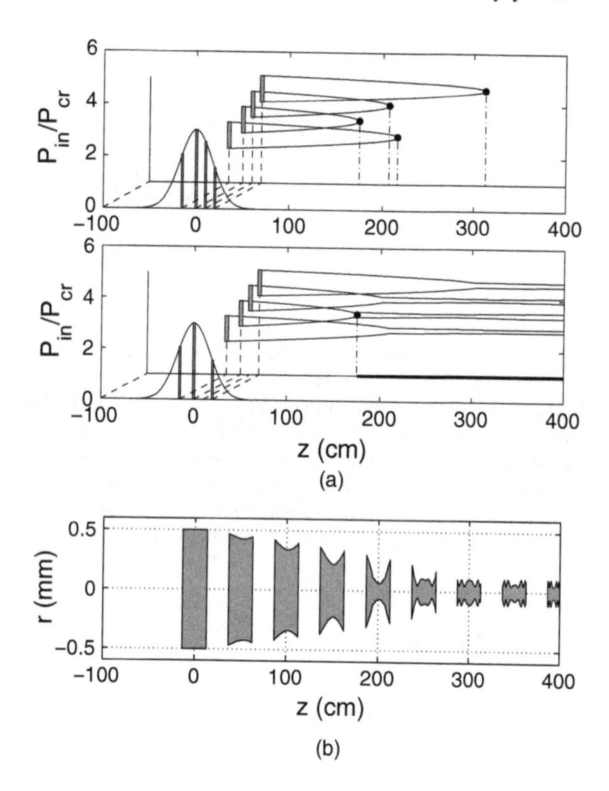

Fig. 12.5 (**a**) Prediction of the moving focus model for the collapse location of the beam associated with a given time slice and input power in the pulse. The time slice with the maximum power collapse at the shortest distance from the laser. The time slices with smaller power, provided $P_{\rm in} \geq P_{\rm cr}$, collapse farther on the propagation axis (adapted from [26]). (**b**) Prediction of the beam evolution beyond the collapse for each time slice, leading to the illusion of a filament. A high-order saturation effect was assumed. (**c**) Evolution of the filamenting wavepacket as predicted by the moving focus model, (For clarity only the time slices with power between 1.3 and 3 $P_{\rm cr}$ are shown.)

[28,29]. According to this model, the propagation of the wavepacket can be simply obtained by considering the time slices with power between $P_{\rm cr}$ and the peak power, having initially the same beam width. Because the slices with larger power collapse faster, the wavepacket is deformed into a horn shape around the nonlinear focus, as shown in the successive plots at propagation distances in Fig. 12.5(c). Beyond the nonlinear focus, the time slices with smaller radius correspond to a higher intensity, leading to a recurrent occurence of a temporal splitting of the intensity profile (or intensity peak).

12.2.3 Saturation of Self-focusing, Self-channeling

A collapse actually never occurs because there are saturation mechanisms prevailing over self-focusing when the intensity becomes sufficiently high. Two mechanisms efficiently prevent the collapse in fs filamentation: multiphoton absorption and plasma-induced defocusing. When the intensity becomes larger than $10^{13}\,{\rm W/cm^2}$, multiphoton ionization of air sets in and it becomes necessary to take into account the defocusing action of the electron plasma.

This was proposed by Braun et al. [1] as an extension of the self-trapping model [24] in the form of a balance between self-focusing diffraction and ionization. The electron plasma plays the role of a divergent lens as shown schematically in Fig. 12.4. The proposed balance was expressed in terms of an equilibrium of the index variations corresponding to the three involved mechanisms:

$$n_2 I = \frac{\rho(I)}{2\rho_c} + \frac{(1.22\lambda_0)^2}{8\pi n_0 w_0^2} \tag{12.1}$$

The orders of magnitudes of the peak intensity or electron density estimated from Eq. (12.1) are in good agreement with measurements of the same quantities [1–3, 5, 11, 25].

However, this balance was never identified in numerical simulations accounting only for diffraction, self-focusing and ionization (see model Eqs. (12.2) and (12.3) below), which do not converge toward any stable wave-packet in the form of a stationary nonlinear waveguide. As will be seen below, the scenario that emerged from the numerical simulations is that of a competition between Kerr self-focusing, ionization and diffraction. One of the effects nearly always prevails over the others, leading to a set of successive stages where beam refocusing and beam defocusing alternatively dominate the dynamics. At the transition between two stages, the peak intensity saturates at a value in agreement with Eq. (12.1). According to the geometry and parameters of the initial pulse, this scenario can give a filament with clearly separated plasma channels associated with several cycles of focusing-defocusing, a short filament with a single plasma string or a longer filament where several focusing cycles overlap to give rise to an extended plasma channel.

12.2.4 X-Waves

The universality of the filamentation phenomenon suggests that filaments correspond to attractors of the dynamics of the nonlinear Schrödinger equation that involve cylindrically symmetric eigensolutions. Self-trapped spatial solitons provide a class of localized solutions where diffraction effects are compensated by the optical Kerr effect [24]. This type of mode was shown to be spontaneously generated for laser pulses undergoing filamentation in Kerr media [31]. However, for filaments in air, self-focusing is not compensated by diffraction along the whole propagation distance but competes with high-order nonlinear effects including plasma defocusing and multiphoton absorption. Recently, another type of modes called X-waves were proposed to play a key role in filamentation in water [32–36]. Nonlinear X-waves are non-diffracting and non-dispersive, spatially and temporally extended solutions to the NLS equation with a characteristic X-shaped extension both in the far-field and in the near-field [36–39]. These structures were originally shown to be spontaneously generated in quadratic media [40]. The same property is true for laser pulses undergoing filamentation in condensed media [34, 35, 41]. The wavelength-dependent angle of the conical emission measured in different condensed media or in air was shown to match the slope of the characteristic X-shaped extensions of the laser energy in the region beyond the nonlinear focus [42], in agreement with simulation results [32, 36, 43].

12.2.5 Numerical Simulations

In order to go beyond the validity limits of the straightforward models presented above, it is necessary to resort to numerical simulations. The electric field is decomposed into a carrier wave and an envelope as $\mathbf{E}(x,y,z,t) = \frac{1}{2}\mathcal{E}(x,y,z,t)\exp[i(kz-\omega_0 t)]e_x + c.c.$, where z is the propagation direction, k and ω_0 are the central wavenumber and frequency of the laser pulse. The propagation of the envelope $\mathcal{E}(x,y,z,t)$ is described by an envelope equation of nonlinear Schrödinger type written in the reference frame of the pulse $(z, t = t_{\text{lab}} - z/v_g(\omega_0))$ where $v_g(\omega_0) \equiv \partial\omega/\partial k|_{\omega_0}$ denotes the group velocity. The minimal model for numerical simulations takes into account diffraction [first term on the right-hand side of Eq. (12.2)] self-focusing (second term), and plasma defocusing (third term).

$$\frac{\partial E}{\partial z} = \frac{i}{2k}\Delta_\perp \mathcal{E} + ik_0 n_2 |\mathcal{E}|^2 \mathcal{E} - i\frac{k_0}{2n_0}\frac{\rho}{\rho_c}\mathcal{E} \tag{12.2}$$

where $\Delta_\perp \equiv \partial^2/\partial x^2 + \partial^2/\partial y^2$. Equation (12.2) must be solved simultaneously with the equation describing the generation of an electron plasma of density ρ by photoionization:

$$\frac{\partial \rho}{\partial t} = W(|\mathcal{E}|^2)(\rho_{at} - \rho), \tag{12.3}$$

where $W(|\mathcal{E}|^2)$ denotes the intensity-dependent ionization rate. For intensities below 10^{13} W/cm^2, multiphoton ionization (MPI) is considered with $W(|\mathcal{E}|^2) \sim \sigma_K |\mathcal{E}|^{2K}$ where the quantity σ_K denotes the cross section for MPI, involving K photons, where $K \equiv \langle U_i/\hbar\omega_0 + 1\rangle$, U_i denotes the ionization potential of the medium and ρ_{at}, the density of neutral atoms. When the intensity becomes larger than 10^{13} W/cm^2, the general formulation by Keldysh [44] and Perelomov et al. [45] must be introduced in $W(|\mathcal{E}|^2)$ (see, e.g., [22, 23, 46, 47]). For realistic values of the parameters entering in this model, the simulations show a contradiction between the results and the underlying framework of the slowly varying envelope approximation. Extremely short temporal structures are generated from the competition of plasma defocusing and Kerr self-focusing, that even near to a singular behavior. Thus, it is necessary to extend the model to a more realistic situation, i.e., to take into account additional effects such as group velocity dispersion, multiphoton absorption and self-steepening, as described by the nonlinear envelope equation

$$\frac{\partial \mathcal{E}}{\partial z} = \frac{i}{2k} U^{-1}[\Delta_\perp + D]\mathcal{E} + N(\mathcal{E}) \tag{12.4}$$

where $U \equiv 1 + (1/kv_g)i\partial/\partial t$ accounts for space–time focusing, D denotes the Fourier transform in time of the operator $\hat{D}(\omega) \equiv (n(\omega)\omega/c)^2 - k^2\hat{U}^2(\omega)$, (with $\hat{U}(\omega) \equiv 1 + (\omega - \omega_0)/kv_g$), which accounts for group velocity dispersion (GVD) at all orders, via the frequency-dependent refraction index $n(\omega)$. This can be readily seen from a small $\omega - \omega_0$ expansion of $\hat{D}(\omega)/2k \simeq k''(\omega - \omega_0)^2/2 + k'''(\omega - \omega_0)^3/6 + \cdots$. The temporal counterpart of this term reads as $D/2k \simeq (k''/2)\partial^2/\partial t^2 + i(k'''/6)\partial^3/\partial t^3 + \cdots$.

The nonlinear terms $N(\mathcal{E})$ in eq. (12.4) account for the optical Kerr effect, plasma-induced defocusing and multiphoton absorption (MPA):

$$N(\mathcal{E}) = U^{-1}[T^2 N_{\mathrm{Kerr}}(\mathcal{E}) + N_{\mathrm{Plasma}}(\mathcal{E}) + TN_{\mathrm{MPA}}(\mathcal{E})] \tag{12.5}$$

The Kerr term (12.6)

$$N_{\mathrm{Kerr}}(\mathcal{E}) = ik_0 n_2 \left\{ \int_{-\infty}^{t} R(t-\tau)|\mathcal{E}(x,y,z,\tau)|^2 d\tau \right\} \mathcal{E}(x,y,z,t), \tag{12.6}$$

is split into an instantaneous component due to the electronic response in the polarization and a delayed component, of fraction α, due to stimulated molecular Raman scattering [48]. The function $R(t)$ mimics both the electronic and the molecular response with a characteristic time τ_R and frequency Ω:

$$R(t) = (1-\alpha)\delta(t) + \alpha R_0 \exp(-t/\tau_R)\sin \Omega t \tag{12.7}$$

where $R_0 = (1 + \Omega^2\tau_R^2)/\Omega\tau_R^2$. In air at 800 nm, $\tau_R = 70$ fs and $\Omega = 16$ THz [48, 49].

The plasma term (12.8)

$$N_{\mathrm{Plasma}}(\mathcal{E}) = -\frac{\sigma}{2}(1 + i\omega_0\tau_c)\rho(x,y,z,t)\mathcal{E}(x,y,z,t), \tag{12.8}$$

accounts for plasma absorption (real part) and plasma defocusing (imaginary part). The cross section σ for inverse Bremsstrahlung follows the Drude model [50] and reads:

$$\sigma = \frac{k\omega_0\tau_c}{n_0^2\rho_c(1 + \omega_0^2\tau_c^2)} \tag{12.9}$$

where τ_c is the electron collision time. In air, $\tau_c = 350$ fs and $\sigma = 5.1 \times 10^{-20}$ cm^2. Therefore, $\tau_c \gg \omega_0^{-1}$, and in this limit, the defocusing term can be expressed as a function of the critical plasma density since $\sigma\omega_0\tau_c\rho \simeq k_0\rho/n_0\rho_c$. The MPA term in Eq. (12.5)

$$N_{\mathrm{MPA}}(\mathcal{E}) = -\frac{W(|\mathcal{E}|^2)U_i}{2|\mathcal{E}|^2}(\rho_{at} - \rho)\mathcal{E}, \tag{12.10}$$

accounts for energy absorption due to multiphoton ionization and depends on the ionization rate $W(|\mathcal{E}|^2)$ of the medium, its ionization potential U_i and its neutral atom density ρ_{at}. In the case of multiphoton ionization, simplifications are obtained from the scaling $W(|\mathcal{E}|^2) = \sigma_K|\mathcal{E}|^{2K}$, which leads to $N_{\mathrm{MPA}}(\mathcal{E}) = -\frac{\beta_K}{2}|\mathcal{E}|^{2K-2}$ (in the limit $\rho \ll \rho_{at}$), where the cross section for multiphoton absorption reads $\beta_K = K\hbar\omega_0\rho_{at}\sigma_K$.

The operators U and $T \equiv 1 + i\omega_0^{-1}\partial/\partial t$, account for space–time focusing and self-steepening of the pulse (also called optical shock generation) [51–53]. They generally describe the deviations from the slowly varying envelope approximation in time [46, 51, 54]. With these effects, it is possible to relax the constraint of the slowly varying envelope in time to describe the propagation of laser pulses with durations as short as one optical cycle.

12.2.6 Typical Simulation of Filamentation

A typical numerical simulation shows successive cycles of refocusing and defocusing stages as indicated in Fig. 12.6. If one defines the filament length as the distance between the first and the last focusing event, the filament covering several tens to hundreds of meters in air.

The evolution of the beam width, defined as the half-width at half maximum of the time-integrated intensity, as well as the evolution of the peak intensity and electronic density as functions of the propagation distance are detailed in Fig. 12.7 for 100 fs, UV and infrared laser pulses. Figure 12.8 shows the beam width, using a longer (450 fs) UV (248 nm) laser pulse [17]. Filamentation and intensity clamping occurs earlier than at IR wavelengths during the collapse of the beam on its axis. The width of the resulting filament is larger than for an infrared filament.

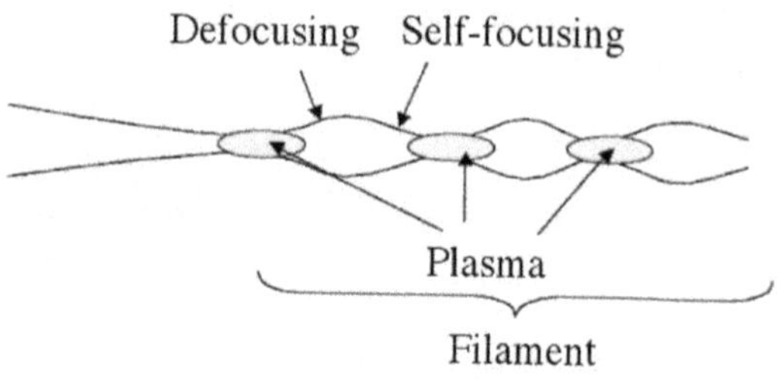

Fig. 12.6 Focusing–defocusing cycles sustaining a long-range propagation of light filaments

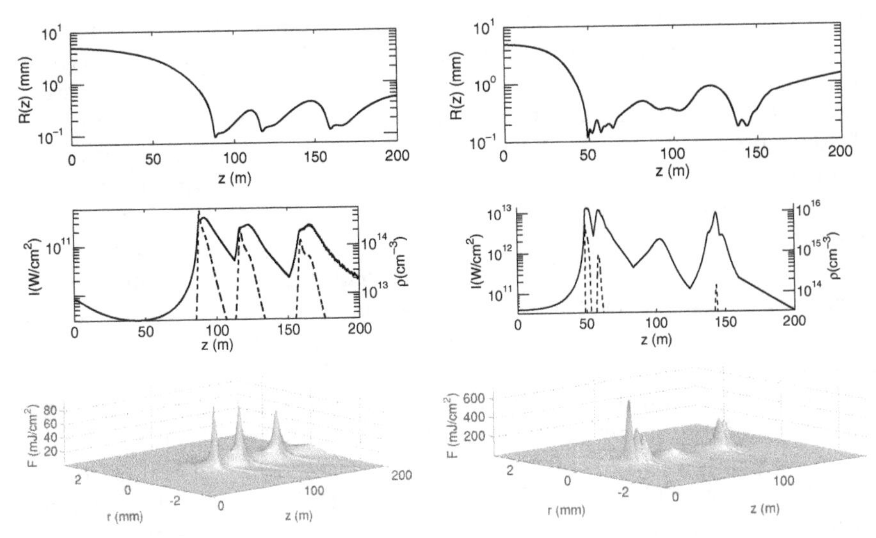

Fig. 12.7 Propagation of a collimated beam (diameter of 1 cm) in air for an ultraviolet laser pulse (first column: $\lambda = 248$ nm, $\tau_{FWHM} = 100$ fs, 1 mJ) and for an infrared laser pulse (second column: $\lambda = 800$ nm, $\tau_{FWHM} = 100$ fs, 3 mJ). The initial power of the pulse is slightly larger than the critical power for self-focusing $P_{\mathrm{cr}}^{UV} = 0.12$ GW, $P_{\mathrm{cr}}^{IR} = 3.2$ GW. The beam radius $R(z)$, the maximum intensity (solid curve, left axis), the density of free electrons on-axis (dashed curve, right axis) and the fluence distribution are plotted as a function of the propagation distance. According to [56]

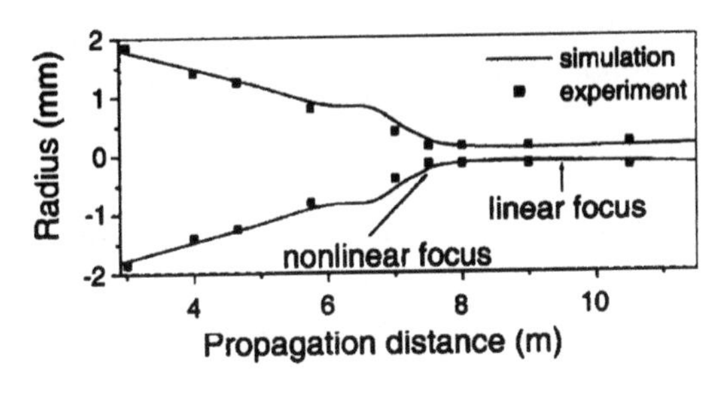

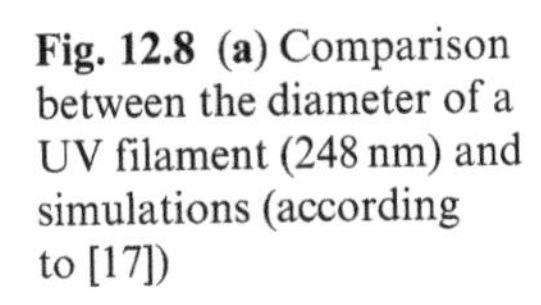
Fig. 12.8 **(a)** Comparison between the diameter of a UV filament (248 nm) and simulations (according to [17])

Depending on the laser wavelength or on other parameters of the propagation (initial beam width, convergence, chirp, etc.), the fluence distribution exhibit either a regular or an irregular periodicity, as illustrated in Fig. 12.7 by the resurgence of a fluence peak for each refocusing stage. When this scenario was first proposed by Mlejnek et al. [55], the role of plasma defocusing in the saturation of the Kerr effect was emphasized and the refocusing events were called *dynamic spatial replenishment of light*. However, it should be stressed that this scenario generally describes any dynamical competition between several physical effects, leading to successive refocusing events [57].

Recently, several teams have shown that the energy reservoir that surrounds the intense core as well as the losses by multiphoton absorption also play a fundamental role in filamentation. The fraction of energy contained in the core of the filament is a few percent of the total beam energy. The typical distance over which this energy is lost via multiphoton absorption is much smaller than the distance characterizing self-focusing of the surrounding part of the beam containing the same amount of energy [15, 58]. A competition between diffraction, self-focusing and multiphoton absorption therefore leads to a spatial replenishment scenario where the defocusing stage is replaced by a fast stage of nonlinear absorption, as shown recently from numerical simulations of filamentation in water [59]. The latter scenario even possesses a limit case for which a stationary regime is likely to sustain filamentation over long distances: rather than having successive cycles of absorption and refocusing stages, multiphoton absorption which takes place in the core induces an energy flow from the periphery towards the center of the beam; Filamentation over long distances is sustained by this energy flow as long as the energy reservoir surrounding the intense core is not exhausted. The underlying idea is again that of a stationary nonlinear mode; the difference with respect to a soliton is that it is weakly localized. The role of the energy reservoir was highlighted by experiments and simulations where a central mask blocked the filament that self-regenerates beyond the mask as in the Poisson spot experiment [33,60]. The complementary experiment where only the filament core passes through a diaphragm shows that the filament is extinguished a Rayleigh length beyond the diaphragm [61]. These real or numerical experiments highlighted the role of the energy reservoir in filamentation and the spontaneous transformation of the Gaussian pulsed beam into a conical wavepacket, i.e., a nonlinear nondispersive nondiffractive stationary solutions with strong space-time couplings [62].

12.3 Self-phase Modulation and Pulse Mode Cleaning

A significant improvement of the beam quality of the pulse emerging from a femtosecond filament in air was discovered recently by analyzing the self-guided pulse and its associated conical emission in the far field [13]. The far field pattern of the emerging beam takes the form of an intense core surrounded by a weaker radiation forming the conical emission shown in Fig. 12.9. A striking feature is the excellent quality of the spatial profile of the conical emission (Fig. 12.9a) compared with the total beam profile at the same distance (Fig. 12.9b). Three parts constitute the total beam, namely, (i) the filament core, (ii) the much weaker conical emission and (iii) the non-filamentary part of the beam. The conical spatial profile shown in Fig. 12.9(a) corresponds to a fundamental transverse radiation mode into which 30% of the incoming beam energy is coupled, while the non-filamentary part of the beam around the central laser wavelength of 400 nm exhibits a poorer beam quality.

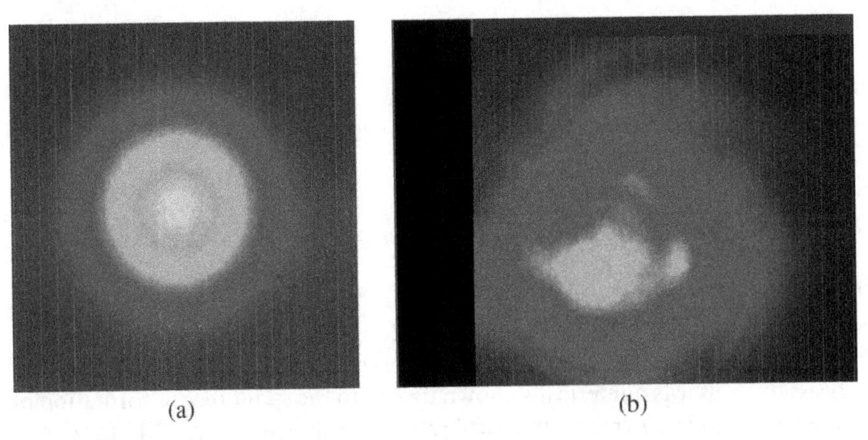

Fig. 12.9 Conical emission and laser pattern measured after 20 m of propagation in air of a 400 nm, 40 fs, 1 mJ laser pulse. (**a**) Surrounding CE only, with the power spectrum close to that of the incident laser pulse removed by a colour filter placed at $z = 10$ m. The wavelengths between 380 and 420 nm have been removed. (**b**) Total laser beam (strongly attenuated) at the same distance

The conical emission was analyzed by measuring with a spectrograph and a CCD camera a quasi-Gaussian profile for each wavelength. Bluer wavelengths were shown to exhibit larger diameters, giving rise to the appearance of the colored rings ,while the conical emission was absent on the visible part of the spectrum. By continuity, one deduces that the filament also assumes a single transverse mode. Indeed, all frequency components showed the striking single spatial mode quality. Around the fundamental frequency, 30% of the fluence can be encompassed in the fundamental spatial mode, which in agreement with numerical simulations [13].

12.4 Single Cycle Pulse Generation by Filamentation

The motivations for generating routinely single optical cycle pulses come from the development of attosecond science. Single-cycle IR pulses constitute one of the key ingredients in the generation of attosecond pulses in the X-UV. Attosecond pulses in turn attract interest because they open up a new field of investigation; for example, ultrafast metrology or measurements of the dynamics of electrons or of molecular orbitals which ultimately control any chemical reaction [63,64]. The only proven way to generate isolated attosecond pulses so far is to convert via harmonics generation from IR to X-UV radiation in a gas. In order to obtain functional attosecond X-UV pulses it is further necessary to control the phase of the IR carrier with respect to the pulse envelope so that the maximum of the oscillations of the electric field coincides with the maximum of the envelope in each laser shot (CEO locking) During filamentation, pulse self-shortening occurs, apparently down to the single cycle limit [65,66]. Moreover, CEO phase locking is preserved during the filamentation process. This opens an interesting prospect to produce well-controlled single-cycle pulses in a simple set-up.

12.4.1 Single-Cycle Pulse Generation in Low Pressure Gas Cells

The set-up shown in Fig. 12.10 corresponds to self-compression experiments performed in Zurich [66,67]. It consists of a low-pressure argon gas cell in which an infrared laser pulse is launched to generate a filament. The filamentation process by itself was numerically shown to lead to the spontaneous formation of a single-cycle pulse. Typical numerical results are shown in Fig. 12.11 for one of the argon gas cells shown in Fig. 12.10, which exhibit a 40 cm long filament with a peak intensity of a few 10^{13} W/cm^2. The electronic density reaches 10^{17} cm^{-3}. As can be seen, the beam profile after filamentation is excellent and highly reproducible. This is another example of pulse mode cleaning discussed in Section 12.3. The dynamics in the filament involves many splitting in time and spatial effects. Several very short temporal subpulses are formed during the propagation but at the end of the filamentation process, an isolated 3-fs pulse is generically obtained for a large range of input conditions. In the experiment, two gas cells were used in order to optimize the compression. After the second cell, a $\sim$ 5 fs pulse (about two cycles) was obtained with 45% of the energy of the input pulse and an excellent spatial quality with a nearly flat phase front. If a CEO pulse was used at the entrance, the phase locking was maintained throughout the two filamentation processes.

12.4.2 Single Cycle Pulse Generation in a Pressure Gradient

A drawback of pulse self-compression via filamentation in a gas of uniform density is the interface with the target. Single-cycle pulses are very fragile; any propagation through a dispersive medium, such as gas beyond the filament end or optical windows-quickly deteriorates the pulse duration. This problem could be partly alleviated by using chirped mirrors to compensate for the linear chirp acquired in the dispersive medium. However, the performances of chirped mirrors become increasingly a challenge when dealing with pulses less than 6 fs in duration. In fact, the results shown in Fig. 12.10, suggest that single-cycle pulses are obtained at the exit of the filament but are deteriorated by the final chirped mirrors. The Fourier transform of the pulse spectrum recorded before the last recompressing mirrors gives a single cycle pulse, whereas the spectrum recorded after the chirped mirrors shows distortions due to the limited band-pass of the mirrors, resulting in a longer pulse.

A control of the filamentation process is possible so as to avoid propagation of the compressed pulse in a dispersive medium. The idea is to use pressure

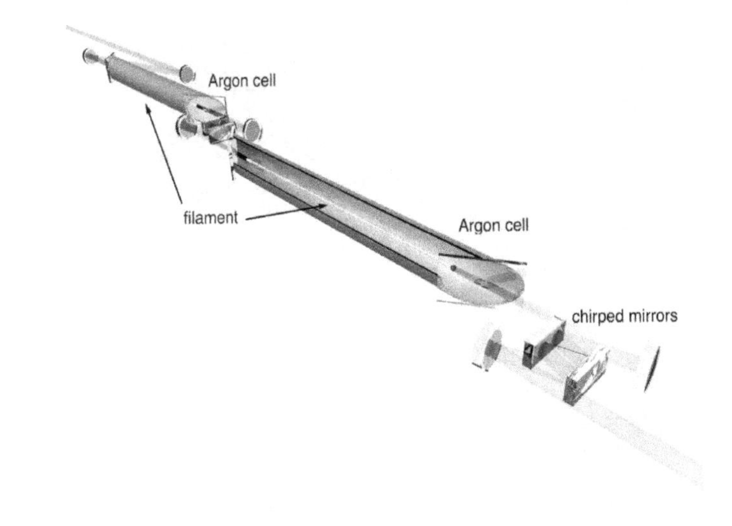

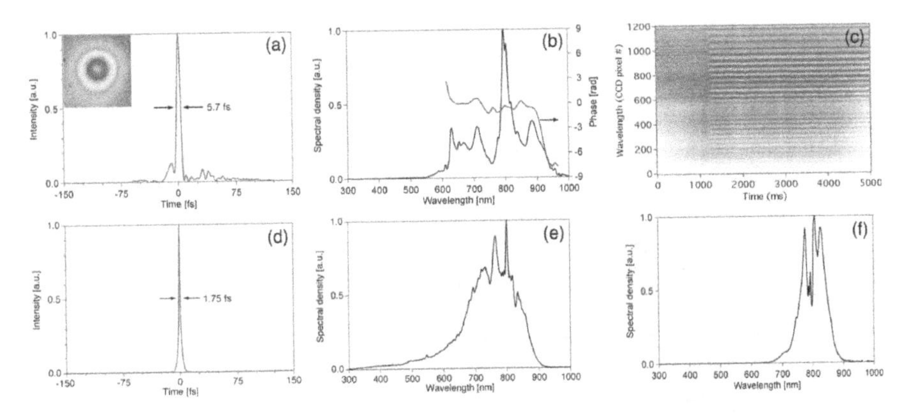

Fig. 12.10 Setup of a self-compression experiment: Two cells are filled with argon at 840 mbar and 700 mbar, respectively. CEO-phase-locked 43-fs IR pulses with 0.84 mJ energy form a 10×15-cm long filament in the middle of the first cell and are compressed with chirped mirrors to 10.5 fs at 790 μJ. Sending those pulses into the second cell results in a filament also 15×20 cm long that, after final recompression, leads to a ~ 5 fs pulse with 45% of the initial pulse energy in an excellent spatial profile shown in (**a**). The associated spectrum and spectral phase are given in (**b**). (**c**) Time series from a f-2f spectral interferometry measurement. The CEO-phase lock is switched on after roughly 1.2 s with the emergence of fringes confirming the CEO phase lock. (**d**) Transform-limited pulse duration supported by the spectrum after the second gas-cell, shown in (**e**), before the chirped mirror compressor, assuming a flat spectral phase, perfect bandpass and phase characteristics of the chirped mirrors. (**f**) Pulse spectrum after the first cell. The pressures in the first and second cell were optimized for maximal spectral broadening and were measured to be 840 mbar and 1060 mbar, respectively. From [66]

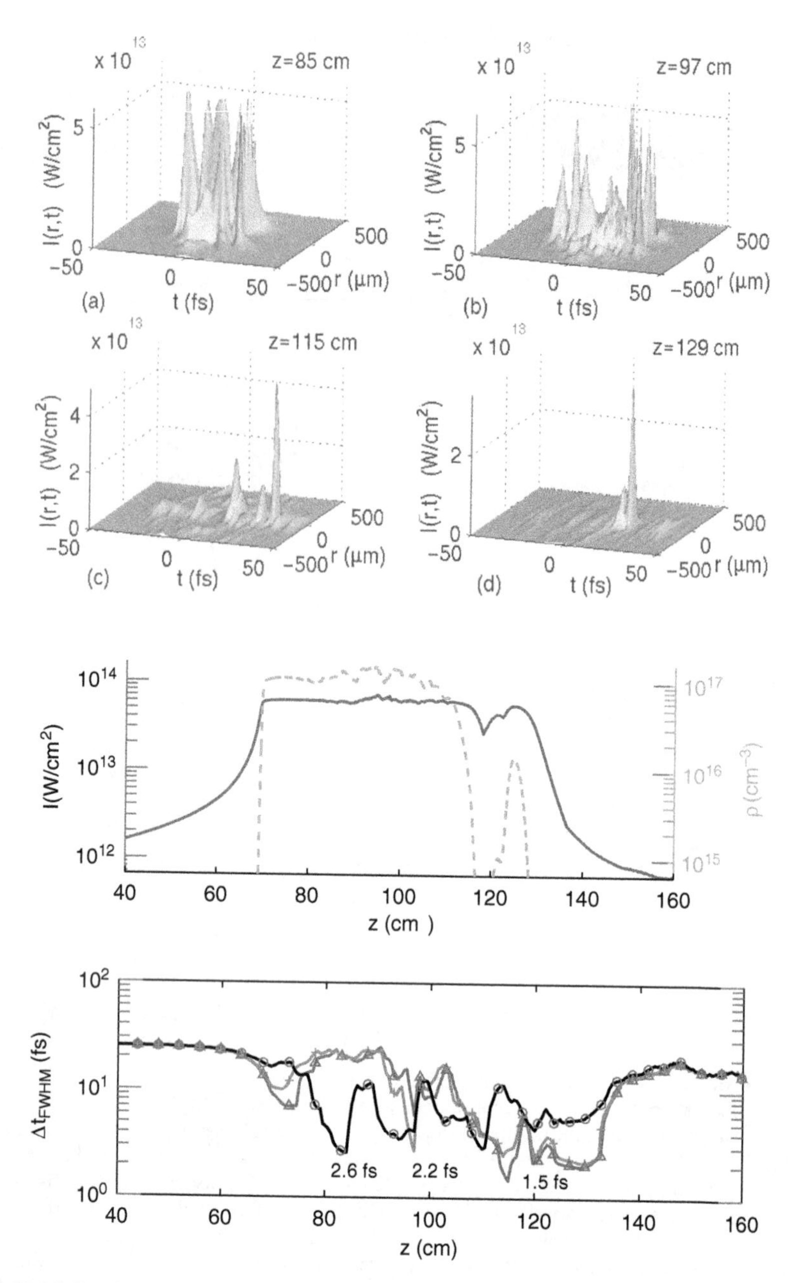

Fig. 12.11 Spatio-temporal intensity distribution of the filamented pulse at (**a**) 85 cm, (**b**) 97 cm, (**c**) 115 cm, and (**d**) 129 cm. (**e**) Peak intensity (continuous curve, scale on the left axis) and electron density (*dashed curve*, scale on the right axis) as a function of propagation distance. The linear focus of the lens is located at $z = 107$ cm. (**f**) FWHM pulse duration as a function of propagation distance. The initial 1 mJ laser pulse with 25 fs duration is focused in a 160- cm-long argon cell at 800 mbar. Three minimal durations reached at 85, 97, and 115 cm are indicated below the minima. The markers indicate that the intensity is averaged over a diameter of 40 μm (*triangles*), 80 μm (*crosses*), and 200 μm (*circles*). The shortest pulse with duration between 2 and 4 fs is obtained between 110 and 130 cm. From [65]

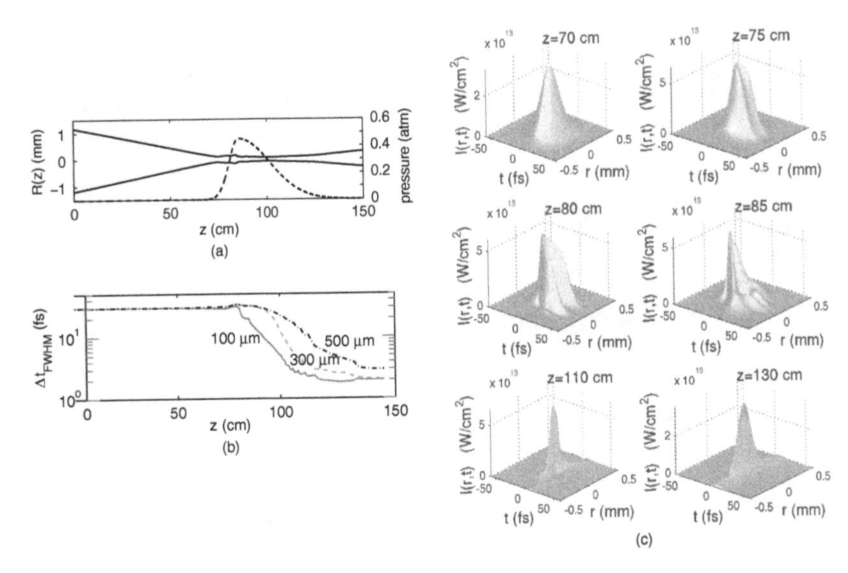

Fig. 12.12 (**a**)Pressure distribution (*dotted lines*, scale on the right axis) and computed beam width (continuous lines, scale on the left axis) along the propagation axis. The linear focus is at $z = 80$ cm. (**b**) Evolution of the pulse duration (FWHM) as a function of the propagation distance. The duration is computed after radial integration of the beam intensity over a beam radius of 100, 300, or 500 μm. (**c**) Pulse dynamics during propagation through the pressure gradient

gradients, i.e., a gas density varying along the propagation distance, so as to end up with a short pulse directly in a vacuum (therefore allowing for transport to the target in a vacuum). The results shown in Fig. 12.12 demonstrate the feasibility of a self-compression scheme via filamentation in a pressure gradient [68]. Numerical simulations were performed by changing three control parameters: the maximum pressure of the gas, the length of the pressure gradient, and its position with respect to the focus of the lens. An excellent control of the filamentation process leading to the generation of single-cycle pulses was achieved for different shapes (parabolic, triangular, and Gaussian) of the pressure gradient. The temporal contrast could even be optimized.

Figure 12.12 displays the case of an asymmetric bi-Gaussian pressure gradient (dashed curve). The corresponding beam width (continuous curve) exhibits a small-scale filament. The pulse duration is computed by integration of the intensity over a 100-μm diameter, which contains the filament. The duration decreases down to 2 fs along the propagation distance. A single-cycle pulse is finally obtained beyond the filament, in vacuum. The complete dynamics of this single-cycle pulse generation, i.e., the evolution of the pulse in space and time is shown on the right figure. The initial beam is focused onto the steep density gradient, thereby minimizing the complications associated with a moving focus. Once the collapse is arrested by MPA, a plasma is generated on the trail of the pulse and defocuses the light on-axis, thus generating a U-shaped pulse. Instead of being followed by a refocusing of the trail on-axis, as is the case in a gas cell with constant pressure, the decrease of the pressure leads to the extinction of the Kerr effect and the pulse finally takes the shape of a pancake that keeps its single-cycle duration and diffracts in a vacuum. Thus, a pressure gradient constitutes a very efficient mechanism to control the filamentation process and to stop it immediately after the generation of a single-cycle pulse. This technique should make it possible to routinely generate single-cycle pulses for applications in attoscience.

12.5 Amplification of Filaments

One key issue in many applications is the amount of energy that can be carried by a filament core. Most studies so far show that for an incident power well beyond the critical power for filamentation, the beam decomposes into a multi-filamentary pattern, each of the sub filaments carrying the same amount of energy. Although multiple filamentation can be organized by suitable amplitude and phase modulation of the incident pulse, it nevertheless represents a major limiting effect in many applications, such as for attosecond pulse production. It has even been argued by some authors that there is a fundamental limit to the amount of energy that can be carried by a filament. Recent works have shown the possibility to increase the filament energy by generating filaments in an amplifying medium [69,70]. In this scheme, the medium consists of a transparent nonlinear host responsible for filamentation by optical Kerr self-focusing and plasma defocusing, and for broadband fluence amplification by the adjunction of a dopant with population inversion. Because the intensity is clamped by multiphoton ionization of the medium, the amplification of a single filament should correspond to an increase of its size. The proof of principle that single filaments can carry significantly more energy in the core than dictated by the critical power for self-focusing has been performed in sapphire doped with Ti^{3+} [70]. This system provides a convenient compact system with a broadband gain that can be easily pumped to population inversion. Filamentation occurs over a distance of the order of 2 cm for an unpumped crystal, provided the input power exceeds a few MW (Fig. 12.13). In Figs. 12.14(a) and (b), the spot at the exit surface of the crystal is shown, displaying the shrinking of the beam size when P_{cr}. If the input power is further increased above several tens of MW, multifilamentation occurs [Fig. 12.14(c)]. On the other hand, propagation of the

Fig. 12.13 Side view photograph of a filament generated in an unpumped Ti:sapphire crystal with a pulse duration of 50 fs, a focusing lens of 20 cm, and an incident energy of 3.7 mJ. The converging laser beam is incident on the left side. A filament with 10-μm diameter is formed over the crystal length, displaying several secondary intensity maxima. From Philip et al. [70]

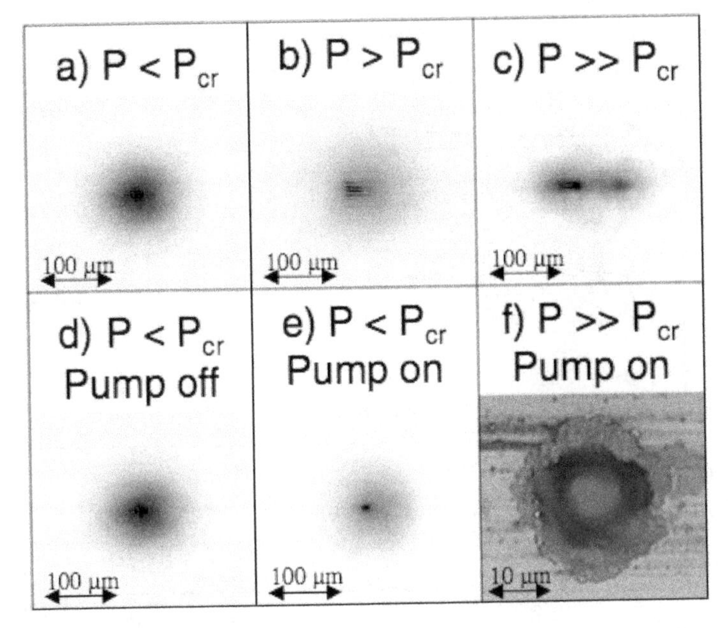

Fig. 12.14 Beam profile measured at the exit face of the crystal for several conditions. (a) $P < P_{\mathrm{cr}}$, unpumped crystal; (b) $E_{in} = 12$ μJ, unpumped crystal; (c) $E_{in} = 47$ μJ, unpumped crystal; (d) same as (a), (e) $E_{in} = 3.6$ μJ, pumped crystal; (f) image of the damage seen through a microscope (note the change of scale). In all cases, the pulse duration was 80 fs and the focus length of the lens was 8 cm. From Philip et al. [70]

infrared pulse in the inverted crystal shows (i) a decrease of the critical power, (ii) an increase of the filament length, and (iii) an increase of the fluence above the value for irreversible damage. All these features are retrieved in the numerical simulations (see Fig. 12.15). Inspection of the damage spot near the exit surface [Fig. 12.14(f)] revealed that no break-up into multiple filaments occurred even if the fluence was larger than in the case of Fig. 12.14(c). This result shows that it is possible to increase the energy contained in a single filament while avoiding multifilamentation. D'Amico et al. have obtained similar results in methanol with a dye [69]. They show that the increase of the fluence of a single filament by a factor of 9 is accompanied by a growth of the filament diameter and they obtained significantly higher energies, up to 80 μJ, inside a single mode.

12.6 Organization of Multiple Filamentation

Several groups have reported on multifilamentary structures propagating over several hundreds of meters [11, 52, 70, 71]. The formation of filaments at kilometric distances in the sky was also reported from the detection of the supercontinuum by a telescope [7, 72]. During horizontal propagation campaign led with the [73, 74], the presence of up to 600 m was clearly demonstrated from

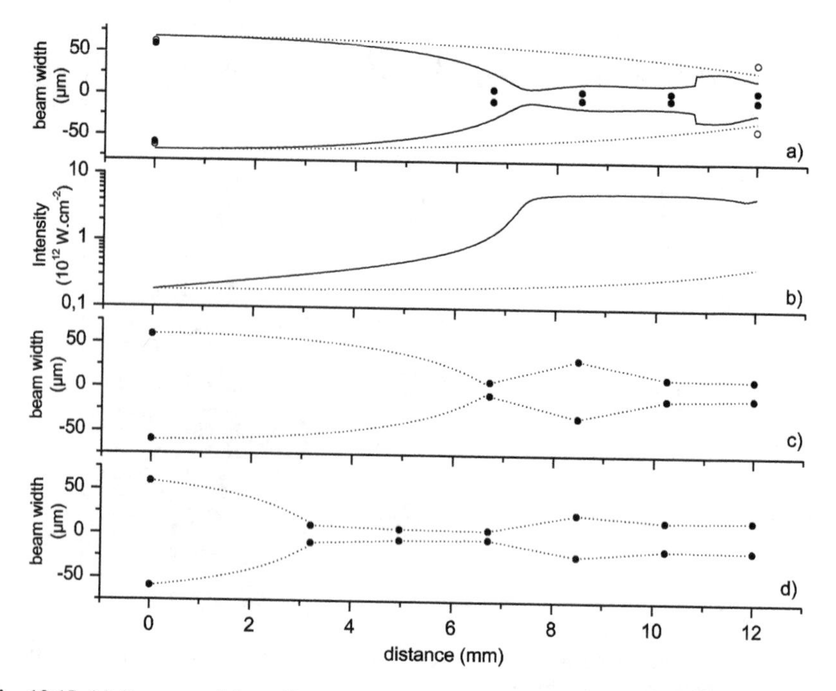

Fig. 12.15 (a) Beam spatial profile as a function of propagation distance inside the crystal. Circles: experimental data with pump at 527 nm off; *black dots*: experimental data with pump on; dotted line: simulation with pump off; *solid line*: simulation with pump on. The incident pulse energy is 3.6 µJ. (**b**) Computed intensity as a function of propagation distance. Dotted line: with pump off; *solid line*: with pump on. The pulse parameters correspond to (**a**). (**c**) Beam spatial profile measured at different depths in the crystal with pump off. The incident pulse energy is 12 µJ. Black dots: experimental data; Dotted line: guide fore the eyes. (**d**) Beam spatial profile measured at different depths in the crystal with pump. The incident pulse energy is 12 µJ. *black dots*: experimental data; *Dotted line*: guide for the eyes. In all cases, the pulse duration was 80 fs and the focus length of the lens was 8 cm. From Philip et al. [70]

step-by-step measurements of the plasma density using three different techniques (conductivity measurements, detection of the luminescence, and detection of the sub-THz EMP). The presence of light channels of millimetric size was demonstrated up to 2.2 km. The domain where multiple filaments were observed with intensities sufficient to ionize air molecules extends to more than 450 m. The intensity of the filaments lies between 10^{10} and 10^{13} W/cm^2.

The origin of multiple filamentation patterns traces back to the works of Bespalov and Talanov [75] on the modulational instability of powerful laser beams. Irregularities in the incident beam profile, even modest, are rapidly reinforced and lead to a breakdown of the beam in several hot spots that act as nuclei for several filaments. Mlejnek et al. [57] performed the first realistic (3 + 1)-dimensional simulations about multiple filamentation and called this propagation regime *optical turbulence*, which seems to describe the self-guided propagation rather well on long distances from an initially collimated beam.

Recently, alternative explanations for multiple filamentation were proposed by Fibich and coworkers [77–79]. They showed that deterministic vectorial effects could prevail over the amplification of noise in the process of multiple filamentation. A small ellipticity of the input beam should also lead to well-determined multiple filamentation patterns [80,81]. Building on this idea, it was shown that multiple filaments can be organized by various control mechanisms acting either on the intensity or on the phase of the input field [76,82]. For instance, large intensity gradients induced by a non-circular diaphragm are shown to lead to beam break-up and multiple filaments regularly located at the periphery of the input beam (see Fig. 12.16). According to the simulations,

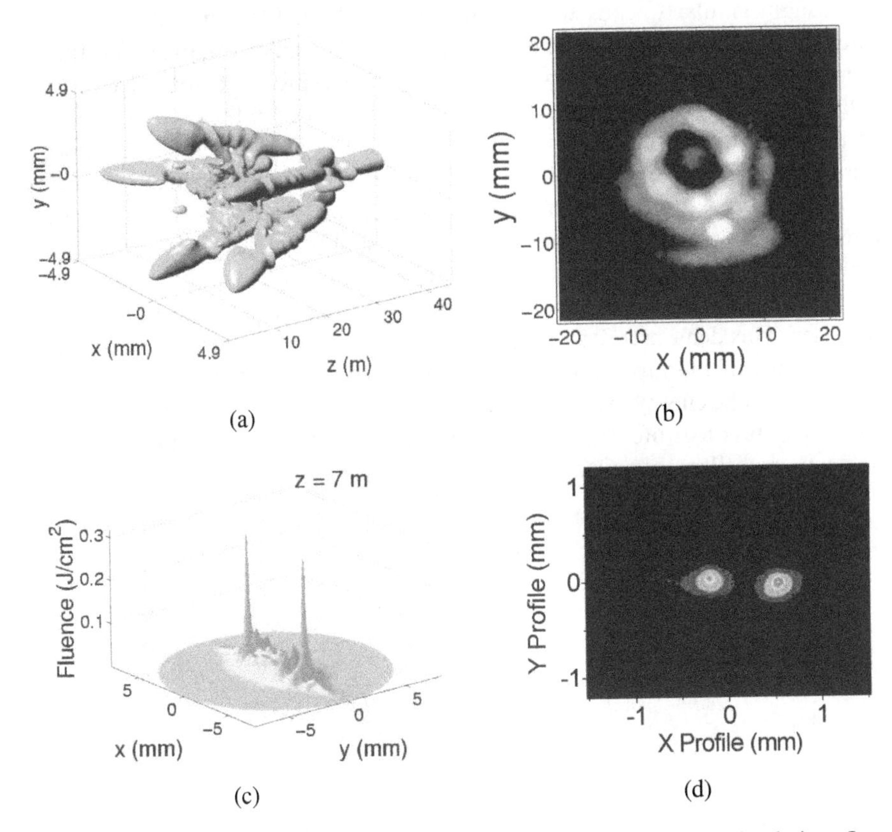

Fig. 12.16 (**a**) Organization of multiple filamentation predicted by (3 + 1)D simulations. Iso-surfaces for the fluence distribution are shown for a 10 mJ, 130 fs laser pulse whose beam is initially reshaped by a five-foil mask (according to Méchain et al. [76]. (**b**) Filamentation pattern measured after propagation over 30 m of a blue (400 nm) laser pulse; the laser beam was diaphragmed by a circular mask. (**c**) Numerically computed filamentation pattern obtained by propagating over 7 m a 7.7-mJ input pulse; the initial curvature of the wave front and its astigmatism was measured and introduced in the numerical simulation. (**d**) Fluence distribution measured for the same conditions (energy 7.7 mJ) at 3.7 m from the tilted defocusing lens of an inverse telescope (according to [76])

the multiple filaments are further expected to coalesce into a single filament on axis, which constitutes a very simple process to enlarge the propagation distance of a multifilamenting pulse and possibly the energy content of a single filament. Such a coalescence has in fact already been observed for twin filaments [30]. Controlled distortions of the spatial phase also leads to organized multifilamentation patterns, as shown in Fig. 12.16, which compares the measured pattern with that obtained from simulations in the case of an input beam with astigmatism [76].

Using large negative initial chirps so as to delay the beginning of the filamentation process, it was shown numerically by means of (3 + 1)D simulations that the effect of azimuthal perturbations on an input beam traversing a circular diaphragm similarly leads to long and intense light filaments that mutually interact via the background energy reservoir [73]. Similar organized multifilamentation patterns, although with a symmetry breaking, were obtained numerically when the measured beam intensity was introduced in the code as an initial condition. The pattern in Fig. 12.17(a) was obtained during a horizontal filamentation campaign [73,74] which led to the observation of weakly ionizing filaments over kilometric distances, connected by a network of moderately intense energy clearly visible on Fig. 12.17(a). Figure 12.17(b) shows good agreement between this measurement and the numerically obtained pattern from (3 + 1)D simulations starting with the beam measured at the output of the laser. This demonstrates that the multifilamentation pattern is governed by the features of the input pulse more than by shot-to-shot fluctuations or air turbulence. The energy exchange between the background energy reservoir and the filaments constitutes the process sustaining the propagation, extinction, and nucleation of filaments over long distances [72–74].

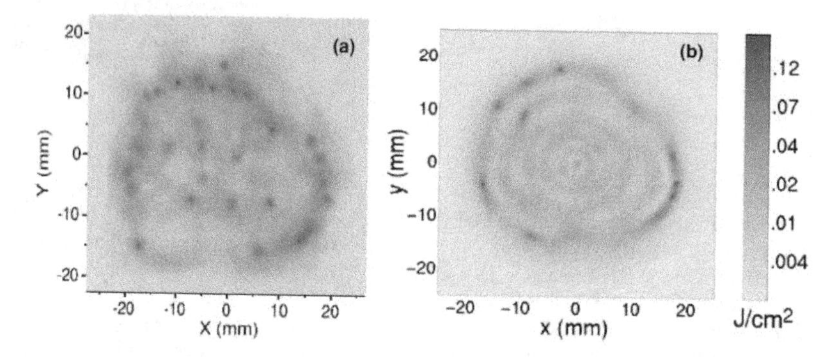

Fig. 12.17 Comparison between the multiple filamentation patterns obtained in (**a**) experiments and (**b**) simulations after propagation over 68 m of a 190 mJ, 800 nm laser pulse stretched to 1.2 ps by a negative chirp (according to Méchain et al. [73,74])

12.7 Conclusion

In summary, the physics of femtosecond filaments involves several self-action effects which can be used for shortening of laser pulses down to a single cycle, for an improvement of the beam quality, or for the amplification of filaments. Multiple filamentation patterns arise for powers $P \gg P_{\text{cr}}$ and result from the growth of modulationally unstable perturbations in the beam. Their nucleation and coalescence are seemingly disordered; yet they can be organized by means of several control processes. They are sustained by the background energy reservoir which can refill a specific filament or reform another light string if it is destroyed by an obscurant on its path. This leads to their observation at considerable distances, reaching several kilometers.

Filaments therefore bear remarkable properties in the time domain (pulse self-compression), frequency domain (spectrum broadening) and spatial domain (mode improvement and controllable multifilamentation patterns), making them a unique source for applications.

Acknowledgments The authors gratefully acknowledge the contribution of Michel Franco, Bernard Prade, Aurélien Houard, Ciro D'Amico and the partners of the Teramobile team. A.C. and A.M. are at the CNRS.

References

1. Braun, A., Korn, G., Liu, X. et al. (1995). Self-channeling of high-peak-power femtosecond laser pulses in air. *Opt. Lett.*, **20**(1), 73–75.
2. Lange, H.R., Grillon, G., Ripoche, J.-F. et al. (1998). Anomalous long-range propagation of femtosecond laser pulses through air: moving focus or pulse self-guiding? *Opt. Lett.*, **23**(2), 120–122.
3. Nibbering, E.T.J., Curley, P.F., Grillon, G. et al. (1996). Conical emission from self-guided femtosecond pulses in air. *Opt. Lett.*, **21**(1), 62–64.
4. Couairon, A., Mysyrowicz, A. (2007). Femtosecond filamentation in transparent media. *Phys. Rep.*, **441**, 47–189.
5. Kosareva, O.G., Kandidov, V.P., Brodeur, A. et al. (1997). Conical emission from laser–plasma interactions in the filamentation of powerful ultrashort laser pulses in air. *Opt. Lett.*, **22**(17), 1332–1334.
6. Kasparian, J., Sauerbrey, R., Mondelain, D. et al. (2000). Infrared extension of the supercontinuum generated by femtosecond terawatt laser pulses propagating in the atmosphere. *Opt. Lett.*, **25**(18), 1397–1399.
7. Wöste, L., Wedekind, C., Wille, H. et al. (1997). Femtosecond atmospheric lamp. *Laser und Optoelektronik*, **29**(5), 51–53.
8. Couairon, A., Franco, M., Méchain, G., Olivier, T., Prade, B., and Mysyrowicz, A. (2006a). Femtosecond filamentation in air at low pressures: Part I: Theory and numerical simulations. *Opt. Commun.*, **259**, 265–273.
9. Méchain, G., Olivier, T., Franco, M. et al. (2006). Femtosecond filamentation in air at low pressures: Part II: Laboratory experiments. *Opt. Commun.*, **261**, 322–326.
10. Kasparian, J., Rodriguez, M., Méjean, G. et al. (2003). White light filaments for atmospheric analysis. *Science*, **301**, 61.

11. La Fontaine, B., Vidal, F., Jiang, Z. et al. (1999). Filamentation of ultrashort pulse laser beams resulting from their propagation over long distances in air. *Phys. Plasmas*, **6**, 1615.
12. Mikalauskas, D., Dubietis, A., Danielus, R. (2002). Observation of light filaments induced in air by visible picosecond laser pulses. *Appl. Phys. B*, **75**, 899–902.
13. Prade, B., Franco, M., Mysyrowicz, A. et al. (2006). Spatial mode cleaning by femtosecond filamentation in air. *Opt. Lett.*, **31**(17), 2601.
14. Couairon, A., Bergé, L. (2002). Light filaments in air for ultraviolet and infrared wavelengths. *Phys. Rev. Lett.*, **88**(13), 135003.
15. Schwarz, J., Rambo, P., Diels, J.-C. et al. (2000). Ultraviolet filamentation in air. *Optics Commun.*, **180**, 383–390.
16. Tzortzakis, S., Lamouroux, B., Chiron, A. et al. (2000). Nonlinear propagation of subpicosecond ultraviolet laser pulses in air. *Opt. Lett.*, **25**, 1270–1272.
17. Tzortzakis, S., Lamouroux, B., Chiron, A. et al. (2001b). Femtosecond and picosecond ultraviolet laser filaments in air: experiments and simulations. *Opt. Commun.*, **197**, 131–143.
18. Dubietis, A., Tamošauskas, G., Diomin, I. et al. (2003). Self-guided propagation of femtosecond light pulses in water. *Opt. Lett.*, **28**(14), 1269–1271.
19. Tzortzakis, S., Sudrie, L., Franco, M., Prade, B., Mysyrowicz, A., Couairon, A., and Bergé, L. (2001c). Self-guided propagation of ultrashort IR laser pulses in fused silica. *Phys. Rev. Lett.*, **87**, 213902 1–4.
20. Tzortzakis, S., Papazoglou, D.G., Zergioti, I. (2006). Long-range filamentary propagation of subpicosecond ultraviolet laser pulses in fused silica. *Opt. Lett.*, **31**, 796.
21. Wille, H., Rodriguez, M., Kasparian, J. et al. (2002). Teramobile: a mobile femtosecond-terawatt laser and detection system. *Eur. Phys. J. Appl. Phys.*, **20**, 183–190.
22. Couairon, A., Sudrie, L., Franco, M. et al. (2005a). Filamentation and damage in fused silica induced by tightly focused femtosecond laser pulses. *Phys. Rev. B*, **71**, 125435.
23. Sudrie, L., Couairon, A., Franco, M. et al. (2002). Femtosecond laser-induced damage and filamentary propagation in fused silica. *Phys. Rev. Lett.*, **89**(18), 186601.
24. Chiao, R.Y., Garmire, E., Townes, C.H. (1964). Self-trapping of optical beams. *Phys. Rev. Lett.*, **13**(15), 479–482.
25. Brodeur, A., Chien, C.Y., Ilkov, F.A. et al. (1997). Moving focus in the propagation of ultrashort laser pulses in air. *Opt. Lett.*, **22**(5), 304–306.
26. Shen, Y. R. (1984). *The Principles of Nonlinear Optics*. Wiley-Interscience, New York.
27. Marburger, J.H. (1975). Self-focusing: Theory. *Prog. Quant. Electr.*, **4**, 35–110.
28. Couairon, A. (2003a). Dynamics of femtosecond filamentation from saturation of self-focusing laser pulses. *Phys. Rev. A*, **68**, 015801.
29. Couairon, A. (2003c). Light bullets from femtosecond filamentation. *Eur. Phys. J. D*, **27**, 159–167.
30. Tzortzakis, S., Bergé, L., Couairon, A. et al. (2001a). Break-up and fusion of self-guided femtosecond light pulses in air. *Phys. Rev. Lett.*, **86**, 5470–5473.
31. Moll, K.D., Gaeta, A.L., Fibich, G. (2003). Self-similar optical wave collapse: Observation of the Townes profile. *Phys. Rev. Lett.*, **90**, 203902.
32. Couairon, A., Gaizauskas, E., Faccio, D., et al. (2006b). Nonlinear X-wave formation by femtosecond filamentation in Kerr media. *Phys. Rev. E*, **73**, 016608.
33. Dubietis, A., Gaižauskas, E., Tamošauskas, G. et al. (2004a). Light filaments without self-channeling. *Phys. Rev. Lett.*, **92**(25), 253903.
34. Faccio, D., Di Trapani, P., Minardi, S. et al. (2005a). Far-field spectral characterization of conical emission and filamentation in Kerr media. *J. Opt. Soc. Am. B*, **22**(4), 862–869.
35. Faccio, D., Matijosius, A., Dubietis, A. et al. (2005b). Near- and far-field evolution of laser pulse filaments in Kerr media. *Phys. Rev. E*, **72**, 037601.
36. Kolesik, M., Wright, E.M., Moloney, J. V. (2004). Dynamic nonlinear X-waves for femtosecond pulse propagation in water. *Phys. Rev. Lett.*, **92**, 253901 1–4.

37. Conti, C., Trillo, S., Di Trapani, P. et al. (2003). Nonlinear electromagnetic X-waves. *Phys. Rev. Lett.*, **90**(17), 170406.
38. Day, C. (2004). Intense X-shaped pulses of light propagate without spreading in water and other dispersive media. *Physics Today*, **57**(10), 25–26.
39. Valiulis, G., Kilius, J., Jedrkiewicz, O. et al. (2001). Space–time nonlinear compression and three-dimensional complex trapping in normal dispersion. In *OSA Trends in Optics and Photonics (TOPS), Technical Digest of the Quantum Electronics and Laser Science Conference (QELS 2001)*, volume 57, pages QPD10-1-2. Optical Society of America, Washington DC, 2001.
40. Di Trapani, P., Valiulis, G., Piskarskas, A., et al. (2003). Spontaneously generated X-shaped light bullets. *Phys. Rev. Lett.*, **91**(9), 093904.
41. Faccio, D., Porras, M., Dubietis, A. et al. (2006b). Conical emission, pulse splitting and X-wave parametric amplification in nonlinear dynamics of ultrashort light pulses. *Phys. Rev. Lett.*, **96**, 193901.
42. Faccio, D., Porras, M.A., Dubietis, A. et al. (2006a). Angular and chromatic dispersion in Kerr-driven conical emission. *Opt. Commun.*, **265**, 672–677.
43. Kolesik, M., Wright, E.M., Moloney, J. V. (2005). Interpretation of the spectrally resolved far field of femtosecond pulses propagating in bulk nonlinear dispersive media. *Opt. Express*, **13**(26), 10729–10741.
44. Keldysh, L.V. (1965). Ionization in the field of a strong electromagnetic wave. *Sov. Phys. JETP*, **20**(5), 1307–1314.
45. Perelomov, A.M., Popov, V.S., Terent'ev, M.V. (1966). Ionization of atoms in an alternating electric field. *Sov. Phys. JETP*, **23**(5), 924–934.
46. Couairon, A., Tzortzakis, S., Bergé et al. (2002). Infrared light filaments: simulations and experiments. *J. Opt. Soc. Am. B*, **19**(13), 1117–1131.
47. Couairon, A., Méchain, G., Tzortzakis et al. (2003). Propagation of twin laser pulses in air and concatenation of plasma strings produced by femtosecond infrared filaments. *Opt. Commun.*, **225**, 177–192.
48. Ripoche, J.-F., Grillon, G., Prade, B. et al. (1997). Determination of the time dependence of n_2 in air. *Opt. Commun.*, **135**, 310–314.
49. Nibbering, E.T.J., Grillon, G., Franco, M.A. et al. (1997). Determination of the inertial contribution to the nonlinear refractive index of air, N_2, and O_2 by use of unfocused high-intensity femtosecond laser pulses. *J. Opt. Soc. Am. B*, **14**(3), 650–660.
50. Yablonovitch, E., Bloembergen, N. (1972). Avalanche ionization and the limiting diameter of filaments induced by light pulses in transparent media. *Phys. Rev. Lett.*, **29**(14), 907–910.
51. Gaeta, A.L. (2000). Catastrophic collapse of ultrashort pulses. *Phys. Rev. Lett.*, **84**, 3582.
52. Zozulya, A.A., Diddams, S.A., Clement, T.S. (1998). Investigations of nonlinear femtosecond pulse propagation with the inclusion of Raman, shock, and third-order phase effects. *Phys. Rev. A*, **58**, 3303–3310.
53. Zozulya, A.A., Diddams, S.A., Van Engen, A.G. et al. (1999). Propagation dynamics of intense femtosecond pulses: Multiple splittings, coalescence, and continuum generation. *Phys. Rev. Lett.*, **82**, 1430.
54. Brabec, T., Krausz, F. (1997). Nonlinear optical pulse propagation in the single-cycle regime. *Phys. Rev. Lett.*, **78**, 3282–3285.
55. Mlejnek, M., Wright, E.M., Moloney, J.V. (1998). Dynamic spatial replenishment of femtosecond pulses propagating in air. *Opt. Lett.*, **23**(5), 382–384.
56. Couairon, A., Mysyrowicz, A. (2006a). Femtosecond filamentation in air. In: K. Yamanouchi, S.L. Chin, P. Agostini et al., eds., *Progress in Ultrafast Intense Laser Science*, chapter 13, pages 235–258, Springer New York.
57. Mlejnek, M., Kolesik, M., Moloney, J.V., et al. (1999). Optically turbulent femtosecond light guide in air. *Phys. Rev. Lett.*, **83**(15), 2938–2941.
58. Couairon, A. (2003b). Filamentation length of powerful laser pulses. *Appl. Phys. B*, **76**, 789–792.

59. Dubietis, A., Couairon, A., Kučinskas, E., et al. (2006). Role of nonlinear losses in self-focusing dynamics. *Appl. Phys. B*.
60. Kolesik, M., Moloney, J.V. (2004). Self-healing femtosecond light filaments. *Opt. Lett.*, **29**(6), 590–592.
61. Dubietis, A., Kučinskas, E., Tamošauskas, G. et al. (2004c). Self-reconstruction of light filaments. *Opt. Lett.*, **29**, 2893–2895.
62. Gaižauskas, E., Dubietis, A., Kudriašov, V. et al. (2007). On the role of conical waves in self-focusing and filamentation of femtosecond pulses, this book. chapter 19.
63. Zewail, A.H. (1996). Femtochemestry: Recent progress in studies of dynamics and control of reactions and their transition rates. *J. Phys. Chem.*, **100**, 12701.
64. Zewail, A.H. (2000). Femtochemestry: atomic-scale dynamics of chemical bonds. *J. Phys. Chem. A*, **104**, 5660–5694.
65. Couairon, A., Biegert, J., Hauri, C.P., et al. (2005c). Self-compression of ultrashort laser pulses down to one optical cycle by filamentation. *J. Mod. Opt.*, **53**(1-2), 75–85.
66. Hauri, C.P., Kornelis, W., Helbing, F.W. et al. (2004). Generation of intense, carrier-envelope phase-locked few-cycle laser pulses through filamentation. *Appl. Phys. B*, **79**, 673–677.
67. Hauri, C.P., Guandalini, A., Eckle, P. et al. (2005). Generation of intense few-cycle laser pulses through filamentation - parameter dependence. *Opt. Express*, **13**(19), 7541.
68. Couairon, A., Franco, M., Mysyrowicz, A., et al. (2005b). Pulse-compression to the single-cycle limit by filamentation in a gaz with a pressure gradient. *Opt. Lett.*, **30**(19), 2657–2659.
69. D'Amico, C., Prade, B., Franco, M. et al. (2006). Femtosecond filament amplification in liquids. *Appl. Phys. B*, **85**, 49–53.
70. Philip, J., DAmico, C., Chériaux, G., et al. (2005). Amplification of femtosecond laser filaments in Ti:Sapphire. *Phys. Rev. Lett.*, **95**, 163901.
71. Chin, S.L., Talebpour, A., Yang, J. et al. (2002). Filamentation of femtosecond laser pulses in turbulent air. *Appl. Phys. B*, **74**, 67–76.
72. Rodriguez, M., Bourayou, R., Méjean, G. et al. (2004). Kilometric-range nonlinear propagation of femtosecond laser pulses. *Phys. Rev. E*, **69**, 036607.
73. Méchain, G., Couairon, A., André, Y.-B. et al. (2004a). Long range self-channeling of infrared laser pulses in air: a new propagation regime without ionization. *Appl. Phys. B*, **79**, 379–382.
74. Méchain, G., D'Amico, C., André, Y.-B. et al. (2005). Length of plasma filaments created in air by a multiterawatt femtosecond laser. *Optics Commun.*, **247**, 171–180.
75. Bespalov, V.I., Talanov, V.I. (1966). Filamentary structure of light beams in nonlinear liquids. *Zh. Eksper. Teor. Fiz. Pis'ma*, **3**, 471–476. [*JETP Lett.* 3 (1966) 307–310].
76. Méchain, G., Couairon, A., Franco, M., et al. (2004b). Organizing multiple femtosecond filamentation in air. *Phys. Rev. Lett.*, **93**(3), 035003.
77. Fibich, G., Ilan, B. (2001a). Deterministic vectorial effects lead to multiple filamentation. *Opt. Lett.*, **26**(11), 840–842.
78. Fibich, G., Ilan, B. (2001b). Vectorial and random effects in self-focusing and in multiple filamentation. *Physica D*, **157**, 112.
79. Fibich, G., Ilan, B. (2002). Multiple filamentation of circularly polarized beams. *Phys. Rev. Lett.*, **89**(1), 013901.
80. Dubietis, A., Tamošauskas, G., Fibich, G. et al. (2004b). Multiple filamentation induced by input-beam ellipticity. *Opt. Lett.*, **29**(10), 1126.
81. Fibich, G., Eisenmann, S., Ilan, B. et al. (2004). Control of multiple filamentation in air. *Opt. Lett.*, **29**(15), 1772.
82. Schroeder, H., Liu, J., Chin, S.L. (2004). From random to controlled small-scale filamentation in water. *Opt. Express*, **12**(20), 4768–4774.
83. Yang, H., Zhang, J., Li, Y. et al. (2002). Characteristics of self-guided laser-plasma channels generated by femtosecond laser pulses in air. *Phys. Rev. E*, **66**(1), 016406.

Chapter 13
Self-organized Propagation of Femtosecond Laser Filamentation in Air

Jie Zhang, Zuoqiang Hao, Tingting Xi, Xin Lu, Zhe Zhang, Hui Yang, Zhan Jin, Zhaohua Wang, and Zhiyi Wei

Abstract A long plasma channel is formed with a length up to a few hundred meters when intense femtosecond laser pulses propagate in air. We find that the propagation of the filaments in the channel shows a very complicated process including the evolution from a single filament into two and three and even more distinct filaments periodically, and the merging of multiple filaments into two filaments that propagate stably and fade away eventually. From the point of view of applications, the lifetime of the plasma channel can be prolonged to the order of microseconds when another sub-ns laser pulse is introduced. The filaments' distribution is optimized using a pinhole with different diameters. Our experiments also demonstrate simultaneous triggering and guiding of large gap discharges in air by laser filaments. A new concept of "laser plasma channel propulsion" is proposed. It is demonstrated that the plasma channel can continuously propel a light paper airplane without complicated focusing optics. As for the long distance propagation of the laser pulses, the filamentation process and the surpercontinuum (SC) emission are closely dependent on the initial negative chirp and the divergence angle of the laser beam. Most of laser energy deposited in the background serves as an energy reservoir for further propagation of the filamentation. We have shown that an energy reservoir over ten times the size of the filament core (mm size) is necessary to feed a single filament undisturbed propagation. At last, the characteristics of the multiple filaments formed by pre-focused and freely propagating fs laser pulses are investigated and compared.

13.1 Introduction

Although the propagation of a laser beam in nonlinear optical media dates back to the 1960s [1], femtosecond (fs) pulses introduce a new concept into the "old" field. Due to the advantages of intense fs pulses, many interesting phenomena

J. Zhang (✉)
Beijing National Laboratory for Condensed Matter Physics, Institute of Physics, Chinese Academy of Sciences, Beijing 100190, China

R.W. Boyd et al. (eds.), *Self-focusing: Past and Present*,
Topics in Applied Physics 114, DOI 10.1007/978-0-387-34727-1_13,

occur and potential applications, such as lightning control and remote sensing attract scientists' attentions all over the world. This chapter reviews the rapid development of the propagation of intense fs laser pulses in air. It includes not only the mechanism of filamentation and many interesting phenomena observed in experiments, but also some efforts toward potential applications including the lifetime prolongation, laser guided discharge, laser propulsion, and so on.

13.2 Mechanism of Filamentation

The mechanism underlying the propagation of intense fs laser pulses has been well studied. Three models were proposed during the past decade. Braun et al. [2] proposed the self-waveguiding model to interpret the self-guided filament formation as the results of balance between self-focusing and plasma defocusing. Brodeur et al. [3] adopted the moving-focus model whereby different time slices of the pulse focus at different nonlinear focal lengths [4–6]. However, the two models do not reveal the dynamic process inside the filamentation and long-range propagation. The moving-focus model cannot explain the persistence of a filament beyond the linear focus, and the self-waveguiding model also presents some problems of interpretation [7]. The dynamic spatial replenishment model proposed by Mlejnek et al. [8] believes that the mechanism should be the periodical energy transformation between the filament and background energy reservoir. That is to say, the focusing–defocusing cycle leads to the long-distance propagation of laser pulses in air. The recurrent collapse mechanism is consistent with almost all recent experimental results.

There are mainly two different nonlinearities affecting the laser propagation in air. One is the Kerr nonlinearity which leads to strong self-focusing if the input power is high enough. The other is the plasma-induced nonlinearity which has a strong defocusing effect to the laser propagation. The dynamic balance between these two effects leads to a self-guided propagation, as indicated in Fig. 13.1.

The propagation of fs laser pulses in air can be described by an extended nonlinear Schrödinger equation, coupled to the equation of electron density induced by multiphoton ionization [9]. The coupled equations can be written as:

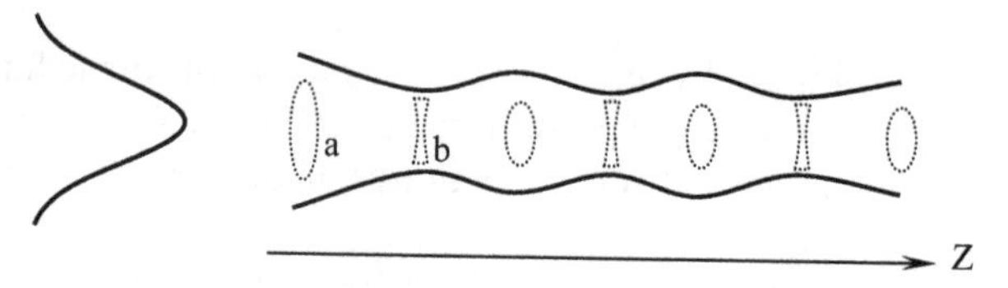

Fig. 13.1 Schematic of a fundamental Gaussian optical beam propagating in air. Left: The Gaussian laser pulse. Right: Self-guided propagation induced by the balance between self-focusing (**a**) with plasma defocusing (**b**)

$$2i\frac{\partial E}{\partial z}+\frac{1}{k_0}\Delta_\perp E-k''\frac{\partial^2 E}{\partial t^2}+k_0n_2(|E|^2+\tau_K^{-1}\int_{-\infty}^{t}e^{-(t-t')/\tau_K}|E(t')|^2dt')E$$
$$-k_0\frac{\omega_{pe}^2(\rho)}{\omega_0^2}E+i\beta^{(K)}|E|^{2K-2}E=0 \tag{13.1}$$

$$\frac{\partial\rho}{\partial\tau}=\frac{\beta^{(K)}}{K\hbar\omega_0}|E|^{2K}\left(1-\frac{\rho}{\rho_{at}}\right). \tag{13.2}$$

Diffraction, group-velocity dispersion, Kerr self-focusing effect (composed of an instantaneous and a delayed part), defocusing due to plasma generation and multiphoton absorption are considered in the equations. The parameters in equations are as follows. $k_0=2\pi/\lambda_0$, $\lambda_0=800\,\text{nm}$ is the central wavelength of the input beam. $K''=0.2\,\text{fs}^2/\text{cm}$ represents the coefficient of group velocity dispersion (GVD). $n_2=3.2\times10^{-19}\,\text{cm}^2/\text{W}$ refers to the nonlinear refraction index of air. $\tau_K=70\,\text{fs}$ denotes the characteristic time of the delayed part of the Kerr effect. The multi-photon ionization coefficient is $\beta^{(K)}=1.27\times10^{-126}\,\text{cm}^{17}/\text{W}^9$ for the number of photons $K=10$. The plasma frequency is $\omega_{pe}=(q_e^2\rho/m_e\varepsilon_0)^{1/2}$ (q_e, m_e and ρ are the electron charge, mass and density, respectively), and the density of neutral atoms is $\rho_{at}=2.7\times10^{19}\,\text{cm}^{-3}$.

Based on this model, we simulated the interaction of the two light bullets generated by the fs laser pulse in air. The evolution of multiple filaments (MF) pattern can be regarded as the propagation of a group of interacting "light bullets" (small-scale light filaments) [10]; thus, the study of the interaction dynamics of light bullets is critical in understanding the characteristics of the propagation of an ultrashort laser beam in air. We have carried out simulations on the interaction between two light filaments. The effects of their relative phase shift, crossing angle and the initial position on the bullets interaction are considered. The attraction, fusion, repulsion, and collision are observed in the simulations. The stability of the channel formed by two interacting light bullets strongly depends on the relative phase shift and the crossing angle between the two light bullets [11].

13.3 Diagnostics of Filamentation

It is difficult to directly measure the intensity distribution of laser pulses during filamentation. Some methods have been used to diagnose the generated plasma channels, such as shadowgraphy [12], interferometry [13], fluorescence detection [14], THz detection [15], electromagnetic pulse measurement [16], and so on. These methods have their own advantages and drawbacks. In this section, we use five techniques to diagnose the filaments formed by intense fs laser pulses in air. They all reveal the evolution of filaments in detail. The five methods are based on the different properties of filaments. Because of the high intensity of filaments, we can image the cross-section of the laser beam using a simple experimental setup.

Using the conductivity of the filaments, the resistivity measurement is used to detect the filaments. By detecting the acoustic wave radiation, the length and the gross intensity distribution in the channels as well as the electron density can be obtained. The gross intensity distribution of the filaments can be reflected by measuring the fluorescence radiation. From the shift of the interference fringes in the interferometry, we can obtain the electron density. Furthermore, a comparison is made among the five diagnostics, indicating that these diagnostic methods have many advantages and drawbacks compared with each other.

The laser system we used in our experiments is a home-made fs laser system called Xlite–II (XL–II) with an output energy up to 640 mJ in 30-fs pulses at a central wavelength of 800 nm. The repetition rate is 10 Hz. At the output of the compressor chamber, the initial laser beam diameter is about 2.5 cm (FWHM).

13.3.1 Imaging of Beam Cross-Section

A glass plate is inserted at a 33° angle with the aim of sampling the cross section of the channels. A lens images the channels onto a charged-coupled device (CCD) camera (512 × 512 pixels) with a pixel size of 24 μm. The wavelength response of the CCD used in our experiments is about 200–1100 nm. A high-speed shutter is used to take single shot image of the filaments on the plate. The glass plate is placed on a translation stage, which can move in parallel with the CCD. This ensures that each laser pulse irradiates a new site of the plate. The imaging system is set up on a small stage that moves along the laser propagation axis to obtain the images at different positions. The setup is shown in Fig. 13.2(3). Because of the high intensity in filaments, white light emits from the glass plate when the filaments shoots the plate. By measuring the light radiation, the filaments patterns are obtained. We can also get some information about filaments, such as the diameter and length of filaments.

Using the imaging measurements, the spatial evolution of the filaments can be studied. The processes of filamentation, splitting, and fusion of plasma filaments are observed. Therefore, the imaging technique is suitable to study the filaments evolution in detail. The detailed results can be found in the following sections.

13.3.2 Resistivity Measurement

Many experiments have demonstrated that the electron density in the plasma channels is about 10^{16}–10^{18} cm^{-3}. The plasma channel is conductive, like a wire. Based on this property, the lightning control using plasma channels in air has been studied at different laboratories. On the other hand, the conductivity can

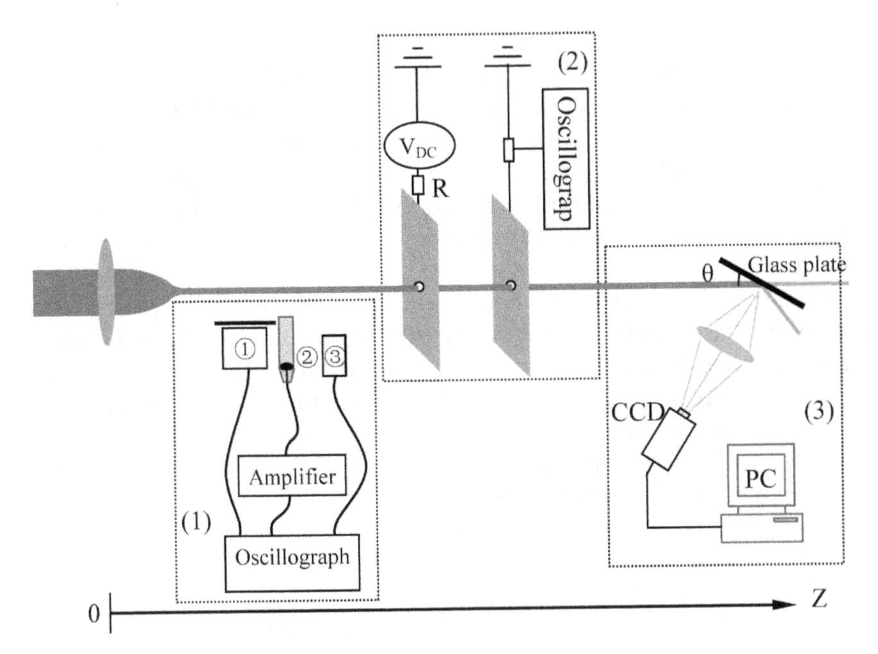

Fig. 13.2 The experimental setup. (1) acoustic detection and fluorescence measurement setup. ① a PMT to detect the fluorescence, ② a microphone to measure the sound, and ③ a photodiode to trigger the oscillograph; (2) resistivity measurement; (3) imaging setup

be used to diagnose plasma channels. Through measuring the resistivity of the channels, the length, the electron density and the evolution of the channels can be retrieved. We can optimize the condition of filamentation and its resistivity, for some potential applications, such as guiding a very long electrical discharge.

Two parallel copper plates are placed on the beam axis. The plasma channels traverse through the two plates. A dc voltage of $V_{DC} = 880$ V is applied between both electrodes. The electrical signals are recorded by a digital oscilloscope. An $R = 10$ kΩ resistance is used to limit the current and the signal voltage is obtained from a 200-Ω resistance by an oscilloscope. The experimental setup is shown in Fig. 13.2(2). The diameter of the pinhole drilled in the plate center is 0.5 mm, and the laser energy is 50 mJ.

The method of the resistivity measurement is simple in principle, which can be used to estimate the plasma channels quickly. However, the error is introduced through the contact resistivity between the filaments and the copper electrodes. The detail analysis about the contact resistivity can be found in Ref. [17].

13.3.3 Acoustic Diagnostics

In the plasma channels, air molecules are partially ionized through the multi-photon ionization instantly by intense laser pulses. Plasma shock waves are formed and then decay to plasma sound wave subsequently, which are the sound signals we detect in the experiments. The intensity of the acoustic wave is directly proportional to the amount of laser energy absorbed in the filaments, and thus proportional to the initial free electron density. It is demonstrated that by detecting the sound signals along the channels, the length and the electron density of the channels can be obtained [18, 19]. Therefore, the acoustic diagnostics method is based on the laser intensity and electron density in the filaments.

In our experiments, a microphone (bandwidth, 16–20 kHz together with its amplifier) is installed inside a shielding tube oriented perpendicularly toward the filament at a distance of 7 cm from it. The tube has a length of 3 cm and an inner diameter of 4 mm, restricting the directly measured filament to a length of 1.5 cm. A pulsed acoustic signal is detected by the microphone and recorded by a digital oscilloscope synchronized to a laser pulse using a photodiode that detects scattered light from the lens.

The experimental setup is shown in Fig. 13.2(3). The detailed results and discussions of the method can be found in Refs. [18, 20]. By measuring the sound signals along the channels, the channels length and distribution of electron density along the channels can be obtained conveniently. Only moving the microphone along the channels, we can measure many parameters of the plasma channels. And thus, the acoustic diagnostics method is a simple and convenient approach to detect the plasma channels nondestructively.

13.3.4 Fluorescence Detection

During the propagating of the fs laser pulse in air, it will ionize air molecules through the process of multiphoton ionization (MPI). In the plasma induced by the laser pulse, there are many ions and molecules with high excited states through dynamic multiphoton absorption in the strong laser field. It has been demonstrated that the photo-emission of N_2 and N_2^+ can be used to reflect the gross intensity distribution of the filament.

The laser pulse with a 40 mJ laser energy is focused through a 8-m lens in air. Two methods, fluorescence measurement and acoustic diagnostics, are used to probe the plasma channel simultaneously. The experimental setup is shown in Fig. 13.2(1). The PMT and the microphone scan along the plasma channel. The fluorescence radiation from the filaments is analyzed spectrally, There is no detectable continuum radiation in the spectrum. The detail discussion can be found in Refs. [14, 21]. The advantage of the simultaneous measurements by the two methods is that we can get more information from the different sides at the same time. Furthermore, both methods are nondestructive for the plasma channel.

13.3.5 Interferometry

In these experiments, the laser beam was focused in air and formed a plasma channel. A small portion of the laser beam was frequency doubled by a BBO crystal to serve as a probing laser beam. The delay between the main laser beam and the probing laser beam could be adjusted. After transmitting through the plasma channel, the probing beam was focused through a Wollaston prism and a polarizer. Finally the interferometric fringes were recorded by a charge-coupled device (CCD) camera. At the same time, we measured the length and diameter of the plasma channel with a CCD camera (512 × 512 pixels) with a pixel size of 24 μm.

In the experiment, when the delay between the main laser and the probing beam is about 600 ps, the shift of the interference fringes was measured. The detailed results can be found in Ref. [13]. The average shift number can be measured to be $\bar{D} = 1/8$. We can then obtain the average electron density

$$\bar{n}_e = 2.7 \times 10^{18}\text{cm}^{-3} \tag{13.3}$$

13.4 Nonlinear Interactions in the Filamentation

In this section, we will demonstrate some experimental studies on the spatial evolution of the filaments and the spectral analysis of the white light emission for the plasma channel. Several major filaments and small-scaled additional filaments are detected in the plasma channel. The complicated interaction process of filaments as splitting, fusion, and spreading is observed. The major filaments propagate stably, and the small scaled additional filaments can be attracted to the major filaments and merged with them. As for the supercontinuum generation, the conversion efficiency from the 800-nm fundamental to the white light has been observed to be higher for the circular polarization than for the linear polarization, when the laser intensity exceeds the threshold of the breakdown of the air.

13.4.1 Spatial Evolution of Filamentation

The dynamics of filamentation is complicated, and the physics mechanism of multiple filaments (MF) evolution in air is still not well understood. Some experimental results indicate that the filament formed by a free propagating laser pulse has millimeter-size diameter and the intensity inside the filament is around the ionization threshold of air. However, as to the prefocused laser pulses in air with a high intensity level the nonlinear effects and the interaction between filaments are extremely stronger, and consequently, the filamentation and its evolution are complicated. The behaviors of the filaments should be different in some respects from the case of free propagation of collimated laser pulse. In this part, we present our experimental investigations on the spatial evolution of filaments in air generated by prefocused fs laser pulse [22].

In our experiments, 22 mJ and 50 mJ of laser pulse energy are used, corresponding to peak power values of about 0.7 and 1.7 TW, respectively. These are significantly higher than the critical power for self-focusing. We track the profile of filaments in detail along the direction of propagation. The plasma channel is stable, and the filaments image is reproducible from shot to shot. We owe the reproducible MF pattern to the input beam ellipticity. The propagation of the filaments in the channel shows a very complicated process including the evolution from a single filament into two and three and even more distinct filaments periodically, and the multiple filaments merge into two filaments and propagate stably and fade away eventually.

Furthermore, in the beam cross section, the onset position of the filament deviates from the beam center. The input laser profile is inhomogeneous, which is amplified by modulational instability during the propagation, and the filament forms at a position where the power exceeds the critical power for self-focusing.

It is noted that the random nucleation found in the case of the collimated laser beam free propagation [23] is not observed in our experiments. The reason may be the great differences of the diameter, intensity, and the degree of modulational instability of laser beam between the two cases of prefocused and collimated laser pulses (see Section 13.7). The size of the background reservoir is also different, about several millimeters in our experiments versus tens of centimeters in the case of free-propagating laser pulses. As a result, the nonlinear effects including the self-focusing, plasma defocusing, self-phase modulation, and modulational instablility, have different degree and different influences. Therefore, the random nucleation has no chance to develop itself in our experimental condition.

13.4.2 Third-Harmonic Generation (THG)

The THG is a key process to understand important features in the propagation of the self-channeled femtosecond laser pulses, such as conical emission and spectral supercontinuum. Although the peak intensity in the plasma channel is clamped down to 10^{14} W/cm^2, it is sufficient to generate strong third-harmonic emission. In this part, we will exploit the relationship between the TH generation and the self-guiding propagation of ultra-intense laser pulses in air. It is demonstrated that the energy conversion efficiency of the fundamental wave to TH wave is very high, with maximum efficiency reaching 1.2×10^{-3}, because the interaction length is largely prolonged due to the self-guiding propagation of ultra-short laser pulse in air.

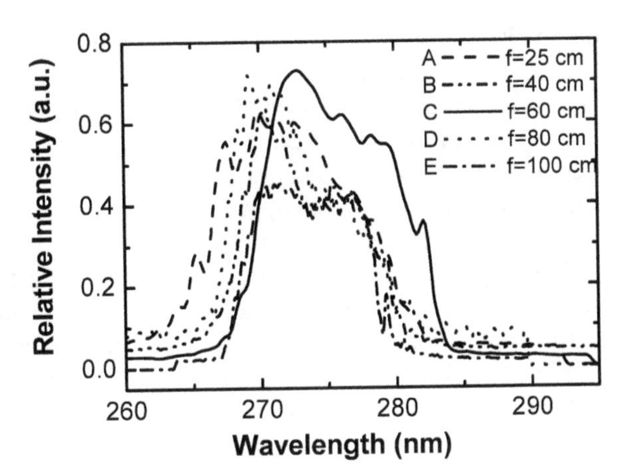

Fig. 13.3 The spectrum of the TH versus laser energy

In the experiment, the characteristics of the spectra of the TH wave are studied in various conditions. We find that there exists an optimum condition, under which maximum conversion efficiency from the fundamental wave into the TH emission can be obtained.

For a fixed energy (28 mJ) of laser pulse, the measured TH spectral profile versus various focal lengths of the lens is shown in Fig. 13.3 With increasing focal length, the TH spectra are red shifted, which is largest at $f = 60$ cm as shown by curve C in Fig. 13.3 When the focal length is longer than this focal length, the TH spectra become blue shifted. The reason for this phenomenon may be self-phase modulation. That is to say, relative red shift or blue shift may be decided by product of the laser intensity and effective interaction length, because the product of laser pulse intensity with effective interaction length is maximum at around $f = 60$ cm, which corresponds to the maximum conversion efficiency 1.2×10^{-3} [24–26].

13.5 Experimental Studies for Potential Applications

The propagation of ultra-intense femtosecond laser pulses in air has been extensively investigated recently in view of many potential applications such as in lightning control and remote sensing of the atmosphere. For potential applications, we devote our attention to lifetime prolongation, optimization of the filamentation, electric discharge, and, especially, laser propulsion.

13.5.1 Lifetime Prolongation of the Filaments

The idea to prolong the plasma channel by adding a delayed long laser pulse has been suggested by our earlier work [13] and also discussed in the context of high-voltage discharge control by Zhao et al. [27]. In this part, we present our systematic investigations on the prolongation of the lifetime of plasma channels in air generated by femtosecond laser pulses. In the experiments, a succeeding sub-nanosecond (sub-ns) laser pulse is injected into the plasma channel. We find that the lifetime of the plasma is prolonged by several tens of times than the case without the succeeding pulse. As far as we know, this is the first experimental study on the prolongation of the lifetime of the plasma channels [28].

In order to postpone electron recombination and detach electrons from O^- and O_2^-, we launch another succeeding sub-ns laser pulse to prolong the lifetime of the plasma channel and find that the lifetime is greatly prolonged.

Figure 13.4 shows the dependence of the lifetime of the plasma on the energy of the sub-ns laser pulse. The energy of the fs laser pulse is 38 mJ. The FWHM of the signal is prolonged to 200 ns when we add the succeeding sub-ns laser pulse with energy of 42 mJ, as shown in Fig. 13.4(a). More interestingly, the tail width of the signal is greatly extended to about 1.5 microseconds (μs) as shown in Fig. 13.4(b). We can see that the more the energy of the sub-ns laser pulse is injected, the longer the lifetime of plasma is prolonged. The error bar of the signals measured in our experiments includes both the standard deviation of measurements and the uncertainty in the reading of the tail. The dotted curves in Fig. 13.4 represent the exponential fitting. The fitting results show an increase of $\sim\exp(0.08 \times E_{\text{sub-ns}})$ for the FWHM and $\sim\exp(0.07 \times E_{\text{sub-ns}})$ for the tail width respectively, where $E_{\text{sub-ns}}$ is the energy of the sub-ns pulse in mJ. They are quite similar.

We also measure the lifetime of the plasma by changing the delay time between the fs and the sub-ns laser pulse to 20 ns. The experimental results demonstrate that the 10 ns delay time is much better than 20 ns. The characteristic relaxation time of the electron attachment is about 16 ns in the standard air conditions [29], so the electrons in the plasma channel have no time to attach to

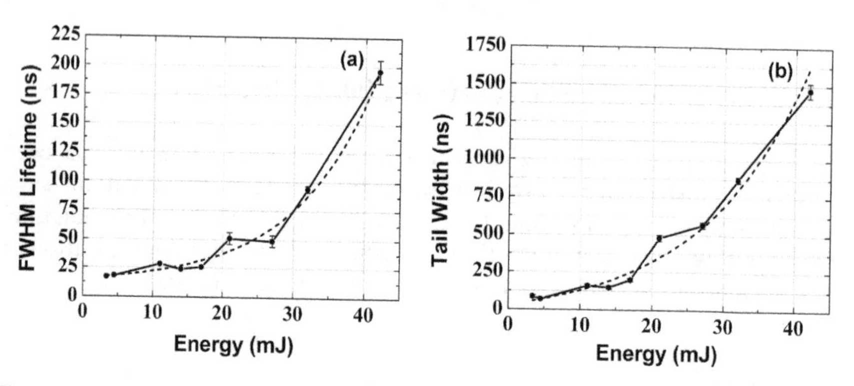

Fig. 13.4 The lifetime of plasma versus the energy of a sub-ns laser pulse. The fs laser pulse energy is 38 mJ and the time delay is 10 ns. Part (**a**) represents the lifetime at the FWHM; part (**b**) represents the lifetime at the tail width of the fluorescence signal. The dotted curves represent the exponential fitting

neutral molecules when the succeeding laser pulse arrives with a 10 ns delay. Otherwise, when the delay time is greater than 16 ns, for example 20 ns, the attachment of electrons to neutral molecules causes the electron density to decay exponentially, so that the electron density is too low to maintain a long lifetime [29].

13.5.2 *Optimization of Multiple Filaments (MF)*

There are several studies for modulational instability (MI) of MF. In this part, we propose an alternative approach to form more stable and longer filaments. We demonstrated that when the cross section of the plasma channel is restricted by a pinhole, remarkably robust and prolonged filaments can be attained. The intensity and propagation distance of the filament are controllable by varying the aperture of the pinhole. Our simulations retrieve the phenomena that the filaments can be prolonged and stabilized, which are in qualitative agreement with experimental results.

In the experiments, a laser pulse of energy 15 mJ is focused by an $f = 200$ cm convex lens. A CCD camera records the beam cross-section at 520 cm distance from the lens. Another CCD camera records the beam along the laser axis. Filters with a frequency range of 350 nm $< \lambda <$ 500 nm are used at both CCD cameras to reduce the scattering noise from the pump laser. A pinhole in a copper plate 4 mm thick is centered on the beam axis at $Z_0 = 190$ cm. Thus, the axial and transverse profiles of the plasma filaments are measured directly by the two CCD cameras.

To control the diameter of the channel, a pinhole with variable apertures is placed at the center of the laser beam axis. In our experiments condition, the pinhole position $Z_0 = 190$ cm is the optimized choice to study the filaments. We find that if the distance between the lens and the pinhole is smaller, there is no chance for filaments to develop into mature ones. The filaments will decay fast due to the lack of laser energy. The reason should be the energy background surrounding filaments includes too much energy, whereas the pinhole blocks too much energy. However, if it is too large, some dynamic interactions have already harmfully influenced the MF. As a result, the MF cannot be controlled effectively by the pinhole setup. Therefore, we chose $Z_0 = 190$ cm of the pinhole position, which results in a remarkable phenomenon that the intensity of the fluorescence from the filaments becomes brighter and the propagation distance longer when a pinhole between 0.8 and 1.8 mm in diameter is inserted.

Figure 13.5 shows the filaments patterns for different diameters of the pinhole. The first image is for a free propagation without any pinhole. We see that the beam is dispersed into a relatively large area showing several bright spots. The positions of these spots are unfixed from shot to shot. As we limit the plasma channel by pinholes of diameters from 2.0 to 0.8 mm, the laser beam cross section becomes smaller, and the number, position, and intensity of the

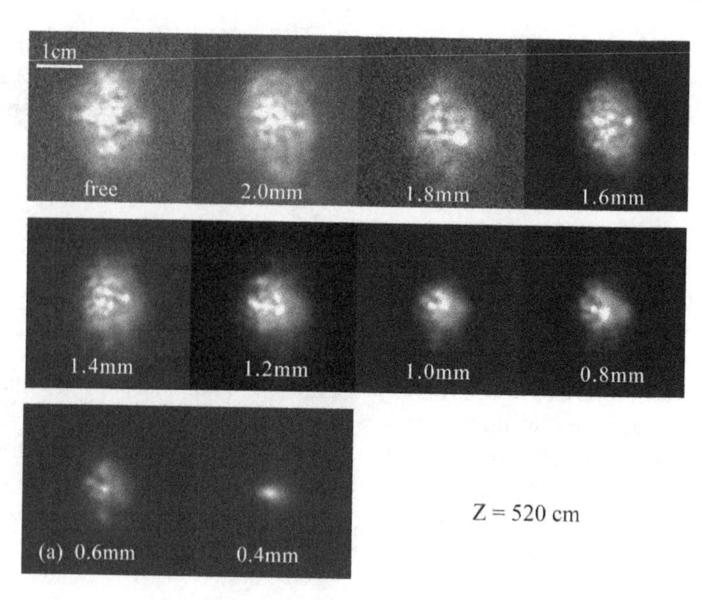

Fig. 13.5 Typical transverse filamentation patterns at $Z = 520$ cm versus the diameter of the pinhole

filaments change. When the diameter is 1.6 mm, the filaments are robust and distinct from one another. With further decrease of the diameter, the number of filaments becomes less. When the pinhole diameter is 0.8 mm, there are only two filaments, one large and one small. All MF patterns are reproducible. Finally, for $D < 0.6$ mm we observe only a dispersed light spot. We can conclude that the MF patterns for $D = 1.6$–0.8 mm are highly stable and reproducible. The number and locations of the filaments hardly change from shot to shot.

The pinhole aperture plays an important role in controlling the filaments. On the one hand, it reduces the turbulence in the periphery of the plasma channel and leads to relatively stable MF. On the other hand, it should allow enough energy through to feed the remaining filaments. In our experiments, the optimum diameter of the pinhole is 1.2 mm, and a deterministic MF pattern is obtained. This method enables one to achieve optimized or otherwise desired MF patterns, which are important for many research and practical applications.

13.5.3 Laser-Guided Discharge

The laser plasma channel is continuous and has low resistivity, and so it will be more efficient to trigger and guide a long gap electric discharge. Many laboratories have reported air discharges triggered by laser plasma channels [27–34]. In this part, we present our investigation of fs laser pulses triggering and guiding quasi-stationary high-voltage discharges in 3–23 cm gaps. We studied the relation of the discharges breakdown voltage and the length of gaps, and get the speed of the leader by some simple measurements.

For the experiment reported here the pulse duration is 50 fs, and the energy is only 40 mJ. The high-voltage system can produce a positive quasi-stationary high voltage up to 180 kV. The fs laser pulses are focused with an f = 4 m lens to produce plasma channels. A circular plane of 55 cm diam is placed vertically as a high-voltage anode. And a ground sphere of 1 cm diam is placed tens of centimeters away from the plane. There are apertures of 2 mm diam both in the center of the plane and the sphere. The laser plasma channel connects the plane and the sphere to trigger the high-voltage discharge.

The discharges are shown in Fig. 13.6. The bright channels in the photos are the strokes of discharges. Fig. 13.6(a) is a random irregular natural discharge channel. Fig. 13.6(b) is a straight laser triggered discharge channel, which is along the path of the laser beam.

First we investigated the triggering and guiding ability of a laser plasma channel, and measured the discharge breakdown voltage both of natural discharges and laser-triggered discharges. We adjust the gap from 3 up to 23 cm. The breakdown voltage is shown in Fig. 13.7. Laser plasma channels have lowered the breakdown voltage a lot. As with a 5-cm gap, the laser plasma channel reduced the breakdown voltage from 110 to 45 kV by about 60%. And at a breakdown voltage of 110 kV, the laser plasma channel has increased the discharge length from 5 to 15 cm.

We also investigated the time characteristics of the discharges by measuring the white light irradiated from the discharges. We use two photodiodes to measure the laser pulses and the white light simultaneously. We find that there are delays of several microseconds between the laser pulse and the white light. And the delay time increases with the prolongation of the discharge gaps. It means the laser-triggered discharges also have four steps: inception, stepped leader, connection and return stoke.

The difference between a natural discharge and a laser-triggered discharge is when the laser has triggered the inception, the stroke will propagate along the laser plasma channel. Figure 13.8 shows the delay signals between the laser

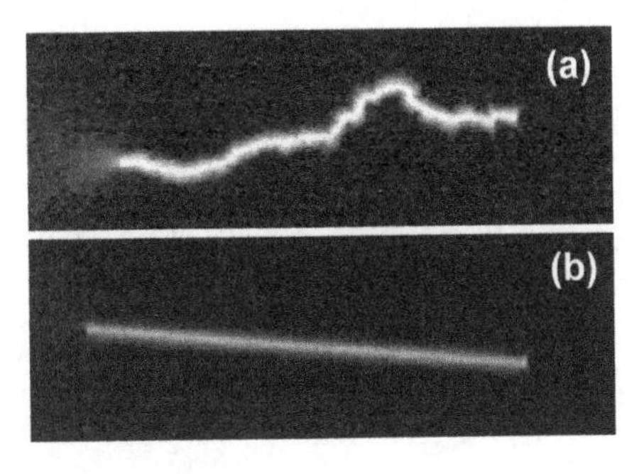

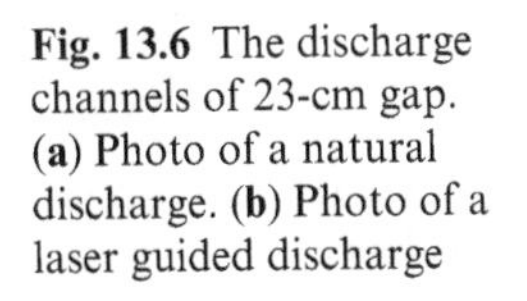
Fig. 13.6 The discharge channels of 23-cm gap. (**a**) Photo of a natural discharge. (**b**) Photo of a laser guided discharge

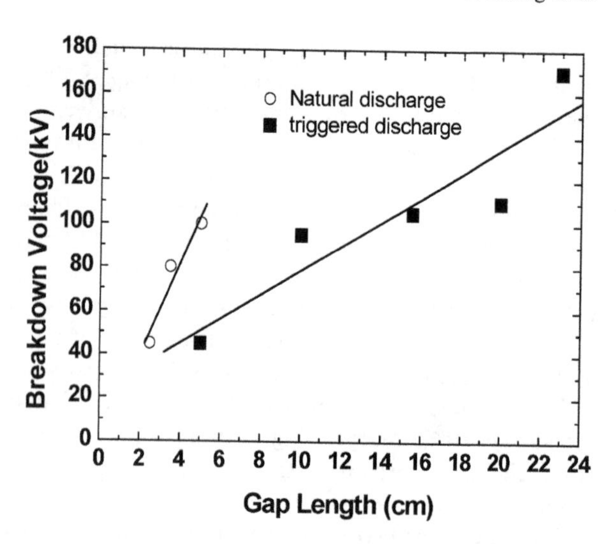

Fig. 13.7 Breakdown voltage as a function of the gap length. The solid squares represent breakdown voltage of laser guided discharges. The open circles represent the breakdown voltage of natural discharges

pulse and the discharge white light in the oscilloscope. Figure 13.8(a) is the signal of a 3-cm gap discharge, and Fig. 13.8(b) is the signal of a 11-cm gap discharge. Because the photodiodes gather light signals from different positions, there is an inceptive delay between the signals; this delay is 92 ns by our measuring. This 92 ns should be decreased from the delay time shown in the oscilloscope. The delay time increases from 464 to 901 ns, when the gap is lengthened from 3 to 15 cm.

The peaks at the ascent of the white light signals are the disturbance caused by the return stroke electric irradiation. These peaks show the time when the return stroke happens and also indicate that the duration of the electric current is about 0.5 μs.

The stepped leader needs longer time to reach the ground when the gap is longer. So the delay between the laser pulse and the white light increases. We

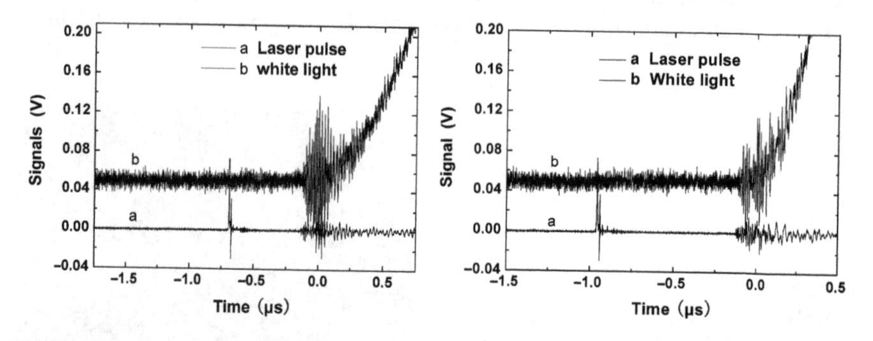

Fig. 13.8 The delay between laser pulses and discharge white light shown in the oscilloscope. (**a**) In a 3-cm gap, the delay is 737 ns. (**b**) In a 11-cm gap, the delay is 1130 ns

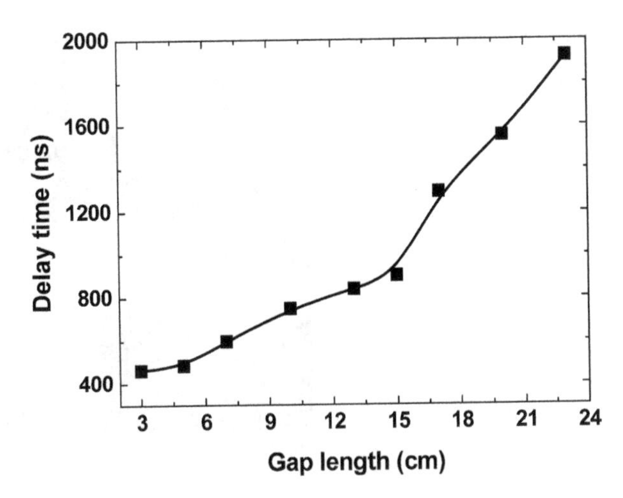

Fig. 13.9 The delay time of different gap length

have measured the delays of different gap length, as shown in Fig. 13.9. The delays increase along with the increase of the gap length. By analyzing the delays and the gap length, we can estimate the speed of the stepped leader to be about 10^7 cm/s. This is consistent with the result in Ref. [30].

13.5.4 Laser Propulsion

Research in recent years on laser plasma propulsion has gained much attention due to its many advantages over other conventional methods of producing thrust [35–38]. Although the concept of laser propulsion is not new, we believe that the idea to use self-guided filaments is very elegant. In this part, we propose a new concept of "laser plasma channel propulsion" in which a long plasma channel generated by intense fs laser pulses in air is served as a propulsive power. Preliminary experiments demonstrate that the plasma channel can continuously propel a paper airplane in a long distance without complicated optical focus.

It is demonstrated that such a plasma channel can continuously propel a light paper airplane without complicated focusing optics, as shown in Fig. 13.10. The paper airplane is located on an air-cushion track and sustained by the airflow from an air-compressor. For 36-mJ laser energy, a paper airplane weighing 1.32 g can be accelerated to 2.05 cm/s^2.

One advantage of using plasma channels to propel a craft is that within a long distance the craft can be steadily propelled without complicated focusing optics. Another advantage of the plasma channel propulsion is that in the plasma channel the air is served as the propellant. Furthermore, it is demonstrated that the laser plasma comes from the air, not from the target material. In this case, the propulsive source only comes from the detonation wave generated with the air ionization. From this point of view, the plasma channel is a very

Fig. 13.10 Flight trajectories of the paper airplane on the air-cushion track

useful source to realize a long propulsion in the atmosphere. With the development of this laser technology and further research into laser propagation, a very long distance propulsion with high coupling coefficient can be expected.

13.6 Long-Distance Filamentation

In this section, light filaments greater than 158 m long are generated by intense fs laser pulses in air. At the same time, the "rainbow-like" SC generation is observed when a pulse with an appropriate negative chirp propagates in air. The conversion efficiency from the 800-nm fundamental to white light is observed to be the highest for 257-fs pulses with an initial negative chirp. Furthermore, the effects of an energy reservoir on the filaments core are investigated by inserting a diaphragm into a selected single filament path to block the propagation of the energy reservoir. The evolution of filamentation is studied experimentally and numerically as a function of diaphragm size. Both experimental and numerical results show that for light filaments that are hundreds of meters long with millimeter size, the energy reservoir can be extended to about 10 mm from the filament core.

13.6.1 Chirp-Dependent Propagation of Filamentation

Despite the importance of controlling SC generation in air [39–43], the subject has not been systematically studied experimentally. In this chapter, we investigate the effect of the initial laser pulse chirp on laser filamentation and SC generation. We find that there exists an optimal pulse duration for maximizing SC generation at a long distance, and that the onset and length of the filaments are closely related to the initial laser pulse chirp.

In the experiments, a laser pulse of energy 50 mJ propagates freely in air and forms a long plasma channel in several filaments that can be observed directly by the naked eye, and is monitored by a digital camera at 150 m. Please note: the distance of 158 m is the maximum length for our measurements, not the maximum length of the filamentation. The actual length of the filamentation is estimated to be around 500 m.

Photos were taken from a white screen positioned in the plane orthogonal to the beam path. Figure 13.11 shows the growth of the filaments and the generated white light over a 158-m distance for three representative laser durations of + 200, –257, and –390 fs, respectively (the sign of the pulse duration indicates the sign of the applied laser chirp). The filament patterns change greatly as the pulse duration changes; so do the filament number and the SC emission. Furthermore, the onset and length of the filaments are also affected by the initial pulse duration. It is found that an increase in the initial pulse duration leads to an increase in the distance between the filament formation point and the output aperture of the laser system [44].

The effect of the initial chirp on the formation of filaments and SC generation are determined mainly by two factors [40, 41]. On the one hand, the peak power decreases if pulse duration is continuously increased at constant pulse energy. On the other hand, because of the initial negative chirp and group velocity dispersion (GVD) in air, the power in the pulse time layer will increase with the distance [40].

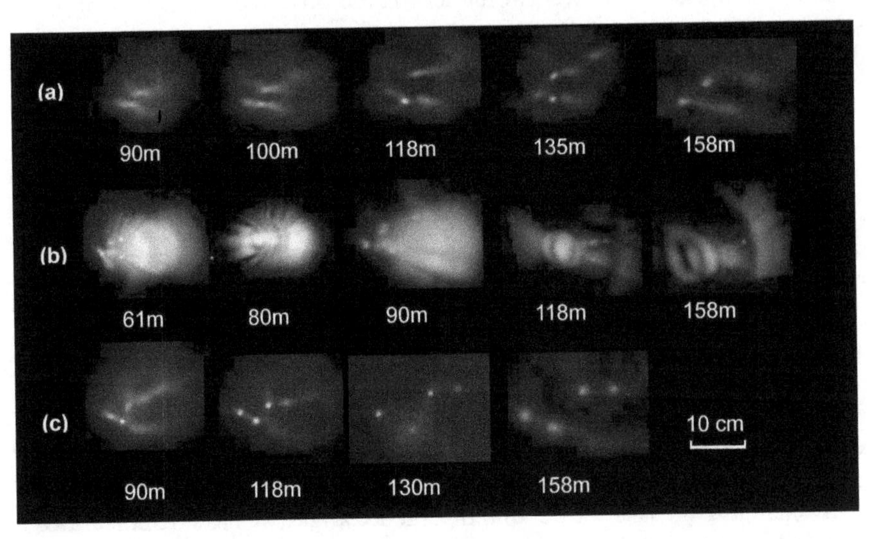

Fig. 13.11 Measured beam profile at various distances with three representative pulse durations of + 200 fs (**a**), –257 fs (**b**), and –390 fs (**c**), respectively. The sign of pulse duration means the sign of applied laser chirp

By adjusting the initial pulse duration, we can achieve the desired filamentation and SC for different applications, such as in LIDAR, where the filament pattern or the white light is needed. Furthermore, we can also expand the laser beam, using an adaptive optics system for example [45], to control the initial focus or divergence of the beam, as well as to control the initial chirp of the pulses to study the propagation of intense fs laser pulses in the atmosphere.

13.6.2 Divergence Angle–Dependent Propagation of Filamentation

Most of the applications on laser plasma channels require long and continuous filaments. The spatial position where the filaments start to form is also an important parameter and should be controllable. In this part, we studied the effects of the energy and divergence angle of fs laser pulses on laser self-focusing and filamentation. It is shown that by controlling the laser energy and divergence angle, one can realize a long-distance propagation of filaments starting at precise position.

We measured the positions (z_f) where the filaments were formed for different laser power and divergence angle. A rapid increase of z_f was observed when we increased the divergence of the laser beam. One can control the position of the filamentation by adjusting the laser divergence angle and the laser power because z_f is very sensitive to the laser divergence angle and laser power, especially when the former is positive.

We now present a theoretical model for the observed behavior [46]. Considering a Gaussian laser beam propagating in air, we obtain the focusing distance

$$z_m \sim -\left(gb_0 + \sqrt{g^2 b_m^2 - 2\alpha_1}\right) / \left(g^2 + 2\alpha_1 b_0^2\right), \tag{13.4}$$

where $g(= \tan\theta)$ is the divergence of the beam, b_0 is initial beam size, b_m is the focal spot size, $\alpha_1 = 1 - P_0/P_N$, P_0 is the laser power and P_N is the critical power for self-focusing (~2 GW at 800 nm). The space is normalized by λ_0, where λ_0 is the laser central wavelength. Assuming that the position where the filaments appeared was at the self-focusing position, we can compare the calculation results with the experiment.

The solid line in Fig. 13.12 shows the calculated z_m when for $P_0 = 40$ GW and $b_0 = 2.4$ mm. The squares are the filamentation positions of the 300-GW laser at different divergences. The calculation well fits the experimental results.

By changing the laser power and divergence angle, we have demonstrated a new way to control the positions of the filaments besides GVD and pulse duration control. A much longer filamentation is possible by simultaneous control of spatial and temporal characteristics of the laser beam.

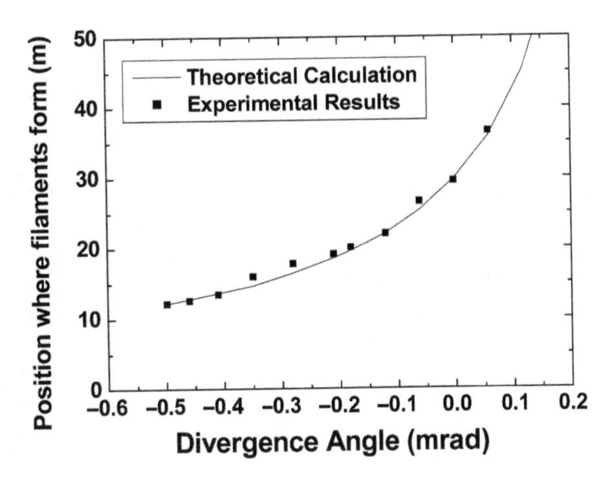

Fig. 13.12 Comparison of the calculation and experimental results. The solid line shows the calculated result for $P_0 = 40$ GW and $b_0 = 2.4$ mm. The squares are the measured filamentation position of a 300-GW laser beam

13.6.3 Energy Reservoir

Several experiments have shown that the filament itself contains only about 6–10% of the total pulse energy, while most of the energy is located in the background. The energy background, that is, the energy reservoir, plays a crucial role in the filament formation and long-distance propagation. In this section, we experimentally and numerically investigate the properties of background reservoir of mm-size light filaments generated by free-propagating fs laser pulses. We demonstrate that the spatial size and energy of the reservoir strongly affects the filament's evolution. The background reservoir with about ten times more than the energy of the filament core is necessary to feed a single filament propagation stably, as compared with an undisturbed free filament. Numerical simulations are also performed to investigate the effect of the energy reservoir on filament formation, and the results of numerical simulations are in good agreement with our experimental observations.

We adjust the grating pair in the compressor of the XL-II laser system to obtain 350-fs pulses with a negative chirp, in order to be for the filaments long-distance propagation in air. In the experiments, a laser pulse with an energy of 60 mJ propagates freely in air and forms a long plasma channel containing several filaments that can be observed directly by the naked eye. Filaments begin to form at about 39 m after the compressor of the laser system. A diaphragm is used at 46 m to single out a filament and block energy reservoir. The diameter of diaphragm can be varied from 2 to 20 mm. We measure the beam profiles using a scientific CCD camera (512×512 pixels) with a pixel size of 24 μm, from a white screen positioned in the beam path at several propagation distances. In our experiments, the multiple filamentation is stable, and the beam pattern is well repeated from shot to shot.

Figure 13.13 details the evolution of the filament singled by the diaphragm with different diameters. The position of pinholes is taken as $Z = 0$ m. When

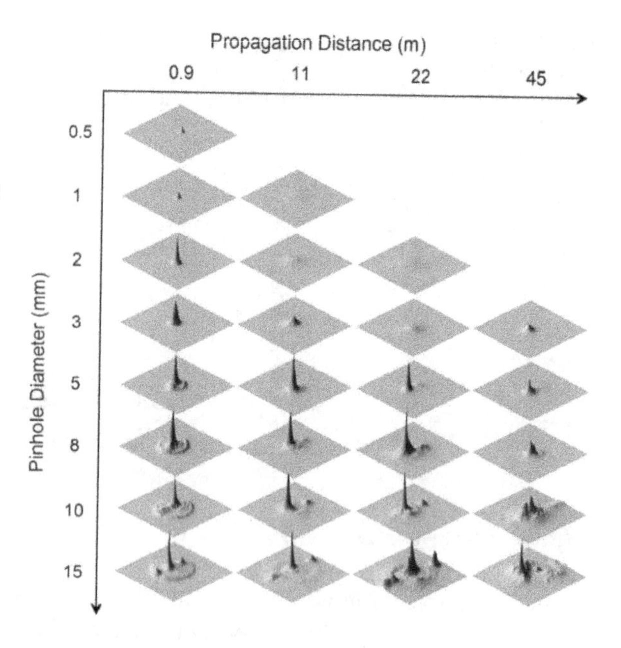

Fig. 13.13 Three-dimensional intensity profiles of the clipped filament for different diameter pinholes and propagation distances. The position of pinholes is signed $Z = 0$ m. Frame transverse size: $20 \times 20\ \text{mm}^2$, the intensity units are the same in all insets

the diameter is relatively small, 2 mm for example, the filament is quickly terminated beyond the pinhole. At 11 m, only a large weak laser spot is observed, whose peak intensity is two orders of magnitude smaller than that of free propagating filament. When a 3-mm pinhole is used, the clipped filament decays and then tries to refocus at $Z = 45$ m.

The energy to support the refocusing of the filament comes from the low-energy reservoir. When the diameter varies from 0.5 to 15 mm, the filament beyond pinholes becomes more stable and can survive longer distance. The filament cannot propagate as the free-filament until the diameter of the diaphragm increases to 20 mm. Figure 13.14 shows the comparison between two beam profiles at 45 m, where (a) is the filament clipped by a pinhole of 20 mm in diameter, and (b) is the profile of the whole free propagating beam. The filament with a maximum intensity in Fig. 13.14 is what we are interested in. The filament is not significantly affected by the diaphragm with 20-mm diameter. Their peak intensities are almost the same. Thus, there is almost no difference of the filament between a free propagation and clipped by a 20-mm pinhole. Therefore, the energy reservoir around 10 mm far away from the filament core can support the propagation of filaments.

Furthermore, our experiments and simulation results (not shown here) indicate that it is impossible to propagate for a single filament without the support of an energy reservoir, and the useful energy reservoir can be extended to about 10 mm away from the filament core.

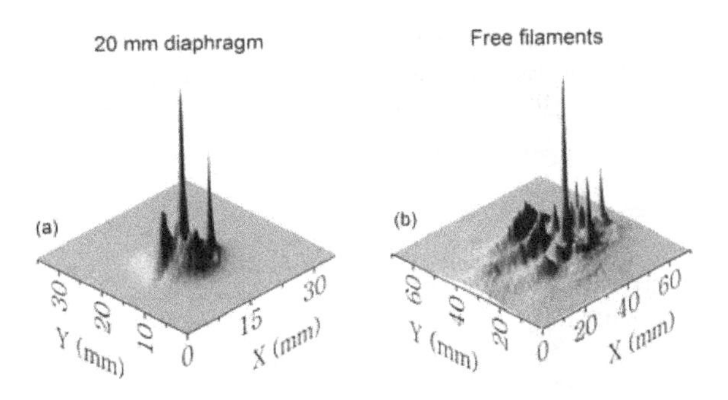

Fig. 13.14 Comparison between two beam profiles at 45 m (after the pinhole): (**a**) with a pinhole of 20 mm diameter, (**b**) free filaments propagation

13.7 Comparison of Filamentation in Focused and Unfocused Laser Beams

The traditional method for investigating MF in the laboratories is to focus the fs laser using a convex lens. MF are formed before the geometrical focus of the lens. For convenience, this type of filaments shall be called pre-focused-MF. However, recent experiments demonstrate that freely propagating fs laser pulses without initial focusing can form MF with very long propagation distance [23, 42, 47]. We shall call such filaments free-MF. The characteristics of this type of MF are quite different from those of the pre-focused-MF. The understanding of MF is still not quite clear. In this section, we experimentally compare the properties of the pre-focused-MF and free-MF. Here, the free-MF propagation implies relatively wide-beam, long-distance propagation for several hundreds of meters up to kilometers, instead of the propagation of a small-size transverse laser beam (or contracted beam) using a telescope system consisting of convex and concave lens [48]. We have found that almost all the characteristics of the MF are quite different for the two cases. In particular, free-MF show unique and unexpected properties.

For investigating prefocused propagation, tens of millijoule (22 mJ in our experiments) is sufficient. Nonlinear effects such as SC generation, dynamic MF interactions, and fluorescence and acoustic radiation, are significant. The laser pulses are focused by an $f = 4$ m lens with geometrical focal length about 5.05 m in air because the laser beam used has a divergence angle in our experiments. For studying free propagation, the laser system runs with less than 200 mJ of laser energy. The grating pair in the compressor chamber is adjusted so that the fs pulse has a negative chirp, favorable for long-distance propagation of the MF in air. The effect of chirp on the free-MF has been studied by our recent experiments and other studies [47, 49]. The MF evolution is recorded by a charged-coupled device (CCD) camera (512 × 512 pixels) with a pixel size of 24 μm and from a white screen positioned in the beam path at different propagation distances. The SC generation is analyzed by a spectrometer. Acoustic and fluorescence measurements are also used in the MF diagnostics.

Table 13.1 Characteristics of multiple filaments generated by the pre-focused and free propagation laser pulse. D is diameter; I is intensity; ρ is electron density; L is filament length; *F-B* denotes the filament–background interaction, and *F-F* denotes the filament–filament interaction

	D(μm)	I(W/cm^2)	P(cm^{-3})	L(m)	Stability	Interactions
Prefocused-MF	10^2	$10^{13\sim14}$	$10^{16\sim18}$	$10^{0\sim1}$	Reproducible	F-B, F-F
Free-MF	10^3	$10^{12,a}$	$10^{12,b}$	$10^{1\sim3}$	Not good	F-B

[a]References [42, 47].
[b]Reference [47].

Recent experimental results indicate that the free-MF have mm size diameter, and the intensity is near the ionization threshold of air, corresponding to very low density plasma generation. However, for the prefocused-MF, many experiments have shown that the typical diameter of a filament is about 100 μm and the light intensity inside is about 10^{13}–10^{14} W/cm^2, corresponding to an electron density of 10^{16}–10^{18} cm^{-3}. The nonlinear effects and the interaction among the filaments are extremely strong, so that the filamentation process and its evolution are complex. The behaviors of the prefocused-MF are thus different in many respects from that of the free-MF.

The detailed results and discussions will appear in other place. Here we just show the final comparisions.

Table 13.1 summarizes the characteristics of MF. The differences in the diameter, intensity, electron density, length, stability, interaction type, and SC generation for the two cases are pointed out.

We have also used acoustic and fluorescence measurements to detect the MF. We can get strong sound and fluorescence signals in the pre-focused-MF case. However, in the free-MF case, we cannot get obvious signals from both methods. The reason could be the too-low electron density inside the free-MF because of the weak ionization of the air. It seems so far there is no promising method for measuring the precise laser intensity as well as the electron density in the free-MF. Furthermore, the mechanism underlying the free-MF including its interaction and evolution also needs further studies.

Each type of MF has its own advantages. The pre-focused-MF is stable from shot to shot with tiny-scaled intensity and strong SC generation. The free-MF can deliver high energy to remote desired destinations. Thus, one can choose different laser energies, chirp conditions, and focal lengths of the lens (if any), to achieve different purposes.

13.8 Conclusions

The self-organized propagation of femtosecond laser filamentation in air has been investigated. Many nonlinear phenomena and their underlying physics have been revealed. Starting from the diagnostic techniques, the complicated interactions among filaments, strong SC and TH generation, and long-distance filamentation have been reviewed. Especially, the lifetime prolongation and optimal control of the filaments, the study on the energy background, the chirp-dependent SC generation, and long-distance filamentation open the promising perspectives for many applications.

It should be noted that our results presented in this paper cannot cover all scopes of this scientific field. Many other potential applications, such as white light LIDAR, femto-LIBS, and propagation in adverse conditions can be found in the corresponding literatures.

Acknowledgements This work was supported by the National Natural Science Foundation of China under Grant Nos. 10390161, 60321003, 60478047, 10734130, 10634020, and 60621063, National Basic Research Programme of China (No. 2007CB815101), and the National Hi-tech ICF Programme.

References

1. R.Y. Chiao, E. Garmire, C.H. Townes: Self-trapping of optical beams, *Phys. Rev. Lett.* **13**(15), 479–482 (1964).
2. A. Braun, G. Korn, X. Liu et al.: Self-channeling of high-peak-power femtosecond laser pulses in air, *Opt. Lett.* **20**(1), 73–75 (1995).
3. A. Brodeur, C.Y. Chien, F.A. Ilkov et al.: Moving focus in the propagation of ultrashort laser pulses in air, *Opt. Lett.* 22(5), 304–306 (1997).
4. J.H. Marburger, W.G. Wagner: Self-focusing as a pulse-sharpening mechanism, *IEEE J. Quantum Electronics* **QE-3**, 415–416 (1967).
5. V.N. Lugovoi, A.M. Prokhorov: A possible explanation of the small-scale self-focusing filaments, *JETP Lett.* **7**, 117–119 (1968).
6. M.M. Loy, Y.R. Shen: Small-scale filaments in liquids and tracks of moving foci, *Phys. Rev. Lett.* **22**, 994–997 (1969).
7. M. Mlejnek, E.M. Wright, and J.V. Moloney: Moving-focus versus self-waveguiding model for long-distance propagation of femtosecond pulses in air, *IEEE J. Quantum Electronics* **35**(12), 1771–1776 (1999).
8. M. Mlejnek, E.M. Wright, J.V. Moloney: Dynamic spatial replenishment of femtosecond pulses propagating in air, *Opt. Lett.* **23**(5), 382–384 (1998).
9. S. Tzortzakis, L. Bergé, A. Couairon et al.: Breakup and fusion of self-guided femtosecond light pulses in air, *Phys. Rev. Lett.* **86**(24), 5470–5473 (2001).
10. Y. Silberberg: The collapse of optical pulses, *Opt. Lett.* **15**, 1282 (1990).
11. T.T. Xi, X. Lu, J. Zhang: Interaction of light filaments generated by femtosecond laser pulses in air, *Phys. Rev. Lett.* **96**(2), 025003 (2006).
12. S. Talebpour, B. Prade, M. Franco et al.: Time-evolution of the plasma channels at the trail of a self-guided IR femtosecond laser pulse in air, *Opt. Commun.* **181**(1–3), 123–127(2000).
13. H. Yang, J. Zhang, Y. J. Li et al.: Characteristics of self-guided laser plasma channels generated by femtosecond laser pulses in air, *Phys. Rev. E* **66**(1), 016406 (2002).
14. A. Talebpour, S. Petit, S.L. Chin: Refocusing during the propagation of a focused femtsecond Ti: sapphire laser pulse in air, *Opt. Commun.* **171**(4–6), 285–290 (1999).
15. S. Tzortzokis, G. Méchain, G. Patalano et al.: Coherent subterahertz radiation from femtosecond infrared filaments in air, *Opt. Lett.* **27**(21), 1944–1946 (2002).
16. S.A. Hosseini, B. Ferland, S.L. Chin: Measurement of filament length generated by an intense femtosecond laser pulse using electromagnetic radiation detection, *Appl. Phys. B* **76**(5), 583–586 (2003).

17. Z. Zhang, J. Zhang, Y.T. Li et al.: Measurements of electric resistivity of plasma channels in air, *Acta Phys. Sin.* **55**(1), 357–361 (2006).
18. Z.Q. Hao, J.Yu, J. Zhang et al.: Acoustic diagnostics of plasma channels induced by intense femtosecond laser pulses in air, *Chin. Phys. Lett.* **22**(3), 636–639 (2005).
19. J. Yu, D. Mondelain, J. Kasparian et al.: Sonographic probing of laser filaments in air, *Appl. Opt.* **42**(36), 7117–7120 (2003).
20. Z.Q. Hao, J. Zhang, J. Yu et al.: The comparison study of diagnostics of light filaments in air, *Sci. China, Ser. G. Phys., Mechan. & Astron.* **49**(2), 228–235 (2006).
21. Z.Q. Hao, J. Zhang, J. Yu et al.: Fluorescence measurement and acoustic diagnostics of plasma channels in air, *Acta Phys. Sin.* **55**(1), 299–303 (2006).
22. Z.Q. Hao, J. Zhang, X Lu et al.: Spatial evolution of multiple filaments in air induced by femtosecond laser pulses, *Opt. Exp.* **14**(2), 773–778 (2006).
23. L. Bergé, S. Skupin, F. Lederer et al.: Multiple filamentation of terawatt laser pulses in air, *Phys. Rev. Lett.* **92**(22), 225002 (2004).
24. R.R. Alfano (ed.): *The Supercontinuum Laser Source*, Springer-Verlag, New York (1989).
25. P.B. Corkum, C. Rolland: Femtosecond continua produced in gases, *IEEE J. Quantum Electron.* **25**(12), 2634–2639 (1989).
26. Z.Q. Hao, J. Zhang, Z. Zhang et al.: Third harmonic generation in plasma channels in air induced by intense femtosecond laser pulses, *Acta. Phys. Sin.* **54**(7), 3173–3177 (2005).
27. X.M. Zhao, J.-C. Diels, C.Y. Wang et al.: Femtosecond ultraviolet laser pulse induced lightning discharges in gases, *IEEE J. Quantum Electron.* **31**(3), 599–612 (1995).
28. Z.Q. Hao, J. Zhang, Y.T. Li et al.: Prolongation of the fluorescence lifetime of plasma channels in air induced by femtosecond laser pulses, *App. Phys. B* **80**(4–5), 627–630 (2005).
29. X. Lu, T.T. Xi, Y.J. Li et al.: Lifetime of the plasma channel produced by ultra-short and ultra-high power laser pulse in the air, *Acta Phys. Sinica* **53**(10), 3404–3408 (2004).
30. B. La Fontaine, D. Comtois, C.-Y. Chien et al.: Guiding large-scale spark discharges with ultrashort pulse laser filaments, *J Appl. Phys.* **88**(2), 610–615 (2000).
31. N. Khan, N. Mariun, I. Aris et al.: Laser-triggered lightning discharge, *New J Phys* **4**, 61. 1–61.20 (2002).
32. D. Comtois, C.Y. Chien, A. Desparoi et al.: Triggering and guiding leader discharges using a plasma channel created by an ultrashort laser pulse, *Appl. Phys. Lett.* **76**(7), 819–821 (2000).
33. R. Ackermann, K. Stelmaszczyk, P. Rohwetter et al.: Triggering and guiding of megavolt discharges by laser-indued filaments under rain conditions, *Appl. Phys. Lett.* **85**(23), 5781–5783 (2004).
34. G. Méjean, R. Ackermann, J. Kasparian et al.: Improved laser triggering and guiding of meqavolt discharges with dual fs-ns pulses, *Appl. Phys. Lett.* **88**(2), 021101 (2006).
35. S.A. Metz: Impulse loading of targets by subnanosecond laser pulses, *Appl. Phys. Lett.* **22**(5), 211–213 (1973).
36. T. Yabe, C. Phipps, M. Yamaguchi et al.: Microairplane propelled by laser-driven exotic target, *Appl. Phys. Lett.* **80**(23), 4318–4320 (2002).
37. Z.Y. Zheng, X. Lu, J. Zhang et al.: Experimental study on the meomentum coupling efficiency of laser plasma, *Acta Phys. Sin.* **54**(1), 192–196 (2005).
38. Z.Y. Zheng, J. Zhang, Z.Q. Hao et al.: Paper airplane propelled by laser plasma channels generated by femtosecond laser pulses in air, *Opt. Exp.* **13**(26), 10616–10621 (2005).
39. J. Kasparian, R. Sauerbrey, D. Mondelain et al.: Infrared extension of the super continuum generated by femtosecond terawatt laser pulses propagating in the atmosphere, *Opt. Lett.* **25**(18), 1397–1399 (2000).
40. I.S. Golubtsov, V.P. Kandidov, O.G. Kosareva: Initial phaser modulation of a high-power femtosecond laser pulse as a tool for controlling its filamentation and generation of a supercontinuum in air, *Quantum Electron.* **33**(6), 525–530 (2003).

41. V.P. Kandidov, I.S. Golubtsov, O.G. Kosareva: Supercontinuum sources in a high-power femtosecond laser pulse propagating in liquids and gases, *Quantum Electron.* **34**(4), 348–354 (2004).
42. G. Méchain, C.D. Amico, Y.-B. André et al.: Range of plasma filaments created in air by a multi-terawatt femtosecond laser, *Opt. Comm.* 247(1–3). 171–180 (2005).
43. F. Théberge, W. Liu, Q. Luo and S.L. Chin: Ultra broadband continuum generated in air (down to 230nm) using ultrashort and intense laser pulses, *Appl. Phys. B* **80**(2), 221–225 (2005).
44. Z.H. Wang, Z.Q. Hao, Z. Zhang et al.: Effects of temporal chirp on laser filamentation in air, *Acta Phys. Sin.* **56**(30), 1434–1438 (2007).
45. Z. Jin, J. Zhang, M.H. Xu et al.: Control of filamentation induced by femtosecond laser pulses propagating in air, *Opt. Express* **13**(25), 10424–10430 (2005).
46. W. Yu, M.Y. Yu,J. Zhang et al.: Long-distance propagation of intense short laser pulse in air, *Phys. Plasmas* **11**(11), 5360–5363 (2004).
47. G. Méchain, A. Couairon , Y.-B. André et al.: Long-range self-channeling of infrared laser pulses in air: a new propagation regime without ionization, *Appl. Phys. B* **79**(3), 379–382 (2004).
48. Q. Luo, S.A. Hosseini, W. Liu et al.: Effect of beam diameter on the propagation of intense femtosecond laser pulses, *Appl. Phys. B* **80**(1), 35–38 (2004).
49. R. Nuter, S. Skupin, L. Bergé: Chirp-induced dynamics of femtosecond filaments in air, *Opt. Lett.* **30**(8), 917–919 (2005).

Chapter 14
The Physics of Intense Femtosecond Laser Filamentation

See Leang Chin, Weiwei Liu, Olga G. Kosareva, and Valerii P. Kandidov

Abstract When a powerful femtosecond laser pulse propagates in a transparent optical medium, be it a gas or a condensed medium, the pulse will self-focus into a series of self-foci, giving rise to the perception of a filament. This universal nonlinear propagation phenomenon is currently an interesting research topic at the forefront of applied physics and attracts more and more people into this field. In this chapter some fundamental physical concepts underlying femtosecond laser filamentation are discussed. These include slice-by-slice self-focusing, intensity clamping, self-transformation into a white light laser pulse (supercontinuum generation), background reservoir and multiple filamentation competition. Some important potential applications are also briefly mentioned.

14.1 Introduction

The discovery of filamentation dates back to the 1960s [1]. During the subsequent decade, active studies were carried out in condensed matters by using nanosecond (ns) and picosecond (ps) laser pulses [2,3]. New interest to study this phenomenon was rekindled by the first observation of filamentation in air [4], which may reach many kilometers [5,6]. Since then, femtosecond laser pulse filamentation has become an attractive topic of research at the forefront of applied physics [7].

The current physical understanding of filamentation is that the major underlying mechanism of the filamentation process is the same in all transparent optical media. It is the dynamical balance of the laser beam self-focusing and the defocusing effect of the self-generated weak plasmas [4,7,8]. The main difference lies in the detail of free electrons generation. In gases, it is mainly tunnel ionization (TI) of the gas molecules inside the self-focal volume resulting in the plasma [9]. In condensed matters, it is the multiphoton excitation (MPE) of free

S.L. Chin (✉)
Centre d'Optique, Photonique et Laser & Département de Physique, de Génie Physique et d'Optique, Université Laval, Québec, GIV OA6, Canada
e-mail: slchin@phy.ulaval.ca

R.W. Boyd et al. (eds.), *Self-focusing: Past and Present*,
Topics in Applied Physics 114, DOI 10.1007/978-0-387-34727-1_14,

electrons from the valence to the conduction bands [10] followed by inverse Bremstrahlung and electron impact ionization [11] before the short pulse is over. The well-known type of optical breakdown of the medium (generation of a spark) by longer laser pulses would not occur in the case of femtosecond filamentation. This is to be contrasted with those studies in the 1960s and 1970s when laser-induced breakdown with nanosecond and picosecond pulses masked many interesting filamentation phenomena that were not foreseen at that time.

Filamentation also involves the interplay of many other physical processes, namely, self-phase modulation (SPM), group velocity dispersion, self-steepening [12,13], third harmonic generation (THG) [14], and so on. The consequence of the filamentation is that the laser pulse will self-transform into a new light source with amazing characteristics, such as ultra-broad spectral bandwidth spanning from the ultraviolet (UV) to the infrared (IR) [13,15–17], fairly high intensity at the self-focal regions (filament) [18,19], and the possibility of self-transforming into a few-cycle (even single) pulse [20,21]. In this chapter, these phenomena will be discussed and experimental confirmation will be given whenever possible.

14.2 Slice-by-Slice Self-focusing of Laser Pulse

14.2.1 Basic Idea

Filamentation is initiated by self-focusing. The dynamic of the laser pulse self-focusing can be described by the slice-by-slice self-focusing scenario according to the moving focus model [22–24] or similar spatial replenishment model [8]. In this scenario, the temporal shape of the laser pulse is visualized as sub-divided into many power slices in time, as shown in Fig. 14.1(a). The propagation of each slice is approximated as that of a continue wave (CW) beam according to the self-focusing theory, where the self-focusing distance z_f is given as [2]:

$$z_f = \frac{0.367ka_0^2}{\left\{\left[(P/P_{crit})^{1/2} - 0.852\right]^2 - 0.02198\right\}^{1/2}}. \tag{14.1}$$

Here, ka_0^2 indicates the diffraction length of the beam and P is the laser peak power. P_{crit} represents the critical power for self-focusing given by $P_{\text{crit}} = 3.77\lambda_0^2/8\pi n_2 n_0$ [2], where, λ_0 is the laser central wavelength, n_0 is the linear index of refraction, and n_2 is the coefficient of the Kerr nonlinear index of refraction. It is clear that the central slice has the highest power; it will self-focus at the shortest distance as illustrated in Fig. 14.1(a). While the slice in front of the central slice would self-focus at a later position because of its lower power.

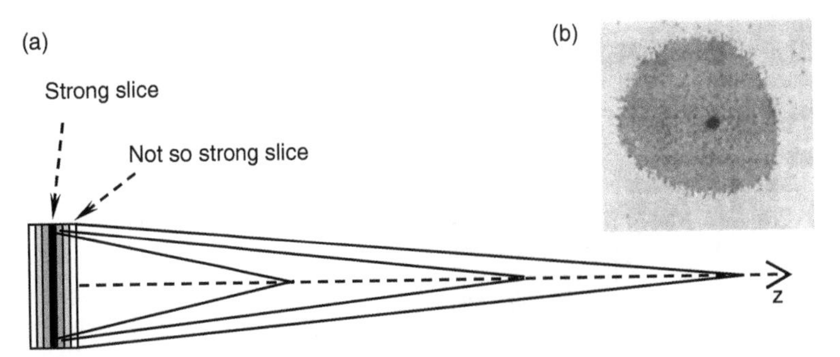

Fig. 14.1 (**a**) Slice-by-slice self-focusing model. (**b**) A typical laser beam profile recorded by burn paper in the path of a filament

Consequently, a series of self-foci is produced on the propagation axis, giving rise to the perception of a light filament.

The slice-by-slice self-focusing model implies that the laser pulse does not degenerate or self-stretch into a thin and long line of intense light in the case of filamentation. It is not the propagation of the self-focus along the axis of propagation either. At any time, there is only one pulse propagating in space. Inside this pulse, there is only one most intense hot spot (self-focus) at the front part of the pulse This most intense hot spot changes its position inside and toward the front of the pulse during propagation.

14.2.2 Experimental Proofs

Implicit evidence of the slice-by-slice self-focusing model can be found in the laser beam's cross-sectional profile in the course of filamentation. If one puts a piece of burn paper in the path of the filament, the slice-by-slice self-focusing model will predict a central strong burn spot surrounded by a larger but weaker burn pattern. This is indeed observed as shown by Fig. 14.1(b), which is a laser burn pattern obtained when a filament was formed by a 800-nm Ti:sapphire laser pulse (200 fs and 30 mJ) [25]. Experimental measurement also revealed that there was less than 10% of the total energy contained within the strong spot's area [24]. This small fraction of the energy can be intuitively seen from Fig. 14.1(a) in the sense that it contains mainly the energy of the already self-focused slice. The other slices form the weak background as shown in Fig. 14.1(b). This weak background is popularly referred to as the energy background reservoir [8,26].

Another more quantitative way to justify the slice-by-slice self-focusing model was demonstrated by Brodeur et al. [24]. They experimentally examined the ending point of a filament produced in a 100-m-long hallway by 200-fs laser pulses with various energies (2.3–8.1 mJ) and found that the filament ending position was independent of the laser pulses energy used therein. This position was proven to be coincident with the diffraction length of the incident

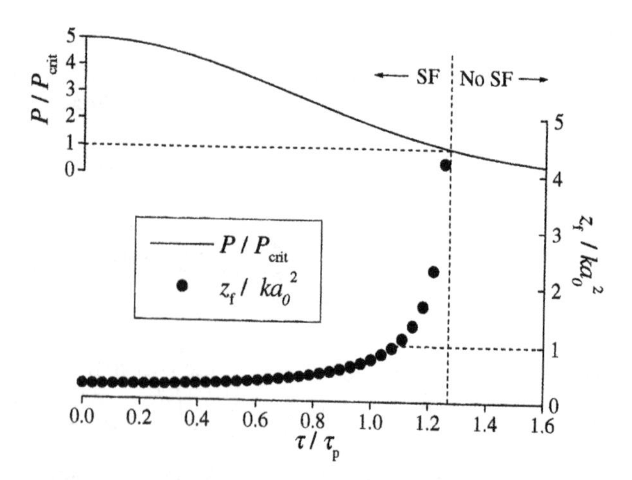

Fig. 14.2 Self-focal distance (*solid circle*) normalized to the diffraction length ka_0^2 for various slices of the front part of the pulse The horizontal scale is normalized to the pulse duration τ_p. No self-focusing should be expected at $P < P_{\text{crit}}$

beam, ka_0^2. In order to interpret this practical observation, a straightforward theoretical analysis based on the slice-by-slice self-focusing model is depicted in Fig. 14.2 [24]. The self-focal distance (circle) is plotted according to Eq. (14.1) for some equally spaced discrete slices of the front part of a Gaussian pulse (solid line). The distance between the self-foci of two adjacent slices starts to increase rapidly from the position where $z_f = ka_0^2$. In the experiment, it is reflected as the significant decrease of the energy constrained around the axis. Thus, the filament will end from that distance onward.

14.2.3 Pulse Duration Dependence of the Critical Power

We have known that the critical power for self-focusing is inversely proportional to n_2 [27,28]. It is necessary to point out that for a non-centrosymmetric molecular medium, the effective n_2 includes both the "instantaneous" electronic response to the light and nuclear responses involving Raman transition [29]. This phenomenon can be expressed as [30]:

$$n_{2,eff}(t) = n_{2,inst}[(1 - R) + \frac{1}{I(t)} R \int_{-\infty}^{t} H(t - t')I(t')dt']. \qquad (14.2)$$

where $n_{2,\text{inst}}$ is the instantaneous nonlinear refractive index coefficient contributed by the electronic response. $H(t)$ accounts for a time-dependent molecular response function introducing a delayed nonlinear effect. R is the fraction of the CW nonlinear optical response that has its origin in the delayed component. In

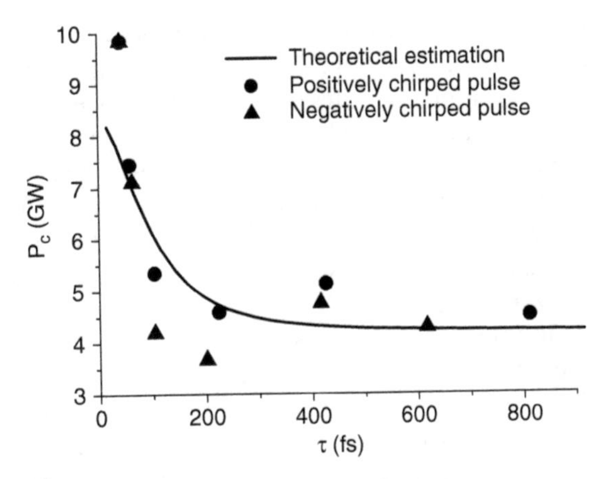

Fig. 14.3 The critical power changes in terms of the pulse duration. *Solid line*: calculated; *triangles*: experimentally measured with positively chirped pulse; *circles*: experimentally measured with negatively chirped pulse

this case, the critical power for self-focusing will have strong pulse duration dependence if the pulse length is comparable to the characteristic delay time of the nuclear response.

For instance, in air, $H(t)$ can be modeled by a damped oscillator function [8]:

$$H(t) = \Omega^2 e^{-\frac{\Gamma t}{2}} \frac{\sin(\Lambda t)}{\Lambda}, \tag{14.3}$$

where $\Omega = 20.6$ THz; $\Gamma = 26$ THz, and $\Lambda = \Omega^2 - \Gamma^2/4$ to mimic the response function given in [30]. In Fig. 14.3, the effective critical power estimated by Eq. (14.2)–(14.3) is plotted (solid line) versus laser pulse duration (full width at half maximum, FWHM) [31]. We take $n_{2,\,\mathrm{inst}} = 2.3 \times 10^{-19}\,\mathrm{cm}^2/\mathrm{W}$ and $R = 0.5$ [30]. It is in good agreement with the recent measurements (symbols) [32]. Figure 14.3 demonstrates that P_{crit} changes from about 10 GW for pulse durations shorter than 50 fs to about 5 GW for pulses longer than 200 fs.

14.3 Intensity Clamping

14.3.1 The Physics of Plasma Generation

As we mentioned in the introduction, plasma generation will take place as soon as the laser intensity at the self-focus is sufficiently high. However, the physics of plasma generation in the case of femtosecond filametation is distinctly different from the long pulse induced optical breakdown.

The well-known optical breakdown in gases using long laser pulses follows a three-step process: (1) Multiphoton ionization (MPI) of impurity molecules with low ionization potentials would easily provide a few free electrons with low initial kinetic energy in the focal volume at the front part of the pulse [33]. (2) The free electrons in the strong laser field can absorb n photons ($n = 0, 1, 2, 3, \ldots$) while colliding (scattering) with a much heavier particle (e.g., atom, molecule, or ion). The heavy particle is to conserve momentum during the interaction. This process is called inverse Bremsstrahlung or free–free transition. (3) After one or more inverse Bremsstrahlung processes, the free electron will acquire a kinetic energy E_e higher than the ionization potential of the molecule/atom. Subsequent collision will give rise to the ejection of an extra electron from the molecule/atom. This will result in two low-energy electrons. They will undergo the same processes as before, each giving rise to two electrons, and so on, until the gas is fully ionized. This is called cascade or avalanche ionization, i.e. optical breakdown [34].

But this three-step process cannot be accomplished because TI dominates all the interaction physics of gas ionization if a femtosecond laser pulse is used [9]. Because the intensity of the femtosecond laser pulse rises very quickly from very low to the peak where TI occurs, the probability of free electrons generation through TI will be next to zero at the front part [9]. The same thing will take place at the falling part of the pulse. The intensity drops so fast that there is no enough time for collision to happen, which is required to reach the step (3) of the breakdown process. For example, the collision time in air is about 1 ps, which is much longer than the femtosecond pulse duration.

Thus, when femtosecond filamentation occurs in gases at pressures of, say, one atmosphere, only partial ionization will take place because of the absence of cascade ionization. In condensed matters, some cascade ionization will contribute to the total number of free electrons [7,35]. Its contribution strongly depends on the geometrical focusing condition [35]. However, even in the case of extremely strong external focusing, not many cycles of collision can be involved, and it is too little to induce total ionization [7]. That is to say, in both gases and condensed matters, the plasma density induced by femtosecond laser pulses during self-focusing is only a tiny fraction (about 10^{-3}) of the neutral density [13,36].

14.3.2 Plasma Defocusing Effect Cancels Self-Focusing

The primary consequence of plasma generation is that its defocusing effect will counteract the self-focusing and seize the spatial collapse of the laser beam. Finally, the balance between the self-focusing and plasma defocusing will set a limit for the minimum beam diameter and the highest intensity. This phenomenon is now known as intensity clamping [19,37,38] and occurs when the effective total nonlinear refractive index is equal to zero,

$$\Delta n = \Delta n_{kr} + \Delta n_p = 0 \tag{14.4}$$

with $\Delta n_{kr} = n_2 I$ being the Kerr effect–induced nonlinear refractive index and $\Delta n_p = -e^2 N_e / 2\varepsilon_0 m_e \omega_0{}^2$ corresponds to the plasma contribution to the refractive index (ω_0 is the central frequency of the pulse, e and m_e denote the charge and mass of the electron, and N_e stands for the electron density).

However, due to the high intensity in the filament, precise measurement of the intensity distribution is difficult. Alternative ways of determining the information about the intensities inside the filament can be offered by the observation of quantities outside the interaction volume. In air, the measurement can be performed by detecting the strength of the fluorescence spectra emitted by nitrogen molecules and ions inside the filament. As shown in Fig. 14.4, the peak intensities of the two strongest fluorescence lines, namely, 337 nm (open squares) and 391 nm (solid circles) [39], are plotted in log–log scale as a function of the input pulse energy [37]. The data are obtained at atmospheric pressure with 250-fs laser pulses. A characteristic energy can be clearly noticed. Below this energy, the steep rise of the fluorescence signal indicates that a highly nonlinear process takes place as the intensity increases.

The nonlinear process is the tunnel ionization of the nitrogen molecules into the ionic ground and excited states which, after relaxation and recombination, emit the above-mentioned fluorescence [39]. After excluding the possibility of the depletion of neutral molecules, the deviation from the steep rise therefore implies that the increase of the peak intensity at high energy is restrained [37]. It indicates the presence of intensity clamping.

In condensed matters, further experimental proofs of intensity clamping can be found in Fig. 14.5 [38]. It is well known that due to SPM, the laser spectrum will be greatly broadened during the filamentation [40]. A few representative spectra for different input pulse energies taken at the output of a water cell are shown in the inset of Fig. 14.5 (pulse duration: 170 fs). When we plot the maximum anti-Stokes (blue) frequency shift of these spectra as a function of the incident pulse energy, the plot reaches a distinct constant if the laser energy is more than 2 μJ in Fig. 14.5. To explain this observation, it is necessary to recall that in the case of filamentation, the maximum spectral broadening introduced by SPM is determined by the pulse peak intensity, the greatest temporal intensity gradient and the filament length; see Eqs. (14.7)–(14.8) in Section 14.4. In

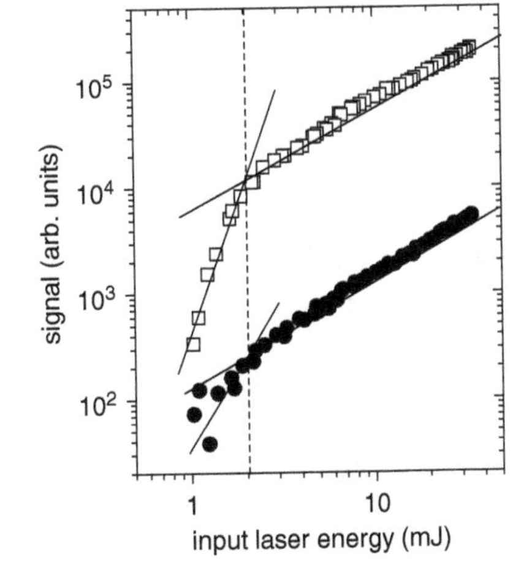

Fig. 14.4 Strengths of the nitrogen fluorescence 337 nm (*open squares*) and 391 nm (*solid circles*) as a function of the input laser pulse energy at 760 Torr

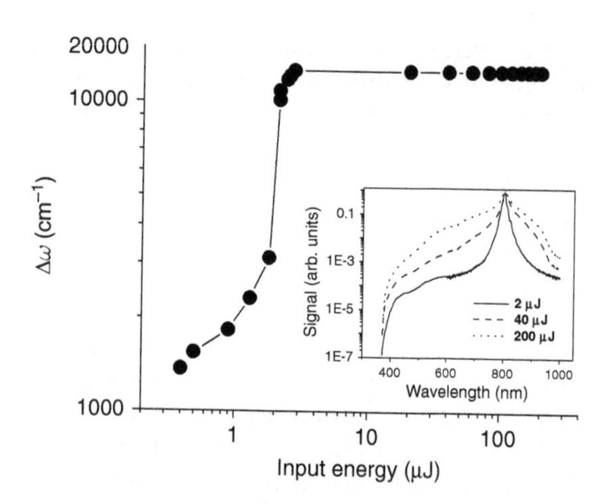

Fig. 14.5 Maximum anti-Stokes shift in water as a function of the input energy. Inset: a few representative white light spectra at different energies

general, the largest temporal intensity gradient is induced by the self-steepenning at the back part of the pulse, giving rise to an anti-Stokes shift; see Eq. (14.8). Taking into account that self-steepening also critically depends on the pulse peak intensity and the filament length is almost constant in condensed matters according to the previous work [17], the constant anti-Stokes frequency shift manifests a nonvariable peak intensity, i.e., intensity clamping. The same phenomenon has been observed in other media, namely, chloroform and soda lime glass [38].

Intensity clamping is a profound physical phenomenon of self-focusing and filamentation. It sets an upper limit to the intensity at the self-focus in all optical media. In air it is about 5×10^{13} W/cm^2 [18,19], the filament diameter is about 100 μm. In condensed matters, the diameter can reach a few microns [17,41].

We also note that the clamped intensity in air (or gases) is independent of pressure. This is because when intensity clamping occurs, the nonlinear Kerr index and the nonlinear index due to plasma generation is equal: $n_2 I = -e^2 N_e / 2\varepsilon_0 m_e \omega_0^2$ [19]. Both n_2 and N_e are linearly proportional to the gas density because N_e comes from tunnel ionization of the individual molecules. Hence, the gas density cancels out at the two sides of the equation, leaving behind an equation for the solution of the same clamped intensity I at any pressure. Thus, when filamentation occurs at a high altitude in the atmosphere, the clamped intensity is always the same as that at sea level.

14.4 White Light Laser (or Supercontinuum Generation) and Conical Emission

14.4.1 White Light Laser

Another signature of filamentation is called the supercontinuum generation: the output spectrum after filamentation has dramatically broadened. In particular, due to the contribution of THG [14], the reported supercontinuum in air spans from 230 nm to 4 μm [16,42]. On the other hand, the supercontinuum spectrum is generally asymmetrical with much larger broadening at the blue shift side as seen in Fig. 14.5. The supercontinuum has been identified as a white light laser pulse because it has been proved that each frequency component within the supercontinuum features the same relative coherence length as that of the input fundamental light by comparing to an incoherent white light source [43].

In this case, SPM is responsible for the supercontinnum generation [40]. The induced pulse frequency change in an optical medium (using gases as an example for simplicity) is equal to:

$$\Delta\omega = \frac{\partial}{\partial t}\left(-\frac{\omega_0 \Delta n(t)}{c} z\right) = -\frac{\omega_0}{c} z \frac{\partial[\Delta n(t)]}{\partial t}, \tag{14.5}$$

where z is the propagation distance and the nonlinear refractive index is given by: $\Delta n(t) = n_2 I(t) - \frac{e^2 \sigma N_0}{2\varepsilon_0 m_e \omega_0^2} \int_{-\infty}^{t} I^m(t)\mathrm{d}t$ after substituting $N_e(t) \approx \sigma N_0 \int_{-\infty}^{t} I^m(t)\mathrm{d}t$ into Eq. (14.4). The dependence of $N_e(t)$ on I^m is an empirical representation of the experimentally observed tunnel ionization of molecules or atoms in which the number of ions (or electrons) generated by the laser follows an effective power law [9]. Here, N_0 represents the neutral material density, σ is a proportional constant, and m is an constant (not necessary an integer). We note that the electron–ion recombination time is normally much longer than the femtosecond time scale of the pulse. Hence, the generated plasma can be considered as static during the interaction with the pulse. Because the front part of the pulse always sees the neutral, from Eq. (14.5) without the plasma contribution:

$$\Delta\omega = -\frac{\omega_0 z}{c}\frac{\partial[\Delta n(t)]}{\partial t} = -\frac{\omega_0 z}{c} n_2 \frac{\partial I(\text{front part})}{\partial t} < 0. \tag{14.6}$$

The last inequality in Eq. (14.6) arises because the front part of the pulse has a positive temporal slope; hence, the front part of the pulse contributes to red (Stokes) shift/broadening. But at the back part of the pulse, much stronger blue shift/broadening will be induced because of the SPM in the plasma [Eq. (14.7)] and the very steep descent of the back part of the pulse duo to self-steepening [Eq. (14.8)] [12,13]:

$$\Delta\omega = -\frac{\omega_0 z}{c}\frac{\partial[\Delta n_p(t)]}{\partial t} = \frac{z e^2 \sigma N_0}{2c\varepsilon_0 m_e \omega_0} I^m(t) > 0 \qquad \text{(SPM in plasma)} \tag{14.7}$$

$$\Delta\omega = -\frac{\omega_0 z}{c} n_2 \frac{\partial I(\text{very steep back part with negative slope})}{\partial t} > 0 \tag{14.8}$$

The propagation distance z also plays a role in both the red and blue broadening ; see Eqs. (14.6)–(14.8). Thus, during experimental observations, the spectral broadening of the pulse develops progressively as the propagation distance increases [44].

14.4.2 Conical Emission

Moreover, the supercontinuum will appear on a screen as a concentric rainbow with colors ranging outward from red to blue. This phenomenon is called conical emission (CE) [15,40]. In Fig. 14.6, we demonstrate some spectrally resolved CE patterns in air. The input laser pulse is 45 fs, 5 mJ, with 6-mm diameter at FWHM. The single filament starts at a propagation distance of 12 m and the pictures are taken at 15 m (a)–(d) and 30 m (e)–(i). It can be seen from Fig. 14.6 that the anti-Stokes light propagates in the form of CE [Fig. 14.6(a)(b) and (e)(f)], while no CE can be observed for Stokes light [Fig. 14.6(d) and (i)]. The shorter the anti-Stokes wavelength is, the larger its diameter is.

If we superpose all these patterns, it will become one with a white pattern at the center surrounded by concentric rings with a continuous rainbow color ranging outward from red to blue [7]. The physical explanation of CE lies in that it is an analog to the temporal SPM and comes from the phase modulation of the beam in the transverse direction. The radial phase variation in the front part $\partial\phi_{nl}/\partial r \propto -\partial I/\partial r$ is positive due to the intensity profile $I(r)$ up to the on-axis maximum. Therefore, the radiation in the front of the pulse is frequency-shifted toward the infrared and converges toward the filament axis. The result is confined propagation of the Stokes components. Later in the pulse, plasma contributes to the nonlinear phase growth: $\partial\phi_{nl}/\partial r \propto \partial N_e(r)/\partial r$ is negative due to the rapid decrease of electron density in the off-axis direction. As a result, the anti-Stokes components of the supercontinuum experience a nonlinear (plasma-induced) divergence. In fact, based on the same arguments, the simulation model, which includes the ionization term into the nonlinear Schrödinger equation, has successfully predicted the CE divergence angles obtained in the experiments (Fig. 14.7) [11,40,45,74]. In both the experiment and simulation, pulse duration $\tau_0 = 250$ fs (FWHM), pulse energy is 10 mJ, and input beam radius $a_0 = 3.5$ mm (FWHM). The propagation distance is 40 m. This theoretical model has become

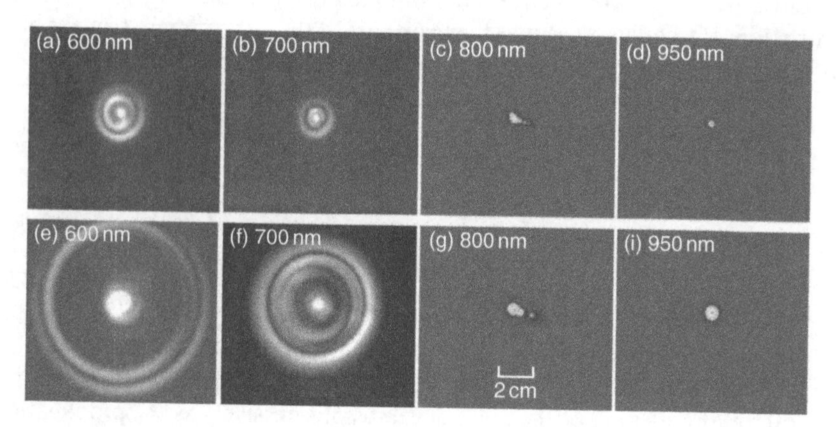

Fig. 14.6 CCD-camera-recorded conical emission patterns at various wavelengths

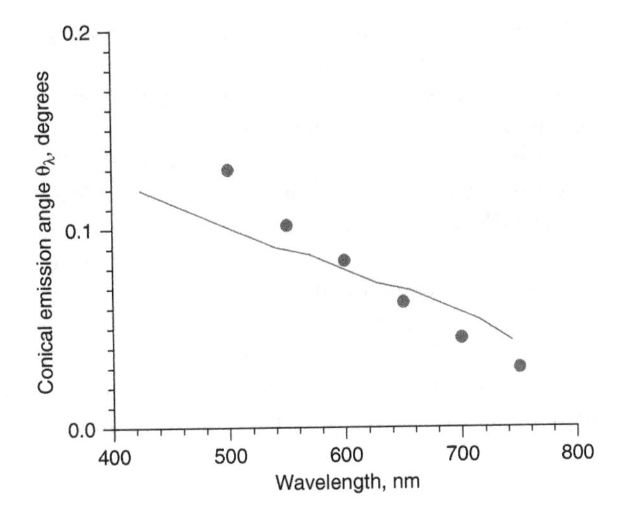

Fig. 14.7 Dependence of the conical emission angle on wavelength. Experimental data are shown by the *black dots* and the simulation results by the *solid curve*

the widely accepted method to simulate the physical process of femtosecond filamentation.

14.4.3 Band Gap Dependence of Supercontinuum in Condensed Matters

Guided by the above idea, Brodeur and Chin found a band gap energy dependence of supercontinnum width in condensed matters [10]. Their measurements are illustrated in Fig. 14.8, which shows the maximum anti-Stokes broadening, $\Delta\omega_+$ as a function of their band gap energies (E_{gap}) in various media. First, in Fig. 14.8 a threshold band gap is found below which there is no continuum

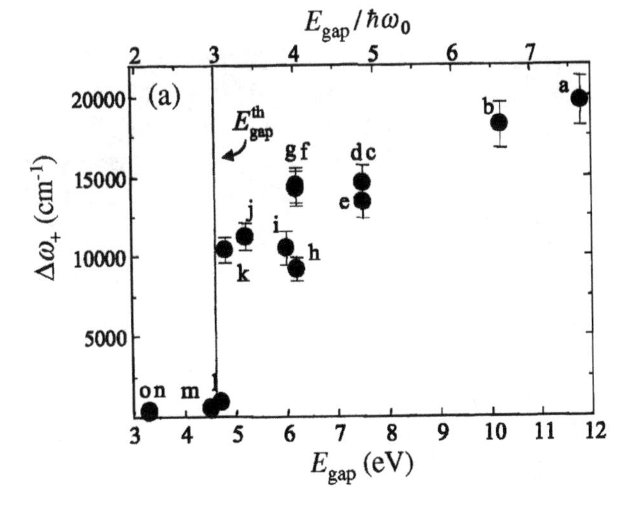

Fig. 14.8 The anti-Stokes broadening vs. band gap in 15 different condensed optical media. (a) LiF; (b) CaF_2; (c) UV-grade fused silica; (d) water; (e) D_2O; (f) 1-propanol; (g) methanol; (h) NaCl; (i) 1,4-dioxane; (j) chloroform; (k) CCl_4; (l) C_2HCl_3; (m) benzene; (n) CS_2; (o) SF-11 glass

generation. Above that threshold, the higher band gap energy tends to give rise to larger spectral broadening.

During self-focusing, an intensity spike corresponding to the self-focusing of a slice of the pulse develops. This generates a sufficiently high density of free electrons in the conduction band by MPE. The defocusing effect of this electron "gas" will cancel the self-focusing effect. For low-band gap material such as CS_2 even if it has a large n_2 to favor an easy self-focusing of the pulse, the relative ease of exciting free electrons into the conduction band will produce a large self-focal diameter. The intensity at the self-focus is clamped down to a low value. According to the previous analyses, low clamped intensity on the SPM of the pulse results in a negligible $\Delta\omega_+$. Figure 14.8 shows that only when the band gap is larger than $3h\omega_0$ (i.e., four photon excitation or higher) $\Delta\omega_+$ will become significantly large and qualified as supercontinuum.

14.5 Background Energy Reservoir

14.5.1 Proof of the Existence of Energy Reservoir

We mentioned earlier that accompanying the occurrence of a filament, in the transverse pattern of the laser pulse, the tiny strong hot spot is embraced by the energy reservoir [Fig. 14.1(b)]. Recently, this concept of energy reservoir has become a central topic in the field of filamentation study [46,–49].

The crucial role of energy reservoir for the whole filamentation process is revealed by the experiment shown in Fig. 14.9 [49]. In the experiment, a 800-nm, 45-fs, 2.5-mJ, 2-mm-diam (FWHM) pulse propagates freely in air and self-focuses into a filament. The nitrogen fluorescence emitted from the plasma column is imaged by an ICCD camera from the side to represent the filament in

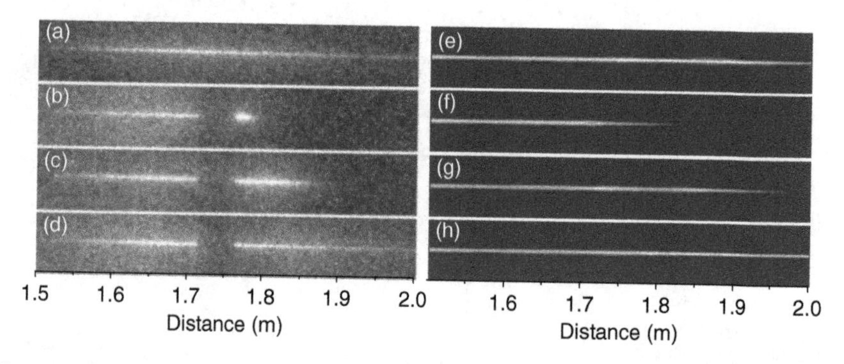

Fig. 14.9 (*Left column*): Images of the nitrogen fluorescence signal recorded by an ICCD camera. (*Right column*): Electron density distribution from numerical simulations. (a),(e) free propagation; (b),(f): pinhole diameter 220 μm; (c),(g): pinhole diameter 440 μm; (d)(h) pinhole diameter 2 mm

air. Figure 14.9(a) shows the free propagation of the pulse. A filament over almost half a meter is seen along the propagation axis.

The other panels in Fig. 14.9 correspond to the images when pinholes of different diameters were inserted co-axially with the filament. The diameters of pinholes are from 220 μm [Fig. 14.9(b)] to 2 mm [(Fig. 14.9(d)]. It is seen that for the smallest diameter [Fig. 14.9(b)], the filament is terminated by the pinhole. As the diameter of the pinhole is doubled [Fig. 14.9(c), 440 μm], the filament partially survives. When the pinhole diameter is increased to 2 mm [Fig. 14.9(d)], the filament formation looks unchanged compared to the case of free propagation within the field of view of the camera. Corresponding simulations with the same initial experimental conditions give rise to quantitative agreements as shown in Fig. 14.9(e)–(h).

Figure 14.9 can be understood qualitatively in the following way. The slice already focused produces a strong hot spot on the propagation axis, while the background arises from the majority of slices that focus at shorter or longer distances. According to the slice-by-slice self-focusing scenario [Fig. 14.1(a)], blocking the background at a certain position should end the propagation of those slices that have not yet self-focused, and hence prevent the production of the next self-foci. The filament will be interrupted. Conversely, the filamentation process will not be markedly influenced if the hot spot is blocked at one position. In principle, only the propagation of the already self-focused slice is terminated, while the rest of the slices will continue to self-focus with slightly reduced power. These arguments deduced from the slice-by-slice self-focusing scheme explain well the phenomenon that the filament can penetrate the cloud in the atmosphere [50].

It is also important to emphasize that the background is not stationary. Essentially, every slice, be it the one from the front or the back, after being de-focused by the self-induced plasma in the self-focal region will "release" its energy back into the pulse outside the self-focal zone. By this repetition, the energy at the self-focus comes always from the outside reservoir.

14.5.2 *Multiple Refocusing*

The impact of the existence of the energy reservoir is extensive. First, it can induce multiple refocusing of the pulse [24,37,51]. Figure 14.10(a) shows an experiment that demonstrates the refocusing phenomenon when we focus a 38-fs, 2.15-μJ, and spatially filtered laser pulse into a 10-cm-long glass cell. The cell contains a very dilute solution of Coumarin 440 in methanol [51]. The on-axis high-intensity zone (filament) induces three-photon fluorescence of coumarin. The background reservoir at much lower intensity does not excite the dye molecules. Hence, each line represents the filament core. Over the entire length of the sample cell, up to six refocusing filaments are observed along the direction of the pulse propagation.

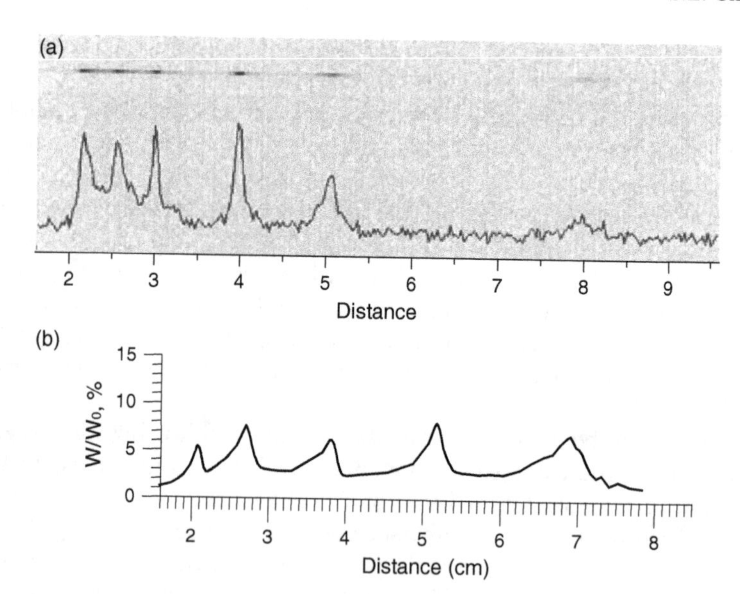

Fig. 14.10 (**a**) Three-photon fluorescence signal registered by the CCD camera in the experiment. (**b**) The change of energy as a function of propagation distance in the near-axis region $0 < r < 10$ μm; input pulse energy is 1.6 μJ (simulation)

More details obtained from the numerical simulation may greatly assist us in understanding this multiple-refocusing phenomenon [51]. In Fig. 14.10(b), we present the simulated dependence of the pulse energy contained in the near-axis region with the radius of 10 μm as a function of the propagation distance. At the beginning of propagation ($z < 2$ cm), the background low-intensity light field converges from the peripheral regions toward the central region because of self-focusing. Soon it leads to the first fluorescence peak at $z = 2.1$ cm. After that, the energy in the central region decreases due to the defocusing effect of plasma. Therefore, the energy flows from the near-axis regions to the outer region. However, the laser power is still high and the divergence is not too large, self-focusing will take over again afterwards. Then, further in the propagation ($z \approx 2.75$ cm) the local maximum is attained in the near-axis region. The next "cycle" of the energy interchanges start at $z \approx 2.8$ cm. The result of this repeated energy interchange is the multiple refocusing picture in the fluorescence signal [Fig. 14.10(a)].

In brief, the refocusing of the pulse channels energy back into the propagation axis, forming the hot spot of the beam, and thus represents the process in which energy is exchanged between the hot spot and the energy reservoir. This is one of the important physical mechanisms of the femtosecond laser pulse long-range propagation and filamentation.

Moreover, when one slice is diffracted by the plasma from the self-focus, it will interfere with the light field contained in the background energy reservoir.

Since the background field (wave) is a quasi-plane wave and the diverging wave is cylindrically symmetrical along the filamentation axis, their interference will lead to ring structures [52].

14.6 Multiple Filamentation Competition

So far, the above description pertains to a single filament arising from one smooth beam profile with only one intensity maximum ("warm zone"). In the real experimental condition, multiple filaments are very often observed when the laser peak power is much higher than the critical power for self-focusing. The explanation of the formation of multiple filaments is a nonuniform spatial distribution in the transverse beam section [53], leading to more than one "warm" zone across the beam profile. Each warm zone will undergo self-focusing when the pulse propagates in air as if it were an independent pulse. But these warm zones are not independent from one another because they all try to feed energy from the whole pulse's back ground reservoir into their own self-foci. This constitutes a competition for energy [54].

The underlying physics of multiple filament competition is essentially field re-distribution inside the pulse during propagation. It consists of two inter-related scenarios. One scenario is linear field interference inside the pulse during propagation and self-deformation through the optical medium. The other is nonlinear field re-distribution due to nonlinear propagation effects.

In the first scenario, consider first each warm zone as being a single pulse. During the propagation, each gives rise to concentric rings around the filament [52]. When two adjacent warm zones self-focus into two nearby filaments, the two sets of rings (or rather, the two conical waves and the background quasi-plane wave) will interfere, giving rise to a star-like pattern as shown Fig. 14.11(a) [52]. Note that in order to clearly show the ring structures, the hot spot zones have been saturated and appear as white color. In this experiment, the laser characteristics are 14 mJ/pulse, 45 fs with diameter of 1.2 cm (FWHM) and the propagation distance was 87 m.

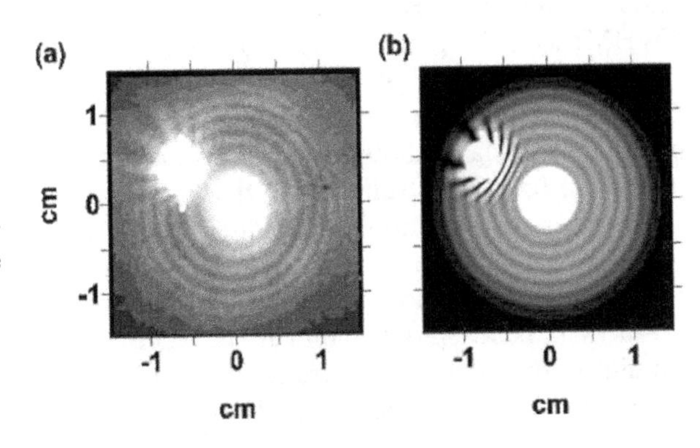

Fig. 14.11 (**a**) Experimentally measured transverse fluence distributions for two filaments' interference. (**b**) Beam pattern theoretically calculated according to the interference idea

Theoretical calculation based on the above idea of the interference of conical waves and the background quasi-plane wave has indeed excellently reproduced the experimental observation [Fig. 14.11(b)] [52]. When more than two nearby filaments interfere, the resultant field would give rise to more complicated structures with more new warm zones [45]. Numerical simulations tell us that these newly generated warm zones will undergo self-focusing again during further propagation. New "children" filaments are thus formed at new positions [26].

However, if the warm zones or filaments are far apart, interference will be too weak to form new and sufficiently warm zones for self-focusing. This will constitute the second scenario in which the initial warm zones each go their own ways as if they were independent. During propagation, the nonlinear self-focusing effect will help each of the warm zones to "pull" the field toward its own self-focus as if each hot spot "sucks" energy from the background reservoir. The consequence of this competition for energy from the same background reservoir will be such that the filaments do not have enough energy to develop fully into maturity.

Such competition for energy will also take place in the first scenario with filaments adjacent to one another but in a more constructive way, because apart from creating "children" filaments, the central bunch of hot spots will collectively "suck" energy from the background reservoir toward them as if they were single one. The consequence of this latter case is that all these filaments become mature almost at the same time over a short distance of propagation. In air, in such a short distance, the nitrogen fluorescence signal appears very strong. After this short distance, the "children" filaments take over but will not be as strong as before [54].

These two scenarios were observed in our recent experiments [54,55]. We measured the nitrogen back-scattered fluorescence (BSF) from long filaments in air using a LIDAR (laser radar) technique. We found that with a beam diameter of about 1.2 cm (FWHM) over which the multiple warm zones were sufficiently far apart, the BSF signal from the generated multiple filaments had a huge fluctuation. For the same input laser peak power of about 1 terawatt (TW), the fluorescence signal sometimes came from the full propagation length of 100 m, but sometimes there was very little after about 20 m of filamentation, and sometimes nothing at all. However, if the beam diameter is made smaller (0.6 cm at FWHM) while keeping the distribution of 'warm' zones roughly the same; i.e. forcing the generated hot spots to be closed to one another, we could detect BSF from the full 100 m propagation length and beyond up to 500 m through extrapolation for all laser shots; also, in the first 10 m, the fluorescence intensity was more than 100 times stronger than the case of larger diameter beam [55].

In order to give more sight into the multiple filamentation competition behavior, a comprehensive simulation in water was carried out [54]. The input beam is a superposition of two sets of Gaussian profile. Their diameters are $a_1 = 0.18$ mm and $a_2 = 0.045$ mm at $1/e$, respectively. While keeping all the other conditions the same [pulse duration 42 fs (FWHM), $P = 11P_{crit}$], two situations were considered: long separation distance between the centers of two Gaussian

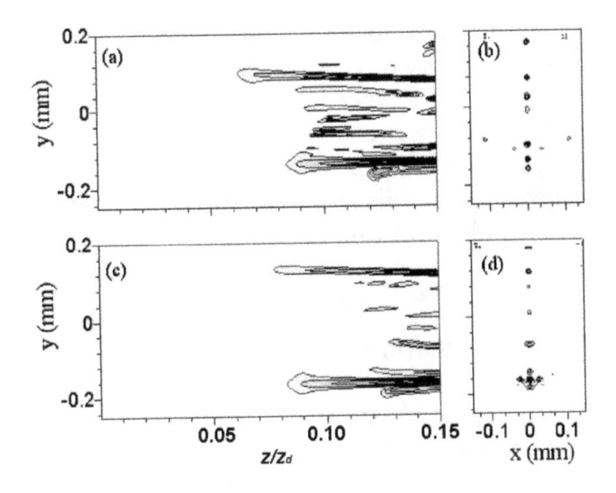

Fig. 14.12 Electron density distribution for the shorter $d_{\text{short}} = 1.15a_1 = 0.2$ mm (**a**)(**b**) and the longer $d_{\text{long}} = 1.6a_1 = 0.29$ mm (**c**)(**d**) separation distances between the maxima of the initial perturbations in a 42-fs pulse propagating in air. (**a**)(**c**) The electron density $N_e(x = 0, y, z)$ at the end of the pulse. (**b**)(**d**) The transverse distribution of the electron density $N_e(x, y)$ at $z = 0.15z_d$

profiles ($d = 0.29$ mm); and short separation distance ($d = 0.2$ mm). The resulting plasma evolution along the propagation distance is presented in Fig. 14.12(a) and (c) as well as their cross sections at $z = 0.15\ z_d\ (z_d = ka_1^2)$ [Fig. 14.12(b),(d)].

In panels (a) and (b), corresponding to the shorter separation distance, the plasma density is higher and there is a larger number of children plasma columns in comparison with the case shown in panels (c) and (d), especially in the range $0.08\ z_d \sim 0.11\ z_d$. Thus, the larger amount of free electrons in the beam cross section is obtained at nearly all distances z. Notably, the shorter separation distance case will induce more significant fluorescence signals in experiments.

The idea of competition between multiple filaments emphasizes that the processes of producing and maintaining multiple filaments are governed by the energy competition among the hot spots and interferences within the background reservoir. In fact, we believe some earlier theories, such as the optical turbulence effect [56] or "filament fusion and breakup" [57] can all be enclosed in the framework of this scenario. It also can give the explanation to the recent proposal that an elliptical beam may naturally lead to multiple filamentation [7,58,59].

In a laboratory experiment, we also observed that the diameter of a 45-fs, 1-TW laser pulse's pattern at the original 800-nm wavelength remains practically constant over 100 m of propagation [45]. The initial pulse pattern contained a lot of warm zones and the final pattern still contained a lot of hot spots at the end of the propagation path in the corridor of 100 m. However, when the pulse length was lengthened (positively chirped) to about 200 ps, everything else being identical, the diameter at 100 m was more than three times larger than the initial diameter.

The reason for the quasi-maintenance of the beam diameter is due to slice-by-slice self-focusing and multiple filamentation competition. When there are many self-foci (multiple filamentation), each of the self-foci contributes to "sucking" energy from the background reservoir toward its respective focal zone and then releasing it back to interfere with the reservoir after being diffracted by plasma. New filaments are formed afterward. As long as the power in a warm zone is above the critical power for self-focusing, this "sucking" process will take place and slow down the linear diffraction of the pulse. We call this pulse a "self-guided hot light pulse" [45]. This is perhaps another indication that the pulse may propagate very far in air and still give rise to high intensity in the pulse. This optimism is indicated by recent experiments showing that the filament end at a distance of about 2 km in the atmosphere using 3-TW pulses [6]. A long pulse laser cannot do so.

14.7 Applications

The potential applications of the femtosecond filamentation phenomenon include making use of the back-scattered white light as a source to measure pollutant absorption using the LIDAR technique in the atmosphere [60]. It is conceivable also to use the LIDAR technique to measure the back-scattered fluorescence from the ionized and fragmented molecules inside the filaments in the path of propagation [44,61]. Furthermore, one can take advantage of the resulting plasma "channel" left behind by filamentation and apply it to lightning control [62,63] and probably even artificial rainmaking [7].

On the other hand, the filamentation of powerful femtosecond laser pulses in bulk glass (fused silica or quartz) leads to the melting of glass [64,65]. Corresponding studies have indicated that melting are responsible for the writing of good waveguide by filamentation [41,66].

We have elucidated that successive slices in front of the central slice will go through self-focusing, intensity clamping, de-focusing, and so on as long as their peak powers are higher than the critical power. The back slices symmetrical to the front slices would in principle also self-focus at a position slightly behind the self-foci of the front slices. However, this would never happen because it would be diffracted by the plasma left behind by the central and successive front slices. Thus, the front part of the pulse would become thinner and thinner as the pulse propagates. This is what could be interpreted as pulse self-compression. The thinner front part of the pulse evolves continuously as the pulse propagates. If somehow one could extract this thin part of the pulse and eliminate the rest of the pulse, the consequence would be a clean few-cycle pulse [20]. This is currently an experimental challenge to efficiently and very simply generate few-cycle down to single-cycle pulses [21].

14.8 Summary

In conclusion, the filamentation phenomenon is a manifestation of the dynamical interplay between the self-focusing and the defocusing effect of the self-generated plasma. The physics of this dynamic has been described on the basis of the slice-by-slice self-focusing scheme. According to this scenario, during the propagation of an intense femtosecond laser pulse, it will self-transform into a new light source having ultra-broad spectral bandwidth and clamped high intensity over a long distance. The main fundamental reason for this long-distance filamentation is because of the existence of the energy reservoir, which is in fact a naturally embedded concept in the slice-by-slice self-focusing model.

Acknowledgments This chapter represents the fruit of many years of research supported by the Natural Sciences and Engineering Research Council of Canada (NSERC), Canada Research Chairs, Canada Foundation for Innovation (CFI), Defence Research and Development Canada-Valcartier (DRDC-Valcartier), Canadian Institute for Photonics Innovation (CIPI), Le Fonds Quebecois de Recherche sur la Nature et les Technologies (FQRNT), Femtotech of Vallorisation de Recherche du Quebec (VRQ), and NATO linkage grants. O.G. Kosareva and V.P. Kandidov thank the support of the European Research Office of the U.S. Army through contract No. W911NF-05-1-0553, and the Russian Foundation for Basic Research through grant No. 06-02-17508-a. W. Liu acknowledges the support of the 973 Program (grant No. 2007CB310403.), NSFC (grants No. 10804056 and No. 60637020), NCET, SRFDP and Fok Ying Tong Education Foundation.

References

1. M. Hercher: Laser–induced damage in transparent media, *J. Opt. Soc. Am.* **54**, 563 (1964).
2. J.H. Marburger: Self-focusing: theory, *Prog. Quant. Electr.* **4**, 35–110 (1975).
3. Y.R. Shen: Self-focusing: experimental, *Prog. Quant. Electr.* **4**, 1–34 (1975).
4. A. Braun, G. Korn, X. Liu et al.: Self-channeling of high-peak-power femtosecond laser pulses in air, *Opt. Lett.* **20**, 73–75 (1995).
5. L. Wöste, C. Wedekind, H. Wille et al.: Femtosecond atmospheric lamp, *Laser Optoelektronik* **29**, 51–53 (1997).
6. M. Rodriguez, R. Bourayou, G. Méjean et al.: Kilometer-range nonlinear propagation of femtosecond laser pulses, *Phys. Rev. E* **69**, 036607 (2004).
7. S.L. Chin, A. Brodeur, S. Petit et al.: Filamentation and supercontinuum generation during the propagation of powerful ultrashort laser pulses in optical media (white light laser), *J. Nonlinear Opt. Phys. Mater.* **8**, 121–146 (1999).
8. M. Mlejnek, E.M. Wright and J.V. Moloney: Dynamic spatial replenishment of femtosecond pulses propagating in air, *Opt. Lett.* **23**, 382–384 (1998).
9. S.L. Chin: From multiphoton to tunnel ionization. In: S.H. Lin, A.A. Villaeys, and Y. Fujimura, (Eds.): *Advances in Multiphoton Processes and Spectroscopy*, World Scientific, Singapore, pp. 249–272 (2004).
10. A. Brodeur, S.L. Chin: Band-gap dependence of the ultrafast white-light continuum, *Phys. Rev. Lett.* **80**, 4406–4409 (1998).
11. V.P. Kandidov, O.G. Kosareva, I.S. Golubtsov et al.: Self-transformation of a powerful femtosecond laser pulse into a white-light laser pulse in bulk optical media (or supercontinuum generation), *Appl. Phys. B* **77**, 149–165 (2003).
12. A.L. Gaeta: Catastrophic collapse of ultrashort pulses, *Phys. Rev. Lett.* **84**, 3582–3585 (2000).
13. N. Aközbek, M. Scalora, C.M. Bowden et al.: White light continuum generation and filamentation during the propagation of ultra-short laser pulses in air, *Opt. Commun.* **191**, 353–362 (2001).
14. N. Aközbek, A. Iwasaki, A. Becker et al.: Third-harmonic generation and self-channeling in air using high-power femtosecond laser pulses, *Phys. Rev. Lett.* **89**, 143901 (2002).

15. E.T.J. Nibbering, P.F. Curley, G. Grillon, et al.: Conical emission from self-guided femtosecond pulses in air, *Opt. Lett.* **21**, 62–64 (1996).
16. J. Kasparian, R. Sauerbrey, D. Mondelain et al.: Infrared extension of the supercontinuum generated by femtosecond terawatt laser pulses propagating in the atmosphere, *Opt. Lett.* **25**, 1397–1399 (2000).
17. A. Brodeur, S. L. Chin: Ultrafast white-light continuum generation and self-focusing in transparent condensed media, *J. Opt. Soc. Am.* **B 16**, 637–650 (1999).
18. H.R. Lange, A. Chiron, J.-F. Ripoche et al.: High-order harmonic generation and quasiphase matching in xenon using self-guided femtosecond pulses, *Phys. Rev. Lett.* **81**, 1611–1613 (1998).
19. J. Kasparian, R. Sauerbrey, S.L. Chin: The critical laser intensity of self-guided light filaments in air, *Appl. Phys. B* **71**, 877–879 (2000).
20. C.P. Hauri, W. Kornelis, F.W. Helbing et al.: Generation of intense, carrier-envelope phase-locked few-cycle laser pulses through filamentation, *Appl. Phys. B* **79**, 673–677 (2004).
21. A. Couairon, M. Franco, A. Mysyrowicz et al.: Pulse self-compression to the single-cycle limit by filamentation in a gas with a pressure gradient, *Opt. Lett.* **30**, 2657–2659 (2005).
22. J. Marburger, W. Wagner: Self-focusing as a pulse-sharpening mechanism, *IEEE J. Quant. Electron.* **3**, 415–416 (1967).
23. M.M.T. Loy, Y.R. Shen: Small-scale filaments in liquids and tracks of moving foci, *Phys. Rev. Lett.* **22**, 994–997 (1969).
24. A. Brodeur, C.Y. Chien, F.A. Ilkov et al.: Moving focus in the propagation of powerful ultrashort laser pulses in air, *Opt. Lett.* **22**, 304–306 (1997).
25. S.L. Chin, A. Brodeur, S. Petit et al.: Filamentation and supercontinuum generation during the propagation of powerful ultrashort laser pulses in optical media (white light laser), *J. Nonlinear Opt. Phys. Mater.* **8**, 121–146 (1999).
26. V.P. Kandidov, O.G. Kosareva, A.A. Koltuna: Nonlinear–optical transformation of a high-power femtosecond laser pulse in air, *Quantum Electron.* **33**, 69–75 (2003).
27. Y.R. Shen: *The Principles of Nonlinear Optics*, Wiley, New York (1984).
28. R.W. Boyd: *Nonlinear Optics*, Academic Press, Boston (2003).
29. E.T.J. Nibbering, G. Grillon, M.A. Franco et al.: Determination of the inertial contribution to the nonlinear refractive index of air, N_2. and O_2 by use of unfocused high-intensity femtosecond laser pulses, *J. Opt. Soc. Am. B* **14**, 650–660 (1997).
30. J.-F. Ripoche, G. Grillon, B. Prade et al.: Determination of the time dependence of n_2 in air, *Opt. Commun.* **135**, 310–314 (1997).
31. V.Y. Fedorov, V.P. Kandidov, O.G. Kosareva et al.: Filamentation of a femtosecond laser pulse with the initial beam ellipticity, *Laser Physics* **16**, 1227–1234 (2006).
32. W. Liu, S.L. Chin: Direct measurement of the critical power of femtosecond Ti:sapphire laser pulse in air, *Opt. Express* **13**, 5750–5755 (2005)
33. S.L. Chin, N.R. Isenor: Multiphoton ionization in atomic gases with depletion of neutral atoms, *Can. J. Phys.* **48**, 1445–1447 (1970).
34. C.G. Morgan, Laser-induced breakdown of gases, *Rep. Prog. Phys.* **38**, 621–665 (1975).
35. W. Liu, O.G. Kosareva, I.S. Golubtsov et al.: Femtosecond laser pulse filamentation versus optical breakdown in H_2O, *Appl. Phys. B* **76**, 215–229 (2003).
36. X. Mao, S.S. Mao, R.E. Russo: Imaging femtosecond laser–induced electronic excitation in glass, *Appl. Phys. Lett.* **82**, 697–699 (2003).
37. A. Becker, N. Aközbek, K. Vijayalakshmi et al.: Intensity clamping and re-focusing of intense femtosecond laser pulses in nitrogen molecular gas, *Appl. Phys. B* **73**, 287–290 (2001).
38. W. Liu, S. Petit, A. Becker et al.: Intensity clamping of a femtosecond laser pulse in condensed matter, *Opt. Commun.* **202**, 189–197 (2002).
39. A. Talebpour, M. Abdel-Fattah, S.L. Chin: Focusing limits of intense ultrafast laser pulses in a high-pressure gas: road to new spectroscopic source, *Opt. Commun.* **183**, 479–484 (2000).

40. O.G. Kosareva, V.P. Kandidov, A. Brodeur et al.: Conical emission from laser plasma interactions in the filamentation of powerful ultrashort laser pulses in air, *Opt. Lett.* **22**, 1332–1334 (1997).
41. K.M. Davis, K. Miura, N. Sugimoto et al.: Writing waveguides in glass with a femtosecond laser, *Opt. Lett.* **21**, 1729–1731 (1996).
42. F. Théberge, W. Liu, Q. Luo et al.: Extension of ultra-broadband continuum generated in air up to 230 nm using ultra-short and intense laser pulse, *Appl. Phys. B* **80**, 221–225 (2005).
43. S.L. Chin, S. Petit, F. Borne et al.: The white light supercontinuum is indeed an ultrafast white light laser, *Jpn. J. Appl. Phys.* **38**, L126–128 (1999).
44. F. Théberge, W. Liu, S.A. Hosseini et al.: Long-range spectrally and spatially resolved radiation from filaments in air, *Appl. Phys. B* **81**, 131–134 (2005).
45. W. Liu, S.A. Hosseini, Q. Luo et al.: Experimental observation and simulations of the self-action of white light laser pulse propagating in air, *New J. Phys.* **6**, 6 (2004).
46. A. Dubietis, E. Kucinskas, G. Tamosauskas et al.: Self-reconstruction of light filaments, *Opt. Lett.*, **29**, 2893–2894 (2004).
47. S. Skupin, L. Bergé, U. Peschel et al.: Interaction of femtosecond light filaments with obscurants in aerosols, *Phys. Rev. Lett.*, **93**, 023901 (2004).
48. M. Kolesik, J.V. Moloney: Self-healing femtosecond light filaments, *Opt. Lett.* **29**, 590–592 (2004).
49. W. Liu, F. Théberge, E. Arevalo et al.: Experiment and simulations on the energy reservoir effect in femtosecond light filaments, *Opt. Lett.* **30**, 2602–2604 (2005).
50. F. Courvoisier, V. Boutou, J. Kasparian et al.: Ultraintense light filaments transmitted through clouds, *Appl. Phys. Lett.* **83**, 213–215 (2003).
51. W. Liu, S.L. Chin, O.G. Kosareva et al.: Multiple re-focusing of a femtosecond laser pulse in a dispersive liquid (methanol), *Opt. Commun.* **225**, 193–209 (2003).
52. S.L. Chin, S. Petit, W. Liu et al.: Interference of transverse rings in multi-filamentation of powerful femtosecond laser pulses in air, *Opt. Commun.* **210**, 329–341 (2002).
53. V.I. Bespalov, V.I. Talanov: Filamentary structure of light beams in nonlinear liquids, *JETP Lett.*, **3**, 307 (1966).
54. S.A. Hosseini, Q. Luo, B. Ferland et al.: Competition of multiple filaments during the propagation of intense femtosecond laser pulses, *Phys. Rev. A* **70**, 033802 (2004).
55. Q. Luo, S.A. Hosseini, W. Liu et al.: Effect of beam diameter on the propagation of intense femtosecond laser pulses, *Appl. Phys. B* **80**, 35–38 (2005).
56. M. Mlejnek, M. Kolesik, J.V. Moloney et al.: Optically turbulent femtosecond light guide in air, *Phys. Rev. Lett.* **83**, 2938–2941 (1999).
57. S. Tzortzakis, L. Bergé, A. Couairon et al.: Breakup and fusion of self-guided femtosecond light pulses in air, *Phys. Rev. Lett.* **86**, 5470–5473 (2001).
58. Dubietis, G. Tamosauskas, G. Fibich et al.: Multiple filamentation induced by input-beam ellipticity, *Opt. Lett.* **29**, 1126 (2004).
59. S. Carrasco, S. Polyakov, H. Kim et al.: Observation of multiple soliton generation mediated by amplification of asymmetries, *Phys. Rev. E* **67**, 046616 (2003).
60. J. Kasparian, M. Rodriguez, G. Méjean et al.: White-light filaments for atmospheric analysis, *Science* **301**, 61–63 (2003).
61. J.-F. Gravel, Q. Luo, D. Boudreau et al.: Sensing of halocarbons using femtosecond laser–induced fluorescence, *Anal. Chem.* **76**, 4799–4805 (2004).
62. H. Pépin, D. Comtois, F. Vidal et al.: Triggering and guiding high-voltage large-scale leader discharges with sub-joule ultrashort laser pulses, *Phys. Plasma* **8**, 2532–2539 (2001).
63. M. Rodriguez, R. Sauerbrey, H. Wille et al.: Triggering and guiding high-voltage discharge by ohmic connection through ionized filaments created by femtosecond laser pulses, *Opt. Lett.* **27**, 772–774 (2002).

64. V. Koubassov, J.F. Laprise, F. Théberge et al.: Ultrafast laser–induced melting of glass, *Appl. Phys. B* **79**, 499–505 (2004).
65. M.R. Kasaai, V. Kacham, F. Théberge et al.: The interaction of femtosecond and nanosecond laser pulses with the surface of glass, *J. Non-Cryst. Solids* **319**, 129–135 (2003).
66. N.T. Nguyen, A. Saliminia, S.L. Chin et al.: Control of femtosecond laser–written waveguides in silica glass, *Proc. SPIE* **5578**, 665–676 (2004).

Chapter 15
Self-focusing and Filamentation of Powerful Femtosecond Laser Pulses

V.P. Kandidov, A.E. Dormidonov, O.G. Kosareva, S.L. Chin, and W. Liu

Abstract The physical picture of the phenomenon of filamentation of a powerful femtosecond laser pulse in the bulk of a transparent dielectric is considered on the basis of the experimental results at Laval University (Quebec, Canada), and the computer simulations performed at M.V. Lomonosov Moscow State University (Russia).

In this chapter the dynamic moving-focus model is discussed. The quasi-stationary model is offered for the analysis of filament origin. The quasi-stationary estimation of critical power for self-focusing is received with consideration of the delayed nonlinear response in gases. Generalization of the known Marburger formula is given to femtosecond pulses with an elliptic intensity distribution and to the initially chirped pulses. The influence of turbulence and scattering particles in the atmosphere on chaotic pulse filamentation is considered. Spatial regularization of filament by means of the regular intensity and phase perturbations introduced into the transverse section of the pulse is investigated. Chirped pulse propagation is studied in a single filament regime. The scenario of filament competition is considered on the basis of three-dimensional nonstationary propagation model.

15.1 Introduction

The phenomenon of filamentation of laser radiation, which has been known since the 1960 s, has again attracted the attention of scientists. Tightening of a powerful laser pulse in a filament that causes concentration of energy and intensification of nonlinear optical processes is of interest from the points of view of both fundamental science and its applications.

A.E. Dormidonov (✉)
International Laser Center, Physics Department, MV. Lomonosov Moscow State University, Moscow 119992, Russia
e-mail: adorm@list.ru

R.W. Boyd et al. (eds.), *Self-focusing: Past and Present*,
Topics in Applied Physics 114, DOI 10.1007/978-0-387-34727-1_15,

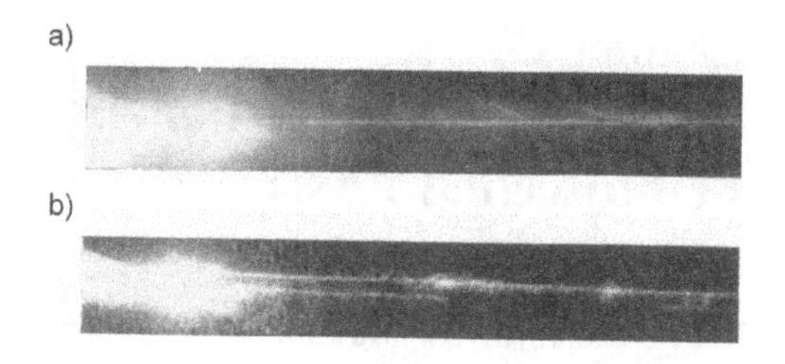

Fig. 15.1 The first self-focusing experiment with a 20-MW pulse focused into a cell with organic liquids. (**a**) Side view of one channel in cyclohexane and (**b**) of two glowing filaments in orthoxylol [1].

In 1965, post-graduate and undergraduate M.V. Lomonosov Moscow State University students N. Pilipetsky and A. Rustamov were the first to demonstrate filament images in the course of laser beam focusing [1]. A glowing filament, which contained about 1% of radiation energy, appeared in the laser pulse with the power of 20 MW focused into a cell with organic liquids (Fig. 15.1). In some cases two or three filaments appeared. The first report about self-focusing observation was made in [2]. Earlier, the self-focusing effect had been predicted in [3]. In the same paper a waveguide mode of the electromagnetic beam propagation had been suggested. It had been shown that self-focusing was possible in the beams only if their power exceeded the so-called "critical" power for self-focusing in the medium [4]. For the first time the self-focusing in air of focused and collimated beams was registered in [5] and [6], respectively.

At high density of laser radiation power a small-scale self-focusing develops, at which the beam breaks up as a result of modulational instability of an intense light field in a medium with the Kerr nonlinearity. The theory of spatio-temporal instability of intense light field in a cubic medium was developed in [7] and then was confirmed experimentally in [8]. Small-scale self-focusing was a serious barrier on the way of creation of powerful solid-state lasers [9]. The peak power in nonlinear foci reached the damage threshold of a material and led to the destruction of the optical elements [10]. To overcome this obstacle, large-beam-diameter laser systems began to be used in which the peak intensity in cascades of amplification was decreased by the scaling of the transverse size of a beam. Now this principle, being transferred on the temporal scale of radiation [11], allows experimenters to create powerful femtosecond laser systems.

The development of powerful femtosecond lasers made it possible to receive the filamentation of pulses in gaseous media, particularly in air at atmospheric pressure. In the first experiments [12–14] with pulses centered at 0.8 μm, peak power of 5–50 GW, duration of 150–230 fs, 10-m filaments were observed. In a thin filament of about 100 μm in diameter, 8–10% of the pulse energy was localized. The parameters of a filament are stable enough along its length and do not depend on the initial pulse energy and duration.

Color rings of the conical emission were formed around a filament. It is essential, that the filamentation take place without geometrical focusing of the pulse. The main parameters for the formation of filaments are femtosecond duration and central wavelength of the laser pulse. In the 1970 s–1980 s, powerful laser systems of atmospheric optics were created on the basis of CO_2 lasers, which generated millisecond pulses. When such pulses propagate in air, delocalization of their power occurs, which is caused by thermal self-influence, optical breakdown, and other nonlinear-optical effects with the power threshold lower than that of self-focusing [15].

The phenomenon of femtosecond filamentation of a laser pulse is accompanied by the formation of plasma channels, generation of white light and conical emission [16–19]. On the basis of these properties of filamentation systems for broadband laser probing, fluorescent and emission spectroscopy of the atmosphere, and remote control of the high-voltage discharge in air, methods of creation of microoptics devices are now under development [20–24]. The phenomenon of femtosecond filamentation in air, liquids and solid dielectrics, and its possible applications, are discussed in detail in [25].

In the present chapter the phenomenon of filamentation of a high-power femtosecond laser pulse constructed on the basis of the moving-focus model, as well as the quasi-stationary model, are both represented.

15.2 Femtosecond Filamentation of a Laser Pulse and the Moving-Focus Model

15.2.1 The Dynamic Moving-Focus Model

The formation of filament during the propagation of a femtosecond laser pulse can be explained in the following way. In a pulse with peak power, which exceeds the critical power for self-focusing, the intensity increases as it approaches the nonlinear focus. When the photoionization threshold of the medium is achieved, the laser-produced plasma is generated, in which the defocusing limits subsequent growth of the intensity in nonlinear focus.[1] The dynamic balance between the Kerr self-focusing and the plasma defocusing leads to localization of energy and stabilization of parameters in a lengthy filament. Nevertheless, the balance of the Kerr and the plasma nonlinearity in a filament does not mean *sensu stricto* the formation of a waveguide mode and self-channeling of a pulse in the medium.

The physical picture of filamentation can be visually demonstrated on the basis of the moving-focus model [29]. First, moving foci were registered in [5]. Review [30] describes other experimental confirmations of the model. According to the model a laser pulse is represented by a sequence of thin temporal slices, which are transformed in the course of nonlinear-optical interaction with the medium (Fig. 15.2).

---

[1] In air the photoionization threshold $I_{\mathrm{Thr}} = 10^{13} \div 10^{14}\,\mathrm{W/cm^{-2}}$ [26], in water and methanol $I_{\mathrm{Thr}} \approx 10^{13}\,\mathrm{W/cm^{-2}}$ [27,28].

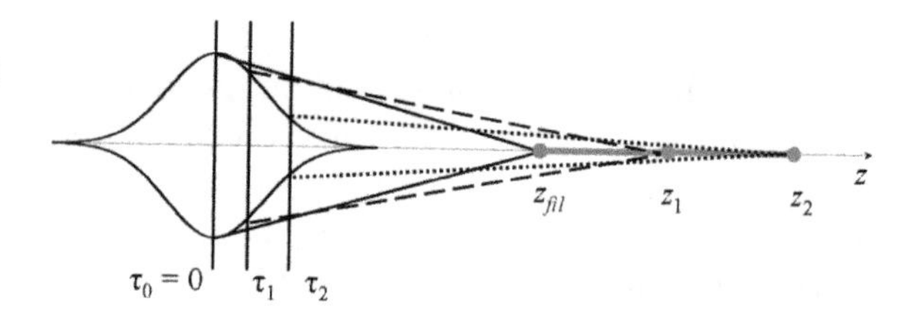

Fig. 15.2 The moving-focus model. Three temporal slices of the pulse focused at different distances z. The *bold curve* indicates the continuous succession of foci forming the filament

The moving-focus model explaining beam self-focusing in the medium with instantaneous Kerr response and the two-photon absorption is modified in [14] and [31] for the analysis of dynamic filamentation of a femtosecond pulse. As a result, the model of moving foci, which is used for the filamentation, can be naturally called the dynamic model.

The dynamic moving-focus model reflects the influence of one temporal slice of a pulse on subsequent slices, which originates from nonlinearity of laser plasma accumulated with time. According to the dynamic model [14,31] the filamentation of a high-power femtosecond laser pulse can be presented as follows. The temporal slice with the peak power is focused at the minimal distance, defining the starting point of the filament. With increasing intensity in this slice up to the threshold of multiphoton ionization, the laser plasma is produced. Defocusing in plasma stops the subsequent growth of intensity.

In other temporal slices of the pulse the power decreases continuously at their displacement from the peak slice. Therefore, slices in the front of the pulse are focused at the distance, which increases with their removal from the slice with the peak power. The continuous succession of nonlinear foci, which arises in a pulse from the slice with the peak power and then, consequently, in slices of pulse up to the slice with the critical power for self-focusing, represents a lengthy filament. The electron generation caused by photoionization of a medium in nonlinear foci forms a plasma channel, where defocusing occurs of all subsequent temporal slices of the pulse (Fig. 15.3). Therefore, in the case of femtosecond filamentation, unlike the quasi-stationary self-focusing and filamentation of a long pulse, the back of a femtosecond pulse is exposed to the aberrational defocusing and as a result the intensity ring distribution is formed around the filament [18,32,33].

The ring structure was first observed in [34] in the course of self-focusing of a ruby laser pulse with power up to 100 kW into a cell with CS_2. The appearance

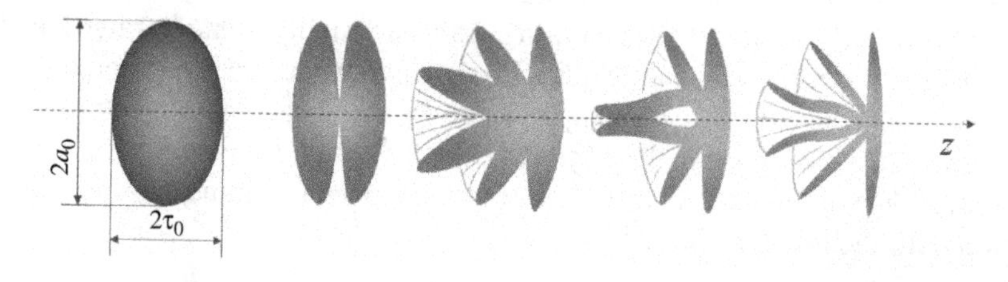

Fig. 15.3 The dynamic moving-focus model. Temporal slices at the back of the femtosecond pulse are exposed to the aberrational defocusing in the laser plasma. The tone images of spatio-temporal intensity distributions are presented

of rings is explained by interference between a spherical wave diffracting from a point in the liquid and the untrapped beam. The beam self-focusing with saturating nonlinearity was numerically considered in [35]. In the course of saturation the decreasing of the Kerr-lens focal power takes place at the beam axis and gives rise to the rings.

The ring formation, surrounding the femtosecond filament in air, was registered in [36] by measurements of silicate sample thickness at ablation under exposure to a focused femtosecond 85-mJ laser pulse. Typical ring structure, received experimentally and numerically in the course of ~90 m filamentation of a collimated pulse in air, is given in Fig. 15.4. Physically the formation of intensity rings in the pulse transverse section is explained by a simple model [37], evolving the assumptions stated in [34,35]. In [37] it is shown that rings are the result of superposition of the background light field with a plane wavefront and divergent field of the filament, the wavefront of which is the surface of a cone.

In the journal *Optics Letters* there was a terminological discussion about models of moving foci and self-channeling for interpretation of the phenomenon of femtosecond filamentation. In an experiment [38] with laser pulses focused in a cell with gases at various pressures, the filamentation of a pulse was observed behind the geometrical focus of a lens. The authors are of the opinion that this result is unexplained from the point of view of the moving foci model [14], and regard it as a confirmation of the formation of a self-guided mode. The opposite point of view is stated in [39] on the basis of laboratory and numerical experiments with the femtosecond pulse focused in a cell with water. It is shown,that a light filament does not behave as a self-channeled wave packet. The authors of [32] used the term "dynamic spatial replenishment" for explaining long-range filament formation by femtosecond laser pulses.

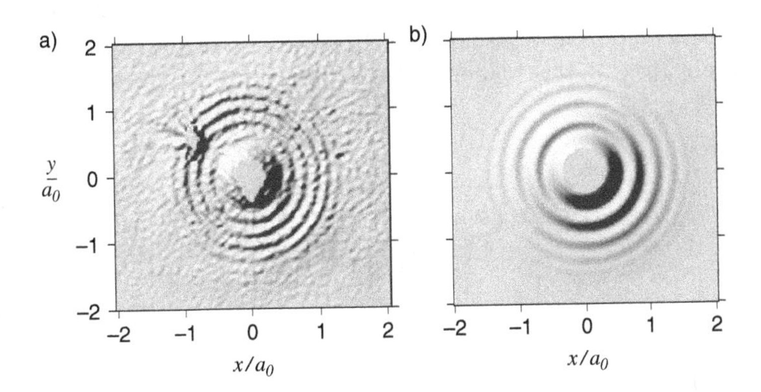

Fig. 15.4 Typical pictures of the interference of ring structures produced by two developed filaments at $z = 87$ m in a 14-mJ pulse. (a) Fluence distribution obtained from the experiment. (b) Intensity distribution obtained in the simulations is plotted in the relative units [37,44]

15.2.2 Refocusing and the Multiple Foci Model

In pulses with peak power of 6–10 times as much as the critical power for self-focusing in the medium, refocusing arises and becomes apparent through the formation of fluence maxima along the femtosecond filament. This effect, which first was found out for femtosecond filamentation in [14], is theoretically investigated in [15,31–33]. Later, the nonmonotonic character of intensity change in the filament was confirmed in [40] by measurements of a photoemission signal of molecules and ions of nitrogen. The influence of air pressure on the refocusing of a focused pulse was considered in [41].

Physically, refocusing is explained by recurring self-focusing of pulse temporal slices, which were defocused in the laser plasma. An increase in intensity at the beam axis toward the back of the pulse — simultaneously with continued self-focusing at the pulse front — leads to the fluence growth registered in experiments. This is schematically explained in Fig. 15.3. Computer experiments [31] visually illustrate how divergent rings of aberrational defocusing at the back of the pulse tighten again to the beam axis (Fig. 15.5).

At the beginning of refocusing the dispersion of the pulse plays an essential role. Strong material dispersion of the medium can result in multiple refocusing, at which a filament breaks up into sequence of "hot" spots with high localization of energy [29]. With an increase of group velocity dispersion the number of hot spots along the filament decreases, and the distance between them reduces.

The dynamic picture of refocusing of a femtosecond pulse has some different aspects in comparison with the multi-focus model proposed in [42] on the basis of numerical investigation of stationary self-focusing in a medium with two-photon absorption. According to this model, to the first nonlinear focus the power flows from a circular area on the beam axis. Then this power comes out from the "game" due to absorption in a medium and diffraction. To the second focus the power flows from the ring enveloping the circular area; to the third focus the power flows from the following ring; and so on [43]. At femtosecond filamentation in transparent dielectric energy losses are insignificant, and the collapse of the nonlinear focus stops due to a strong nonlinear

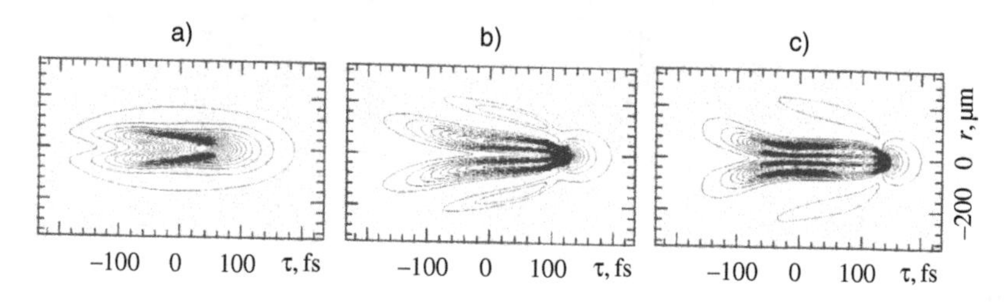

Fig. 15.5 Spatio-temporal evolution of the intensity distribution of the pulse in the filament: $z = 27\,\mathrm{m}$ (a); 33 m, (b); 40 m (c). The contour interval is $0.25 \times 10^{13}\,\mathrm{W/cm^{-2}}$, $a_0 = 3.5\,\mathrm{mm}$, $P_{\mathrm{peak}} = 5P_{cr}$ [31]

refraction of the pulse in the laser plasma, instead of absorption in the medium. Formation of the maxima of filament fluence at refocusing occurs from simultaneous self-focusing at different temporal slices of the pulse.

15.2.3 Filamentation and Transverse Energy Flows

A narrow filament with a high concentration of light field is surrounded by the background reservoir of energy that provides its existence [32]. From this reservoir the sequence of nonlinear foci in the front of the pulse is supplied by energy, forming a filament. In the pulse trailing part, the energy in the rings provides a subsequent refocusing. In the course of filamentation there is a continuous exchange of energy in the transverse section of the pulse between its paraxial area and its periphery [44].

During nonlinear focus formation, power from the periphery of the transverse section flows toward the axis in a temporal slice of a pulse (Fig. 15.6). When the intensity of focus reaches the threshold of photoionization, laser plasma is generated. Defocusing in plasma causes a power flow from the paraxial area to the periphery of transverse section of temporal slice. At the distance z where refocusing arises, the power flows down again to the paraxial area, increasing intensity on the axis. After refocusing the power again flows to the periphery of the transverse section of the pulse temporal slice. Thus, a lengthy filament exists due to the background reservoir of energy with which it continuously exchanges.

The determining role of the energy reservoir in formation of lengthy filament is confirmed visually with laboratory experiments. In [39] a section of filament was completely blocked by a small circular screen with diameter 55 μm.

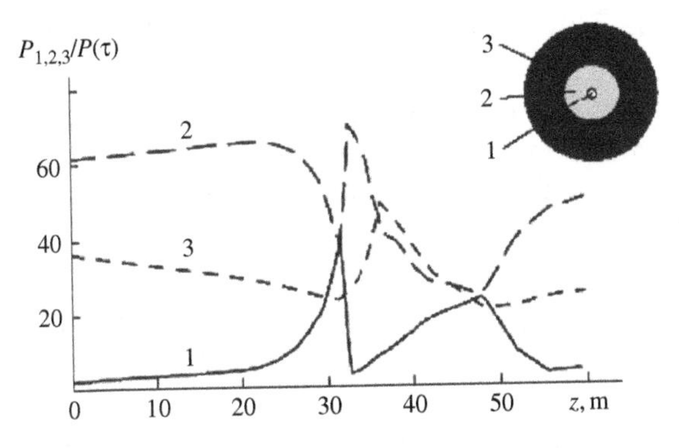

Fig. 15.6 Dependences of power in different spatial regions of the pulse temporal slice $\tau = -20$fs on the distance z: in the axial region of radius 0.5 mm (1), in a ring with radii of 0.5 and 3.5 mm (2), and in a ring with radii of 3.5 and 8 mm (3). Pulse energy $W = 10$ mJ, duration $\tau_0 = 140$fs and $a_0 = 3.5$ mm. The insert shows the positions of the regions in a pulse transverse section [44]

Nevertheless, the filament was interrupted only a small distance behind the screen, but then reappeared again on all its length as in the absence of the screen.

In one experiment [45] thin aluminum foil screen was located on the trace of the filament. The intensive field of the filament burned a hole through the foil. The screen with the hole burnt completely blocked the filament behind the screen. Energy that transferred from the periphery of the transverse section of the pulse was cut off by the screen, and in the absence of a background reservoir of energy the filament was eliminated.

15.3 Quasi-Stationary Model of Filament Origination

15.3.1 Initial Stage of Filamentation

The process of filamentation can be separated into two stages: the initial stage, at which the formation of nonlinear focus occurs, and the dynamic stage, at which an aberrational defocusing dominates the self-induced laser plasma. The initial stage, i.e., the prefilamentation stage, consists of the pulse transformation, with intensity under the threshold of medium photoionization. At this stage. plasma is not yet generated, and the light field varies only due to the Kerr self-focusing. For pulses with duration of some tens of femtoseconds, the filamentation in air and, as a rule, in optical glasses begins on distances considerably smaller than the dispersive length, and the influence of material dispersion on the origin of filament can be neglected [46,47]. Thus, at the initial stage of filamentation the complex amplitude $E(x, y, z, t)$ of the light field in temporal slice of a pulse submits to the equation:

$$2ik\frac{\partial E}{\partial z} = \frac{\partial^2 E}{\partial x^2} + \frac{\partial^2 E}{\partial y^2} + \frac{2k^2}{n_0}\Big(\Delta n_k(|E|^2) + \Delta\tilde{n}(x, y, z)\Big)E, \qquad (15.1)$$

where $\Delta n_k(|E(x, y, z, t)|^2)$ is the nonlinear correction to the refractive index due to the Kerr effect, and $\Delta\tilde{n}(x, y, z)$ is the random fluctuations of the refractive index in the medium. The electric field envelope $E(x, y, z = 0, t)$ at the laser system output is usually represented in the form:

$$E(x, y, z = 0, t) = E_0 \exp\left\{-\frac{t^2}{2\tau_0^2}\right\} \exp\left\{-\frac{x^2 + y^2}{2a_0^2}\right\}. \qquad (15.2)$$

In the central slice of the pulse $t = 0$ the power achieved the peak value $P_{\text{peak}} = \pi a_0^2 I_0$, where $I_0 = \frac{cn}{8\pi}|E_0|^2$ is the peak intensity.

The solution to problem (15.1) for the central pulse slice $t = 0$, which contains the peak power, allows us to investigate the prefilamentation, i.e., the origin of filament due to the Kerr self-focusing. The distance z_{fil} on which the intensity in the central pulse slice achieves the threshold of photoionization of a medium is the distance of the filamentation start.

15.3.2 Quasi-Stationary Estimation of Critical Power

For solids and liquids the increment Δn_k is caused only by the electronic component, and the nonlinearity can be considered as an instantaneous $\Delta n_k = n_2 I$, where n_2 is the nonlinearity coefficient of a medium. In gases, Δn_k is defined by the electronic nonlinearity and the stimulated Raman scattering of radiation on rotational transitions of molecules, which leads to delay of the nonlinear response. For air this delay is $\tau_{nl} \approx 70$fs [48], and it is necessary to consider it for pulses of femtosecond duration. According to [32] in air the following approximation for increment Δn_k can be used:

$$\Delta n_k(t) = \frac{1}{2} n_2 \left(|E(t)|^2 + \int_{-\infty}^{t} H(t-t')|E(t')|^2 dt' \right), \tag{15.3}$$

where $H(t)$ is the response function, and $n_2 = (1.5 \div 5.6) \times 10^{-19} \mathrm{cm}^2/\mathrm{W}$ is the coefficient of air nonlinearity, which was established for pulses with duration $\tau_0 >> \tau_{nl}$ [48].

It is possible to introduce the effective coefficient of nonlinearity $n_2^{qst}(t)$ for the temporal slice t of pulse with the known profile $E(t)$ [49]:

$$n_2^{qst}(t) = n_2 \cdot \eta(t), \text{ where } \eta(t) = \frac{1}{2} \left(1 + \int_{-\infty}^{t} H(t-t')|E(t')|^2 dt' / |E(t)|^2 \right). \tag{15.4}$$

The coefficient of nonlinearity received in this way corresponds to the quasi-stationary estimation in approximation of the given profile of the pulse. In particular, for the Gaussian pulse the quasi-stationary model leads to following expression for the function $\eta(t=0)$ in the central pulse slice:

$$\eta(0) = \frac{1}{2} \left(1 + \int_{-\infty}^{0} H(t-t') \exp\left\{ -\frac{t'^2}{\tau_0^2} \right\} dt' \right). \tag{15.5}$$

The coefficient of nonlinearity determines the critical power P_{cr} [48]:

$$P_{cr} = 3.77 \frac{\pi n_0}{2k^2 n_2}. \tag{15.6}$$

The quasi-stationary estimation of the critical power $P_{cr}^{qst}(t)$ for the temporal slice t of femtosecond pulse in air follows from (15.4, 15.6):

$$P_{cr}^{qst}(t) = P_{cr}/\eta(t). \tag{15.7}$$

The function $\eta(t) \leq 1$ and $P_{cr}^{qst}(t) \geq P_{cr}$. For example, for a Gaussian pulse (15.1) with duration $\tau_0 = 27$fs the effective coefficient of nonlinearity for the central temporal slice equals $n_2^{qst}(t=0) = 0.56n_2$ [49]. The quasi-stationary critical power for self-focusing of the central slice of the pulse $P_{cr}^{qst}(t=0)$ is 1.8 times as mush as the critical power for self-focusing of a long pulse ($P_{cr}^{qst}(t=0) = 1.8P_{cr}$).

Direct measurements of critical power for self-focusing of Ti:sapphire laser pulses with energy of 0.42 mJ and the pulse duration from 42 to 800 fs were executed in [50]. Critical power for self-focusing was determined by displacement of a focal spot of the focused laser pulse, the duration of which changed by means of the initial phase modulation. The experimental results received with positive and negative initial phase modulation are in agreement with an analytical estimation.

15.3.3 Generalized Marburger Formula

The concept of critical power for self-focusing was introduced at the analysis of the axially symmetric Gaussian beam [3]. With the change of intensity distribution in a beam the value of critical power varies, reaching a minimum for a beam whose profile coincides with the Townes mode [51]. For beams without axial symmetry the critical power for self-focusing increases because of significant outflow of its power at formation of the axisymmetric Townes mode. For elliptical beams the critical power $P_{cr}(a/b)$ depending on the axial ratio a/b of initial intensity distribution was determined by means of numerical simulation [52–54]. As the ratio a/b increases the critical power for self-focusing of elliptical beam rises several times (Fig. 15.7).

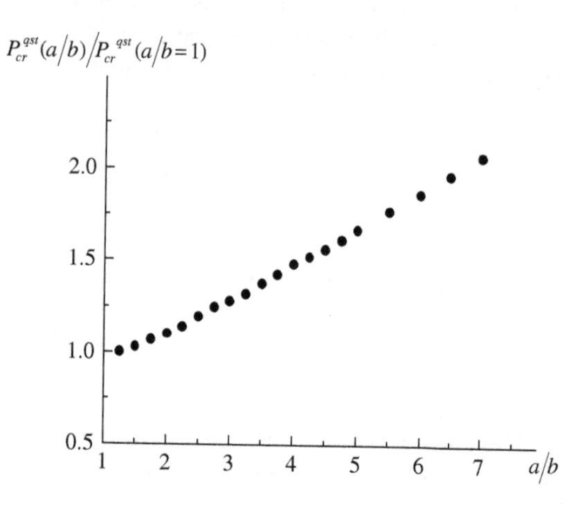

Fig. 15.7 Dependences of the critical power for self-focusing P_{cr}^{qst} on the axial ratio a/b of the initial elliptic transverse section of a collimated Gaussian beam: circles are numerical simulation [54]; the solid curve is the empirical expression [52]

The quasi-stationary model allows one to analytically estimate the distance z_{fil} before the start of filamentation. From the computer simulation it follows that the Marburger formula [55] for the distance of the beam self-focusing is generalized on the quasi-stationary model of the initial stage of filamentation of the pulse with elliptical intensity distribution. According to this generalization the distance before the filament start z_{fil} is calculated with the generalized Marburger formula [55,56]:

$$z_{fil} = \frac{0.367kab}{\left\{\left[\left(\frac{P_{\mathrm{peak}}}{P_{cr}^{qst}(t=0,a/b)}\right)^{1/2} - 0.852\right]^2 - 0.0219\right\}^{1/2}}. \qquad (15.8)$$

where a, b are the axes of elliptic distribution of intensity in the transverse section of a pulse, P_{peak} is the pulse peak power reached in the central ($t = 0$) temporal slice, $P_{cr}^{qst}(t = 0, a/b)$ is the quasi-stationary estimation (15.7) of the critical power for self-focusing of the central temporal slice of the femtosecond pulse with elliptic intensity distribution (vide supra Fig. 15.7). Application of formula (15.8) is confirmed with the computer simulation results of pulse filamentation in air with duration $\tau_0 = 27$ and 140 fs and peak power $P_{\mathrm{peak}} = 10 - 64\mathrm{GW}$ at a various axial ratio a/b of elliptic intensity distribution [56].

15.4 Modulational Instability and Multifilamentation

During the propagation of femtosecond pulses, the power of which exceeds the critical power for self-focusing, multifilamentation occurs as the pulse breaks up into multiple filaments. The break-up of a pulse is the unavoidable consequence of transverse modulational instability of an intense light field in a medium with the cubic nonlinearity. The initial stage of multifilamentation at which nonlinear foci are formed is defined by small-scale self-focusing in temporal slices of a pulse. The centers of origins of nonlinear foci and, hence, of filaments in the transverse section of the pulse are chaotic perturbations of intensity which can be caused by quality of output laser beam, fluctuations of a refractive index, and scattering by particles of a medium.

15.4.1 Origin of Filaments from Initial Intensity Perturbations

The formation of nonlinear foci at the initial stage of filamentation causes the redistribution of power in the transverse section of a pulse. The character of this redistribution depends on pulse energy and geometry of intensity perturbation at the laser system output. In experiments [57] with the pulses centered at

1053 μm, the peak power 30 GW, and duration 500 fs, it was observed that at small distances there were 3–4 maxima of energy density formed in the transverse section of a beam.

However, later on not all of these maxima were transformed to filaments. In [58] the complex picture of the formation of filaments in a pulse with nonunimodal distribution of energy density in the transverse section was investigated. It was noted that at the beginning, due to the modulational instability, the initial laser beam breaks up into two maxima in transverse section, which then merge, forming one filament.

In experiments [49] in air with collimated pulses with duration of 45 fs (FWHM), filaments were formed independently from large-scale perturbations of the initial profile [Fig. 15.8(a)]. In the numerical simulation of this experiment, the quasi-stationary model of the filament origin for the pulse with an intensity profile coincided with the measured fluence on the laser system output. Intensity distributions in the transverse section of the central slice of the pulse were received numerically for the distance, where filamentation began, are similar to the fluence registered in the experiment [Fig. 15.8(b)]. At increasing pulse energy up to 40 mJ, a bunch containing many filaments is formed. These filaments are irregularly located in the transverse section of the beam and begin at different distances from the output aperture of the laser [Fig. 15.8(c)]. As follows from estimations [5], in such a pulse [Fig. 15.8(a)] the intensity perturbations, with transverse section size equal to 0.1 cm, have the greatest increment of growth. Thus, with the pulse energy increasing. filaments arise from small-scale perturbations of intensity of the initial profile, which generally vary irregularly from shot-to-shot.

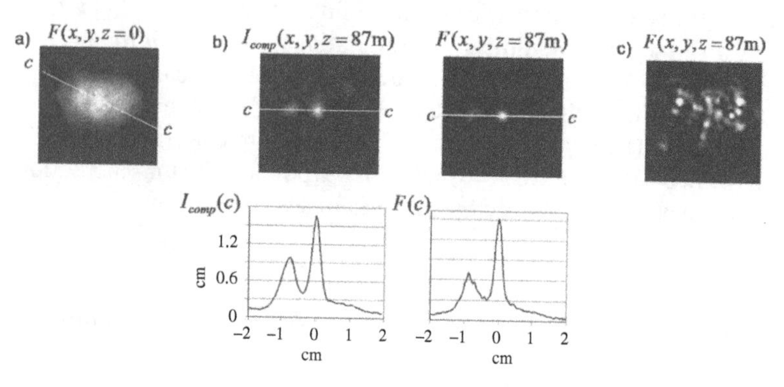

Fig. 15.8 (a) Experimentally measured initial fluence distributions $F(x, y, z = 0)$. (b) Measured fluence $F(x, y, z)$ and computer intensity distributions $I_{\text{comp}}(x, y, z)$ and fluence $F(c)$ and intensity $I_{\text{comp}}(c)$ profiles. Pulse energy $W = 10\,\text{mJ}$, distance $z = 87\,\text{m}$. (c) Fluence $F(x, y, z)$ at pulse energy of $W = 40\,\text{mJ}$ [49]

15.4.2 "Energy" Competition between Initial Perturbations

Detailed research of the formation of several filaments in a pulse with initial intensity perturbations are performed in [59] on the example of superposition of two partial coherent beams whose centers are displaced with respect to each other in the transverse section plane.

$$E(x,y,z=0,\tau)=e^{-\tau^2/2\tau_0^2}E(x,y).$$
$$E(x,y))=E_0\exp\left[-\frac{(y-d/2)^2+x^2}{2a_0^2}\right]+E_0\exp\left[-\frac{(y+d/2)^2+x^2}{2a_0^2}\right], \quad (15.9)$$

where d is the distance between the partial beams of radius a_0. From computer simulations on the basis of the quasi-stationary model it follows that for the pulse considered there are two critical powers $P_{cr}^{(1)}$ and $P_{cr}^{(2)}$, which define borders of areas of various modes of filamentation depending on distance d [Fig. 15.9(a)]. The area $P_{cr}^{(1)} < P_{\text{peak}} < P_{cr}^{(2)}$ corresponds to the mode of the single-filament formation, the area $P_{\text{peak}} > P_{cr}^{(2)}$ to the mode of two and more filaments. At the peak power $P_{\text{peak}} < P_{cr}^{(1)}$ the filamentation is absent.

The distance to the start of filamentation z_{fil} changes nonmonotonically with an increase of the peak power at the constant distance between partial beams [Fig. 15.9(b)]. Such a dependence $z_{fil}(P_{\text{peak}})$ is explained by a "power" competition between initial perturbations. For the peak power $P_{\text{peak}} > P_{cr}^{(1)}$, the initial perturbations at the beginning of the propagation merge into one, and then self-focusing occurs. In this case, the distance to the filament origin decreases with increasing P_{peak}. For $P_{\text{peak}} \leq P_{cr}^{(2)}$, initial perturbations increase independently at beginning. Then, the extended competition occurs between enhanced perturbations, and the power accumulated in them transfers to the beam axis, where

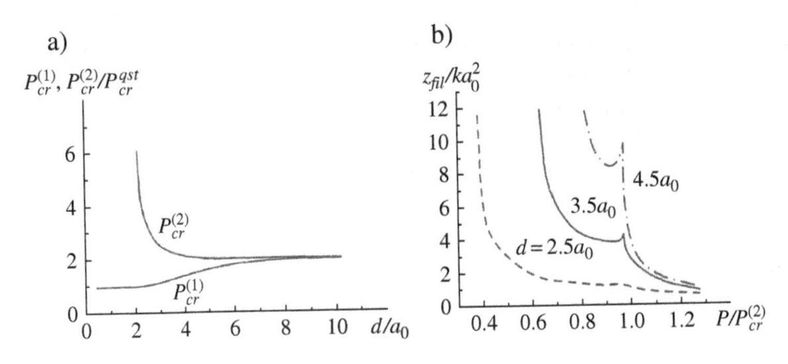

Fig. 15.9 Dependences of the critical powers $P_{cr}^{(1)}$ and $P_{cr}^{(2)}$ on the distance d between the maxima of perturbations in the initial intensity distribution (a) and dependences of the distance z_{fil} to the filament start on the peak power P_{peak} (b). P_{cr}^{qst} is the critical power for self-focusing of axisymmetric collimated Gaussian beam; $P_{cr}^{(2)}$ is the critical power for two-foci formation [59]

one nonlinear focus forms. The redistribution of power in the pulse transverse section increases the distance to the filament starts. For $P_{\text{peak}} > P_{cr}^{(2)}$, the distance $z_{fil}(P_{\text{peak}})$ again monotonically decreases with increasing power.

15.4.3 Spatial Regularization of Filaments

The stochastic character of the multifilamentation, which is caused by modulation instability of the pulse with random perturbations of the initial profile, decreases the effectiveness of the femtosecond laser systems, for example, of atmospheric lidars because of an unstable back-scattering signal [60]. In recent years, various methods for spatial regularization of a bunch of filaments in high-power pulses have been offered. The opportunity for filament regularization by means of introducing a circular diaphragm, the mask with apertures, and introducing phase astigmatism in a pulse at the laser output are considered in [61]. It can be noted that a regular bunch of filament whose centers are located on a circle was observed in 1973 [62].

The general method for spatial regularization of multifilamentation consists in formation in the pulse transverse section of a regular system of perturbations of the light field, capable of suppressing the influence of random fluctuations on the origin and formation of filaments. In this case the modulation instability of the light field develops on the preset array of perturbations, creating the centers of origin of regular filaments. For the creation of amplitude perturbations in the pulse transverse section for example a wire mesh can be used, and phase perturbations can be created by lens arrays.

The method of spatial regularization of multiple filaments by means of a mesh imposed in a plane of the pulse transverse section is experimentally and numerically investigated in [63–66]. For the "stochastic" multifilamentation achieved when only the random mask was located in front of the methanol cell entrance window, filaments were arranged chaotically in the pulse transverse section. For the "periodic" filamentation, when only the mesh was in front of the cell, a regular system of filaments was formed. For "regularization" mode both the mask and the mesh were placed before the cell entrance window, an ordered spatial arrangement of filaments was observed, which is in contrast to the "stochastic" filamentation.

For computer simulation of spatial regularization of multiple filaments the quasi-stationary model (15.1) was used with $\Delta\tilde{n}(x, y, z) = 0$ [67]. At the "stochastic" filamentation, the light field $\tilde{E}(x, y, z = 0, t)$ for the ith laser shot was represented with the additive perturbations $\tilde{\xi}_i(x, y)$:

$$\tilde{E}_i(x, y, z = 0, t) = C_i\big(1 + \tilde{\xi}(x, y)\big) \cdot E(x, y, z = 0, t), \qquad (15.10)$$

where C_i is the normalizing factor. The random function $\tilde{\xi}(x, y)$ obeys the normal distribution law with zero mean value and variance σ^2. The spatial correlation of the perturbations is given by the Gaussian function.

The correlation radius $R_{\rm cor}$ of perturbations was chosen so, that in near-axial region of a pulse the power $P_{R_{\rm cor}}$ contained in the perturbation with this radius R_{cor} met the condition: $P_{R_{\rm cor}} \geq P_{cr}^{qst}(t=0)$. In this case, perturbations with the radius $R_{\rm cor}$ are most "dangerous" because they have the greatest increment of growth [5]. At research of the periodic filamentation the amplitude of the light field $E_{\rm per}(x,y,z=0,t)$ after a mesh equals:

$$E_{\rm per}(x,y,z=0,t) = E(x,y,z=0,t) \cdot T_{\rm array}(x,y). \tag{15.11}$$

The mesh transmission factor $T_{\rm array}(x,y)$ periodically changed with period d in the XY plane. The width h of the opaque part of the mesh was small compared with the mesh period d, so that the resulting energy loss was within 10–20% of the initial pulse energy calculated before the mesh.

At simulation of the "regularization" mode, the light field $E_{\rm reg}$ was present as follow:

$$E_{\rm reg}(x,y,z=0,t) = \tilde{E}_i(x,y,z,=0,t) \cdot T_{\rm array}(x,y). \tag{15.12}$$

From the numerical simulation it was learned [67] that at the stochastic filamentation the average number of filaments monotonously increases with distance z, as the "periodic" filaments appear in groups. In the mode of regularization there is a significant suppression of the contribution of chaotic perturbations, and filaments origin on average close to the group formation (Fig. 15.10). For the spatial regularization it is necessary that the modulation instability on the regular perturbations, which are created by a mesh, developed on smaller distance, than on random fluctuations of the light field. It is

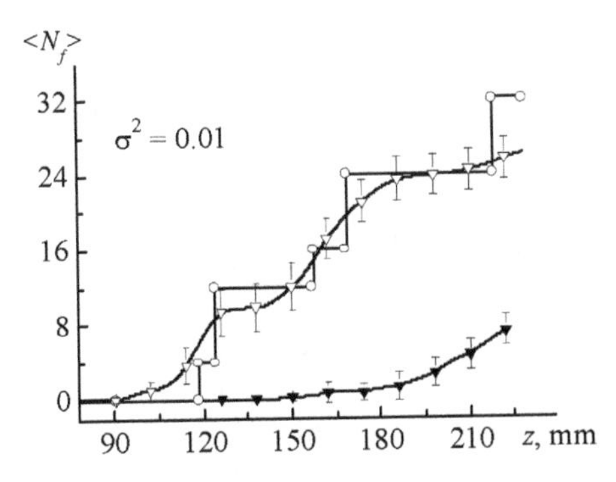

Fig. 15.10 Average number of filaments as a function of the distance. Open circles indicate periodic filamentation; the curves marked by closed triangles indicate the stochastic mode for $\sigma^2 = 0.01$; the curves marked by open triangles indicate the regularization mode. The mesh period $d = 0.2a_0$, the ratio $P_{\rm peak}/P_{cr}^{qst} = 370$, $P_{\rm unit}/P_{cr}^{qst} = 3.7$, $P_{R_{\rm cor}}/P_{cr}^{qst} = 1.5$. Laser pulse parameters: $\lambda = 800$nm, duration 42 fs FWHM, energy 130 μJ, and radius $a_0 = 2.4$ mm [67]

established, that the filament regularization by means of a mesh is most effective if a mesh cell contains the power P_{unit} equals $(3.1 \div 3.2)P_{cr}^{qst}$. Therefore, the optimum period d_{opt} of regularizing mesh follows the estimation:

$$d_{\text{opt}} = (1.76 \div 1.79)\sqrt{P_{cr}^{qst}(t=0)/I_0}. \tag{15.13}$$

15.5 Femtosecond Pulse Filamentation in the Atmosphere

In the first experiments on filamentation of femtosecond laser pulse in atmosphere [68] a focusing lens with a 30-m focal length was used for positioning of filament in space. Theoretically problems of stochastic filamentation, influences of dispersion at the propagation of femtosecond pulses in atmosphere are discussed in [65].

15.5.1 Filament Wandering

In laboratory experiments [12] it was observed that in the plane of observation the position of lengthy filament wanders from shot-to-shot. In subsequent works [46,69,70], random displacements of filament at different distances from the laser system output was measured, and statistical investigations of the distribution function of these displacements were performed.

The quasi-stationary model is applicable for the stochastic problem about random displacements of filament in the turbulent atmosphere. In the computer simulation, fluctuations of the refractive index $\Delta\tilde{n}(x,y,z)$ were represented by a chain of δ-correlated phase screens located along the propagation axis [71]. For the spatial spectrum of the refraction index $F_n(\kappa_x,\kappa_y,\kappa_z)$ the following expression was used [72]:

$$F_n(\kappa_x,\kappa_y,\kappa_z) = 0.033C_n^2(\kappa^2+\kappa_0^2)^{-11/6}\exp\{-\kappa^2/\kappa_m^2\}, \tag{15.14}$$

where $\kappa_0 = 2\pi/L_0$, $\kappa_m = 5.92/l_0$, L_0 and l_0 are the outer and inner scales of turbulence, respectively, and C_n^2 is the structure constant.

The statistical analysis of the results of laboratory experiments and of numerical simulations has shown that in the turbulent atmosphere distribution function of filament wanderings obeys Rayleigh's law. It is necessary to note that statistical investigations [49] have shown that for a pulse with random fluctuation of intensity at the laser system output the distribution function of displacement of the filament does not obey Rayleigh's law.

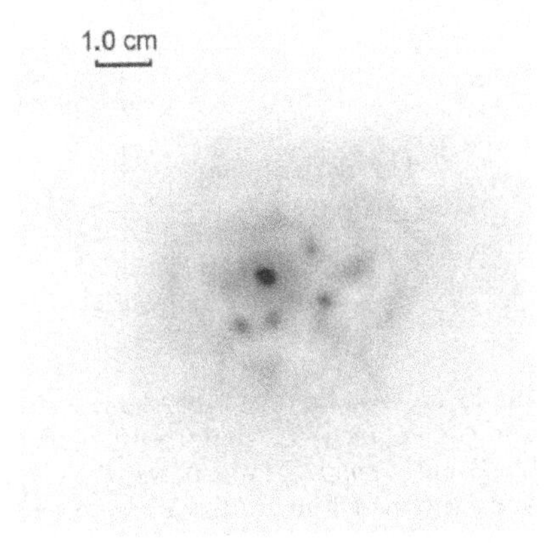

Fig. 15.11 Typical tone picture of the multifilamentation of a 40-mJ pulse with duration 42 fs (FWHM) [60]

15.5.2 Bunch of Chaotic Filaments in the Turbulent Atmosphere

Pulses of an atmosphere terawatt-power laser break up into multiple filaments. In experiments [73,74] the phase-modulated pulses with duration of 100–600 fs, energy of 230 mJ were used. In experiments [25,60] with spectral-limited 45-fs pulses with energy of 50 mJ, the formation of multiple "hot" spots at multifilamentation have been registered (Fig. 15.11). Such a picture of multifilamentation explains that the modulation instability of a powerful pulse develops both for inhomogeneities of the initial profile of intensity and from irregular perturbations caused by atmospheric turbulence.

At multifilamentation, the quasi-stationary model describes the origin of filaments that are initiated by fluctuations of a refractive index of a medium. Statistical characteristics of a bunch of filaments were determined from a series of 100 pulses, any one of which was resolved by problem (15.1) with statistically independent chains of phase screens. The average distance $< z_{fil}(N) >$ at which the pulse can have N "primary" filaments is reduced with an increase in the structural constant C_n^2 — characterizing the "force" of turbulence — and weakly depends on the internal scale l_0 (Fig. 15.12). The average width of a bunch of N chaotic filaments increases with the growth of C_n^2.

15.5.3 Scattering in Aerosols and Formation of Filaments

In laboratory experiments on filamentation of laser pulses in an aerosol it has been established [76,77] that with an increase of the optical thickness of the medium, the number of filaments is reduced. The first natural experiment [47] has shown that filaments can be formed and propagate in a rain. In pulse transverse section, the formation of a diffraction pattern (typical for scattering

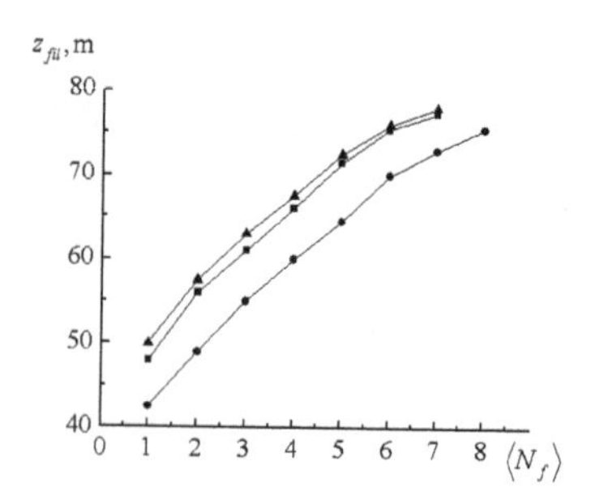

Fig. 15.12 Distance z_{fil} to the filament start in the atmosphere for: $C_n^2 = 15 \times 10^{-15}\,\text{cm}^{-2/3}$, $l_0 = 1$ mm (*circles*); $C_n^2 = 3.0 \times 10^{-15}\,\text{cm}^{-2/3}$, $l_0 = 1$ mm (*squares*); $C_n^2 = 3.0 \times 10^{-15}\,\text{cm}^{-2/3}$, $l_0 = 5$ mm (*triangles*). Pulse power $P_{\text{peak}} = 1.2 \times 10^{11} =$ W and $a_0 = 0.82$ cm [75]. On the trace with a length of 80 m, statistically independent screens with the spectrum (15.14) were located, each 10 m long

by spherical particles) was observed. Originating intensity perturbations may become self-focusing initiators sites and thus initiators of random filaments.

The influence of a separate particle on the filamentation of a pulse was investigated in the experiment [78] where water droplets with a diameter from 30 to 100 μm were placed on the trace of the filament propagation. It was revealed that particles comparable with the filament diameter had an insignificant influence on the existence of the filament behind the drop.

"Survivability" of the filament while hitting an aerosol particle is explained by preservation of the surrounding energy reservoir, which provides its subsequent existence. The simple model in which the particle is replaced with an absorbing disk was offered in [79]. On the basis of this model, numerical investigations of the propagation of the femtosecond pulse through separate opaque particles displaced from the pulse axis were performed [80]. The model of absorbing disks for aerosol particles is used in [77] for interpretation of laboratory experiments on pulse filamentation in clouds of high optical density. The influence of dense aerosol, which is represented by a continuous strongly absorbing layer, on the length of filament is numerically considered in [81].

At the same time, water particles of rain, clouds, and fog absorb weakly at 800 nm and mainly cause the scattering. The stratified model of pulse propagation in an aero-disperse medium, which describes coherent scattering by particles, diffraction, and Kerr and plasma nonlinearity in air, was offered in [82]. The stratified model consists of a chain of screens with particles, at which takes place coherent pulse scattering leading to the small-scale intensity redistribution in its transverse section. At free sections between aerosol screens, diffraction and nonlinear-optical interaction of a pulse with gas components of air takes place.

On the basis of the stratified model the pattern of intensity distribution in the transverse section of a subterawatt pulse that propagated in a drizzling rain was received in [83]. The distribution contains randomly located "hot" spots, and

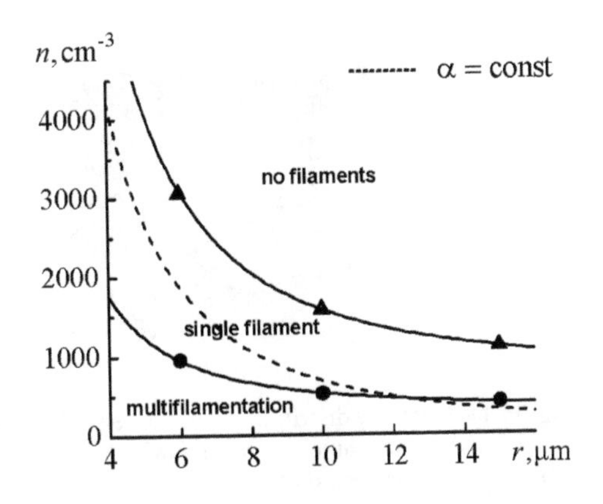

Fig. 15.13 Different modes of filamentation versus concentration n and radius r of aerosol particles. Circles mark the upper board of the multifilamentation regime; triangles represent the upper board of a single filamentation regime; the dashed curve corresponds to the constant value of the attenuation $\alpha \sim nr^2$

ring formations such as in [49], which are caused by the diffraction of a light field on the droplets. The statistical analysis performed in [84] has shown that the size of perturbations with the greatest increment of growth [5] and does not depend on drops concentration, and is determined only by the beam power.

From numerical simulation it has been established that scattering by particles with a radius of 2–5 μm does not practically influence the filamentation of a pulse with peak power $P_{\text{peak}} \sim 100P_{cr}$, whereas large particles with a radius of 10–15 μm produce in pulses the formation of intensity perturbations that stimulate the origin of filaments, and the distance before the filaments start is reduced. Depending on the concentration and the size of atmospheric aerosol particles, the different modes are possible: the formation of many filaments, the formation of a single filament in a pulse, and, at last, the absence of filamentation at the large optical thickness (Fig. 15.13).

15.5.4 Regularization of Filaments in the Atmosphere by a Lens Array

In atmosphere it is reasonable to use the regular phase modulation in the transverse section of a pulse for suppression of the influence of turbulent fluctuations of a refractive index. Moreover, the phase regularization does not bring energy losses, unlike amplitude mask. In [85] the lens array of the spatial regularization of a stochastic bunch of filament is numerically investigated (Fig. 15.14). The analysis was performed on the basis of the quasi-stationary models (15.1) for the light field amplitude of the pulse $E_{\text{reg}}(x, y, z = 0, t)$, represented in the form of (15.12). Each array element has the square form with size d and corresponds to a lens with focal length R_f. Lens focusing radius in array $R_f = 16.75$m, lenses size $d/a_0 = 0.4 \div 1.0$. Upon regularization by means of a homogeneous lens, array filaments are formed by groups. Optimization of the lens array elements allows formation of all filaments at the same distance [86].

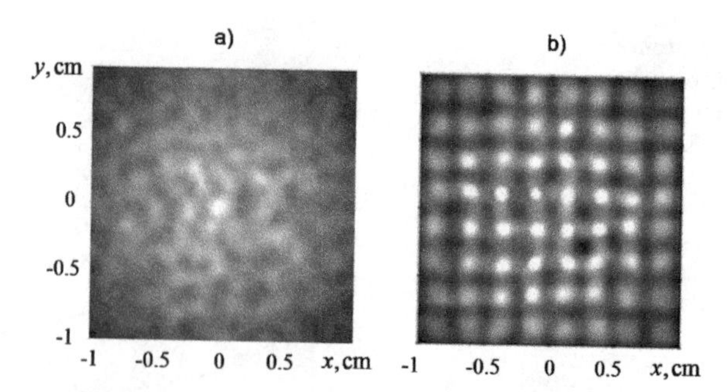

Fig. 15.14 Intensity distributions in the central slice of a pulse in a turbulent atmosphere after $z = 4.7$ m: (a) without lens array, (b) with lens array. The beam radius $a_0 = 1$ cm and the ratio $P/P_{cr} = 135$. The period of the lens array is $d = 2.5$ mm. The turbulence structure constant is $C_n^2 = 5 \times 10^{-13}\,\mathrm{cm}^{-2/3}$, the inner scale is $l_0 = 1$ mm and the outer scale is $L_0 = 0.16$ m [85]

15.6 Spatio-Temporal Picture of Femtosecond Filamentation

15.6.1 Dynamic Model

The spatio-temporal scenario of femtosecond pulse filamentation is reproduced completely by the dynamic model which covers both the initial and the dynamic stages of the process. The dynamic model includes diffraction, pulse dispersion, non-stationary nonlinear-optical response of medium and induced laser plasma [19,25,33,87,88]:

$$2ik\frac{\partial E}{\partial z} = \Delta_{\perp}E - kk'\frac{\partial^2 E}{\partial t^2} + \frac{2k^2}{n_0}\left[(\Delta n_k + \Delta n_p + \Delta\tilde{n})E\right] - ik\alpha E. \qquad (15.15)$$

Equation (15.15) leaves out the self-steepening of the pulse. This effect is investigated in [87–90]. The Kerr nonlinearity $\Delta n_k(x, y, z, t)$ is determined by (15.3). In air the plasma increment of the refractive index $\Delta n_p(x, y, z, t)$ is described by the expression:

$$\Delta n_p(x, y, z, t) = -\omega_p^2(x, y, z, t)/(2n_0\omega_0^2), \qquad (15.16)$$

where $\omega_p(x, y, z, t) = \sqrt{4\pi e^2 N_e(x, y, z, t)/m}$ is the plasma frequency, ω_0 is the central frequency of pulse spectrum, and m and e are the mass and charge of the electron. The free-electron density $N_e(x, y, z, t)$ is described by the kinetic equation:

$$\frac{\partial N_e}{\partial t} = R(|E|^2)(N_0 - N_e), \qquad (15.17)$$

where $R(|E|^2)$ is the ionization rates, determined from the accepted process model.

On the basis of the dynamic model (15.3, 15.17–15.19) of the pulse filamentation (2) the spacio-temporal scenarios of formation of filaments and plasma channels were received in [25,31–33]. The processes of energy exchange between a filament and the background reservoir, which explains refocusing of the pulse and existence of lengthy robust filament [28,31,32,44], and mechanisms of supercontinuum generation and conical emission, are also investigated [17,19].

15.6.2 Chirped Pulse

For femtosecond pulses the spatial compression of a beam at self-focusing is inseparably connected with redistribution of power. At self-focusing the normal dispersion of a medium leads to essential transformation of the temporal profile of a laser pulse, and to its breakdown on some spikes [91–93].

The pulse–phase modulation allows controlling the distance z_{fil} before filamentation starts. The influence of the initial pulse-phase modulation on the distance before filamentation starts was noted in [94]. It is revealed therein that the distance before the beginning of the filament increases at the detuning of the outlet compressor, independently of a sign of the initial phase modulation. In subsequent experiments [95] the pre-compensation of group velocity dispersion of a pulse in a medium with a normal dispersion was investigated.

Physically, the chirp influence on the formation of filaments is determined by two factors [96]. The first one, quasi-stationary, consists of a reduction of initial power in temporal slices at pulse–phase modulation and does not depend on the sign of the phase modulation. The second, dynamic, factor, consists of the pre-compensation of group velocity dispersion and depends on the sign of phase modulation. In a medium with a normal dispersion, recompensation of the pulse with negative chirp occurs, and the power in pulse temporal slices increases with distance z. The duration of a phase-modulated pulse is $\tau_p(\delta) > \tau_0$, and the peak power of a pulse is $P_{\text{peak}}(\delta) < P_{\text{peak}}$:

$$P_{peak}(\delta) = P_{peak}\tau_0/\tau_p(\delta). \tag{15.18}$$

At negative chirp, the pulse in a medium with normal dispersion undergoes a temporal compression; at positive chirp the duration of the pulse increases. The distances z_{fil} before the beginning of a filament were obtained numerically in [96] on the basis of the quasi-stationary (15.1) and dynamic (15.15–15.17) models (Fig. 15.15). The analytical evaluation of z_{fil} by the generalized Marburger formula (15.8) — in which instead of P_{peak} the peak power $P_{\text{peak}}(\delta)$ of the chirp pulse is substituted — is close to the numerical results.

The initial pulse–phase modulation allows us to control the distance and length before the start of a filament, and also to increase the effectiveness of the

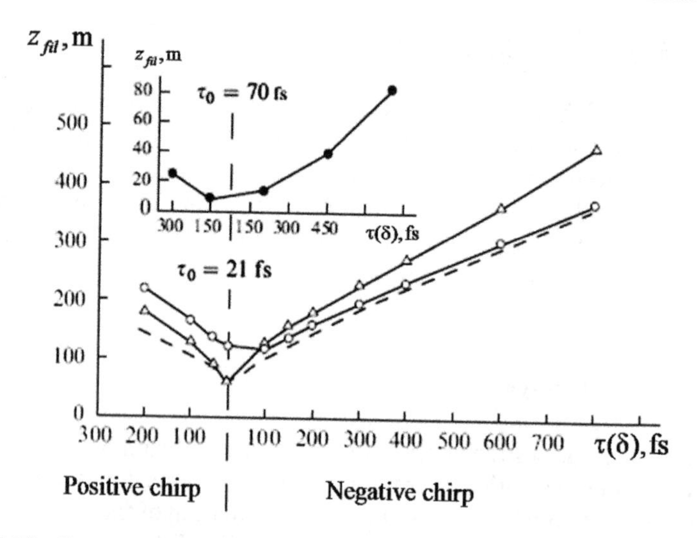

Fig. 15.15 The distance of a filament start z_{fil} calculated from both the quasi-stationary model (Δ) and from the dynamical model (○), as a function of the duration $\tau(\delta)$ of chirp pulse. The estimate by generalized Marburger formula (15.8) is represented by the dashed curve). The energy of the transform-limited pulse is $W_0 = 60\,\text{mJ}$, its duration is $\tau_0 = 21\,\text{fs}$, and $a_0 = 1.5\,\text{cm}$ [96]. The inset shows the experimental results [95]

supercontinuum generation [96]. In natural experiments [20] the advantages of the phase-modulated pulses for receiving filaments that are kilometers long are evidently shown. Detailed research of the propagation in air of the phase-modulated pulses with energy of 190 mJ was performed in [97]. At phase modulation the pulse duration $\tau_p(\delta)$ varied from 0.2 to 9.6 ps, which corresponded to a change of the peak power in the range $P_{\text{peak}}(\delta) = (190 \div 4)P_{cr}$. At duration $\tau(\delta) = 2.4$ps, phase modulation was the optimum, and the laser plasma was registered at distances up to 400 m.

15.6.3 Dynamic Multiple Filament Competition

The filament number in terawatt and subterawatt pulses increases with distance [25]. The arrangement of the "hot" spots defining the position filament in a transverse section changes. Upon side visualization of filaments in rhodamine B a merging and birth of new filaments along the propagation axis are visible [98]. Changes of number and arrangement of filaments with distance are confirmed by supercontinuum patterns [99]. Its frequency-angular distribution becomes more complex due to interference of radiation at an increased number of filaments with distance and, hence, of supercontinuum sources.

The basic laws of dynamic multifilamentation are based on the example of a simple model of a pulse with two perturbations at the initial intensity

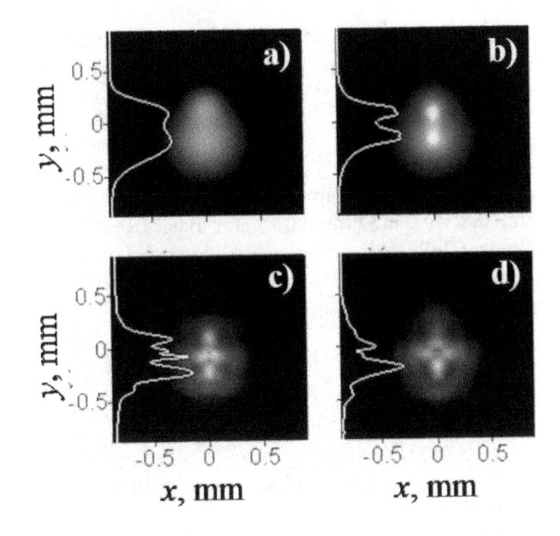

Fig. 15.16 (**a**) Fluence distributions at different distances z: initial conditions ($z = 0$); (**b**) independent development of two "parent" filaments ($z = 7.5$ cm); (**c**) birth of a "daughter" ($z = 16.9$ cm); (**d**) "survival" of one of them ($z = 20.6$ cm) [100]

distribution [60,100]. The system of equations (15.3), (15.18), (15.19), and (15.23) is numerically investigated for a pulse (15.9). The dynamic scenario of multifilamentation is evidently reproduced by fluence distributions $F(x, y, z)$ in the pulse in water (Fig. 15.16). In the beginning of propagation from each perturbation so-called "parent" filaments were formed [Fig. 15.16(a),(b)]. At the interference of the ring structures from "parent" filaments [37], some local maxima are formed. These maxima become the center of origin of "daughter" filaments [Fig. 15.16(c),(d)]. The development of "daughter" filaments is accompanied by energy flow in a pulse transverse section. As a result of dynamic energy competition between filaments only one filament survives in the pulse.

The dynamic competition of the filament is the cause of instability of a back-scattering fluorescent signal, which arises at excitation of molecules of nitrogen in plasma channels of competing filaments [100]. It was shown [84] that at reduction of the transverse sizes of a pulse, plasma channels arrange more densely, and the registered signal of fluorescence increases.

15.7 Conclusions

Self-focusing is the necessary condition for filament initiation. In the present chapter we have considered only one aspect of the many-sided phenomenon of femtosecond laser pulse filamentation in the bulk of transparent medium, namely, the formation of a filament that is directly connected to the phenomenon of self-focusing. The questions of plasma channel formation, supercontinuum, and conical emission generation influence of various factors on pulse filamentation — all these issues are subjects of independent reviews. Research aspects associated with the filamentation in dielectric solids and applications of this research to the solution of microoptics problems are of significant interest as well.

Acknowledgment V.P. Kandidov, A.E. Dormidonov, and O.G. Kosareva thank the support of the European Research Office of the U.S. Army through contract No. W911NF-05-1-0553, and the Russian Foundation for Basic Research through grants No. 06-02-17508-a and No. 06-02-08004.

References

1. N.F. Pilipetskii, A.R. Rustamov: Observation of self-focusing of light in liquids, *JETP Lett.* **2**, 55–56 (1965).
2. M. Hercher: Laser-induced damage in transparent materials, *J. Opt. Soc. Am.* **54,** 563 (1964).
3. G.A. Askar'yan: Effects of the gradient of a strong electromagnetic beam on electrons and atoms, *Sov. Phys. JETP* **15**, 1088–1090 (1962).
4. R.Y. Chiao, E. Garmire, C.H. Townes: Self-trapping of optical beams, *Phys. Rev. Lett.* **13**, 479 (1964).
5. V.V. Korobkin, A.J. Alcock: Self-focusing effects associated with laser-induced air breakdown, *Phys. Rev. Lett.* **21**, 1433 (1968).
6. N.G. Basov, P.G. Kryukov, Yu.V. Senatskii et al.: Production of powerful ultrashort light pulses in a neodymium glass laser, *Sov. Phys. JETP* **30,** 641–645 (1970).
7. V.I. Bespalov, V.I. Talanov: Filamentary structure of light beams in nonlinear liquids, *Sov. Phys. JETP Lett.*, **3**, 307–310 (1966).
8. A.J. Campillo, S.L. Shapiro, B.R. Suydam: Relationship of self-focusing to spatial instability modes, *Appl. Phys. Lett.*, **24**, 178–180 (1974).
9. A.N. Zherikhin, Yu.A. Matveets, S.V. Chekalin: Self-focusing limitation of brightness in amplification of ultrashort pulses in neodymium glass and yttrium aluminum garnet, *Quantum Electron.* **6**, 858–860 (1976).
10. I.A. Fleck, Jr., C. Layne: Study of self-focusing damage in a high-power Nd:glass–rod amplifier, *Appl. Phys. Letts.* **22**, 467–469 (1973).
11. D. Stricland, G. Mourou: Compression of amplified chirped optical pulses, *Optics Commun.* **56**, 219–221 (1985).
12. A. Braun, G. Korn, X. Liu et al.: Self-channeling of high-peak-power femtosecond laser pulses in air, *Opt. Lett.* **20**, 73 (1995).
13. E.T.J. Nibbering, P.F. Gurley, G. Grillon et al.: Conical emission from self-guided femtosecond pulses in air, *Opt. Lett.* **21**, 62 (1996).
14. A. Brodeur, O.G. Kosareva, C.Y. Chien et al.: Moving focus in the propagation of ultrashort laser pulses in air, *Opt. Lett.* **22**, 304 (1997).
15. V.P. Kandidov, O.G. Kosareva, A. Brodeur et al.: State-of-the-art of investigations into the filamentation of high-power subpicosecond laser pulses in gases, *Atmos. Oceanic Opt.* **10**, 966–973 (1997).
16. H. Schillinger, R. Sauerbrey: Electrical conductivity of long plasma channels in air generated by self-guided femtosecond laser pulses, *Appl. Phys. B.* **68**, 753–756 (1999).
17. O.G. Kosareva, V.P. Kandidov, A. Brodeur et al.: Conical emission from laser–plasma interactions in the filamentation of powerful ultrashort laser pulses in air, *Opt. Lett.* **22**, 1332 (1997).
18. I.S. Golubtsov, V.P. Kandidov, O.G. Kosareva: Conical emission of high–power femtosecond laser pulse in the atmosphere, Atmos. Oceanic. Opt. **14**, 303 (2001).

19. V.P. Kandidov, O.G. Kosareva, I.S. Golubtsov et al.: Self-transformation of a powerful femtosecond laser pulse into a white-light laser pulse in bulk optical media (or supercontinuum generation), *Appl. Phys. B.* **77**, 149 (2003).
20. J. Kasparian, M. Rodrigues, G. Mejean et al.: White-light filaments for atmospheric analysis, *Science* **301**, 61 (2003).
21. F. Theberge, W. Liu, S.A. Hosseini et al.: Long-range spectrally and spatially resolved radiation from filaments in air, *Appl. Phys. B.* **81**, 13–34 (2005).
22. S. Tzortzakis, B. Prade, M. Franco et al.: Femtosecond laser-guided electric discharge in air, *Phys. Rev. E.* **64**, 057401 (2001).
23. K.M. Davis, K. Miura, N. Sugimoto et al.: Writing waveguides in glass with a femtosecond laser, *Opt. Lett.* **21**, 1729 (1996).
24. K. Yamada, W. Watanabe, K. Kintaka et al.: Volume grating induced by a self-trapped long filament of femtosecond laser pulse in silica glass, *Jpn. J. Appl. Phys.* **42**, 6916–6919 (2003).
25. S.L. Chin, S.A. Hosseini, W. Liu et al.: The propagation of powerful femtosecond laser pulses in optical media: physics, applications, and new challenges, *Can. J. Phys.* **83**, 863–905 (2005).
26. J. Kasparian, R. Sauerbrey, S.L. Chin: The critical laser intensity of self-guided light filaments in air, *Appl. Phys. B.* **71**, 877–879 (2000).
27. W. Liu, O.G. Kosareva, I.S. Golubtsov et al.: Femtosecond laser pulse filamentation versus optical breakdown in H_2O, *Appl. Phys. B* **76**, 215 (2003).
28. W. Liu, S.L. Chin, O.G. Kosareva et al.: Multiple refocusing of a femtosecond laser pulse in a dispersive liquid (methanol), *Opt. Commun.* **225**, 193–209 (2003).
29. V.N. Lugovoi, A.M. Prokhorov: A possible explanation of the small-scale self-focusing filaments, *JETP Lett.*, **7**, 117–119 (1968).
30. V.V. Korobkin.: Experimental investigation of the propagation of powerful radiation in nonlinear media, *Sov. Phys. Uspekhi*, **15**, 520–521 (1973).
31. O.G. Kosareva, V.P. Kandidov, A. Brodeur et al.: From filamentation in condensed media to filamentation in gases, *Nonlinear Opt. Phys. Mater.* **6**, 485 (1997).
32. M. Mlejnek, E.M. Wright, J.V. Moloney: Dynamic spatial replenishment of femtosecond pulses propagating in air, *Opt. Lett.* **23**(5), 382–384 (1998).
33. A. Chiron, B. Lamoroux, R. Lange et al.: Numerical simulation of the nonlinear propagation of femtosecond optical pulses in gases, *Eur. Phys. J. D* **6**, 383–396 (1999).
34. E. Garmire, R.Y. Chao, C.Y. Townes: Dynamics and characteristics of the self-trapping of intense light beams, *Phys. Rev. Lett.* **19**, 347–349 (1966).
35. E.L. Dawes, J.H. Marburger: Computer studies in self-focusing, *Phys. Rev. Lett.* **179**, 862 (1969).
36. S.L. Chin, N. Akozbek, A. Proulx et al.: Transverse ring formation of a femtosecond laser pulse propagating in air, *Opt. Commun.* **188**, 181–186 (2001).
37. S.L. Chin, S. Petit, W. Liu et al.: Interference of transverse rings in multi-filamentation of powerful femtosecond laser pulses in air, *Opt. Commun.* **210**, 329–341 (2002).
38. H.R. Lange, G. Grillon J.-F.Repoche et al.: Anomalous long-range propagation of femtosecond laser pulses through air: moving focus or pulse self-guiding, *Opt. Lett.* **23**, 120–122 (1998).
39. A. Dubietis, E. Gaizauskas, G. Tamosauskas et al.: Light filaments without self-channeling, *Phys. Rrev. Lett.* **29**, 253903-1-4 (2004).
40. A. Talebpour, S. Petit, S.L. Chin: Re-focusing during the propagation of a focused femtosecond Ti:sapphire laser pulse in air, *Opt. Commun.* **171**, 285 (1999).
41. A. Becker, N. Akozbek, K. Vijayalakshmi et al.: Intensity clamping and re-focusing of intense femtosecond laser pulses in nitrogen molecular gas, *Appl. Phys. B.* **73**, 287 (2001).
42. A.L. Dyshko, V.N. Lugovoi, A.M. Prokhorov: Self-focusing of intense light beams, *JETP Lett.* **6**, 146 (1967).

43. V.N. Lugovoi, A.M. Prokhorov: Theory of the propagation of high-power laser radiation in a nonlinear medium, *Sov. Phys.-Usp.* **16**, 658–679 (1974).
44. V.P. Kandidov, O.G. Kosareva, A.A. Koltun: Nonlinear-optical transformation of a high-power femtosecond laser pulse in air, *Quant. Electron.* **33**, 69–75 (2003).
45. W. Liu, J.-F. Cravel, F. Theberge et al.: Background reservoir: its crucial role for long-distanse propagation of femtosecond laser pulse in air, *Appl. Phys. B* **80**, 857–860 (2005).
46. S.L. Chin, A. Talebpour, J. Yang et al.: Filamentation of femtosecond laser pulses in turbulent air, *Appl. Phys. B.* **74**, 67 (2002).
47. G. Méchain, G. Méjean, R. Ackermann et al.: Propagation of fs TW laser filaments in adverse atmospheric conditions, *Appl. Phys. B* **80**, 785–789 (2005).
48. E.T.J. Nibbering, G. Grillon, M.A. Franco et al.: Determination of the inertial contribution to the nonlinear refractive index of air, N_2, and O_2 by use of unfocused high-intensity femtosecond laser pulses, *J. Opt. Soc. Am. B* **14**, 650–660 (1997).
49. K.Yu. Andrianov, V.P. Kandidov, O.G. Kosareva et al.: The effect of the beam quality on the filamentation of high-power femtosecond laser pulses in air, *Bull. Russ. Acad. Sci. Phys.* **66**(8), 1091–1102 (2002).
50. W. Liu, S.L. Chin: Direct measurement of the critical power of femtosecond Ti:sapphire laser pulse in air, *Opt. Express.* **13**, 5750–5755 (2005).
51. G. Fibich, A.L. Gaeta: Critical power for self-focusing in bulk media and in hollow waveguides, *Opt. Lett.* **25**, 335–337 (2000).
52. G. Fibich, B. Ilan: Self-focusing of elliptic beams: an example of the failure of the aberrationless approximation, *Opt. Soc. Am. B.* **17**, 1749–1758 (1999).
53. A. Dubietis, G. Tamosauskas, G. Fibich et al.: Multiple filamentation induced by input-beam ellipticity, *Opt. Lett.* **29**, 1126–1128 (2004).
54. V.P. Kandidov, V.Yu. Fedorov: Properties of self-focusing of elliptic beam, *Quant. Electron.* **34**, 1163 (2004).
55. J.H. Marburger: Self-focusing: theory. *Prog. Quant. Electr.* (Printed in Great Britain: Pergamon Press) **4**, 35 (1975).
56. V.Yu. Fedorov, V.P. Kandidov, O.G. Kosareva et al.: Filamentation of a femtosecond laser pulse with the initial beam ellipticity, *Laser Phys.* **16**(8), 1227–1234 (2006).
57. B. La Fontaine, F. Vidal, Z. et al.: Filamentation of ultrashort pulse laser beams resulting from their propagation over long distances in air, *Phys. Plasmas* **6**(3), 1615–1621 (1999).
58. S. Tzortzakis, L. Berge, A. Couairon et al.: Breakup and fusion of self-guided femtosecond light pulses in air, *Phys. Rev. Lett.* **86**, 5470–5473 (2001).
59. V.P. Kandidov, O.G. Kosareva, S.A. Shlenov et al.: Dynamic small-scale self-focusing of a femtosecond laser pulse, *Quant. Electron.* **35**, 59–64 (2005).
60. S.A. Hosseini, Q. Luo, B. Ferland et al.: Competition of multiple filaments during the propagation of intense femtosecond laser pulses, *Phys. Rev. A* **70**(3), 033802-1-12 (2004).
61. G. Mechain, A. Couairon, M. Franco et al.: Organizing multiple femtosecond filaments in air, *Phys. Rev. Lett.* **93**, 035003-1 (2004).
62. A.J. Campillo, S.L. Shapiro, B.R. Suydam: Periodic breakup of optical beams due to self-focusing, *Appl. Phys. Lett.*, **23**, 628–630 (1973).
63. V.P. Kandidov, N. Akozbek, M. Scalora et al.: A method for spatial regularization of a bunch of filaments in a femtosecond laser pulse, *Quant. Electron.* **34**, 879 (2004).
64. V.P. Kandidov, N. Akozbek, M. Scalora et al.: Towards a control of multiple filamentation by spatial regularization of a high-power femtosecond laser pulse, *Appl. Phys. B.* **80**, 267 (2005).
65. O.G. Kosareva, T. Nguyen, N.A. Panov et al.: Array of femtosecond plasma channels in fused silica, *Opt. Commun.* **267**, 511 (2006).
66. H. Schroeder, S.L. Chin: Visualisation of the evolution of multiple filaments in methanol, *Opt. Commun.* **234**, 399–406 (2004).

67. V.P. Kandidov, A.E. Dormidonov, O.G. Kosareva et al.: Optimum small-scale management of random beam perturbations in a femtosecond laser pulse, *Appl. Phys. B.* **87**(1), 29–36 (2007).
68. L. Woste, C. Wedekind, H. Wille et al.: Femtosecond atmospheric lamp, *Laser Optoelectron.* **29**, 51 (1997).
69. V.P. Kandidov, O.G. Kosareva, E.I. Mozhaev et al.: Femtosecond nonlinear optics of the atmosphere, *Atmos. Oceanic. Opt.* **13**, 394–401 (2000).
70. V.P. Kandidov, O.G. Kosareva, M.P. Tamarov et al.: Nucleation and random movement of filaments in the propagation of high–power laser radiation in a turbulent atmosphere, *Quant. Electron.* **29**, 911 (1999).
71. V.P. Kandidov: Monte Carlo method in nonlinear statistical optics, *Physics-Uspekh I.* **39**, 1243–1272 (1996).
72. A. Ishimaru: Wave propagation and scattering in random media, vol. 2. Academic Press, New York (1978).
73. L. Berge, S. Skupin, F. Lederer et al.: Multiple filamentation of terawatt laser pulses in air, *Phys. Rev. Lett.* **92**, 225002-1-4 (2004).
74. S. Skupin, L. Berge, U. Peschel et al.: Filamentation of femtosecond light pulses in the air: Turbulent cells versus long-range clusters, *Phys. Rev. E* **70**, 046602-1-15 (2004).
75. S.A. Shlenov, V.P. Kandidov: Filament bunch formation upon femtosecond laser pulse propagation through the turbulent atmosphere. Part1. Method; Part2. Statistical characteristics, *Atmos. Oceanic Opt.* **17**, 565–575 (2004).
76. N.N. Bochkarev, A.A. Zemlyanov, Al.A. Zemlyanov et al.: Experimental investigation into interaction between femtosecond laser pulses in aerosol, *Atmos. Oceanic Opt.* **17**, 861–867 (2004).
77. G. Mejean, J. Kasparian, J. Yu et al.: Multifilamentation transmission through fog, *Phys. Rev. E.* **72**(2), 026611-1-7 (2005).
78. F. Courvoisier, V. Boutou, J. Kasparian et al.: Ultraintense light filaments transmitted through clouds, *Appl. Phys. Lett.* **83**(2), 213–215 (2003).
79. M. Kolesik, J.V. Moloney: Self-healing femtosecond light filaments, *Opt. Lett.* **29**(6), 590–592 (2004).
80. S. Skupin, L. Berge, U. Peschel et al.: Interaction of femtosecond light filamentation with obscurants in aerosols, *Phys. Rev. Lett.* **93**, 023901-1-4 (2004).
81. A.A. Zemlyanov, Yu.E. Geints: Filamentation length of ultrashort laser pulse in presence of aerosol layer, *Opt. Commun.* **259**(2), 799–804 (2006).
82. V.O. Militsin, L.S. Kuzminsky, V.P. Kandidov: Stratified-medium model in studying propagation of high-power femtosecond laser radiation through atmospheric aerosol, *Atmos. Oceanic Opt.* **17**, 630–641 (2005).
83. V.P. Kandidov, V.O. Militsin: Computer simulation of laser pulses filament generation in rain, *Appl. Phys. B.* **83**, 171–174 (2006).
84. V.P. Kandidov, V.O. Militsin: Multiple filaments formation in high-power femtosecond laser pulse in rain, *Atmos. Oceanic Opt.* **19**(9), 765–772 (2006).
85. N.A. Panov, O.G. Kosareva, V.P. Kandidov et al.: Controlling the bunch of filaments formed by high-power femtosecond laser pulse in air, *Proc. SPIE.* **5708**, 91–101 (2005).
86. N.A. Panov, O.G. Kosareva, I.N. Murtazin: Arranged femtosecond light filaments in the bulk of transparent medium, *J. Opt. Technol.* **73**(11), (2006).
87. N. Akozbek, M. Scalora, C.M. Bowden et al.: White-light continuum generation and filamentation during the propagation of ultrashort laser pulses in air, *Opt. Commun.* **191**, 353 (2001)
88. I.S. Golubtsov, O.G. Kosareva: Influence of various physical factors on the generation of conical emission in the propagation of high-power femtosecond laser pulses in air, *J. Opt. Technol.* **69**, 462–467 (2002).
89. J.K. Ranka, A.L. Gaeta: Breakdown of the slowly varying envelope approximation in the self-focusing of ultrashort pulses, *Opt. Lett.* **23**, 534 (1998).

90. Y.R. Shen: *The Principles of Nonlinear Optics*, John Wiley, New York (1984).
91. J.K. Ranka, R.W. Schirmer, A.L. Gaeta: Observation of pulse splitting in nonlinear dispersive media, *Phys. Rev. Lett.* **77**, 3783 (1996).
92. M. Mlejnek, E.M. Wright, J.V. Moloney: Power dependence of dynamic spatial replenishment of femtosecond pulses propagating in air, *Opt. Express.* **4**, 223 (1999).
93. V.P. Kandidov, O.G. Kosareva, E.I. Mozhaev et al.: Femtosecond nonlinear optics of the atmosphere. *Atmos. Oceanic. Opt.* **13**, 394–401 (2000).
94. S.L. Chin: A study of the fundamental science underling the transport of intense femtosecond laser pulses in the atmosphere, Final Report for Grant No. DAAG55-97-1-0404 (1999).
95. H. Wille, M. Rodriguez, J. Kasparian et al.: A mobile femtosecond–terawatt laser and detection system, *Eur. Phys. J. AP.* **20**, 183 (2002).
96. I.S. Golubtsov, V.P. Kandidov, O.G. Kosareva: Initial phase modulation of a high-power femtosecond laser pulse as a tool for controlling its filamentation and generation of supercontinuum in air, *Quant. Electron.* **33**(6), 525–530 (2003).
97. G. Mejean, C. D'Amico, Y.-B.Andre et al.: Range of plasma filaments created in air by a multi-terawatt femtosecond laser, *Opt. Commun.* **247**, 171–180 (2005).
98. H. Schroeder, S.L. Chin: Visualisation of the evolution of multiple filaments in methanol, Opt. Commun. 234, 399–406 (2004).
99. W. Liu, S.A. Hosseini, Q. Luo et al.: Experimental observation and simulations of the self-action of white light laser pulse propagating in air, *New J. Phys.* **6**, 6.1–6.22 (2004).
100. O.G. Kosareva, N.A. Panov, V.P. Kandidov: Scenario of multiple filamentation and supercontinuum generation in a high-power femtosecond laser pulse, *Atmos. Oceanic Opt.* **18**(3), 204–211 (2005).

Chapter 16
Spatial and Temporal Dynamics of Collapsing Ultrashort Laser Pulses

Alexander L. Gaeta

Abstract In nonlinear optics, a wave collapse manifests itself as the self-focusing of light. There exist a number of universal features associated with wave collapse, such as self-similar evolution, pulse splitting, filamentation, and shock formation, in which nonlinear optics has offered an unparalleled opportunity for study and comparison of theory with experiment. This chapter reviews recent theoretical and experimental work in which such phenomena have been observed.

16.1 Introduction

One of the most dramatic phenomena in nonlinear wave physics is the collapse of a wave leading to a divergence in its peak amplitude. Such singular behavior occurs in plasma physics, hydrodynamics, Bose–Einstein condensates, and nonlinear optics, although many different sets of nonlinear equations exhibit wave collapse, and there are universal features shared by these systems.

In nonlinear optics, wave collapse is equivalent to self-focusing of the light field. Because the highly nonresonant interactions in transparent materials such as air, water, and glass are very well modeled by a third-order optical nonlinearity over a dynamic range of many orders of magnitude in the intensity of the electromagnetic field, nonlinear optics has provided an unmatched system for studying collapse dynamics, which can be accurately modeled by the nonlinear Schrödinger-type wave equation.

Studies of self-focusing began soon after the discovery of the laser when researchers determined [1–3] that a light beam could self-focus when the power in the beam was greater than a certain critical power P_{cr}. The origin of this effect is the third-order nonlinear susceptibility, which for most optically transparent media, be it a gas, liquid or solid, gives rise to an index of refraction n that

A.L. Gaeta (✉)
Cornell University, School of Applied and Engineering Physics, Ithaca, NY, USA
e-mail: a.gaeta@cornell.edu

R.W. Boyd et al. (eds.), *Self-focusing: Past and Present*,
Topics in Applied Physics 114, DOI 10.1007/978-0-387-34727-1_16,

increases with local intensity I of the light wave (i.e., $n = n_0 + n_2 I$, where n_0 is the linear refractive index and n_2 is the nonlinear index coefficient). Above the critical power P_{cr}, the intensity dependent refractive index gives rise to a nonlinear curvature of the wavefront that overcomes the effects of linear diffraction and produces runaway focusing of the beam. Eventually, the peak intensity of the collapsing wave becomes sufficiently high that other processes occur, such as nonlinear absorption and plasma formation. In most cases self-focusing imposes an upper limit on the power that can be effectively transmitted through a medium. For example, for solid-state materials such as glass, $P_{cr} \sim 10^6$ W, whereas for gases such as air, $P_{cr} \sim 10^{10}$ W. Although self-focusing may be viewed solely as a deleterious process, there are situations where self-focusing can be beneficial, such as in the production of femtosecond pulses from solid-state lasers via Kerr-lens mode-locking [4], supercontinuum generation for spectroscopy [5,6], and remote sensing in the lower atmosphere [7].

16.2 Spatial Collapse Dynamics

For a beam propagating along the z-axis with a wavevector amplitude $k = n_0\omega/c$, where ω is the central frequency of the field, the fundamental model for describing the self-focusing of laser beams is the 2-D nonlinear Schrödinger equation (NLSE),

$$\frac{\partial u}{\partial \zeta} = \frac{i}{4}\nabla_\perp^2 u + i\frac{L_{df}}{L_{nl}}|u|^2 u, \tag{16.1}$$

where $u(r, z, t) = A(r, z, t)/A_0$ is the normalized amplitude of the electric-field amplitude, $L_{df} = kw_0{}^2/2$ is the diffraction length, $\zeta = z/L_{df}$, $L_{nl} = (c/\omega\, n_2 I_0)$ is the nonlinear length, ω is the frequency of the wave, and $I_0 = n_0 c|A_0|^2/2\pi$ is the peak input intensity. The ratio L_{df}/L_{nl} is proportional to P/P_{cr}, where P is the total power of the input field, and the critical power P_{cr} is given by [8],

$$P_{cr} = \alpha\frac{\lambda^2}{4\pi n_0 n_2}, \tag{16.2}$$

where λ is the vacuum wavelength. The value of the parameter α depends on the initial spatial profile of the beam, and for a Gaussian beam numerical analysis [8] shows that $\alpha \sim 1.9$.

16.2.1 *Self-similar Collapse*

The profile that exhibits the lowest critical power for collapse is known as the Townes profile (see Fig. 16.1). This profile also corresponds to the situation in which the diffractive and nonlinear effects are in a precise balance, and the

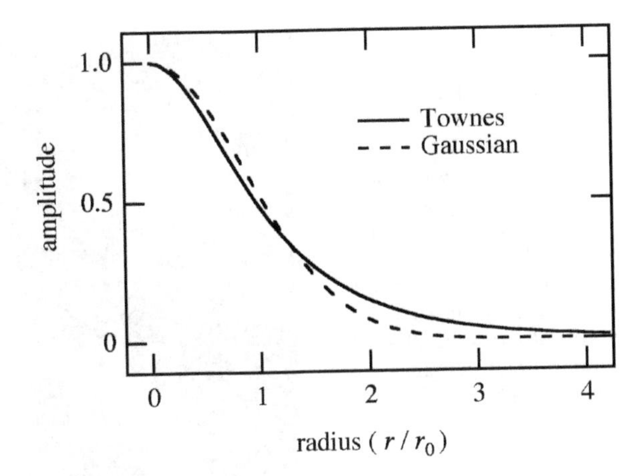

Fig. 16.1 Plot of the Townes profile. Also shown is a Gaussian profile with the same peak intensity and power [8]

beam propagates without change [1] and was termed the "waveguide" solution. Nevertheless, as a waveguide solution, the Townes profile is unstable in the sense that for powers above P_{cr} the beam undergoes collapse, whereas for P less than P_{cr} the beam diffracts.

It was thought for decades that this waveguide solution was simply a physically irrelevant solution to the 2-D cylindrically symmetric NLSE. However, it was suggested [9–12] that the shape of a collapsing beam always evolves to the radially symmetric Townes profile as it approaches the point of total collapse and that the collapsing beam always contains the Townes critical power (i.e., $\alpha \sim 1.86$), with the remaining energy shed as it undergoes linear diffraction. This behavior is known as self-similar collapse in which the profile maintains the same functional form (i.e., the Townes profile), but is not stationary in the sense that the spatial profile become narrower with a corresponding higher peak intensity as it approaches the collapse singularity.

In fact, it can be shown numerically that the functional form to which it evolves is precisely the Townes profile, regardless of the shape of the input profile [see Fig. 16.2(a),(c)]. These predictions were verified experimentally for both randomly distorted [see Fig. 16.2(b), (d)] and elliptically shaped input beams in glass [13]. This evolution to a self-similar symmetric profile is actually a universal feature of many different types of nonlinear equations that exhibit collapse, and these observations represent the first experimental verification of this behavior.

Although the full temporal dynamics are not included in this self-similar analysis, it does provide deeper insight into other optical self-focusing experiments, even for ultrashort pulses. For example, it offers a compelling argument for why high-intensity beams propagating in air [14] break up into multiple filaments of equal energy and why for ultrashort pulses with peak powers more than 1000 times the critical power, only a small fraction of the power (and pulse energy) is contained in the collapsing region. This also explains why very high-power femtosecond pulses that undergo self-focusing do not produce damage in materials.

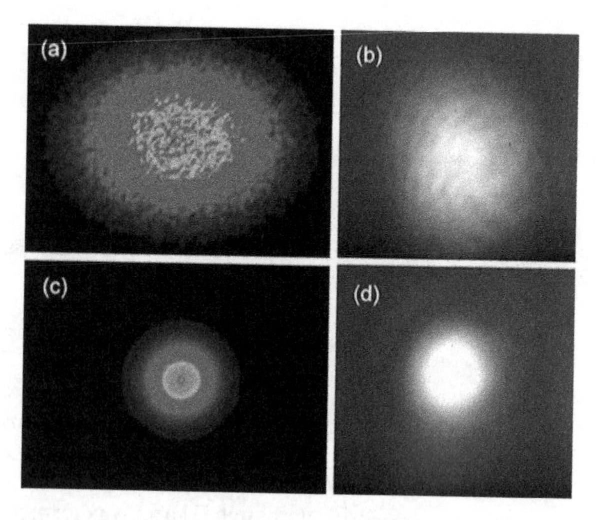

Fig. 16.2 (**a**) Theoretical input intensity profile with added noise and (**c**) the predicted evolution to the smooth, circularly symmetric Townes profile as the beam undergoes collapse. (**b**) Experimentally generated distorted beam of a 50-fs pulse using a weakly etched microscope slide and (**d**) the observed beam after propagating through a 30-cm-long block of glass [13]

16.2.2 Modulational Instability versus Townes Collapse

As discussed in the previous section, the presence of noise on a near-Gaussian beam does not necessarily lead to multiple filamentation via modulation instability [15], as had been commonly believed, even for powers many times the critical power for self-focusing. This issue can be understood as a competition between self-focusing to a Townes profile and the modulational instability. Analysis shows that the transition from Townes collapse to multiple filamentation occurs for powers $\sim 100P_{cr}$ [16]. This transition is also accompanied by a change in the dependence of the collapse distance on power from $P_{cr}^{-1/2}$ to P_{cr}^{-1} [16,17]. However as discussed in the next section, an alternative route to multiple filamentation can also occur via self-similar evolution to ring profiles, which are unstable and can undergo break up.

16.2.3 Self-focusing with Non-Gaussian Beams

Further research has shown that other self-similar profiles can exist and that evolution to these profiles can occur at sufficiently high powers and for certain shapes of the input beam. Theoretical analysis showed that for super-Gaussian [18,19] or vortex [20] profiles, the beams evolve to flat-phase and vortex, respectively, self-similar ring profiles. For example for a super-Gaussian input beam, the beam will collapse to a ring profile for powers greater than 10 P_{cr}, rather than to the Townes profile. However, in the presence of azimuthal noise this ring profile become unstable as it approaches the collapse point and undergoes multiple filamentation along the ring. By performing an azimuthal modulation stability analysis, the number of filaments can be accurately predicted as a function of the input power and topological charge (see Fig. 16.3 for the case

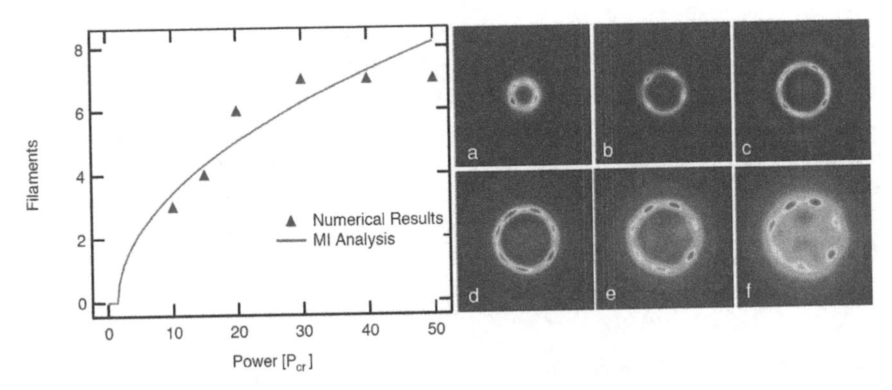

Fig. 16.3 *Left*: Comparison of a modulational instability analysis with the approximate number of filaments observed in numerical simulations of Eq. (16.1) with a super-Gaussian input beam. *Right*: Numerically simulated profiles illustrating multiple filamentation of an initially super-Gaussian beam with 10% amplitude noise for input powers of (**a**) 10 P_{cr}, (**b**) 15 P_{cr}, (**c**) 20 P_{cr}, (**d**) 30 P_{cr}, (**e**) 40 P_{cr}, and (**f**) 50 P_{cr} (Grow, et al. 2006)

of a flat-phase profile). After the profile breaks up, the shape of these individual filaments evolves to the Townes profile. This route to multiple filamentation occurs at powers that are a factor of 10 times lower than for multiple filamentation via the modulational instability with Gaussian beams, as discussed in the previous section, which requires powers in excess of 100 P_{cr}. In addition to experiments performed in water [19], multiple filamentation in a ring shape has been observed in experiments with terawatt femtosecond laser pulses in air [7,21].

16.3 Spatio-Temporal Collapse Dynamics

For initial pulse durations in the femtosecond regime, the role of dispersion can play an important role and can dramatically alter the self-focusing dynamics. By including dispersion into the wave equation and by not making the slowly varying envelope equation in time, it is possible to derive a modified 3-D NLSE, which is also known as the nonlinear envelope equation (NEE) [22]. Assuming the pulse propagates along the z-axis, the NEE describing the evolution of the normalized amplitude $u(r, z, t) = A(r, z, t)/A_0$ of the pulse is given by [23],

$$\frac{\partial u}{\partial \zeta}=\frac{i}{4}\left(1+\frac{i}{\omega\tau_p}\frac{\partial}{\partial \tau}\right)^{-1}\nabla_{\perp}^2 u - i\,\mathrm{sgn}(\beta_2)\frac{L_{df}}{2L_{ds}}\frac{\partial^2 u}{\partial \tau^2}+i\left(1+\frac{i}{\omega\tau_p}\frac{\partial}{\partial \tau}\right)\frac{L_{df}}{L_{nl}}|u|^2 u, \quad (16.3)$$

where $L_{ds} = {\tau_p}^2/|\beta_2|$ is the dispersion length, τ_p is the pulse duration, β_2 is the group-velocity dispersion (GVD), $\zeta = z/L_{ds}$ is the normalized propagation distance, and $\tau = (t - z/v_g)/\tau_p$ is the normalized retarded time for the pulse traveling at the group-velocity v_g. The normal- (anomalous-) GVD regime corresponds to the case in which β_2 is greater (less) than 0. The presence of the operator $1 + i\partial/\omega\tau_p\partial\tau$ in the diffraction and in the nonlinear-index terms gives rise to space–time focusing [24] and self-steepening [25–27], respectively, and allows for the modeling of pulses with spectral widths comparable to the optical frequency ω. For an 80-fs laser pulse $\omega\tau_p = 140$, which suggests that the presence of this operator is unnecessary; however, as discussed below, after the pulse undergoes self-focusing and spatio-temporal collapse occurs, these terms can play an important, if not critical, role.

16.3.1 Self-Focusing in the Normal-GVD Regime

The 3-D NLSE without the self-steepening and space–time focusing terms has been extensively studied in the normal-GVD regime. One of the early and most important predictions [28–32] based on the NLSE is that even for input powers significantly above the critical power for self-focusing, the effects of normal dispersion halt catastrophic self-focusing through the process of pulse splitting. As the pulse collapses it undergoes self-phase modulation, which spectrally broadens and produces a nonlinear temporal chirp in which red-shifted and blue-shifted frequencies are generated to the front and back of the pulse, respectively.

The presence of normal dispersion then acts to split the collapsing pulse into red-shifted and blue-shifted wavepackets. Experiments [33,34] verified this pulse-splitting scenario. An example of the temporal behavior above the splitting threshold is shown in Fig. 16.4. However, unlike the predictions of the

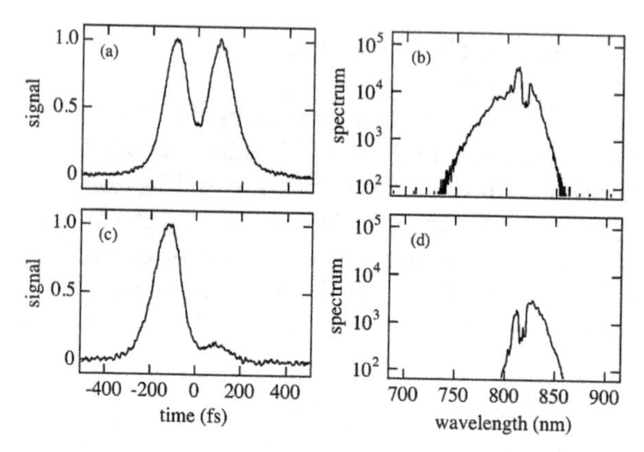

Fig. 16.4 (**a**) Cross-correlation measurement of temporal pulse splitting of an initially 80-fs pulse at 800 nm via self-focusing in a fused-silica glass sample. The corresponding spectrum is shown in part (**b**). (**c**) Cross-correlation and (**d**) spectral measurements under the same conditions as (a) and (b), but the transmitted pulsed is passed through a Schott long-pass RG-785 filter. This illustrates that the front pulse is red-shifted and the rear pulse is blue-shifted [37]

NLSE, the observed splitting and spectral broadening were asymmetric, which was determined [23,35] to be primarily a result of a breakdown in the slowly varying envelope approximation and could be explained by including the self-steepening and space–time focusing terms. At first glance, this result is surprising because the pulses under consideration were all much longer than an optical cycle; however, because the pulse is undergoing nonlinear spectral broadening as it undergoes collapse, these terms become increasingly important. Although multiple pulse splitting was observed [33], theoretical simulations [36] indicate that such phenomena is not an intrinsic feature of the NLSE and is most likely a result of higher-order nonlinear effects such as multiphoton absorption and plasma formation.

16.3.2 Optical "Shock" Formation and Supercontinuum Generation

At powers higher than those used in the pulse-splitting studies described above, the action of the self-focusing process can overcome dispersive effects that lead to pulse splitting, and the peak intensity grows explosively. Experimentally, it is found that as the input power is increased above a certain threshold power $P_{th} > P_{cr}$ an extremely broad pedestal appears to the blue side of the transmitted pulse spectrum. This process is termed supercontinuum generation (SCG) or "white-light" generation [6], because by eye the transmitted beam appears white, even for an infrared input pulse. This phenomenon was first observed [38] in 1970, and because then it has been observed in many different solids [39–41], liquids [42,43] and gases [44–46] under a wide variety of experimental conditions.

The shape of the spectra for various media are similar, which indicates that SCG is a universal feature of the laser–matter interaction. For spectroscopic applications [6], SCG has proven to be a useful source of broadly tunable ultrafast pulses from the near-ultraviolet to the far-infrared, and is commonly used as a seed for optical parametric amplifiers. Experimental measurements [41] in solids show that the cutoff wavelength of the supercontinuum spectrum on the short-wavelength side roughly scales with the band-gap of the material and that as long as the ratio of the band-gap energy to the photon energy is roughly equal to or greater than four, SCG can occur. Despite numerous studies, the basic underlying process responsible for SCG had resisted explanation via any of the standard one-dimensional models that incorporate self-phase modulation and self-steepening [26], which led Corkum et al. [44] to suggest that the mechanisms responsible for SCG were connected to the full 3-D self-focusing dynamics.

Using the NEE [Eq. (16.3)] to investigate the dynamics of self-focusing of femtosecond laser pulses both near and above the point at which the peak intensity grows explosively [47], reveals that as the pulse approaches the collapse point, a steep edge is formed at the back of the pulse (i.e., an optical

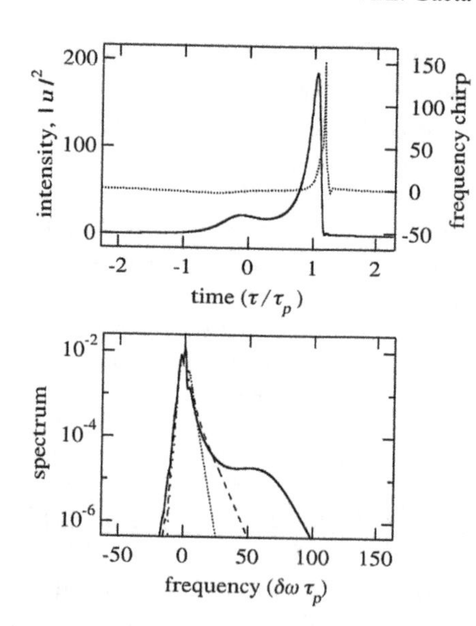

Fig. 16.5 (*Top*): Theoretical simulation showing the intensity and temporal chirp of a 70-fs pulse at 800 nm undergoing self-focusing collapse in sapphire for $P/P_{cr} = 1.8$ at a propagation distance of $\zeta = 1.63$, which shows the formation of an optical shock wave at the rear of the pulse. (*Bottom*): The corresponding spectrum exhibits a long blue pedestal similar to SCG observed in experiments [47]

"shock wave"), which is accompanied by a large phase jump. Figure 16.5 shows plots of the on-axis intensity and the temporal chirp just beyond the point at which the intensity peaks and dispersion acts to halt collapse of the pulse. The back edge of the pulse becomes steep and the narrow peak in the frequency chirp at the back of the pulse is indicative of the abrupt phase jump that accompanies the optical shock wave. The resulting spectrum of the pulse exhibits a long blue tail with a sharp cut-off, which is in excellent qualitative agreement with experimental observations [41,44] and thus provides an understanding for the underlying mechanisms for SCG.

16.3.3 Filamentation and Light Strings

At still higher powers above the power for critical collapse, the peak intensity of the wave can become sufficiently high that multi-photon absorption (MPA) and avalanche ionization can result in the formation of a plasma, which produces a defocusing effect on the laser pulse [48], such that the pulse forms an apparent light filament over distances that are larger than the diffraction length associated with the filament width. A remarkable example of this effect in the femtosecond regime has been the observation of filamentation in air [14,49,50] for pulse energies greater than 10 mJ. The lengths of these filaments can be on the order of meters, and for higher energy beams, it is possible to observe the continual formation of filaments over distances as long as 12 km [51]. Such filaments have been proposed for use in remote sensing of chemical species [7] or for guiding of lightning [52] and electrical [53] discharges. In addition, recent studies show that spatial cleaning related to self-similar collapse occur [54] and that under carefully controlled conditions, self-focusing and filamentation in noble gases can lead to strong pulse compression down to few-cycle pulses [55,56]. This compression can be explained as a natural consequence of the self-focusing, shock-formation process, and plasma interaction [47,57].

Theoretical simulations [58,59] of the femtosecond filamentation have been able to explain a number of the features of the process, but ultimately it is necessary [47,60,61] to include space–time focusing and self-steepening to accurately simulate the spectral content of the generated radiation and the structure of the conical emission. However unlike a true spatial soliton, the peak intensity and the accompanying plasma density in the extended filaments is not constant [62], and at sufficiently high powers multiple collapse/filamentation occurs in the longitudinal direction. This process has been termed spatial replenishment [57]; the beam undergoes collapse, defocuses as a result of the plasma formation, and then collapses at another point along the beam axis within the medium [63].

As a result of the excitation of the broad frequency and spatial frequencies near the collapse point, it is also possible to excite linear eigenmodes, known as "X-waves" [64,65], which are nondiffracting waves corresponding to Bessel-type solutions of the wave equation in a dispersive medium. These waves contain a small fraction of the energy within the collapsing region, but they do offer intuition for understanding the observed conical emission in this system.

At still higher powers, multiple filamentation is predicted to occur in the transverse direction, which may be a result of modulational instability, and it has been suggested that the resulting behavior is turbulent [66]. Extensive transverse multiple filamentation is observed with peak powers that are in the terawatt regime [7,67]. Similar filamentation and multiple collapse effects have also been observed in solids [68–70] and liquids [65,71,72] in the normal-dispersion regime. The primary distinction with the behavior in gases is that dispersive effects and the contribution of avalanche ionization to the production of the plasma density play a critical role in condensed materials.

16.3.4 Self-focusing in the Anomalous-Dispersion Regime

Only a few theoretical [73–76] and experimental studies have investigated pulse collapse in the anomalous-GVD (i.e., $\beta_2 < 0$) regime, and the dynamics are found to be completely different from that of the normal-GVD regime. In the anomalous-GVD regime, no pulse splitting occurs and the pulse can undergo full three-dimensional collapse, even for powers less than the critical power P_{cr}, which ultimately depends on the initial focusing conditions.

When plasma formation occurs, the resulting filaments are far more extended than in the normal-GVD regime, with the spatial replenishment occurring at distances that can be many diffraction lengths away from the initial collapse point [75]. This extended re-formation distance is a result of the anomalous

GVD, which allows for recompression of the pulse even after plasma defocusing. The possibility that "light bullets" (i.e., 3-D solitons) might be formed that are completely localized in space and time and can propagate without change were investigated [74]; however, these pulses are unstable unless another nonlinear process such as a saturating nonlinear refractive index is present to stabilize it [77].

Although the observation of full 3-D spatio-temporal solitons in the normally dispersive regime of glass and of gases was claimed [78], it was pointed out [79] that the pulses in those experiments traveled only a small fraction of a dispersion length and that it was not possible to draw any conclusions about the formation of a spatio-temporal soliton. Spatio-temporal compression and guiding has been observed [80] in a single transverse dimension in a planar silica waveguide. Under suitable conditions, materials with a quadratic nonlinearity can possess an effective saturating nonlinearity, which has allowed for the formation of spatio-temporal solitons in one transverse dimension [81].

16.4 Conclusions

The nonlinear dynamics associated with self-focusing has proven to be remarkably complex and rich in its behavior. Since the interaction is very well modeled by the nonlinear Schrödinger equation, experiments have been able to reproduce numerous features generic to wave collapse. There are a number of promising directions for future study. For example, all studies of self-focusing in the presence of dispersion have assumed that the initial pulse is Gaussian; we expect that novel dynamics will occur at high powers and for the case in which the initial pulses are non-Gaussian. Another potential area of study is self-focusing interactions with multiple pulses or with complicated initial spatial profiles, such as "necklace" beams [82].

References

1. R.Y. Chiao, E. Garmire, C.H. Townes: Self-trapping of optical beams, *Phys. Rev. Lett.* **13**, 479–481 (1964).
2. V.I. Talanov: Self-focusing of wave beams in nonlinear media, *JETP Lett.-USSR* **2**, 138 (1965).
3. P.L. Kelley: Self-focusing of optical beams, *Phys. Rev. Lett.* **15**, 1005–1007 (1965).
4. D.E. Spence, P.N. Kean, W. Sibbett: 60-fsec pulse generation from a self-mode-locked Ti:sapphire laser, *Opt. Lett.* **16**, 42–44 (1991).
5. J.H. Glownia, J. Misewich, P.P. Sorokin: Ultrafast ultraviolet pump probe apparatus, *J. Opt. Soc. Am. B-Opt. Phys.* **3**, 1573–1579 (1986).
6. R.R. Alfano, *The Supercontinuum Laser Source*. Springer-Verlag, New York (1989).
7. J. Kasparian, M. Rodriguez, G. Mejean et al.: White-light filaments for atmospheric analysis, *Science* **301**, 61–64 (2003).
8. G. Fibich, A.L. Gaeta: Critical power for self-focusing in bulk media and in hollow waveguides, *Opt. Lett.* **25**, 335–337 (2000).

9. G.M. Fraiman: Asymptotic stability of manifold of self-similar solutions on self-focusing, *Sov. Phys. JETP* **61**, 228–233(1985).
10. M.J. Landman, G.C. Papanicolaou, C. Sulem et al.: Rate of blowup for solution of the nonlinear Schrodinger equation at critical dimension, *Phys.Rev. A* **38**, 3837–3843 (1988).
11. B.J. Mesurier, G.C. Papanicolaou, C. Sulem et al.: Local structure on the self-focusing singularity of the cubic Schrödinger equation, *Physica D. Nonlinear Phenomena* **32**, 210–226 (1988).
12. G. Fibich, B. Ilan: Self-focusing of elliptic beams: an example of the failure of the aberrationless approximation, *J. Opt. Soc. Am. B-Opt. Phys.* **17**, 1749–1758 (2000).
13. K.D. Moll, A.L. Gaeta, G. Fibich: Self-similar optical wave collapse: observation of the townes profile, *Phys. Rev. Lett.* **90**, 203902 (2003).
14. A. Braun, G. Korn, X. Liu et al.: Self-channeling of high-peak-power femtosecond laser pulses in air, *Opt. Lett.* **20**, 73–75 (1995).
15. V.I. Bespalov, V.I. Talanov: Filamentary structure of light beams in nonlinear liquids, *JETP Lett.-USSR* **3**, 307–309 (1966).
16. G. Fibich, S. Eisenmann, B. Ilan et al.: Self-focusing distance of very high power laser pulses, *Opt. Express* **13**, 5897–5903 (2005).
17. A.J. Campillo, S.L. Shapiro, B.R. Suydam: Relationship of self-focusing to spatial instability modes, *Appl. Phys. Lett.* **24**, 178–180 (1974).
18. G. Fibich, N. Gavisha, X.P. Wang: New singular solutions of the nonlinear Schrodinger equation, *Physica D-Nonlinear Phenomena* **211**, 193–220 (2005).
19. T.D. Grow, A.A. Ishaaya, L.T. Vuong et al.: Collapse dynamics of super-Gaussian beams, *Opt. Express* **14**, 5468–5475 (2006).
20. L.T. Vuong, T.D. Grow, A. Ishaaya et al.: Collapse of optical vortices, *Phys. Rev. Lett.* **96**, 133901 (2006).
21. A.L. Gaeta: Collapsing light really shines, *Science* **301**, 54–55 (2003).
22. T. Brabec, F. Krausz: Nonlinear optical pulse propagation in the single-cycle regime, *Phys. Rev. Lett.* **78**, 3282–3285 (1997).
23. J.K. Ranka, A.L. Gaeta: Breakdown of the slowly varying envelope approximation in the self-focusing of ultrashort pulses, *Opt. Lett.* **23**, 534–536 (1998).
24. J.E. Rothenberg: Space–time focusing: breakdown of the slowly varying envelope approximation in the self-focusing of femtosecond pulses, *Opt. Lett.* **17**, 1340–1342 (1992).
25. T.K. Gustafson, P.L. Kelley, R.Y. Chiao et al.: Self-trapping in media with saturation of nonlinear index, *Appl. Phys. Lett.* **12**, 165–167 (1968).
26. G.Z. Yang, Y.R. Shen: Spectral broadening of ultrashort pulses in a nonlinear medium, *Opt. Lett.* **9**, 510–512 (1984).
27. J.E. Rothenberg, D. Grischkowsky: Observation of the formation of an optical-intensity shock and wave breaking in the nonlinear propagation of pulses in optical fibers, *Phys. Rev. Lett.* **62**, 531–534 (1989).
28. N.A. Zharova, A.G. Litvak, T.A. Petrova et al.: Multiple fractionation of wave structures in a nonlinear medium, *JETP Lett.-USSR* **44**, 13–17 (1986).
29. J.E. Rothenberg: Pulse splitting during self-focusing in normally dispersive media, *Opt. Lett.* **17**, 583–585 (1992).
30. P. Chernev, V. Petrov: Self-focusing of light-pulses in the presence of normal group-velocity dispersion, *Opt. Lett.* **17**, 172–174 (1992).
31. G.G. Luther, A.C. Newell, J.V. Moloney et al.: Short-pulse conical emission and spectral broadening in normally dispersive media, *Opt. Lett.* **19**, 789–791 (1994).
32. G. Fibich, V.M. Malkin, G.C. Papanicolaou: Beam self-focusing in the presence of a small normal time dispersion, *Phys. Rev. A* **52**, 4218–4228 (1995).
33. J. K. Ranka, R. W. Schirmer and A. L. Gaeta: Observation of pulse splitting in nonlinear dispersive media, Phys. Rev. Lett. **77**, 3783-3786 (1996).
34. S.A. Diddams, H.K. Eaton, A.A. Zozulya et al.: Amplitude and phase measurements of femtosecond pulse splitting in nonlinear dispersive media, *Opt. Lett.* **23**, 379–381 (1998).

35. H.K. Eaton, T.S. Clement, A.A. Zozulya et al.: Investigating nonlinear femtosecond pulse propagation with frequency-resolved optical gating, *IEEE J. Quantum Electron.* **35**, 451–458 (1999).
36. G. Fibich, W.Q. Ren, X.P. Wang: Numerical simulations of self-focusing of ultrafast laser pulses, *Phys. Rev. E* **67**, 056603 (2003).
37. J.K. Ranka, A.L. Gaeta, unpublished.
38. R.R. Alfano, S.L. Shapiro: Observation of self-phase modulation and small-scale filaments in crystals and glasses, *Phys. Rev. Lett.* **24**, 592–594 (1970).
39. P.L. Baldeck, P.P. Ho, R.R. Alfano: Effects of self-induced and cross-phase modulations on the generation of picosecond and femtosecond white-light supercontinua, *Revue De Physique Appliquee* **22**, 1677–1694 (1987).
40. P.B. Corkum, P.P. Ho, R.R. Alfano et al.: Generation of infrared supercontinuum covering 3–14 μm in dielectrics and semiconductors, *Opt. Lett.* **10**, 624–626 (1985).
41. A. Brodeur, S.L. Chin: Band-gap dependence of the ultrafast white-light continuum, *Phys. Rev. Lett.* **80**, 4406–4409 (1998).
42. W.L. Smith, P. Liu, N. Bloembergen: Superbroadening in H_2O and D_2O by self-focused picosecond pulses from a YAlG:Nd laser, *Phys. Rev. A* **15**, 2396–2403 (1977).
43. R.L. Fork, C.V. Shank, C. Hirlimann et al.: Femtosecond white-light continuum pulses, *Opt. Lett.* **8**, 1–3 (1983).
44. P.B. Corkum, C. Rolland, T. Srinivasanrao: Supercontinuum generation in gases, *Phys. Rev. Lett.* **57**, 2268–2271 (1986).
45. J.H. Glownia, J. Misewich, P.P. Sorokin: Ultrafast ultraviolet pump probe apparatus, *J. Opt. Soc. Am. B-Opt. Phys.* **3**, 1573–1579 (1986).
46. T.R. Gosnell, A.J. Taylor, D.P. Greene: Supercontinuum generation at 248 nm using high-pressure gases, *Opt. Lett.* **15**, 130–132 (1990).
47. A.L. Gaeta: Catastrophic collapse of ultrashort pulses, *Phys. Rev. Lett.* **84**, 3582–3585 (2000).
48. E. Yablonovitch, N. Bloembergen: Avalanche ionization and limiting diameter of filaments induced by light-pulses in transparent media, *Phys. Rev. Lett.* **29**, 907–909 (1972).
49. E.T.J. Nibbering, P.F. Curley, G. Grillon et al.: Conical emission from self-guided femtosecond pulses in air, *Opt. Lett.* **21**, 62–64 (1996).
50. A. Brodeur, C.Y. Chien, F.A. Ilkov et al.: Moving focus in the propagation of ultrashort laser pulses in air, *Opt. Lett.* **22**, 304–306 (1997).
51. P. Rairoux, H. Schillinger, S. Niedermeier et al.: Remote sensing of the atmosphere using ultrashort laser pulses, *Appl. Phys. B-Lasers Opt.* **71**, 573–580 (2000).
52. J.C. Diels, R. Bernstein, K.E. Stahlkopf et al.: Lightning control with lasers, *Sci. Am.* **277**, 50–55 (1997).
53. S. Tzortzakis, B. Prade, M. Franco et al.: Femtosecond laser–guided electric discharge in air, *Phys. Rev. E* **6405**, 057401 (2001).
54. B. Prade, M. Franco, A. Mysyrowicz et al.: Spatial mode cleaning by femtosecond filamentation in air, *Opt. Lett.* **31**, 2601–2603 (2006).
55. C.P. Hauri, W. Kornelis, F.W. Helbing et al.: Generation of intense, carrier-envelope phase-locked few-cycle laser pulses through filamentation, *Appl. Phys. B-Lasers Opt.* **79**, 673–677 (2004).
56. G. Stibenz, N. Zhavoronkov, G. Steinmeyer: Self-compression of millijoule pulses to 7. 8 fs duration in a white-light filament, *Opt. Lett.* **31**, 274–276 (2006).
57. A. Couairon, J. Biegert, C.P. Hauri et al.: Self-compression of ultra-short laser pulses down to one optical cycle by filamentation, *J. Mod. Opt.* **53**, 75–85 (2006).
58. M. Mlejnek, E.M. Wright, J.V. Moloney: Dynamic spatial replenishment of femtosecond pulses propagating in air, *Opt. Lett.* **23**, 382–384 (1998).
59. O.G. Kosareva, V.P. Kandidov, A. Brodeur et al.: Conical emission from laser–plasma interactions in the filamentation of powerful ultrashort laser pulses in air, *Opt. Lett.* **22**, 1332–1334 (1997).

60. N. Akozbek, M. Scalora, C.M. Bowden et al.: White-light continuum generation and filamentation during the propagation of ultra-short laser pulses in air, *Opt. Commun.* **191**, 353–362 (2001).
61. A. Couairon, S. Tzortzakis, L. Berge et al.: Infrared femtosecond light filaments in air: Simulations and experiments, *J. Opt. Soc. Am. B-Opt. Phys.* **19**, 1117–1131 (2002).
62. A. Talebpour, S. Petit, S.L. Chin: Re-focusing during the propagation of a focused femtosecond Ti:sapphire laser pulse in air, *Opt. Commun.* **171**, 285–290 (1999).
63. S. Tzortzakis, L. Berge, A. Couairon et al.: Breakup and fusion of self-guided femtosecond light pulses in air, *Phys. Rev. Lett.* **86**, 5470–5473 (2001).
64. C. Conti, S. Trillo, P. Di Trapani et al.: Nonlinear electromagnetic X waves, *Phys. Rev. Lett.* **90**, 170406 (2003).
65. D. Faccio, M.A. Porras, A. Dubietis et al.: Conical emission, pulse splitting, and X-wave parametric amplification in nonlinear dynamics of ultrashort light pulses, *Phys. Rev. Lett.* **96**, 193901 (2006).
66. M. Mlejnek, M. Kolesik, J.V. Moloney et al.: Optically turbulent femtosecond light guide in air, *Phys. Rev. Lett.* **83**, 2938–2941 (1999).
67. J. Kasparian, R. Sauerbrey, D. Mondelain et al.: Infrared extension of the supercontinuum generated by femtosecond terawatt laser pulses propagating in the atmosphere, *Opt. Lett.* **25**, 1397–1399 (2000).
68. S. Tzortzakis, L. Sudrie, M. Franco et al.: Self-guided propagation of ultrashort IR laser pulses in fused silica, *Phys. Rev. Lett.* **87**, 213902 (2001).
69. Z.X. Wu, H.B. Jiang, L. Luo et al.: Multiple foci and a long filament observed with focused femtosecond pulse propagation in fused silica, *Opt. Lett.* **27**, 448–450 (2002).
70. K.D. Moll, A.L. Gaeta: Role of dispersion in multiple-collapse dynamics, *Opt. Lett.* **29**, 995–997 (2004).
71. A. Matijosius, J. Trull, P. DiTrapani et al.: Nonlinear space–time dynamics of ultrashort wave packets in water, *Opt. Lett.* **29**, 1123–1125 (2004).
72. M.A. Porras, A. Dubietis, E. Kucinskas et al.: From X- to O-shaped spatiotemporal spectra of light filaments in water, *Opt. Lett.* **30**, 3398–3400 (2005).
73. S.N. Vlasov, L.V. Piskunova, V.I. Talanov: Three-dimensional wave collapse in a nonlinear Schrodinger-equation model, *Sov. Phys. JETP* **68**, 125–1128 (1989).
74. Y. Silberberg: Collapse of optical pulses, *Opt. Lett.* 15, 1282–1284 (1990).
75. K.D. Moll, A.L. Gaeta: Role of dispersion in multiple-collapse dynamics, *Opt. Lett.* **29**, 995–997 (2004).
76. L. Berge, S. Skupin: Self-channeling of ultrashort laser pulses in materials with anomalous dispersion, *Phys. Rev. E* **71**, 065601 (2005).
77. R. McLeod, K. Wagner, S. Blair: (3 + 1)-dimensional optical soliton dragging logic, *Phys. Rev. A* **52**, 3254–3278 (1995).
78. I.G. Koprinkov, A. Suda, P.Q. Wang et al.: Self-compression of high-intensity femtosecond optical pulses and spatiotemporal soliton generation, *Phys. Rev. Lett.* **84**, 3847–3850 (2000).
79. A.L. Gaeta, F. Wise: Comment on “Self-compression of high-intensity femtosecond optical pulses and spatiotemporal soliton generation,” *Phys. Rev. Lett.* **87**, 229401 (2001).
80. H.S. Eisenberg, R. Morandotti, Y. Silberberg et al.: Kerr spatiotemporal self-focusing in a planar glass waveguide, *Phys. Rev. Lett.* **87**, 043902 (2001).
81. X. Liu, K. Beckwitt, and F. Wise: Two-dimensional optical spatio-temporal solitons in quadratic media, *Phys. Rev. E* **62**, 1328–1340 (2000).
82. M. Soljacic, S. Sears, M. Segev: Self-trapping of “Necklace” Beams in self-focusing Kerr media, *Phys. Rev. Lett.* **81**, 4851–4854 (1998).

Chapter 17
Some Modern Aspects of Self-focusing Theory

Gadi Fibich

Abstract In this chapter, we give a brief summary of the present status of self-focusing theory, while trying to highlight the fascinating evolution of this theory.

17.1 Introduction

During the 1960s and early 1970s, there was intense theoretical and experimental research on self-focusing of intense laser beams in bulk media. In 1975, the results of this research activity were summarized in two long review papers: "Self-Focusing: Experimental," by Shen [1], and "Self-Focusing: Theory, by Marburger [2].

Around that time, mathematicians started to become interested in self-focusing theory. A large part of this research effort concerned collapsing (i.e., singular) solutions of the 2D cubic nonlinear Schrodinger equation (NLS)

$$i\psi_z(z,x,y) + \Delta\psi + |\psi|^2\psi = 0, \qquad \Delta = \partial_{xx} + \partial_{yy}. \tag{17.1}$$

Here, ψ is the complex-valued electric field envelope, z is the direction of propagation, and x and y are the transverse coordinates. Let the Kerr medium be located at $z>0$. Then the NLS is solved for $z>0$, subject to the initial condition

$$\psi(z=0,x,y) = \psi_0(x,y), \qquad -\infty < x,y < \infty,$$

where ψ_0 is the envelope of the electric field that impinges on the Kerr medium interface at $z=0$. Analysis of singular solutions of the NLS (17.1) turned out to be a hard mathematical problem for several reasons. First, the NLS is genuinely nonlinear, so linearization-based techniques are not applicable. Second, unlike

G. Fibich (✉)
School of Mathematical Sciences, Tel Aviv University, Tel Aviv, Israel
e-mail: fibich@tau.ac.il

R.W. Boyd et al. (eds.), *Self-focusing: Past and Present*,
Topics in Applied Physics 114, DOI 10.1007/978-0-387-34727-1_17,

the 1D cubic NLS, Eq. (17.1) is not integrable, so one cannot use inverse scattering theory. Finally, it turned out that the 2D cubic NLS is a "borderline case" for collapse in the following sense. Consider the d-dimensional focusing NLS with nonlinearity exponent p, i.e.,

$$i\psi_z(z, x_1, \ldots, x_d) + \Delta\psi + |\psi|^{p-1}\psi = 0, \qquad \Delta = \partial_{x_1x_1} + \cdots + \partial_{x_dx_d}. \tag{17.2}$$

Then the NLS (17.2) has collapsing solutions if $(p-1)d > 4$, which is the *supercritical* case. When, however, $(p-1)d < 4$, the *subcritical* case, there are no singular solutions. Therefore, $(p-1)d = 4$, the *critical* NLS, is a "borderline case" for collapse, in particular, the NLS (17.1) for which $p = 3$ and $d = 2$ is critical. As a result, self-focusing in the NLS (17.1) is characterized by a delicate balance between the focusing Kerr nonlinearity and diffraction. Hence, for example, collapse in the critical NLS is highly sensitive to small perturbations, which can arrest the collapse even when they are still small [3].

The present mathematical theory of self-focusing is very different from the one in 1975, because it is mostly based on results that were obtained after 1975.[1] For various reasons, some of these results have not become part of the common knowledge of the nonlinear optics community. The goal of this chapter is, thus, to provide a short survey of the current status of the mathematical theory of self-focusing, and to express it in a nonlinear optics context. Obviously, we focus on the mathematical results which we believe are of most relevance to the nonlinear optics applications. For more extensive reviews, see [3,4].

17.2 Some Pre-1975 Results

We begin with some pre-1975 results. The NLS (17.1) has several conservation laws. Most important are the conservation of *power*

$$P(z) \equiv P(0), \qquad P(z) = \int |\psi|^2\, dxdy,$$

and of the *Hamiltonian*

$$H(z) \equiv H(0), \qquad H(z) = \int |\nabla\psi|^2\, dxdy - \frac{1}{2}\int |\psi|^4\, dxdy.$$

Another identity that plays an important role in NLS theory is the *variance identity* [5]. Let

---

[1] The year 1975 was chosen as the "borderline" between the "past" and "present" in self-focusing theory, because of the review papers of Marburger and Shen [2,1] that appeared in that year. Clearly, this choice is somewhat arbitrary.

$$V(z) = \int r^2 |\psi|^2 \, dxdy, \qquad r = \sqrt{x^2 + y^2}.$$

Then,

$$\frac{d^2}{dz^2} V = 8H(0). \tag{17.3}$$

Because vanishing of variance can only occur if the whole solution collapses into the singularity, the variance identity leads to the following result:

Theorem 1. *Let ψ be a solution of the NLS (17.1) with initial condition ψ_0. Assume that one of the following three conditions holds:*

1. $H(0) < 0$.
2. $H(0) = 0$ *and* $\mathrm{Im} \int \psi_0^*(x, y) \cdot \nabla \psi_0 \, dxdy < 0$.
3. $H(0) > 0$ *and* $\mathrm{Im} \int \psi_0^*(x, y) \cdot \nabla \psi_0 \, dxdy \leq -\sqrt{H(0)V(0)}$,

where $\nabla = (\partial_x, \partial_y)$ and ψ_0^* is the complex conjugate of ψ_0. Then, ψ becomes singular at a finite propagation distance.

An important symmetry of the critical NLS (17.1), known as the *lens transformation*, is as follows [6]. Let ψ be a solution of the NLS (17.1) with initial condition $\psi_0(x, y)$, and let $\tilde{\psi}$ be the solution of the NLS (17.1) with initial condition $\tilde{\psi}(0, x, y) = e^{-i\frac{r^2}{4F}} \psi_0(x, y)$. Then, $\tilde{\psi}$ is given by

$$\tilde{\psi}(z, x, y) = \frac{1}{L(z)} \psi(\zeta, \xi, \eta) e^{i\frac{L_z}{L}\frac{r^2}{4}},$$

where

$$L(z) = 1 - \frac{z}{F}, \quad \zeta = \int_0^z L^{-2}(s) \, ds, \quad \xi = \frac{x}{L}, \quad \eta = \frac{y}{L}.$$

17.2.1 *Effect of a Lens*

Because

$$\frac{1}{z} = \frac{1}{F} + \frac{1}{\zeta}, \tag{17.4}$$

and because the addition of the quadratic phase term $e^{-i\frac{r^2}{4F}}$ corresponds to adding a focusing lens with focal length F at $z = 0$, the lens transformation shows that in a bulk Kerr medium, the effect of a lens at $z = 0$ is the same as in linear geometrical optics.

In particular, let Z_c be the collapse distance of a laser beam. If we now add a lens with focal length F at $z = 0$, then the new collapse distance, denoted by Z_c^F, follows immediately from Eq. (17.4), and is given by

$$\frac{1}{Z_c^F} = \frac{1}{F} + \frac{1}{Z_c}. \tag{17.5}$$

Therefore, the collapse distance decreases when the lens is focusing ($F > 0$), increases when the lens is defocusing ($F < 0$), and collapse is arrested by the defocusing lens if $-Z_c < F < 0$.

17.2.2 The R (Townes) Profile

The NLS (17.1) has waveguide solutions of the form $\psi_{w-g} = e^{i\lambda^2 z} R_\lambda(r)$, where $R_\lambda(r) = \lambda R(\lambda r)$, and R is the solution of

$$R''(r) + \frac{1}{r}R' - R + R^3 = 0, \qquad R'(0) = 0, \qquad R(\infty) = 0.$$

This equation turns out to have an infinite number of solutions. Of most interest, however, is the ground-state solution, known as the R *profile* or the *Townes profile*, which is positive and monotonically decreasing to zero.

The R-based waveguide solutions were originally considered by Chiao et al. [7] in the context of *self-trapping*, i.e., the observation of long and narrow filaments in experiments. It later turned out, however, that the NLS waveguide solutions ψ_{w-g} cannot be used to explain self-trapping, because they are unstable. Indeed, if we perturb the initial condition as $\psi_0 = (1+\epsilon)R(r)$, where $0 < \epsilon \ll 1$, then the corresponding solution will collapse after a finite propagation distance [8]. Nevertheless, the R profile is extremely important in NLS theory, because it has the critical power for collapse (Section 17.3) and because it is the universal, self-similar profile of collapsing beams (Section 17.4).

17.3 Critical Power

Kelley [9] was the first to predict that the key quantity that determines whether a beam would collapse is its input power (and not, e.g., its initial radius or focusing angle). During the sixties and seventies, however, there was some confusion regarding the value of the critical power.

Weinstein [8] proved that a necessary condition for collapse is that the input power will exceed the power of the R profile, i.e., that $P \geq P_{cr}$, where

$$P_{\text{cr}} = \int R^2\, dxdy \approx 11.70.$$

Merle [10,11] proved that the only collapsing solutions with the *critical power* $P = P_{\text{cr}}$ are those for which the input profile is given by the R profile. Therefore, for any other input profile the threshold power for collapse P_{th} is strictly above P_{cr}. For example, the threshold power for a Gaussian profile $\psi_0 = c \cdot \exp(-r^2)$ is $\approx 2\%$ above P_{cr}, and for a super-Gaussian profile $\psi_0 = c \cdot \exp(-r^4)$ is $\approx 9\%$ above P_{cr} [12]. In general, the threshold power for collapse of an input profile which is "close" to the R profile will be lower (i.e., closer to P_{cr}) than of an input profile which is "less similar" to the R profile.

Because the condition $H(0) < 0$ implies collapse (see Section 17.2), some researchers estimated the threshold power for collapse from the condition of a zero Hamiltonian. Although $P_{\text{cr}} = \int R^2\, dxdy$ is only a *lower bound* for the threshold power, in the case of cylindrically symmetric initial conditions, it provides a much better estimate for the threshold power than the one obtained from the condition of a zero Hamiltonian ([12], and see also Section 17.6.1).

The threshold power for collapse increases with input beam ellipticity. Fibich and Ilan [13] showed that for an elliptic input profile $\psi_0 = cF(\sqrt{(x/a)^2 + (y/b)^2})$, the threshold power for collapse can be approximated with

$$P_{\text{th}}(e) \approx \left[0.4\frac{e + 1/e}{2} + 0.6\right] P_{\text{th}}(e = 1),$$

where $e = b/a$ and $P_{\text{th}}(e = 1)$ is the threshold power for the circular profile $\psi_0 = cF(r)$. In this case, the threshold power obtained from the condition of a zero Hamiltonian is also highly inaccurate [13].

It is important to realize that while the threshold power P_{th} for collapse is above P_{cr}, the amount of power that eventually collapses into the blowup point is always equal to P_{cr} (see Section 17.6).

17.3.1 Hollow Waveguides (Bounded Domains)

The propagation of laser beams in a hollow-core waveguide with radius r_a can be modeled by the NLS (17.1) on the bounded domain $0 \le x^2 + y^2 \le r_a^2$, subject to the boundary condition $\psi(z, r = r_a) = 0$. Fibich [14] proved that in this case, P_{cr} is still a *lower bound* for the threshold power. In addition, the numerical simulations of Fibich and Gaeta [12] suggest a stronger result, namely, that in this case the threshold power for collapse is *equal* to P_{cr} for "any" input profile. The reason for this behavior is that, unlike in bulk media, the reflecting walls prevent the shedding of power and keep the power localized in the transverse domain. Therefore, unlike in bulk medium, the collapsing core does not lose power due to radiation as it "rearranges" itself in the form of the self-similar ψ_R profile (see Section 17.4).

17.4 The Universal Blowup Profile ψ_R

The early analytical studies of collapsing solutions in the NLS (17.1) assumed that the blowup profile is a self-similar Gaussian or *sech* profile (see, e.g., [15]). Numerical simulations of collapsing NLS solutions that were carried out during the 1980s and early 1990s (see, e.g., [16]) showed, however, that regardless of the initial profile, near the singularity, collapsing NLS solutions approach the universal, self-similar profile ψ_R, which is a modulated R profile. In other words, $\psi \sim \psi_R$ near the collapse point Z_c, where

$$|\psi_R| = \frac{1}{L(z)} R\left(\frac{r}{L(z)}\right),$$

and $\lim_{z \to Z_c} L(z) = 0$. Therefore, in particular, even if the initial condition is elliptically shaped and/or noisy, near the singularity the blowup profile becomes smooth and symmetric.

The fact that collapsing NLS solutions approach the universal blowup profile ψ_R was crucial for the derivation of the reduced system (17.7) that was used to find the NLS blowup rate (see Section 17.7), as well as for the development of *modulation theory* for the asymptotic analysis of the effect of small perturbations (see Section 17.10.2). *Proving* the convergence to ψ_R, however, turned out to be a very hard mathematical problem. Indeed, this result was proved by Merle and Raphael [17] only in 2003, and its proof required developing new analytical tools. The proof of Merle and Raphael is valid for any initial condition whose power P satisfies $P_{\text{cr}} < p < P_{\text{cr}} + \alpha$, where α is a universal constant whose value is less than P_{cr}. In particular, the proof holds for initial conditions which are elliptically shaped or randomly distorted.

In 2003, Moll et al. [18] showed experimentally that the spatial profile of a collapsing beam evolves to the cylindrically symmetric Townes profile, for elliptically shaped as well as for randomly distorted Gaussian input beams that propagated in glass. Subsequently, in 2006, Grow et al. [19] observed experimentally the evolution to the Townes profile for collapsing Gaussian input pulses that propagated in water.

17.4.1 New Blowup Profiles

Although the proof of Merle and Raphael does not hold for solutions whose power is above $2P_{\text{cr}}$, it was widely believed, based on numerous numerical simulations since the late 1970s, that any singular solution of the NLS collapses with the ψ_R profile. Even when the solution broke into multiple filaments (see Section 17.9), each filament was found to collapse with the ψ_R profile.

Recently, however, Fibich et al. [20] showed that high-power super-Gaussian beams collapse with a self-similar *ring profile*, which is different from the R profile; see Fig. 17.1. Similarly to the ψ_R profile, these solutions are stable under radial perturbations. Therefore, the radially symmetric NLS

$$i\psi_z(z,r) + \psi_{rr} + \frac{1}{r}\psi_r + |\psi|^2\psi = 0, \tag{17.6}$$

does have stable collapsing solutions whose blowup profile is not given by ψ_R. Unlike the ψ_R profile, however, these ring solutions are unstable under azimuthal perturbations. Therefore, the ring solutions are unstable as solutions of the NLS (17.1).

17.5 Super-Gaussian Input Beams

The self-focusing dynamics of super-Gaussian input beams is very different from the dynamics of Gaussian beams. Here we briefly discuss some aspects of this topic. For more information and references, see the chapter in this book by Lukishova et al.

In the linear regime (i.e., for beam power $\ll P_{cr}$), the super-Gaussian evolves into a ring shape. This linear effect is due to Fresnel diffraction, and it occurs at propagation distances of the order of the diffraction (Fresnel) length, which is defined as $L_{\text{diff}} = k_0 r_0^2$, where k_0 is the wavenumber and r_0 is the input beam width. For high peak-power Nd:glass lasers, it has been shown, both experimentally and numerically, that super-Gaussian profiles evolve into a ring profile [21,22].

Grow et al. [19] showed theoretically that *ring formation of high-power super-Gaussian beams is a nonlinear phenomena which is due to ray bending as a result of nonlinear self-phase modulations* (see Fig. 17.2). Therefore, ring formation of high-power super-Gaussian beams is a nonlinear geometrical optics effect, rather than a linear Fresnel diffraction one. This implies that it can occur over distances much shorter than a single diffraction length, and that this phenomena occurs for any $N > 2$ in the initial profile $\psi_0 = c \cdot \exp(-r^N)$ (see Fig. 17.2).

In the absence of azimuthal noise, high-power super-Gaussian beams collapse with the self-similar ring profile described in Section 17.4.1 (see, e.g., Fig. 17.1). These collapsing rings are, however, highly unstable under azimuthal perturbations. As a result, the ring quickly disintegrates into a ring of filaments, each of which collapses with the ψ_R profile.

In [19], both the ring formation and its subsequent propagation dynamics as it breaks up into a ring of filaments were observed experimentally for high-power super-Gaussian beams propagating in water. The diffraction length in

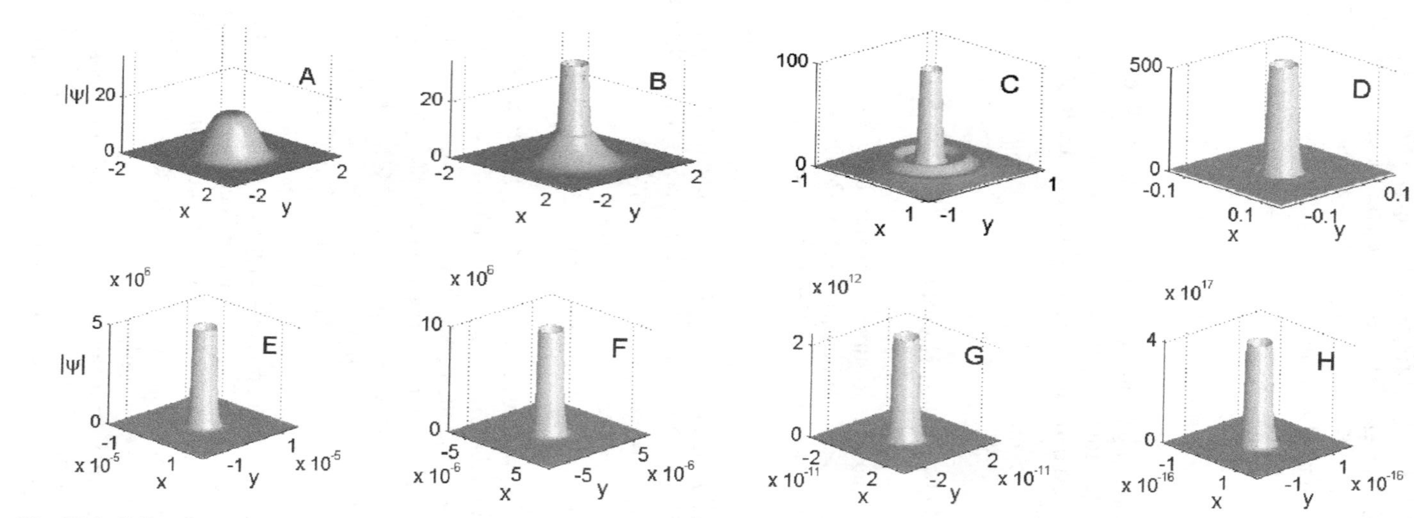

Fig. 17.1 Collapsing self-similar ring solution of the NLS (17.6) with $\psi_0 = 15e^{-r^4}$. From Ref. [20]

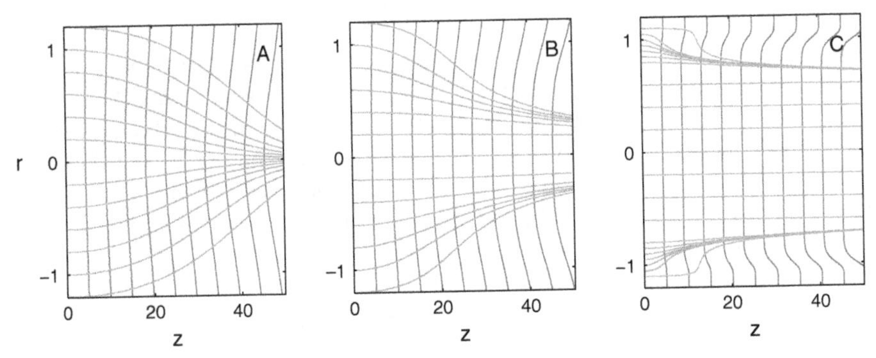

Fig. 17.2 Nonlinear propagation of rays (*horizontal lines*) and phase fronts (*vertical lines*) for high-power beams $\psi_0 = c \cdot \exp(-r^N)$. (**A**) Gaussian $N = 2$, (**B**) super-Gaussian $N = 4$, and (**C**) super-Gaussian $N = 20$. Graphs (A) and (B) are from Ref. [19]

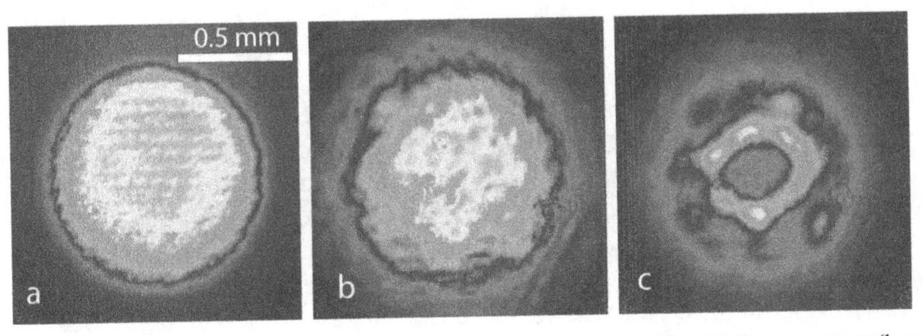

Fig. 17.3 Experimental intensity distributions for: (**a**) input profile, (**b**) linear output (low power) after a 7-cm cell, and (**c**) nonlinear output with $E = 12.2\ \mu$J after a 7-cm cell. Data supplied by T.D. Grow and A.L. Gaeta

these experiments was approximately 40 cm, and the distances at which the rings were observed were usually less than 10 cm. Therefore, the experimental data agrees with the ring developing at a fraction of a diffraction length. Figure 17.3 shows data from the same set of experiments that clearly shows the ring formation at high powers and its absence at low powers. This further confirms that ring formation for high-power super-Gaussian beams is a nonlinear phenomena.

Finally, we note that the collapse dynamics of high-power super-Gaussians implies that they can undergo multiple filamentation during the initial collapse at much lower powers ($P \sim 10P_{\mathrm{cr}}$) than Gaussian beams (see Section 17.9) [19,23].

17.6 Partial Beam Collapse

The fact that the blowup profile is given by ψ_R implies that

$$|\psi|^2 \longrightarrow P_{\mathrm{cr}} \cdot \delta(r), \qquad z \to Z_c.$$

In other words, the amount of power that collapses into the singularity is independent of the initial condition, and is always given by P_{cr}.[2] We already have seen that for any input profile different from the R profile, collapse can occur only if $P > P_{\text{cr}}$. Therefore, the "outer part" of the beam, whose power is equal to $P - P_{\text{cr}}$, does not collapse into the singularity, but rather continues to propagate forward. This shows that generically, only part of the beam power collapses into the singularity (*partial beam collapse*). Hence, *the variance at the collapse point is generically positive and not zero*, i.e.,

$$\lim_{z \to Z_c} V(z) > 0$$

17.6.1 Common Misinterpretations of the Variance Identity

There are some common *misinterpretations* of the variance identity, all of which follow from the wrong assumption that the variance vanishes at the blowup point: (1) The variance identity can be used to predict the collapse distance. (2) The threshold power can be well approximated from the condition of a zero Hamiltonian, (3) Solutions with positive Hamiltonian undergo partial-beam collapse whereas those with negative Hamiltonian undergo *whole-beam collapse* (i.e., all the beam power collapses into the focal point), and so on. We stress that all of these statements are false, as they are based on the wrong assumption that the variance vanishes at the blowup point. See [13] for more details.

17.7 Blowup Rate

In the rigorous mathematical theory of the NLS, the blowup rate is usually defined as $L(z) = (\int |\nabla\psi|^2 \, dxdy)^{-1/2}$. However, up to a multiplicative constant, the blowup rate can be defined as $L = 1/\max_{x,y} |\psi(z, x, y)|$. A question that has been open for many years is what is the blowup rate of NLS solutions. In other words, does $L(z) \sim c(Z_c - z)^p$ for some p as $z \to Z_c$?

17.7.1 The Loglog Law

Finding the blowup rate of the NLS turned out to be a very difficult problem, and over the years various power-law relations were proposed [24]. In retrospect, the mathematical difficulties had to do with the fact that collapse in the

---

[2] This explains why in experiments it is often found that all filaments have the same power.

critical NLS is "only" *quasi self-similar*, i.e., the collapsing core approaches the self-similar profile ψ_R, but the "outer part" of the beam has a completely different dynamics. Moreover, the coupling between these two components of the solution is exponentially weak. Eventually, Fraiman [25], and independently (and in a different way) Landman, et al. [26,27] showed that the NLS dynamics can be approximated with the following reduced ODE system for $L(z)$:

$$L_{zz}(z) = -\frac{\beta}{L^3}, \qquad \beta_z(z) = -\frac{\nu(\beta)}{L^2}, \tag{17.7}$$

where $0 < \beta \ll 1$,

$$\nu(\beta) = c_\nu e^{-\pi/\sqrt{\beta}}, \qquad c_\nu = \frac{2A_R^2}{M}, \tag{17.8}$$

$A_R = \lim_{r\to\infty} e^r r^{1/2} R(r) \approx 3.52$, and $M = \frac{1}{4}\int_0^\infty r^2 R^2 \, rdr \approx 0.55$. Asymptotic analysis of the reduced Eqs. (17.7) showed that the blowup rate of the critical NLS is a square root with a loglog correction (the *loglog law*) [25–27].

$$L \sim \left(\frac{2\pi(Z_c - z)}{\ln\ln(1/(Z_c - z))}\right)^{1/2}, \qquad z \longrightarrow Z_c. \tag{17.9}$$

Subsequent numerical simulations with specialized codes that could reach very high focusing levels (e.g., $1/L = O(10^{10})$) have confirmed that the blowup rate is slightly faster than a square root, but failed to detect the loglog correction. The reason for this "failure" was explained by Fibich and Papanicolaou [3,28], who showed that the loglog law does not become valid even after the solution has focused by 10^{100}. Because the validity of the NLS model breaks down at much lower focusing levels, the loglog law turned out to be more of mathematical interest than of real physical value. However, Malkin [29] and Fibich [28] showed that the same reduced equations that lead to the loglog law, Eq. (17.7), can be solved differently, yielding the *adiabatic laws of collapse*, which become valid after moderate focusing levels [3]. More importantly, Fibich and Papanicolaou [3] showed that a similar approach can be used to analyze the effect of small perturbations on the collapse dynamics at physically-relevant focusing levels (see Section 17.10.2).

Although the loglog law (17.9) was derived in the late 1980s, a rigorous proof was obtained only in 2003 by Merle and Raphael [17]. The proof holds for all initial conditions whose power P satisfies $P_{\text{cr}} < p < P_{\text{cr}} + \alpha$, where α is a universal constant which is smaller than P_{cr}.

17.7.2 A Square-Root Law

Although the rigorous proof of Merle and Raphael does not hold for input powers $P > 2P_{\mathrm{cr}}$, it was widely believed that generically all NLS solutions collapse according to the loglog law. However, in 2005, Fibich et al. showed asymptotically and numerically that collapsing self-similar *ring* solutions of the NLS (see, e.g., Fig. 17.1) blowup at a square-root rate, with no loglog correction [20]. At present, it is still an open question whether the self-similar ring profile, hence the square-root blowup rate, is maintained all the way up to the singularity. Indeed, the numerical simulations of [20] become unreliable after focusing by $\approx 10^{16}$. Therefore, it is possible that at higher focusing levels the self-similar ring profile would change to the Townes profile, in which case the blowup rate would change from a square root to the loglog law. This open question is, however, only of mathematical interest, as the validity of the NLS model breaks down at much smaller focusing levels.

17.8 Self-focusing Distance

It would have been very useful to have an exact analytical formula for the location of the singularity as a function of the input beam. Unfortunately, such a formula does not exist. Some researchers estimated the location of the blowup point by using the variance identity (17.3) to calculate the location where the variance should vanish. As we pointed out in Section 17.6.1, this approach usually leads to very inaccurate results. Fibich [28] used the *adiabatic law of collapse* to derive the following asymptotic formula for real initial conditions (i.e., for collimated beams)

$$Z_c \sim \sqrt{\frac{MP_{\mathrm{cr}}}{P/P_{\mathrm{cr}} - 1}} \left(\int |\nabla \psi_0|^2 \right)^{-1}, \tag{17.10}$$

which gives reasonable predictions for $P \le 2P_{\mathrm{cr}}$.

In the absence of an analytical formula, the only way to find the location of the singularity is through numerical simulations. For example, Dawes and Marburger [30] used the results of numerical simulations to derive the following *curve-fitted* formula for the location of the singularity of collimated *Gaussian* input beams $\psi_0 = ce^{-r^2/2}$:

$$Z_c = 0.367[(\sqrt{P/P_{\mathrm{cr}}} - 0.852)^2 - 0.0219]^{-1/2}. \tag{17.11}$$

Kelley [9] was the first to show that the collapse distance Z_c scales as $1/\sqrt{P}$ for $P \gg P_{\mathrm{cr}}$. In theory, the $1/\sqrt{P}$ relation should become more and more accurate as P increases. This is, indeed, the case when the input beam is noiseless. In practice, however, input beams are always noisy. At input powers that are roughly above$100P_{cr}$, the propagation becomes highly sensitive to the effect of input beam noise. As a result, the beam becomes modulationally unstable and breaks into multiple filaments at a distance that scales as $\sim 1/P$. Campillo et al. [31] predicted theoretically, and observed experimentally for continuous-wave (cw) beams propagating in CS_2, that this noise-induced multiple filamentation implies that the collapse distance scales as $1/P$. This result was rediscovered by Fibich et al. [32], who also observed numerically, and experimentally for femtosecond pulses propagating in air, that the collapse distance scales as $1/\sqrt{P}$ for input powers that are moderately above the critical power for self-focusing, but that at higher powers the collapse distance scales as $1/P$.

17.8.1 Effect of a Lens

Let Z_c be the collapse distance of a laser beam. As we have seen before, if we add a lens with focal length F at $z = 0$, then the new collapse distance, denoted by Z_c^F, is given by Eq. (17.5). Using this lens relation, one can predict the collapse point of focused beams based on predictions for the collapse point of collimated beams, such as Eqs. (17.10) or (17.11).

Recently, Fibich et al. [33] showed that this lens relation is in excellent agreement with experimental measurements of the collapse distance in atmospheric propagation of femtosecond pulses. This showed that the relatively simple NLS model (17.1) can be used to predict the collapse point, because all other mechanisms (multiphoton absorption, plasma formation, Raman scattering, etc.) become important only *after* the pulse has collapsed.

17.9 Multiple Filamentation

Since the early 1960s, it has been observed experimentally that when the laser power is significantly larger than the critical power P_{cr}, the beam can break into several long and narrow filaments, a phenomenon known as *multiple filamentation* (MF) or as *small-scale self-focusing*. Because MF involves a complete breakup of the beam cylindrical symmetry, it must be initiated by a symmetry-breaking mechanism.

17.9.1 Noise-Induced Multiple Filamentation

For many years, the standard (and only) explanation for MF in the literature, due to Bespalov and Talanov [34], has been that it is initiated by input beam noise. Briefly, the noise leads to a *modulational instability* (MI), which ultimately results in MF. Although the analysis of Bespalov and Talanov was based on linear stability of (infinite power) plane waves solutions of the NLS (17.1), it was believed to hold for beams whose power is roughly above $10P_{cr}$. This is, however, not the case. For example, Fibich and Ilan [35] solved numerically the NLS (17.1) with an input Gaussian beam with $P = 15P_{cr}$ and 10% random noise, and observed that it did not break into multiple filaments, but rather collapsed at a single location. Indeed, noise would lead to MF of Gaussian beams in the NLS (17.1) only when the distance for noise-induced MF, L_{MF}, is smaller than the self-focusing distance Z_c. Because $L_{MF} \sim 1/P$ and $Z_c \sim 1/\sqrt{P}$, this would occur for powers above a second power threshold $P_{th}^{(2)}$, which is roughly of the order of $100P_{cr}$. Therefore, input beam noise would lead to MF in the NLS (17.1) only for $P > P_{th}^{(2)}$ [31,32].

Until recently, the only known way to observe noise-induced MF numerically at input powers $\ll P_{th}^{(2)}$ was to add to the NLS a collapse-arresting mechanism. Consider, for example,the 2D NLS with a nonlinear saturation

$$i\psi_z(z,x,y) + \Delta\psi + \frac{|\psi|^2}{1+\epsilon|\psi|^2}\psi = 0, \qquad \epsilon > 0, \qquad \Delta = \partial_{xx} + \partial_{yy}. \quad (17.12)$$

Solutions of this equation do not become singular. Rather, when the input power is above P_{cr}, the solution initially self-focuses, but then the collapse is arrested (see Figs. 17.4–17.5). Subsequently, the solution undergoes focusing–defocusing cycles (a *multifocus structure*). During the defocusing stage of each cycle, the solution has a ring (or crater) shape. Because a ring is an unstable shape for a beam, it can disintegrate into multiple filaments in the presence of noise, as was first demonstrated numerically in 1979 by Konno and Suzuki [36].

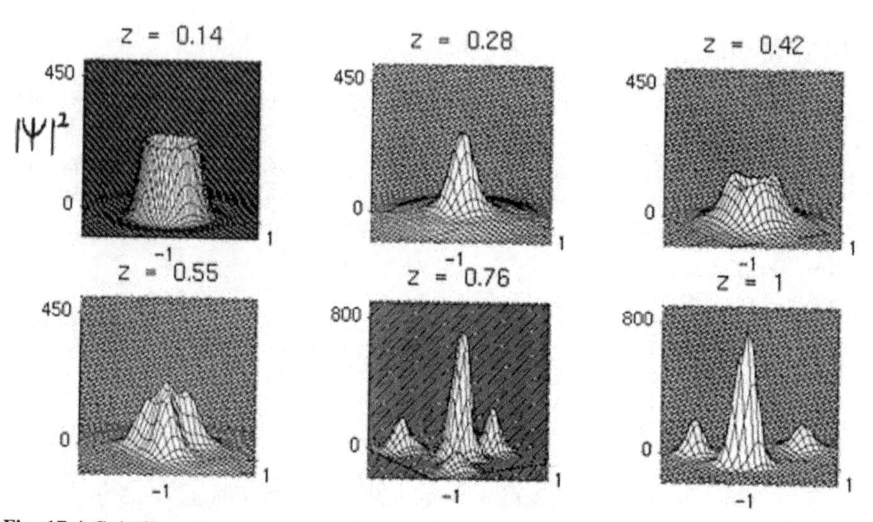

Fig. 17.4 Solution of Eq. (17.12) with a noisy input Gaussian beam with power $P \approx 15P_{cr}$. From Ref. [35]

Soto-Crespo et al. [37] developed an approximate stability analysis that explains the instability of the rings and predicts the number of filaments.

It is important to note that the two ingredients needed for MF at input powers $\ll P_{\text{th}}^{(2)}$ are:

1. A "mechanism" that arrests the collapse, so that as the beam defocuses it will assume the unstable ring shape. This mechanism does not have to be nonlinear saturation. For example, in [38] the collapse-arresting mechanisms were nonparaxiality and vectorial effects.
   Recently, it turned out that this mechanism does not even have to be collapse-arresting, as the only real requirement is that it will make the beam assume the unstable ring shape. Moreover, the ring does not have to be defocusing, because a collapsing ring is also unstable. Indeed, this is the case in MF of high-power super-Gaussian beams [19], where the ring is formed as the beam is collapsing (see Section 4.1), and also in MF of collapsing vortices [39].
2. A "mechanism" that breaks the symmetry of the ring. This can be input-beam noise, but also deterministic input-beam astigmatism [38,40,41] or vectorial effects [35,42] (see Section 17.9.2).

There are several important differences between noise-induced multiple filamentation for $P \ll P_{\text{th}}^{(2)}$ and for $P \geq P_{\text{th}}^{(2)}$ [32]. When $P \ll P_{\text{th}}^{(2)}$, the beam initially collapses as a single filament, then undergoes a few focusing–defocusing cycles and only then breaks into MF (see, e.g., Fig. 17.5). In this case, the distance where MF occurs scales as $1/\sqrt{P}$. When, however, $P \geq P_{\text{th}}^{(2)}$, MF occurs during the initial collapse, and the distance where MF occurs scales as $1/P$. Indeed, in the experiments in [32], it was observed that the beam initially collapses as a single filament in the $1/\sqrt{P}$ "low-power" regime, but as multiple filaments in the $1/P$ "high-power" regime.

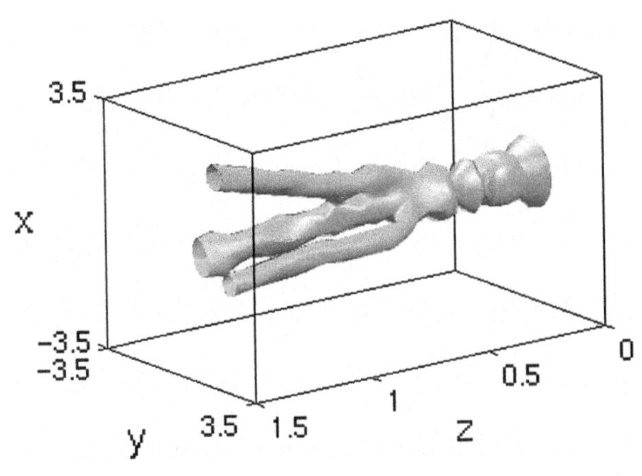

Fig. 17.5 Iso-intensity surface of the solution of Fig. 17.4. Note that the beam propagates from right to left. From Ref. [35]

17.9.2 Deterministic Multiple Filamentation

Because noise is by definition random, the MF pattern of noise-induced MF would be different from shot to shot; i.e., the number and location of the filaments is unpredictable. This constitutes a serious drawback in applications in which precise localization is crucial.

Recall that the NLS is only the leading-order model for propagation of linearly polarized beams in a Kerr medium, and that a more comprehensive model is given by the vectorial Helmholtz equations. In the latter model, a linear polarization state breaks up the cylindrical symmetry by inducing a prefered direction, which is the direction of the linear polarization state. Fibich and Ilan [35,42] showed

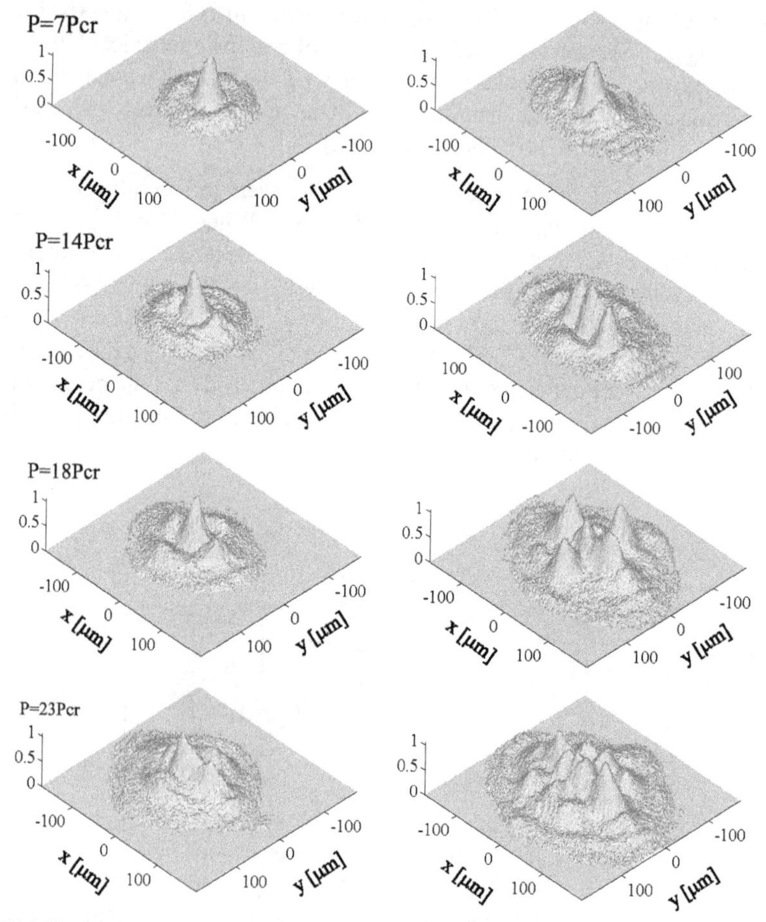

Fig. 17.6 Normalized 3D views of filamentation patterns after propagation of 31 mm in water, for a near-circular incident beam ($e = b/a = 1.09$, *left panel*) and elliptical incident beam ($e = 2.2$, *right panel*). The major axis of the ellipse lies along the x-axis of the plots. From Ref. [43]

numerically that the deterministic breakup of cylindrical symmetry by a linear polarization state can lead to a deterministic multiple filamentation. However, vectorial-effects–induced MF has not been observed in experiments [43]. The reason for this is probably as follows. In order that the vectorial coupling between the electric fields component will lead to MF, the beam radius should self-focus down to approximately two wavelengths. In experiments, however, self-focusing is arrested at a much earlier stage, due to plasma effects.

In [38,40,41] it was predicted theoretically and observed for laser pulses propagating in sodium vapor [41] and in water [43] that input beam astigmatism can also lead to a deterministic MF pattern, i.e., a pattern that is reproducible from shot to shot. Dubietis et al. [43] pointed out that when the input beam is elliptically shaped, e.g., $\psi_0 = c \cdot \exp(-(x/a)^2 - (y/b)^2)$, the MF pattern can only consist of a combination of (1) a single on-axis central filament, (2) pairs of identical filaments located along the ellipse major axis at $(\pm x, 0)$, pairs of identical filaments located along the minor axis at $(0, \pm y)$, and (4) quadruples of identical filaments located at $(\pm x, \pm y)$. Moreover, in that study all the above four filament types were observed experimentally in water (see Fig. 17.6) and numerically. Subsequent approaches for deterministic MF which is induced by a deterministic breakup of the input beam symmetry include a titled lens and a phase mask [44,45]. In [44], it was shown experimentally that sufficiently large astigmatism can dominate noise in the determination of the MF pattern in atmospheric propagation. Hence, rather than trying to eliminate noise, one can control the MF pattern by adding sufficiently large astigmatism.

17.10 Perturbation Theory: Effect of Small Mechanisms Neglected in the NLS Model

The 2D cubic NLS (17.1) is the leading-order model for propagation of intense laser beams in a Kerr medium. As we have seen, in the NLS model the electric field intensity becomes infinite at the blowup point. Because physical quantities do not become infinite, this indicates that near the blowup point, some physical mechanisms that were neglected in the derivation of the NLS become important. Moreover, since the NLS (17.1) is *critical* (see Section 17.1), these mechanisms can have a large effect even when they are still small compared with the Kerr nonlinearity or diffraction.

17.10.1 Unreliability of Aberrationless Approximation and Variational Methods

Because a direct analysis of NLS equations with additional perturbations is hard, the standard approach has been to approximate these equations with reduced equations that do not depend on the transverse (x, y) coordinates, and then to analyze the (much simpler) reduced equations. The key issue, naturally, has been how to derive the "correct" reduced equations.

Starting with [15], for many years all the derivations used the *aberrationless approximation*, i.e., the assumption that during its propagation, the beam maintains a known self-similar profile

$$|\psi| \sim \frac{1}{L(z)} G\left(\frac{r}{L(z)}\right),$$

where G is a Gaussian, *sech*, etc. In the *aberrationless approximation method*, this self-similar profile is substituted in the NLS, and the reduced equations follow from balancing the leading-order terms. In the *variational method*, the self-similar profile is substituted in the NLS Lagrangian. Integration over the (x, y) coordinates gives the averaged Lagrangian, whose variational derivative gives the reduced equations.

There has been considerable research effort on the optimal way to derive the reduced equations using either of these two methods. However, as was pointed out and explained in [13], both methods are unreliable, in the sense that they sometimes lead to predictions that are quantitatively inaccurate, or even qualitatively wrong. It is important to note that the unreliability of the aberrationless approximation and variational methods is related to the fact that the NLS (17.1) is *critical*. Indeed, these methods have been successful in analysis of perturbed non-critical NLS, such as the perturbed one-dimensional cubic NLS.

The reason for the unreliability of these methods is that they make use of the *aberrationless approximation*. One problem with this approximation is that it implicitly implies that when collapse occurs, all the beam collapses into the focal point (*whole beam collapse*), whereas, in fact, collapsing beams undergo *partial beam collapse* (see Section 17.6). Another problem with this approximation is that it has been usually applied with the wrong profile, i.e., with a profile G different from the R profile. This may not seem like a serious issue, because a well-fitted Gaussian can be quite close to the R profile. However, the R profile is the *only* profile that has (1) the critical power for collapse, and (2) a zero Hamiltonian. Hence, it is a borderline case for the two conditions for collapse $P > P_{\text{cr}}$ and $H < 0$. These two properties are at the heart of the sensitivity of the ψ_R profile to small perturbations, which cannot be captured by any self-similar profile that is not based on the R profile.

It may seem, therefore, that one can use the aberrationless approximation, so long as it is being used with the ψ_R profile. Even this is not true, however, because the ψ_R profile represents a complete balance between diffraction and nonlinearity. Hence, the collapse dynamics is determined by the *small difference* between ψ and ψ_R. Therefore, the derivation of the reduced equations should be based on balancing the leading-order deviations from ψ_R.

17.10.2 Modulation Theory

As noted, in order to derive the "correct" reduced equations, one should take into account that

1. As the solution collapses, its profile approaches the ψ_R profile.
2. The collapse dynamics is determined by the *small differences* from ψ_R.

Fibich and Papanicolaou [3] used these observations to derive a systematic method for deriving reduced equations for the effect of additional small mechanisms on critical self-focusing, known as *modulation theory*.[3] Consider the NLS with a general small perturbation ϵF

$$i\psi_z(z,x,y) + \Delta\psi + |\psi|^2\psi + \epsilon F(\psi) = 0, \qquad \Delta = \partial_{xx} + \partial_{yy}. \tag{17.13}$$

As the solution collapses, its profile becomes closer to ψ_R. When that happens, self-focusing in the perturbed NLS (13) is given, to leading order, by the reduced ODE system

$$\beta_z(z) + \frac{\nu(\beta)}{L^2} = \frac{\epsilon}{2M}(f_1)_z - \frac{2\epsilon}{M}f_2, \qquad L_{zz}(z) = -\frac{\beta}{L^3}, \tag{17.14}$$

where

$$f_1(z) = 2L(z)\cdot \mathrm{Re}\int F(\psi_R)e^{-iS}[R(\rho) + \rho R'(\rho)]\,dxdy, \tag{17.15}$$

$$f_2(z) = \mathrm{Im}\int F(\psi_R)\psi_R^*\,dxdy, \tag{17.16}$$

and

$$S(z,r) = \zeta(z) + \frac{L_z}{L}\frac{r^2}{4}, \qquad \rho = \frac{r}{L}, \qquad \zeta = \int_0^2 L^{-2}(s) \qquad ds.$$

As expected, when $\epsilon = 0$, Eqs. (17.14) become the reduced equations (17.7) for the unperturbed NLS. The effect of the perturbation enters through the auxiliary functions f_1 and f_2, which correspond to the conservative and non-conservative components of the perturbation, respectively.

The reduced Eqs. (17.14) have been used to analyze the effect of various small perturbations, such as a weak defocusing quintic nonlinearity, saturating nonlinearities, small normal time dispersion, nonparaxiality, vectorial effects, linear and nonlinear damping and Debye relaxation [14,35,38,46–48]. In all

[3] Not to be confused with modulational instability, which refers to the destabilizing effect of small perturbations in the input profile, see Section 17.9.1.

cases, the predictions of the reduced equations were found to be in full agreement with numerical simulations of the corresponding perturbed NLS.

The reduced equations (17.14) show that additional mechanisms can have a large effect on the self-focusing dynamics even when they are still small compared with the Kerr nonlinearity and diffraction. This "sensitivity" property is unique to the *critical* NLS (see Section 17.1), and it reflects the fact that in critical collapse, the Kerr nonlinearity is nearly balanced by diffraction. Hence, a small mechanism can shift the delicate balance between these two competing (much larger) effects, and even arrest the collapse [3].

17.11 Effect of Normal Group Velocity Dispersion

The basic model for propagation of *ultrashort pulses* in a bulk Kerr medium is given by the NLS with group velocity dispersion (GVD)

$$i\psi_z(z,x,y,t) + \Delta\psi - \gamma_2\psi_{tt} + |\psi|^2\psi = 0, \qquad \Delta = \partial_{xx} + \partial_{yy}, \qquad (17.17)$$

where

$$\gamma_2 = \frac{r_0^2 k_0 k_0''(\omega_0)}{T_0^2}$$

is the dimensionless GVD parameter, and r_0, T_0, k_0 and ω_0 are the input pulse radius, temporal duration, wavenumber and carrier frequency, respectively.

When dispersion is *anomalous* ($\gamma_2 < 0$) the pulse undergoes temporal and transverse compression. Indeed, in that case Eq. (17.17) is the 3D [i.e., (x, y, t)] supercritical NLS (see Section 17.1), which has solutions that become singular in finite distance z. The dynamics in the case of *normal GVD* ($\gamma_2 > 0$) is much more complicated, however, because of the opposite signs of diffraction and dispersion.

The question of whether small normal dispersion can arrest singularity formation has defied research efforts for many years. In 1986, Zharova et al. [49] derived a reduced ODE for the evolution (in z) of the pulse amplitude at t_m, the time of the initial peak amplitude. Analysis of this ODE showed that small normal GVD arrests the collapse at t_m. As a result, the pulse splits into two temporal peaks that continue to focus. The numerical simulations of Zharova et al. confirmed the predicted pulse splitting, and also showed what was interpreted as a secondary splitting. Therefore, Zharova et al. conjectured that the new peaks would continue to split into "progressively smaller-scale," and hence that small normal dispersion would arrest self-focusing through multiple splitting.

Numerical simulations of Eq. (17.17) carried during the 1990s [47,50–54] showed that self-focusing of the two peaks leads to the formation of temporal shocks at the peaks edges. As a result, in all the above studies, which used "standard" numerical methods, the simulations could not go beyond the shock formation and were thus unable to determine whether secondary splittings occur and whether the solution ultimately becomes singular. In 1995, Fibich et al. [47] used *modulation theory* (see Section 17.10.2) to derive a reduced system of PDEs for self-focusing in (17.17) which is valid for all t cross-sections, and not just for t_m.

Analysis of the reduced system showed that while self-focusing is arrested in the near vicinity of t_m, it continues elsewhere, i.e., that small normal GVD does not arrest the collapse. However, the validity of the reduced system breaks down as the shock edges of the two peaks form. Therefore, one cannot use the reduced system to predict whether multiple splitting would occur and/or whether the solution ultimately becomes singular. Analysis of the reduced system did reveal, however, that temporal splitting is associated with the transition from independent 2D collapse of each t cross-section to a full 3D dynamics. Therefore, it was suggested in [47] that the two peaks would not necessarily split again.

Temporal splitting of ultrashort pulses was first observed experimentally in 1996 by Ranka and Gaeta [53] and later by Diddams et al. [51]. In these experiments secondary splittings were also observed at even higher input powers. Nevertheless, these observations do not imply that solutions of Eq. (17.17) undergo multiple splittings, because these secondary splittings were observed at such high powers where the validity of Eq. (17.17) breaks down, as additional physical mechanisms become important.

In 2001, Germaschewski et al. [55] used an adaptive mesh refinement method to solve Eq. (17.17) beyond the pulse splitting. These simulations show that after the pulse splitting the two peaks do not undergo a similar secondary splitting. Rather, the collapse of the two peaks is arrested by dispersion. Similar results were also obtained by Fibich et al. using the iterative grid distribution method [56]. In that study, the authors solved Eq. (17.17) with the same initial conditions used by Zharova et al. in [49] and observed no secondary splitting, thus confirming that the numerical observation of secondary splitting in that study was in fact a numerical artifact.

Based on the numerical simulations of [55–57], it is now believed that after the pulse splitting the two peaks do not undergo a secondary splitting, and that small normal GVD arrests the collapse of the two peaks. However, a rigorous proof that solutions of Eq. (17.17) cannot become singular is still not available, and is considered a hard analytical problem.

17.12 Nonparaxiality and Backscattering

In nonlinear optics, the NLS model is derived from the more comprehensive scalar nonlinear Helmholtz equation (NLH)

$$E_{xx}(x,y,z) + E_{yy} + E_{zz} + k_0^2\left(1 + 4\epsilon_0 c n_2 |E|^2\right)E = 0. \qquad (17.18)$$

In general, $E = e^{ik_0z}\psi(z,x,y) + e^{-ik_0z}B(z,x,y)$, where ψ and B are the slowly varying envelopes of the forward-propagating and backscattered waves, respectively. The NLS is derived by *neglecting the backscattered wave* (i.e., setting $B \equiv 0$) and then applying the *paraxial approximation*

$$\psi_{zz} \ll k_0\psi_z.$$

Because there are no singularities in nature, a natural question is whether initial conditions that lead to blowup in the NLS, correspond to global (i.e., non-singular) solutions of the corresponding NLH. In other words, do nonparaxiality and backscattering arrest the collapse, or is the collapse arrested only in a more comprehensive model than the NLH (17.18)?

The observation that the paraxial approximation breaks down near the singularity had already been noted in 1965 by Kelley in [9]. Vlasov [58] derived a perturbed NLS that includes the leading-order effect of nonparaxiality. He solved this equation, numerically and observed that collapse is arrested in this "nonparaxial NLS". Feit and Fleck [59] used numerical simulations of the NLH to show that nonparaxiality can arrest the blowup for initial conditions that lead to singularity formation in the NLS (17.1). After the arrest of collapse in the NLH, the beam undergoes focusing– defocusing oscillations (*multiple foci*). In these simulations, however, they did not solve a true boundary value problem for the NLH. Instead, they solved an initial value problem for a "modified" NLH that describes only the forward-going wave (while introducing several additional assumptions along the way). Akhmediev and collaborators [60,61] analyzed an initial-value problem for a different "modified" NLH; their numerical simulations also suggested that nonparaxiality arrests the singularity formation. All of the above numerical approaches ([58–61]), however, did not account for the effect of backscattering. Fibich [46] used *modulation theory* (see Section 17.10.2) to derive a reduced ODE (in z) for self-focusing in the presence of small nonparaxiality. His analysis suggests that nonparaxiality indeed arrests the singularity formation, resulting instead in decaying focusing-defocusing oscillations. Moreover, it showed that nonparaxiality arrests collapse while it is still small compared with the focusing Kerr nonlinearity. However, backscattering effects were neglected in this asymptotic analysis.

In [62,63], Fibich and Tsynkov developed a novel numerical method for solving the NLH as a true boundary-value problem. The key issue has been to develop a *two-way absorbing boundary condition* that allows for the impinging electric field to enter the Kerr medium, while allowing the backscattered wave to be fully transmitted in the opposite direction. This method allowed for the first quantitative calculation of backscattering due to the nonlinear Kerr effect [64]. Unfortunately, so far the method can only compute solutions whose input power is below P_{cr}, leaving open the issue of global existence of solutions with power above P_{cr} in the NLH model. Some progress in that direction was made in [65], when Fibich et al. used the same numerical method to compute global solutions of the linearly damped NLH for initial conditions that lead to collapse in the corresponding linearly damped NLS. For these solutions, therefore, the arrest of collapse has to be due to nonparaxiality and backscattering. Recently, Sever proved that solutions of the NLH exist globally [66]. However, Sever's proof holds only for real boundary conditions, whereas the correct physical radiation boundary conditions that allow for power propagation are complex-valued. Therefore, all the results so far suggest that nonparaxiality and backscattering arrest the collapse, but a rigorous proof of this result is still an open problem.

Finally, we note that the NLH (17.18) is derived under the assumption of linear polarization, and that a more comprehensive model is the vectorial NLH for the three components of the vector electrical field [67]. The effects of the coupling to the two other components of the electric field are of the same order as nonparaxiality, and have the same qualitative effect on the arrest of collapse, but are ≈ 7 times stronger [35].

17.13 Final Remarks

At present, there is a fairly good understanding of self-focusing of "low-power" beams, i.e., those beams whose power is moderately above the critical power P_{cr}. Indeed, since they collapse with the ψ_R profile, one can use *modulation theory* to analyze the effects of most perturbations. There are still some open mathematical questions, such as proving rigorously that normal dispersion or nonparaxiality arrest the collapse (Sections 17.11 and 17.12). However, the lack of rigorous proofs for these open problems is probably a "mathematical issue," which need not concern the nonlinear optics community.

In recent years, there has been growing evidence that the self-focusing dynamics of "high-power" beams, i.e., those whose power is many times the critical power P_{cr}, is very different from the one of "low power" beams. Indeed, "high power" beams can undergo multiple filamentation prior to the initial collapse (Section 17.9), have a different scaling law for the collapse distance (Section 17.8), and collapse with self-similar ring profile which is different from ψ_R (Section 17.4.1). At present, there is no good understanding of self focusing of "high power" beams, and most results are based either on numerical simulations or on elementary analysis. For example, there is no good theory for multiple filamentation that can predict the number of filaments of an input beam, whether two close filaments will merge into a single filaments, and so on. The lack of a good theory for "high-power beams" is because, unlike "low-power" beams, there is no universal attractor ψ_R. Developing new analytical tools for the "high power" regime is probably one of key challenges for the future of self-focusing theory.

References

1. Y.R. Shen. Self-focusing: Experimental. *Prog. Quant. Electr.* **4**, 1–34 (1975).
2. J.H. Marburger. Self-focusing: theory. *Prog. Quant. Electr.* **4**, 35–110 (1975).
3. G. Fibich, G.C. Papanicolaou. Self-focusing in the perturbed and unperturbed nonlinear Schrödinger equation in critical dimension. *SIAM J. Applied Math.* **60**, 183–240 (1999).
4. C. Sulem, P.L. Sulem. *The Nonlinear Schrödinger Equation.* Springer, New-York (1999).
5. S.N. Vlasov, V.A. Petrishchev, V.I. Talanov. Averaged description of wave beams in linear and nonlinear media. Izv. Vuz Radiofiz (in Russian), **14**, 1353–1363 (1971) Radiophys. and Quantum Electronics **14**, 1062–1070 (1971) (in English)
6. V.I. Talanov. Focusing of light in cubic media. *JETP Lett.* **11**, 199–201 (1970).
7. R.Y. Chiao, E. Garmire, C.H. Townes. self-trapping of optical beams. *Phys. Rev. Lett.* **13**, 479–482 (1964).
8. M.I. Weinstein. Nonlinear Schrödinger equations and sharp interpolation estimates. *Comm. Math. Phys.* **87**, 567–576 (1983).
9. P.L. Kelley. Self-focusing of optical beams. *Phys. Rev. Lett.* **15**, 1005–1008 (1965).
10. F. Merle. On uniqueness and continuation properties after blow-up time of self-similar solutions of nonlinear Schrödinger equation with critical exponent and critical mass. *Comm. Pure Appl. Math.* **45**, 203–254 (1992).
11. F. Merle. Determination of blow-up solutions with minimal mass for nonlinear Schrödinger equation with critical power. *Duke Math. J.* **69**, 427–454, (1993).
12. G. Fibich, A. Gaeta. Critical power for self-focusing in bulk media and in hollow waveguides. *Opt. Lett.* **25**, 335–337 (2000).
13. G. Fibich, B. Ilan. Self focusing of elliptic beams: An example of the failure of the aberrationless approximation. *JOSA B* **17**, 1749–1758 (2000).
14. G. Fibich. Self-focusing in the damped nonlinear Schrödinger equation. *SIAM. J. Appl. Math* **61**, 1680–1705 (2001).
15. S.A. Akhmanov, A.P. Sukhorukov, R.V. Khokhlov. Self-focusing and self-trapping of intense light beams in a nonlinear medium. *JET* **23**, 1025–1033 (1966).
16. M.J. Landman, G.C. Papanicolaou, C. Sulem et al. Stability of isotropic singularities for the nonlinear Schrödinger equation. *Physica D* **47**, 393–415, (1991).
17. F. Merle, P. Raphael. Sharp upper bound on the blow-up rate for the critical nonlinear Schrödinger equation. *Geom. Funct. Anal.* **13**, 591–642 (2003).
18. K.D. Moll, A.L. Gaeta, G. Fibich. Self-similar optical wave collapse: Observation of the Townes profile. Phys. Rev. Lett. **90**, 203902 (2003).
19. T.D. Grow, A.A. Ishaaya, L.T. Vuong et al. Collapse dynamics of super-gaussian beams. *Opt. Express* **14**, 5468–5475 (2006).
20. G. Fibich, N. Gavish, X.P. Wang. New singular solutions of the nonlinear schrodinger Schrödinger equation. *Physica D* **211**, 193–220 (2005).
21. V.R. Costich, B.C. Johnson. Apertures to shape high-power laser beams. *Laser Focus* **10**, 43–46 (1974).
22. I. Kryzhanovskiĭ, B.M. Sedov, V.A. Serebryakov et al. Formation of the spatial structure of radiation in solid-state laser systems by apodizing and hard apertures. *Sov. J. Quant. Electron.* **13**, 194–198 (1983).
23. L. Bergé, C. Goutéduard, J. Schjodt-Erikson, H. Ward. Filamentation patterns in Kerr media vs. beam shape robustness, nonlinear saturation, and polarization states. *Physica D* **176**, 181–211, (2003).
24. V.M. Malkin. Dynamics of wave collapse in the critical case. Phys. Lett. A **151**, 285–288 (1990).
25. G.M. Fraiman. Asymptotic stability of manifold of self-similar solutions in self-focusing. *Sov. Phys. JETP* **61**, 228–233 (1985).

26. M.J. Landman, G.C. Papanicolaou, C. Sulem et al. Rate of blowup for solutions of the nonlinear Schrödinger equation at critical dimension. *Phys. Rev. A* **38**, 3837–3843, (1988).
27. B.J. LeMesurier, G.C. Papanicolaou, C. Sulem, P.L. Sulem. Local structure of the self-focusing singularity of the nonlinear Schrödinger equation. Physica D **32**, 210–226 (1988).
28. G. Fibich. An adiabatic law for self-focusing of optical beams. *Opt. Lett.* **21**, 1735–1737 (1996).
29. V.M. Malkin. On the analytical theory for stationary self-focusing of radiation. *Physica D* **64**, 251–266, (1993).
30. E.L. Dawes, J.H. Marburger. Computer studies in self-focusing. *Phys. Rev.* **179**, 862–868 (1969).
31. A.J. Campillo, S.L. Shapiro, B.R. Suydam. Relationship of self-focusing to spatial instability modes. *Apple. Phys. Lett.* **24**, 178–180 (1974)
32. G. Fibich, S. Eisenmann, B. Ilan et al. Self-focusing distance of very high power laser pulses. *Opt. Express* **13**, 5897–5903 (2005).
33. G. Fibich, Y. Sivan, Y. Ehrlich et al. Control of the collapse distance in atmospheric propagation. *Opt. Express* **14**, 4946–4957 (2006).
34. V.I. Bespalov, V.I. Talanov. Filamentary structure of light beams in nonlinear media. *Hz. Eksper. Tenor. Fizz. - Pis'ma Redact. (U.Sis.R. JET)*, **3**, 471–476, 1966. Transf. in *JET Lett.* **3**, 307–310 (1966).
35. G. Fibich, B. Ilan. Vectorial and random effects in self-focusing and in multiple filamentation. *Physica D* **157**, 112–146 (2001).
36. K. Konno, H. Suzuki. Self-focusing of a laser bean in nonlinear media. *Physica Scripta* **20**, 382–386 (1979).
37. J.M. Soto-Crespo, E.M. Wright, N.N. Akhmediev. Recurrence and azimuthal-symmetry breaking of a cylindrical Gaussian beam in a saturable self-focusing medium. *Phys. Rev. A* **45**, 3168–3174 (1992).
38. G. Fibich, B. Ilan. Self-focusing of circularly polarized beams. *Phys. Rev. E* **67**, 036622 (2003).
39. L.T. Vuong, T.D. Grow, A. Ishaaya et al. Collapse of optical vortices. *Phys. Rev. Lett.* **96**, 133901 (2006).
40. G. Fibich, B. Ilan. Multiple filamentation of circularly polarized beams. *Phys. Rev. Lett.* **89**, 013901 (2002)
41. J. W. Grantham, H.M. Gibbs, G. Khitrova et al. Kaleidoscopic spatial instability: Bifurcations of optical transverse solitary waves. *Phys. Rev. Lett.* **66**, 1422–1425 (1991).
42. G. Fibich, B. Ilan. Vectorial effects in self-focusing and multiple filamentation. *Opt. Lett.* **26**, 840–842 (2001).
43. A. Dubietis, G. Tamošauskas, G. Fibich, et al. Multiple filamentation induced by input-beam ellipticity. *Opt. Lett.* **29**, 1126–1128 (2004).
44. G. Fibich, S. Eisenmann, B. Ilan et al.: Control of multiple filamentation in air. *Opt. Lett.* **29**, 1772–1774 (2004).
45. G. Mechain, A. Couairon, M. Franco et al. Organizing multiple femtosecond filaments in air. *Phys. Rev. Lett.* **93**, 035003 (2004).
46. G. Fibich. Small-beam nonparaxiality arrests self-focusing of optical beams. *Phys. Rev. Lett.* **76**, 4356–4359 (1996).
47. G. Fibich, V.M. Malkin, G.C. Papanicolaou. Beam self-focusing in the presence of small normal time dispersion. *Phys. Rev. A* **52**, 4218–4228 (1995).
48. G. Fibich, G.C. Papanicolaou. Self-focusing in the presence of small time dispersion and nonparaxiality. *Opt. Lett.* **22**, 1379–1381 (1997).
49. N.A. Zharova, A.G. Litvak, T.A. Petrova et al. Multiple fractionation of wave structures in a nonlinear medium. *JETP Lett.*, **44** 13–17 (1986).
50. P. Chernev, V. Petrov. Self-focusing of light pulses in the presence of normal group-velocity dispersion. *Opt. Lett.* **17**, 172–174 (1992).
51. S.A. Diddams, H.K. Eaton, A.A. Zozulya, et al. Amplitude and phase measurements of femtosecond pulse splitting in nonlinear dispersive media. *Opt. Lett.* **23**, 379–381 (1998).

52. G.G. Luther, A.C. Newell, J.V. Moloney. The effects of normal dispersion on collapse events. *Physica D* **74**, 59–73 (1994).
53. J.K. Ranka, R.W. Schirmer et al. A.L. Gaeta. Observation of pulse splitting in nonlinear dispersive media. *Phys. Rev. Lett.* **77**, 3783–3786 (1996).
54. J.E. Rothenberg. Pulse splitting during self-focusing in normally dispersive media. *Opt. Lett.*, **17** 583–585, (1992).
55. K. Germaschewski, R. Grauer, L. Berge et al. Splittings, coalescence, bunch and snake patterns in the 3D nonlinear Schrödinger equation with anisotropic dispersion. *Physica D* **151**, 175–198 (2001).
56. G. Fibich, W. Ren, X.P. Wang. Numerical simulations of self-focusing of ultrafast laser pulses. *Phys. Rev. E* **67**, 056603 (2003).
57. J. Coleman, C. Sulem. Numerical simulations of blow-up solutions of the vector nonlinear Schrödinger equation. *Phys. Rev. E* **66**, 036701 (2002).
58. S.N. Vlasov. Structure of the field of wave beams with circular polarization near a nonlinear focus in a cubic medium. *Sov. J. Quantum Electron.* **17**, 1191–1193 (1987).
59. M.D. Feit, J.A. Fleck. Beam nonparaxiality, filament formation, and beam breakup in the self-focusing of optical beams. *J. Opt. Soc. Am. B* **5**, 633–640 (1988).
60. N. Akhmediev, J.M. Soto-Crespo. Generation of a train of three-dimensional optical solitons in a self-focusing medium. *Phys. Rev. A* **47**, 1358–1364, (1993).
61. N. Akhmediev, A. Ankiewicz, J.M. Soto-Crespo. Does the nonlinear Scharödinger equation correctly describe beam propagation? *Opt. Lett.* **18**, 411–413 (1993).
62. G. Fibich, S.V. Tsynkov. High-order two-way artificial boundary conditions for nonlinear wave propagation with backscattering. *J. Comput. Phys.* **171**, 1–46 (2001).
63. G. Fibich, S.V. Tsynkov. Numerical solution of the nonlinear Helmholtz equation using nonorthogonal expansions. *J. Comput. Phys.* **210**, 183–224 (2005).
64. G. Fibich, B. Ilan, S.V. Tsynkov. Computation of nonlinear backscattering using a high-order numerical method. *J. Sci. Comput.* **17**, 351–364 (2002).
65. G. Fibich, B. Ilan, S.V. Tsynkov. Backscattering and nonparaxiality arrest collapse of nonlinear waves. *SIAM. J. Appl. Math.* **63**, 1718–1736 (2003).
66. M. Sever. An existence theorem for some semilinear elliptic systems. *J. Differnetial Equations* **226**, 572–593 (2006).
67. S. Chi, Q. Guo. Vector theory of self-focusing of an optical beam in Kerr media. *Opt. Lett.* **20**, 1598–1560 (1995).

Chapter 18
X-Waves in Self-Focusing of Ultra-Short Pulses

Claudio Conti, Paolo Di Trapani, and Stefano Trillo

Abstract X waves are emerging as a universal concept that allows us to interpret experiments of filamentation and trapping in a normally dispersive medium. We briefly review the main experimental and theoretical results behind this paradigm.

18.1 Introduction

The process of self-focusing of short pulses is well known to be affected by the linear dispersive properties of dielectric media. In particular, different phenomena are expected depending on the sign of the group-velocity dispersion (GVD) $k'' = d^2k/d\omega^2|_{\omega_0}$ [1]. When the GVD is anomalous, a focusing nonlinearity can balance both the diffractive (spatial) and the dispersive (temporal) spreading of the wavepacket at once. Under perfect balance, in principle one can have a spatio-temporal soliton, the so-called "light bullet," as a self-confined propagation-invariant solution of the wave equation. While for ideal Kerr media the observation of such entities is generally hampered by the occurence of collapse-type instabilities, the phenomenon of trapping can occur under different configurations that involve stabilizing mechanisms [2] (for further details on this topic we refer the reader to the other chapters of this book).

Conversely, in this chapter, we are concerned with media that exhibit normal GVD. In this case the elementary consideration that focusing nonlinearities enhance the linear temporal broadening of a pulse seem to imply that spatio-temporal trapping is no longer possible. Such an argument is enforced by the prediction and observation of the phenomenon of pulse splitting [3] and by the fact that modulational instability (the amplification of noise at the expense of a strong cw plane-wave) is no longer a narrow bandwidth process, leading instead

C. Conti (✉)
Research Center "Enrico Fermi," Via Panisperna 89/A, 00100 Roma, Italy and
Research Center SOFT INFM-CNR, University "La Sapienza," P.le Aldo 5, 00185 Roma, Italy
e-mail: claudio.conti@phys.uniroma1.it

R.W. Boyd et al. (eds.), *Self-focusing: Past and Present*,
Topics in Applied Physics 114, DOI 10.1007/978-0-387-34727-1_18,

to conical emission [4,5]. This simple point of view can not explain why both real and numerical experiments [6] show that the evolution of a Gaussian (both in time and space) wave-packet (WP) propagating in a normally dispersive medium with nonlinearity that mimics a focusing Kerr nonlinearity via second-harmonic generation (SHG), exhibits remarkable features of self-confinement. The explanation becomes possible, however, by assuming that new entities, namely weakly (because the decay is slower than exponential characteristic of solitons) localized WPs with a conical envelope, are spontaneously generated during the process. These entities are strictly related to the hyperbolic nature of the dispersion relationship [see Fig. 18.1(a)] and can be termed nonlinear X waves because of the characteristic shape of their axial cross section (resembling the letter X), as illustrated in Fig. 18.1(b),(c). Since then they have been the object of several studies devoted to assess their role in parametric conversion (quadratic media) [7–15] and self-action phenomena in Kerr-like media [16–27], which have also further stimulated the investigation of the linear limit [28–35].

Historically, X waves have been introduced in the context of linear wave propagation [36] as suitable non-monochromatic generalizations of ideal (infinite aperture) non-diffracting Durnin beams , namely plane waves evenly distributed on a cone, giving rise to a field described by the Bessel function J_0 [37]. The

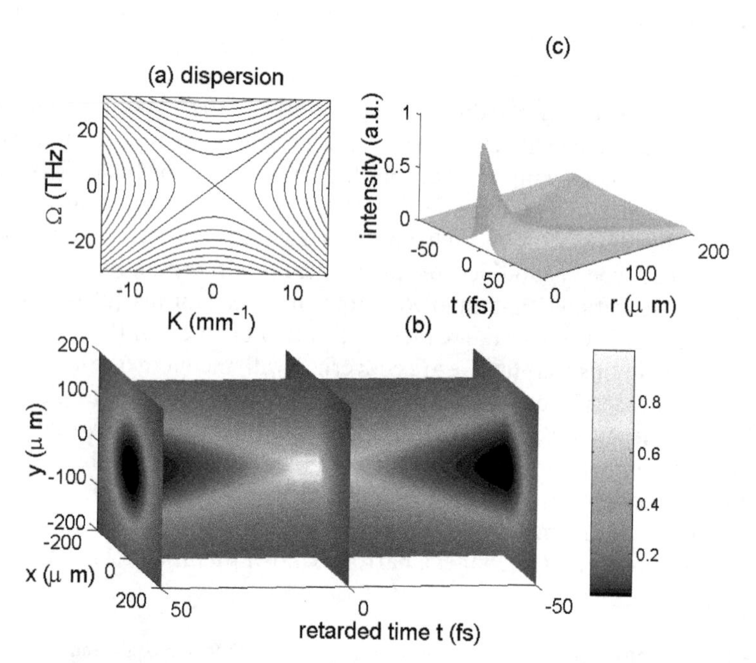

Fig. 18.1 (**a**) Dispersion relation in the plane of frequency $\Omega = \omega - \omega_0$ and transverse wavenumber K for a normally dispersive medium. (**b**) Isointensity surfaces $|A(x, y, \tau)|^2$ const. of an axially symmetric envelope X wave; (**c**) Intensity profile of the same wave in the plane of radial and retarded time coordinates

propagation features of X waves have been experimentally tested in acoustics [38,39] and optics [40–45]. Although they encompass a more general class of weakly localized WPs, those that are relevant in the present context are *envelope* X waves , so-called because it is the slowly varying (narrow bandwidth) envelope that has the form of an X wave. Indeed, nonlinear X waves are nothing but suitable deformations of linear envelope X waves, and the scenario that emerges from experiments is that the nonlinearity, regadless of its order, acts as a driving mechanism for reshaping for conventional bell-shaped beams into X waves. In the following two sections we briefly review the main experimental results and the theoretical achievements that have led to the emergence of this new area of nonlinear optics.

18.2 Experimental Results

18.2.1 Historical Preamble: The Key Role of Angular Dispersion

The interpretation of the self-focusing dynamics of ultrashort pulses in terms of X waves arose from investigations into parametric wave mixing in $\chi^{(2)}$ materials, the birefringence of which makes the impact of chromatic dispersion typically more dramatic than for the $\chi^{(3)}$ (Kerr) regime. In this context, X waves had been proposed in alternative to spatial solitons for describing self-channeling, in conditions where time-domain effects were seen to play a crucial role. The concept of "filament," on the other hand, is historically linked to self-focusing in Kerr-driven processes. In spite of the fact that quite different models and scenarios have been proposed for describing $\chi^{(2)}$- and $\chi^{(3)}$- driven self-trapping, experiments indicate indeed a great number of analogies, both regimes being featured by self-focusing, apparent diffraction-free propagation over some transient region, spectral broadening, pulse splitting, angular dispersion, nonlinear losses, and so on, the overall behavior being more "extreme" for the $\chi^{(3)}$ case due to the higher intensities involved. The fact that nonlinear X waves were quite successful in interpreting both $\chi^{(2)}$ and $\chi^{(3)}$ dynamics should be taken as a further indication of the relevant analogy existing between these two regimes.

The key feature that distinguishes X waves and, more generally, tilted pulses, from conventional Gaussian WP is the angular dispersion (AD), i.e., the dependence of the temporal frequency on an angle. AD adds a geometrical contribution to axial phase rotation of each wavelength mode, whose amount depends both on wavelength and angle. As a result, the overall WP effective phase velocity, group velocity (GV), and GVD depend not only on WP spectrum and on material dispersive properties, but also on the WP geometry. Notably, WP shapes exist for which phase modulation arising from diffraction, chromatic dispersion, nonlinear phase shift, nonlinear losses, etc., all cancel one another, thus leading to the achievement of a stationary profile.

In a linear regime, AD is generally established by optical elements that disperse wavelengths in a given plane, e.g, prisms or diffraction gratings. Notably, a planar AD produces an amplitude front that is "tilted" with respect to the phase front, leading to the formation of the so-called "tilted pulse" [46].

If the dispersive element has radial symmetry (as for circular diffraction gratings or conical prisms, e.g., the so-called "axicones"), the front tilt spans the 2π azimuthal-coordinate domain, producing the clepsydra (or double cone) shaped WP that is usually called "X-wave." Tilted pulses and X waves exhibit identical effective dispersion [39]. By imposing an appropriate AD over the entire spectrum within the typical transparent-material transmission window, linear waves could be obtained with sub-cycle duration that propagate without spreading in spite of material dispersion [47].

The control of the effective "first-order dispersion" (i.e., GV) by means of tilted-pulses AD was demonstrated in the late 1980s by Martinez [46] and by Szabo [48], who obtained "achromatic phase matching" in SHG. In the early nineties, investigations in the field of femtosecond (fs) optical-parametric amplification (OPA) outlined the key role of AD in this regime too. On one hand conical (off-axis) emission was shown to be connected to spontaneous GV self-matching in non-collinear regime. On the other hand, non-collinear interaction [49,50] as well as the use of pre-tilted pulses [51] were successfully adopted for enhancing conversion efficiency when chromatic dispersion plays indeed a major role (e.g. in the case of fs UV pump pulses). Note that tilted fronts and AD are naturally produced in non-collinear interactions, so to ensure overlap among ultrashort pulses that propagate in different directions. Nowadays the non-collinear and tilted-pulse scheme are adopted in all modern commercial optical-parametric amplifiers. The control of the effective "second-order dispersion" (i.e., GVD) by means of AD was demonstrated few years later by Szatmari for the tilted-pulse case [53], and by Sönajalg for X waves [40]. These WP were shown to propagate without spreading in linear media, in spite of the presence of GVD. By combining the two aforementioned features, i.e., by controlling simultaneously group-velocity mismatch (GVM) and GVD, temporal solitons were finally formed in the regime of phase-mismatched SHG with tilted pulses [54]. Notably, the latter was the first AD experiment performed in a genuine nonlinear regime, i.e., in a condition where nonlinearly induced phase modulation plays a dominant role in counteracting phase modulation induced by AD and chromatic-dispersion.

18.2.2 Spontaneous X Waves in Second-Harmonic Generation

In the SHG temporal-soliton experiment, as well as in all the $\chi^{(2)}$ experiments of above, the crystals were carrying angular dispersion (i.e., a dependence of refraction index on an angle), leading to lateral walkoff between interacting waves. This material property required using planar AD (i.e., tilted pulses) and fairly large beams. As a consequence, the beam showed transverse modulational instability just above the threshold for temporal trapping, causing the immediate quenching of the stationary regime [55]. In fact, transverse localization is compatible with AD only for X-type wave, which require radial symmetry to be supported. For this reason, investigation were started in the radial symmetric "SHG non-critical phase-matching" configuration, i.e., by propagating light along one of the index-ellipsoid optical axes. In this context, the linear generation of X-type WP with suitable AD for compensating GVM turned out to be a much more complex task than for the previous tilted-pulse case, so much so that the problem has yet to be solved. Quite surprisingly, experiments performed with a standard (i.e., Gaussian), AD-free, launching input condition led to the same spatio-temporal localization that was expected for the X wave regime. In fact spatio-temporal trapping between the first harmonic (FH) and the second harmonic (SH) was observed over several diffraction and GVM lengths in lithium triborate crystal (LBO), this regime being by far non-compatible with soliton-like (bullet) WP owing to the huge GVM and the wrong regime (normal) of GVD. Relying upon numerical simulations, which showed the emergence of the expected X-type structure, the results were interpreted by conjecturing the existence of a stationary, nonlinear, X-like waves that act, to some extent, as an attractor for the observed dynamic [6].

Figure 18.2 summarizes the results obtained in a first series of experiments [6,10] performed with Gaussian input (45 μm waist) and pulse durations 100–200 fs: the spatial beam profiles and the autocorrelation traces (left frames) show clear evidence of space–time focusing at the output of a 22-mm LBO

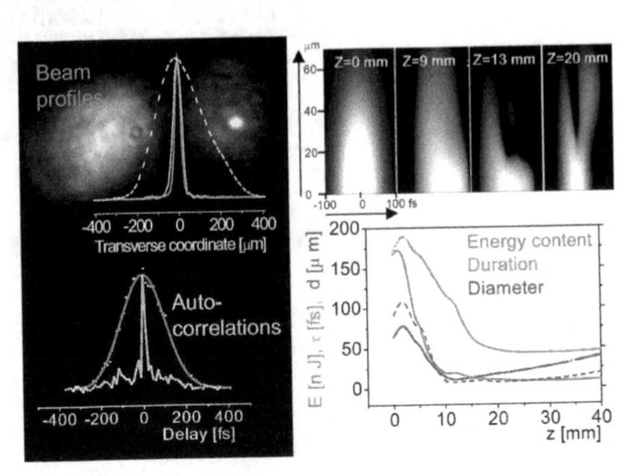

Fig. 18.2 *Left*, experimental data: (*top*): output transverse beam profiles in the linear (*dashed white*) and nonlinear (*solid green*: SH; red: FF) regime. The diffracted (*left*) and localized (*right*) spots recorded at the output of the crystal are also shown. *Bottom*: temporal autocorrelation traces (collinear technique) showing the transition from the linear regime (*dashed red*) to the compressed WP (*solid green*). *Right*: numerical simulations showing the transformation of the beam into an X wave, with quasi-stationary characteristics after 20 mm propagation (along z-axis). After Ref [10], color online

crystal, while numerical results (right frames) show that such a process is accompanied by a dramatic beam reshaping with formation of conical tails (corresponding to off-axis pulse splitting), after which the beam remains nearly stationary.

However, verifying the correctness of this claim and deepening the knowledge of the nonlinear dynamics require sizable work, both on the experimental and theoretical sides. Experimentally, the main complication to be overcome was the absence of an appropriate diagnostic set-up to characterize in space and time this novel type of WP. Although several different experimental techniques have been applied to characterize the process [11], ultimately the capture of the weak WP tails skewed along the conical space–time surface has required to set-up a new tomographic optical (20 fs) gating technique. The apparatus was based on the cross-correlation via sum-frequency mixing of the object to be characterized (the X wave) with a delayed probe made by a wide (quasi-plane-wave) beam and short pulse (for further details see [10,13]). Measurements made by means of this technique are reported in Fig. 18.3. As shown, the WP exhibits a clear conical structure that is particularly well formed along the trailing edge (corresponding to positive delays) of the pulse. The problem of exciting a doubly conical symmetric structure is still open.

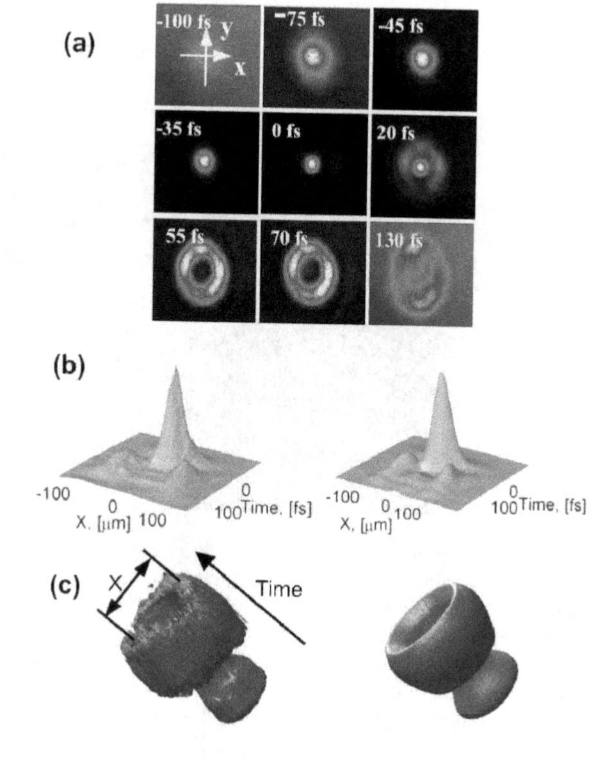

Fig. 18.3 (**a**) Time-resolved transverse profiles of a WP after propagation through a 22-mm LBO crystal (data obtained by means of cross-correlation nonlinear gating technique; see [23]). (**b**) Experimental (*left*) and simulated (*right*) spatio-temporal intensity profile of the X wave as captured after 5 mm of free-space propagation, and (**c**) experimental (*left*) and simulated (*right*) iso-intensity surface at 7% of peak intensity

18.2.3 X Waves in Kerr-Like Media

The issue of X waves in Kerr media is closely connected with the well-known phenomenon of filamentation induced by ultra-short pulses, especially because recent experiments have observed the phenomenon in condensed matter (water or transparent solids) where GVD is expected to play a more determinant role compared to air. In particular, water seems to be the best candidate because of the accessibility of the propagation along the sample and virtually no damage threshold. In this case, on the other hand, one operates above (from few to several times) the ideal (cw) collapse threshold power, so that energies and intensities are higher than in the SHG case, thus forcing one to explore also the impact of additional (other than diffraction, GVD and nonlinear phase shift) phenomena such as nonlinear losses, plasma-induced defocusing, self-steepening, and so on.

The picture that emerges from recent investigations of the phenomenon [18–22,24,26,27] is that filaments formed in condended matter can by no means be thought of as self-guided soliton-like structures stabilized by an internal mechanism (e.g., plasma defocusing action). Rather, the filaments have conical oscillating tails that carry a substantial fraction of the energy [19] and a complicated spatio-temporal structure [21]. Although interpretation of the filaments with conical (Bessel-like) structure have been proposed also in the framework of cw models due to the key action of nonlinear losses [19,20] (see also the chapter by Gaizauskas et al. in this book), the work of Kolesik and coworkers [18] has explicitly pointed out the importance of GVD in the dynamics of filament formation in water. In particular, the simulations show that the dynamics is governed by cycles of pulse splitting and subsequent temporal replenishment of the on-axis pulse that is perhaps at the origin of the illusion that the filament (in time-integrated measurements) is self-guided. Past the first splitting point the field pattern shows evidence of X waves. Because the latter is highly dynamical the X features are best seen in the spectral domain (spectrally resolved far field) where the wave exhibits characteristic tails that follow the X-shaped dispersion curve [see Fig. 18.1(a)]. A typical experimental spectrum in the angle wavelength plane obtained at 2 μJ input energy in a water sample of 3 cm is shown in Fig. 18.4. X-shaped tails are clearly evident.

The spectral arms in Fig. 18.4 are due to conical emission [4,5,16], i.e., amplification of weak perturbations whose transverse wave-number K or angle $\theta \simeq K/k_0$ with propagation axis increases with frequency detuning $\Omega = \omega - \omega_0$. While one expect that at large angles and detunings the arms follow the asymptotic lines $K = \sqrt{k_0 k''}\Omega$ characteristic of the linear dispersion relationship [see Fig. 18.1(a)], the spectral content at low frequencies suggests a complicated scenario, as suggested in [27]. The key observation is that the locus of points θ, Ω corresponding to maximum intensity in the arms [see circles in Fig. 18.5(a)] fits better with two hyperbolic curves (solid and dashed curves) with a gap in frequency [as visibile in Fig. 18.5(b)], rather than a single curve (dotted) with

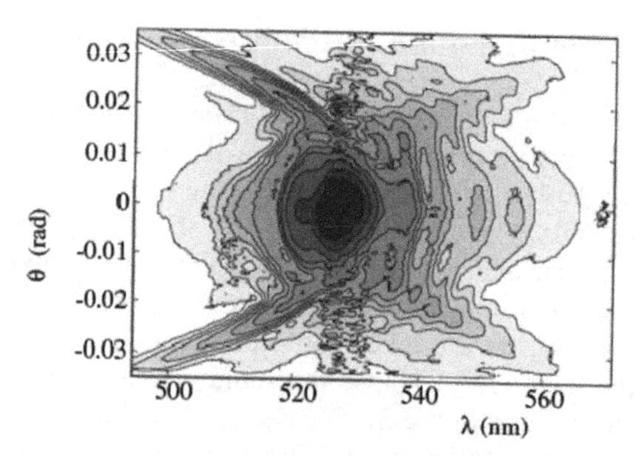

Fig. 18.4 Contour plot (in logarithmic scale over two decades) of a typical measured angle-wavelength ($\theta - \lambda$) spectrum after propagation in a 3-cm cell of water at input energy level of 2 μJ (similar spectral arms are obtained in the energy range 1.8–4 μJ, and also at other carrier wavelengths). Data obtained with laser wavelength $\lambda_0 = 527$ nm, pulse duration of 200 fs, and 100μm input waist (FWHM)

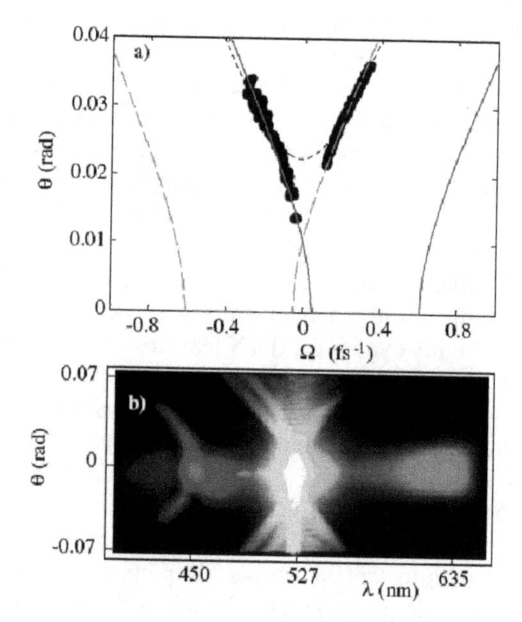

Fig. 18.5 (**a**) *Solid circles,* distribution of peak fluence in Fig. 18.4 in the plane of angle θ and frequency detuning Ω; *black dotted line,* best fit of the entire set of experimental data with the locus of frequencies amplified through conical emission at the expense of a cw plane wave pump; *dashed red and blue solid curves* are best fit for hyperbolas that describe the spectrum of X-wave with red-shift and a blue-shift carrier, respectively. (**b**) Low resolution spectrum showing measured tails around input frequency and two frequency shifted localized modes (input energy is 3 μJ). After Ref. [27], color online

gap in angle as it would have been obtained on the basis of calculations on conical emission by a cw plane pump. This suggests the presence of two weak X waves (each one corresponding to a pair of hyperbolas with a gap in frequency—see also Fig. 18.1(a)—which are amplified through a dominant phase-matchable four-photon process at the expense of two strongly localized pump modes. Pulse temporal splitting emerges in this model as the necessary temporal dynamics for preserving group matching among the interacting waves [27].

We conclude this section by saying that this is a very dynamical area, and though it is clear that X waves have an important role in filamentation processes in Kerr-like media, many details certainly require further experiments and analysis.

18.3 Theory

The aforementioned experiments involve WPs with a narrow bandwidth usually propagating in the paraxial (small-angle) regime. As such, their theoretical description can be based on envelope wave equations, whose propagation invariant solutions are envelope X waves. Here we show in what sense nonlinear X waves can be regarded as linear solutions suitably dressed by the nonlinearity. To this end we start with the most simple linear formulation.

18.3.1 Linear X Waves

In the linear case, the slowly varying envelope $A = A(r, \phi, z, t)$ of the electric field $E(r, \phi, z, t) = A(r, \phi, z, t) \exp[i(k_0 z - \omega_0 T)]$, where ω_0 is the central frequency, $k_0 = k(\omega_0)$ is the central wave-number, and $r = (x^2 + y^2)^{1/2}, \phi = \tan^{-1}(y/x)$ are polar coordinates in the transverse plane x, y, is ruled by the equation

$$i\partial_z A + \frac{1}{2k_0} \nabla_\perp^2 A - \frac{k''}{2} \partial_t^2 A = 0 \tag{18.1}$$

where $t \equiv T - k'z$ is the retarded time in the frame traveling at light GV $1/k' = (dk/d\omega|_{\omega_0})^{-1}$, and T is the time in the lab frame. Assuming a superposition of modes $A = \sum_m A_m(r, t) \exp(im\phi)$ whose angular dependence is fixed by integer values of m, the total field turns out to be propagation invariant or diffraction-free ($\partial_z = 0$) provided that the temporal spectrum of the m-th mode, $\tilde{A}_m(r, \Omega) = \mathcal{F}[A_m(r, t)]$, where $\Omega = \omega - \omega_0$, obeys the Bessel equation

$$\left(\partial_r^2 + \frac{1}{r}\partial_r\right)\tilde{A}_m + \left(k_0 k'' \Omega^2 - \frac{m^2}{r^2}\right)\tilde{A}_m = 0, \tag{18.2}$$

which, for $k''>0$, has non-singular solutions $\tilde{A}_m(r,\Omega) = J_m\left(\sqrt{k_0 k''}|\Omega|r\right)$. An axially symmetric solution that does not spread in space and time can be constructed by superimposing $m = 0$ modes $J_0\left(\sqrt{k_0 k''}|\Omega|r\right)$ at different frequency Ω according to a spectral distribution $f(\Omega)$ (centered around $\Omega = 0$, i.e., $\omega = \omega_0$)

$$A(r,t) = \mathcal{F}^{-1}[\tilde{A}(r,\Omega)] = \frac{1}{2\pi}\int_{-\infty}^{\infty} d\Omega f(\Omega) J_0\left(\sqrt{k_0 k_0''}|\Omega|r\right)\exp(-i\Omega t). \quad (18.3)$$

Equation (18.3) constitutes a (sufficiently general) expression of an envelope X wave. A closed-form expression is obtained, e.g., with a spectrum $f(\Omega) = (\pi/\Delta\Omega)e^{-|\Omega|/\Delta\Omega}$

$$A(r,t) = Re\left[\frac{1}{\sqrt{k_0 k'' \Delta\Omega^2 r^2 + (1 + i\Delta\Omega t)^2}}\right], \quad (18.4)$$

which corresponds to the wave envelope represented in Fig. 18.1(b),(c).

It is worth pointing out that Eq. (18.3) is very similar to the formal general expression of an X wave (see, e.g., [36, 40]) that exists also in a non-dispersive medium (even in a vacuum), with the *total field E* determined by the superposition of Durnin J_0 beams of the same angular semi-aperture θ and different frequency (in an arbitrary bandwidth) as

$$E(r,z,t) = \frac{1}{2\pi}\int_{-\infty}^{\infty} d\omega f(\omega) J_0(Kr)\exp[i(k_z z - \omega T)], \quad (18.5)$$

where the transverse wave-number $K = k\cos\theta$, is fixed by the free parameter, so-called *cone angle*, θ, and the longitudinal wave-number $k_z = k\sin\theta$ $\left[K^2 + k_z^2 = k(\omega)^2\right]$ must be a linear function of ω for the wave to be dispersion-free at any order. Despite this formal similarity, however, there are important physical differences. In a non-dispersive medium $k = k(\omega) = \omega/c$ and, assuming $k_z = \omega/V_X$, one has that the wave (18.5) fulfills the constraint $\sin\theta = k_z/k(\omega) = c/V_X < 1$, which leads us to conclude that the wave is superluminal $(dk_z/d\omega)^{-1} = V_X > c$. Conversely, envelope X waves (18.3) propagate at the group velocity of light $1/k'$, while their cone angle θ is not fixed. Rather, because in this case $K = \sqrt{k_0 k_0''}|\Omega|$, one has cone angle dispersion $\theta = \theta(\Omega) = \sin^{-1}[\sqrt{k_0 k_0''}|\Omega|/k(\omega)]$ with cone angle vanishing at carrier frequency $\Omega = 0$ (note that for envelope X waves one can define instead an *envelope cone angle* $\theta_e = \tan^{-1}\sqrt{k_0 k''}/k'$ fixed by the dispersive properties of the medium [28]).

From the physical viewpoint, it is clear that stationariety of the envelope X wave arises from a mechanism of mutual compensation of diffractive and dispersive phase shifts (the non-uniformity of phase shift with frequency is indeed at the origin of WP spreading). This can be easily seen by considering the general solution of Eq. (18.1) written in terms of the Fourier–Bessel spectrum $\tilde{A}(K, \Omega)$

$$A(r,z,t) = \frac{1}{2\pi}\int_{-\infty}^{\infty} d\Omega \int_{0}^{\infty} dKK\tilde{A}(K,\Omega)J_0(Kr)e^{-i\frac{K^2}{2k_0}z}e^{i\frac{k''\Omega^2}{2}z}e^{-i\Omega t}. \tag{18.6}$$

Equation (18.6) reduces to Eq. (18.3) when $\tilde{A}(K,\Omega) = f(\Omega)\delta(K - \sqrt{k_0 k''}|\Omega|)$, i.e., when the envelope spectrum lies on the two crossing spectral lines in Fig. 18.1(a). In this case the two phase shift terms in Eq. (18.6) that originate from dispersion and diffraction, respectively, cancel each other.

We end up this section by warning the reader that envelope X wave solutions more general than those discussed so far can be generated, possessing a phase anomaly (central wave-number differing from the plane-wave one) and/or group-velocity different from $1/k'$, with spectrum lying on hyperbola of Fig. 18.1(a) [17,29,30]. Moreover, one can account for higher-order dispersive properties of the medium [28], and extend the definitions to non-scalar [33] and one transverse dimension (in a slab waveguide) [35,46] cases. We refer the reader to the literature for further details on this topic.

18.3.2 Nonlinear Models

At high intensities, Eq. (18.1) must be generalized to account for the nonlinear response. In a Kerr medium, the governing model is the well-known 1 + 3D nonlinear Schrödinger (NLS) equation, which in terms of dimensionless units reads as

$$(i\partial_z + \nabla_\perp^2 - \partial_{tt}^2)u + \chi|u|^2 u = 0; \nabla_\perp^2 \equiv \partial_{rr}^2 + \frac{1}{r}\partial_r \tag{18.7}$$

whereas SHG in a quadratic medium is ruled by the following two NLS equations that couple the envelopes at fundamental frequency (u_1) and second harmonic (u_2)

$$(i\partial_z + \nabla_\perp^2 - \partial_{tt}^2 u_1 + \chi u_2 u_1^* e^{i\delta k z} = 0), \tag{18.8}$$

$$(i\partial_z + \sigma\nabla_\perp^2 + iv\partial_t - d\partial_{tt}^2)u_2 + \chi\frac{u_1^2}{2}e^{-i\delta k z} = 0. \tag{18.9}$$

Here, in order to introduce normalized variables, we make use of the substitutions $z/z_d, r/w_0, t/t_0 \rightarrow z, r, t$, where w_0 is the beam waist, $z_d = 2k_0 w_0^2$ is the diffraction length at carrier frequency ω_0, $t_0 = (|k''|z_d/2)^{1/2}$, and $\chi = z_d/z_{nl}$ is the ratio between the diffraction and the nonlinear length, the latter depending on the peak intensity I_p as $z_{nl} = (\chi_3 I_p)^{-1}$ in Kerr media and as $z_{nl} = (\chi_2\sqrt{I_p})^{-1}$

in SHG (χ_3 [m/W] and χ_2 [W$^{-1/2}$] are standard nonlinear coefficients). Moreover, in Eqs. (18.8) $d = k''(2\omega_0)/|k''(\omega_0)$ stands for the GVD ratio, $\delta k = (k_2 - 2k_0)z_d$ with $k_2 = k(2\omega_0)$ is the wave-number mismatch, $\sigma = k_0/k_2 \simeq 1/2$, and v accounts for the GVM.

Localized solutions of Eq. (18.7) can be sought in the form $u(r,t,z) = \hat{u}(r,t)e^{-i\beta z}$, which yields the following equation for the spatio-temporal profile $\hat{u}(r,t)$

$$\partial^2_{rr}\hat{u} + \frac{1}{r}\partial_r\hat{u} - \partial^2_{tt}\hat{u} + b\hat{u} + \gamma\hat{u}^3 = 0 \qquad (18.10)$$

where, in order to keep only one parameter, we set $b = \beta/|\beta| = 0, \pm 1$, with $\gamma = \chi/|\beta|$ and $r/\sqrt{|\beta|}, t/\sqrt{|\beta|} \to r, t$ (the latter positions are not necessary for the case $b = 0$). Numerical solutions of Eq. (18.10) clearly exhibit the structure of envelope X waves, as shown by the examples reported in Fig. (18.6). Interestingly enough nonlinear X waves can be regarded as linear envelope X waves dressed by the nonlinearity. This is illustrated in Fig. (18.7), which shows the continuation of the linear [$\gamma = 0$ in Eq. (18.10)] profile (18.3), when the intensity (and hence the value of γ) increases. As shown the effect of a moderate nonlinearity ($\gamma = 1$) is to enhance the localization of the wave, while strong nonlinearities (see $\gamma = 10, 100$ cases) introduces in the field envelope additional oscillation without, however, modifying the basic X-like structure.

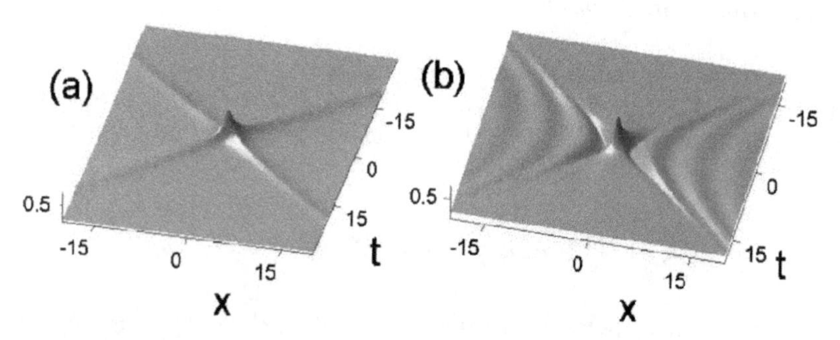

Fig. 18.6 Stationary profile $\hat{u} = \hat{u}(x, y = 0, t)$ of envelope X waves in Kerr media obtained from Eq. (18.10) with parameters: **(a)** $b = 0$, $\gamma = 1$; **(b)** $\mathbf{b} = 1$, $\gamma = 10$

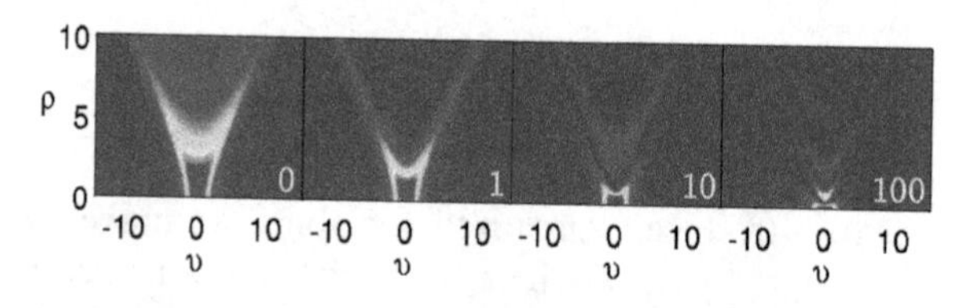

Fig.18.7 Level plots of the intensity of the fundamental X wave for different values of γ ($\Delta\Omega t_0 = 1, b = 0$), showing how the wave is "dressed" by the Kerr nonlinearity, when the latter is progressively increased

Similar results can be obtained in the SHG case by means of the substitution $u_1 = \hat{u}_1(r,t)e^{-i\beta z}$; $u_2 = \hat{u}_2(r,t)e^{-i(2\beta+\delta k)z}$, where the invariant profiles $\hat{u}_1$ and $\hat{u}_2$ can have both the structure of an envelope X wave.

18.3.3 Instability and Generation

From the previous discussion it is clear that the non-linearity is not a necessary ingredient for the existence of X waves. However, in a linear system excitation of X waves requires non-trivial input shaping techniques that involve the use of dispersive elements such as the axicon. Conversely the nonlinearity plays a major role in driving spontaneously the evolution of a standard (bell-shaped, e.g. Gaussian) light beam and pulse into a WP that closely resembles an X wave. The mechanism behind this transformation is spatio-temporal modulational instability that, in a normally dispersive medium, gives rise to conical emission, i.e., amplification of perturbations that correspond to off-axis frequency detuned components.

Exponential amplification at the expense of a plane wave pump has been investigated and is known to occur both in Kerr media [4,5,16] and in SHG [8]. The amplified disturbance is in fact composed of Fourier–Bessel or X wave modes [8,16], and in the case of a realistic input the process can be favored by the fact that such components are already present in the input spectrum. As a result, under proper input conditions, the numerical integration of the non-linear models shows the formation of envelope X waves. As an example, we show in Fig. 18.8 the spatio-temporal profile of the FH beam propagating in the absence and in the presence of GVM, respectively (typical parameters are those characteristics of the experimental results in Fig. 18.2). The case of the Kerr

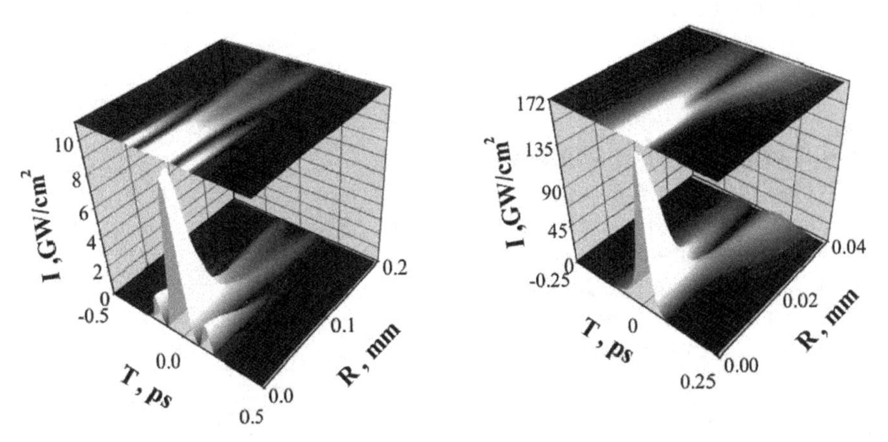

Fig. 18.8 Spatio-temporal output profiles of intensity at FH as obtained from integration of Eqs. (18.8): The *left frame* shows the GVM-matched case after propagation through a 4-cm sample of LBO. The *right frame* is relative to non-zero GVM after 1.5 cm of propagation

media is more complicated because the formation of X waves can be highly dynamical and involve complicated process of mixing as discussed before. Moreover, the dynamics is affected by several ingredients, whose relative strength is not entirely known with good accuracy. The reader can find a detailed report on numerical calculations in Ref. [26].

An important point is that the final state cannot be represented by the solutions illustrated in the previous section because it turns out that they possess infinite energy regardless of the fact they are linear or nonlinear. In turn, the circumstance that stationary solution are found only with infinite energy implies, that physical (finite-energy) beams cannot be strictly stationary, or in other words that a single X wave cannot be a strict attractor for the dynamical evolution of the field. Nevertheless one can have quasi-stationary situations [9] or the formation of the dynamical X wave through cycles of depletion and replenishments as discussed with reference to experiments (see also [18]).

Although the general problem of theoretically describing such scenarios and finding families of finite-energy solutions is still an open problem, different approaches has been already proposed. For instance in the liner regime finite energy beams can be constructed by superposition of X waves with different velocities [17,57]. The same approach can hardly be applied unchanged to the nonlinear regime where the superposition principle ceases to be valid. Nevertheless, a succesful attempt for extending the formalism to deal with the nonlinear case has been proposed in [17], where it is shown that the intrinsic 1 + 3D dynamics with normal GVD is accounted for by an averaged NLS model where the effective dispersion is anomalous. Breather solutions of the reduced equation describes the initial stage of pulse compression followed by splitting and replenishment stages.

Finally, we emphasize that other mechanisms of generation of X waves have been proposed that make use of interference of a Gaussian WP with cw plane wave [25].

18.4 Perspectives and Other Systems

It is worth mentioning that the existence of conical waves in the nonlinear regime has implications that go certainly beyond the specific area of self-focusing dynamics. For instance, X waves have been predicted to play a role in a Bose–Einstein condensed gas where it is well known that kinetic spreading of the atom cloud can be counteracted by the nonlinearity induced by atom–atom collisions. In a one-dimensional lattice, the effective mass along the lattice direction can be negative, playing a role analogous to normal GVD in self-focusing of light pulses (in the transverse direction, the kinetic term with positive mass is equivalent to diffraction). In this situation one can envisage, as an unusual manifestation of atomic coherence, the formation of a matter X waves even in the absence of conventional atomic traps [57]. A similar concept applied to optics allows us to predict that X waves arise in periodic structures that exhibit bandgaps. For instance, both arrays of evanescently coupled parallel waveguides that exhibit properties of anomalous diffraction [56,58] and 2D or 3D photonic crystals [59–60] support the existence of X waves. In these structures one can envisage intriguing phenomena such as tunneling in the bandgap mediated by nonlinear X waves which, however, will need future investigations.

Another interesting perspective is that of X waves in dissipative systems of the driven-damped type. Preliminar investigations [61] let us envisage that X waves can have an interesting role in cavity nonlinear optics.

Finally, we mention also that, in the anomalous dispersion regime where the dispersion relationship has an onion-like structure, an other type of non-diffractive and non-dispersive conical waves exist, the so-called "O waves". As shown recently [62–64] these type of waves were also shown to have a role in the non-linear dynamics of ultrashort pulses when operating in the anomalous dispersion regime.

18.5 Conclusions

We have presented an overview of the recent results obtained in the context of nonlinear X waves. Experiments suggest that, in media with normal GVD, the nonlinearity acts as the driving mechanism that reshapes a narrow beam and ultra-short pulse into conical filaments with features of X waves. From a theoretical viewpoint, the phenomenon is supported by the existence of weakly localized self-trapped solutions of envelope wave equations, which, unlike solitons, constitute the continuation (or dressing) of linear envelope X waves.

Acknowledgments The authors thank all colleagues who have collaborated with them over the years, without whom it would have not been possible to reach the present status of knowledge in this exciting field. They acknowledge also financial support from MIUR in the framework of PRIN and FIRB 2001 projects.

References

1. H.S. Eisenberg, R. Morandotti, Y. Silberberg, et al.: Kerr spatio-temporal self-focusing in a planar glass waveguide. *Phys. Rev. Lett.* **87**, 043902 (2001).
2. F. Wise, P. Di Trapani: Spatio-temporal solitons: the hunt for light bullets. *Opt. Photon. News*, Feb. 2002, 28–32 (2002).
3. J.K. Ranka, R.W. Schirmer, A.L. Gaeta: Observation of pulse splitting in nonlinear dispersive media. *Phys. Rev. Lett.* **77**, 3783–3786 (1996).
4. L.W. Liou, X.D. Cao, C.J. McKinstrie, et al.: Spatio-temporal instabilities in dispersive nonlinear media. *Phys. Rev. A* **46** 4202–4208 (1992).
5. G.G. Luther, A.C. Newell, J.V. Moloney, et al.: Short-pulse conical emission and spectral broadening in normally dispersive media. *Opt. Lett.* **19** 789–791 (1994).

6. G. Valiulis, J. Kilius, O. Jedrkiewicz, et al.: In *Quantum Electronics and Laser Science Conference* (Optical Society of America), paper QPD10-1; arXiv:physics/311081.
7. S. Orlov, A. Piskarskas, A. Stabinis: Focus wave modes in optical parametric generators. *Opt. Lett.* **27**, 2103 (2002).
8. S. Trillo, C. Conti, P. Di Trapani, et al.: Colored conical emission by means of second-harmonic generation. *Opt. Lett.* **27**, 1451–1453 (2002).
9. C. Conti, S. Trillo, P. Di Trapani, et al.: Nonlinear electromagnetic X waves. *Phys. Rev. Lett.* **90**, 170406 (2003).
10. P. Di Trapani, G. Valiulis, A. Piskarskas, et al.: Spontaneously generated X-shaped light bullets. *Phys. Rev. Lett.* **91**, 093904 (2003).
11. O. Jedrkiewicz, J. Trull, G. Valiulis et al.: Nonlinear X waves in second-harmonic generation: Experimental results. *Phys. Rev. E* **68**, 026610 (2003).
12. C. Conti, S. Trillo: X waves generated at second-harmonic. *Opt. Lett.* **28**, 1251 (2003).
13. J. Trull, O. Jedrkiewicz, P. Di Trapani et al.: Spatio-temporal three-dimensional mapping of nonlinear X waves. *Phys. Rev. E* **69**, 026607 (2004).
14. S. Longhi: Parametric amplification of spatiotemporal localized envelope waves. *Phys. Rev. E* **69** 016606 (2004).
15. R. Butkus, S. Orlov, A. Piskarskas et al.: Phase matching of optical X-waves in nonlinear crystals. *Opt. Commun.* **244** 411–421 (2005).
16. C. Conti: X-wave-mediated instability of plane waves in Kerr media. *Phys. Rev. E* **68**, 016606 (2003).
17. C. Conti: Generation and nonlinear dynamics of X-waves of the Schrodinger equation. *Phys. Rev. E* **70**, 046613 (2004).
18. M. Kolesik, E.M. Wright, J.V. Moloney: Dynamic nonlinear X waves for femtosecond pulse propagation in water. *Phys. Rev. Lett.* **92** 253901 (2004).
19. A. Dubietis, A. Gaizauskas, G. Tomosauskas et al.: Light filaments without self-guiding. *Phys. Rev. Lett.* **92**, 253903 (2004).
20. M. A. Porras, A. Parola, D. Faccio et al.: Nonlinear unbalanced Bessel beams: stationary conical waves supported by nonlinear losses. *Phys. Rev. Lett.* **93**, 153902 (2004).
21. A. Matijosius, J. Trull, P. Di Trapani et al.: Nonlinear space–time dynamics of ultrashort wave packets in water. Lett. **29**, 1123–1125 (2004).
22. D. Faccio, A. Matijosius, A. Dubietis et al.: Near- and far-field evolution of laser pulse filaments in Kerr media. *Phys. Rev. E* **72**, 037601 (2005).
23. P. Polesana, D. Faccio, P. Di Trapani et al.: High localization, focal depth, and contrast by means of nonlinear Bessel beams. *Opt. Express* **13**, 6160 (2005).
24. D. Faccio, P. Di Trapani, S. Minardi et al.: Far-field spectral characterization of conical emission and filamentation in Kerr media. *J. Opt. Soc. Am.* B **22**, 862 (2005).
25. Y. Kominis, N. Moshonas, P. Pagagiannis et al.: Bessel X waves in 2D and 3D bidispersive optical systems. *Opt. Lett.* **30**, 2924 (2005).
26. A. Couairon, E. Gaizauskas, D. Faccio et al.: Nonlinear X-wave formation by femtosecond filamentation in Kerr-media. *Phys. Rev. E* **73**, 016608 (2006).
27. D. Faccio, M. Porras, A. Dubietis et al.: Conical emission, pulse splitting and X-wave parametric amplification in nonlinear dynamics of ultrashort light pulses. *Phys. Rev. Lett.* 96, 193901 (2006).
28. M. A. Porras, C. Conti, S. Trillo et al.: Paraxial envelope X waves. *Opt. Lett.* **28**, 1092 (2003).
29. M. Porras, G. Valiulis, P. Di Trapani: Unified description of Bessel X waves with cone dispersion and tilted pulses. *Phys. Rev. E* **68**, 016613 (2003).
30. M. Porras, P. Di Trapani; Localized and stationary light wave modes in dispersive media. *Phys. Rev. E* **69**, 066606 (2004).
31. S. Longhi: Localized subluminal envelope waves in dispersive media. *Opt. Lett.* **29** 147 (2004).
32. A. Ciattoni, C. Conti, P. Di Porto: Universal space-time properties of X waves. *J. Opt. Soc. Am. A* **21** 451 (2004).

33. A. Ciattoni, C. Conti, P. Di Porto: Vector electromagnetic X waves. *Phys. Rev. E* **69** 036608 (2004).
34. P. Saari, K. Reivelt: Generation and classification of localized waves by Lorentz transformation in Fourier space. *Phys. Rev. E* **69**, 036612 (2004).
35. A. Ciattoni, P. Di Porto: One-dimensional nondiffracting pulses. *Phys. Rev. E* **69** 056611 (2004).
36. J. Lu, J.F. Greenleaf: Nondiffracting X-waves exact solutions to free-space scalar-wave equation and their finite aperture realization. *IEEE Trans. Ultrason. Ferrelec. Freq. Contr.* **39**, 19–31 (1992).
37. J. Durnin, J.J. Miceli, J.H. Eberly: Diffraction-free beams. *Phys. Rev. Lett.* **58**, 1499–1502 (1987).
38. R.W. Ziolkowski, D.K. Lewis, B.D. Cook: Evidence of localized wave transmission. *Phys. Rev. Lett.* **62**, 147–150 (1989).
39. J. Lu, J.F. Greenleaf: Experimental verification of nondiffracting X-waves. *IEEE Trans. Ultrason. Ferrelec. Freq. contr.* **39**, 441–446 (1992).
40. H. Sönajalg, P. Saari: Suppression of temporal spread of ultrashort pulses in dispersive media by Bessel beam generators. *Opt. Lett.* **21**, 1162–1164 (1996).
41. H. Sönajalg, M. Rtsep, P. Saari: Demonstration of the Bessel–X pulse propagation with strong lateral and longitudinal localization in a dispersive medium. *Opt. Lett.* **22**, 310–312 (1997).
42. P. Saari, K. Reivelt: Evidence of X-shaped propagation invariant localized light waves. *Phys. Rev. Lett.* **79**, 4135–4138 (1997).
43. Z. Jiang, X.C. Zhang: 2D measurement and spatio-temporal coupling of few-cycle THz pulses. *Opt. Express* **5**, 243 (1999).
44. D. Mugnai, A. Ranfagni, R. Ruggeri: Observation of superluminal behaviors in wave propagation. *Phys. Rev. Lett.* **84**, 4830 (2000).
45. R. Grunwald, V. Kebbel, et al,: Generation and characterization of spatially and temporally localized few-cycle optical wave packets. *Phys. Rev. A* **67**, 063820 (2003).
46. O.E. Martinez: Pulse distortions in tilted pulse schemes for ultrashort pulses. *Opt. Comm.* **59**, 229 (1986).
47. S. Orlov, A. Piskarskas, A. Stabinis, Localized optical subcycle pulses in dispersive media. *Opt. Lett.* **27**, 2167–2169 (2002).
48. G. Szabo, Zs. Bor: Broadband frequency doubler for femtosecond pulses. *Appl. Phys. B* **50**, 51 (1990).
49. P. Di Trapani, A. Agnesi, G. P. Banfi et al.: Off-axis parametric generation in the femtosecomd regime. *Lith. J. Phys.* **33** 324–327 (1993).
50. P. Di Trapani, A. Andreoni, G.P. Banfi et al., Group-velocity self-matching of femtosecond pulses in noncollinear parametric generation. *Phys. Rev. A.* **51** 3164–3168 (1995).
51. P. Di Trapani, A. Andreoni, P. Foggi et al.: Efficient conversion of femtosecond blue pulses by travelling-wave parametric generation in non-collinear phase matching. *Opt. Commun.* **119** 327–332 (1995).
52. R. Danielius, A. Piskarskas, P. Di Trapani et al.: Matching of group velocities by spatial walkoff in collinear three-wave interaction with tilted pulses. *Opt. Lett.* **21**, 973 (1996).
53. S. Szatmari, P. Simon, M. Feuerhake: GVD-compensated propagation of short pulses in dispersive media. *Opt. Lett.* **21**, 1156 (1990).
54. P. Di Trapani, D. Caironi, G. Valiulis et al.: Observation of temporal solitons in second-harmonic generation with tilted pulses. *Phys. Rev. Lett.* **81**, 570 (1998).
55. X. Liu, K. Beckwitt, F. Wise: Transverse instability of optical spatiotemporal solitons in quadratic media. *Phys. Rev. Lett.* **85**, 1871 (2000).
56. D.N. Christodoulides, N.K. Efremidis, P. Di Trapani et al.: Bessel X waves in 2D and 3D bidispersive optical systems. *Opt. Lett.* **29**, 1446 (2004).
57. C. Conti, S. Trillo: Nonspreading wave packets in three dimensions formed by an ultracold Bose gas in an optical lattice. *Phys. Rev. Lett.* **92**, 120404 (2004).

58. S. Droulias, K. Hizanidis, J. Meier et al.: X-waves in nonlinear normally dispersive waveguide arrays. *Opt. Exp.* **13**, 1827 (2005).
59. S. Longhi, D. Janner: X-shaped waves in photonic crystals. *Phys. Rev. B* **70**, 235123 (2004).
60. K. Staliunas, R. Herrero: Nondiffractive propagation of light in photonic crystals. *Phys. Rev. E* 73, 016601 (2006).
61. C.T. Zhou, M.Y. Yu, X.T. He: X-wave solutions of complex GL equation. *Phys. Rev. E* 73, 026209 (2006).
62. M. Porras, A. Dubietis, E. Kucinskas et al.: From X- to O-shaped spatiotemporal spectra of light filaments in water. *Opt. Lett.* **30**, 3398 (2005).
63. M.A. Porras, A. Parola, P. Di Trapani: Nonlinear unbalanced O waves: nonsolitary conical light bullets in nonlinear dissipative media. *J. Opt. Soc. Am. B* **22**, 1406 (2005).
64. M.A. Porras et al.: Characterization of O-shaped conical emission of light filaments in fused silica, *J. Opt. Soc. Am. B* **24**, 581 (2007).

Chapter 19
On the Role of Conical Waves in Self-focusing and Filamentation of Femtosecond Pulses with Nonlinear Losses

Eugenijus Gaižauskas, Audrius Dubietis, Viačeslav Kudriašov, Valdas Sirutkaitis, Arnaud Couairon, Daniele Faccio, and Paolo Di Trapani

Abstract This chapter concerns with the experimental observations and theoretical investigations on the propagation of intense femtosecond pulses in water and fused silica. It emphasizes spontaneous transformation of a beam into a conical (Bessel-like) wave during the filamentary propagation in media with nonlinear losses. This transformation constitutes an interpretation of the energy reservoir surrounding the high intensity central core of the filament. The adopted model is shown as being able to explain related phenomena such as the formation of multiple filaments and that of X-waves, observed experimentally in both water and fused silica.

19.1 Introduction

Since the advent of the chirped pulse amplification (CPA) technique [1], considerable attention has been paid to the nonlinear propagation of intense femtosecond pulses in various media. The discovery of pulse self-channeling in air [2] triggered a renewed interest in the phenomenon which was known for more than four decades in the regime of longer pulses, namely, the self-focusing of optical beams. In contrast to the longer pulses, femtosecond pulses can propagate in air over km-range distances by forming one or several narrow structures with a hot core (diameter less than 100 μm). This regime was called femtosecond filamentation and is the subject of an intense and growing research activity. It should be pointed out that the experimental observation of femtosecond light filaments exhibit substantially the same features in all the investigated media: gases [2,3], liquids [4,5] and solids [6,7] at various wavelengths and pulse durations or energies, and under diverse experimental conditions.

Many features of femtosecond filaments were initially interpreted based on the extensive investigation of self-focusing developed in the regime of longer

E. Gaižauskas (✉)
Department of Quantum Electronics, Vilnius University, Vilnius, Lithuania
e-mail: eugenijus.gaizauskas@ff.vu.lt

R.W. Boyd et al. (eds.), *Self-focusing: Past and Present*,
Topics in Applied Physics 114, DOI 10.1007/978-0-387-34727-1_19,

pulse durations (for review, see, e.g., [8,9]). The complexity of the phenomena involved in the regime of femtosecond pulses, however, required further investigations both experimenally and with numerical simulations; this led to several interpretations of the physics of femtosecond filaments, which enriched the picture obtained by simply considering self-focusing as the main nonlinear effect responsible for filamentation.

The self-trapping of intense optical beams due to the intensity-dependent refraction index increase was predicted in the pioneering work by Askar'yan [10] in 1962. Shortly afterwards the radial distribution of the electric field in the monochromatic self-trapped beam was computed, and is now known as the Townes profile [11]. In particular, the power of this self-trapped solution is exactly equal to one critical power for self-focusing, P_{cr}. A beam with larger power collapses at a finite distance on-axis. The dependence of the collapse distance on beam power was predicted by Kelley [12] and verified experimentally by Wang [13]. Explaining striking features of the optical damage as observed in glasses [14], Chiao and his co-workers conjectured the occurrence of multiple filamentation in 1964 [11]: "...if a broad beam is considerably above threshold, presumably it breaks up into several beams of threshold power." This prediction was supported by Bespalov and Talanov, who demonstrated that the modulational instability (MI) of plane waves propagating in nonlinear media leads to a breakup into filaments, each of which contain power comparable to the critical power for self-focusing [15]. Experimental confirmations were provided by Lallemand and Bloembergen [16], Pilipetskii and Rustamov [17], Campillo [18], and in many papers on small-scale focusing in high-peak-power Nd:glass laser systems; see, e.g., [19,20].

The question of whether experimental and theoretical investigations from the past are transposable to the regime of femtosecond filamentation in transparent media remained open. In particular, it was well established that the Townes mode constituted an unstable self-trapping mode. It can be stable either in planar systems or in media with a saturation of nonlinearity. For longer pulses, self-trapping (spatial solitons) were observed in planar waveguides (carbon disulfide, glass, semiconductor). In bulk media self-trapping was observed in atomic (sodium) vapors and other media with saturation of nonlinearity or other stabilization mechanisms [21]. The stabilization mechanism of femtosecond filaments beyond the distance at which they are formed, remained to be determined.

There are several viewpoints on the modeling, description, and understanding of the femtosecond filamentation. A first model is based on the assumption that an electromagnetic beam can produce its own dielectric waveguide and propagate without spreading [2,10,11]. In this approach, the filament is interpreted as a genuine soliton-like beam, whereas the role of the non-trapped radiation is marginal. A second approach is based on the moving focus model [22–25] in which the wavepacket is stacked into different *temporal* slices considered as independently giving rise to nonlinear foci located at distances depending on the power contained in each slice. Under such treatment, the filament appears as an optical illusion related to the use of time-integrated detection. The dynamic spatial replenishment model [26] was proposed as an extension of the moving focus model interpreting the results of numerical simulations: in this case, the filament is continuously absorbed and regenerated from the refocusing of the non-focused radiation that occurs at different distances for different *temporal* slices of the wavepacket. This approach allows us to interpret the occurrence of several focusing–defocusing cycles by considering the plasma-induced refractive index changes. The recurrence of the focusing–defocusing cycles in the replenishment scenario appears justified by the fact that the core of the beam contains a small fraction of the wave-packet energy. Multiphoton absorption occurring at each focusing action, leads to a step-like but overall slow decrease of the pulse energy. Therefore, the beam power stays above the critical value over a long propagation distance, thus permitting a large number of cycles to occur. Measurements of the energy losses confirm slow energy decay with propagation [27].

Nonlinear losses (NLL) associated with multiphoton ionization were properly accounted in most of the numerical investigations related to the femtosecond beam filamentation. The above-quoted contexts were inherently associated with the free-carrier excitation, plasma formation, plasma defocusing, and soon. The interpretation of the role played by NLL in filament dynamics remained a concern. Notably, NLL were mentioned to play a role as a collapse-arresting mechanism in several investigations concerning self-focusing (see, e.g., [9,12,15,28,29]), also for models that do not account for the contribution of plasma or of any other beam defocusing effect [30]. Finally, NLL have been considered as contributing to the quenching of the filament regime.

Our interpretation of light filaments in terms of conical waves arose from the merging of two main results: (i) numerical modeling of Sirutkaitis and his co-workers involving NLL [7,31] and (ii) extension of the unique property of the spatiotemporal linear X-waves, which were known to support stationary propagation in dispersive media [33], to the nonlinear realm by Di Trapani and co-workers [34,35].

In 2003, Sirutkaitis and his co-workers proposed an interpretation of light filaments dynamics formed in condensed media by means of a model that accounts only for self-focusing (i.e., Kerr response), diffraction, and NLL [7,31]. Notably, the proposed numerical results on beam propagation did not account for chromatic dispersion, plasma defocusing, and saturation of the nonlinearity, which were usually considered as counteracting the optical Kerr effect. The numerical results outlined a scenario featured by a *long-range stationarity* in propagation of the central hot core. This indicated a possible *active* role of NLL not only counteracting the collapse but also contributing to the self-focusing over long distances, leading to a (locally) balanced regime. The achievement of this balance can be interpreted in terms of the phase shift produced by NLL, via the intermediate action of beam flattening due to NLL and distortion by linear diffraction of the beam regions where the flat parts end and the intensity drops quickly [32]. Linear diffraction leads to a nonlinear phase, generating a flux of radiation towards the center of the beam. Obviously, this interpretation accounts only for a local, transient effect. Over large distances, it requires a loss-replenishment mechanism and leaves open the question of the possible stationarity of the entire transverse beam profile.

Detailed experimental and numerical investigations have evidenced the transition from the input Gaussian to the output X-wave. The distinguishing feature of a nonlinear X-wave is its conical (Bessel-like) structure. In fact, a conical wave can be composed of a central hot core, which experiences nonlinear interaction with matter, and of slowly decaying Bessel-like tails, which propagate in a virtually linear regime and provide a large energy reservoir. The name "conical" indicates the fact that energy does not flow over the propagation direction, as in the case of conventional Gaussian beams, but over a cone-shaped surface (i.e., non-collinearly). This conical flux creates the central hot spot as an interference effect. Here we intend to demonstrate how the concept of a "nonlinear conical wave" supports a scenario for describing light-filaments dynamics and, particularly, the key role played by NLL in the apparent stationary regime.

This chapter presents experimental observations and theoretical investigations of the propagation of intense femtosecond pulses in water and fused silica at power levels several times above the critical power for self-focusing. Our objective is to underline the spontaneous transformation of a beam into a conical wave during the filamentary propagation in the media with NLL. In particular, this transformation constitutes an interpretation of the role of the energy reservoir surrounding the high-intensity central core of the filament. We will show by means of a simple model retaining the minimal necessary set of physical effects that this transformation may result from the combined action of nonlinear losses, optical Kerr effect, and diffraction. The adopted model will be shown to also capture related phenomena such as the formation of multiple filaments and (with addition of chromatic dispersion) that of X-waves, observed experimentally in both considered media (water and fused silica).

19.2 Light Filaments Supported by a Conical Wave

19.2.1 Model Equations

We start by briefly presenting the physical model used for performing numerical simulations of filamentation with a minimal or a larger set of physical effects. We model the linearly polarized beam around the propagation axis z by the envelope $\mathcal{E}$ of the electric field $\mathbf{E}$, written as $\mathbf{E} = \mathrm{Re}[\mathcal{E}\exp(ik_0 z - i\omega_0 t)]\mathbf{e}_x$, where $k_0 = n_0\omega_0/c$ and ω_0 are the wavenumber and frequency of the carrier wave and n_0 denotes the refraction index of water at ω_0. The input pulses are Gaussian with energy E_{in} and a temporal half width t_p (the FWHM duration $\tau_{FWHM} \equiv t_p\sqrt{2\log(2)}$ is the quantity given in the following):

$$\mathcal{E}(x,y,t,0) = \mathcal{E}_0 \exp\left(-\frac{x^2}{w_x^2} - \frac{y^2}{w_y^2} - \frac{t^2}{t_p^2}\right), \tag{19.1}$$

The peak input power is computed from the energy and pulse duration $P_{\text{in}} = E_{\text{in}}/t_p\sqrt{\pi/2}$ and the peak input intensity is computed from the input power and the transverse waists $w_{x,y}$ of the (possibly elliptic) beam $\mathcal{E}_0^2 = 2P_{\text{in}}/\pi w_x w_y$. The scalar envelope $\mathcal{E}(x,y,t,z)$ evolves along the propagation axis z according to the nonlinear envelope equation proposed by Brabec and Krausz [36]. In order to simplify the notations, we consider dispersive terms up to the third-order only with the second- and third-order dispersive coefficients given by $k_0'' \equiv \partial^2 k/\partial\omega^2|_{\omega_0}$ and $k_0''' \equiv \partial^3 k/\partial\omega^3|_{\omega_0}$. By using the retarded time $t \equiv t_{\text{lab}} - z/v_g$, this equation can be written in the time domain as follows:

$$U\frac{\partial \mathcal{E}}{\partial z} = \frac{i}{2k_0}\nabla_\perp^2 \mathcal{E} + U\left(-i\frac{k_0''}{2}\frac{\partial^2 \mathcal{E}}{\partial t^2} + \frac{k_0'''}{6}\frac{\partial^3 \mathcal{E}}{\partial t^3}\right) + N(\mathcal{E}). \tag{19.2}$$

Equation (19.2) accounts for diffraction in the transverse plane, space–time focusing via the operator $U \equiv (1 + \frac{i}{k_0 v_g}\frac{\partial}{\partial t})$ [36], group velocity dispersion up to the third-order and nonlinear sources $N(\mathcal{E})$:

$$N(\mathcal{E}) = i\frac{\omega_0}{c} n_2 T^2 |\mathcal{E}|^2 \mathcal{E} - T\frac{\beta_K}{2}|\mathcal{E}|^{2K-2}\mathcal{E}. \tag{19.3}$$

The nonlinear terms describe the optical Kerr effect with possible optical shock terms (self-steepening is described via the operator $T \equiv 1 + \frac{i}{\omega_0}\frac{\partial}{\partial t})$ and nonlinear losses. Self-focusing related to the Kerr effect occurs for pulses with P_{in} above P_{cr}.

Numerical simulations were performed for water and for fused silica, which are typical Kerr and dispersive media. The coefficient β_K for multiphoton absorption, where K is the number of photons involved in the process, may be estimated from the Keldysh formulation [37]. The accuracy of this estimation, however, should be considered with care. For fused silica, for example, values for the multiphoton absorption cross section found in the literature can differ from each other by four orders of magnitude [6,38]. Alternatively, the multiphoton absorption coefficient can be deduced from experiments. For instance, transmission measurements were recently performed for green laser pulses (527 nm) undergoing filamentation in water [27]. Our measurements could be perfectly reproduced by numerical simulations of the model Eq. (19.2,19.3) and for a multiphoton absorption coefficient that differs from the value computed from Keldysh's model. Table 19.1 below indicates the values we have used for water and fused silica at different wavelengths.

With the values given in this table (here U_i denotes an ionization potential) our model shows that NLL alone are sufficient to arrest the collapse [38]. The minimal set of physical effects taken into account (diffraction, optical Kerr

Table 19.1 Parameters for water and fused silica used in the numerical simulations

@ 527 nm	n_2 (cm$^{2/W}$)	U_i (eV)	K	β_K (cm^{2K-3}W^{1-K})
Water	2.7×10^{-16}	6.5	3	2×10^{-25}
Water	2.7×10^{-16}	7.1	4	2×10^{-34}
@ 800 nm	n_2	U_i (eV)	K	β_K (cm^{2K-3}W^{1-K})
Water	4.1×10^{-16}	6.5	5	3.5×10^{-50}
Fused silica	3.5×10^{-16}	7.5	5	5×10^{-50}

effect and NLL) allows us to reproduce many features of femtosecond filamentation. In particular, these effects are sufficient to show the natural tendency of the wave-packet to readjust its shape in the form of a conical wave beyond the nonlinear focus. The addition of group velocity dispersion, space–time focusing and self-steepening is necessary for a realistic description of the generation of polychromatic conical waves as shown in Section 19.2.4.

19.2.2 Single-Filament Formation

The experiment was performed by launching into a water-filled cuvette the 527nm, $\sim 3\mu J$, 200-fs, spatially filtered beam with an approximately 0.1 mm FWHM waist at the input facet of the cuvette. The laser pulse was provided by an SHG-compressed, CPA Nd:glass laser (TWINKLE, Light Conversion Ltd.), operated at a 33 Hz repetition rate. The cuvette was made of 1-mm-thick, syringe-shaped quartz, which allowed us to continuously tune the sample length in the range $z = 5$–40mm and to record the beam evolution along the propagation distance. In order to monitor the fluence distribution (the time- integrated intensity in the transverse dimension) on the output facet of the cuvette, the output beam was imaged onto the CCD camera (8-bit Pulnix TM-6CN) by an $f = +50$ mm achromatic objective, with $8\times$ magnification.

The results displayed in Fig. 19.1 highlight the appearance of a single filament with almost constant $\sim 20\,\mu$m FWHM diameter in the investigated $z = 15$–40 mm range. Measurements show that the central spike (filament) contains $\simeq$ 20 percent of the transmitted energy. The filament length was found to increase with the incident power. Three transient stages showing the transformation of the transverse fluence distribution are plotted in Fig. 19.1. Note that the beam at 15 mm exhibits a weak ring-like structure, containing the major part of the input energy. For comparison, numerical simulations performed with the parameters of the experiment are shown in Fig. 19.1. Both the numerical simulations and the experiments indicate that the top part of the beam profile flattens in the first millimeters of propagation, which occurs due to the competition between the slow self-focusing of the beam and the high-order NLL process. As for super Gaussian beams, which form ring structures under the action of self-focusing [39], the flattened radial profile generated by the action of NLL self-focuses by developing a sharp and intense ring-shaped

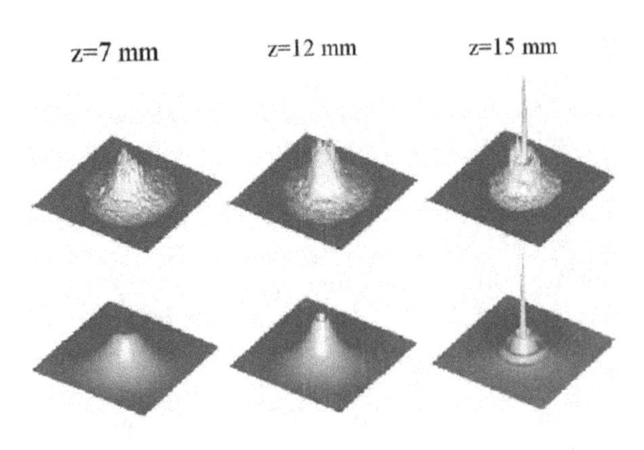

Fig. 19.1 Transient stages of the filament formation in water: Fluence profiles measured from 200 fs pulses propagating in water over 7 (*first column*), 12 (*second column*) or 15 mm (*third column*). (*top*); numerical simulations according to Eqs. (19.2)–(19.3), (*bottom*). (Reprinted with permission from [31]. Copyright 2004 by the American Physical Society.)

modulation which further shrinks and results in the very intense spike in the center of the beam [40], still surrounded by the ring-shaped structures. Due to its very high intensity, the spike in turn leads to strong on-axis NLL.

Note that even though the above-presented experimental results and numerics were performed by propagating femtosecond pulses, the proposed model (see below) does not rely on dispersive effects. The first observations of persistent ring formation during self-focusing were reported by Garmire et al. [44]. In the regime of femtosecond pulses, NLL are enhanced and might prevail over self-focusing and diffraction in the beam transformation. It is therefore worth emphasizing the role of NLL.

Consequently, the formation of filaments can be interpreted as due to the on-axis sink to which flows the energy from the surrounding reservoir. In order to highlight the role of the reservoir, we recall here the generalized solutions $\mathcal{E}(r_\perp, z, \omega) = e^{ik_0 z} u(r_\perp, z, \omega)$ to the wave equation (19.4) that are invariant along the optical axis z in the presence of a point-like sink at $\mathbf{r} = 0$. These solutions must satisfy Eq. (19.4):

$$(\nabla_\perp^2 + k_\perp^2) u = -4\pi\delta(\mathbf{r}), \qquad (19.4)$$

where $k_\perp^2 = k^2 n^2(r_\perp, \omega, \mathcal{E}) - k_0^2$, $\mathbf{r} = (x, y)$ and n is a refractive index of the medium. Equation (19.4) has two independent solutions: (1) $u(\mathbf{r}) = i\pi H_0^{(1)}(k_\perp r)$ which corresponds to an outgoing wave propagating with the wave vector $\mathbf{k} = k_\perp \hat{r} + k_0 \hat{z}$, and (2) $u(\mathbf{r}) = i\pi H_0^{(2)}(k_\perp r)$ which corresponds to an ingoing wave, for which $\mathbf{k} = -k_\perp \hat{r} + k_0 \hat{z}$. $H_0^{(1,2)}$ are the zero-order Hankel functions of the first and second kind. Accordingly, the general solution outside the intense core of the filament can be written as a sum of ingoing and outgoing conical waves:

$$\mathcal{E}(r_\perp, z, \omega) = e^{ik_0 z}\{a_+ H_0^{(1)}(k_\perp r) + a_- H_0^{(2)}(k_\perp r)\} \qquad (19.5)$$

In the absence of any on-axis sink (or source), this solution transforms into the well-known zero-order Bessel beam: $\mathcal{E}(r_\perp, z, \omega) = \mathcal{E}_0 e^{ik_0 z} J_0(k_\perp r)$ [41].

Solutions (19.5) constitute conical waves with a cone angle $\simeq k_\perp/k_0$ and, in general, an unbalance between the ingoing and outgoing components ($a_+ \neq a_-$) that characterize the energy flow from the periphery to the core of the beam. The interpretation of filamentation proposed here relies on the trend of the beam to spontaneously transform into a conical wave under the action of nonlinear losses. The nonlinear losses generate the unbalance in the core of the filament and drive the energy flow from the periphery of the beam towards the hot spike, precisely determined by the unbalance ratio $a_+ \neq a_-$. The underlying assumption is that the dynamics of the beam attempt to reach stationarity [31]. In this respect, one should note that the existence of fully nonlinear, stationary, monochromatic, conical waves was demonstrated numerically [42], which constitute the generalization of solutions (19.5) in the presence of Kerr self-focusing and NLL in the hot core. Moreover, the long-lived self-localized matter waves in Bose–Einstein condensates with a nonlinear dissipative mechanism have been discovered very recently [43]. This strongly supports the above assumption, i.e., the existence of stationarity supported by the refilling process which drives the energy from the reservoir to the intense core where it is absorbed. Due to the finite-energy content of the beam, stationarity can only be approached until the central spike eventually decays when the energy reservoir is exhausted.

Therefore, the quasi-stationary propagation results from the combined action of linear and non-linear effects: a purely linear contribution comes from the periphery of the beam, the conical shape of which establishes an inward power flux, which permits the compensation of the NLL occurring in the center; non-linear (Kerr) self-focusing action shrinks the hot central spot and acts in support of stationarity by limiting radially the overall losses to the hot core in the beam.

19.2.3 Filament Reconstruction

The proposed filamentation scenario raises two relevant questions. (i) The first one concerns the possible occurrence of a non-solitary, yet stationary filamentation regime in the presence of NLL. This regime could be reached not only in water (or in condense-matter) but also in gases. The experimental evidence of the robustness of filaments in air generated by terawatt laser pulses in adverse atmospheric conditions while suffering collisions with water droplets in clouds [45,46] already provides a compelling argument in support of the generality of the process. This was clearly stated by Kolesik and Moloney [47], who interpreted the process as the *non-linear replenishment* caused by the expulsion of light from the ring-shaped plasma channel beyond the droplet. The theoretical determination of nonlinear unbalanced Bessel beams, which are non-solitary stationary beams undergoing Kerr and NLL, provides another compelling argument. (ii) The second question concerns the possible role of *linear replenishment* in the filament reconstruction and, more generally, in the filament dynamics [48]. The self-reconstruction property is indeed an inherent feature of conical, non-diffracting waves, i.e., Bessel beams [49]. A set of real and numerical experiments gives an answer to these questions.

Figure 19.2 (top) shows the fluence distribution measured for the free propagation of a filament in water. In order to show that the filament propagates over several Rayleigh lengths because it is continuously refilled by the surrounding energy, we blocked the central spike with a 55-μm beam stopper printed on a 100-μm-thick BK7 glass plate and inserted into the beam path at

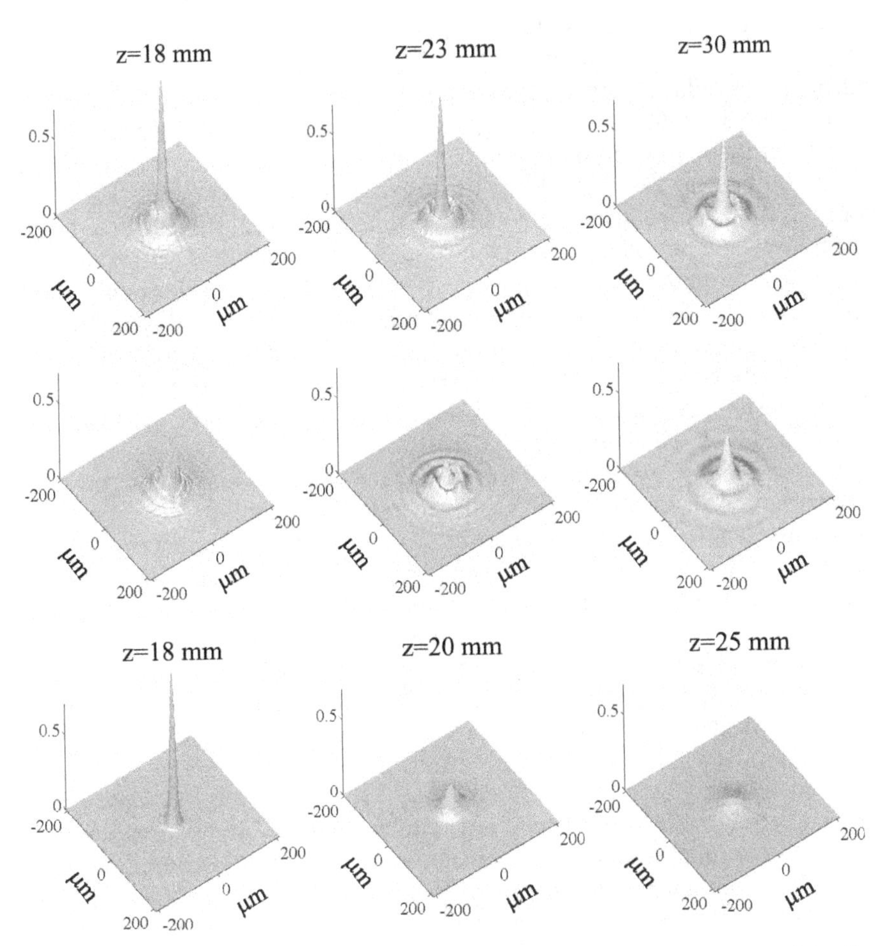

Fig. 19.2 Measurements of free filamentation in water (*top*), self reconstruction of the filament after blocking the central spike with a 55-μm beam stopper (*middle*), and filament extinction after inserting a 55-μm-diameter pinhole at $z = 18$mm (*bottom*)

$z = 18$mm. Figure 19.2 (middle) depicts the reappearance of a central spike at $z = 25$mm, gaining power while propagating, and with transverse dimensions equal to the original filament diameter. Comparison of Figs. 19.2 (top and middle) shows that the effect of the beam stopper is barely detectable after only 12mm of propagation. We performed a complementary experiment in the same range of distances by inserting a 55-μm-diameter pinhole in the water cuvette at $z = 18$mm. The energy transmitted by the pinhole was 20% of the total incident energy. In this case, Fig. 19.2 (bottom) shows the rapid extinction of the filament. Although the transmitted beam attempted to focus after a short distance, (i.e., within one diffraction length; not shown here), the filament did not survive, and a rapid decay was observed with a divergence roughly two times larger than that of a Gaussian beam of the same FWHM diameter.

These results unequivocally show that the observed filament can by no means be described in terms of solitons, if one accept the definition of a soliton as a localized stationary nonlinear solution for which, in the present case, the diffraction is balanced by the focusing nonlinear effects. Figure 19.3 displays the calculated evolution (vs. the propagation distance z) of the fluence distribution under the same conditions as in the experiments discussed in Fig. 19.2. The results outline the cases of (i) free propagation and filamentation (Fig. 19.3, top), (ii) filament reconstruction after blocking the central spike with a 55-μm beam stopper at $z = 18$mm (Fig. 19.3, middle) and (iii) filament extinction after inserting a 55-μm-diameter pinhole at $z = 18$mm (Fig. 19.3, bottom). These results clearly exhibit excellent agreement with their experimental counterparts, which demonstrates that filamentation is sustained over many Rayleigh lengths owing to the energy flow from the periphery toward the core of the beam. Similar results showing the role of the energy reservoir have been recently reproduced with filaments generated in air [50].

In the second experiment, we investigated the propagation and self-reconstruction in air of the filament generated in water. Because the beam power was below the critical power for self-focusing in air (about 3 GW) and the propagation beyond the water cell is linear, the experiment allowed us to determine the role of linear effects in the reconstruction process. The new experiment as well as the corresponding numerical simulations were performed under identical input conditions except for the stopper which was placed *outside* the water cell. The first contribution to the filament reconstruction is evidenced by the filament propagation in free space (air), show in Fig. 19.4 (left). This contribution is a linear, geometrical effect, which we attribute to the conical structure of the wave. It leads to the fast (i.e., within one Rayleigh range) reconstruction of the central spot and to its further spreading, as expected in the case of a finite-power linear conical wave. The other contribution is, in contrast, a non-linear effect that prevents the central-spot spreading owing to the optical Kerr effect and simultaneously drives the energy flow from the periphery towards the center of the beam owing to NLL. In Fig. 19.4 (right) we summarize our findings showing the calculated (solid curves) and measured (circles) FWHM radii of the reconstructed filament in water (a) and air (b).

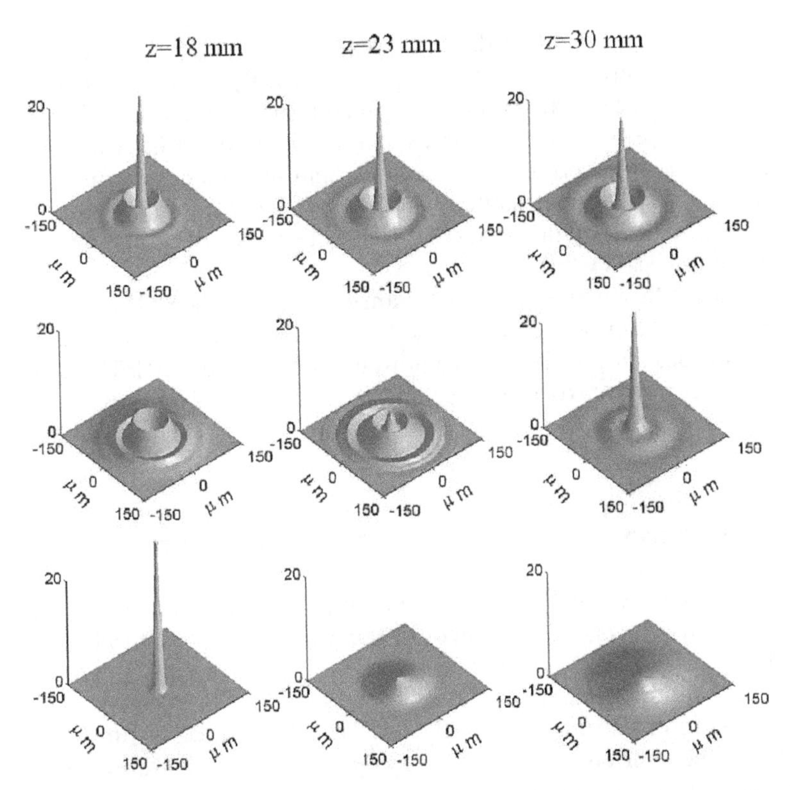

Fig. 19.3 Numerical simulations of the experiments of filamentation, regeneration, and extinction of filaments in water. *Top*: free propagation of the filament. *Middle*: filament reconstruction after blocking the central spike with a 55-μm beam stopper. *Bottom*: filament extinction after inserting a 55-μm-diameter pinhole at $z = 18$mm

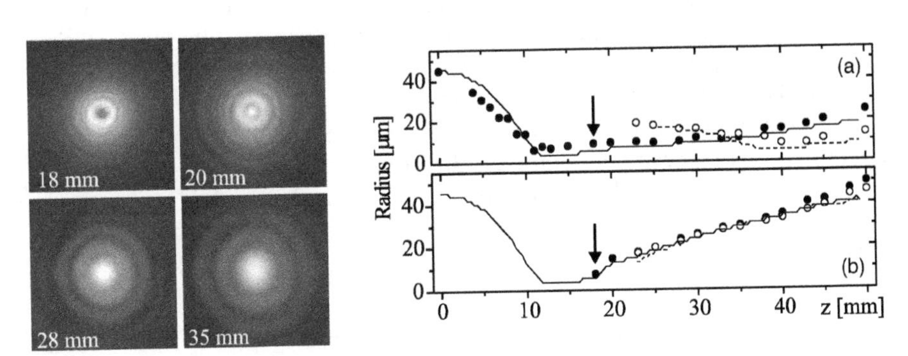

Fig. 19.4 *Left*: Self-reconstruction of the central spike in free space in air. *Right*: summary of the calculated and measured FWHM radii of free-propagating (*solid curves* and *filled circles*) and self-reconstructed (*dashed curves* and *open circles*) filament in water (**a**) and in free space (**b**). *Arrow* indicates the location of the beam stopper at the boundary between water and air

These results shed light on the linear replenishment mechanism that lead to filament over a long distance in non-linear media. By placing a small circular stopper on the (linear) path of a large beam, one introduces a perturbation that evolves, within few diffraction lengths, into a bright point (the Arago spot) on the optical axis. The spot is surrounded by light and dark rings similar to the Airy diffraction pattern generated by a lens. NLL can be viewed as a smooth stopper extended along the optical axis which continuously adds conical components to the beam while it propagates.

Concluding this section, note that filamentation of Bessel–Gauss pulses was found recently to produce damage lines extending over hundreds of micrometers and consisting of discrete, equidistant damage spots [51]. These discrete damage traces were explained by self-regeneration of Gauss–Bessel beams during propagation, which is consistent with the above-described mechanism.

19.2.4 X-Waves Generated from Femtosecond Filaments

It is clear from the discussion so far that filament formation and dynamics may be interpreted in terms of spontaneous beam shaping into a conical wave. The model presented above [Eqs.(19.2) and (19.3)] and experiments are able to indicate the full spatiotemporal nature of the spontaneously generated conical waves. The monochromatic conical beam may be identified with the Bessel beam, i.e., the interference pattern resulting from the superposition of infinite plane waves propagating along the surface of a cone with a well-defined cone angle θ with respect to the propagation direction. On the other hand, ultrashort, broadband conical-wave pulses may have a wavelength-dependent cone angle, $\theta = \theta(\lambda)$. The further requirement of stationarity implies that the actual geometry of the pulse in (θ, λ) space will depend on the material through the refractive index $n = n(\lambda)$. In normal dispersion the stationary conical wave is often referred to as the X-wave and in anomalous dispersion as the O-wave. In this section we shall focus attention on the case of ultrashort pulse filamentation in normally dispersive media and the spontaneous formation of X-waves (see also the chapter by Conti et al. in this book).

X-waves possess a bi-conical shape (an X) both in the space–time, (r, t), domain (for the near-field) and in the transverse wavenumber-frequency, $(k_\perp, \Omega)$, domain (equivalent to the (θ, λ) space). In the far-field it is possible to write out the form of the X spectrum explicitly in a simple analytical form [52,53]:

$$k_\perp = \sqrt{k_0 k_0''(\Omega - \tilde{\Omega})^2 + 2k_0\beta} \tag{19.6}$$

where third- and higher-order dispersion terms are neglected. $\Omega \equiv 2\pi c/\lambda - \omega_0$ denotes the frequency shift from ω_0. $\tilde{\Omega}$ and β and are two parameters related to the X-wave such that the carrier frequency is given by $(\omega_0 + \tilde{\Omega})$, the phase velocity is given $v_p = (\omega_0 + \tilde{\Omega})/[k(\omega_0 + \tilde{\Omega}) - \beta]$ and the group velocity is $v_g = 1/k'(\omega_0 + \tilde{\Omega})$.

The link between conical-wave modes and filament dynamics was first suggested by Dubietis et al. in connection to the measurements of filamentary propagation in water [5], and conical emission was suggested to be the energy-stabilizing factor sustaining the apparent stationarity of the filament. These measurements and the whole filamentation process in water were then successfully interpreted on the basis of numerical simulations by Kolesik et al. as a dynamical interaction between X-waves spontaneously generated within the nonlinear medium with chromatic dispersion [54].

Figure 19.5(a) shows a typical angular spectrum measured for a $\tau_{FWHM} = 200$ fs, green laser pulse ($\lambda_0 = 527$ nm) with energy $E_{in} = 3$ μJ, at the propagation distance of $z = 2$ cm. Specific remarkable features should be noted: (i) there is a strong on-axis emission for small transverse wave vectors or angles, extending both into the blue and red regions. (ii) there is an X pattern with long arms, extending at specific angles. (iii) The angular spectra are modulated, with fringes exhibiting parabolic-like dependence on the wavelength. These features were found in various media, both liquid and solid, under different pulse and focusing conditions and can therefore be considered as a signature of filamentation in condensed media.

Numerical simulations of pulse filamentation in water with the model Eqs. (19.2) and (19.3) clearly indicate a signature in the form of X-waves in the wavenumber-frequency domain. Figure 19.5(b) shows the angular spectrum obtained from numerics performed with the same parameters as in the experiment discussed in Fig. 19.5(a). It shows that the physical effects that contribute to the beam reshaping into a conical wave, i.e., the optical Kerr effect, NLL and diffraction are also driving the spontaneous generation of polychromatic conical waves in the dispersive medium [38].

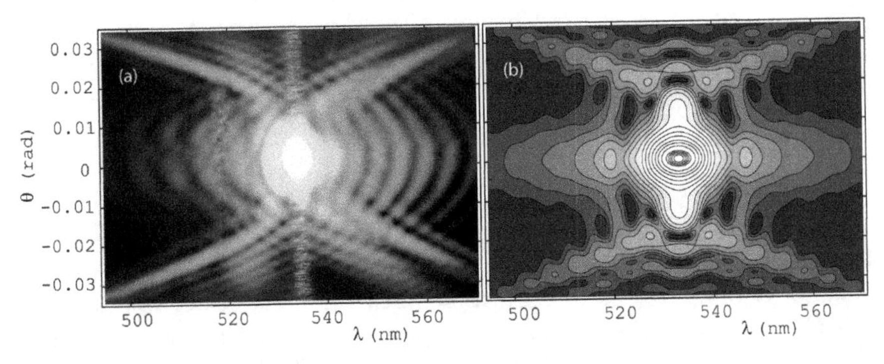

Fig. 19.5 Far-field spectra of a single filament generated in 2 cm of water with an input Gaussian pump with 200 fs FWHM duration, 130 μm FWHM diameter, 527 nm wavelength, input energy of 2 μJ. (**a**) Experimental, (**b**) numerical result using model Eqs. (19.2)–(19.3). Both are in logarithmic scale.

Laser pulse filamentation is characterized by an apparently complex dynamics accompanied by specific features such as temporal pulse splitting, formation of shock fronts, or multiple refocusing, thus inducing one to consider a complete near-field (r, t) characterization as the most appropriate. However, the same amount of information, if not more, may be retrieved from a proper analysis of the far-field $(k_\perp, \Omega)$ spectrum [55–57]. For example pulse splitting is clearly highlighted by the interference fringes observable in the axial part of the spectrum ([Fig. 19.5(a)]) and the spacing of the fringes is directly related to the temporal separation of the split pulses [58]. Going into more detail, it is possible to fit the experimental spectra using Eq. (19.6) with β and $\tilde{\Omega}$ taken as free parameters. However, a satisfactory fit cannot be obtained by using a single X-wave. On the other hand excellent agreement may be found under the assumption that the spectrum is composed from two X-modes, i.e., we obtain one set of values for β and $\tilde{\Omega}$ corresponding to a red-shifted X-wave ($\tilde{\Omega} > 0$) and one set corresponding to a blue-shifted X-wave ($\tilde{\Omega} < 0$). The general picture that emerges from this analysis is that of a parametric interaction, mediated by a four-wave mixing process in the Kerr medium, between the input pump and the two daughter X-waves. Furthermore these are characterized by distinct carrier frequencies $(\omega_0 + \tilde{\Omega})$ and consequently distinct group velocities. Group matching among the interacting waves is then better attained if the two pumps, breaking their initial degeneracy, split also to co-propagate with the X waves. In this view, pulse splitting is not a mere collapse-arresting mechanism but rather a further aspect of the same parametric process [57].

Finally we note that although the above analysis has been conducted on filaments generated in water, recent results highlight the spontaneous formation of X-waves in Kerr-induced filaments in a wide range of media including solids, but also air [55]. In all cases investigated the measured far-field spectra exhibit the expected dependence for X-waves with the slope of the tails determined solely by the material dispersion, as predicted by Eq. (19.6) for large frequency shifts. The filamentation process is therefore characterized by a series of well-defined but very general features, all of which are well described in terms of the spontaneous generation of conical waves and within the framework of the model outlined in Eqs. (19.2) and (19.3).

19.3 Multiple Filaments Supported by a Conical Wave

The self-focusing of optical beams in Kerr media leads either to catastrophic collapse for $P_{in} > P_{cr}$ or to diffraction when the input power is below the critical. Nonlinear losses arrest the collapse, and generate ring structures. However, the rings are azimuthally unstable. Beams that carry powers exceeding P_{cr} are likely to break transversally into multiple filaments. This phenomenon was interpreted as due to the growth of perturbations or noise present in the input beam, which are modulationally unstable [15]. In this case, the multifilamentation (MF) pattern characterized by the number and the location of the filaments in the transverse diffraction plane must undergo shot-to-shot fluctuations because it is determined by the input noise. In this respect, the beam break-up of a high-power femtosecond pulse into optically turbulent light filaments in air was demonstrated by Mlejnek et al. [59].

It was pointed out that vectorial-induced symmetry breaking can lead to MF even for cylindrically symmetric input beams [60]. In the latter case, deterministic effects prevail over noise in the generation of the MF pattern. Another deterministic process for the generation of MF in air was recently demonstrated both experimentally and using numerical simulations. It consists of imposing either strong field gradients or phase distortions in the input-beam profile of an intense femtosecond laser pulse so as to obtain a prescribed regular filamentation pattern [61]. Elliptic beams also induce MF patterns, which are organized due to the prevailing effect of beam ellipticity over noise [62,63]. In this section, we will show that multiple filaments induced by the beam ellipticity can be interpreted by the means of the model presented in Section 19.2.1. In particular, we will connect the multiple filamentation pattern to the generation of conical waves.

19.3.1 Multiple Filaments from Femtosecond Pulses in Water

The experiments reported in this subsection provide evidence that ultrashort pulses with sufficiently large input beam ellipticity undergo multiple filamentation and generate a pattern featured by the ellipticity, in contrast to the random nature of the patterns that should result from the growth of modulationally unstable perturbations.

A 170-fs, 527-nm pulse was provided by the second-harmonic compressed Nd:glass laser system (TWINKLE, Light Conversion Ltd., Lithuania) operated at a 33-Hz repetition rate. A spatially filtered beam was focused into a $\sim 85 - \mu$m FWHM beam waist at the entrance of a water cell by means of an $f = +500$ mm lens. The incident energy was varied by means of a half-wave plate and a polarizer. The focused beam had a small intrinsic ellipticity $e = w_y/w_x = 1.09$. Highly elliptical beams ($e = 2.2$) were formed by inserting a slightly off-axis iris into the beam path. The output face of the water cell was imaged on the CCD camera (8-bit dynamic range, Pulnix TM-6CN and a frame grabber from Spiricon, Inc., Logan, Utah) with $7\times$ magnification by means of an achromatic objective ($f = +50$ mm).

First, by increasing the ellipticity of the input beam, we found that the threshold power for the formation of a single filament is larger for an elliptic beam than for a circular beam. However, the threshold power for MF is lower for an elliptic beam than for a circular beam. For instance at, $z = 31$ mm, a single filament was observed from $6P_{cr}$ for $E = 1.09$ and from $4.9P_{cr}$ for $e = 2.2$. This $\sim 20\%$ increase is in good agreement with a recent theoretical prediction for this threshold [64].

Second, we recorded transverse distribution patterns at the fixed propagation length $z = 31$ mm by increasing peak power of the input beam. Figure 19.6 displays the case of a of an elliptic beam ($e = 2.2$) which highlight the scenario for multiple filamentation: at moderate powers above P_{cr} ($P_{in} = 7P_{cr}$), the beam self-focusing generates a ring, i.e., following the interpretation in the previous section, a conical wave that constitutes the energy reservoir. It contains the power that is not trapped in the central filament. For powers exceeding a threshold value ($P_{in} = 10\,P_{cr}$), hereafter called the threshold for multiple filamentation, two filaments are generated along the major axis of the input elliptic beam, in addition to the central filament.

At higher powers ($P_{in} = 14\,P_{cr}$), two additional filaments appear in the perpendicular direction leading to the observation of a quadrupolar filamentation pattern. At even higher powers ($P = 23\,P_{cr}$), a quadrupole of filaments was observed along the bisectors of the major and minor axes. Generally, an increase of the input power led to an increase of the number of filaments. Because the MF patterns shown in Fig. 19.6 were reproducible from shot to shot and followed the orientation of the elliptic input beam, we conclude that they were induced by the beam-ellipticity, thus ruling out the effect of the noise inherent to the input conditions. We checked that the orientation of the MF pattern is not due to a symmetry breaking induced by the linear polarization of the laser which we changed without observing any influence on the MF pattern. Polarization effects are actually predicted to occur for much narrower filaments (with radius comparable to the wavelength) than those we measured in our experiments with FWHM diameter of $\sim 20\mu m$ [60].

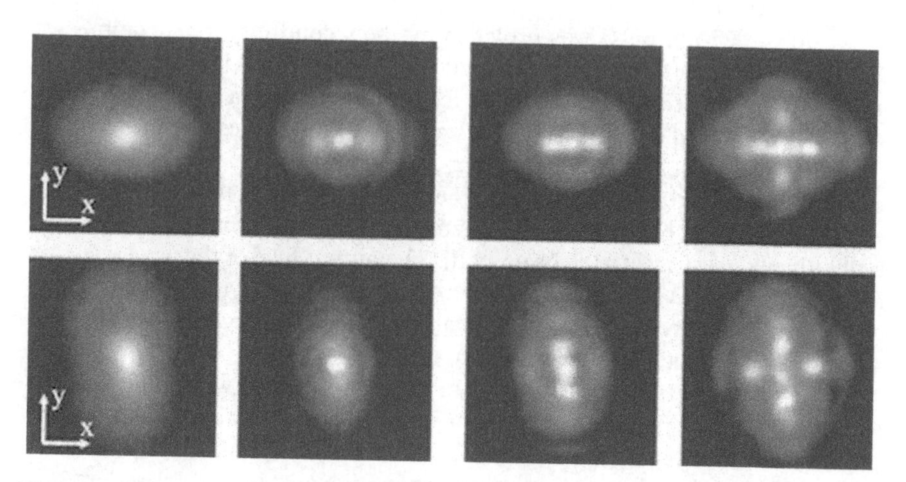

Fig. 19.6 CCD camera images of the filamentation patterns generated from elliptic beams ($e = 2.2$) with different transverse orientation, denoted by x (*top*) and y-axes (*bottom*) at the propagation distance of $z = 31$ mm in water. The image dimensions are $330 \times 330\mu m^2$, the incident power is 5, 7, 10, and $14P_{cr}$ from left to right

Simple symmetry arguments explain the location of the multiple filaments along preferential directions (x- and y- axes) corresponding to the major and/or minor axis of the elliptic input beam; we interpret the quadrupole distribution of filaments located symmetrically along the bisectors of the major and minor axes as an effect of the beam transformation into a conical wave and its ensuing modulational instability.

This interpretation is founded by the linear stability analysis of the stationary conical waves supported by Kerr self-focusing and NLL, i.e., the nonlinear unbalanced Bessel beams (NLUBB) which we identified as the attractors for single filamentation [42]. NLUBB constitute a family of solutions, each member of which is characterized by the strengths of the Kerr effect and of NLL. The growth rate of radial perturbations on the NLUBB solutions with the same Kerr nonlinearity was shown to decrease when the strength of NLL is increasing. The role of NLL is therefore to drive the system towards stability. The conical, Bessel-like solutions in pure Kerr media are unstable; when the strength of NLL exceeds a certain threshold, the corresponding Bessel-like solutions become stable. The same analysis was performed for dipolar and higher order azimuthal perturbations. Solution emerged stable for dipolar perturbations both in the pure Kerr case and in the presence of NLL. In contrast, both systems turned out to be unstable for quadrupole and higher-order perturbations, a result that can be related to the fact that these perturbations involve modulation far from the central spot where intensity is weak and so NLL cannot play its stabilizing role. This result is in keeping with the interpretation of the MF patterns as driven by the modulational instability of the conical wave generated by the competition of NLL and the optical Kerr effect.

19.3.2 Multiple Filaments from Femtosecond Pulses in Fused Silica

The propagation of ultrashort laser pulses in solid dielectrics leads to filamentation with specific features compared to non-solid media. For solids and liquids, the critical power for self-focusing is in the MW range and NLL are generally larger than in gases; both type of media therefore lead to filamentation over reduced scales. However, the most important feature distinguishing these media is permanent modification of the material resulting from the interaction of a femtosecond pulse with solids. This permanent modification appear as a local change of the refraction index [63,65,66], and might not only be due to the self-actions occuring during the passage of the pulse, but also to thermal effects occuring over longer time scales and accumulating from shot to shot. Although the origin of this change in the refraction index is still an open question, this property is promising for the production of photonic structures in the bulk of optical materials. For instance, waveguides can be produced either by transverse illumination of an optical material or by forming a filament for longitudinal writing [6,67,68].

In the following, we show from the experimental and simulation results that the scenario we have outlined for filamentation in water applies also to filamentation in fused silica.

A CPA laser system was used in the experiments (Ti:sapphire oscillator (Tsunami) and Ti:sapphire amplifier (Spitfire), both from Spectra-Physics). The laser system delivers 800nm, 1 mJ, 130 fs pulses at the repetition rate of 1 kHz. Figure 19.7 shows the beam profiles on the output face of a 1-cm-long fused silica sample at different pulse energies.

All the features obtained for filaments in water are visible in Fig. 19.7. Figure 19.7(a) displays beam diffraction for a pulse energy well below the threshold for continuum generation. Figure 19.7(b)(c) display the filament (with a core diameter of 30μm) obtained when the power exceeds the critical threshold and the non-trapped part of the beam energy which formed the surrounding ring [25,31,45,63]. According to our interpretation, this ring is a conical component of the wave-packet, and refills the hot core so as to sustain filamentation. Note that filamentation is accompanied by the supercontinuum generation, which happens with a simultaneous conical emission of colored rings. We interpret this observation as another manifestation of the beam transformation into a conical wave. Figure 19.7(d) shows the appearance of two additional lateral filaments when the energy of the input pulse is increased ($\sim 3\mu$J or $10P_{cr}$). These filaments are located along the major axis of the elliptic

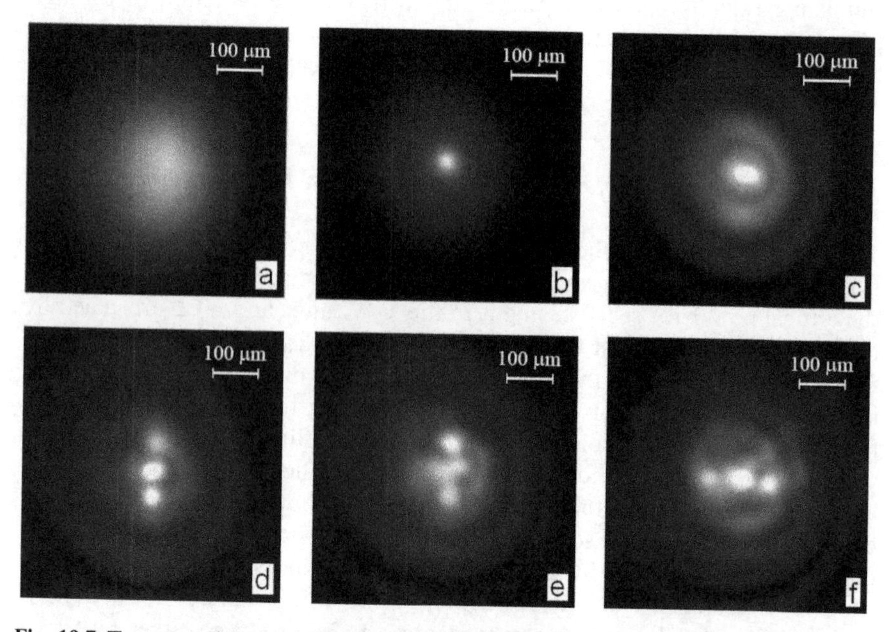

Fig. 19.7 Transverse beam profiles recorded after propagation over 1 cm in fused silica at different pulse energies: 0.7 (**a**), 1.4 (**b**), 2.7 (**c**), 3.4 (**d**), 4.0 (**e**) and 4.8μJ (**f**). (An energy of 0.3 μJ corresponds to the critical power for self-focusing P_{cr})

input beam. By further increasing the incident beam power, filamentation along the minor axis is observed [Fig. 19.7(e)]. At higher pulse energies exceeding 4μJ (13P_{cr}), the quadrupole structure deteriorates [Fig. 19.7(f)]. This sequence of features is interpretable in terms of the beam transformation into a conical wave owing to NLL which can be viewed as a continuous Arago spot experiment. Diffraction of the Arago spot is suppressed by self-focusing which competes with NLL so as to drive the system towards stationarity. However, ideal stationarity of the cylindrically symmetric nonlinear beam supported by these competing effects breaks up into filaments along the major and minor axis of the elliptic input beam.

To capture the propagation dynamics of the beam, numerical simulations were performed by using model Eqs. (19.2) and (19.3) where no axial symmetry was imposed. Our results displayed in Fig. 19.8 are in excellent agreement with the experimental results. The filamentation patterns are shown at a fixed propagation distance of 1 cm for increasing beam powers, from 2P_{cr} to 12P_{cr}. They exhibit a conical beam with a strong central filament [Fig. 19.8(b), 4P_{cr}], from which a pair of weaker filaments are nucleated along the major axis of the elliptic input beam [Fig. 19.8(c), 8P_{cr}], or the quadrupole distribution of filaments with an uneven strength of each lateral pair at larger incident beam power [Fig. 19.8(d), 12P_{cr}]. The comparison of Figs. 19.8(c) and (d) seems to indicate 90-degree rotation of the pair of lateral filaments, which we interpret as an effect of the astigmatism corresponding to the ellipticity of the input beam [61]. Irregular multiple filamentation patterns similar to those obtained in the experiments could be retrieved in numerical simulations performed by adding a small fraction of noise in the input beams [63].

Concluding this section, it is important to stress that filamentation and continuum generation processes in fused silica possibly depend on permanent changes in the properties of the propagating medium induced by successive laser shots. The measurements presented in Fig. 19.7 were performed by monotonically increasing the beam energy while the beam position remained unchanged. Returning to the lower pulse energies (from maximum at 5μJ) revealed that spatial beam structure does not recover [63] for the same values

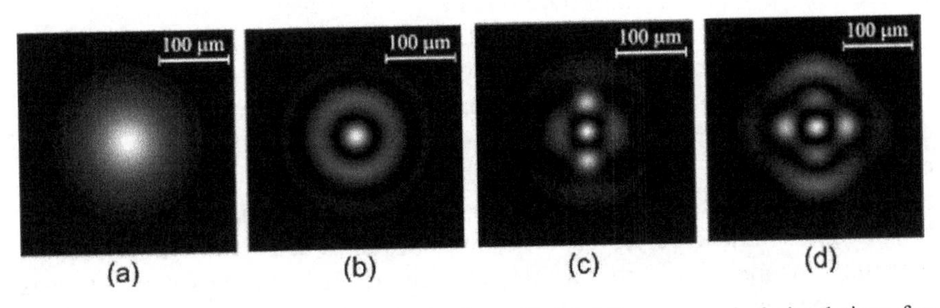

Fig. 19.8 Filamentation patterns in fused silica obtained from numerical simulations for elliptic input beams for different incident beam powers: (a) 2P_{cr}, (b) 4P_{cr}, (c) 8P_{cr}, and (d) 12P_{cr}. The propagation distance was $z = 1$ cm

of the pump energy. The observed effect should be attributed to the formation of a waveguide inside the bulk material.

19.4 Conclusions

While femtosecond lasers have a great potential for processing dielectric materials, the underlying physics both of femtosecond wave-packet propagation and media response (its modification) are far from fully understood, although significant progress has been achieved in the past few years. In this chapter we have focused attention on what we believe is a relevant contribution to light-filament dynamics, i.e., the contribution of nonlinear losses (NLL). In fact, besides quenching the collapse and supporting the excitation of free electrons, NLL should actively participate to a filament regime by means of unique contribution to the wave-packet phase modulation, which they produced in conjunction with propagation and Kerr response.

In order to clarify how the effect comes into play, we have here considered the simplest possible model where filaments are supported in the presence of NLL. We found that this model contains, besides NLL, only diffraction and Kerr response. The object that we propose for the best characterization of the model is its stationary solution, which was found and identified as a nonlinear, unbalanced Bessel beam. Indeed, this object is important not only in respect to the observed filament stationarity, or because it behaves as an attractor for the transient beam-profile dynamics. More relevantly, the stationary solution identifies the model itself, different models leading obviously to different profiles.

For what concerns numerical beam-propagation experiments, here we have shown that their results fully support the emergence of long-living filaments (i.e., fluence profiles featured by a central peak that propagate without apparent diffraction) from input Gaussian beams. The result confirms that plasma defocusing, chromatic dispersion, nonlinearity saturation, and all other phase-modulating mechanisms that are usually considered to counteract Kerr-induced phase modulation are indeed not strictly necessary for the establishment of an apparent stationary regime.

While the proposed analysis identifies a relevant contribution related to NLL, it does not pretend that other physical effects play a negligible role in real settings. In order to clarify this issue, we have devoted a small section of this discussion to the X-wave dynamics, for which a different chapter of this book is entirely devoted. Note that nonlinear X-waves were introduced as the stationary solution of a different, but analogously oversimplified, model that contains, besides diffraction and Kerr response, only chromatic dispersion (GVD). As the first model is aimed at outlining the key role of NLL, so the second one has clarified the key role of GVD. At present, a stationary solution accounting both for NLL and GVD, as well as for all the other effects mentioned above, has not been found yet. However, what one should reasonably expect on the basis of the analysis here proposed is that, if such a solution exists, it should be definitely shaped as a nonlinear conical wave. In fact, the conical nature emerge as necessary key fact for energy replenishment when NLL come into play.

References

1. D. Strickland, G. Mourou Compression of amplified chirped optical pulses. *Opt. Commun.* **56**, 219–221 (1985).
2. A. Braun, G.Korn, X. Liu et al.: Self-channeling of high-peak-power femtosecond laser pulses in air. *Opt. Lett.* **20**, 73–75 (1995).
3. H.R. Lange, A. Chiron, J.-F. Ripoche et al.: High-order harmonic generation and quasiphase matching in xenon using self-guided femtosecond pulses. *Phys. Rev. Lett.* **81**(8), 1611-1613 (1998).
4. W. Liu, O. Kosareva, I.S. Golubtsov et al.: Femtosecond laser pulse filamentation versus optical breakdown in H_2O. *Appl. Phys. B* **76**(3), 215–229 (2003).
5. A. Dubietis, G.Tamošauskas, I.Diomin et al.: Self-guided propagation of femtosecond light pulses in water *Opt. Lett.* **28**, 1269–1271 (2003).
6. L.Sudrie, A. Couairon, M. Franco et al.: Femtosecond laser-induced damage and filamentary propagation in fused silica. *Phys. Rev. Lett.* **89** 186601 (2002).
7. V. Sirutkaitis, E. Gaižauskas, V. Kudriašov et al.: Self-guiding, supercontinuum generation and damage in bulk materials induced by femtosecond pulses. *SPIE Proc.* **4932**, 346–357 (2003).
8. Y.R. Shen: Self-focusing: experimental, *Prog. Quant. Electr.* **4**, 1–34 (1975).
9. J.H. Marburger: Self-focusing: theory, *Prog. Quant. Electr.* **4**, 35–110 (1975).
10. G.A. Askar'yan: Effects of a gradient of a strong electromagnetic beam on electron and atoms. *Soviet Phys.-JETP* **15**,1151 (1962).
11. R.Y. Chiao, E. Garmire, C. H. Townes: Self-trapping of optical beams. *Phys. Rev. Lett.* **13**, 479–482 (1964).
12. P. L. Kelley: Self-focusing of optical beams. *Phys. Rev. Lett.* **15**, 1005–1008 (1965).
13. C.C. Wang: Length-dependent threshold for stimulated Raman effect and self-focusing of laser beams in liquids. *Phys. Rev. Lett.* **16**, 344–346 (1966).
14. M. Hercher: Laser-induced damage in transparent media, *J. Opt. Soc. Am.* **54**, 563 (1964). (See also current book).
15. V.I. Bespalov, V.I. Talanov: Filamentary structure of light beams in nonlinear liquids. *JETP Lett.* **3**, 307–310 (1966).
16. P. Lallemand, N. Bloembergen: Self-focusing of laser beams and stimulated Raman gain in liquids, *Phys. Rev. Lett.* **15**(26), 1010-1012 (1965).
17. N.F. Pilipetskii, A.R. Rustamov: Observation of self-focusing of light in liquids, *JETP Lett.* 2, 55–56 (1965).
18. A.J. Campillo, S.L. Shapiro, B.R. Suydam: Periodic breakup of optical beams due to self-focusing. *Appl. Phys. Lett.* **23**, 628–630 (1973).
19. D.C. Brown: High-peak-power lasers, Springer, New York (1981).
20. A.A. Mak, L.N. Soms, V.A. Fromzel et al.: *Nd:Glass Lasers*, Nauka, Moscow (1990).
21. G.I. Stegeman, M. Segev: Optical spatial solitons. *Science* **286**, 1518–1523 (1999).
22. J.H. Marburger, W.G. Wagner: Self-focusing and pulse sharpening mechanism. *IEEE J. Quant. Electron.* **QE-3**, 415–416 (1967).
23. V.N. Lugovoi A.M. Prokhorov: A possible explanation of the small-scale self-focusing filaments, *JETP Lett.* 7, 117–119 (1968).
24. M.M. Loy Y.R. Shen: Small-scale filaments in liquids and tracks of moving foci, *Phys. Rev. Lett.* 22, 994–997 (1969).

25. A. Brodeur, C.Y. Chien, F.A. Ilkov et al.: Moving focus in the propagation of ultrashort laser pulses in air. *Opt. Lett.* **22**, 304–306 (1997).
26. M. Mlejnek, E.M. Wright, J.V. Moloney: Dynamic spatial replenishment of femtosecond pulses propagating in air. *Opt. Lett.* **23**, 382–384 (1998).
27. A. Dubietis, A. Couairon, E. Kučinskas et al.: Measurement and calculation of nonlinear absorption assotiated to femtosecond filaments in water. *Appl. Phys.* B **84**, 439–446 (2006).
28. S.N. Vlasov, V.A. Petrishchev, V.I. Talanov: Application of the method of moments to certain problems in the propagation of partially coherent light beams. *Radiophys. Quant. Electro.* **14**, 1062–1070 (1971).
29. J.H. Marburger, E. Dawes: Dynamical formation of a small-scale filament. *Phys. Rev. Lett.* **21**, 556–558 (1968).
30. B.J. LeMesurier: Dissipation at singularities of the nonlinear Schrödinger equation through limits of regularisations. *Physica D* **138**, 334–343 (2000)
31. A. Dubietis, E. Gaižauskas, G. Tamošauskas et al.: Light filaments without self-channeling. *Phys. Rev. Lett.* **92**, 253903 (2004).
32. S. Polyakov, F. Yoshino, G. Stegeman: Interplay between self-focusing and high-order multiphoton absorption. *JOSA B* **18**, 1891–1895 (2001).
33. P. Saari, K. Reivelt: Evidence of X-shaped propagation-invariant localized light waves. *Phys. Rev. Lett.* **79**, 4135–4138 (1997).
34. G. Valiulis, J. Kilius, O. Jedrkiewicz et al.: Space–time nonlinear compression and three-dimensional complex trapping in normal dispersion. In: OSA Trends in Optics and Photonics (TOPS), Technical Digest of the Quantum Electronics and Laser Science Conference vol. 57. Optical Society of America, Washington DC, pp. QPD10 1–2 (2001).
35. P. Di Trapani, G. Valiulis, A. Piskarskas et al.: Spontaneously generated X-shaped light bullets. *Phys. Rev. Lett.* **91**, 093904 (2003).
36. T. Brabec, F. Krausz: Nonlinear optical pulse propagation in the single-cycle regime. *Phys. Rev. Lett* **78**, 3282–3285 (1997).
37. L.V. Keldysh: Ionization in the field of a strong electromagnetic wave. *Sov. Phys. JETP* 20, 1307–1314 (1965).
38. A. Couairon, E. Gaižauskas, D. Faccio et al.: Nonlinear X-wave formation by femtosecond filamentation in Kerr media. *Phys. Rev. E* **73**, 016608 (2006).
39. S. Tzortzakis, B. Lamouroux, A. Chiron et al.: Femtosecond and picosecond ultraviolet laser filaments in air: experiments and simulations. *Opt.Commun.* **197**, 131–143 (2001).
40. M. Soljačić, S. Sears, M. Segev: Self-trapping of "necklace" beams in self-focusing media. *Phys. Rev. Lett.* **81**, 4851–4854 (1998).
41. J. Durnin, J.J. Miceli, J.H. Eberly: Diffraction-free beams. *Phys. Rev. Lett.* **58**, 1499–1501 (1987).
42. M.A. Porras, A. Parola, D. Faccio et al.: Nonlinear unbalanced bessel beams: Stationary conical waves supported by nonlinear losses. *Phys. Rev. Lett.* **93**, 153902 (2004).
43. A. Alexandrescu, V.M. Prez-Garcia: Matter-wave solitons supported by dissipation. *Phys.Rev.A* **73**, 053610 (2006).
44. E. Garmire, R.Y. Chiao, C.H. Townes: Dynamics and characteristics of the self- of intense light beams. *Phys. Rev. Lett.* **16**, 347–349 (1966).
45. F. Courvoisier, V. Boutou, J. Kasparian et al.: Ultraintense light filaments transmitted through clouds. *Appl. Phys. Lett.* **83**, 213–215 (2003).
46. G. Méchain, G. Méjean, R. Ackermann et al.: Propagation of fs-TW laser filaments in adverse atmospheric conditions. *Appl. Phys.* B 80, 785–789 (2005).
47. M. Kolesik, J.V. Moloney: Self-healing femtosecond light filaments. *Opt. Lett.* **29**(6), 590–592 (2004).
48. A. Dubietis, E. Kučinskas, G. Tamošauskas, et al. Self-reconstruction of light filaments. *Opt. Lett.* **29**, 2893–2895 (2004).

49. D. Mc Gloin, K. Dholakia: Bessel beams: diffraction in a new light, *Contemporary Phys.* **46**(1) 15–28 (2005).
50. W. Liu, F. Théberge, E. Arévalo et al.: Experiments and simulations on the energy reservoir effect in femtosecond light filaments. *Optics Lett., year,* **30**(19) 2602–2604 (2005).
51. E. Gaižauskas, E. Vanagas, V. Jarutis et al.: Discrete damage traces from filamentation of femtosecond pulses. *Opt. Lett.* 31, 80–82 (2006).
52. M. A. Porras, P. Di Trapani: Localized and stationary light wave modes in dispersive media. *Phys. Rev.* E **69**, 066606 (2004).
53. D. Faccio M.A. Porras, A. Dubietis et al.: Conical emission, pulse splitting, and X-wave parametric amplification in nonlinear dynamics of ultrashort light pulses. *Phys. Rev. Lett.* **96**, 193901 (2006).
54. M. Kolesik, E.M. Wright, J.V. Moloney: Dynamic nonlinear X waves for femtosecond pulse propagation in water. *Phys. Rev. Lett.* **92**, 253901 (2004).
55. D. Faccio, A. Matijosius, A. Dubietis et al.: Near- and far-field evolution of laser pulse filaments in Kerr media. *Phys. Rev.* E **72**, 037601 (2005).
56. M. Kolesik, E.M. Wright, J.V. Moloney: Interpretation of the spectrally resolved far field of femtosecond pulses propagating in bulk nonlinear dispersive media, *Opt. Expr.*, 13(26), 10729–10741 (2005).
57. D.Faccio, M.A. Porras, A. Dubietis et al.: Angular and chromatic dispersion in Kerr-driven conical emission, *Opt. Commun.* **265**, 672–677 (2006).
58. D. Faccio, P. Di Trapani, S. Minardi et al.: Far-field spectral characterization of conical emission and filamentation in Kerr media. *J. Opt. Soc. Am. B* **22**(4), 862–869 (2005).
59. M. Mlejnek, M. Kolesik, E.M. Wright et al.: Optically turbulent femtosecond light guide in air. *Phys. Rev. Lett.* 83, 2938–2941 (1999).
60. G. Fibich, B. Ilan: Deterministic vectorial effects lead to multiple filamentation. *Opt. Lett.* **58**, 840–842 (2001).
61. G. Méchain, A. Couairon, M. Franco et al. Organizing multiple femtosecond filaments in air. *Phys. Rev. Lett.* **93**, 035003 (2004).
62. A. Dubietis, G. Tamošauskas, G. Fibich et al.: Multiple filamentation induced by input-beam ellipticity, *Opt. Lett.* **29**, 1126–1128 (2004).
63. V. Kudriašov, E. Gaižauskas, V. Sirutkaitis: Beam transformation and permanent modification in fused silica induced by femtosecond filaments. *J. Opt. Soc. Am. B* **22**, 2619–2627 (2005).
64. V.P. Kandidov, V. Yu Fedorov: Properties of self-focusing of elliptic beams. *Quant Electron.* **34**(12),1163–1168 (2004).
65. K. M. Davis, K. Miura, K. Sugimoto et al.: Writing waveguides in glass with a femtosecond laser. *Opt. Lett.* **21**, 1709–1711 (1996).
66. L. Sudrie, M. Franco, B. Prade et al.: Writing of permanent birefringent microlayers in bulk fused silica with femtosecond laser pulses. *Opt. Commun.* **171**, 279–284 (1999).
67. K. Y Yamada, W. Watanabe, T. Toma et al.: In situ observation of photoinduced refractive-index changes in filaments formed in glasses by femtosecond laser pulses. *Opt. Lett.* 26, **19** (2001).
68. M. Kamata, M. Obara: Control of the refractive index change in fused silica glasses induced by a loosely focused femtosecond laser. *Appl. Phys. A* **78**, 85–88 (2004).

Chapter 20
Self-focusing and Self-defocusing of Femtosecond Pulses with Cascaded Quadratic Nonlinearities

Frank W. Wise and Jeffrey Moses

Abstract Nonlinear phase shifts, either self-focusing or self-defocusing, can be impressed on pulses that propagate in quadratic nonlinear media. The issues that arise in the extension of these phase shifts to the femtosecond regime are outlined in this chapter, and applications to femtosecond pulse generation and propagation are reviewed. Quadratic media appear to be unique in offering a means of impressing self-defocusing nonlinear phase shifts on ultrashort pulses. Ongoing work that extends the investigation of controllable nonlinear phase shifts to pulses with a duration of a few optical cycles will be introduced.

20.1 Nonlinear Phase Shifts in Quadratic Media

Quadratic optical nonlinearities are usually associated with frequency-conversion applications. However, the production of nonlinear phase shifts through the interactions of light beams in quadratic nonlinear media was identified, if not appreciated, long ago [1]. Isolated studies of the nonlinear phase shifts produced in quadratic processes appeared between 1970 and 1990 [2,3], and in the last decade there has been a surge of interest in this area [4,5]. Nonlinear phase shifts can be produced in any 3-wave mixing process [6]. The simplest case, which is most pertinent here, is phase-mismatched second-harmonic generation (SHG): the fundamental field acquires a nonlinear phase shift in the process of conversion to the second-harmonic field and back-conversion to the fundamental (Fig. 20.1). (Both fields acquire a nonlinear phase shift, but we will focus on the fundamental field here.) Because it involves sequential or cascaded quadratic processes, this is referred to as the cascaded quadratic nonlinearity. This phase shift has the following properties:

F.W. Wise (✉)
Department of Applied and Engineering Physics, Cornell University, Ithaca, New York, USA
e-mail: fwise@ccmr.cornell.edu

R.W. Boyd et al. (eds.), *Self-focusing: Past and Present*,
Topics in Applied Physics 114, DOI 10.1007/978-0-387-34727-1_20,

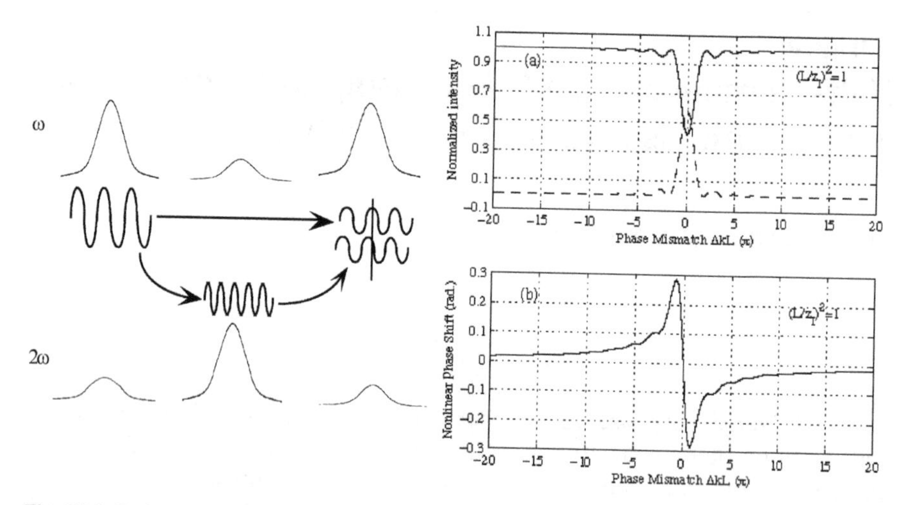

Fig. 20.1 *Left panel*: Origin of the cascaded quadratic phase shift. Some of the fundamental field undergoes one or more cycles of conversion and back-conversion. When the energy is in the second-harmonic field, it propagates with the phase velocity of the second-harmonic frequency, so when it returns to the fundamental frequency it is out of phase with light that was not converted. The resulting phase shift depends on the phase-velocity mismatch and the intensity of the input fundamental field. *Right panel, top*: Normalized peak intensities of the fundamental (*solid*) and second-harmonic (*dashed*) fields. *Bottom*: nonlinear phase shift on the fundamental field

- it can be large (~10 radians in practical situations)
- its sign is controllable through the wave-vector mismatch (Fig. 20.1); the nonlinearity can be self-focusing or self-defocusing.
- its magnitude is controllable through the wave-vector mismatch (Fig. 20.1)
- the magnitude saturates with increasing intensity

When the wave-vector mismatch is large, the phase shifts follow a simple proportionality

$$\Delta\phi_{NL} \propto \frac{d_{eff}^2 IL}{\Delta k} \tag{20.1}$$

where d_{eff} sums all relevant components of the quadratic nonlinear susceptibility, I is the intensity, L is the propagation length, and Δk is the wave-vector mismatch.

The cascaded quadratic phase shifts can be controlled through the wave-vector mismatch by simply adjusting the orientation or temperature of a crystal. For pulse-shaping applications, the ability to control the sign of the nonlinear phase shift, and specifically the ability to generate negative phase shifts without excessive loss, is probably the feature that will underlie the most-revolutionary new possibilities. The present book attests to the enormous effort that has been paid to investigation of self-focusing, while very little work on self-defocusing nonlinear processes has been reported. As this review will demonstrate, the cascaded quadratic nonlinearity has important pulse-shaping applications when either self-focusing or self-defocusing nonlinearities are employed.

The cascaded quadratic nonlinearity saturates when a large fraction of the total energy is converted to the second harmonic field during a single conversion-back-conversion cycle. Saturation of the nonlinear phase shift is undesirable in most pulse-shaping applications because it distorts the temporal variation of the phase shift (and frequency chirp) from the shape produced in cubic nonlinear media, where the phase shift is proportional to intensity for all intensities. However, saturation is crucial to stabilizing multi-dimensional solitons against the collapse that occurs in cubic nonlinear media. This will be discussed further below.

Nonlinear phase shifts are associated with a nonlinear index of refraction n_2, which arises from the cubic nonlinearity, $\chi^{(3)}$. Below saturation, where the nonlinear phase shift is proportional to the intensity of the fundamental field, it can be useful to define an effective nonlinear index for the cascaded quadratic process. Continuing the analogy to cubic nonlinear processes, the residual second-harmonic light that is generated in the phase-mismatched process can be likened to 2-photon absorption, which arises from the imaginary part of $\chi^{(3)}$.

Further discussion of pulse propagation in quadratic nonlinear media is presented in the chapter by P. DiTrapani and co-workers in this volume.

20.1.1 Self-focusing Versus Self-defocusing

Of course, the net phase shift impressed on a pulse will also contain a contribution from the cubic nonlinearity present in all materials, which is generally positive. We may then think of the $\chi^{(2)}$ and $\chi^{(3)}$ processes as contributing to a *total* effective n_2. Figure 20.2 plots the results for a typical quadratic material. Because the real n_2 due to $\chi^{(3)}$ has negligible dependence on crystal orientation and temperature, it acts as a bias to the total nonlinear index. As a result, multiple self-focusing and self-defocusing regions can exist. For the typical case, when cubic phase shifts are positive, self-defocusing is only possible within a finite region on one side of $\Delta k = 0$. This result can be significant when other considerations constrain the magnitude of Δk in a given application, as will be discussed below.

20.1.2 Analytical Framework

The propagation equations for the cascaded quadratic process and definitions of parameters and quantities used throughout this chapter are given below for

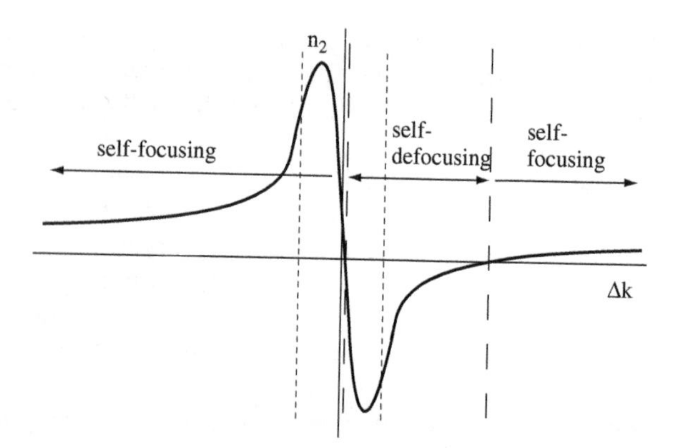

Fig. 20.2 The total effective n_2 resulting from both quadratic and cubic contributions plotted versus Δk for the typical case when $\chi^{(3)}$ phase shifts are positive. Self-focusing and self-defocusing regions are indicated. The short-dashed lines represent a possible location of the stationary boundaries, which will be introduced and defined in Section 20.3

reference. Gaussian units are used. Generally, these details are unnecessary for following the main ideas of the text.

The equations that govern the propagation of fundamental frequency (FF) and second-harmonic (SH) field envelopes in a quadratic nonlinear medium in the presence of group-velocity dispersion (GVD) and diffraction are

$$i\frac{\partial a_1}{\partial \xi} + \frac{\rho_1}{2}\nabla_\perp^2 a_1 - \frac{\alpha_1}{2}\frac{\partial^2 a_1}{\partial s^2} + a_1^* a_2 e^{-i\beta\xi} = 0, \tag{20.2}$$

$$i\frac{\partial a_2}{\partial \xi} + \frac{\rho_2}{2}\nabla_\perp^2 a_2 - \frac{\alpha_2}{2}\frac{\partial^2 a_1}{\partial s^2} - i\frac{\partial a_2}{\partial s} + a_1^2 e^{i\beta\xi} = 0. \tag{20.3}$$

An instantaneous quadratic nonlinearity is assumed. The equations are written in dimensionless coordinates in the frame of the FF group velocity, $\upsilon_{g,1} = c/n_{g,1} = \left(\partial k_1/\partial\omega|_{\omega_0}\right)^{-1}$, where k_j are the electric field wave vectors along the propagation axis, z. They are normalized to the initial FF pulse duration τ_0 and arbitrary beam width ρ_0, and the propagation coordinate is normalized to the mismatch between FF and SH group velocities: $s = t/\tau_0 - z/\upsilon_{g,1}\tau_0$, $x' = x/\rho_0$, $y' = y/\rho_0$ and $\xi = z/L_{GVM}$, where $L_{GVM} = c\tau_0/(n_{g,1} - n_{g,2})$. In these coordinates, the transverse Laplacian $\nabla_\perp^2 = \partial^2/\partial x'^2 + \partial^2/\partial y'^2$. Dependent variables a_1 and a_2 are the FF and SH field complex amplitudes in dimensionless units, which can be multiplied by carrier waves at ω_0 and $2\omega_0$, respectively, to recover the electric fields. They are defined as $a_1 = \Gamma_1(2k_1/k_2)^{1/2}L_{GVM}A_1$ and $a_2 = \Gamma_2(2k_2/k_1)L_{GVM}A_2$, where the actual field amplitudes A_1 and A_2 are normalized to $|A_0|$, the peak value of the FF field amplitude. Parameters Γ_i measure the quadratic nonlinear response, where $\Gamma_1 = 8\pi\omega_0^2 d_{eff}/c^2k_1$ and $\Gamma_2 = 16\pi\omega_0^2 d_{eff}/c^2k_2$. Parameters $\rho_j = L_{GVM}/k_j\rho_0^2$ measure diffraction and $\alpha_j = (\partial^2 k_j/\partial\omega^2|j\omega_0)L_{GVM}/\tau_0^2 \equiv L_{GVM}/L_{GVD,j}$ measure GVD. Finally, $\beta = L_{GVM}\Delta k$ is the normalized phase mismatch, where $\Delta k = 2k_1 - k_2$ is the wave-vector mismatch between fields. Numerical solutions of these equations (often with cubic nonlinear polarization terms also included) provide the calculated results shown in this chapter.

In terms of these parameters we can define both the magnitude of the nonlinear phase shifts generated by the cascaded quadratic process, $\Delta\phi_{NL}$, and a characteristic nonlinear length, L_{NL}, as follows. In the limit of large Δk, $\Delta\phi_{NL} = 128\pi^2\omega_0^4 d_{eff}^2 |A_0|^2 L/c^4 k_1 k_2 \Delta k = \Gamma_1\Gamma_2|A_0|^2 L/\Delta k$, where L is the propagation length. Note, $|A_0|^2$ can be equated to the intensity in the usual way. From this we derive the important relation, Eq. (20.1). Finally, the nonlinear length $L_{NL} = \Delta k/\Gamma_1\Gamma_2|A_0|^2$. From these definitions we may also define an effective nonlinear index of refraction, n_{2eff}.

20.2 Ultrashort Pulse Shaping

The propagation of a light pulse in a transparent medium is governed by the interplay between linear and nonlinear phase accumulations. The linear phases underlie diffraction and GVD in the limits of monochromatic and plane waves, respectively; more generally, these processes are intertwined. Diffraction converts the phase shifts to changes in the transverse spatial profile of a laser beam. If self-focusing and diffraction are precisely balanced, the result is a spatial soliton—a beam that propagates indefinitely without spreading laterally. In cubic nonlinear media, spatial solitons are stable in one transverse dimension, but unstable against collapse in two dimensions.

Phase modulations dominate the generation and propagation of intense ultrashort (picosecond and femtosecond) optical pulses. As an example, consider temporal soliton formation: the positive (or self-focusing) nonlinear phase shift induced on a pulse through the process of self-phase modulation (SPM) is balanced by that arising from anomalous GVD. All materials have cubic nonlinearities, and in the absence of (linear or nonlinear) absorption these are almost always self-focusing, $n_2 > 0$. Anomalous GVD generally cannot be obtained in materials without absorption, so prism pairs, grating pairs, or chirped mirrors are employed to produce it. Soliton-like pulse shaping is ubiquitous in modern femtosecond-pulse lasers, pulse compressors, and related devices.

If the signs of the nonlinearity and dispersion are interchanged in the nonlinear Schrödinger equation that governs plane-wave pulse propagation in cubic nonlinear media, an identical soliton will still form. Therefore, in media with fast negative (self-defocusing) nonlinear refraction, soliton-like pulse shaping can be implemented with normal GVD. In this case, lasers and other devices can be built with transparent material replacing the usual prism pairs or grating pairs. This can lead to obvious simplifications and performance advances. This approach was not pursued prior to 1999, largely because media with self-defocusing nonlinearities ($n_2 < 0$) and low loss have not been identified.

Nonlinear phase shifts are required to generate ultrashort pulses, but excessive nonlinear phase shifts are a ubiquitous problem in the generation of high-energy pulses. These become significant at the nanojoule level in fiber lasers and amplifiers, and at the microjoule or milllijoule level in typical bulk-optics devices. In the spatial domain these phase shifts are manifested as self-focusing and as instabilities, which adversely affect the beam profile. In the temporal domain uncontrolled nonlinear phase shifts lead to excessive spectral bandwidth and phase distortions, which reduce pulse quality or even destroy the pulse through optical wave-breaking.

20.3 Extension of the Cascaded Quadratic Nonlinearity to Femtosecond Pulses

Although the cascaded quadratic nonlinearity offers some remarkable properties, prospects for exploiting cascaded quadratic phase shifts for short-pulse generation initially were considered poor. To be efficient, nonlinear processes must be phase-matched, and nonlinear optics with ultrashort pulses requires group-velocity matching as well. Because the fundamental and second-harmonic pulses generally have different group velocities, they move away from each other in time, as illustrated in Fig. 20.3. To quantify this issue we define the group-velocity mismatch (GVM) length L_{GVM}, the distance over which the fundamental and harmonic pulses move apart by approximately one pulse duration. GVM reduces the magnitude of the nonlinear phase shift, and distorts the temporal variation of the nonlinear phase across the pulse. For a 120-fs pulse at 800 nm in the nonlinear crystal barium metaborate (BBO), $L_{GVM} = 0.6\,\text{mm}$. This situation is representative of the problems that arise in quadratic nonlinear optics with femtosecond pulses.

Liu and Qian presented an intuitive solution to the problems caused by GVM [7]. The cascaded quadratic phase shift is produced by the cycles of conversion and back-conversion between the harmonics. If one arranges for these cycles to occur before the pulses separate in time (i.e., in a distance shorter

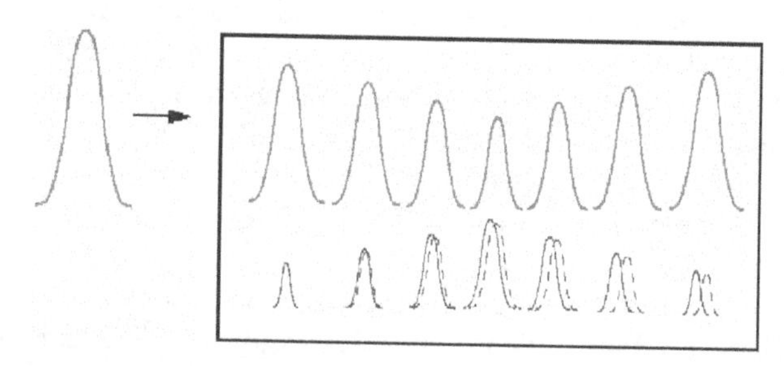

Fig. 20.3 Illustration of FF and SH pulses propagating in a quadratic nonlinear crystal. The *dashed lines* represent the propagation in the absence of GVM

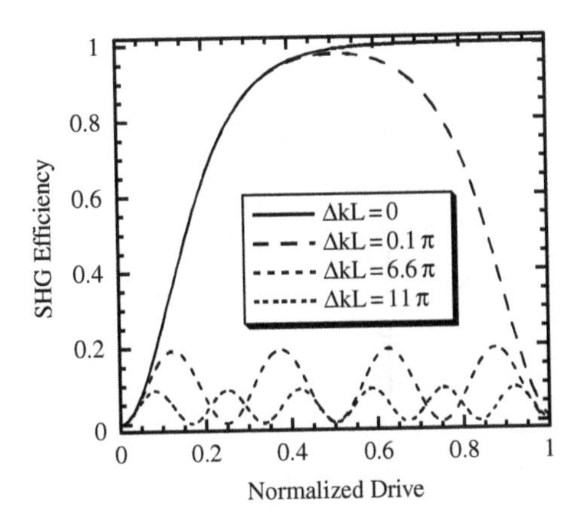

Fig. 20.4 SHG efficiency plotted versus nonlinear drive, which can be taken as proportional to propagation distance, for the indicated values of the phase mismatch. With larger mismatch, the conversion cycles have a smaller period, and less energy resided in the SH field

than L_{GVM}), the resulting phase shift should not suffer excessively from the GVM. This is achieved if the wave-vector mismatch $\Delta k > 4\pi/L_{GVM}$ (Fig. 20.4).

The success of this approach is illustrated in Fig. 20.5. The instantaneous frequency (the time derivative of the nonlinear phase) produced by the cascaded quadratic process is plotted for varying amounts of GVM and small or large phase mismatch. To mimic the electronic Kerr nonlinearity of cubic media, the frequency would follow the peak-valley shape shown as a dashed line. (A negative nonlinear phase shift is assumed.) The bottom two rows of the figure illustrate the improvement in phase-shift quality with increased phase mismatch when GVM is significant. In Fig. 20.5(e), for example, the FF and SH pulses separate by ~4 times the input pulse duration before back-conversion is completed. The distortion of the instantaneous frequency is extreme, and would preclude the production of stable pulses. By increasing the phase mismatch, increased phase-shift quality is obtained at the expense of magnitude [Fig. 20.5(f)]. Thus, the utility of the cascaded quadratic phase shift for a given application ultimately depends on the nonlinearity of available and appropriate SHG crystals.

With a large phase mismatch, the FF and SH pulses effectively overlap despite their short duration. This is referred to as the stationary limit of the cascaded process, and $\Delta k = 4\pi/L_{GVM}$ is called the stationary boundary. Under these conditions, the cascaded quadratic nonlinearity mimics the bound electronic Kerr nonlinearity. In the limit of large phase mismatch, the single-field equation that can be derived for the FF by perturbation methods approaches the ordinary nonlinear Schrodinger equation that governs pulse propagation in cubic nonlinear media [8]. Thus, in that limit the effective nonlinearity is just the instantaneous Kerr nonlinearity. This will be discussed further below.

Phase modulations impact virtually all aspects of the generation and propagation of intense ultrashort pulses. The availability of a simple, adjustable, low-loss way to impress nonlinear phase shifts of either sign on a pulse is therefore a significant new degree of freedom for the generation and propagation of short

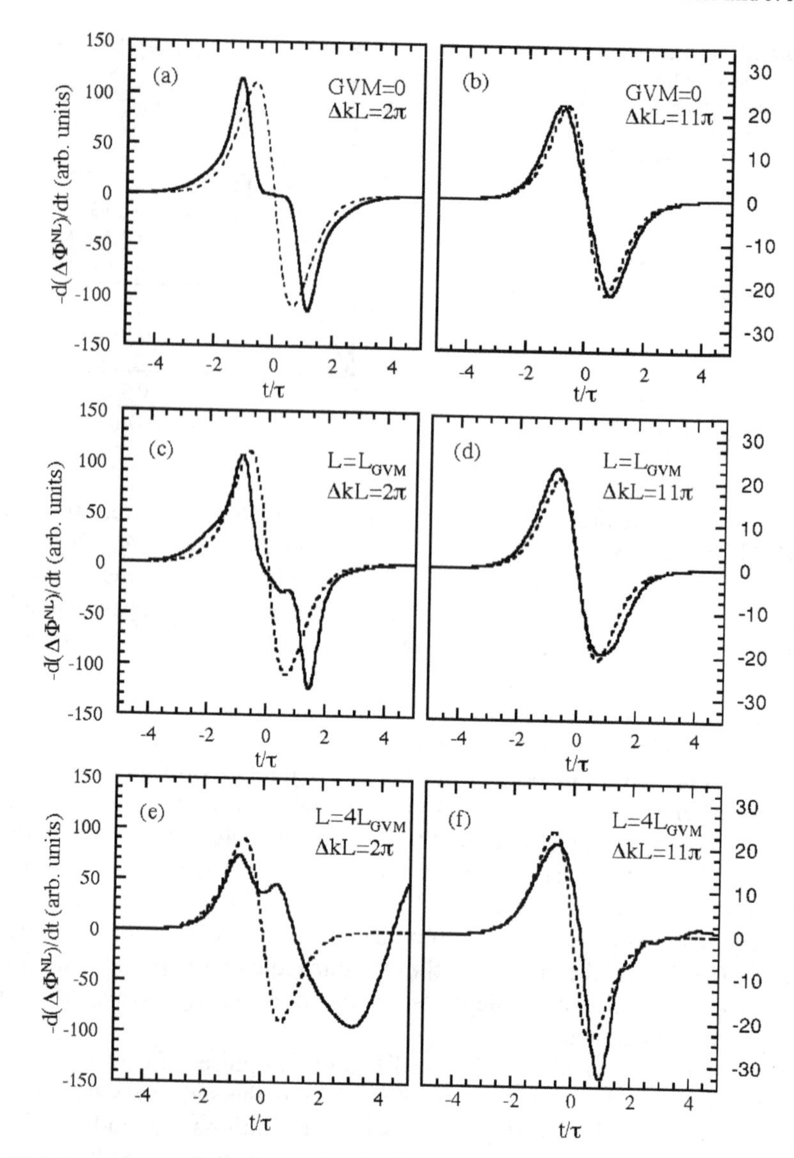

Fig. 20.5 Instantaneous frequency shifts of the fundamental pulse resulting from the cascaded quadratic process with the indicated values of GVM and phase mismatch. The *dashed curve* in each figure represents the ideal Kerr-like shifts. The *top row* shows that the effects of saturation can be reduced by increasing the phase mismatch, as expected. The *bottom two rows* illustrate the tradeoff between magnitude and quality of the nonlinear phase shift in the presence of significant GVM

pulses. In the last few years, initial applications of cascaded quadratic nonlinearities to short-pulse generation have appeared. The remainder of this paper will review some of the highlights of this effort, with an emphasis on the new capabilities that can be obtained.

20.4 Saturable Self-focusing: Space–Time Solitons

Diffraction and GVD cause a pulse to spread in space and time, respectively, and nonlinearities will generally accelerate the decay of a pulse. However, under special conditions self-focusing can counter diffraction, and the phase modulation that underlies the self-focusing can simultaneously balance GVD to produce a soliton in space and time. The production of such spatio-temporal solitons (STS), sometimes called light bullets to convey their particle-like nature (Fig. 20.6), is one of the major goals in the field of nonlinear waves.

Related to STS are nonlinear X-waves, which generalize the "diffraction-free" propagation of Bessel beams to the polychromatic case. X-waves are only weakly localized, but they do manifest a kind of spatio-temporal trapping. Of particular interest, X-waves can emerge spontaneously from unstable propagation of a short pulse in a quadratic medium. The interested reader is referred to the chapter by P. DiTrapani and co-workers in this volume.

The requirements for the generation of stable, localized (i.e., finite-energy) STS are

- a self-focusing nonlinearity
- anomalous GVD
- one or more processes that stabilize the pulse against perturbations.

In contrast to the temporal pulse-shaping described above, it is not sensible to talk about changing the signs of nonlinearity and diffraction; the latter always causes a beam to spread in homogeneous media, so a self-focusing nonlinearity is required to form the soliton. Diffraction and self-focusing can be perfectly balanced, in principle. In cubic nonlinear media, that balance is an unstable equilibrium: if the intensity decreases at all, diffraction makes the

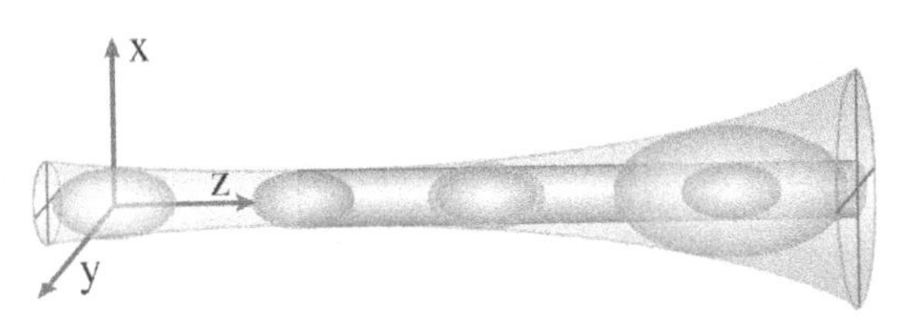

Fig. 20.6 Intuitive picture of a spatio-temporal soliton. The energy of the pulse is represented by the ellipsoid, initially at the origin. In linear propagation the pulse broadens in space and time, to produce the larger ellipsoid. A spatio-temporal soliton propagates stably, without any broadening

beam spread, and if it increases at all, self-focusing reinforces itself and the beam collapses. The collapse can be arrested by a variety of phenomena, such as nonlinear loss, non-paraxiality, or higher-order dispersion, e.g., but the same processes that arrest collapse generally also preclude soliton formation. A convincing demonstration would consist of theoretical modeling that predicts truly stable propagation, along with experimental observation of stable propagation over at least several dispersion/diffraction lengths.

It is well-known that STS in two (one transverse spatial dimension plus time) and three dimensions are unstable against collapse in cubic nonlinear media [9–11], but solutions may be stabilized if the nonlinearity is saturable (i.e., there is a 5th-order contribution to the nonlinearity) [12–14]. As mentioned above, the process of phase-mismatched SHG produces an effective saturable nonlinear index for the FF. In fact, it has been known theoretically for some time that multi-dimensional solitons could be stable in quadratic media [15]. However, it is important to keep in mind that solitons in quadratic media consist of two fields that mutually couple and trap each other, and thus exhibit drastically different properties from the solitons of the nonlinear Schrödinger equation.

An experimental challenge to the generation of STS is the apparent requirement of large anomalous GVD at both fundamental and second-harmonic wavelengths, without loss. An environment with such strong anomalous GVD can be created through a principle that is well-known in the femtosecond-optics community: angular dispersion (e.g., from a diffraction grating) is generally accompanied by anomalous GVD [16]. The effective dispersion occurs because the pulse amplitude front is tilted with respect to its phase fronts.

The generation of solitons in one transverse spatial dimension and time [7,17] was a major step toward the production of light bullets. The saturable self-focusing nonlinearity was obtained from phase-mismatched SHG in lithium iodate or BBO. Tilting of the pulse fronts produced large anomalous GVD and allowed the GVM to be varied from zero to large values. The input beam was taken in the form of a narrow ellipse or stripe (Fig. 20.7). In the direction of the major axis of the ellipse, diffraction was negligible. Numerical simulations predicted that STS would form for moderate values of the GVM, $L_{DS}/L_{GVM} < 3$ (Fig. 20.8). With large GVM, STS form with wave-vector

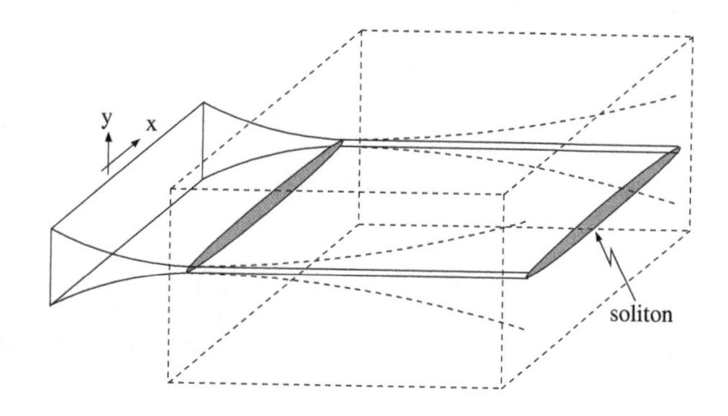

Fig. 20.7 Schematic of experimental arrangement used to generate STS

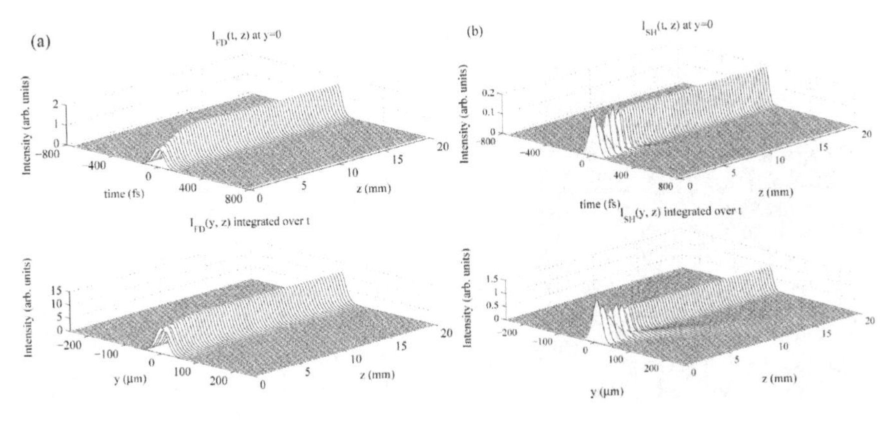

Fig. 20.8 Numerical simulation of the formation of STS. Temporal (*upper panels*) and spatial (*lower*) profiles of the FF (*left*) and SH (*right*) are shown

mismatch larger than a threshold value determined by the stationary limit (see Section 20.3), in agreement with the acceptance range determined by analytic theory [18].

Experimental results obtained with a wide range of input intensities and wave-vector mismatches [17] agree well with the numerical simulations. Clear evidence of STS is summarized in Fig. 20.9. The insets show the pulse measurements and beam profiles recorded with the optimal mismatch and intensity for STS. In this case the propagation covered five characteristic lengths. The main graphs show the constant pulse duration and beam size, contrasted with those

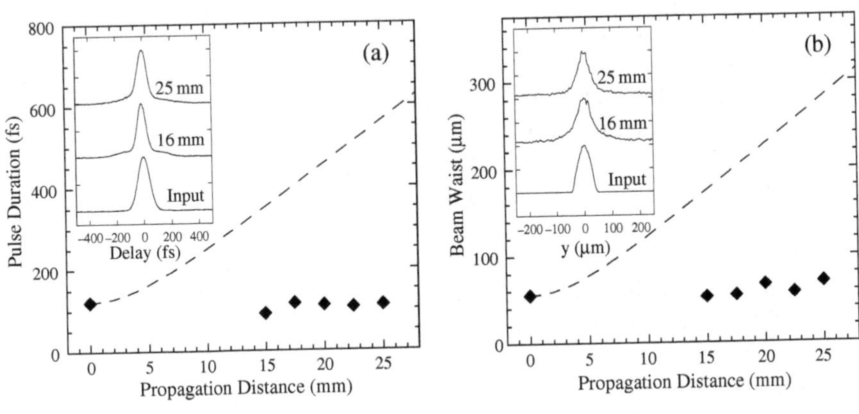

Fig. 20.9 Measured pulse duration (*left*) and beam waist (*right*) as a function of propagation distance for spatio-temporal solitons in barium metaborate. The pulse duration and beam waist were obtained from experimental traces as shown in the insets. The *dashed lines* in each panel show the variation observed at low intensity

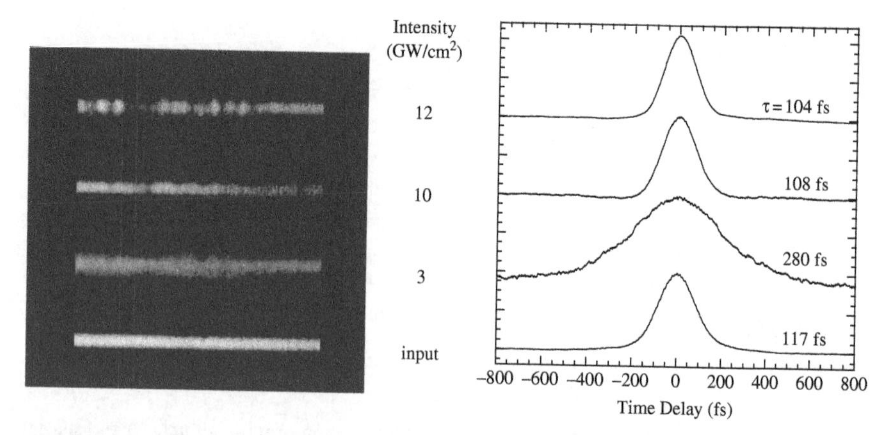

Fig. 20.10 Experimental observations of STS and spatial modulation instability. Spatial (*left*) and temporal (*right*) profiles were recorded with the indicated intensities. At 12 GW/cm^2 (and above), the beam breaks into filaments. The temporal profile of a single filament is shown

of linear propagation. Observations of the output beam and pulse for fixed propagation length and wave-vector mismatch and intensity varying up to 10 GW/cm^2 are presented in Fig. 20.10 (lower three traces on each side).

An important point is that the quadratic soliton has components at both FF and SH frequencies, but only the FF is launched. As is clear in the numerical solution (Fig. 20.8), pulses launched with suitable combinations of intensity and phase mismatch quickly evolved to stable spatial and temporal profiles. Approximately 10% of initial energy is radiated away as the pulse adjusts to its final form.

The ranges of intensity and phase-mismatch for which STS form are small. If the intensity is increased from the threshold value to produce STS by ~20%, the stripe spontaneously breaks into noise-induced filaments along its major axis (Fig. 20.10) [19]. Further increases in the initial intensity cause the stripe to break into denser filaments, which is a signature of a modulation instability (MI), or small-scale self-focusing. The MI would eventually prevent formation of STS if the propagation could be monitored over longer distances. On the other hand, MI could be a way to achieve confinement in the remaining transverse dimension; numerical simulations show that the resulting filaments should in fact be light bullets. Unfortunately, the angular dispersion used to control GVD in the experiments of Liu et al. caused the bullets to disperse soon after their formation.

Enormous progress has been made in the study of optical solitons in one and two dimensions. The fact that STS in all three space dimensions and time have not been observed experimentally leaves a major piece of the puzzle unfilled. Recent theoretical results indicate that experimental conditions for STS formation may not be as stringent as once thought. It is very difficult to find materials with anomalous GVD and acceptably low loss at both FF and SH frequencies. Beckwitt et al. showed theoretically that temporal solitons can form with normal GVD at the SH frequency [20], and Towers et al. found a well-defined region of parameters in which stable STS exist with normal GVD at the SH frequency [21]. These findings may relax the experimental requirements to some degree. Given the difficulty of creating an environment that will support light bullets launched directly, the possibility that bullets might emerge spontaneously, through an instability, remains attractive.

20.5 Self-defocusing Nonlinearities: Applications to Ultrashort Pulse Generation

As mentioned in Sections 20.1–20.3 above, a way to impress controllable self-defocusing nonlinear phase shifts on ultrashort pulses offers several advantages. Soliton-like pulse shaping can be done with normal GVD, which implies that ordinary transparent materials can replace low-loss devices that produce anomalous GVD, such as pairs of prisms or diffraction gratings. This can result in substantial simplification of instruments. The nonlinear phase shift that can be impressed on a pulse via propagation through a homogeneous self-focusing medium (i.e., without any waveguiding) is limited to $\sim \pi/2$, at which point beam distortion becomes significant. Self-focusing collapse is obviously not a concern with a self-defocusing nonlinearity. In addition to avoiding whole-beam self-focusing, small-scale self-focusing does not occur; there is no modulation-instability gain when $n_2 < 0$ and diffraction is normal. Thus, self-defocusing nonlinearities open a route to impressing nonlinear phase shifts on arbitrarily intense pulses. This can be expected to find application in high-energy pulse compression.

In applications of negative nonlinear phase shifts, it is important to keep in mind that the net nonlinear phase shift will be the sum of nonlinear phases from the quadratic and cubic nonlinearities. As mentioned earlier, the nonlinear index of transparent media is positive, and one can think of the cubic nonlinearity as a positive bias added to the nonlinear phase shift that makes it more difficult to achieve a net negative nonlinearity (Fig. 20.2). With femtosecond pulses and in the stationary limit, the cubic nonlinear phase shift can be an appreciable fraction (~5–30%) of the quadratic nonlinear phase shift.

Self-defocusing nonlinearities have been exploited in modelocking of a femtosecond laser [22], pulse compression [23], nonlinear polarization rotation [24], and compensation of self-focusing [25]. A couple highlights of these applications will be reviewed to illustrate the new capabilities.

20.5.1 Compensation for Self-focusing

Self-focusing causes numerous problems in high-energy lasers, ranging from pulse or beam distortion to catastrophic damage of optical components. Many high-power lasers and amplifiers are largely designed around the limitations imposed by self-focusing, so a means of compensating the effects of self-focusing is highly desirable. When the ability to generate negative nonlinear phase shifts was demonstrated, perhaps the most obvious application of the cascaded quadratic process was the compensation of self-focusing; nonlinear phase shifts accumulated in propagation could simply be reversed, in principle.

Beckwitt et al. showed that self-focusing in cubic nonlinear media can be conveniently compensated with the cascaded quadratic nonlinearity [25]. Millijoule-energy pulses from a Ti:sapphire regenerative amplifier were passed through 6 cm of fused silica to generate nonlinear phase shifts as large as 1.5π. The beam profile before passage through the fused silica is shown in the top left panel of Fig. 20.11. When the beam emerges from the fused silica, whole-beam self-focusing is evident as a narrowing of the beam, while small-scale self-focusing is evident as increased modulation depth between the noise peaks and background. A 2.5-cm-long BBO crystal was used to compensate for the self-focusing. As a control experiment, the BBO was oriented so that there were no quadratic nonlinear effects. Here, the pulse experienced additional self-focusing owing to the Kerr nonlinearity of the BBO (bottom left panel). When the BBO crystal was oriented for optimal compensation, the beam was restored to nearly its original quality (middle right panel). Increasing the phase mismatch resulted in undercompensation of the self-focusing (lower right panel), and decreasing the phase mismatch produced overcompensation: whole-beam self-defocusing was observed along with filamentation from small-scale self-focusing (top right panel). Beckwitt et al. actually demonstrated separate optimizations of whole-beam and small-scale self-focusing.

In a separate experiment, Beckwitt et al. showed that cascaded quadratic phase shifts can compensate for the temporal/spectral consequences of SPM in a short-pulse amplifier [25]. Picosecond pulses with millijoule energies are desired for applications that require narrower spectra than accompany femtosecond pulses, such as time-resolved vibrational spectroscopy. The spectral broadening that accompanies accumulation of a nonlinear phase shift is therefore highly undesirable in these systems, but it is also difficult to avoid.

Transform-limited pulses 5 ps in duration and with 0.5 mJ energy were passed through a 2-cm Ti:sapphire crystal to simulate the conditions of an amplifier. The pulses accumulated a nonlinear phase shift of π, and as a result the spectrum broadened to 3.5 times its original transform-limited width. Subsequent passage of the pulse through a 1.7-cm BBO crystal adjusted to compensate the Kerr nonlinearity of the Ti:sapphire rod restored the spectrum to within 1.5 times the transform limit (Fig. 20.12).

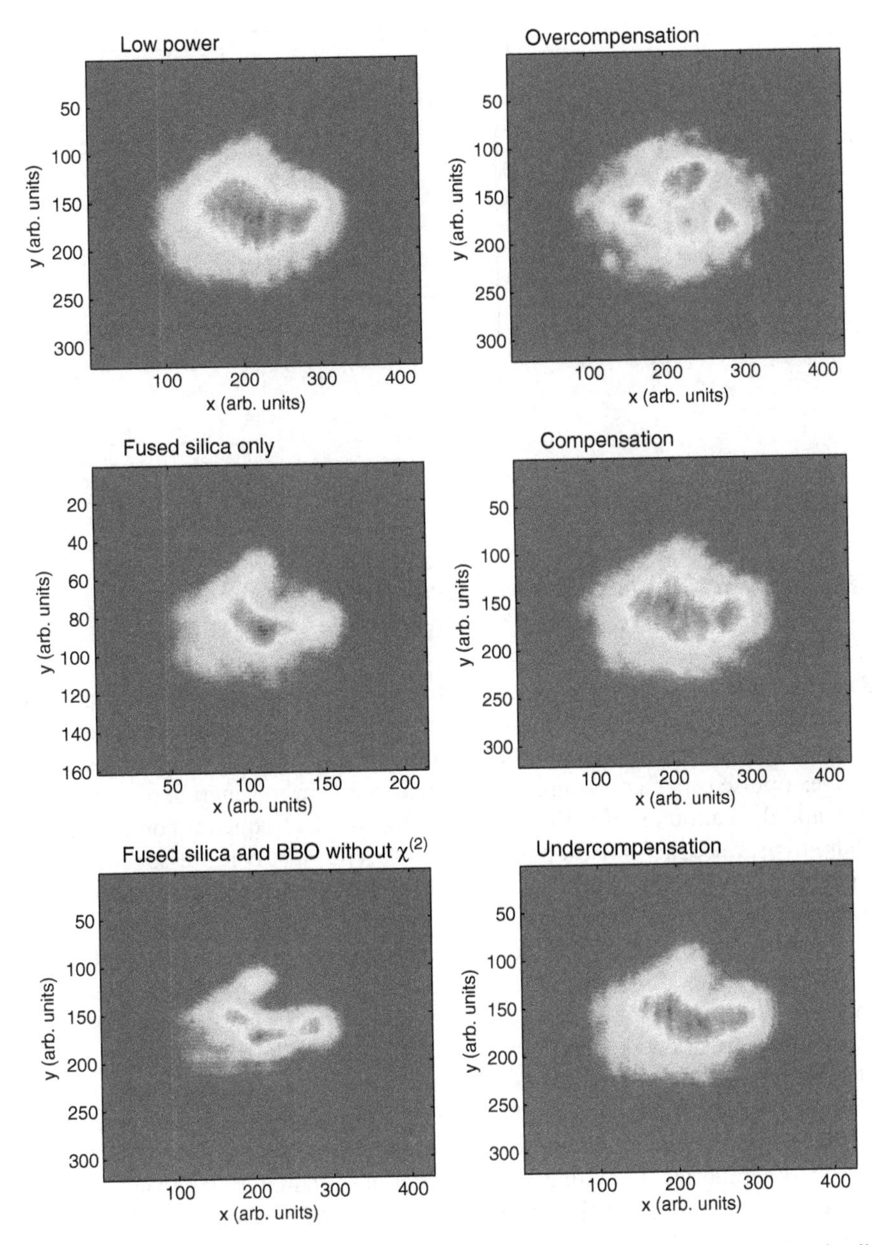

Fig. 20.11 Beam profiles that demonstrate compensation for self-focusing. See text for details

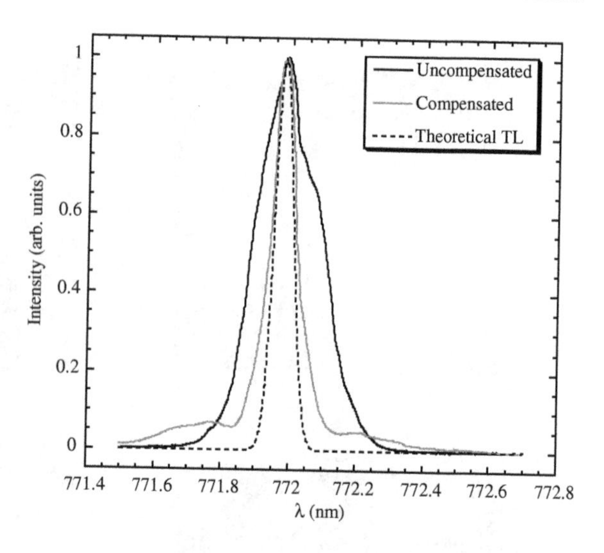

Fig. 20.12 Spectral compression by impression of negative nonlinear phase shift on a millijoule-energy picosecond pulse

20.5.2 Pulse Compression with Self-defocusing Nonlinear Phase Shifts

Scientists often desire optical pulses that are shorter than can be produced with a laser or amplifier, so pulse compression techniques have been widely investigated. The most effective techniques use the soliton-like pulse shaping described above: a pulse is passed through a nonlinear medium to generate new frequencies, and then anomalous GVD is used to align all the frequency components in phase to produce the shortest pulse. The prototype pulse compressor consists of a length of optical fiber followed by a pair of prisms or gratings. Compressors based on single-mode fiber are limited to nanojoule pulse energies by excessive nonlinearity and eventually damage to the fiber. Propagation of a pulse through a hollow capillary filled with a high-pressure noble gas produces substantial spectral broadening of energetic pulses [26], which can be compressed in subsequent prism or grating pairs. This approach is limited to pulse energies of 1 mJ by ionization of the gas as well as self-focusing. Recent demonstrations of compression in filaments [27,28], where the physical waveguiding is replaced by the self-guiding of filamentation in gas, suggest that it may supplant the use of hollow capillaries. The cubic nonlinearities of solid materials can also be used for compression [29], but this approach is hampered severely by (whole-beam and small-scale) self-focusing, which distorts the beam and precludes effective compression.

As a result of the limitations described above, an effective means of compressing pulses with energies much above 1 mJ has not been reported. Although existing compressors are effective, they are all rather inefficient: the compressed-pulse energy is always less than 50% of the input energy, and 20–30% is typical. For the many applications that benefit from the maximum peak power, this inefficiency essentially cuts into the compression ratio. Finally, pulse compressors are fairly large devices that are sensitive to optical alignment.

Pulse compression based on self-defocusing nonlinearities and normal GVD can address these limitations [23]. The use of a self-defocusing nonlinearity eliminates the problems that arise from Kerr self-focusing, and the compressor consists of only two components: an SHG crystal and a piece of glass, e.g., (Fig. 20.13).

Pulses from a Ti:sapphire regenerative amplifier were employed in the first experimental demonstration of the cascaded quadratic compressor. The input pulses were 120 fs in duration. A 1.7-cm BBO crystal was used for the first stage of the compressor, oriented so that $\Delta kL = 200\pi/\mathrm{mm}$ to satisfy the criterion for high-quality phase shifts. With incident intensity $\sim$50 GW/cm^2, the spectrum broadened and developed the doubly peaked structure that is expected theoretically for a nonlinear phase shift of approximately π. The phase-modulated pulses then traversed $\sim$2 cm of calcite or sapphire, where they were compressed to $\sim$30 fs (Fig. 20.11). The energy of the compressed pulse was 85% of that of the input pulse.

Recent work has shown that the compressor can be simplified even further, although some performance may be sacrificed. Most SHG crystals have normal GVD at the wavelengths at which they are used. Thus, the SHG crystal itself, rather than a separate piece of transparent material, can provide the required GVD [30]. In this case the compression would be analogous to soliton-type compression that occurs in fibers.

Pulse compression based on self-defocusing nonlinearities is an excellent example of the new capabilities that are possible with the cascaded quadratic process. The cascaded quadratic compressor is extremely simple and compact, and can be scaled to higher energies simply by increasing the aperture of the crystal. The only losses are $\sim$5% conversion to SH and residual reflections, so the device can easily have 90% energy efficiency. We expect that this approach to pulse compression will be attractive for many future applications. In particular, the cascaded quadratic compressor may offer a unique capability for high-field science.

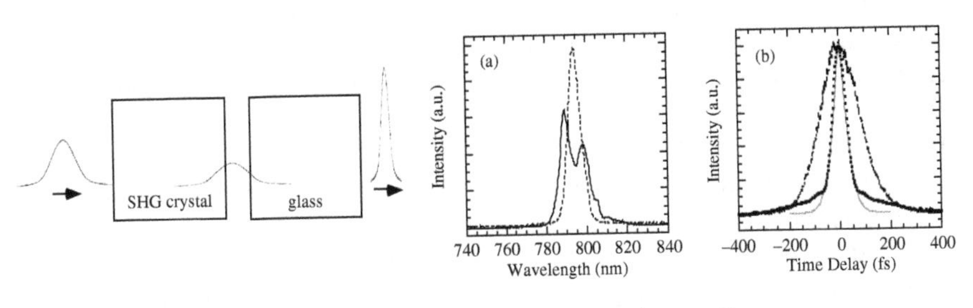

Fig. 20.13 Schematic of quadratic compressor and compression results

20.6 Few-Cycle Pulses and the Non-stationary Regime of the Cascaded Quadratic Nonlinearity

The generation and propagation of pulses consisting of only a few optical cycles ($< \sim 10$ fs) has become an active frontier of nonlinear optics. While much effort has been made to understand few-cycle pulse propagation in cubic media, the quadratic case is largely unexplored. The temporal walk-off inherent in parametric interactions has been the largest impediment to progress in this area. Nonstationary effects become harder to avoid as pulses reach few-cycle durations. These effects disrupt the simple nonlinear-Schrödinger equation framework used in the stationary cascaded quadratic applications described above. The development of an understanding of nonstationary-type propagation has been necessary to tap into the field of quadratic few-cycle propagation, where the ability to produce either self-focusing or self-defocusing nonlinearities of controllable magnitude create a rich field deserving of study.

The definition of the nonstationary regime given above can be expanded as

$$|\Delta k|_{\text{nonstat}} < \frac{4\pi}{L_{GVM}} = \frac{4\pi(n_{g,1} - n_{g,2})}{c\tau_0}. \tag{20.4}$$

As the pulse duration approaches a few cycles, the smaller values of $|\Delta k|$ (where the effective non-linear index is usefully large) correspond to non-stationary propagation. Moreover, due to the biasing effect of $\chi^{(3)}$, the entire self-defocusing region can be non-stationary (see Fig. 20.2). Even $(n_{g,1} - n_{g,2})/c$ as small as 10 fs/mm is significant for several-cycle pulses at near-infrared and shorter wavelengths. The appearance of non-stationary propagation is well known: temporal amplitude and phase profiles become asymmetric (see Fig. 20.5) and the spectral power density is distorted, often resulting in wave breaking. Such distortions become prohibitive when unperturbed nonlinear Schrödinger-type propagation is desired.

20.6.1 Controllable Raman-Like Nonlinearity

Recent progress has made non-stationary propagation more understandable and has allowed for the discovery of practical few-cycle self-focusing and self-defocusing applications. An approximate single-field equation for the FF provides a useful framework, where perturbation methods can be used to uncover the first-order effects of FF/SH coupling on the FF [31,32]. When Δk is large enough that only up to a few percent of the total energy is converted to the SH before back-conversion occurs (Fig. 20.4), a_2 of Eqs. (20.2) and (20.3) can be expanded in powers of $\beta = \Delta k L_{GVM}$. Then using the method of repeated substitution and keeping only the two largest effects due to three-wave mixing, the equation for the FF is

$$i\frac{\partial a_1}{\partial \xi} - \frac{\alpha_1}{2}\frac{\partial^2 a_1}{\partial s^2} - \frac{1}{\beta}|a_1|^2 a_1 - 2i\frac{1}{\beta^2}|a_1|^2\frac{\partial a_1}{\partial s} = 0. \tag{20.5}$$

The first term proportional to β describes the controllable nonlinearity of cubic power analogous to the instantaneous Kerr effect in $\chi^{(3)}$. The next term is the first-order effect of temporal walk-off, and has been recognized as a non-instantaneous nonlinearity analogous to the nuclear or Raman contribution to nonlinear refraction in $\chi^{(3)}$ [32]. It is also comparable to self-steepening in $\chi^{(3)}$. Figure 20.14 illustrates the connection between these terms, and shows that both analogies are close but not exact. Because the coefficient of the walk-off term in Eq. (20.5) is imaginary, like self-steepening, it will have a direct effect on

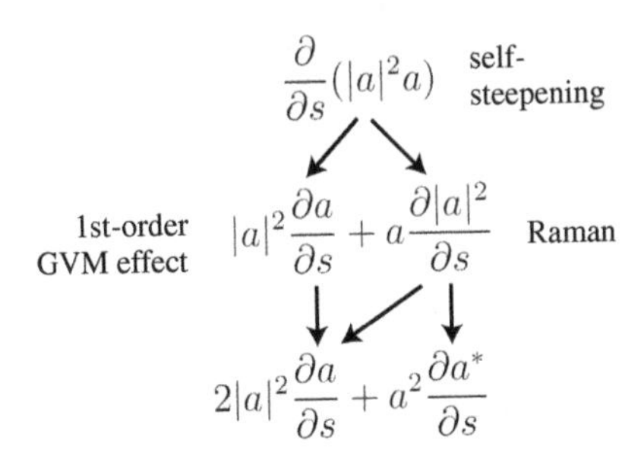

Fig. 20.14 The cubic self-steepening term contains the Raman term and first-order GVM term as its component derivatives. Each successive line is an implementation of the product rule, with cubic self-steepening at the top. Direction of arrows indicate parent derivative to component derivatives. Coefficients are neglected

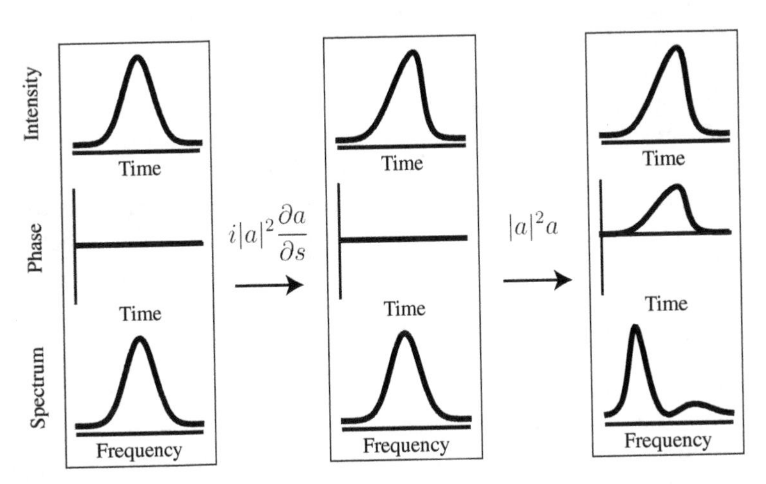

Fig. 20.15 The first-order GVM effect tilts the temporal intensity profile, but does not alter the phase (first to second pane), like self-steepening. SPM converts the amplitude tilt to an asymmetric phase, causing Raman-like spectral shifting (second to third pane)

the amplitude of the pulse: acting as an intensity-dependent group velocity, energy is moved from the peak of the pulse to one of its edges, which makes the pulse asymmetric. However, the instantaneous non linear term converts amplitude asymmetry to phase asymmetry, so the net effect is a frequency shift analogous to Raman scattering (Fig. 20.15).

Like the effective cubic nonlinear term, the Raman-like term has a coefficient controllable through the wave-vector mismatch, because it is proportional to $1/\beta^2 \sim 1/\Delta k^2$. Thus, frequency shifts of controllable magnitude and sign can be impressed on short pulses. These effective Stokes and anti-Stokes processes extend the analogy between cascaded quadratic nonlinearities and true cubic nonlinearities, and provide another significant degree of freedom in ultrashort pulse propagation.

The comparison of cascaded quadratic propagation to nonlinear Schrödinger-type propagation perturbed by Raman scattering provides a powerful tool for predicting and understanding pulse evolution in the nonstationary regime. Equation (20.5) can be rewritten as

$$i\frac{\partial u_1}{\partial \xi} - \frac{sgn(u_1)}{2}\frac{\partial^2 u_1}{\partial \tau^2} - sgn(n_2)N^2|u_1|^2 u_1 = -sgn(n_2)N^2\tau_R|u_1|^2\frac{\partial u_1}{\partial \tau}, \qquad (20.6)$$

where $N = (L_{GVD}/L_{NL})^{1/2}$ and n_2 is the effective nonlinear index of refraction due to $\chi^{(2)}$. We have used a slightly different normalization than elsewhere in the text, such that $\tau = t/\tau_0 - z/\tau_0 v_{g,1}$, $\xi = z/L_{GVD}$, and $u_1 = \Gamma_1(2k_1/k_2)^{1/2}L_{GVD}A_1$. We can define an effective Raman response time, $\tau_R = 2i(n_{g,1} - n_{g,2})/c\Delta k\tau_0$. This parameter together with N^2 determines the coefficient of the Raman-like perturbation.

When α_1 and n_2 have the same sign, Eq. (20.6) with $\tau_R = 0$ has soliton solutions $u_1(\tau) = Nsech(\tau)$, where N is the characteristic soliton order. Of course, this equation is only an approximation of a two-field system, and the solitons will have the usual characteristics that differentiate them from cubic ones, such as having a continuous distribution of possible N "eigenvalues." Still, the FF soliton will experience effects analogous to the cubic case with a Raman pertubation, such as self-frequency shifts and splitting. In light of Eq. (20.6), it is clear that higher-order solitons feel the effective Raman term more strongly. For a given τ_R, therefore, N becomes the critical parameter that determines whether the Raman perturbation plays a significant role in the pulse evolution.

This is evident in the case of soliton-effect compression (Section 20.5.2), where the propagation effects of the Raman-like term act as unwanted distortions. In soliton compression, one uses the initial spectral broadening and pulse narrowing that occurs during the first N-soliton period (for $N > 1$) to compress a pulse, and the compression ratio is proportional to N. For N greater than some optimum, N_{opt}, spectral broadening will cause the effective Raman gain to become too large before propagation ends. As a result, increases in peak power due to compression become outweighed by loss of energy from the soliton spike, due to Raman-like frequency-shifting effects (Fig. 20.16). Recent work has

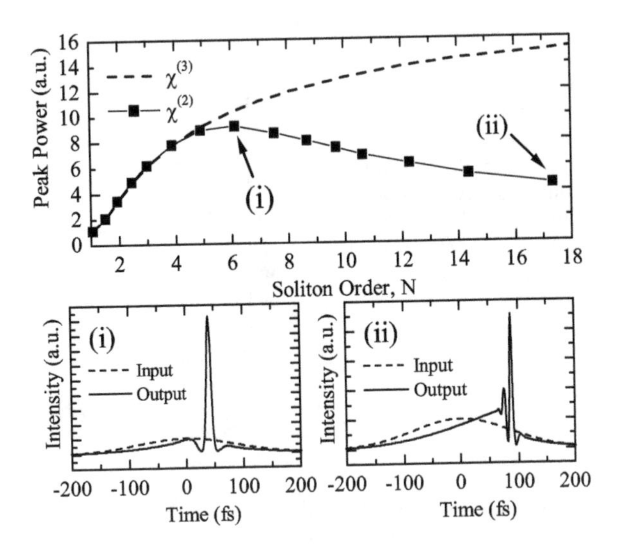

Fig. 20.16 Peak power versus N for soliton compression in $\chi^{(2)}$ and $\chi^{(3)}$. The $\chi^{(3)}$ case is unperturbed, whereas the $\chi^{(2)}$ case feels the perturbation due to GVM. The temporal profiles of compressed pulses at points (i) and (ii) reveal the effect of the Raman-like perturbation in $\chi^{(2)}$. Final durations are ~2 cycles

shown that compression to few-cycle durations is achievable with existing bulk quadratic crystals when $N < N_{\text{opt}}$, demonstrating compression to three cycles (12 fs) for pulses initially ~100 fs at 1250-nm wavelength, and compressed-pulse bandwidths as large as half an octave [33]. Efficient compression to under two cycles (~5 fs) is predicted. Moreover, because the sign of the nonlinear phase shift is negative (or self-defocusing), compression ratios as large as 100× are possible in this scheme without a wave-guiding structure. This compressor is extremely simple, consisting of only a single frequency-doubling crystal, and is a promising device for the generation of high-energy near-single-cycle pulses.

It is also possible to exploit the Raman-like effect, which can be substantial for large values of the effective response time τ_R. Frequency shifting has been demonstrated where the sign and magnitude of the Raman gain is controllable through Δk [32]. Initial experiments demonstrated frequency shifts on the order of $\Delta\omega/\omega_0 = 0.02$ (about one full-width half-maximum of the spectral profile, making the effect useful for switching applications). These experiments were limited by an effective saturation: as the center frequency shifts, Δk also changes, eventually causing the effective Raman response $\tau_R \propto 1/\Delta k$ to vanish. To counter the saturation, Beckwitt et al. proposed the use of an aperiodically poled quasi-phase matched structure [34]. The poling period is tailored to keep Δk constant as ω_0 shifts. In this structure, the analog of a Raman soliton may propagate. For example, when $N = 1$ the frequency of the pulse continuously shifts while the temporal profile remains unchanged. Figure 20.17 shows a predicted frequency shift of $\Delta\omega/\omega_0 = 0.15$, or 200-nm, for a 100-fs pulse at $\lambda_0 = 1550$ nm using aperiodically poled lithium niobate. When $N > 1$, the cascaded quadratic analog to Raman-soliton compression occurs, which has been both predicted and realized [32,35].

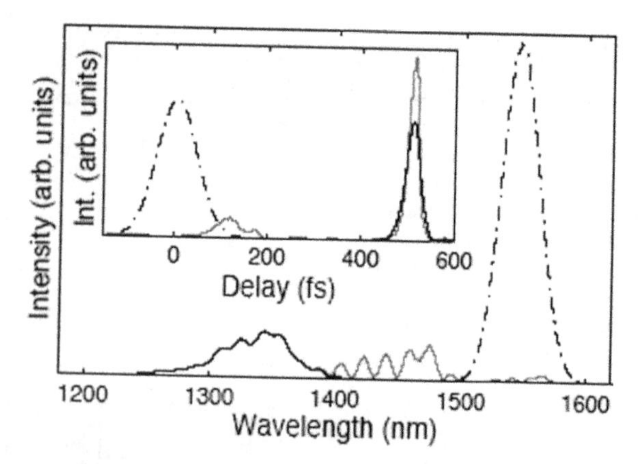

Fig. 20.17 Numerical calculations of pulse propagation through a chirped periodically-poled lithium niobate crystal 4.6 cm long. Spectra of the input pulse (*dash-dotted* and shifted pulse before (*light line*) and after (*heavy line*) filtering. Inset: temporal profiles

Finally, as mentioned above, the first-order GVM term in Eq. (20.5) has the form of an intensity-dependent group velocity. Marangoni et al. have directly measured the group delay occuring in nonstationary cascaded quadratic propagation, the conjugate effect to the frequency-shifting process [36]. The group delay is likewise controllable via Δk.

20.6.2 Beyond the Slowly Varying Envelope Approximation

As seen above, the use of controllable self-focusing and self-defocusing nonlinearities can be extended to the nonstationary regime. However, as pulse bandwidths approach the carrier frequency and changes in the temporal envelope become comparable to the carrier-wave oscillations, one must question the slowly varying envelope approximation (SVEA) employed to model the propagation.

To address this issue, Moses et al. [37] studied cascaded quadratic propagation using the slowly evolving wave approximation (SEWA) [38]. In this model, the temporal portion of the SVEA, $|\partial_\tau A_j| \ll \omega_j |A_j|$, which directly limits pulse duration and bandwidth, is replaced by the condition that the group and phase velocities are approximately equal, $v_g \simeq v_p$. The SEWA allows for accurate modeling down to shorter pulse durations but requires the use of more complicated propagation equations: the coupled SVEA-model equations for three-wave mixing in quadratic media, [Eqs. (20.2) and (20.3)] are replaced by

$$\begin{aligned} i\frac{\partial a_1}{\partial \xi} + \frac{\rho_1}{2}\left(1 + \frac{i}{\omega_0 \tau_0}\frac{\partial}{\partial s}\right)^{-1} \nabla_\perp^2 a_1 - \frac{\alpha_1}{2}\frac{\partial^2 a_1}{\partial s^2} \\ + \left(1 + \frac{i}{\omega_0 \tau_0}\frac{\partial}{\partial s}\right) a_1^* a_2 e^{-i\beta\xi} = 0, \end{aligned} \tag{20.7}$$

$$
\begin{aligned}
&i\frac{\partial a_2}{\partial \xi} + \frac{\rho_2}{2}\left(1 + \frac{i}{2\omega_0\tau_0}\frac{\partial}{\partial s}\right)^{-1}\left[\nabla_\perp^2 a_2 + \frac{\nu}{\rho_2}\frac{\partial^2 a_2}{\partial s^2}\right] \\
&- i\frac{\partial a_2}{\partial s} - \frac{\alpha_2}{2}\frac{\partial^2 a_2}{\partial s^2} + \left(1 + \frac{i}{2\omega_0\tau_0}\frac{\partial}{\partial s}\right)a_1^2 e^{i\beta\xi} = 0.
\end{aligned} \tag{20.8}
$$

These contain several new terms: the $\partial/\partial s$ prefactors to nonlinear terms, which are the quadratic analog to cubic self-steepening (SS); the $(1 + ()\partial/\partial s)^{-1}$ prefactors to diffraction ($\nabla_\perp^2$) terms, known as space–time focusing; and a new correction term to SH linear dispersion resulting from GVM, $(\nu/2)(1 + i\partial/\partial s/2\omega_0\tau_0)^{-1}\partial^2 a_2/\partial s^2$, where $\nu = L_{GVM}(n_{g,1} - n_{g,2})^2/\tau_0^2 k_2 c^2$.

Through the same perturbation methods used to derive Eq. (20.5), one may derive an approximate 1D single-field equation for the FF within the SEWA,

$$
\begin{aligned}
&i\frac{\partial a_1}{\partial \xi} - \frac{\alpha_1}{2}\frac{\partial^2 a_1}{\partial s^2} + \frac{1}{\beta}|a_1|^2 a_1 - 2i\frac{1}{\beta^2}|a_1|^2\frac{\partial a_1}{\partial s} \\
&+ i\frac{1}{\beta}\frac{1}{\omega_0\tau_0}\left(3|a_1|^2\frac{\partial a_1}{\partial s} + a_1^2\frac{\partial a_1^*}{\partial s}\right) = 0,
\end{aligned} \tag{20.9}
$$

where only the lowest-order correction term resulting from the SEWA has been kept. This new term also has a form similar to cubic SS, with only a slight difference in the relative coefficients between its component derivatives [in $\chi^{(3)}$, SS is $\partial/\partial s(|a|^2 a) = 2|a|^2\partial a/\partial s + a^2\partial a^*/\partial s$]. This term, like the Raman-like term due to temporal walk-off (Section 20.6.1), has a coefficient that is controllable through $\beta \sim \Delta k$. In this case, because the dependency is in $1/\Delta k$, the term is controllable in both magnitude and sign.

The repercussions of this "controllable SS" term are several-fold. First, there is a SS effect even when temporal walk-off is negligible, $(n_{g,1} - n_{g,2})/c = 0$. Second, the ability to control the sign of the term allows for shock-front formation on either the front or rear edge of a pulse (see Fig. 20.18). Third, because the two SS-like terms in Eq. (20.9) share a mathematical form, one may be used to enhance or cancel the other. This allows either the enhancement of the Raman-like effect used in applications such as frequency shifting or the removal of the largest perturbation to NLSE-type propagation in $\chi^{(2)}$ for applications such as soliton compression.

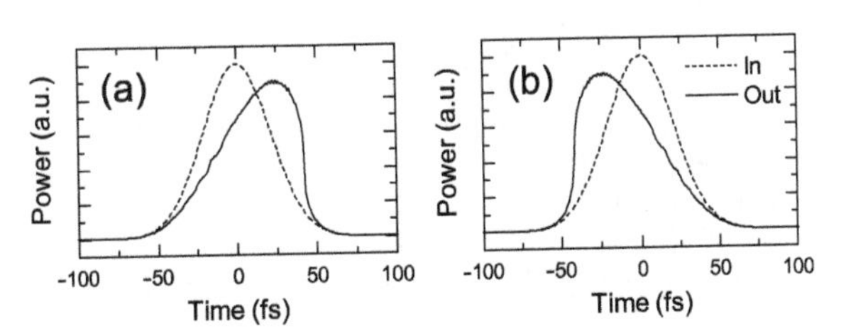

Fig. 20.18 When GVD is negligible, self-steepening causes shock formation that can be on either the leading or trailing pulse edge, depending on the sign of Δk

This last point is rather remarkable. The two effects, one due to temporal walk-off and the other due to nonlinearly dependent group velocity terms in the coupled propagation equations, cause decay or collapse of an ultrashort pulse independently. However, together they can cancel in a manner comparable to the compensation of linear and nonlinear terms in soliton propagation. Furthermore, the compensation works for arbitrary pulse shape. Thus, the first-order correction to propagation beyond the SVEA in quadratic media (within the SEWA model) may under the right conditions act to stabilize the propagation of few-cycle pulses. This finding increases the plausibility of the cascaded nonlinearity serving an important role in few-cycle pulse propagation and pulse-shaping applications.

20.7 Conclusion

Control of the sign and magnitude of nonlinear phase shifts via cascaded quadratic processes enables a variety of new ultrafast phenomena. The first femtosecond applications of cascaded quadratic nonlinear phase shifts demonstrate advantageous properties for a range of sources, from modelocked fiber lasers to terawatt systems. In addition, the cascaded quadratic nonlinearity supports the formation of optical spatio-temporal solitons. Current work suggests that the cascaded quadratic nonlinearity will underlie a variety of novel pulse propagation effects in the future, particularly in the area of few-cycle pulses, along with more applications in ultrafast science.

Acknowledgments The work described here was supported by the National Science Foundation (grants PHY-0099564 and ECS-0217958) and the National Institutes of Health. FW thanks all the students whose work constitutes the advances described here.

References

1. J.A. Armstrong, N. Bloembergen, J. Ducuing et al.: Interactions between light waves in a nonlinear dielectric, *Phys. Rev.* **127**, 1918 (1962).
2. J.M.R. Thomas, J.P.E. Taran: Pulse distortions in mismatched second harmonic generation, *Opt. Commun.* **4**, 329 (1972).
3. N.R. Belashenkov, S.V. Gagarskii, M.V. Inochkin: On the nonlinear light refraction under the 2nd-harmonic generation, *Opt. Spectrosc.* **66**, 806 (1989).
4. H.J. Bakker, P.C.M. Planken, L. Kuipers et al.: Phase modulation in 2nd-order nonlinear-optical processes, *Phy. Rev. A* **42**, 4085 (1990).
5. R. DeSalvo, D.J. Hagan, M. Shiek-Bahae et al.: Self-focusing and self-defocusing by cascaded 2nd-order effects in KTP, *Opt. Lett.* **17**, 28 (1992).
6. G. Stegeman, R. Schiek, L. Torner et al.: Cascading: a promising approach to nonlinear optical phenomena, *Novel Optical Materials and Applications*, John Wiley, New York, pp. 49–76 (1997).
7. X. Liu, L. Qian, F. W. Wise: Generation of optical spatiotemporal solitons, *Phys Rev. Lett.* **82**, 4631 (1999).

8. C.R. Menyuk, R. Schiek, L. Torner: Solitary waves due to $\chi^{(2)}/\chi^{(2)}$ cascading, *J. Opt. Soc. Am. B* **11**, 2434 (1994).
9. E.A. Kuznetsov, A.M. Rubenchik, V.E. Zakharov: Soliton stability in plasmas and hydrodynamics, *Phys. Rep.* **142**, 105 (1986).
10. J.J. Rasmussen, K. Rypdal: Blow-up in nonlinear Schroedinger equations. 1. A general review, *Phys. Scr.* **33**, 481 (1986).
11. Y. Silberberg: Collapse of optical pulses, *Opt. Lett.* **15**, 1282 (1990).
12. J.H. Marburger, E. Dawes: Dynamical formation of a small-scale filament, *Phys. Rev. Lett.* **21**, 556 (1968).
13. R.H. Enns, S.S. Rangnekar, A.E. Kaplan: Bistable-soliton pulse-propagation: stability aspects, *Phys. Rev. A* **35**, 466 (1987).
14. D. Edmundson, R.H. Enns: Robust bistable light bullets, *Opt. Lett.* **17**, 586 (1992).
15. Y.N. Karamzin, A.P. Sukhorukov: Mutual focusing of high-power light beams in media with quadratic nonlinearity, *Soviet Phys. JETP* **41**, 414 (1975).
16. O.E. Martinez: Magnified expansion and compression of subpicosecond pulses from a frequency-doubled Nd:YLF laser, *IEEE J. Quantum Electron.* **25**, 2464 (1989).
17. X. Liu, K. Beckwitt, F.W. Wise: Two-dimensional optical spatiotemporal solitons in quadratic media, *Phys. Rev. E* **62**, 1328 (2000).
18. S. Carrasco, J.P. Torres, L. Torner et al.: Walk-off acceptance for quadratic soliton generation, *Opt. Commun.* **191**, 363–370 (2001).
19. X. Liu, K. Beckwitt, F. W. Wise: Transverse instability of optical spatiotemporal solitons in quadratic media, *Phys. Rev. Lett.* **85**, 1871 (2000).
20. K. Beckwitt, Y.-F. Chen, F. W. Wise et al.: Temporal solitons in quadratic nonlinear media with normal group-velocity dispersion at the second harmonic, *Phys. Rev. E* **68**, 0575601 (2003).
21. I.N. Towers, B.A. Malomed, F.W. Wise: Light bullets in quadratic media with normal dispersion at the second harmonic, *Phys. Rev. Lett.* **90**, 123902 (2003).
22. L.J. Qian, X. Liu, F.W. Wise: Femtosecond Kerr-lens modelocking with negative non-linear phase shifts, *Opt. Lett.* **24**, 166 (1999).
23. X. Liu, L. Qian, F.W. Wise: High-energy pulse compression using negative phase shifts produced by the cascade $\chi^{(2)} : \chi^{(2)}$ nonlinearity, *Opt. Lett.* **24**, 1777 (1999).
24. X. Liu, F.O. Ilday, K. Beckwitt et al.: Femtosecond nonlinear polarization evolution based on quadratic nonlinearities, *Opt. Lett.* **25**, 1394 (2000).
25. K. Beckwitt, F.W. Wise, L. Qian, et al.: Compensation of self-focusing by use of the cascade quadratic nonlinearity, *Opt. Lett.* **26**, 1696 (2001).
26. M. Nisoli, S. De Silvestri, O. Svelto: Generation of high energy 10 fs pulses by a new pulse compression technique, *Appl. Phys. Lett.* **68**, 2793–2795 (1996).
27. C.P. Hauri, W. Kornelis, F.W. Helbing et al.: Generation of intense, carrier-envelope phase-locked few-cycle laser pulses through filamentation, *Appl. Phys. B* **79**, 673 (2004).
28. G. Stibenz, N. Zhavoronkov, G. Steinmeyer: Self-compression of millijoule pulses to 7.8 fs duration in a white-light filament, *Opt. Lett.* **31**, 274 (2006).
29. C. Rolland, P.B. Corkum: Compression of high-power optical pulses, *J. Opt. Soc. Am. B* **5**, 641 (1988).
30. S. Ashihara, J. Nishina, T. Shimura et al.: Soliton compression of femtosecond pulses in quadratic media, *J. Opt. Soc. Am. B* **19**, 2505 (2002).
31. J.P. Torres, L. Torner: Self-splitting of beams into spatial solitons in planar waveguides made of quadratic nonlinear media, *Opt. Quantum Electron.* **29**, 757 (1997).
32. F.O. Ilday, K. Beckwitt, Y.-F. Chen et al.: Controllable Raman-like nonlinearities from nonstationary, cascaded quadratic processes, *J. Opt. Soc. Am. B* **21**, 376 (2004).
33. J. Moses, F.W. Wise: Soliton compression in quadratic media: high-energy few-cycle pulses with a frequency-doubling crystal, *Opt. Lett.*, **31**, 1881 (2006).
34. K. Beckwitt, F.O. Ilday, F.W. Wise: Frequency shifting with local nonlinearity management in nonuniformly poled quadratic nonlinear materials, *Opt. Lett.* **29**, 763 (2004).

35. K. Beckwitt, J.A. Moses, F.O. Ilday et al.: "Cascade-Raman" soliton compression with 30-fs, terawatt pulses, *Nonlinear Guided Waves and Their Applications* 2003, Toronto, ON (2003).
36. M. Marangoni, C. Manzoni, R. Ramponi et al.: Group-velocity control by quadratic nonlinear interactions. *Opt. Lett.* **31**, 534 (2006).
37. J. Moses, F.W. Wise: Controllable self-steepening of ultrashort pulses in quadratic nonlinear media. *Phys. Rev. Lett.* **97**, 073903 (2006).
38. T. Brabec, F. Krausz: Nonlinear optical pulse propagation in the single-cycle regime. *Phys. Rev. Lett.* **78**, 3282 (1997).

Chapter 21
Effective Parameters of High-Power Laser Femtosecond Radiation at Self-focusing in Gas and Aerosol Media

G.G. Matvienko, S.N. Bagaev, A.A. Zemlyanov, Yu.E. Geints, A.M. Kabanov, and A.N. Stepanov

Abstract We consider the regularities of the femtosecond laser pulse propagation from the viewpoint of evolution of its effective parameters (energy transfer coefficient, effective beam radius, effective pulse duration, limiting angular divergence, and effective intensity). Based on the concepts of the spatial self-action zones we present a thorough analysis of beam effective parameters variation along the propagation path, discussing the regime of formation of a single axial filament. Filamentation behavior of femtosecond laser pulses in the air in the presence of aerosol on the optical path is considered also. The effect of a localized attenuating layer modeling an aerosol medium, on spatial extension and evolution of the light filament, is investigated.

21.1 The Evolution of Effective Parameters of High-Power Femtosecond Laser Radiation in the Air

Propagation of high-power femtosecond laser radiation through a gaseous and condensed medium occurs in the nonlinear regime and results in significant changes of temporal, spatial, and spectral characteristics of a light beam. It is known that for stationary self-action of laser radiation the transformation of its energy characteristics is favorably studied on the basis of effective parameters such as power (energy) transmission coefficient, effective (energetic) beam radius, angular divergence and mean intensity which characterize the global changes of light beam as a whole. In a number of cases, for example, at stationary self-focusing in a Kerr medium [1], or at a thermal blooming of long laser pulses [2], we can write equations for some effective characteristics that may quantify and qualify the process.

Self-action of ultrashort laser pulses in the air occurs with a wide set of physical factors that determine a radiation propagation process and may be in

G.G. Matvienko (✉)
Institute of Atmospheric Optics SB RAS, Tomsk, Russia
e-mail: ygeints@iao.ru

R.W. Boyd et al. (eds.), *Self-focusing: Past and Present*,
Topics in Applied Physics 114, DOI 10.1007/978-0-387-34727-1_21,

dynamical balance with each other (see, e.g., [3–5]). These are effects of beam diffraction, air chromatic dispersion, Kerr nonlinear refraction, and effects concerned with plasma channel formation within a laser beam during the multiphoton absorption.

In the general case for correct description of femtosecond pulse self-focusing, we need the numerical solution of the 4D-wave equation for the electric field, which is a very difficult problem even for state-of-the-art computing facilities. The description of self-focusing phenomenon on the basis of effective characteristics of light radiation is productive because it follows the fundamental transformations occurring with light beams in the nonlinear medium, and it forecasts its propagation on distances greatly exceeding the diffraction length of the beam.

Below we consider the peculiarities of the femtosecond laser pulse propagation from the viewpoint of evolution of its effective parameters. Based on the concepts of the spatial self-action zones we present a thorough analysis of beam effective parameters variation along the propagation path, discussing the regime of formation of a single axial filament.

21.1.1 Integral Characteristics of a Light Pulse

According to the classical definition [1], the effective (generalized) light beam parameters are the functionals of the optical field intensity $I(\mathbf{r}_\perp, z'; t)$ at every space point $(\mathbf{r}_\perp, z')$ and at every time moment t. The squared effective beam energetic radius R_e can be defined as:

$$R_e^2(z') = \frac{1}{E(z')} \int\limits_{-\infty}^{\infty} dt' \iint\limits_{\mathbf{R}\perp} d^2\mathbf{r}_\perp I(\mathbf{r}_\perp, z'; t') \left|(\mathbf{r}_\perp - \mathbf{r}_{gr})\right|^2. \qquad (21.1)$$

where $E(z')$ stands for the total energy of the optical pulse, and $\mathbf{r}_{gr}(z')$ is the radius-vector of the beam gravity center. The squared effective pulse duration t_{pe} has a similar form:

$$t_{pe}^2(z') = \frac{1}{E(z')} \iint\limits_{\mathbf{R}_\perp} d^2\mathbf{r}_\perp \int\limits_{-\infty}^{\infty} dt' I(\mathbf{r}_\perp, z'; t')(t' - t_0)^2, \qquad (21.2)$$

where $t_0(z')$ is the position of the beam's temporal gravity center. Also, the parameter

$$\theta_e^2(z') = 1/2\left(d^2 R_e^2 / dz^2\right) \qquad (21.3)$$

is important. It determines the square of the limit angular beam divergence $\theta_{e\infty}^2$ at $z' \to \infty$.

Another important integral parameter of the laser pulse is the light energy transfer coefficient that characterizes energy losses after the beam has passed the distance z': $T_e(z') = E(z')/E_0$,where E_0 is the initial value of the total energy of the optical pulse. The use of effective parameters to describe light beam propagation is equivalent to the use, instead of a real spatiotemporal intensity profile, an Gaussian distribution in a circle with the radius $R_e(z')$ rectangular in time $t_{pe}(z')$ with the mean intensity: $I_e(z') = E(z')/\left(\pi R_e^2(z')t_{pe}(z')\right)$.

Now, using the introduced characteristics (21.1)–(21.3), we will analyze the main phases of the nonstationary self-focusing of a femtosecond laser pulse in the atmosphere. The ground for the numerical calculation of the effective parameters is the nonlinear Schrödinger equation (NLSE) describing propagation of the electromagnetic wave in a medium in the approximation of a slowly varying field amplitude and completed with the rate equation for the concentration of free plasma electrons (see, for example, [6]).

21.1.2 Beam Effective Parameters Evolution

In a nondissipative Kerr medium and in the absence of air chromatic dispersion and nonlinear beam aberrations, the characteristics T_e, t_{pe} are invariant along the z'-axis up to the nonlinear focus, which reflects the quadratic dependence of the effective radius R_e along the longitudinal coordinate and the self-similarity of its temporal profile [1]. With self-focusing of ultrashort radiation resulting from multiphoton medium ionization, the regions of strongest intensity produce plasma. The latter determines the nonlinear absorption of the optical wave and its defocusing. A simultaneous manifestation of Kerr focusing and defocusing effects in plasma gives a strong phase self-modulation of radiation; thus, the parameters T_e and t_{pe} lose their invariance property.

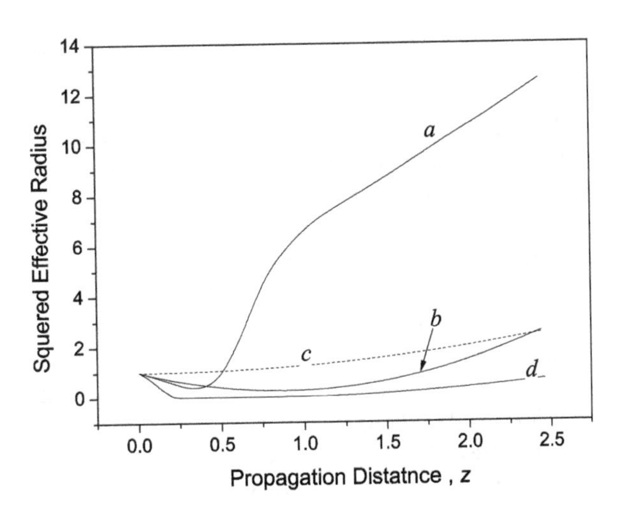

Fig. 21.1 Squared normalized effective radius $\bar{R}_e^2$ of a femtosecond pulse ($t_p = 80$ fs, $F = 1.2\,L_R$, $P_0 = 15P_c$) propagating in air: complete model (a); linear diffraction of focused (b) and collimated (c) beams; squared radius $(R_f/R_0)^2$ of the filamented beam zone (d)

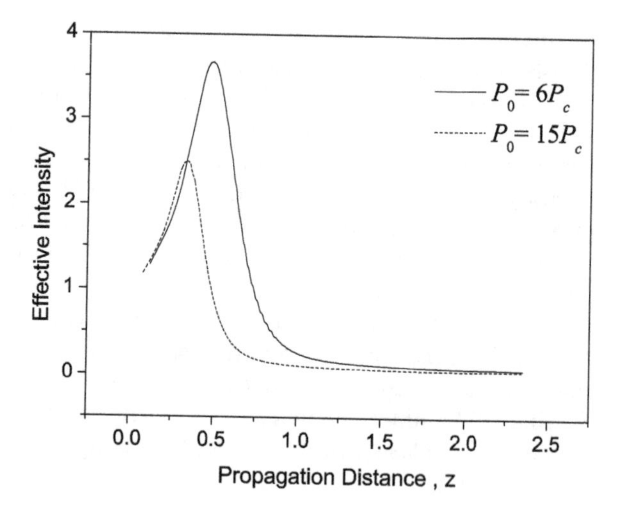

Fig. 21.2 Spatial evolution of the effective intensity $\bar{I}_e(t_p = 80\,\text{fs}, F + 1.2\,L_R)$

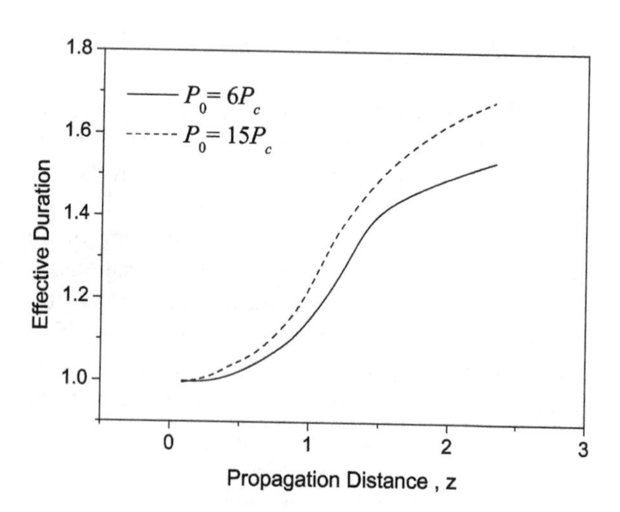

Fig. 21.3 Normalized effective pulse duration $\bar{t}_{pe}$ as the function of z

Figures 21.1–21.3 illustrate the dependences, along the propagation path, of the normalized parameters $\bar{R}_e^2(z) = R_e^2(z)/R_0^2$, $\bar{t}_{pe}(z) = t_{pe}(z)/t_p$ and $\bar{I}_e(z) = I_e(z)/I_0$ obtained by numerically modeling of the self-action of the laser pulse with the Gaussian spatio-temporal profile ($\lambda_0 = 810$nm, pulse duration $t_p = 80\,\text{fs}$, beam initial radius $R_0 = 1$mm, wave front curvature radius $F = 1.2\,L_R$, peak power $P_0 = 15P_c$, critical self-focusing power in the air $P_c = 3.2$ GW). The variation of squared FWHM beam size R_f is plotted in Fig. 21.1 also. All the dependent variables are normalized to their initial values at $z' = 0$, and the variable $z = z'/L_R$ is normalized to the beam Rayleigh length $L_R = 1/2\left(k_0 R_0^2\right)$.

From these figures we see that at the beginning of the self-action, the beam undergoes a fast transverse waist due to the Kerr effect. The growth of the peak intensity (Fig. 21.2) and the resulting strong multiphoton absorption lead to medium ionization and plasma generation. The defocusing effect of plasma together with the energy depletion stop beam collapsing and give rise to an axial waveguide channel (filament) with a quasi-Bessel spatial intensity distribution and a weak angular divergence. The filamentation start point z_f (the local nonlinear focus) depends on the initial pulse peak power P_0 and is determined by the known formula [1].

Immediately behind the nonlinear focus, the effective beam area gets sharply increased. Its divergence around the point z_{sf} (Fig. 21.1) is close to that of the focused Gaussian beam in the linear medium with the initial phase front curvature $F \simeq z_{sf}$. Though a strong diffraction-induced broadening of the energy-intensive part of the beam, the radius of the axial filament generally remains unchanged.

The next spatial zone $z/L_R \simeq 0.6 \div 1.6$ is characterized by the decrease in beam radius growth rate and by a smaller angular divergence (Fig. 21.1). The reason for this is successive refocusing of the beam periphery area, having lower intensities, on concentric toroidal nonlinear lenses formed around the filament in the preceding spatial zone.

Beyond this intermediate zone, the dependence $R_e^2(z)$ becomes quadratic, and there forms the limiting beam divergence $\theta_{e\infty}$. The value of the spatial variable $z' \equiv L_N \simeq 1.6\,L_R$ can therefore be considered a conditional boundary of the nonlinear medium layer, behind which evolution of the effective beam parameters obeys the linear diffraction laws.

21.2 Filamentation of Ultrashort Laser Pulse in the Presence of an Aerosol Layer

A laser pulse while passing through a particle ensemble undergoes energy extinction due to scattering and absorption of radiation by particles. In the linear regime, extinction of laser radiation by an aerosol layer in the single scattering approach can be described by the Bouger law. As the laser pulse shortens, the process of light scattering by particles becomes essentially non-stationary. The extremely high peak intensity of ultrashort radiation leads to additional nonlinear losses of light beam energy due to multiphoton ionization (MPI) of particulate matter and absorption by plasma centers formed inside particles.

There are two aspects of this problem: (1) the study of the filament evolution of a light beam that comes across isolated microparticles, and (2) the study of filamented beam propagation through the aerosol layer as a whole. As to the first aspect, there have been performed both experimental [7] and theoretical studies [8], whose results indicate that the transverse beam fluence distribution is highly resistant to the local extinction and perturbation of its central filamented area by a particle or an absorbing disk screen. This effect known as "filament self-healing" was explained within the dynamical spatial energy replenishment model as a consequence of light energy influx from periphery to beam axis [9].

Here, we have studied a somewhat different situation when the whole beam, rather than its central part only, is screened by aerosol, for example, when radiation propagates through a cloud. In this case, our results presented below clearly show that along with on-axis zones the off-axis beam areas (the "photon bath") can play a key role in light filament self-healing when the laser beam passes through an attenuating aerosol layer. These periphery areas carry the most fraction of the light energy of the beam and are capable of recovering the filament due to Kerr self-focusing.

21.2.1 Experiment

Recently, the first Russian experiments were performed [10] on the propagation of subterawatt femtosecond pulses of the Ti:sapphire laser ($l_0 = 800$ nm, $W_0 < 17$ mJ, $t_p = 80$ fs) in the laboratory air that contained a spatially localized aerosol layer (1.3 cm jet of water droplets with the root-mean-square radius ~ 2.5 μm and particle concentration $< 10^7$ cm^{-3}) placed before the geometrical focus of the beam. The aerosol layer transmittance was measured at two distinct wavelengths: $l_0 = 800$ nm of the pumping high-power laser and $l_0 = 630$ nm of the probing CW He–Ne laser. The experiments show that extinction of the filamented beam by an aerosol layer generally occurs in the quasi-linear regime and obeys the Bouger law, e.g., it exponentially decreases with the increase in particle concentration (see Fig. 21.4). The same results were reported in [11,12].

It has been noted in [10] that the only difference from the linear extinction theory for CW radiation is a somewhat higher ($\sim 35\%$) final extinction coefficient of aerosol for femtosecond pulses. This is obviously the cause of additional

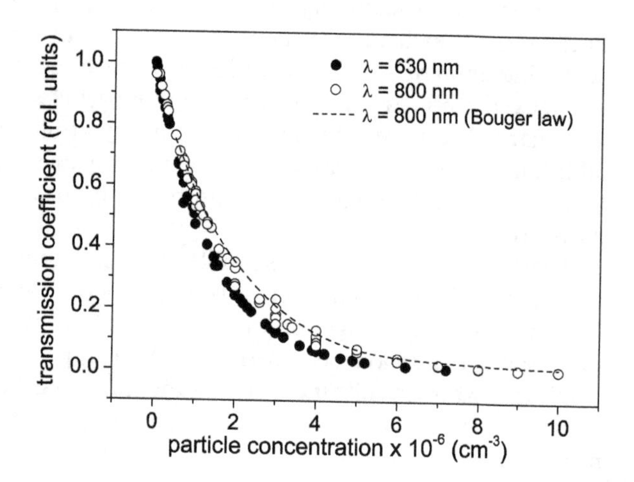

Fig. 21.4 Optical transmission of water aerosol at two wavelengths vs. particle concentration

nonlinear energy losses of the femtosecond pulse due to high ionizing plasma formation in the air.

Thus, we can suggest that aerosol extinction of the ultrashort laser pulse obeys the same laws as in the case of CW radiation. Consequently, in numerical calculations it is possible to simulate propagation of the radiation through an aerosol layer by introducing into the propagation equation (the nonlinear Schrödinger equation) at a given position on the optical path the distributed amplitude screen having the absorption coefficient equal to the coefficient of aerosol extinction.

In this section, such a theoretical model is used to numerically investigate the effect of the localized absorbing layer that simulates an attenuating aerosol medium, on the spatial extension of a light filament formed during the femtosecond laser pulse propagation in the air. The results of numerical simulation presented below indicate that the length of the filament depends not only on the optical density of the aerosol layer, but also on its spatial position on the propagation path and its spatial extension along the beam. To obtain the maximal filament length, one must try and shift the position of nonlinear beam focus behind the aerosol layer rather than before it.

21.2.2 Theoretical Simulations

Figure 21.5 shows two-dimensional pulse fluence profile in the x–z coordinate plane for the laser beam propagating in clear air, and in the presence of an absorbing aerosol layer positioned at different locations on the optical path: (1) before the filamentation zone, and (2) inside it. The laser radiation parameters correspond to the pulse with $l_0 = 800\,\text{nm}$, $t_p = 80\,\text{fs}$, $R_0 = 1.5\,\text{mm}$. The initial peak power P_0 exceeds the critical self-focusing power fifteen times. The aerosol is simulated by an absorbing screen intercepting the whole beam cross-section, with the spatial length $d_a = 0.05\ L_R$ and constant volumetric absorption coefficient α_{a0}. The aerosol optical depth $\tau_0 = \alpha_a d_a$ in the simulation always equaled unity.

From Fig. 21.5 it is clearly seen that the spatial zone of high fluence with the diameter $\sim 0.1 R_0$ is formed on the laser beam axis, and the trace of this zone along the propagation path resembles a light filament. In a clear air, the filament appears much earlier than the diffraction focus of the beam ($z_{f1} \approx 0.2 < F/L_R$) and has the mean length of $L_f/L_R = (z_{f2} - z_{f1}) \sim 0.6$.

As can be seen from this figure, when the absorbing screen simulating the aerosol layer is set on the beam path, the filament length decreases. At the same time, the aerosol influence on the beam filamentation dynamics appears to be different depending on the position of the aerosol layer on the optical path. If the aerosol layer is placed before the nonlinear focus of the beam [Fig. 21.5(b)], the position of this focus shifts towards a larger z because the peak power of the beam P_0 decreases, and SF coordinate z_{f1} is proportional to the fraction $(P_0/$

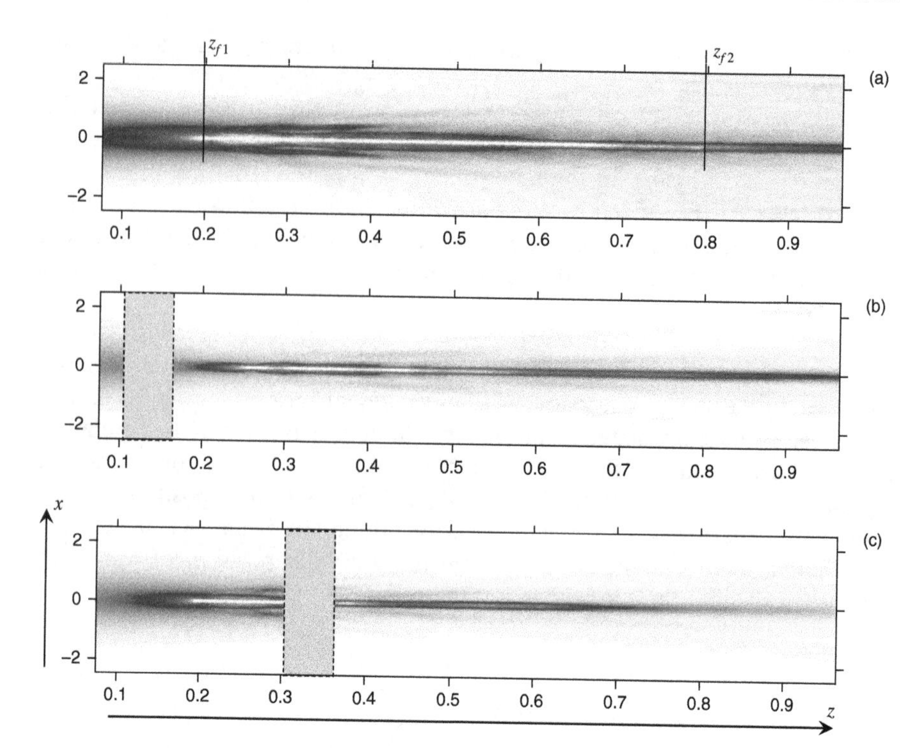

Fig. 21.5 Two-dimensional (x–z) pulse fluence profile (in arbitrary units) of the laser beam ($P_0/P_{cr} = 15$) propagating in the air without aerosol (a) and with aerosol layer (c)–(d) at $\tau_0 = 1$ ($\alpha_{a0} = 2.27m^{-1}$). Position of the aerosol layer is shown by grayed rectangle; the filament is represented by the light band near the beam axis

$P_{cr})^{-1}$ according to the well-known SF-theory. In addition, the radiation energy losses beyond the absorbing screen become critical for further maintenance of the filament due to dynamic influx of the light energy from beam periphery to its axis, which shortens the coordinate z_{f2} as well.

However, the strongest effect on the filament length is observed in the case when the beam that is already filamented is incident on the aerosol layer [see Fig. 21.5(c)]. The aerosol optical depth $\tau_0 = 1$ appears sufficient to block the filament. At the rear boundary of the aerosol layer, the maximal density of the plasma electrons formed on the beam axis decreases sharply to the level of $\rho_e \sim 10^{11}\text{cm}^{-3}$.

Figure 21.6 illustrates the above effect of filament self-healing upon transmission through the aerosol layer. The calculation parameters used here were the same as in Fig. 21.5, but the length of the absorbing layer was fivefold shorter while the aerosol optical depth was kept at the level $\tau_0 = 1$ ($\alpha_{a0} = 11.36\,\text{m}^{-1}$).

As can be seen from this figure, upon leaving the aerosol layer the axial spatially localized zone with high intensity first disappears simultaneously with the decrease in free electron concentration. Then, however, the filament

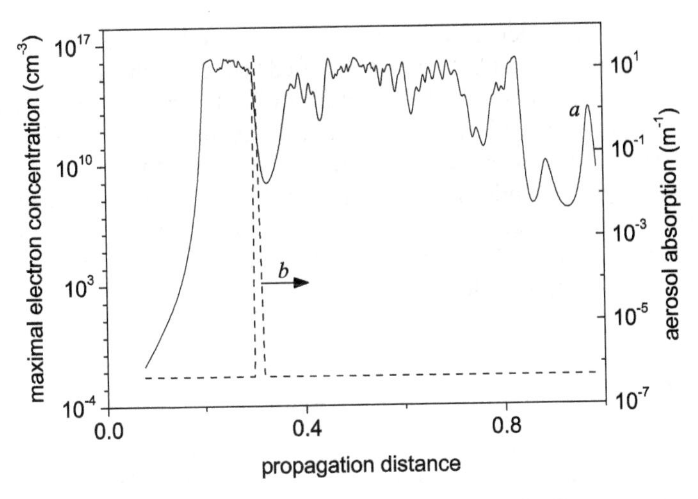

Fig. 21.6 Filament self-healing effect: plasma electrons concentration (curve *a*) on the laser beam axis ($P_0/P_{cr} = 15$), and linear volume absorption coefficient α_α (curve *b*) along the optical path

re-forms due to Kerr self-focusing of the off-axis beam zones, and further filament evolution is similar to that in the case of the clear atmosphere (Fig. 21.5*a*), although the filament itself loses stability (becomes discontinuous) along the propagation direction.

It is interesting to note that if the filament passes through a thicker aerosol layer (curve *b*, Fig. 21.6), it does not self-heal afterwards. The reason for this seems to be the shift in dynamical balance between Kerr self-focusing and beam diffraction in favour of the latter. Indeed, the aerosol distributed along the optical path gradually lowers pulse intensity and thus lowers the energy influx into the filament from beam periphery. This enhances the role of diffraction, which as the beam propagates tends to spread its cross-section and lower its intensity. When the attenuating aerosol layer is very thin, the balance between these effects recovers quickly (of course, if beam power is still above critical one). A lengthy attenuating layer allows the diffraction to significantly spread the beam which then complicates its refocusing.

21.3 Conclusions

On the basis of a numerical solution of the nonlinear Schrödinger equation we have considered the regularities of the spatial evolution of effective parameters of femtosecond laser radiation that undergoes self-focusing and filamentation. We have demonstrated that reasoning from the evolution of the effective parameters on the optical path, it is possible to mark out three spatial zones reflecting different stages of the nonstationary radiation self-focusing: the zone of beam waist toward the nonlinear focus and filament formation around it; the zone of a sharp increase of the effective beam area behind the nonlinear focus, and the zone of free diffraction of the radiation "reformatted" by a nonlinear lens.

The numerical simulations of self-action of femtosecond laser pulses in the air that contains an aerosol layer show that the increase of the optical aerosol thickness leads to a shorter spatial length of the filament formed. In addition, the filament length depends on the position of the aerosol layer on the propagation path and on its geometric thickness (at the constant optical depth). The filament length is maximal if the nonlinear focus of the beam is located after the aerosol layer rather than before it.

References

1. S.N. Vlasov, V.A. Petrishchev, and V.I. Talanov: Averaged description of wave beams in linear and nonlinear media. *Radiophys. Quant. Electron.* **14**, 1353–1363 (1971).
2. V.E. Zuev, A.A. Zemlyanov, Yu.D. Kopytin: *Nonlinear Atmospheric Optics*. Gidrometeoizdat. Leningrad (1989).
3. B. La. Fontaine, F. Vidal, Z. Jiang et al.: Filamentation of ultrashort pulse laser beams resulting from their propagation over long distances in air. *Phys. Plasmas*, **6**, 1615 (1999).
4. M. Rodriguez, R. Bourayou, G. Méjean et al.: Kilometer-range nonlinear propagation of femtosecond laser pulses. *Phys. Rev.* **E69**, 036607 (2004).
5. J. Kasparian, R. Sauerbrey, D. Mondelain et al.: Infrared extension of the supercontinuum generated by femtosecond terawatt laser pulses propagating in the atmosphere. *Opt. Lett.* **25**, 1397 (2000).
6. A.A. Zemlyanov, Yu.E. Geints: Integral parameters of high-power femtosecond laser radiation during filamentation in air. *Atmos. Oceanic Opt.* **18**, 574–759 (2005).
7. F. Courvoisier, V. Boutou, J. Kasparian et al.: Ultraintense light filaments transmitted through clouds. *Appl. Phys. Lett.* **83**, 213 (2003).
8. M. Kolesik, J. Moloney: Self-healing femtosecond light filaments. *Opt. Lett.*, **29**, 590 (2004).
9. M. Mlejnek, M. Kolesik, E.M. Wright et al.: Recurrent femtosecond pulse collapse in air due to plasma generation: numerical results. *Math. Comput. Simul.* **56**, 563 (2001).
10. N.N. Bochkarev, A.A. Zemlyanov, Al.A. Zemlyanov et al.: Experimental investigation into interaction between femtosecond laser pulses and aerosol. *Atmos. Oceanic Opt.* **17**, 861–867 (2004).
11. G. Méchain, G. Méjean, R. Ackermann et al.: Propagation of fs–TW laser filaments in adverse atmospheric conditions. *Appl. Phys.* **B80**, 785 (2005).
12. G. Méjean, J. Kasparian, J. Yu et al.: Multifilamentation transmission through fog. *Phys. Rev.* **E72**, 026611 (2005).

Chapter 22
Diffraction-Induced High-Order Modes of the (2 + 1)D Nonparaxial Nonlinear Schrödinger Equation

Sabino Chávez-Cerda, Marcelo David Iturbe-Castillo, and Jandir Miguel Hickmann

Abstract It is shown that the generation of high-order modes of the nonparaxial nonlinear Schrödinger equation can be induced by a diffracted beam at circular apertures propagating in a self-focusing medium, and that, under perturbations, they may break up into hot spots. Comparison with classic experiments and those patterns due to modulation instability patterns is discussed. A whole analysis of the phenomena involved is done in detail.

22.1 Introduction

Amplitude and phase perturbations of a beam propagating in a self-focusing medium can result in the break-up of the beam and the creation of filaments or hot spots with either regular or irregular transverse patterns. In some cases, *regular* patterns can be explained by the theory of *modulational instability,* which can account for transverse patterns with some degree of symmetry (e.g., annular and rectangular patterns). Within this approach these patterns are such that the power is uniformly distributed among the filaments. However, there exists a second kind of pattern that has no apparent symmetry. These patterns arise as the consequence of random imperfections of the beam or the medium, producing what here will be called *small-scale self-focusing*. It is possible, in some instances, to also attribute such patterns to modulational instability and degeneracy in frequency space. Another physical effect that is responsible for pattern formation is the nonlinear propagation of beams that have been diffracted by apertures.

Despite a very large number of theoretical and experimental studies, there are still some features of transverse spatial structures that arise with a CW beam propagating in nonlinear media and that remain to be explained satisfactorily. It is relevant then to understand such phenomena before theories can be

S. Chávez-Cerda (✉)
Instituto Nacional de Astrofísica, Óptica y Electrónica, Apdo. Postal 51/216, Puebla., Pue, México 72000
e-mail: sabino@inaoep.mx

R.W. Boyd et al. (eds.), *Self-focusing: Past and Present,*
Topics in Applied Physics 114, DOI 10.1007/978-0-387-34727-1_22,

extended to include, for example, the temporal domain. In this contribution some new features of pattern formation arising from the propagation of very intense beams in nonlinear media will be presented. Emphasis will be put on the relation between theory and experiment by studying systems that have been, and could be, implemented in the laboratory. In the first part of the discussion we will deduce analytically some effects concerning the stability of beams and pattern formation. Comparison of these results with numerical simulations of beams propagating in highly nonlinear media will be presented at the final sections of the chapter.

22.2 Spatial Modulational Instability in Nonlinear Media: Analytical Approach

Bespalov and Talanov were the first to present a model of modulational instability for infinite plane waves in a Cartesian frame [1]. We notice that most of the experiments favor cylindrical symmetry around a finite beam. The study of modulational instability of plane waves in different coordinate frames may reveal new and relevant characteristics. In this section we will present some of these features.

We shall consider the linear stability of plane waves propagating along the z-axis and within this context it is possible to neglect any nonparaxial effects. The wave equation, including temporal evolution, can be written as [2, 3]

$$-i\frac{\partial u}{\partial \xi}+\frac{1}{4}\left(\frac{\partial^2 u}{\partial \rho^2}+\frac{1}{\rho}\frac{\partial u}{\partial \rho}+\frac{1}{\rho^2}\frac{\partial^2 u}{\partial \theta^2}\right)+\gamma\frac{\partial^2 u}{\partial \tau^2}+f\left(|u|^2\right)u=0 \tag{22.1}$$

where ρ, θ and ξ are normalized cylindrical coordinates; γ is a measure of the dispersion and f is a function of the local field intensity. The normalization used will be defined below. The first term encompasses the traveling wave nature of the field; the second bracketed term accounts for transverse spatial variations, i.e., diffraction; the third term measures temporal changes, i.e. dispersion; and the last one, is the nonlinear response of the medium, either Kerr or saturable Kerr, and takes the form

$$f\left(|u|^2\right)=\begin{cases}\eta|u|^2 & \text{Kerr}\\ \dfrac{\eta|u|^2}{1+\Gamma|u|^2} & \text{Saturable Kerr}\end{cases} \tag{22.2}$$

where $\eta = P_0/P_C$ and Γ is the reciprocal of a normalized saturation intensity, P_0 is the input power, and P_C is a critical power whose value is given by $\varepsilon_0 c\lambda^2/4\pi n_2$ [4, 5]. We will assume that $\Gamma << 1$, which is a common case for many materials of interest [6]. Note that in previous analyses in the Cartesian frame, mainly the case of Kerr nonlinearity was considered. The field is normalized such that its maximum initial amplitude is unity. It is straightforward to show that a CW plane wave solution satisfies this wave equation and that its phase changes linearly with the input power accordingly to

$$\overline{u}(\xi) = \exp\left(-i\left\{\begin{array}{c}\eta \\ \dfrac{\eta}{1+\Gamma}\end{array}\right\}\xi\right). \tag{22.3}$$

To check the stability of this solution, we add a perturbation which we expect to decay on propagation if the solution is stable. The perturbation evolves both in amplitude and phase and must be small, $|a(\rho,\theta,\xi,\tau,)| << 1$. The perturbed field is taken to be of the form

$$u(\rho,\theta,\xi;\tau) = (1 + a(\rho,\theta,\xi;\tau))\,\overline{u}(\xi) \tag{22.4}$$

substituting (22.4) into (22.1) and linearizing yields for the perturbation

$$-i\frac{\partial a}{\partial \xi} + \frac{1}{4}\left(\frac{\partial^2 a}{\partial \rho^2} + \frac{1}{\rho}\frac{\partial a}{\partial \rho} + \frac{1}{\rho^2}\frac{\partial^2 a}{\partial \theta^2}\right) + \gamma\frac{\partial^2 a}{\partial \tau^2} + \alpha(a + a^*) = 0 \tag{22.5}$$

where $\alpha = \eta$ for the Kerr nonlinearity and $\alpha = \eta(1\text{-}2\Gamma)$ for the saturable Kerr nonlinearity. Equation (22.5) governs the evolution of the perturbation a. If the plane wave solution is unstable, the perturbation grows and, eventually, the condition $|a(\rho,\theta,\xi,\tau,)| << 1$ will not be satisfied. At this point the linear perturbation analysis, and consequently Eq. (22.5), may no longer be valid. To find a solution in analytical form for the small amplitude perturbation, we propose the following ansatz:

$$a(\rho,\theta,\xi;\tau) = \begin{bmatrix} a_1\cos(k_\xi\xi - \Omega\tau - m\theta) \\ + ia_2\sin(k_\xi\xi - \Omega\tau - m\theta) \end{bmatrix} H(k_\rho\rho) \tag{22.6}$$

where, $|a_1|$ and $|a_2| << 1$ are arbitrary constants, H is an unknown complex function of the radial coordinate; $\boldsymbol{k} = (k_x, k_y, k_\xi)$ is the wave vector (its transverse components combine to give $k_\rho^2 = k_x^2 + k_y^2$, the normalized radial frequency), m is the azimuthal frequency and Ω is the normalized temporal frequency. After substitution of Eq. (22.6) in Eq. (22.5), and separating the factor of sines and cosines and, because these functions are linearly independent, one finds

$$\begin{aligned} k_\xi a_2 &= \frac{a_1}{4}\left[\frac{\partial^2 H}{\partial \rho^2} + \frac{1}{\rho}\frac{\partial H}{\partial \rho} - \frac{m^2}{\rho^2}H\right]\frac{1}{H} - \beta\Omega^2 a_1 + 2\alpha a_1 \\ k_\xi a_1 &= \frac{a_2}{4}\left[\frac{\partial^2 H}{\partial \rho^2} + \frac{1}{\rho}\frac{\partial H}{\partial \rho} - \frac{m^2}{\rho^2}H\right]\frac{1}{H} - \beta\Omega^2 a_2. \end{aligned} \tag{22.7}$$

Because the radial dependence is assumed to be separable and included only in the arbitrary function H, then the whole term involving H on the right-hand side must be constant. Defining this constant to be $-k_\rho{}^2$ yields

$$\frac{\partial^2 H}{\partial \rho^2} + \frac{1}{\rho}\frac{\partial H}{\partial \rho} + \left(k_\rho^2 - \frac{m^2}{\rho^2}\right) H = 0. \tag{22.8}$$

This equation is easily recognized as Bessel's equation of order m and with radial frequency k_ρ. Taking the opposite sign for the constant results in the modified Bessel's equation, whose families of modified Bessel solutions diverge either as $\rho \to \infty$ or $\rho \to 0$. This is unphysical for the present situation so these solutions are discarded. Notice as well that, by its definition in (22.6), the azimuthal frequency m is restricted to integer values.

Particular solutions to (22.8) are the Bessel functions of the first kind that are obtained by linear superposition of the two independent Hankel solutions. These particular solutions have the characteristic of representing propagation invariant structures. For this reason we expect that the form of the perturbation will be the Bessel functions of the first kind. Physically, we can argue that the Bessel structure is created by interference between the conical waves traveling inward (towards the z-axis) and those traveling outward (away from the z-axis), as occurs in the cone of light which is necessary for the existence of the Bessel beams in linear media [7].

Seeking a nontrivial solution for the perturbation, the determinant of the set of Eq. (22.7) has to vanish. This results in the following dispersion relation:

$$k_\xi^2 = -\left(\gamma\Omega^2 + \frac{k_\rho^2}{4}\right)\left(2\alpha - \gamma\Omega^2 - \frac{k_\rho^2}{4}\right) \tag{22.9}$$

from which the modulational instability gain coefficient is obtained, $g = 2Im(k_\xi)$. Even though azimuthal variations were allowed in the wave equation, and in the ansatz, the dispersion relation does not depend explicitly on the order m. If a Cartesian frame is used instead, the form of the dispersion relation is exactly the same as Eq. (22.9) [1, 8].

The lack of azimuthal frequency dependence through the values of m is not particularly surprising because, in analyses using Cartesian coordinates, the transverse spatial frequencies (k_x, k_y) appear in a degenerate Pythagorean form. Physically, this is a consequence of the nonlinear medium being homogeneous and isotropic. There are no preferred directions or symmetry axes of the material. Notice also that the nonlinear response of the medium, either Kerr or saturable Kerr, yields the same functional form of the dispersion relation. The only assumption necessary for this to be true is that the saturation intensity is very high compared to the input beam intensity, but this is often the case in real systems. In Eq. (22.9) temporal and spatial effects do not occur in crossed terms, so we can assume a CW beam and center our attention on the spatial aspects of propagation only.

Disregarding the temporal terms, the gain coefficient obtained from (22.9) is

$$g(\alpha, k_\rho) = k_\rho \sqrt{2\alpha - \frac{k_\rho^2}{4}}, \tag{22.10}$$

from which we deduce that perturbations with spatial frequencies within the range $0 < k_\rho < 2\sqrt{2\alpha}$ are subject to exponential growth.

Plots of Eq. (22.10) for a range of values of the power parameter are shown in Fig. 22.1. Laser beams with powers equal to several hundreds or even thousands of critical powers have been used experimentally for more than 40 years [9–11]. Because the value of the critical power, P_C, is inversely proportional to the nonlinear refractive index, n_2, materials with a high n_2 will have a low threshold power for self-trapping. A beam power equal to a few critical powers can be in the range from milliwatts to kilowatts. The frequency at which the gain attains its maximum value is $k_{\rho\max} = 2\sqrt{\alpha}$ and this maximum gain is found to be $g_{\max} = 2\alpha$. This frequency ring is the responsible for the creation of the Bessel modulation on top of the beam [7].

Although $g_{\max}$ can be very large for beams with powers much greater than the critical value, one should recall that the propagation distance to the first focus is approximately inversely proportional to the square root of the input power [4, 5]. The rate of whole-beam self-focusing is expected to affect the effective length over which the modulation will be subject to gain. In addition to local intensity gradients, there will always be a phase profile that accounts for the focusing of the whole beam. These features affect the modulational rings which, once generated, tend to shrink in radius and, in the case of filamentation, result in a decrease in the separation between filaments [10]. The innermost ring is expected to collapse to the center of the beam, creating a very intense filament that will produce the first focal point. Similarly, the coalescence of neighboring filaments may lead to very intense spikes.

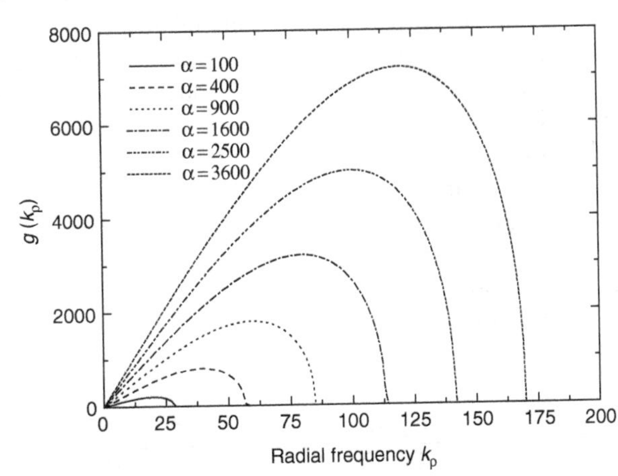

Fig. 22.1 Gain spectra of modulational instability

For simplicity, the stability analysis of this section considered an infinite extent plane wave background field, but it is expected to be applicable to beams whenever the wavelength of the modulation is small compared with the beam width. In Section 22.5, we will present numerical results concerning the modulational stability of beams.

22.3 Nonparaxiality and Filamentation of Intense Beams

To fully describe the process of beam filamentation it is necessary to include nonparaxial propagation effects in the modeling. The reason for this is that a wide spatial spectrum is generated during the evolution of self-focusing filaments. Narrowing of the filaments implies a widening of the corresponding Fourier spectrum. Components of the spectrum with finite transverse wave vector (k_x, k_y) will tend to travel off-axis. The angle between these off-axis components and the z-axis is given by

$$\theta = \sin^{-1}\left(\frac{\sqrt{k_x^2 + k_y^2}}{k}\right), \tag{22.11}$$

where $k = \sqrt{k_x^2 + k_y^2 + k_z^2}$. It has been argued that when the magnitude of the transverse wave vector is bigger than a critical value, such that $\theta \geq \pi/6$, then this value of the wavevector represents the limit for which the paraxial approximation remains valid [6]. The net effect of spectral widening is to decrease the intensity near the z-axis. In nonlinear propagation this can lessen the possibility of catastrophic collapse. By considering only materials that suffer damage when the beam narrows to a very intense filament, almost all theories of optical collapse have assumed the paraxial approximation and have not allowed for nonparaxial effects. However, it has been demonstrated (in liquids and gases) that the beam can propagate beyond the focal point creating a filament of filaments as a result of self-focusing or self-trapping [10, 12–14].

Feit and Fleck first showed [15] that by the inclusion of nonparaxial effects, optical collapse—which is inherent to the paraxial model—can be avoided, allowing further propagation and giving the opportunity to examine beam behavior beyond the predicted catastrophic focus. Other physical effects, such as saturation of the nonlinearity and linear and nonlinear absorption can play a role in avoiding optical collapse. However, nonparaxiality is most fundamental, even in linear propagation.

The normalized nonparaxial wave equation, for waves traveling in the region of space $z > 0$, is

$$\frac{\tan^2\Theta}{4}\frac{\partial^2 u}{\partial \xi^2} - i\frac{\partial u}{\partial \xi} + \frac{1}{4}\left(\frac{\partial^2 u}{\partial x^2} + \frac{\partial^2 u}{\partial y^2}\right) + f\left(|u|^2\right)u = 0. \tag{22.12}$$

where $E = E_0 u, \xi = z/L_D, x = x/w_0, y = y/w_0, k = 2\pi n_0/\lambda, L_D = k w_0^2/2, \tan\Theta = 2/k w_0$ and f is defined in Eq. (22.2). The first term in Eq. (22.12) continuously measures the nonparaxiality of the beam as it propagates along the z-axis in the nonlinear medium. Usually, this term is neglected by assuming the slowly varying envelope approximation. Its physical interpretation is straightforward. Consider an intense beam propagating in a Kerr medium and approaching the predicted focus. As the beam nears this point, its intensity at the center changes very rapidly; see Fig. 22.2. Its beam width follows an inverse relation. The peak intensity grows while its width shrinks. This is accompanied by a substantial increase in the width of the transverse spatial spectrum which renders the paraxial approximation invalid.

It is well known that a Gaussian beam, of waist w_0, and propagating in a linear medium, has an angular beam spread given by [6]

$$\tan\Theta \equiv \frac{\lambda_0}{\pi n_0 w_0} = \frac{2}{k w_0} \tag{22.13}$$

where n_0 is the background refractive index and λ_0 is the wavelength in vacuum. Thus, the constant multiplying the nonparaxial term is a direct geometrical measure of the magnitude of beam divergence at the onset of propagation. For visible optical beams, whose diameters are of the order of millimeters or sub-millimeters, this parameter is small (10^{-4}–10^{-2}). However, if a beam is predicted to focus to submicrometer dimensions, this parameter may approach unity, and hence this term clearly cannot be neglected. This factor also removes the unphysical arbitrariness of the independence of beam collapse on the geometrical features of the transverse profile which can modify the characteristics of

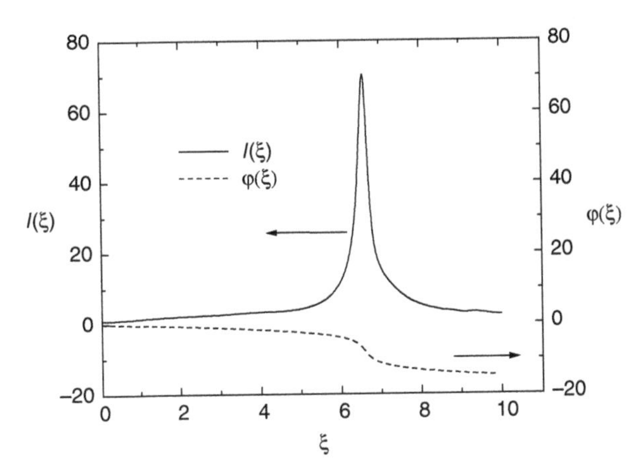

Fig. 22.2 Variation of the peak intensity and the longitudinal phase shift in the nonparaxial regime of an input Gaussian beam with η = 1. The rate of change of the peak intensity is higher than the one of the phase shift near the focus

the propagation. Notice that the geometry of the beam is totally absent in the paraxial approximation.

Either whole-beam self-focusing or rapid local variations of the beam profile can make the nonparaxial term significantly large. Such behavior occurs in a self-focusing Kerr medium, where whole-beam collapse or the formation of rings or filaments occurs. To prove the validity of our argument, in Fig. 22.3 we present results from simulations in which we propagate an input Gaussian beam at the critical power and an initial beam waist which is of the order of one wavelength ($\eta = 1$ and $\tan \Theta \cong 1/\pi$). For comparison, a simulation was performed neglecting nonparaxial effects. The difference between the paraxial and nonparaxial propagation of the same input beam is evident. The nonparaxial solution not only avoids the collapse but also shows that a beam under these conditions is self-trapped. Increasing the power can result in multifoci behavior as shown by Feit and Fleck [15]. In nonparaxial approximation multifocus structure has been suggested by Lugovoi and colleagues [16, 17]. But to arrest the collapse in nonparaxial approximation, nonlinear absorption or another energy dissipation mechanism should be involved.

A different nonparaxiality criterion was proposed by Akhmediev and Soto-Crespo [18, 19] for a single beam (it does not account for filamentation) propagating in Kerr nonlinear media. They assumed that accounting for the rapid changes in the longitudinal on-axis phase is sufficient to describe the nonparaxiality of self-focused beams. Under this assumption they also derived a new energy invariant. Extending their findings, it can be shown, using the variational Rayleigh–Ritz method to Eq. (22.12), that a more general energy invariant can be found showing that transverse characteristics of the beam can be relevant, which is in agreement with our discussion above. As shown in Fig. 22.2, even nonparaxial evolution involves very rapid changes in beam intensity and these changes can be of a magnitude much larger than those of the longitudinal phase.

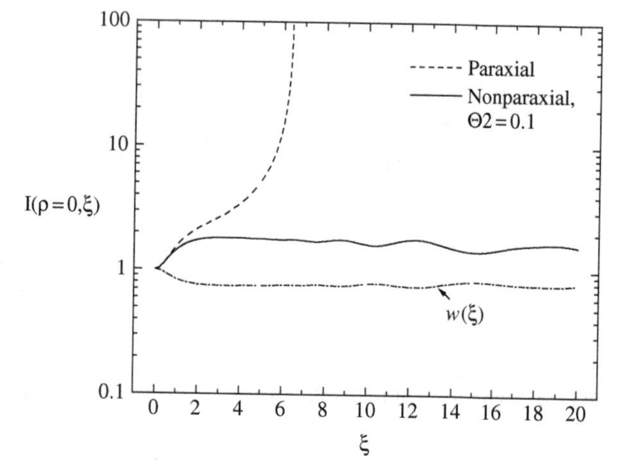

Fig. 22.3 Variation of the peak intensity of a beam with the critical power and a beam width of about one wavelength. Under these conditions the beam is self-confined. The variation of the peak intensity in the paraxial regime is also shown

22.4 Higher-Order Modes of the Nonparaxial Nonlinear Schrödinger Equation

Both the paraxial and the nonparaxial nonlinear Schrödinger equation (NLSE) with Kerr nonlinearity can support a zero-order stationary solution that acts as an attractor in the nonlinear evolution. More general stationary solutions exist that have a central bright spot surrounded by a number of rings that determine the order of the mode. The first-order mode has one ring, the second has two rings, and so on. They represent higher-order modes of the NLSE [20–23]. No analytical form exists for these modes, even in the case of the fundamental zero-order mode; however, it is always possible to find an approximate solution by means of the calculus of variations. For instance, an approximation to the zero-order mode was first obtained by Anderson et al. [24], and later Chen reported analytical approximations to the first three higher-order modes [25]. Here we calculate the exact modes numerically. In this section these higher-order modes will be analyzed because, as we will see, they represent attracting transverse spatial structures in the nonparaxial evolution of an intense beam in a nonlinear media. We shall now formulate their characteristics in the nonparaxial regime.

The mode structure can be found when we separate the radial dependence of the envelope and the longitudinal harmonic behavior of the field through setting $u(\rho,\xi) = U_\beta(\rho)\exp(-i\beta\xi)$. β is a longitudinal wave number, related to k_z through the relation $\beta \equiv k_\xi = L_D\Delta k_z$. Inserting this ansatz function in Eq. (22.12) with the Kerr nonlinearity given in Eq. (22.2) yields

$$\frac{1}{4}\left(\frac{\partial^2 U}{\partial\rho'^2}+\frac{1}{\rho'}\frac{\partial U}{\partial\rho'}\right)-\beta' U+|U|^2U=0 \tag{22.14}$$

in which

$$\beta'=\frac{1}{\alpha}\left(\frac{\tan^2\Theta}{4}\beta^2+\beta\right) \tag{22.15}$$

is the eigenvalue corresponding to the transverse eigenmode $U_\beta(\rho')$. α is the nonlinear parameter as defined above. The transverse coordinate has been renormalized to $\rho'=\sqrt{\alpha}\rho$. The modes satisfy the boundary conditions

$$U(\rho')|_{\rho'\to 0}=1,\;\; U(\rho')|_{\rho'\to\infty}=0,\;\; \left.\frac{\partial U}{\partial\rho'}\right|_{\rho'\to 0}=\left.\frac{\partial U}{\partial\rho'}\right|_{\rho'\to\infty}=0 \tag{22.16}$$

and can be considered real or complex (with a constant phase that can be factored out). If $\beta << 2/\tan\Theta = kw_0$, we recover the expression for the paraxial modes, namely, $\beta' = \beta/\alpha$. The eigenvalues, β'_n, decrease monotonically as the order, n, which corresponds to the shrinking of the longitudinal wavelength related to each mode because the wavenumber, β, augments. This is obvious if

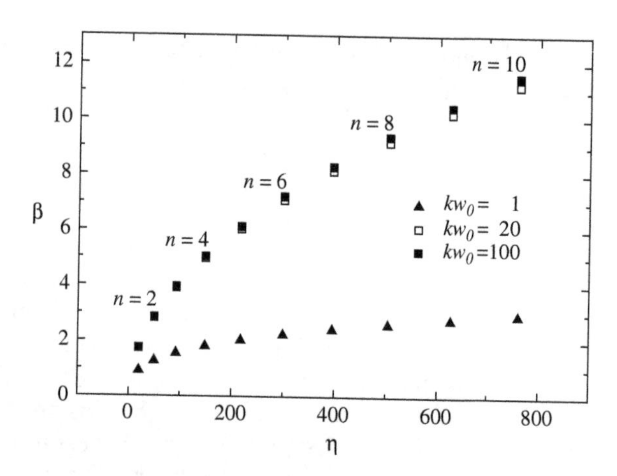

Fig. 22.4 Variation of the wavenumber (not the eigenvalue) with the power parameter for the first ten eigenmodes

one recalls that higher-order modes induce a higher refractive index and that for light propagating in a medium with increasing refractive index the wavelength must decrease. Figure 22.4 shows the dependence of the longitudinal wavenumber, β, on the power parameter, $\alpha = \eta$, for the first ten modes studied here. For $\tan\Theta \sim 0.02$ the paraxial and nonparaxial results are practically identical.

It is known that there are three limit points on the abscissa axis in the phase space $\left(U, \frac{\partial U}{\partial \rho'}\right)$ for the solutions of Eq. (22.14), one negative, one zero and one positive [26, 27]. To which limit point the solution approaches as $\rho' \to \infty$ depends on the value of the eigenvalue β'_n. The solutions that satisfy the conditions of Eq. (22.16) have zero as their point limit in phase space. In Fig. 22.5 the fundamental and the first ten eigenmodes of Eq. (22.14) are shown. The scaling invariance of the NLSE equation leaves the integrated intensities of the modes having the same values as they did prior to the renormalization [27].

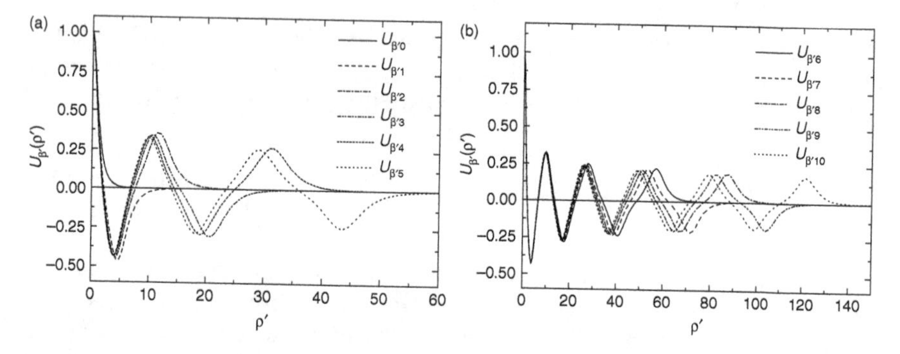

Fig. 22.5 (**a**) Eigenmodes of orders 1–5 of the nonparaxial NLSE. The fundamental mode is also included. (**b**) Eigenmodes of order 6–10

In the next section we will present results that suggest that it is possible to create these modes in realistic experimental configurations. We will support our argument with published experimental results.

22.5 Ring and Filament Formation of Beams in Self-Focusing Media: Numerical Study

In the previous sections, we studied spontaneous regular optical patterns, structures that are, in one way or another, inherent to three-dimensional beam propagation in nonlinear Kerr and saturable Kerr media. Now, we will show that the radially symmetric modulational instability modes derived in Section 22.2 can arise in beam propagation. We shall define their characteristics and the conditions under which they arise. A related instability of propagating beams in focusing nonlinear media is small-scale self-focusing. It also arises from amplitude and phase imperfections of the beam that grow with distance and leads to “narrow beams” or filaments. The emerging structures, under the conditions of small-scale self-focusing, usually form irregular patterns. These will be compared with the symmetric patterns that result from regular modulational instability and we will see that the latter can be destroyed by the presence of the former. Cartesian modulational instability modes are less likely to arise spontaneously in experiments. For radially symmetric beams we will see that the agreement with the predictions in Section 22.2 is outstanding. The radial symmetry predictions are justified by results from numerical experiments, involving full three-dimensional propagation.

A very important result is that we have found for the first time the spontaneous creation of modes of the three-dimensional NLSE. We relate these results to the eigenmodes studied in Section 22.4. Moreover, these modes reproduce the experimental findings of Campillo et al. better that any other previous theoretical study [28, 29]. Filamentation of the eigenmode-rings has been found to occur only after they have almost reached a steady state. It is possible that such filamentation can be due either to a pure one-dimensional modulational instability or to small-scale self-focusing along the circumference of the rings. However, the main point in this section of our results is that the filaments formed on the rings are not necessarily periodic and that the power distribution among the filaments is irregular. These are characteristics of small-scale self-focusing.

22.5.1 Modulational Instability of Optical Beams

In this section we will restrict our study to beams with radial symmetry in order to investigate the modulational instability modes presented in Section 22.2. Full two-dimensional analysis will be discussed in the following sections. Because it is our intention to stay close to realistic experimental situations, Gaussian and

Top-Hat profiles were chosen as the initial conditions for the simulations. We have found that results using apertured Gaussian beams only show slight quantitative differences from those using Top-Hat beams, and for this reason are not included. In Section 22.2 we derived the characteristics of a spatial modulation that would evolve from small perturbations of a plane wave, or wide beam, propagating through a Kerr or saturable Kerr medium in the paraxial regime. In this section we investigate the behavior of very intense beams when radial symmetry is assumed.

The parameter that determines the characteristics of the modulation was found to be the power parameter α. For a Kerr medium, α is equal to the ratio of the input power and the critical power. For a saturable Kerr medium it also depends on the saturation intensity [see the definitions after Eq. (22.2)]. In Section 22.2 we showed that the range of spatial frequencies subject to modulational instability gain is proportional to the square root of α. We also showed that the spatial frequency that will be most favored, i.e., the frequency at which Eq. (22.10) attains its maximum, has a value of $k_{\rho\max} = 2\sqrt{\alpha}$. The analysis was performed for plane waves and we assumed that it could be applicable to beams. This can be true only if the beam is sufficiently wide that a region of this beam (which is large compared to the characteristic scale of the modulation) can be considered as, essentially, plane.

Here we define such a quasi-plane wave region as the central part of the beam bounded by the curve at which the intensity drops to 95% of its peak value. Such a definition is of crucial importance when studying the modulational instability of bell-shaped beams. It allows an understanding of the lower bound frequency that can be favored by the modulational instability gain and overcome the effects of whole-beam self-focusing. This is because as the beam propagates in the self-focusing medium its beam width is decreased and the quasi-plane wave region becomes smaller. Further to this, it can now be seen that modulational instability requires a growth rate faster than the whole-beam self-focusing to evolve.

In Fig. 22.6 we have plotted the input Gaussian intensity profile $\exp(-\rho^2)$ and the Bessel function $J_0(k_{\rho\max}\rho)$ for several spatial frequencies. Also shown is the quasi-plane wave region, which is a disc of radius $\rho \cong 0.23$. From this plot and with the above considerations one can clearly see why bell-shaped beams with powers of the order of 400 critical powers or less may not be susceptible to the development of *spontaneous* modulational instabilities.

Our main interest is the spontaneous formation of modulational patterns from a noisy input beam. One of the reasons for this is that the theory of modulational instability predicts growth for a *range* of frequencies. The standard approach in the literature is to show that an induced modulation of the beam, at some frequency within this range, is subject to growth. What we want to demonstrate is that the physical system chooses, in a natural way, the frequency with maximum gain. Another reason is to confirm that the modulational instability mode has the Bessel-J_0 profile as predicted by the theory.

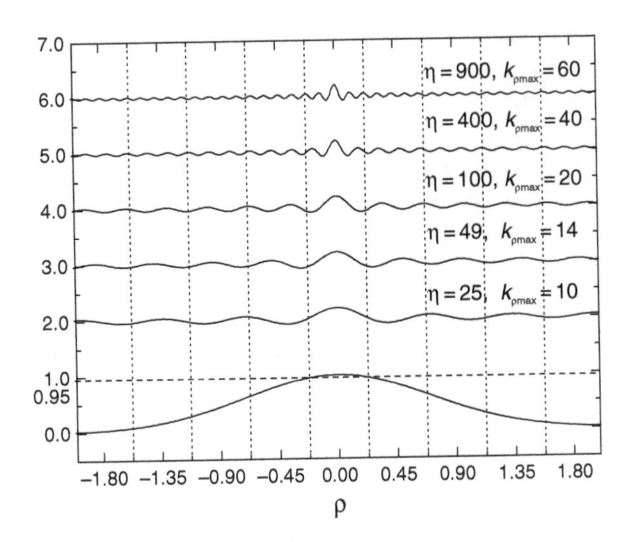

Fig. 22.6 Plots of the zero-order Bessel function for different radial frequencies calculated for the power parameter range $\eta = 25$–900. An intensity Gaussian profile is also plotted; the horizontal dashed line defines the region of the beam with a quasi-plane wave intensity distribution (see text)

Firstly, we shall consider the propagation of a Gaussian beam with $\alpha = 900$ in a Kerr medium. Following the same approach as in Section 22.3, we assume that the beam is wide enough to neglect nonparaxial effects. In Fig. 22.7(a) the evolution of the modulational instability at the center of such a beam is shown. The modulation clearly evolves in the region where the beam is more intense and is symmetric with respect to the propagation axis. This is important to note because it is this region of the beam that propagates most like a plane wave. Thus, it is in the central region where the modulational instability is expected to evolve in agreement with the analysis in Section 22.2. At $\xi = 0.023$ the instability is quite clear, its Fourier–Bessel spectrum at this position is shown in Fig. 22.7(b). Here, we can see that the modulational frequency is very close to

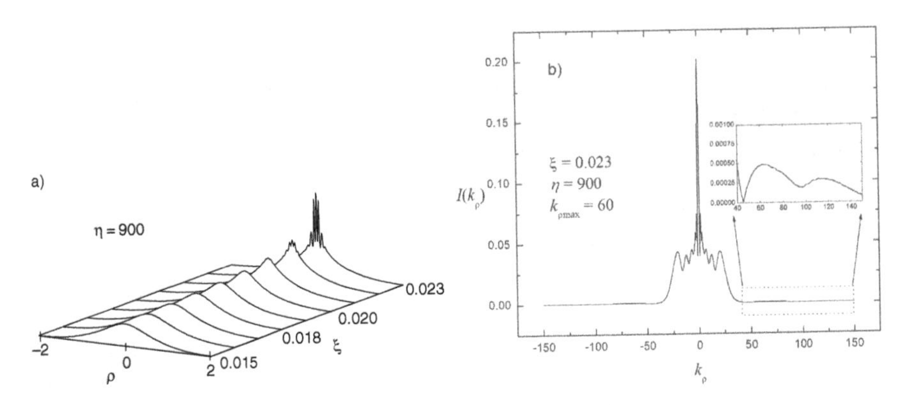

Fig. 22.7 (a) Evolution of the modulational instability on top of a Gaussian beam. The Bessel modulation was not seeded; white noise was used instead. (b) The Fourier–Bessel spectrum of the last profile of (a); in the inset is shown the presence of the predicted modulation frequency of highest gain. Also, the first higher harmonic at twice such a frequency is developing

the predicted value of $k_{\rho\,\max} = 2\sqrt{\alpha} = 60$. Moreover, the first higher harmonic is also present. The center of the spectrum shows the long wavelength components and demonstrates that the spectral content of the background beam is strongly modulated by self-phase modulation. This self-phase modulation effect can be easily understood by referring back to the wave Eq. (22.12) and neglecting, for the moment, diffraction. Doing so yields

$$-i\frac{\partial u}{\partial \xi} + \alpha|u|^2 u = 0. \tag{22.17}$$

It is straightforward to verify that a solution to Eq. (22.17) can be written in the form

$$u(\rho, \xi) = u_0(\rho)\exp\left(-i\alpha\xi|u_0(\rho)|^2\right) \tag{22.18}$$

where $u_0(\rho) = u(\rho, \xi = 0)$. From this equation we observe that the spatially varying phase implies that the longitudinal wave number differs radially over the beam from its value on-axis, k_ξ. Such variation is given approximately by

$$\delta k_\xi\big|_\rho = -\alpha\frac{\partial}{\partial \rho}|u_0(\rho)|^2 \tag{22.19}$$

An intense beam and/or high nonlinear refractive index results in modulation of the Fourier–Bessel spectrum which evolve with ξ, from the region around $k_\rho = 0$. As can be inferred from Fig. 22.1, the value of the power parameter dramatically affects the gain coefficient. When the perturbation is negligible, our simulations show that Gaussian beams with powers $\alpha = 100$ and 400 focus down to a single filament.

When the Gaussian beam is propagating through the nonlinear medium it is the central region which suffers more self-focusing diminishing the region over which the modulational instability could evolve as can be deduced from Fig. 22.6. However, when the power is higher the large value of the gain coefficient is reflected in the rapid formation of *spontaneous* modulational instability.

The evolution of a Gaussian beam with power parameter $\alpha = 3600$ is shown in Fig. 22.8. Details of the modulation at two different stages of propagation are compared with the predicted modulational mode in Fig. 22.9. The modulation profile was obtained by subtracting a Gaussian profile from the numerical data. Figure 22.9(a) shows very good agreement between the spontaneously created modulation and the predicted Bessel profile at the frequency of maximum gain. With further propagation the modulation maintains its shape even while the amplitude increases by a factor of approximately fifty! [see Fig. 22.9(b)].

Comparing results for a Gaussian input beam with those for an apertured plane wave or Top-Hat beam, one finds similar modulation effects occurring at

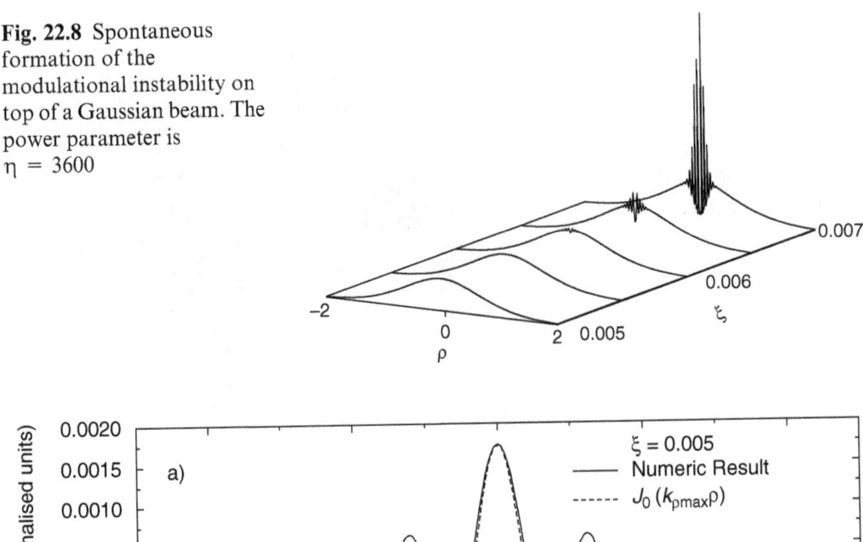

Fig. 22.8 Spontaneous formation of the modulational instability on top of a Gaussian beam. The power parameter is $\eta = 3600$

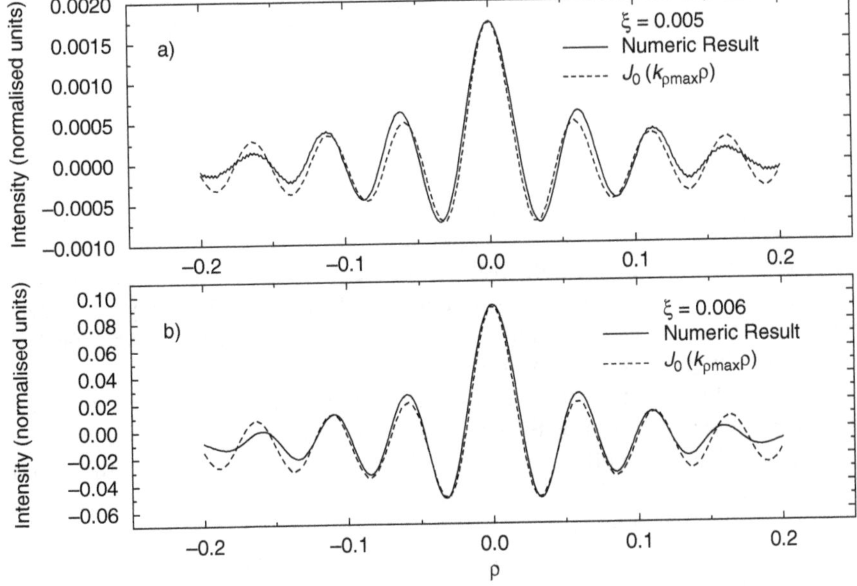

Fig. 22.9 Detail of the center of profiles at (a) $\xi = 0.005$ and (b) $\xi = 0.006$ of Fig. 22.8. The modulation spontaneously developed is contrasted with the predicted zero order Bessel function with the frequency of maximum gain. The agreement is almost perfect within the quasi-plane wave region; out of this the modulation slowly vanishes. At $\xi = 0.006$ the amplitude of the modulation has increased almost 50 times

the center of the beam. Propagation of apertured beams also involves strong diffraction effects that are enhanced by the presence of the nonlinear medium. This is clearly seen in Fig. 22.10, in which propagation of a Top-Hat beam is shown. We observe that, during early stages of propagation, diffraction rings are formed at the outer part of the beam. These rings have large amplitude and cannot be considered as a perturbation. Also observe that they are created where the beam cannot be assumed to approximate a plane wave, even locally.

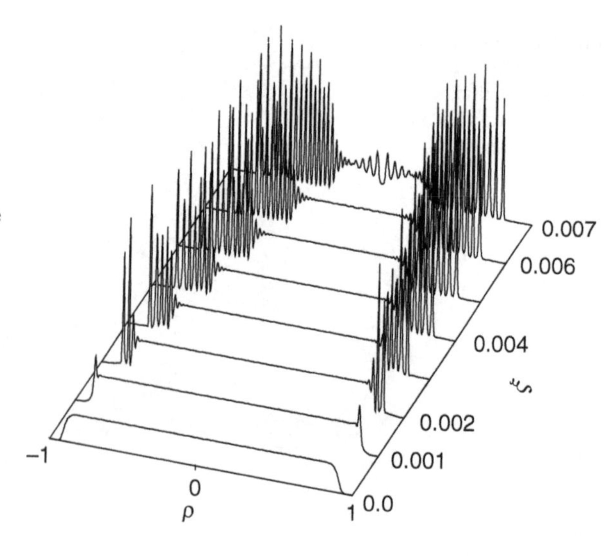

Fig. 22.10 Evolution of a beam with initial Top-Hat profile. The modulational instability is created spontaneously in the center of the beam. The rings formed at the edges of the beam a have different nature (see text)

Hence, it is unlikely that they can be described using the stability theory developed above (their physical explanation will be given in the following section). Nevertheless, in the central region the diffraction-induced oscillations just after the aperture have very small amplitude and variable frequency from which an induced modulational instability can occur which selects the frequency with highest gain before the outer rings take over.

Figure 22.11 extends the results of Fig. 22.9 to include Top-Hat input beams. In this case the agreement, while good, is not as impressive due to the presence of aperture diffraction effects which produce small oscillations near the center of the beam. Even with these aperture effects, agreement is within an acceptable accuracy. From the agreement shown in Figs. 22.9 and 22.11 we can discard the possibility of sinusoidal modulations that are predicted by analysis in an infinite Cartesian frame.

A final simulation with power such that $\alpha = 1600$ is presented in Fig. 22.12. Here there is a smaller value of the gain coefficient and hence the rate of growth of the modulational instability is smaller as well. The growth of the diffraction rings is faster than those of the modulational instability. This large amplitude perturbation does not allow the possibility of the predicted Bessel pattern. Figure 22.13 shows that modulational instability frequencies are present in the spatial spectrum at early stages of propagation, $\xi = 0.004$. However, further propagation allows the diffraction rings to reach the center, annihilating the predicted modulational instability. In Fig. 22.14 we compare the central part of the beam at $\xi = 0.012$ with the Bessel modulation. Certainly there is not agreement. The frequency of the diffraction rings is about 2.5 times bigger than $k_{\rho\max}$ for this power, $\alpha = 1600$, and thus lie out of the range for modulational instability gain.

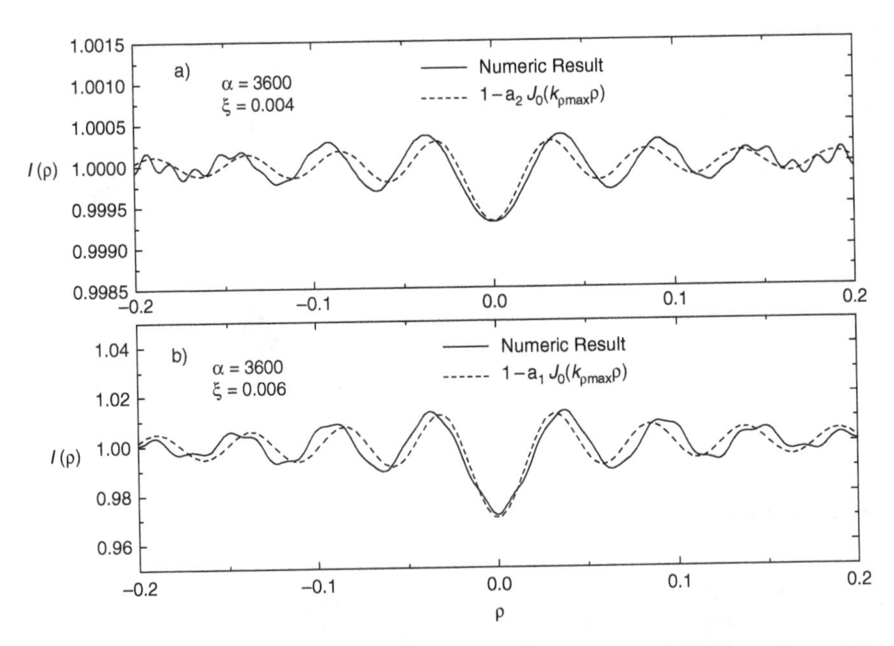

Fig. 22.11 As in Fig. 22.9, but for the case of a Top-Hat input profile. The presence of the Fresnel diffraction ripples makes the agreement not as good as in the previous case

This final result is important because it shows that beams propagating in nonlinear media are *not only* subject to modulational instability. It also demonstrates the distinctiveness of the modulational instability and, by consequence, the validity of our new analysis which describes *spontaneous* spatial structure

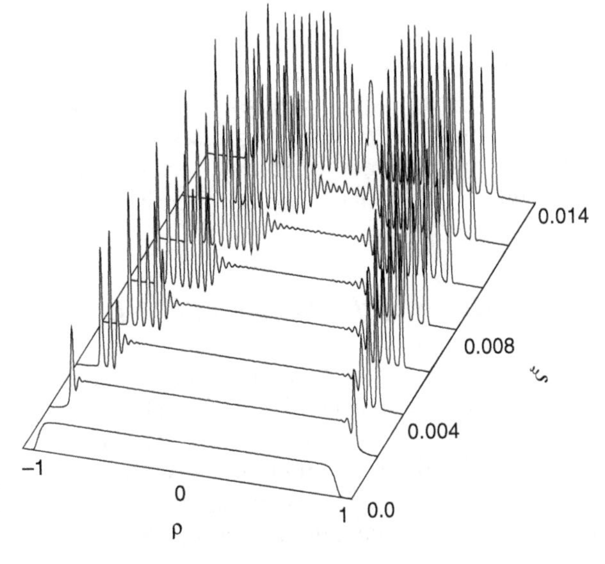

Fig. 22.12 Evolution of a beam with input Top-Hat profile and $\eta = 1600$. In this case the gain of the modulational instability is not large enough for the modulation to evolve before the diffraction-induced rings reach the center

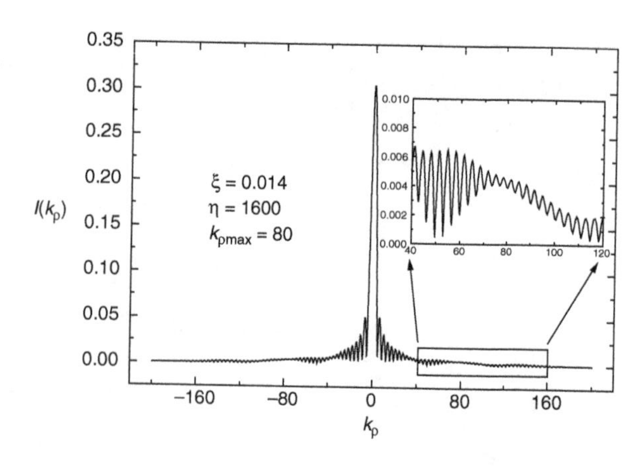

Fig. 22.13 Fourier–Bessel spectrum of the beam profile at $\xi = 0.014$. The inset shows that the predicted modulational frequency is present in the spectrum

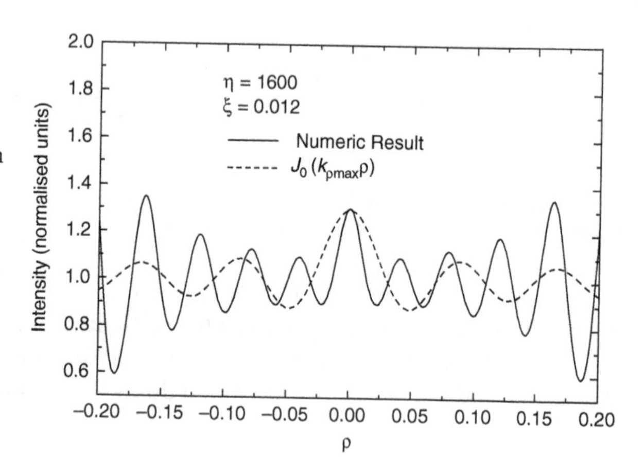

Fig. 22.14 Detail of the profile at $\xi = 0.012$ of Fig. 22.12. The zero-order Bessel function with the frequency of maximum gain is also plotted. The rings induced by diffraction have a different frequency that the one of the modulational instability

arising from nonlinear beam propagation. Patterns with rectangular or other symmetries may arise in single-pass beam propagation and may be due to different and more complex processes that are not yet fully understood. In a following section another process that competes with regular modulational instability, namely small-scale self-focusing, will be considered.

22.5.2 Modes of the Nonparaxial NLSE

In Section 22.4 we found that the nonparaxial NLSE has stationary solutions designated as the higher-order modes. These solutions have a ringed structure and the number of rings defines their order, the first-order mode has one ring, and so on. We also found that the transverse structure of the modes is the same as the paraxial NLSE modes, and that they differ from the paraxial case just in terms of their longitudinal wavenumber. Now, we will show that it is possible to simulate their formation, in Kerr media or saturable Kerr with high saturation intensity, by considering propagation of a smooth input beam.

The parameter that characterized the modes of the nonparaxial NLSE, $U_{\beta'n}$, was their eigenvalue, β'_n, which depends on the longitudinal wavenumber and the input power through the relation (22.15). For bell-shaped beams at powers near the critical power, the balancing of linear and nonlinear effects can lead to a beam with characteristics close to the zero-order mode. We observe that for higher orders, $n \geq 2$, as the radius of the rings is increased, the distance between successive rings grows, see Fig. 22.4. A similar situation occurs in the Fresnel diffraction patterns of a circular aperture. This suggests that Top-Hat profiled beams could be an initial condition that could lead to the formation of the modes of the nonparaxial NLSE. The input power to choose would correspond to that calculated for the modes of Eq. (22.14). These powers are given in Table 22.1.

In the previous section, while studying the propagation of a beam with an initial Top-Hat transverse profile, we observed the formation of rings at the outer part of the beam; see Figs. 22.10 and 22.12. These rings cannot be the result of any type of conventional modulational instability because they are created one-by-one and stop growing at some stage of their propagation. These features are contrary to the expected behavior, and characteristic length scales of the modulational instability modes. The latter are patterns that should appear uniformly across a transverse region and grow exponentially [see Eq. (22.6) and Figs. 22.7, 22.8 and 22.10].

The formation of the rings in our simulations was always found to occur in the same sequence. An external first ring would grow up to between five or seven times the local amplitude at $\xi = 0$; then a second ring would be formed reaching about the same amplitude as the first one. This process evolves until the rings filled the whole beam. The final number of rings on the beam increases with the input power. After each ring is formed it propagates almost with no change in shape, maintaining the radial symmetry.

Table 22.1 Eigenvalues for the first ten eigenmodes

Eigenmode order	Eigenvalue	Integrated intensity
1	0.09007263793937	19.29
2	0.05806094935894	48.75
3	0.04287233472234	91.83
4	0.03399070386553	147.87
5	0.02816013404235	217.12
6	0.02403805305375	299.52
7	0.02096916749454	395.06
8	0.01859541830025	503.06
9	0.01670459482222	625.65
10	0.01516289224172	760.67

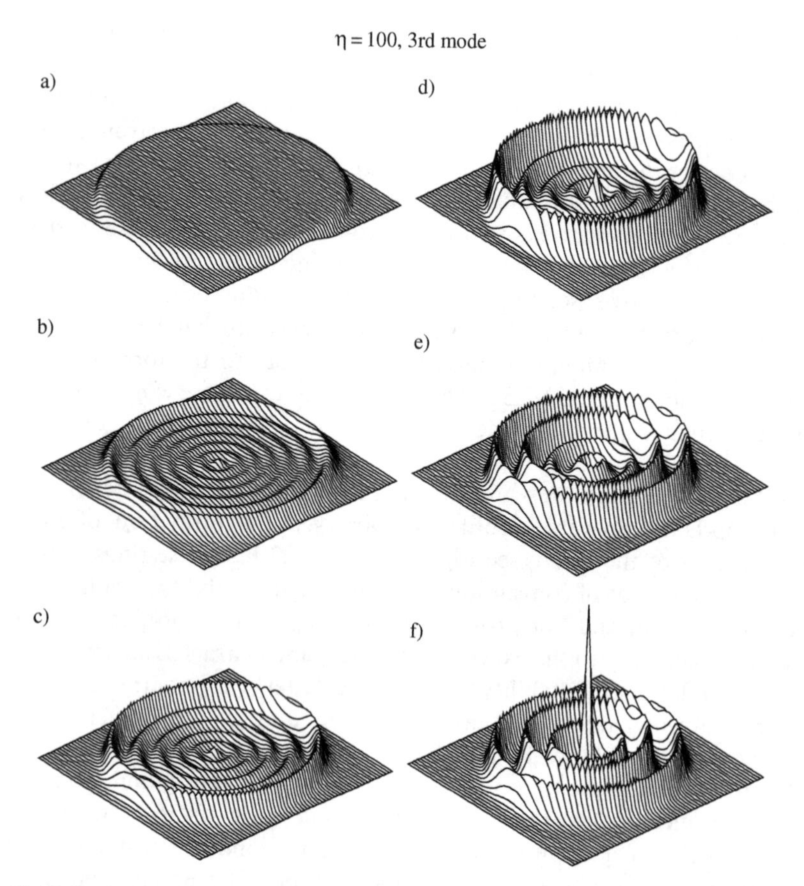

Fig. 22.15 Evolution of an initially Top-Hat beam with $\eta = 100$, this value is close to the integrated intensity of the third-order eigenmode. All plots have the same scale. The mode forms from the edge of the beam toward the center, removing the oscillations due to diffraction

In the simulations that we present we have used beams with $\Theta = 0.04$. Figure 22.15 shows the different stages of propagation of an initial Top-Hat beam with $\alpha = 100$. This value is 10% above the power for the third-order mode. At early stages of propagation the Fresnel diffraction characteristics are evident but, as the beam propagates, a large amplitude ring is formed, and then a second, while the diffractive higher frequency oscillations are disappearing. Finally, a third ring is formed. However, the centre is not stationary in ξ. Linear and nonlinear effects are still competing and the pattern has not reached an exact steady state. Reducing the power to 90 does not removes this oscillation. An important point to observe is the preservation of radial symmetry. We have repeated the calculation for each of the values given in Table 22.1 and the agreement was very good.

22.5.3 Small-Scale Self-Focusing

The two dimensions in the transverse plane of a beam propagating in a homogenous medium allow for an infinite number of possible directions for spatial modulation. With restriction to one degree of freedom, Soto-Crespo et al. found that it is possible to analyze filament structures of paraxial propagation in saturable Kerr media that have low saturation intensity [30]. Although their simulations were three-dimensional, their initial conditions were the steady-state modes which, in principle, should not change in the radial direction. Thus, to some extent, only azimuthal variations were allowed to evolve freely. In our studies, we have considered nonlinear Kerr and saturable Kerr media with high saturation intensity, and the theory developed by Soto-Crespo et al. might not be applicable, in a straightforward manner, to all of our results.

The first experimental reports on filamentation of beams showed that beams with very high powers are unstable and break up into a number of filaments that are not always regularly spaced [14, 28, 31]. In the previous section, we showed that if the system is restricted to be radially symmetric, excellent agreement between the simulations and the theory is obtained. In this section, this restriction is removed and we perturb the otherwise smooth input profiles with white noise of small amplitude. We find that, if the beam is not perturbed deliberately, it maintains radial symmetry.

We have propagated beams with powers in the range 100–900 critical powers. Because catastrophic filamentation was seen to occur in paraxial simulations, we present results for simulations in the nonparaxial regime. The initial transverse beam profiles were, as before, Gaussian and Top-Hat, with beam radius and wavelength such that $\Theta = 0.04$. Figure 22.16 shows the intensity distribution of the central region of a beam bounded by a square of side 2, $|x_{max} - x_{min}| = 2$, for an initially Gaussian profile, but three different perturbations of amplitude that is 1% of the input beam. The input power is such that $\eta = 100$ and the beam has propagated a distance of $\xi = 0.06$. Figure 22.17(a) shows the result when the input Gaussian beam was modulated with the Bessel function J_0 with amplitude 0.01 and at the frequency of maximum gain. In Figs. 22.17(b) and 22.17(c) the Gaussian beam was perturbed with two different noisy masks at the onset of propagation.

Observe that for the noisy beams there is filamentation, but no regular patterns are formed, neither rectangularly nor cylindrically. We also note that different random perturbation leads to a different pattern. In Fig. 22.17(c) we see that two filaments are coalescing into a single filament. Similar conclusions arise for an input power such that $\eta = 400$ (see Fig. 22.17). In this case the three pictures are of the same beam taken at different stages of propagation in which the filaments can be seen growing from the background beam, but the transverse pattern is irregular.

These results are in complete agreement with the experimental report of Pilipetskii and Rustamov [14] who observed that increasing the input beam

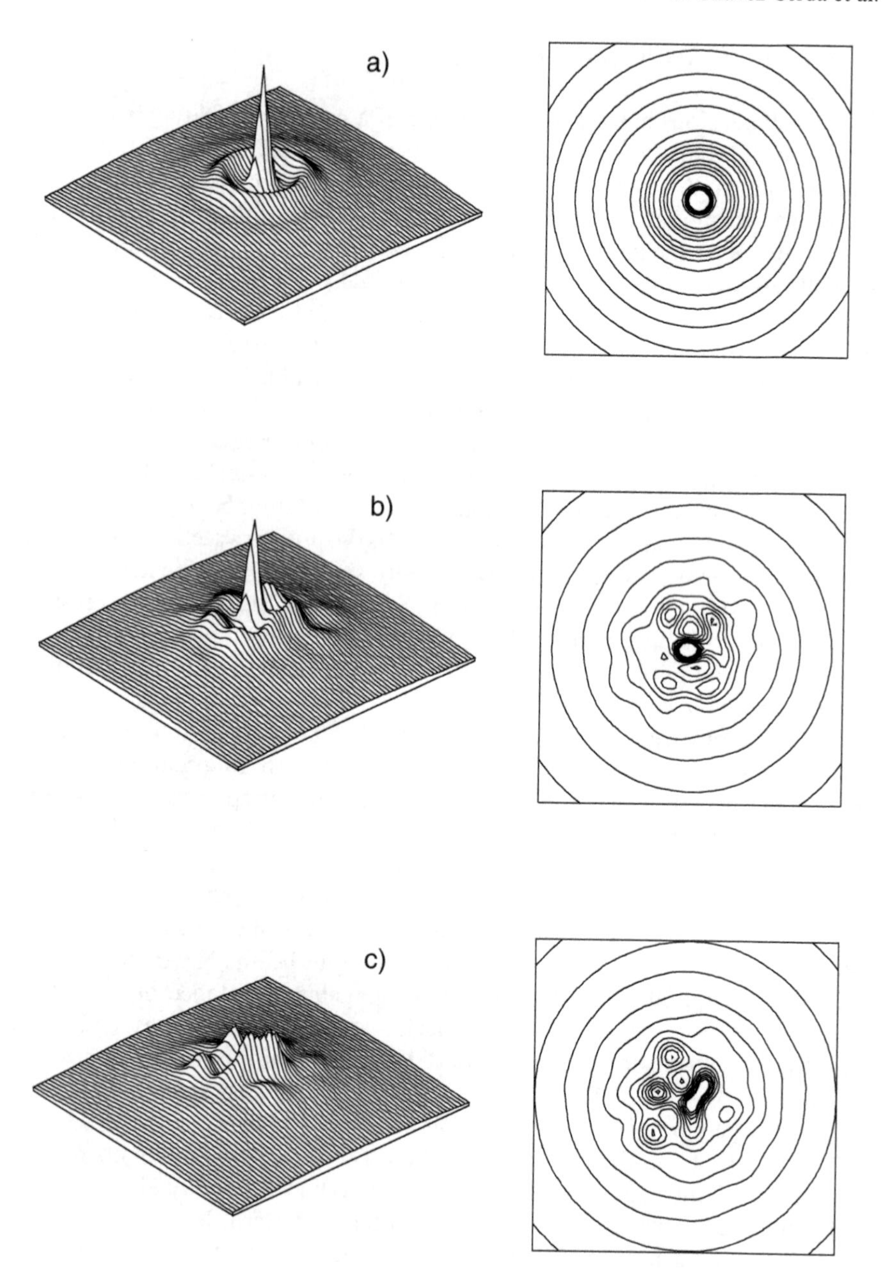

Fig. 22.16 Intensity distribution of an initially Gaussian beam with $\eta = 100$ with three different perturbations: (**a**) $0.005\,J_0(k_{\rho\max}\rho)$; (**b**) and (**c**) random noise generated with different masks. Different noise perturbation produces different pattern; however, the small-scale filaments develop within the region corresponding to the first ring of $J_0(k_{\rho\max}\rho)$

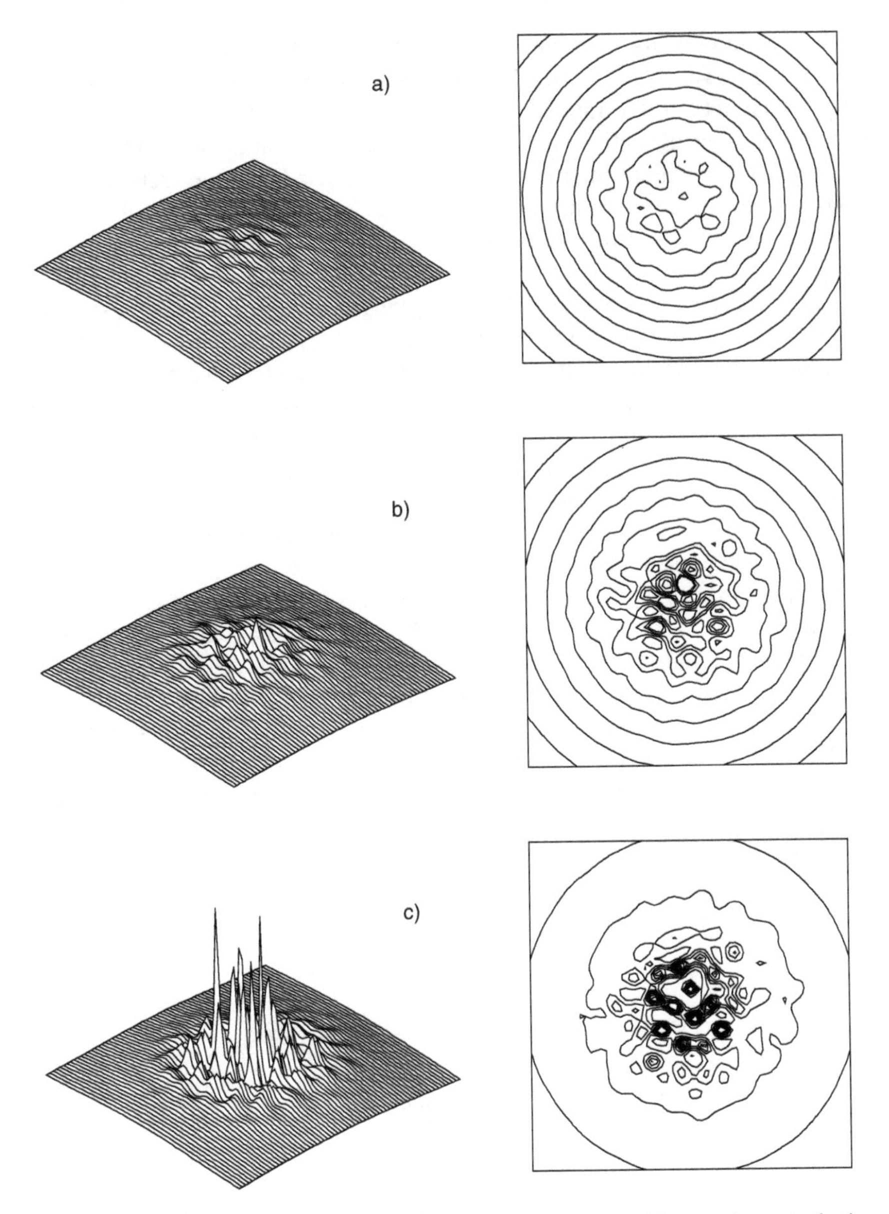

Fig. 22.17 Evolution of small-scale self-focusing filaments generated from noise perturbation on an initial Gaussian beam with $\eta = 400$. No regular pattern is formed by the filaments

power reduces the filaments diameter and that the filament distribution is not necessarily symmetric with respect to propagation axis. These conclusions agree perfectly with the results shown in Figs. 22.16 and 22.17, which are obtained from our nonparaxial simulations.

A very interesting experiment that showed small scale self-focusing was performed by Fleck and Laine [31]. In their experiment the beam formed wide rings which in turn split into small filaments. Although they accompanied their experimental results with simulations, these were done under a radial symmetry assumption. To extend their results, we have modeled just one such diffraction rings by using an annular input profile. To be able to account for higher power, one may change the transverse dimensions in our simulation or increase the value of η.

In Fig. 22.18 we see that the propagating ring splits into a sub-ringed structure and then into a very large number of filaments. This pattern may be compared with their experimental results. We have found, in accord with experiments, that a reduction of the power of the input annulus leads to a corresponding decrease in the number of filaments that are formed. Another feature that is common in both our simulations and this particular experiment is that the initial annulus splits into subrings that subsequently break into filaments. We acknowledge that early experiments performed by Garmire et al. reckon that observed stable-ringed patterns may be related to the modes of the NLSE [12].

The final experiment to be discussed, and the one that we consider to be the most important, is that reported by Campillo et al. [28]. The experimental set-up used involved collimating devices and apertures, and the resulting input beam can be represented by means of an apertured plane wave. The laser that was used delivered input powers of the order of 60–100 critical powers. After our model parameters corresponding to those of the experiment were defined, we performed simulations using Top-Hat beams with powers $\alpha = 60$ to $\alpha = 100$.

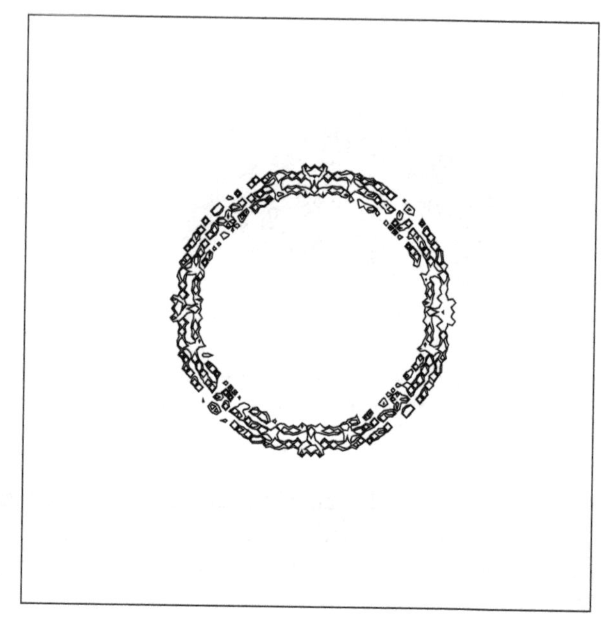

Fig. 22.18 Filament pattern generated from an original ring-beam with the Gaussian profile perturbed with noise. Observe that the pattern consists of several "principal" rings that have a filamented subring structure

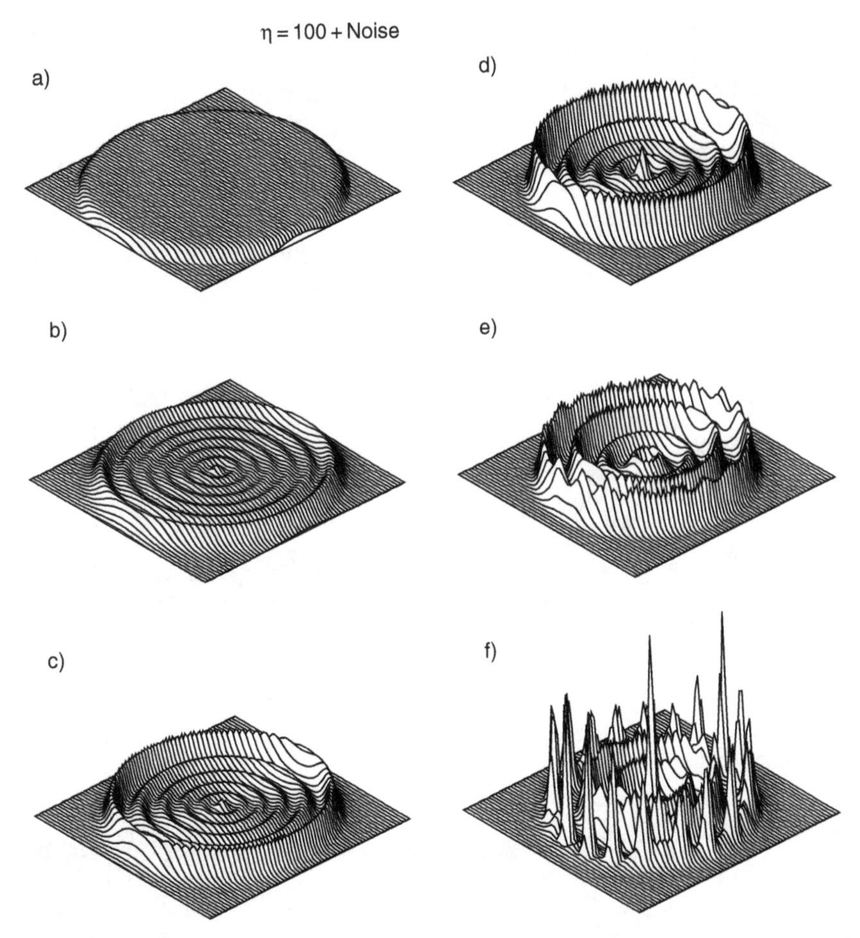

Fig. 22.19 Evolution of a Top-Hat beam with $\eta = 100$ and a noise perturbation at onset. The initial stages are similar to those when the perturbation is absent (Fig. 22.15). When a ring reaches the quasi-steady state, the filaments start to develop (e)

Figure 22.19 shows the intensity characteristics of the beam at several consecutive points of propagation for $\alpha = 100$. White noise, of amplitude 0.005, was added to the input to induce the rings to break up into filaments. Note that the evolution of the beam is very similar to the case in which no noise was added; see Fig. 22.15. Initially high frequency diffraction ripples are formed, and then the predicted eigenmode appears. Thus, the attraction of the solution to the eigenmodes is still a valid conclusion even in the presence of noise. Filamentation occurs, firstly in the outer ring, but only after it has reached a quasi-steady state.

As discussed before, we may consider this ring as an essentially one-dimensional structure. It is possible, however, that either one-dimensional modulational instability or small scale-self-focusing will take place along the circumference of this ring. For modulational instability one expects uniform growth of filaments which are regularly distributed on the circumference. However, the filaments apparently grow independently of each other, and the overall pattern varies as the noise perturbation is changed.

Although there is almost a regular spacing between filaments in the rings, some filaments break the periodicity. Neighboring filaments can also coalesce to form a single filament during propagation. As the inner rings approach a quasi-stable state in the radial direction, filaments start to form on top of them until all the rings have been completely filamented. In Fig. 22.20, we show

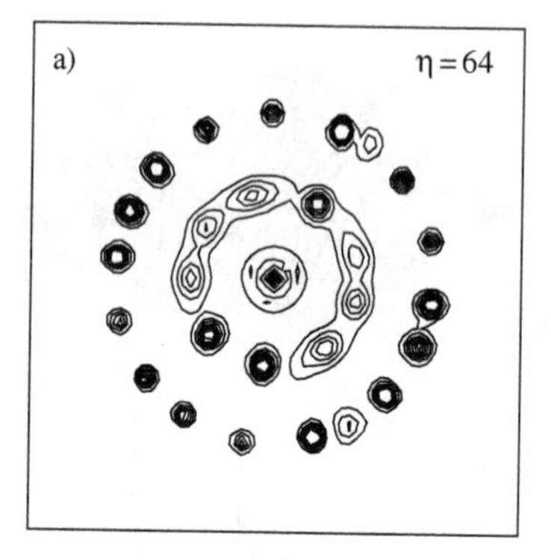

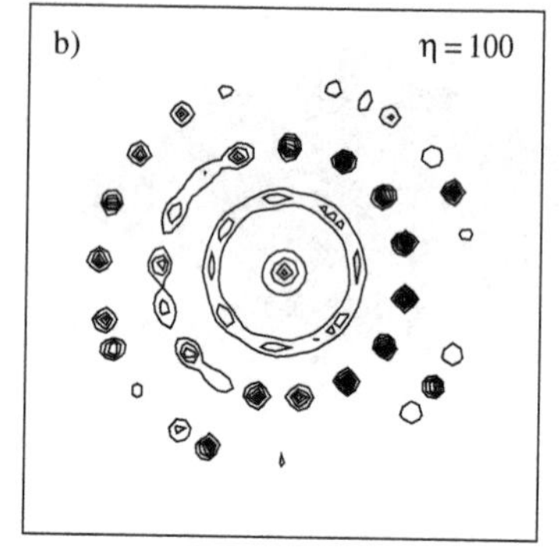

Fig. 22.20 (**a**) Resulting pattern of a perturbed Top-Hat beam with $\eta = 64$. (**b**) Pattern corresponding to the plot (f) in Fig. 5.22 ($\eta = 100$)

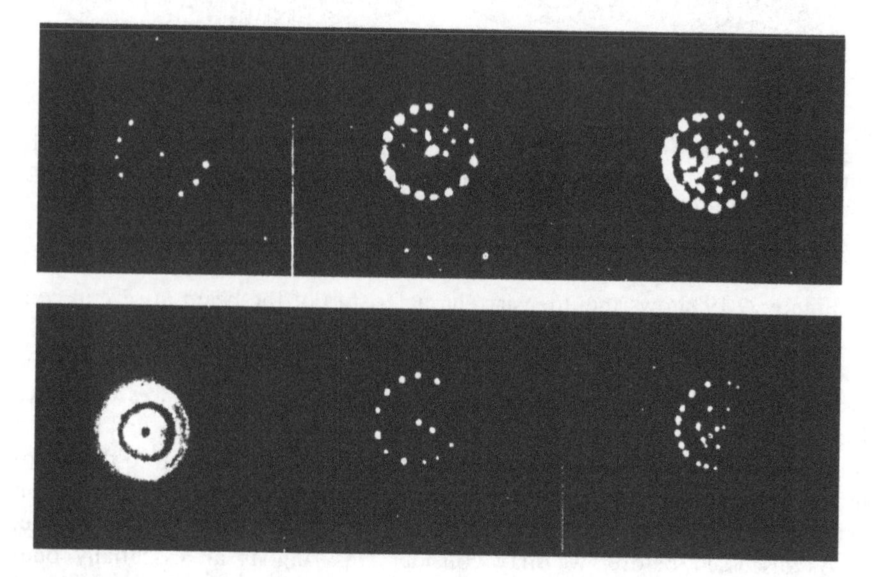

Fig. 22.21 Patterns observed in the experiment published in Ref. [28]. The beam powers used in the experiment correspond to those used in the simulations shown in Fig. 22.20

contour plots of the filamented modes, and we remark on the extraordinary resemblance with the experimental results reported by Campillo et al.; see Fig. 22.21. The similarity is more than approximate. The main impact of our results comes when we remark that the experimental results were explained by means of a simple modulational instability theory. Further to this, the theory assumed a Cartesian frame! Our results suggest that Campillo's results were the NLSE modes which underwent filamentation due to interplay between one-dimensional modulational stability and small-scale self-focusing. One of their results was a uniform ringed pattern, and this observation strongly supports the experimental feasibility of generating the modes. An alternative approach to this kind of filamentation in saturable media is presented in Ref. [32].

22.6 Conclusions

In this chapter we have reviewed the theory of modulation instability and have examined the consequences of allowing cylindrical symmetries. We have proved analytically and numerically, using realistic experimental parameters that, under strict radial symmetry, the modes predicted by the modulational instability theory are in fact described by the zero-order Bessel function of the first kind. When the constraint of radial symmetry is removed, input beam profiles that are perturbed by noise break up into filaments. However, the resulting patterns do not possess any of the symmetries predicted by conventional modulational instability theory. The dominant process that was observed for Gaussian input profiles, which were perturbed with noise, is a type of modulational instability called small-scale self-focusing. The underlying physical mechanism for this type of instability has been verified by the examination of spatial spectra which showed the presence of modulational frequencies bounded by the frequency of maximum gain. This was expected due to the Pythagorean degeneracy of the transverse Cartesian spatial frequencies and because there is no preferred direction, i.e., the nonlinear medium is homogeneous and isotropic.

We have also presented the characteristics of the modes of the nonparaxial NLSE. We have found that their transverse structures coincide with the modes of the paraxial NLSE, and also that they act as attractors in the nonlinear evolution of beams. By propagating collimated beams, each with power content approximately that of a particular mode, we have discovered that ring structures induced by diffraction can be formed in which the number of rings corresponds to the respective mode. These ring structures, once formed, propagate with almost no change either in the radial direction or with respect to their amplitude. When input beams are perturbed with noise, these eigenmodes are still formed, but undergo filamentation due to the interplay between one-dimensional modulational instability and small-scale self-focusing after each ring has reached its quasi-steady state. The transverse patterns created in this way have an extraordinary resemblance with published experimental results. We have also shown that a laboratory realization of the modes of the NLSE is feasible and that they may also have implication in nonparaxil formation of light bullets.

Acknowledgments The authors acknowledge support by CONACYT (SALUD-2005-01-14012), INAOE (Mexico) and the CAPES (Brasil). SCHC deeply acknowledges fruitful discussions and proof reading to G. S. Mcdonald and G. H. C. New.

References

1. V.I. Bespalov, V.I. Talanov: Filamentary structure of light beams in nonlinear liquids, *JETP Lett.* **3**, 307–310 (1966).
2. P. Chernev, V. Petrov: Self-focusing of light-pulses in the presence of normal group-velocity dispersion, *Opt. Lett.* **17**, 172–174 (1992).
3. J.E. Rothenberg: Space–time focusing: breakdown of the slowly varying envelope approximation in the self-focusing of femtosecond pulses, *Opt. Lett.* **17**, 1340–1342 (1992).
4. Y.R. Shen: Self-focusing: experimental, *Progr. Quant. Electr.* **4**, 1–34 (1975).
5. J.H. Marburger: Self-focusing: theory, *Progr. Quant. Electr.* **4**, 35–110 (1975).
6. A.E. Siegman: *Lasers*, University Science Books, Mill Valley, CA (1986).
7. S. Chavez-Cerda: A new approach to Bessel beams, *J. Mod. Opt.* **46**, 923–930 (1999).
8. L.W. Liou, X.D. Cao, C.J. McKinstrie et al.: Spatiotemporal instabilities in dispersive nonlinear media, *Phys. Rev. A* **46**, 4202–4208 (1992).
9. A.J. Campillo, S.L. Shapiro: Relationship of self-focusing to spatial instability modes, *IEEE J. Quant. Electr.* **QE-10**, 705–706 (1974).
10. A.J. Campillo, S.L. Shapiro, B.R. Suydam: Relationship of self-focusing to spatial instability modes, *Appl. Phys. Lett.* **24**, 178–180 (1974).
11. E.S. Bliss, D.R. Speck, J.F. Holzrichter et al.: Propagation of a high-intensity laser pulse with small-scale intensity modulation, *Appl. Phys. Lett.* **25**, 448–450 (1974).
12. E. Garmire, R.Y. Chiao, C.H. Townes: Dynamics and characteristics of the self-trapping of intense light beams, *Phys. Rev. Lett.* **16**, 347–349 (1964).
13. J.E. Bjorkholm, A. Ashkin: CW self-focusing and self-trapping of light in sodium vapor, *Phys. Rev. Lett.* **32**, 129–132 (1974).
14. N.F. Pilipetskii, A.R. Rustamov: Observation of self-focusing of light in liquids, *JETP Lett.* **2**, 55–56 (1965).
15. M.D. Feit, J.A. Fleck, Jr.: Beam nonparaxiality, filament formation, and beam breakup in the self-focusing of optical beams, *J. Opt. Soc. Am. B* **5**, 633–640 (1988).
16. A.L. Dyshko, V.N. Lugovoi, A.M. Prokhorov: Self-focusing of intense light beams, *JETP Lett.* **6**, 146–148 (1967).
17. V.N. Lugovoi, A.M. Prokhorov: Theory of the propagation of high-power laser radiation in a nonlinear medium, *Sov. Phys. Uspekhi* **16**, 658–679 (1974).
18. N. Akhmediev, A. Ankiewicz, J. M. Soto-Crespo: Does the nonlinear Schrödinger equation correctly describe beam propagation? *Opt. Lett.* **18**, 411–413 (1993).
19. J.M. Soto-Crespo, N. Akhmediev: Description of the self-focusing and collapse effects by a modified nonlinear Schrödinger-equation, *Opt. Commun.* **101**, 223–230 (1993).
20. H.A. Haus: Higher order trapped light beam solutions, *Appl. Phys. Lett.* **8**, 128–129 (1966).
21. Z.K. Yankauskas: Radial field distributions in a self-focusing light beam, *Radiophys Quant. Electron.*, **9**, N 2, 261–263 (1966).

22. Y, Chen: TE and TM families of self-trapped beams, *IEEE J. Quant. Electron.* **27**, N5, 1236 (1991).
23. S.N. Vlasov, V.I. Talanov: Wave self-focusing, Institute of Applied Physics of the Russian Academy of Science, Nizhny Novgorod, 220 pp (1997).
24. D. Anderson, M. Bonnedal, M. Lisak: Self-trapped cylindrical laser beams, *Phys. Fluids* **22**, 1838–1840 (1979).
25. Y.J. Chen: Self-trapped beams with cylindrical symmetry, *Opt. Comm.* **82**, 255–259 (1991).
26. R.D. Small: Paraxial self-trapped beams in non-linear optics, *J. Math. Phys.* **22**, 1497–1503 (1981).
27. P.K. Newton, S. Watanabe: The geometry of nonlinear Schrödinger standing waves - pure power nonlinearities, *Physica D* **67**, 19–44 (1993).
28. A.J. Campillo, S.L. Shapiro, B.R. Suydam: Periodic breakup of optical beams due to self-focusing, *Appl. Phys. Lett.* **23**, 628–630 (1973).
29. D.R. Heatley, E.M. Wright, G.I. Stegeman: Spatial ring emission and filament formation in an optical fiber with a saturable nonlinear cladding, *Opt. Lett.* **16**, 291–293 (1991).
30. J.M. Soto-Crespo, D.R. Heatley, E.M. Wright et al.: Stability of the higher-bound states in a saturable self-focusing medium, *Phys. Rev. A* **44**, 636–644 (1991).
31. J.A. Fleck, Jr., C. Layne: Study of self-focusing damage in a high-power Nd-glass-rod amplifier, *Appl. Phys. Lett.* **22**, 467–469 (1973).
32. J.M. Soto-Crespo, E.M. Wright, N.N. Akhmediev: Recurrence and azimuthal-symmetry breaking of a cylindrical Gaussian beam in a saturable self-focusing medium, *Phys. Rev. A*, **45**, 3168–3175 (1992).

Chapter 23
Self-Focusing and Solitons in Photorefractive Media

E. DelRe and M. Segev

Abstract We describe the basic physical mechanisms supporting the formation of spatial solitons in photorefractive crystals, and provide an up-to-date account of the developments in the field.

23.1 Introduction

Diversity and complexity, on one side, and extreme regularity and stability, on the other, are two faces of nonlinearity. Solitons are a paradigm of the second, deriving their scientific and technological importance from a remarkable universality and a specific amenability to application. The phenomenological trait of a soliton is a nonlinear wave that propagates without suffering distortion to the point that, when made to interact with other waves, it maintains its localized identity in a manner analogous to a particle. This striking stability and robustness is a consequence of the action of two counterbalancing effects: linear dispersion and nonlinear self-phase modulation, a dynamic feedback loop that locks the wave into a soliton.

Although soliton science dates more than a century back, the accessible generation and observation of solitons in optics has in the past decade caused a revival of interest. A tassel in this revival is without doubt played by photorefractive spatial solitons, which are micron-size beams that propagate even tens of millimeters without diffracting. Discovered in 1992, they have become arguably the principal playground for soliton studies, pairing relative ease in experiments with a rich diversity of underlying physical mechanisms.

In this chapter we describe how photorefraction can support optical spatial solitons, review some of the principal phenomenology, recall the main implications they have had on soliton science, and discuss some potential applications.

E. DelRe (✉)
Dipartimento di Ingegneria Elettrica e dell'Informazione, Università dell'Aquila and INFM-CNR CRS SOFT, Italy
e-mail: edelre@ing.univaq.it

R.W. Boyd et al. (eds.), *Self-focusing: Past and Present*,
Topics in Applied Physics 114, DOI 10.1007/978-0-387-34727-1_23,

Previous reviews can found in [1,2,3,4,5], whereas an up-to-date account can be found in [6].

23.2 Self-Trapping in Photorefractives

One of the principal characteristics of photorefraction is that optical self-action can build up over time, accumulating into a strong nonlinearity even for low-power continuous-wave laser beams [7,8]. In 1992, Segev et al. proposed the first scheme to use photorefraction to self-focus and ultimately self-trap a low-power beam into a soliton [9]. The clue to their discovery was in how they were able to transform conventional photorefractive nonlinearity from a wave-mixing process to a mutual phase-modulation process. Photorefraction mediates self-action through the redistribution of photoexcited charge, which, forming a light-induced space-charge field, modulates the material index of refraction through electro-optic effects. In the absence of an external bias field, photorefraction is driven by diffusion of charge carriers. This leads to a light-induced change in the index of refraction that scatters light from the original beam into new optical modes in different directions, leading to a process known as beam fanning. Segev et al. suggested that, when the diffusion space-charge field can be neglected with respect to the external bias field, strong self-focusing occurs and self-trapping becomes possible [9,10].

The first experiments, reported in [11] were carried out with the set-up shown in Fig. 23.1. The beam was launched in a zero-cut uniaxial photorefractive sample of rhodium-doped SBN (strontium-barium-niobate) along the ordinary

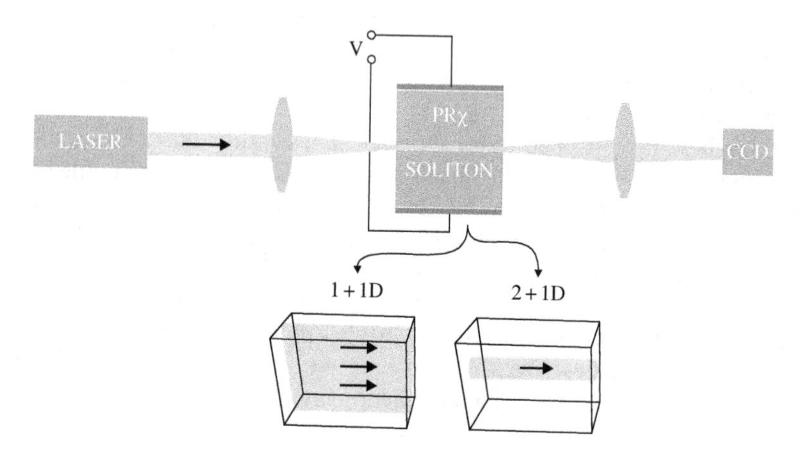

Fig. 23.1 Basic experimental scheme allowing the observation of 1 + 1D or 2 + 1D photorefractive solitons. For the 1 + 1D case, the laser beam is focused by a cylindrical lens onto the input face of the photorefractive crystal (PRχ). The beam is propagating in the crystal, exiting at the output face, and is imaged by a second lens onto a CCD camera. For the 2 + 1D case, the beam is focused by a standard spherical lens. The beam forms a soliton for an appropriate applied V and during a specific time window

axis a, with the external biasing field E_0 applied along the poling optical axis c, through two electrodes brought to a relative potential V. A characteristic electro-optic change in index of refraction $|\Delta n| \simeq (1/2)n^3 r_{33} E_0 \sim 2 - 5 \cdot 10^{-4}$ can be reached for fields of the order of E_0 ~1-3 kV/cm ($r_{33} \simeq 220$ pm/V), a nonlinearity sufficient to produce the self-lensing to compensate the diffraction of a 10 μm wide beam in the visible wavelength range.

The basic result of early experiments was the self-trapping of visible continuous-wave beams into spatial solitons. A 15-μm-wide continuous-wave 457 nm μW beam was observed to propagate without spreading for external fields in the range of 400–500 V/cm. The effect was found to persist even under the influence of considerable noise and under conditions in which the launch was not of the shape thought to be optimal for soliton formation [12]. Within a short time a series of experiments confirmed this finding and provided the phenomenological basis for the field [12–14].

Although the results confirmed the basic qualitative predictions, they reflected a far richer and more complex phenomenology. Two principal and interesting features were observed. First, the self-trapping was transient, occurring during a specific temporal window. This has caused the self-trapping phenomenon to be termed a quasi-steady-state soliton. The second feature observed was that self-trapping can be obtained in both one and two transverse directions, suggesting the existence not only of 1+1D (one-transverse-plus-one-propagation-dimension) photorefractive solitons (envisaged by the first models) (see Fig. 23.4), but also of 2+1D (two-transverse-plus-one-propagation-dimension) solitons (see Fig. 23.3).

In a second set of experiments, it was found that illuminating the photorefractive sample with a second plane-wave background beam, effectively increasing the dark conductivity in the sample (see Fig. 23.2), a particular set of experimental parameters transformed the transient (quasi-steady-state)

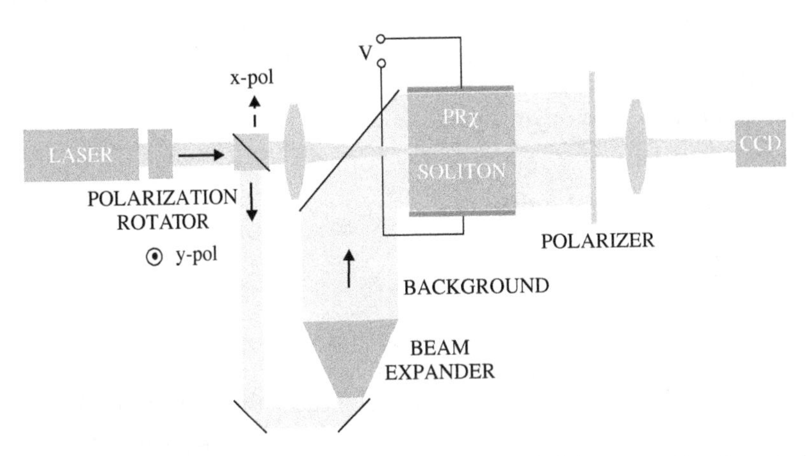

Fig. 23.2 A scheme to generate screening solitons [20–22]. The extraordinarily polarized soliton-forming beam is co-propagating with an ordinarily polarized beam of uniform intensity

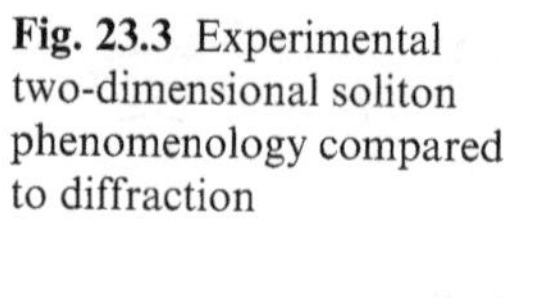

Fig. 23.3 Experimental two-dimensional soliton phenomenology compared to diffraction

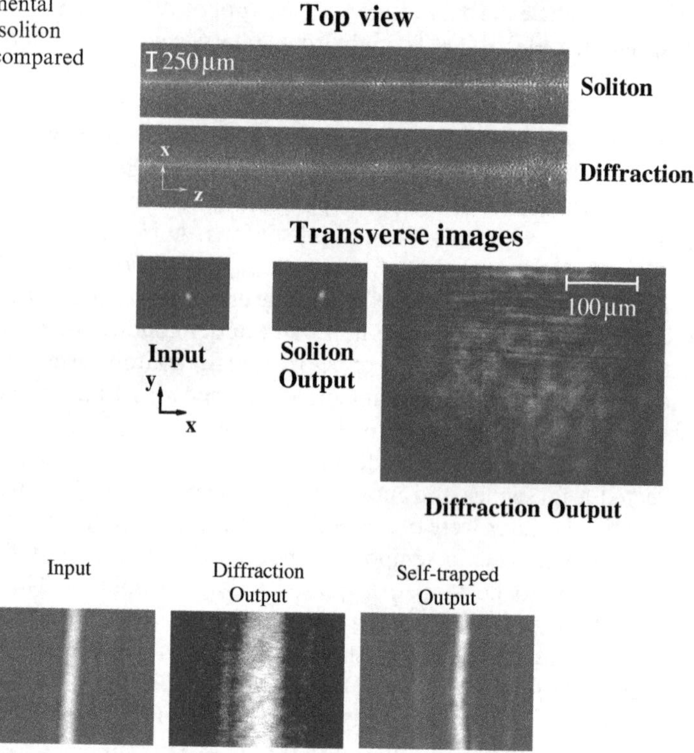

Fig. 23.4 A one-dimensional soliton observed in biased KLTN. From left, input 9 μm FWHM intensity distribution, output 29 μm linear diffraction (no bias), and output 9 μm self-trapped beam

nature of the solitons into stable steady-state effects [15–19]. These steady-state photorefractive solitons, which are termed screening solitons and form the most commonly studied photorefractive self-trapped beams, have since been observed in SBN, BSO, BGO, BTO, $BaTiO_3$, $LiNbO_3$, InP, CdZnTe, KLTN, $KNbO_3$, polymers, and organic glass.

23.3 Nonlinear Mechanism

23.3.1 Photorefraction

Photorefraction is observed in specific doped electro-optic crystals. In the most common case, impurities form deep donor sites and, in a lesser concentration, acceptor sites. At visible wavelengths the crystal is transparent but the donor site can be photoionized. Hence illuminated regions generate an out-of-equilibrium concentration of mobile electrons that rearranges into a space-charge distribution by diffusing to less illuminated regions and by drifting in an externally applied electric field. The charge distribution settles into the donor sites that, at equilibrium, are ionized by nearby acceptor sites, thus rendering the dislocation semi-permanent. Yet the space-charge creates a field, the space-charge field, that changes (locally) the index of refraction through the electro-optic effect. This causes the light beam (which originally generated the charge) to experience changes in its waveform, which again changes the charge distribution and hence the space charge field. The process eventually leads to the formation of a soliton when the light beam induces such a refractive index change that acts as a waveguide, while guiding the light beam itself in its own induced waveguide.

23.3.2 *Light-Induced Space-Charge Field*

To grasp how photorefraction can support solitons, the first step is to simplify the system to a one-dimensional condition, in which the beam depends only on one transverse direction x, i.e., the light intensity is $I(x,z)$, z being the propagation axis, and for conditions in which a time-independent steady-state regime has been reached. The corresponding solitons, for which I is moreover independent of z, are 1+1D solitons. In the conditions of interest, the optical intensity distribution I is such that the resulting concentration of photo-excited electrons N, the concentration of acceptor impurities N_a, and the concentration of donor impurities N_d follow the scaling $N << N_a << N_d$. In this case the x-directed space-charge field E is related to the optical intensity I through the nonlinear differential equation [23]

$$E(I_b + I)\frac{1}{1 + \frac{\epsilon}{N_a q}\frac{dE}{dx}} + \frac{k_B T}{q}\frac{d}{dx}\left((I_b + I)\frac{1}{1 + \frac{\epsilon}{N_a q}\frac{dE}{dx}}\right) = g, \tag{23.1}$$

where ϵ is the sample dielectric constant, q is the electron charge, k_B is the Boltzmann constant, T is the temperature, and g is a constant related to the boundary conditions, i.e., to the voltage V applied on the x-facets L_x apart. I_b is the effective background illumination, the homogeneous optical intensity that allows a finite crystal conductivity.

The structure of Eq. (23.1) even in the one-dimensional case is complicated. Setting $Y = E/E_0$, $Q = (I_b + I)/I_b$, and $\xi = x/x_q = x/[\epsilon E_0/(N_a q)]$, Eq. (23.1) is

$$\frac{YQ}{1 + Y'} + a\left[\frac{Q'}{1 + Y'} - \frac{Q}{(1 + Y')^2}Y''\right] = G, \tag{23.2}$$

with $a = N_a k_B T/\epsilon E_0^2$ and $G = g/E_0 I_b$. The prime stands for $(d/d\xi)$, or equally

$$Y = -a\frac{Q'}{Q} + \frac{G}{Q} + \frac{GY'}{Q} + a\frac{Y''}{1 + Y'}. \tag{23.3}$$

This form is rendered tractable by the fact that the greater part of spatial soliton study involves the trapping of beams with an intensity full-width-half-maximum (FWHM) $\Delta x \sim 10\,\mu\text{m}$. For most configurations, $x_q \sim 0.1\,\mu\text{m}$, and $\eta = x_q/\Delta x \sim 0.01$ represents a smallness parameter, and the evaluation of the various terms indicates that

$$Y^{(0)} = \frac{G}{Q} + o(\eta), \tag{23.4}$$

since $a \sim 2.5$, and $G \simeq -1$ [19]. A first correction is obtained by iterating this solution into Eq. (23.3), and the resulting expression for Y is

$$Y^{(1)} = \frac{G}{Q} - a\frac{Q'}{Q} - \frac{Q'}{Q}\left(\frac{G}{Q}\right)^2 + o(\eta^2). \tag{23.5}$$

The first dominant term is generally referred to as the screening term and is the main agent leading to solitons. It is a local term, in the sense that the field (and hence the index change) at a given location depends on the optical intensity only at that (same) location. This "locality feature" is manifested in the fact that this leading term does not involve spatial derivatives or integration, has the same symmetry of the optical intensity Q, and represents a decrease in E with respect to E_0 as a consequence of charge rearrangement ($G \simeq -1$).

The second term, of first order in η, involves a spatial derivative and can be identified with the diffusion field. The third, again of first order in η, is the coupling of the diffusion field with the screening field, a component sometimes referred to as deriving from charge-displacement [24]. Both these two last terms are nonlocal, in that they involve a spatial derivative, and thus provide an anti-symmetric contribution to the space charge field (Y) for a symmetric beam $I(x) = I(-x)$. That is, these last two terms lead to a beam self-action with symmetry opposite to that required to support solitons. Such terms lead to beam self-bending, which for most configurations amounts to a slight parabolic distortion of the preferentially z-oriented trajectory. The subject has attracted interest over the years [14,21,24] and has helped build an understanding into the limits of the local saturable nonlinearity model [25].

23.3.3 Nonlinear Index Change

In order to identify the nonlinearity, we must now translate the space-charge field E into an index modulation. Screening solitons are observed both in the noncentrosymmetric ferroelectric phase (e.g., room temperature SBN) and in the centrosymmetric paraelectric phase (e.g., room temperature KLTN). The standard configuration for generating screening solitons is such that a zero-cut crystal is positioned so that the x-axis is the direction along which E_0 is applied, parallel to the optical axis for ferroelectrics, the soliton beam of intensity I is extraordinarily-polarized and is propagating along z, while I_b is obtained through a co-propagating ordinarily polarized plane-wave [20]. For a noncentrosymmetric photorefractive crystal, like SBN, $\Delta n = -\frac{1}{2}n^3 r_{33}E$, n being the unperturbed crystal index of refraction, and r_{ij} the linear electro-optic tensor of the sample. Consistent with our iterative scheme of Eq. (23.4), we obtain the nonlinearity [16]

$$\Delta n(I) = -\frac{1}{2}n^3 r_{33}\frac{V}{L_x}\frac{1}{1+I/I_b} = -\Delta n_0\frac{1}{1+I/I_b}, \quad (23.6)$$

which constitutes a saturable nonlinearity. The nature of the self-action evidently depends on the sign of Δn_0, and is self-focusing, when $\Delta n_0 > 0$ and defocusing for $\Delta n_0 < 0$. The sign of Δn_0 is established by the orientation of the external bias with respect to the crystalline axes, having established the sign of r_{33} with respect to the chosen system of reference. For example, in SBN applying E_0 in the direction of the crystalline (ferroelectric) axis we observe a self-focusing nonlinearity. It is possible to apply E_0 in a direction opposite to the ferroelectric axis, thus effectively changing the sign of Δn_0, then E_0 must be smaller than the coercive field, otherwise it may render the ferroelectric crystalline structure unstable and de-pole the crystal.

Analogously, for the centrosymmetric case of KLTN, the electro-optic response is quadratic $\Delta n = -(1/2)n^3 g_{eff}(\epsilon_r - 1)^2\epsilon_0^2 E^2$, where g_{eff} is the effective quadratic electro-optic coefficient, and ϵ_0 and ϵ_r are the vacuum and relative dielectric constants, and the zero order in η solution is [26,27]

$$\Delta n(I) = -\Delta n_0\frac{1}{(1+I/I_b)^2}, \quad (23.7)$$

where $\Delta n_0 = (1/2)n^3 g_{eff}(\epsilon_r - 1)^2\epsilon_0^2 E_0^2$. Here the nonlinearity is either focusing or defocusing, depending on the sign of g_{eff} and not evidently on the orientation of the applied field E_0.

23.3.4 The Soliton-Supporting Nonlinear Equation

Soliton formation is described by the nonlinear wave equation, representing the evolution of the beam in the light-induced index of refraction pattern Δn. Under scalar conditions (i.e., when no relevant polarization dynamics intervene), and for beam sizes much larger than the wavelength of the monochromatic beam, this evolution is described by the nonlinear monochromatic paraxial equation

$$\left[\frac{\partial}{\partial z} - \frac{i}{2k}\frac{\partial^2}{\partial x^2}\right]A(x,z) = -\frac{ik}{n}\Delta n A(x,z) \quad (23.8)$$

where $k = 2\pi n/\lambda$ is the wave-vector, A is the slowly varying optical field, i.e., $E_{opt}(x, z, t) = A(x, z)\exp(ikz - i\omega t)$, $\omega = 2\pi c/n\lambda$, and $I = |A(x, z)|^2$.

Soliton solutions can now be identified through the self-consistency method. The balancing of diffraction (second term on the left-hand side of Eq. (23.8)) by the nonlinearity (the term on the right-hand side) leads to a solution A that has a stationary or non-evolving intensity distribution I, and hence must be of the form $A(x, z) = u(x)e^{i\Gamma z}\sqrt{I_b}$. The transverse spatial scale is normalized to the so-called nonlinear length scale $d = (\pm 2kb)^{-1/2}$, i.e., $\xi = x/d$. For photorefractive solitons in noncentrosymmetric crystals where the electro-optic response is linear in the field, i.e., $\Delta n = -\frac{1}{2}n^3 r_{33}E$, $b = (1/2)kn^2 r_{33}(V/L_x)$ and we obtain [16,19]

$$\frac{d^2u(\xi)}{d\xi^2} = \pm\left(\frac{\Gamma}{b} - \frac{1}{1+u(\xi)^2}\right)u(\xi). \tag{23.9}$$

The plus sign is for $b>0$, the minus for $b<0$. The sign of b corresponds to the sign of Δn_0, and, as mentioned above, implies a self-focusing, for $b>0$, or a self-defocusing, for $b<0$, nonlinearity, having established that E decreases across the beam profile (see Eq. (23.6)). Both defocusing and focusing nonlinearities support solitons. A self-focusing nonlinearity traps a conventional bell-shaped beam into a bright soliton; a self-defocusing nonlinearity gives rise to a dark soliton: a non-broadening dark notch generated by a π phase jump upon an otherwise uniform amplitude wave.

A parallel formulation holds for paraelectrics, where Eq. (23.7) substitutes Eq. (23.6) [26].

23.3.5 Soliton Waveforms and Existence Curve

The basic screening nonlinearity expressed by Eq. (23.6) indicates that what plays a role in the attainment of self-trapping is the ratio I/I_b, but not the actual value of the intensity. This important result is the basis for low-power solitons in photorefractives, the logical consequence of a cumulative effect brought to steady-state. Yet not any bell-shaped beam will necessarily self-trap. Equation (23.9) identifies the specific set of waveforms that can form solitons, and the observation of self-trapping is conditioned to launching a beam that reasonably approximates a given soliton waveform. Furthermore, given the saturable nature of Δn, the self-trapped waveforms of Eq. (23.9) not only do not have an explicit form, such as those of standard Kerr solitons, but more importantly, their shape changes for different values of saturation.

The experimentally accessible parameters are evidently not the actual beam shape, but the nonlinear paramater b, the beam width, or full-width-at-half-maximum Δx, and the intensity, which, for bright solitons, is generally parametrized through the intensity ratio $u_0^2 = I(0)/I_b$, i.e., the beam peak intensity at

$x=0$ normalized to the background intensity. Similarly for dark solitons, the relevant parameter is $u_\infty^2 = I(x \to \infty)/I_b$. The fundamental role of Eq. (23.9) is in providing, for each value of nonlinear response b, the values of u_0 and Δx (or $\Delta\xi$) of the bright soliton solution, the set of these points in the $(u_0, \Delta\xi)$ parameter space being termed the soliton existence curve. Analogously, the dark soliton existence curve will be in the $(u_\infty, \Delta\xi)$ plane. The experimental generation of, for example, a bright soliton will then be achieved by launching a bell-shaped waveform, such as a Gaussian beam from a laser, with a correct value of Δx and u_0, for the given b.

To construct the existence curve the first step is to reduce the number of relevant parameters in Eq. (23.9) by noting that it can be integrated once, giving the relationship $\Gamma/b = \log(1+u_0^2)/u_0^2$ for bright beams, and $\Gamma/b = 1/(1+u_\infty^2)$ for dark, where $u_\infty = u(\infty) = -u(-\infty)$. Thus, for example, for bright solitons Eq. (23.9) is

$$\frac{d^2u(\xi)}{d\xi^2} = \left(\frac{\log(1+u_0^2)}{u_0^2} - \frac{1}{1+u(\xi)^2}\right)u(\xi). \tag{23.10}$$

Next, Eq. (23.10) is integrated once by quadrature, and then the resultant first-order ordinary differential equation can be solved numerically [16,19]. The values of $(u_0, \Delta\xi)$ that correspond to the soliton waveforms are obtained by solving a simple integral numerically, giving rise to the soliton existence curve.

The usefulness of the notion of a soliton existence curve is evidently associated with experiments. As solutions of Eq. (23.10), the different points in the parameter space represent different levels of saturation and hence different waveforms, and u_0 and $\Delta\xi$ are not sufficient to characterize them. In experiments, the launch beam is a focused Gaussian beam from a laser, and for this family of beams, u_0 and FWHM unambiguously identify the waveform. The point is that since the soliton solutions are stable and robust with respect to perturbations, they attract the beam dynamics to the closest self-trapped solution by reshaping the initial launch beam in the first segments of propagation. In turn, this closest self-trapped solution will have, to a good approximation, the very same u_0 and $\Delta\xi$ of the launch.

23.3.6 Experiments and Theory

The basic test for the validity of the approximations leading to Eq. (23.9) is in comparing the experimental conditions leading to solitons with those predicted through the existence curve. The best established method to carry out this test is to keep the actual Gaussian launch beam unchanged, and scan, for each fixed value of u_0, the value of E_0 that causes self-trapping. Because Δx is fixed, changing E_0 changes $\Delta\xi$ through d.

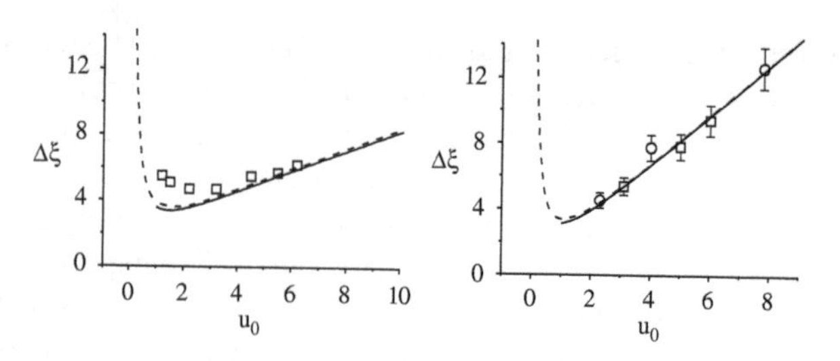

Fig. 23.5 Existence curves (dashed curve) for bright noncentrosymmetric (left) and centrosymmetric (right) solitons compared to results, from Refs. [22] and [28]. The solid line is the explicit asymptotic function describing the existence conditions from [28]

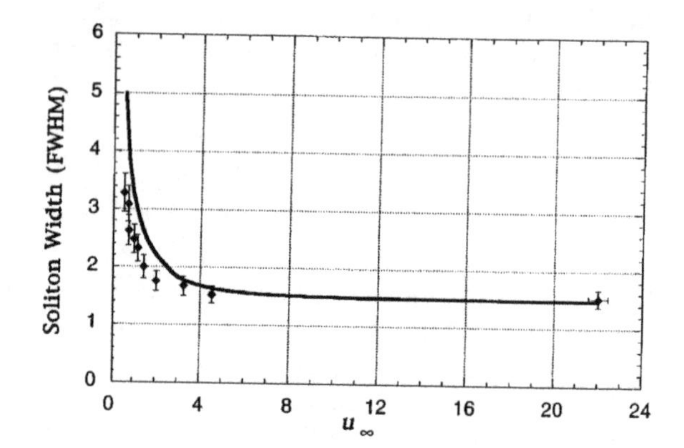

Fig. 23.6 Comparison between experiments and theory for (1 + 1)D dark screening solitons, from [29]

Experimental results compared to theory are shown in Fig. 23.5 for bright solitons and in Fig. 23.6 for dark [19,26,27,29]. Whereas qualitative agreement is full, quantitative agreement in some experiments is weaker. The very small discrepancy in these cases is generally attributed to partial guiding of the background beam, for which also I_b depends on x, and in the evaluation of the actual value of the electro-optic coefficients, which depend on poling, clamping, and temperature.

23.4 Two-Dimensional Solitons

Photorefraction is able to support, both as quasi-steady-state and as steady-state effects, also 2 + 1D solitons (see Fig. 23.3). These are beams whose intensity $I(x, y)$ is well-localized in both transverse directions x and y [11,20,21]. They form in the greater part of photorefractive materials, such as other ferroelectrics [30], semiconductors [31], paraelectrics [32], sillenites [33], and for most types of self-trapping: bright, photovoltaic [34], multimode [35–37], and incoherent [38,39]. Furthermore, even (2 + 1)D dark solitons have been observed in photorefractives, in quasi-steady-state [40] and in steady-state [41] under a bias field,

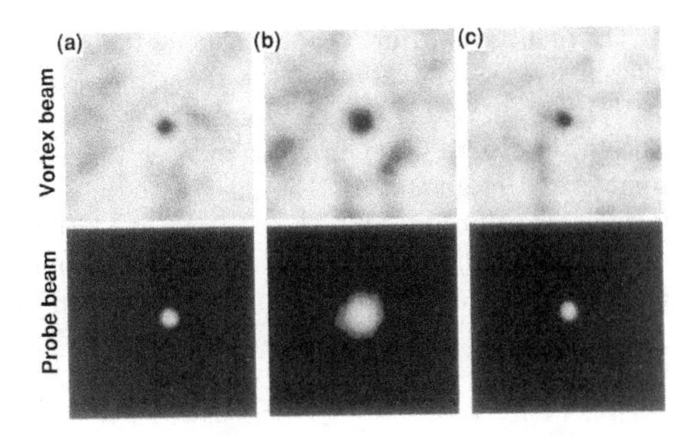

Fig. 23.7 A vortex screening soliton from [41]. (**a**) Input intensity distribution of the vortex; (**b**) Diffracting vortex after linear propagation to the output of the sample; (**c**) Self-trapped output intensity distribution in a biased sample. (*bottom*) Probe beam guided propagation

as well as photovoltaic [42] and incoherent [43] dark "vortex" solitons. An example of a dark vortex screening soliton is shown in Fig. 23.7.

Because the Kerr nonlinearity cannot lead to stable 2 + 1D bright solitons, the effect, which appears similar to the formation of a self-induced optical fiber inside the bulk of the sample, represents an important achievement for optical soliton science, and the associated studies have contributed to understanding of higher-than-one-dimensional nonlinear waves.

The generalized relationship between the light-induced electric field **E** and the optical intensity of the beam I is

$$\nabla \cdot \left[\mathbf{E}(I_b + I) \frac{1}{1 + \frac{\epsilon}{N_a q} \nabla \cdot \mathbf{E}} + \frac{k_b T}{q} \nabla \left((I_b + I) \frac{1}{1 + \frac{\epsilon}{N_a q} \nabla \cdot \mathbf{E}} \right) \right] = 0, \quad (23.11)$$

and the irrotational condition for the dc space-charge field

$$\nabla \times \mathbf{E} = 0, \quad (23.12)$$

along with proper boundary conditions (the field at the electrodes). Although this two-dimensional situation can in general create components of the field **E** both in the x and in the y directions, in most conditions $E_x \gg E_y$ (there is an intrinsic asymmetry in the direction of the externally applied field E_0 along x), and hence the generally tensorial electro-optic response reduces to a scalar response $\Delta n = -(1/2)n^3 r_{\text{eff}} E_x$ analogous to the 1 + 1D case for an x-polarized beam. In this manner, the propagation equation is the scalar (i.e., the optical field remains uniformly x-polarized)

$$\left[\frac{\partial}{\partial z} - \frac{i}{2k} \left(\frac{\partial^2}{\partial x^2} + \frac{\partial^2}{\partial y^2} \right) \right] A(x, y, z) = -\frac{ik}{n} \Delta n A(x, y, z) \quad (23.13)$$

Finding **E** (and hence E_x) through Eq.(23.11) in itself involves a nontrivial three-dimensional, anisotropic, and spatially nonlocal nonlinear problem [44–48].

The first basic feature characteristic of the 2 + 1D process is that the photorefractive response does not follow the shape of the optical intensity (as for the 1 + 1D case of Eq. (23.6)). The combination of the x-oriented external field E_0 with the localized $I(x, y)$ gives rise to a central guiding index pattern, which to some extent recalls the 1 + 1D index pattern, and to two lateral antiguiding lobes in the x direction [49]. These emerge as a basic feature of the response even at zero order in η. Using the very same normalization procedures described for the 1 + 1D case, at zero order in η the nonlinear problem becomes

$$\nabla \cdot (\mathbf{Y}Q) = 0 \tag{23.14}$$

and the irrotational condition

$$\nabla \times \mathbf{Y} = 0. \tag{23.15}$$

From these the lobular structure illustrated in the top right insert of Fig. (23.8) emerges, generated through numerical calculation of a specific solution.

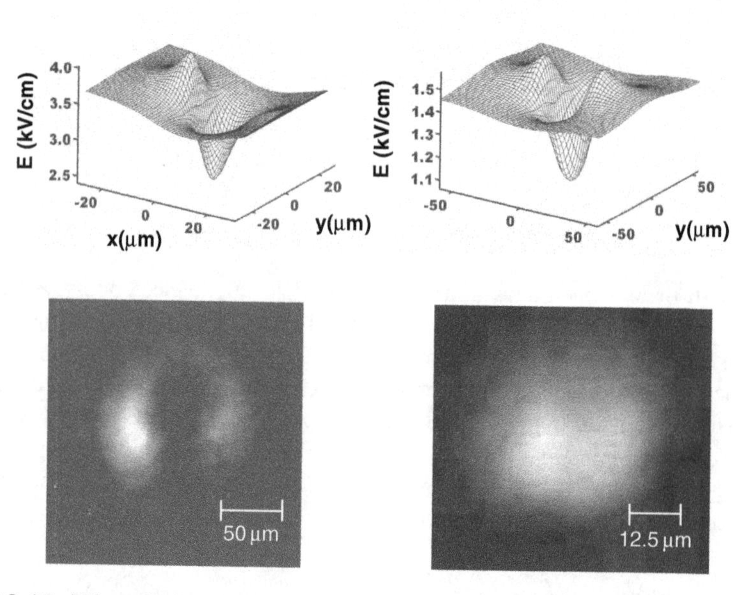

Fig. 23.8 (*Top*) Numerical evaluation of nonlinear index response for a narrow (6.5 μm) beam (*left*) and a wide (14 μm) beam (*right*). In the first case, first-order corrections in η play an important role, whereas in the second, these are negligible. (*Bottom*) Zero field electro-optic read-out of the index pattern underlying a 6.5 μm round soliton (*left*), and the read-out for a 14-μm soliton (elliptic) (*right*). Note how for the smaller beam one of the lateral lobes is almost absent (from [50])

The index pattern resembles the panda-fiber index distribution of polarization maintaining optical fiber. It clearly indicates that the nonlinearity is spatially anisotropic and self-focusing is astigmatic. In most cases more pronounced in the x-direction with respect to the y, where the lobes are absent. The result is that the general response in conditions of small η supports elliptic soliton profiles [44]. Many experiments, including many early discoveries, indicate also that circular solitons can form, i.e., with an approximately circular symmetric propagation invariant I [20,21,32].

The possibility of generating solitons with a round mode is particularly important for optoelectronic applications, because the circular symmetry provides optimal overlap with standard optical fiber. Without considering first corrections in η, circular symmetric 2 + 1D solitons cannot be explained, because the lateral lobes render the self-focusing in the x-direction stronger than in the y. In other words, in the absence of a finite contribution to the nonlinearity of the nonlocal mechanisms such as charge diffusion and displacement, round solitons are excluded by anisotropy. For relatively narrow launch beams η becomes finite and the lobular structure suffers an asymmetric distortion (see Fig. 23.8). In particular conditions, this distortion greatly decreases the effect of one of the two lobes, both bending the soliton trajectory and decreasing the x directed self-focusing power. When astigmatism is sufficiently decreased, round solitons emerge [50]. The apparently simple formation of round solitons from round Gaussian launch beams is in fact the consequence of a rather involved combination of anisotropy and response nonlocality.

The fact that 2 + 1D solitons are supported by this more complex nonlinearity does not substantially modify our soliton picture, other than the fact that we do not have a means to formulate in a straightforward manner an existence curve for (2 + 1)D solitons. Nevertheless, if we phenomenologically build the set of points in which it is possible to observe circular-symmetric self-trapping [21,32], we find a single-valued continuous curve that behaves and looks just like the existence curve of (1 + 1)D solitons (albeit at somewhat higher values of $\Delta\xi$).

23.5 Temporal Effects and Quasi-Steady-State Dynamics

Although the cumulative nature of photorefraction is the basis for strong nonlinear response, actual time dynamics play a negligible role in the physics behind steady-state 1 + 1D and 2 + 1D solitons. Temporal effects become relevant when we ask what happens to the beam during the transient from an initially diffracting Gaussian beam to a steady-state soliton, what occurs if the parameters, such as the external bias field E_0 or the light intensity distribution, are modulated in time, or, simply, what is the physical origin of quasi-steady-state or transient self-trapping?

The time-dependent version of Eq. (23.2) truncated at zero order in η reads

$$\frac{\partial Y}{\partial \tau} + QY = G, \tag{23.16}$$

where $\tau = t/\tau_d$, $\tau_d = \epsilon_0 \epsilon_r \gamma N_a / (q\mu s(N_d - N_a)I_b)$ is the characteristic dielectric time constant, γ is the recombination rate, μ the electron mobility, and s the donor impurity photoionization efficiency. As occurs for most configurations of interest to soliton dynamics, the charge recombination time $\tau_r = 1/(N_a\gamma)$ is much shorter than charge transport time, and no time dependence in the boundary conditions are considered ($G = -1$). If Q is almost constant in space and time, Eq. (23.16) gives an exponential build-up of Y with the time constant τ_d/Q. For solitons, Q is both space- and time-dependent and the continuum of different time constants leads to a highly time-nonlocal response, as can be appreciated by the formally equivalent integral version of Eq. (23.16)

$$Y = Ge^{-\int_0^{\tau} Q d\tau'} \left(1 + \int_0^{\tau} d\tau' e^{\int_0^{\tau'} Q d\tau''}\right), \tag{23.17}$$

indicates [45]. The result is an evolution that presents a number of surprising phenomena.

23.5.1 *The Transition from a Diffracting Wave to a Soliton*

The full complexity of the nonlocality emerges when I undergoes relevant changes in time, i.e., during the very first collapsing stage from a diffracting to a self-trapped beam. This stage occurs for times $\tau \leq \tau_s = 1/(1 + u_0^2)$ and is characterized by a stretched exponential dynamic. For example, if we consider the physically relevant time evolution of the output beam FWHM $\Delta x_{out}(t)$, we find that $\Delta x_{\text{out}}(t) = (\Delta x_{\text{out}}(0) - \Delta x_{\text{in}})e^{-(\tau/\tau_s)^{\beta}} + \Delta x_{\text{in}}$, where Δx_{in} is the input beam FWHM and $\beta < 1$ is the characteristic stretching parameter [51]. Intuitively, since the final size of the beam depends on the distributed self-focusing along the entire propagation axis z and each self-focusing process has a different time constant for each z (owing to the initial diffracting I), the stretching is a direct consequence of the superposition of a continuum of different time scales.

The situation is even more complicated for 2+1D solitons. Here there are generally two coupled dynamics $\Delta x_{\text{out}}(t)$ and $\Delta y_{\text{out}}(t)$, which lead also to an evolution in time of beam ellipticity [52].

23.5.2 *External Modulation of Soliton Parameters*

Although specific experiments have been dedicated to the study of self-trapping with a time-dependent external bias E_0 [53–56], or the spatial self-trapping of a single pulse [57,58], the most widely studied case is when the transverse beam intensity is randomly modulated by having the beam pass through a rotating diffuser before being launched in the photorefractive sample. The effect gives rise to what are generally termed incoherent solitons. Consider Q to be a stochastic process with a characteristic time scale $\tau_r \ll \tau_s$. Defining $\overline{Q}(\tau) = \int_{\tau}^{\tau+\tau_r} Q d\tau'$, this will be a deterministic function of τ, and the entire space-charge formation process is described by Eq. (23.4) with Q substituted with $\overline{Q}$. In particular, the steady state-solution (for $\tau \to \infty$) will simply be, in the tractable 1+1D case, $Y = G/\overline{Q}(\tau \to \infty)$, and hence follows the case of a saturable nonlinearity (see Section 23.10).

23.5.3 *Quasi-Steady-State Solitons*

As described previously, in the absence of background illumination self-trapping can occur during a time window, known as the soliton plateau, after which the beam once again diffracts through the sample. The numerical solution of Eq. (23.17) coupled to the parabolic wave equation confirms experimental findings. To understand the process and the underlying mechanisms, a generalized spatio-temporal soliton self-consistent approach can be implemented, the resulting soliton supporting equation involves an exponential nonlinearity that allows the prediction of the size of the quasi-steady-state soliton Δx as a function of experimental conditions [59]. In particular,

$$\Delta x = \frac{\Delta \xi_{\min} \lambda}{2\pi n^2 a_m} E_0^{-m/2}, \tag{23.18}$$

where $\Delta\xi_{\min} \simeq 3.07$, $a_1 = (r_{\text{eff}})^{1/2}$ and $a_2 = \varepsilon_0 \varepsilon_r (g_{\text{eff}})^{1/2}$, and $m = 1(2)$ for noncentrosymmetric (centrosymmetric) samples, and λ is the beam wavelength.

23.5.4 *Response Change in Beams That Approximately Do Not Evolve in Time*

In very special cases in which the beam does not undergo time evolution, we can considerably simplify the prediction for the build of the space-charge field [60]. In this case Eq. (23.17) is simplified to give

$$Y = e^{-\tau Q}\left(1 + \frac{1}{Q}(e^{\tau Q} - 1)\right), \tag{23.19}$$

This approach can be meaningful and useful for conditions in which the beam initially suffers a negligible amount of diffraction [61–65].

Table 23.1 Principal non-screening self-trapping mechanisms

Mechanism	Solitons	References
Photovoltaic	1 + 1D dark, 2 + 1D vortex	[42,66–68]
Diffusion-driven	1 + 1D and 2 + 1D self-focusing	[69,70]
Resonance enhancement	1 + 1D, 2 + 1D bright	[31,71,72]
Spontaneous	1 + 1D, 2 + 1D self-trapping	[73]

23.6 Non-screening Self-Trapping Mechanisms

In both the 1 + 1D and 2 + 1D self-trapping mechanisms described above, the driving process is the displacement of photoexcited charge so as to screen the externally applied field E_0, the mechanism being generally termed the screening nonlinearity. Further studies have uncovered other self-trapping mechanisms through different photorefractive effects, the main developments being summarized in Table 23.1.

23.7 Materials

Photorefractive solitons can be observed in a number of different materials, these including most photorefractive crystals, polymers, and organic gels. A map of main experimental findings associated to the different materials is contained in the Table 23.2.

23.8 Soliton Interaction-Collisions

A soliton is not simply a propagation invariant wave, but the result of the balancing of diffraction by distributed nonlinear self-lensing. This "dynamic" equilibrium makes the phenomenon stable to perturbations and leads to a characteristic particle-like soliton–soliton phenomenology. Photorefractive crystals form an ideal setting for the observation and study of this multi-beam property, for a number of reasons. First, photorefractives offers a very strong

Table 23.2 Principal experiments on materials supporting solitons

Material	Soliton mechanism	References
SBN, $KNbO_3$, $BaTiO_3$, polymers	Screening	[16,30,74–76]
$LiNbO_3$, KNSBN	Photovoltaic	[34,66–68]
KLTN, organic gels, unpoled SBN	Quadratic screening	[27,32,77,78]
KLTN (near transition)	Diffusion-driven, spontaneous	[70,73]
BGO, BSO, BTO	Screening in optically-active media	[15,33,79]
InP, CdZnTe	Resonantly enhanced	[31,72]

Table 23.3 Principal experimental results on photorefractive soliton collisions

Phenomenon	Year	References
Incoherent collisions between 1 + 1D solitons and fusion *	1996	[80]
Incoherent collisions between 2 + 1D solitons and fusion	1996	[81]
Soliton annihilation and birth (fission) in coherent collisions *	1997–1998	[82,83]
Coherent interaction between 1 + 1D and 2 + 1D solitons	1997–1998	[84,85]
3D soliton spiraling *	1997	[86,87]
Hybrid-dimensional collisions *	2000	[88]
Collisions between counter-propagating solitons *	1999–2004	[89–91]

nonlinearity at very low (microwatt) power levels. Second, launching and detecting different beams propagating through a bulk environment is relatively simple. Third, the saturable nature of the nonlinearity makes the collisional phenomenology richer, including events such as soliton fission and fusion. Last, the possibility of experimenting with 2 + 1D solitons makes previously inaccessible soliton–soliton interaction scenarios observable, such as soliton spiraling and interactions between solitons carrying angular momentum. In fact, the system is so amenable to soliton propagation, that it can simultaneously support a 1 + 1D and a 2 + 1D soliton, allowing the singular study of collisions between solitons of different dimensionality. For all of these reasons, almost all pioneering experiments of soliton interactions in 2 + 1D settings were obtained first with photorefractive solitons, and only many years later were followed up by similar experiments in other soliton-supporting saturable material systems.

The principal experiments on soliton interactions are summarized in Table 23.3. The * marks those cases where these experiments were the first in all soliton studies, including those beyond optics.

23.9 Vector and Composite Solitons

Vector solitons are an important part of basic soliton phenomenology. Vector solitons are self-trapped beams composed of more than one (independent) optical field. In analogy to linear guiding terminology, a scalar (i.e., single component) soliton occupies the lowest mode of its self-induced waveguide. In turn, a vector soliton emerges when this waveguide is the result of the joint action of two (or more) independent optical fields (e.g., when the fields are not coherent with each other or when they are at two orthogonal polarizations), and all fields occupy the lowest mode. The vector soliton is said to be composite or multi-mode when one or more of the independent fields or components occupies higher modes. The components can be independent because they have orthogonal polarizations (Manakov-like solitons), different wavelengths, or simply are from mutually incoherent sources [92]. The principal experimental results on photorefractive vector soliton studies are summarized in Table 23.4.

Table 23.4 Principal experimental results on photorefractive vector solitons

Phenomenon	Year	References
Manakov-like soliton	1996	[93,94]
Bright-dark vector soliton	1996	[95]
Multi-mode multi-hump solitons *	1998	[96]
Dipole-type composite solitons *	2000	[36,37]
Propeller soliton *	2001	[35]
Collisions of Manakov-like solitons *	1999–2001	[97,98]
Collisions of multi-mode solitons *	1999	[99]

The * marks those cases where these experiments were the first in all soliton studies, including those beyond optics.

23.10 Incoherent (Random-Phase) Solitons

As briefly discussed previously, photorefraction is able to also trap beams that have a randomly varying Q, because the cumulative nature of the nonlinear response can in specific conditions be exclusively driven by $\overline{Q}$, which for a static stochastic process is well defined and deterministic. This allows the self-trapping of a spatially incoherent light beam, and even of a spatio-temporally incoherent beam. The result, an "incoherent soliton," is due to the simultaneous guiding of all the underlying independent light fields by a nonlinear index pattern that is generated by the time-average of the intensity resulting from the superposition of all the components. The phenomenon has attracted a considerable amount of interest and opened up a field in its own right. The principal achievements are summarized in Table 23.5. In this case, all the pioneering experimental work was performed in photorefractives.

Table 23.5 Principal experimental and theoretical results on incoherent solitons

Achievement	Year	References
Self-trapping of a partially incoherent beam	1996	[38]
White-light soliton	1997	[39]
Coherent-density and modal theory	1997	[100,101]
Dark incoherent solitons	1998	[43,102]
Mutual coherence theory	1998	[103]
Anti-dark incoherent states	2000	[104]
Elliptic incoherent solitons	2000–2004	[105,106]
Interaction of incoherent solitons	1998–2000	[105,107,108]
Incoherent modulation instability	2000–2004	[109,110]
Arresting transverse instabilities via incoherence	2000	[111,112]
White-light soliton theory	2003	[113,114]
Modulation instability of white light	2002–2005	[115]
Modulation instability of white incoherent light	2004	[116]

Table 23.6 Principal results on applications of photorefractive solitons

Achievement	Year	References
Guiding a beam through a transient soliton	1995	[117]
Guiding a beam through a steady-state soliton	1996	[118]
Soliton-based Y junction	1996	[119–121]
Solitons at telecommunication wavelengths	1997	[31,72,122]
Soliton-based directional coupler	1999	[123]
Second-harmonic generation in a soliton	1999–2004	[124–126]
Soliton electro-optic effects	2000	[127]
Image transmission through waveguides induced by incoherent solitons	2001	[128]
Permanent fixing of multiple soliton-based devices	2001	[129]
Soliton electro-activation	2002	[130]
Optical parametric oscillation in solitons	2002	[131]
Coupling of fibers to soliton waveguides	2004	[132]
Low-voltage solitons through top electrodes	2004	[133]
Soliton waveguides in organic glass	2005	[134]
Soliton-based fiber-slab couplers	2005	[135]

23.11 Applications

Photorefractive solitons are instruments not only for a substantial expansion of our understanding of nonlinear physics, but also for their role in developing new applicative designs, concepts, and devices. A soliton in itself is a beam that propagates through a bulk medium in a guided fashion, i.e., without losing its spatial definition. In turn, because photorefraction is wavelength dependent (long wavelengths are not able to photoactivate impurities) but the electro-optic response is much lesser so, a photorefractive soliton can guide passively infrared beams—these undergoing a purely linear propagation. In other words, in conditions in which the photorefractive charge does not redistribute, a photorefractive soliton imprints in the bulk a waveguide for non-photorefractively active light. The purely nonlinear nature of soliton interaction allows also an all-optical type of elaboration of light signals. Finally, the photorefractive crystal is both electro-optic and generally has strong electronic nonlinearity. The electro-optic response allows a fast, versatile, and multi-functional optical manipulation technique based on solitons also at low voltage levels (see Fig. 23.9), known as soliton electro-activation, whereas the nonlinear response for wavelength mixing and conversion can be strongly enhanced when combined with self-trapping. A brief list of principal experimental achievements is reported in Table 23.6.

23.12 Concluding Remarks

The study of photorefractive solitons has, in the past decade, played a major role in the development of the present understanding of optical solitons, and has had an important impact on soliton science and nonlinear waves in general,

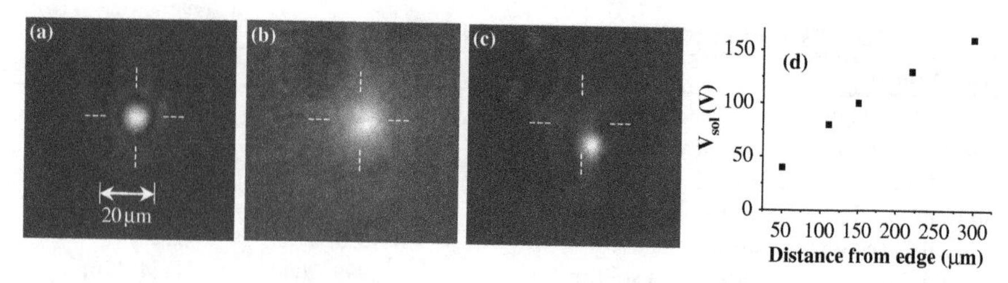

Fig. 23.9 2 + 1D 7 μm photorefractive soliton at quasi-digital voltages generated through a top-sided electrode geometry. Input intensity distribution (**a**); diffraction at output face (**b**); and self-trapped soliton for 40 V (**c**). Measured values of required bias voltage to achieve self-trapping vs. distance from the crystal edge is plotted in (**d**). From [133]

in a variety of systems beyond optics. The drive continues to this day. Recent years (2002–2006) have witnessed another important breakthrough obtained with solitons in photorefractives: the invention of the optical induction method to make nonlinear photonic lattices [136,137,138], which has become the main experimental scheme to explore spatial soliton phenomena in periodic systems. The ability to induce 1D or 2D photonic lattices of any structure, and to tune the polarity and strength of the nonlinearity, have led to the observation of a series of soliton phenomena, many of which have been the first observation in any system in nature. Examples include the observations of 2D lattice ("discrete") solitons [138], spatial gap solitons [137], vortex lattice solitons [139,140], random-phase lattice solitons [141], solitons in quasi-crystals [142], as well as closely related phenomena of Brillouin-zone spectroscopy of photonic lattices [143], Zener tunneling in 2D photonic lattices [144], dynamics of polarons in photonic lattices [145], and much more. These are just a small sample group from the recent experiments with solitons in periodic structures, and they were all observed by employing the photorefractive screening nonlinearity while taking advantage of its inherent nonlinear anisotropy [136].

Acknowledgments The work of the authors has been supported over the years by a number of agencies. In particular, we wish to thank the Istituto Italiano di Fisica della Materia, the Ministero Italiano della Ricerca through the FIRB funding initiative, the U.S. Army Research Office, the U.S. National Science Foundation, and the Israeli Science Foundation.

References

1. M. Segev, G. Stegeman: Self-trapping of optical beams: Spatial solitons, *Phys. Today* **51**, 42–48 (1998).
2. M. Segev: Optical spatial solitons, *Opt. Quant. Electron.* **30**, 503–533 (1998).
3. G.I. Stegeman, M. Segev: Optical spatial solitons and their interactions: Universality and diversity, *Science* **286**, 1518–1523 (1999).
4. See Chapter 4 by E. DelRe, B. Crosignani, P. Di Porto, and Chapter 5 by M. Segev, D.N. Christodoulides, in S. Trillo and W. Torruellas (Eds.), *Spatial Solitons* (Springer-Verlag, Berlin 2001).

5. G.I. Stegeman, D.N. Christodoulides, and M. Segev: Optical spatial solitons: historical perspectives, Millennium Issue of the *IEEE J. Selected Topics Quant. Electron.* **6**, 1419–1427 (2000).
6. See Chapter 11 by E. DelRe, M. Segev, D.N. Christodoulides, B. Crosignani, and G. Salamo in P. Gunter and J.P. Huignard (Eds.), *Photorefractive Materials and Their Applications* (Springer-Verlag, Berlin, 2006).
7. P. Yeh, Introduction to photorefractive nonlinear optics (Wiley, New York, 1993).
8. L. Solymar, D. Webb, and A. Grunnet-Jepsen, *The Physics and Applications of Photorefractive Materials*, (Claredon Press, Oxford, 1996).
9. M. Segev, B. Crosignani, A. Yariv et al.: Spatial solitons in photorefractive media, *Phys. Rev. Lett* **68**, 923–926 (1992).
10. B. Crosignani, M. Segev, D. Engin et al.: Self-trapping of optical beams in photorefractive media, *J. Opt. Soc. Am B* **10**, 446–453 (1993).
11. G.C. Duree, J.L. Shultz, G.J. Salamo et al.: Observation of self-trapping of an optical beam due to the photorefractive effect, *Phys. Rev. Lett* **71**, 533–536 (1993).
12. M. Segev, B. Crosignani, P. Diporto et al.: Stability of photorefractive spatial solitons, *Opt. Lett.* **19**, 1296–1298 (1994).
13. G. Duree, G. Salamo, M. Segev et al.: Dimensionality and size of photorefractive spatial solitons, *Opt. Lett.* **19**, 1195–1197 (1994).
14. D.N. Christodoulides, M.I. Carvalho: Compression, self-bending, and collapse of gaussian beams in photorefractive crystals, *Opt. Lett.* **19**, 1714–1716 (1994).
15. M.D.I. Castillo, P.A.M. Aguilar, J.J. Sanchez-Mondragon et al.: Spatial solitons in photorefractive Bi12Tio20 with drift mechanism of nonlinearity, *Appl. Phys. Lett.* **64**, 408–410 (1994).
16. M. Segev, G.C. Valley, B. Crosignani et al.: Steady-state spatial screening solitons in photorefractive materials with external applied-field, *Phys. Rev. Lett.* **73**, 3211–3214 (1994).
17. S.R. Singh, D.N. Christodoulides: Evolution of spatial optical solitons in biased photorefractive media under steady-state conditions, *Opt. Commun.* **118**, 569–576 (1995).
18. D.N. Christodoulides, M.I. Carvalho: Bright, dark, and gray spatial soliton states in photorefractive media, *J. Opt. Soc. Am. B* **12**, 1628–1633 (1995).
19. M. Segev, M.F. Shih, G.C. Valley: Photorefractive screening solitons of high and low intensity, *J. Opt. Soc. Am. B* **13**, 706–718 (1996).
20. M.F. Shih, M. Segev, G.C. Valley et al.: Observation of two-dimensional steady-state photorefractive screening solitons, *Electron. Lett.* **31**, 826–827 (1995).
21. M.F. Shih, P. Leach, M. Segev et al.: Two-dimensional steady-state photorefractive screening solitons, *Opt. Lett.* **21**, 324–326 (1996).
22. K. Kos, H.X. Meng, G. Salamo et al.: One-dimensional steady-state photorefractive screening solitons, *Phys. Rev.E* **53**, R4330–R4333 (1996).
23. E. DelRe, A. Ciattoni, B. Crosignani, et al.: Approach to space-charge field description in photorefractive crystals, *J. Opt. Soc. Am. B* **15**, 1469–1475 (1998).
24. M.I. Carvalho, S.R. Singh, D.N. Christodoulides: Self-deflection of steady-state bright spatial solitons in biased photorefractive crystals, *Opt. Commun.* **120**, 311–315 (1995).
25. E. DelRe, A. Ciattoni, E. Palange: Role of charge saturation in photorefractive dynamics of micron-sized beams and departure from soliton behavior, *Phys. Rev. E* **73**, 017601-1–017601-4 (2006).
26. M. Segev, A.J. Agranat: Spatial solitons in centrosymmetric photorefractive media, *Opt. Lett.* **22**, 1299–1301 (1997).
27. E. DelRe, B. Crosignani, M. Tamburrini et al.: One-dimensional steady-state photorefractive spatial solitons in centrosymmetric paraelectric potassium lithium tantalate niobate, *Opt. Lett.* **23**, 421–423 (1998).
28. E. DelRe, A. D'Ercole, A.J. Agranat: Emergence of linear wave segments and predictable traits in saturated nonlinear media, *Opt. Lett.* **28**, 260–262 (2003).

29. Z. Chen, M. Mitchell, M.F. Shih et al.: Steady-state dark photorefractive screening solitons, *Opt. Lett.* **21**, 629–631 (1996).
30. S. Lan, M.F. Shih, M. Segev: Self-trapping of one-dimensional and two-dimensional optical beams and induced waveguides in photorefractive KNbO3, *Opt. Lett.* **22**, 1467–1469 (1997).
31. M. Chauvet, S.A. Hawkins, G.J. Salamo et al.: Self-trapping of two-dimensional optical beams and light-induced waveguiding in photorefractive InP at telecommunication wavelengths, *Appl. Phys. Lett.* **70**, 2499–2501 (1997).
32. E. DelRe, M. Tamburrini, M. Segev et al.: Two-dimensional photorefractive spatial solitons in centrosymmetric paraelectric potassium-lithium-tantalate-niobate, *Appl. Phys. Lett.* **73**, 16–18 (1998).
33. E. Fazio, W. Ramadan, A. Belardini et al.: (2 + 1)-dimensional soliton formation in photorefractive Bi12SiO20 crystals, *Phys. Rev. E* **67**, 026611 (2003).
34. W.L. She, K.K. Lee, W.K. Lee: Observation of two-dimensional bright photovoltaic spatial solitons, *Phys. Rev. Lett.* **83**, 3182–3185 (1999).
35. T. Carmon, R. Uzdin, C. Pigier et al.: Rotating propeller solitons, *Phys. Rev. Lett.* **87**, 143901 (2001).
36. W. Krolikowski, E.A. Ostrovskaya, C. Weilnau et al.: Observation of dipole-mode vector solitons, *Phys. Rev. Lett.* **85**, 1424–1427 (2000).
37. T. Carmon, C. Anastassiou, S. Lan et al.: Observation of two-dimensional multimode solitons, *Opt. Lett.* **25**, 1113–1115 (2000).
38. M. Mitchell, Z.G. Chen, M.F. Shih et al.: Self-trapping of partially spatially incoherent light, *Phys. Rev. Lett.* **77**, 490–493 (1996).
39. M. Mitchell, M. Segev: Self-trapping of incoherent white light, *Nature* **387**, 880–883 (1997).
40. G. Duree, M. Morin, G. Salamo et al.: Dark photorefractive spatial solitons and photorefractive vortex solitons, *Phys. Rev. Lett.* **74**, 1978–1981 (1995).
41. Z. Chen, M.F. Shih, M. Segev et al.: Steady-state vortex-screening solitons formed in biased photorefractive media, *Opt. Lett.* **22**, 1751–1753 (1997).
42. Z. Chen, M. Segev, D.W. Wilson et al.: Self-trapping of an optical vortex by use of the bulk photovoltaic effect, *Phys. Rev. Lett.* **78**, 2948–2951 (1997).
43. Z. Chen, M. Mitchell, M. Segev et al.: Self-trapping of dark incoherent light beams, *Science* **280**, 889–892 (1998).
44. A.A. Zozulya, D.Z. Anderson, A.V. Mamaev et al.: Self-focusing and soliton formation in media with anisotropic nonlocal material response, *Europhys. Lett.* **36**, 419–424 (1996).
45. B. Crosignani, P. DiPorto, A. Degasperis et al.: Three-dimensional optical beam propagation and solitons in photorefractive crystals, *J. Opt. Soc. Am. B* **14**, 3078–3090 (1997).
46. S. Gatz, J. Herrmann: Anisotropy, nonlocality, and space-charge field displacement in (2 + 1)-dimensional self-trapping in biased photorefractive crystals, *Opt. Lett.* **23**, 1176–1178 (1998).
47. M.R. Belic, D. Vujic, A. Stepken et al.: Isotropic versus anisotropic modeling of photorefractive solitons, *Phys. Rev. E* **65**, 066610 (2002).
48. G.F. Calvo, F. Agullo-Lopez, M. Carrascosa, et al.: Two-dimensional soliton-induced refractive index change in photorefractive crystals, *Opt. Commun.* **227**, 193–202 (2003).
49. E. DelRe, A. Ciattoni, A.J. Agranat: Anisotropic charge displacement supporting isolated photorefractive optical needles, *Opt. Lett.* **26**, 908–910 (2001).
50. E. DelRe, G. De Masi, A. Ciattoni et al.: Pairing space-charge field conditions with self-guiding for the attainment of circular symmetry in photorefractive solitons, *Appl. Phys. Lett.* **85**, 5499–5501 (2004).
51. C. Dari-Salisburgo, E. DelRe, E. Palange Molding and stretched evolution of optical solitons in cumulative nonlinearities *Phys. Rev. Lett.* **91**, 263903 (2003).
52. C. Denz, W. Krolikowski, J. Petter et al.: Dynamics of formation and interaction of photorefractive screening solitons, *Phys. Rev. E* **60**, 6222–6225 (1999).

53. G.M. Tosi-Beleffi, M. Presi, E. DelRe et al.: Stable oscillating nonlinear beams in square-wave-biased photorefractives, *Opt. Lett.* **25**, 1538–1540 (2000).
54. G.M. Tosi-Beleffi, F. Curti, D. Boschi et al.: Soliton-based Y-branch in photorefractive crystals induced through dispersion-shifted optical fiber, *Opt. Lett.* **28**, 1561–1563 (2003).
55. C.A. Fuentes-Hernandez, A.V. Khomenko: Beam collapse and polarization self-modulation in an optically active photorefractive crystal in an alternating electric field, *Phys. Rev. Lett.* **83**, 1143–1146 (1999).
56. M.N. Frolova, S.M. Shandarov, M.V. Borodin: Self-action of a light beam in a photorefractive crystal in an alternating electric field upon synchronous intensity modulation, *Quantum. Electron.* **32**, 45–48 (2002).
57. D. Wolfersberger, N. Fressengeas, J. Maufoy et al.: Self-focusing of a single laser pulse in a photorefractive medium, *Phys. Rev. E* **62**, 8700–8704 (2000).
58. D. Wolfersberger, F. Lhomme, N. Fressengeas et al.: Simulation of the temporal behavior of one single laser pulse in a photorefractive medium, *Opt. Commun.* **222**, 383–391 (2003).
59. E. DelRe, E. Palange: Optical nonlinearity and existence conditions for quasi-steady-state photorefractive solitons, *J. Opt. Soc. Am. B* **23**, 2323–2327 (2006).
60. N. Fressengeas, J. Maufoy, G. Kugel: Temporal behavior of bidimensional photorefractive bright spatial solitons, *Phys. Rev. E* **54**, 6866–6875 (1996).
61. N. Fressengeas, J. Maufoy, D. Wolfersberger et al.: Experimental transient self-focusing in Bi12TiO20 crystal, *Ferroelectrics* **202**, 193–202 (1997).
62. N. Fressengeas, D. Wolfersberger, J. Maufoy et al.: Build-up mechanisms of (1 + 1)-dimensional photorefractive bright spatial quasi-steady-state and screening solitons, *Opt. Commun* **145**, 393–400 (1998).
63. N. Fressengeas, D. Wolfersberger, J. Maufoy et al.: Experimental study of the self-focusing process temporal behavior in photorefractive Bi12TiO20, *J. Appl. Phys* **85**, 2062–2067 (1999).
64. D. Wolfersberger, N. Fressengeas, J. Maufoy et al.: Experimental study of the behaviour of narrow nanosecond laser pulses in biased photorefractive Bi12TiO20, *Ferroelectrics* **223**, 381–388 (1999).
65. J. Maufoy, N. Fressengeas, D. Wolfersberger et al.: Simulation of the temporal behavior of soliton propagation in photorefractive media, *Phys. Rev. E* **59**, 6116–6121 (1999).
66. G.C. Valley, M. Segev, B. Crosignani et al.: Dark and bright photovoltaic spatial solitons, *Phys. Rev. A* **50**, R4457–R4460 (1994).
67. M. Taya, M.C. Bashaw, M.M. Fejer et al.: Observation of dark photovoltaic spatial solitons. *Phys. Rev. A* **52**, 3095–3100 (1995).
68. M. Segev, G.C. Valley, M.C. Bashaw et al.: Photovoltaic spatial solitons, *J. Opt. Soc. Am. B* **14**, 1772–1781 (1997).
69. B. Crosignani, E. DelRe, P. Di Porto et al.: Self-focusing and self-trapping in unbiased centrosymmetric photorefractive media, *Opt. Lett.* **23**, 912–914 (1998).
70. B. Crosignani, A. Degasperis, E. DelRe et al.: Nonlinear optical diffraction effects and solitons due to anisotropic charge-diffusion-based self-interaction, *Phys. Rev. Lett.* **82**, 1664–1667 (1999).
71. M. Chauvet, S.A. Hawkins, G.J. Salamo et al.: Self-trapping of planar optical beams by use of the photorefractive effect in InP:Fe, *Opt. Lett.* **21**, 1333–1335 (1996).
72. T. Schwartz, Y. Ganor, T. Carmon et al.: Photorefractive solitons and light-induced resonance control in semiconductor CdZnTe, *Opt. Lett.* **27**, 1229–1231 (2002).
73. E. DelRe, M. Tamburrini, M. Segev et al.: Spontaneous self-trapping of optical beams in metastable paraelectric crystals, *Phys. Rev. Lett.* **83**, 1954–1957 (1999).
74. M.F. Shih, F.W. Sheu: Photorefractive polymeric optical spatial solitons, *Opt. Lett.* **24**, 1853–1855 (1999).
75. E. DelRe, M. Tamburrini, G. Egidi: Bright photorefractive spatial solitons in tilted BaTiO3, presented at the Eleventh Annual Meeting of the [IEEE Lasers and Electro-Optics Society] (LEOS 98), Orlando, Fla., 3–4 December 1998.

76. J. Andrade-Lucio, M. Iturbe-Castillo, P. Marquez-Aguilar et al.: Self-focusing in photorefractive BaTiO3 crystal under external DC electric field, *Opt. Quantum Electron.* **30**, 829–834 (1998).
77. Z.G. Chen, M. Asaro, O. Ostroverkhova, et al.: Self-trapping of light in an organic photorefractive glass, *Opt. Lett.* **28**, 2509–2511 (2003).
78. M. Chauvet, A. Guo, G. Fu and G. Salamo, Electrically switched photoinduced waveguide in unpoled strontium barium niobate, *J. Appl. Phys.* **99**, 113107-1–113107-5 (2006).
79. M.D. Castillo, J.J. Sanchezmondragon, S.I. Stepanov et al.: Probe beam wave-guiding induced by spatial dark solitons in photorefractive Bi12TiO2 crystal, *Rev. Mex. Fis* **41**, 1–10 (1995).
80. M.F. Shih, Z.G. Chen, M. Segev et al. Incoherent collisions between one-dimensional steady-state photorefractive screening solitons, *Appl. Phys. Lett.* **69**, 4151–4153 (1996).
81. M.F. Shih, M. Segev: Incoherent collisions between two-dimensional bright steady-state photorefractive spatial screening solitons, *Opt. Lett.* **21**, 1538–1540 (1996).
82. W. Krolikowski, S.A. Holmstrom: Fusion and birth of spatial solitons upon collision, *Opt. Lett.* **22**, 369–371 (1997).
83. W. Krolikowski, B. Luther-Davies, C. Denz et al.: Annihilation of photorefractive solitons, *Opt. Lett.* **23**, 97–99 (1998).
84. H.X. Meng, G. Salamo, M.F. Shih et al.: Coherent collisions of photorefractive solitons, *Opt. Lett.* **22**, 448–450 (1997).
85. A.V. Mamaev, M. Saffman, A.A. Zozulya: Phase-dependent collisions of (2+1)-dimensional spatial solitons, *J. Opt. Soc. Am. B* **15**, 2079–2082 (1998).
86. M.F. Shih, M. Segev, G. Salamo: Three-dimensional spiraling of interacting spatial solitons, *Phys. Rev. Lett.* **78**, 2551–2554 (1997).
87. A.V. Buryak, Y.S. Kivshar, M.F. Shih et al.: Induced coherence and stable soliton spiraling, *Phys. Rev. Lett.* **82**, 81–84 (1999).
88. E. DelRe, S. Trillo, A.J. Agranat: Collisions and inhomogeneous forces between solitons of different dimensionality, *Opt. Lett.* **25**, 560–562 (2000).
89. E. DelRe, A. Ciattoni, B. Crosignani et al.: Nonlinear optical propagation phenomena in near-transition centrosymmetric photorefractive crystals, *J. Nonlinear. Opt. Phys.* **8**, 1–20 (1999).
90. D. Kip, C. Herden, M. Wesner: All-optical signal routing using interaction of mutually incoherent spatial solitons, *Ferroelectrics* **274**, 135–142 (2002).
91. C. Rotschild, O. Cohen, O. Manela et al.: Interactions between spatial screening solitons propagating in opposite directions, *J. Opt. Soc. Am. B* **21**, 1354–1357 (2004).
92. D.N. Christodoulides, S.R. Singh, M.I. Carvalho et al.: Incoherently coupled soliton pairs in biased photorefractive crystals, *Appl. Phys. Lett.* **68**, 1763–1765 (1996).
93. Z.G. Chen, M. Segev, T.H. Coskun et al.: Observation of incoherently coupled photorefractive spatial soliton pairs, *Opt. Lett.* **21**, 1436–1438 (1996).
94. Z. Chen, M. Segev, T. Coskun et al.: Coupled photorefractive spatial soliton pairs, *J. Opt. Soc. Am. B* **14**, 3066–3077 (1997).
95. Z. Chen, M. Segev, T. Coskun et al.: Observation of incoherently coupled dark-bright photorefractive spatial soliton pairs, *Opt. Lett.* **21**, 1821 (1996).
96. M. Mitchell, M. Segev D.N. Christodoulides: Observation of multi-hump multi-mode solitons, *Phys. Rev. Lett.* **80**, 4657–4660 (1998).
97. C. Anastassiou, M. Segev, K. Steiglitz et al.: Energy-exchange interactions between colliding vector solitons, *Phys. Rev. Lett.* **83**, 2332–2335 (1999).
98. C. Anastassiou, J.W. Fleischer, T. Carmon et al.: Information transfer through cascaded collisions of vector solitons, *Opt. Lett.* **26**, 1498 (2001).
99. W. Krolikowski, N. Akhmediev, B. Luther-Davies: Collision-induced shape transformations of partially coherent solitons, *Phys. Rev. E* **59**, 4654–4658 (1999).
100. D.N. Christodoulides, T.H. Coskun, M. Mitchell et al.: Theory of incoherent self-focusing in biased photorefractive media, *Phys. Rev. Lett.* **78**, 646–649 (1997).

101. M. Mitchell, M. Segev, T.H. Coskun et al.: Theory of self-trapped spatially incoherent light beams, *Phys. Rev. Lett.* **79**, 4990–4993 (1997).
102. D.N. Christodoulides, T. Coskun, M. Mitchell et al.: Theory of dark incoherent solitons, *Phys. Rev. Lett.* **80**, 5113–5116 (1998).
103. V.V. Shkunov, D.Z. Anderson: Radiation transfer model of self-trapping spatially incoherent radiation by nonlinear media, *Phys. Rev. Lett.* **81**, 2683–2686 (1998).
104. T.H. Coskun, D.N. Christodoulides, Y.R. Kim et al.: Bright spatial solitons on a partially incoherent background, *Phys. Rev. Lett.* **84**, 2374–2377 (2000).
105. E.D. Eugenieva, D.N. Christodoulides, M. Segev: Elliptic incoherent solitons in saturable nonlinear media, *Opt. Lett.* **25**, 972–974 (2000).
106. O. Katz, T. Carmon, T. Schwartz et al.: Observation of elliptic incoherent spatial solitons, *Opt. Lett.* **29**, 1248–1250 (2004).
107. T.H. Coskun, A.G. Grandpierre, D.N. Christodoulides et al.: Coherence enhancement of spatially incoherent light beams through soliton interactions, *Opt. Lett.* **25**, 826–828 (2000).
108. T.H. Coskun, D.N. Christodoulides, M. Mitchell et al.: Dynamics of incoherent bright and dark self-trapped beams and their coherence properties in photorefractive crystals, *Opt. Lett.* **23**, 418–420 (1998).
109. M. Soljacic, M. Segev, T. Coskun et al.: Modulation instability of incoherent beams in noninstantaneous nonlinear media, *Phys. Rev. Lett.* **84**, 467–470 (2000).
110. D. Kip, M. Soljacic, M. Segev et al.: Modulation instability and pattern formation in spatially incoherent light beams, *Science* **290**, 495 (2000).
111. C. Anastassiou, M. Soljacic, M. Segev et al.: Eliminating the transverse instabilities of Kerr solitons, *Phys. Rev. Lett.* **85**, 4888–4891 (2000).
112. C.C. Jeng, M. Shih, K. Motzek et al.: Partially incoherent optical vortices in self-focusing nonlinear media, *Phys. Rev. Lett.* **92**, 043904 (2004).
113. H. Buljan, M. Segev, M. Soljacic et al.: White light solitons, *Opt. Lett.* **28**, 1239–1241 (2003).
114. H. Buljan, A. Siber, M. Soljacic et al.: White light solitons in logarithmically saturable nonlinear media, *Phys. Rev. E* **68**, 036607 (2003).
115. H. Buljan, A. Siber, M. Soljacic et al.: Propagation of incoherent "white" light and modulation instability in non-instantaneous nonlinear media, *Phys. Rev. E* **66**, R35601 (2002).
116. T. Schwartz, T. Carmon, H. Buljan et al.: Spontaneous pattern formation with incoherent "white" light, *Phys. Rev. Lett.* **93**, 223901-1–223901-4 (2004).
117. M. Morin, G. Duree, G. Salamo et al.: Wave-guides formed by quasi-steady-state photorefractive spatial solitons, *Opt. Lett.* **20**, 2066–2068 (1995).
118. M.F. Shih, M. Segev, G. Salamo: Circular waveguides induced by two-dimensional bright steady-state photorefractive spatial screening solitons, *Opt. Lett.* **21**, 931–933 (1996).
119. M. Taya, M.C. Bashaw, M.M. Fejer et al.: Y-junctions arising from dark-soliton propagation in photovoltaic media, *Opt. Lett.* **21**, 943–945 (1996).
120. Z.G. Chen, M. Mitchell, M. Segev: Steady-state photorefractive soliton-induced Y-junction waveguides and high-order dark spatial solitons, *Opt. Lett.* **21**, 716–718 (1996).
121. J. Petter, C. Denz: Guiding and dividing waves with photorefractive solitons, *Opt. Commun.* **188**, 55–61 (2001).
122. M. Wesner, C. Herden, D. Kip et al.: Photorefractive steady state solitons up to telecommunication wavelengths in planar SBN waveguides, *Opt. Commun.* **188**, 69–76 (2001).
123. S. Lan, E. DelRe, Z.G. Chen et al.: Directional coupler with soliton-induced waveguides, *Opt. Lett.* **24**, 475–477 (1999).
124. S. Lan, M.F. Shih, G. Mizell et al.: Second-harmonic generation in waveguides induced by photorefractive spatial solitons, *Opt. Lett.* **24**, 1145–1147 (1999).
125. C. Lou, J. Xu, H. Qiao et al. Enhanced second-harmonic generation by means of high-power confinement in a photovoltaic soliton-induced waveguide, *Opt. Lett.* **29**, 953–955 (2004).

126. J.R. Salgueiro, A.H. Carlsson, E. Ostrovskaya et al.: Second-harmonic generation in vortex-induced waveguides, *Opt. Lett.* **29**, 593–595 (2004).
127. E. DelRe, M. Tamburrini, A.J. Agranat: Soliton electro-optic effects in paraelectrics, *Opt. Lett.* **25**, 963–965 (2000).
128. D. Kip, C. Anastassiou, E. Eugenieva et al.: Transmission of images through highly nonlinear media by gradient-index lenses formed by incoherent solitons, *Opt. Lett.* **26**, 524–526 (2001).
129. A. Guo, M. Henry, G.J. Salamo et al.: Fixing multiple waveguides induced by photorefractive solitons: directional couplers and beam splitters, *Opt. Lett.* **26**, 1274–1276 (2001).
130. E. DelRe, B. Crosignani, P. Di Porto et al.: Electro-optic beam manipulation through photorefractive needles, *Opt. Lett.* **27**, 2188–2190 (2002).
131. S. Lan, J.A. Giordmaine, M. Segev et al.: Optical parametric oscillation in soliton-induced waveguides, *Opt. Lett.* **27**, 737–739 (2002).
132. E. DelRe, E. Palange, A.J. Agranat: Fiber-launched ultratight photorefractive solitons integrating fast soliton-based beam manipulation circuitry, *J. Appl. Phys.* **95**, 3822–3824 (2004).
133. A. D'Ercole, E. Palange, E. DelRe et al.: Miniaturization and embedding of soliton-based electro-optically addressable photonic arrays, *Appl. Phys. Lett.* **85**, 2679–2681 (2004).
134. M. Asaro, M. Sheldon, Z. Chen et al.: Soliton-induced waveguides in an organic photorefractive glass, *Opt. Lett.* **30**, 519–521 (2005).
135. E. DelRe, A. D'Ercole, E. Palange et al.: Observation of soliton ridge states for the self-imprinting of fiber-slab couplers, *Appl. Phys. Lett.* **86**, 191110-1–191110-3 (2005).
136. N.K. Efremidis, S. Sears, D.N. Christodoulides et al.: Discrete solitons in photorefractive optically induced photonic lattices, *Phys. Rev. E* **66**, 046602 (2002).
137. J.W. Fleischer, T. Carmon, M. Segev et al.: Observation of discrete solitons in optically induced real time waveguide arrays, *Phys. Rev. Lett.*, **90**, 023902 (2003).
138. J.W. Fleischer, M. Segev, N.K. Efremidis et al.: Observation of two-dimensional discrete solitons in optically induced nonlinear photonic lattices, *Nature* **422**, 147–150 (2003).
139. J.W. Fleischer, G. Bartal, O. Cohen et al.: Observation of vortex-ring "discrete" solitons in 2D photonic lattices, *Phys. Rev. Lett.*, **92**, 123904 (2004).
140. D.N. Neshev, T.J. Alexander, E.A. Ostrovskaya et al.: Observation of discrete vortex solitons in optically induced photonic lattices, *Phys. Rev. Lett.*, **92**, 123903 (2004).
141. O. Cohen, G. Bartal, H. Buljan et al.: Observation of random-phase lattice solitons, *Nature* **433**, 500–503 (2005).
142. B. Freedman, G. Bartal, M. Segev et al.: Wave and defect dynamics in nonlinear photonic quasicrystals, *Nature* **440**, 1166–1169 (2006).
143. G. Bartal, O. Cohen, H. Buljan et al.: Brillouin-zone spectroscopy of nonlinear photonic lattices, *Phys. Rev. Lett.* **94**, 163902 (2005).
144. H. Trompeter, W. Krolikowski, D.N. Neshev et al.: Bloch oscillations and Zener tunneling in two-dimensional photonic lattices, *Phys. Rev. Lett.* **96**, 053903 (2006).
145. H. Martin, E.D. Eugenieva, Z. Chen et al.: Discrete solitons and soliton-induced dislocations in partially coherent photonic lattices, *Phys. Rev. Lett.* **92**, 123902 (2004).

Chapter 24
Measuring Nonlinear Refraction and Its Dispersion

Eric W. Van Stryland and David J. Hagan

Abstract We describe methods for measuring the nonlinear refraction of nominally transparent materials that involve propagation from the near to the far field, which changes a phase distortion into an amplitude redistribution. These methods include beam distortion methods and Z-scan. We also look at methods to determine the spectral dependence of these changes in refractive index. Recent advances here include using femtosecond white-light continua as the source for Z-scan. The types of nonlinear refractive mechanisms are also briefly discussed including bound-electronic, excited state or free-carrier generation, reorientation, electrostrictive, and thermal nonlinear refraction as well as cascaded second-order nonlinearities.

24.1 Introduction

Nonlinear refraction (NLR) is the general name ascribed to phenomena that give rise to an intensity-dependent refractive index. A wide variety of mechanisms can give rise to NLR, and magnitudes and response times can vary by many orders of magnitude for the different mechanisms; however in many cases, the nonlinear refractive index may be adequately characterized by:

$$n(I) = n_0 + \Delta n(I) = n_0 + n_2 I, \tag{24.1}$$

where n_0 is the linear refractive index, I is the irradiance, and n_2 is the nonlinear refractive index. Nonlinear absorption (NLA) and NLR were among the very first nonlinear optical effects reported [1–4]. The experimental study of NLR is typically more complex than for NLA, due to the fact that its effects are usually only observed after some amount of propagation. Essentially, the input beam to a nonlinear material sets up a phase mask whose amplitude profile mimics the

E.W.Van Stryland (✉)
CREOL, The College of Optics and Photonics, University of Central Florida, Orlando, FL 32817, USA
e-mail: ewvs@creol.ucf.edu

R.W. Boyd et al. (eds.), *Self-focusing: Past and Present*,
Topics in Applied Physics 114, DOI 10.1007/978-0-387-34727-1_24,

irradiance profile of the incoming beam. The "lens" thus created with this spatially graded index then either focuses or defocuses the beam upon propagation with the imposed aberrations.

If the sample is thick compared to the focusing length, the nonlinear phase shift may significantly alter the irradiance distribution within the sample. For self-focusing nonlinearities this can have catastrophic consequences (optical damage). If the sample is thin (to be defined later) the alterations in the irradiance distribution occurs only after propagation outside the sample and can be measured, usually in tshe far field. This method of measuring the beam in the far field and comparing to the input beam can determine the sign and magnitude of the nonlinear refraction. Z-scan is a form of this type of measurement. Before describing Z-scan, however, we first look at the beam propagation in Section 24.2. Section 24.3 describes Z-scan and its variants, while Section 24.4 introduces a relatively new technique for measuring the dispersion of nonlinear refraction. This method, referred to as the "white-light continuum Z-scan," relies on high spectral energy density broadband femtosecond continua. Because any discussion of measurement techniques must address the physical processes being measured, in Section 24.5 we briefly discuss several physical mechanisms leading to nonlinear refraction. This leads to questions of how to determine which physical mechanisms are present in any given sample, and ways of unraveling the physics are suggested. An important part of determining the physical mechanisms is determining the dispersion of the nonlinear refraction.

24.2 Beam Propagation

While it is possible to solve the wave equation to calculate the energy distribution at any position along a beam within a nonlinear material, [5–9] for the purpose of quantitatively measuring nonlinear refraction, it is far simpler to use a "thin" sample. By "thin" we mean that the input beam does not change size or shape within the length of the sample, L. This is often referred to as the "external self-action" regime [10]. For this to be valid, neither diffraction nor nonlinear refraction may cause any change of beam profile within the nonlinear sample. The diffraction criterion is simply that the thickness of the sample $L << Z_0$ where Z_0 is the Rayleigh range or depth of focus of the beam. The criterion that NLR does not change the beam shape is $L << Z_0/\Delta\Phi_0$ where $\Delta\Phi_0$ is the maximum nonlinearly-induced phase distortion. This latter requirement simply states that the effective focal length of the induced nonlinear lens in the sample should be much longer than the sample thickness itself [10].

The external self-action limit simplifies the problem considerably, because the amplitude and relative phase, $\Delta\varphi$, of the electric field E inside the nonlinear material are then separately governed by the following pair of simple equations:

$$\frac{d\Delta\phi}{dz'} = \frac{2\pi}{\lambda}\Delta n(I) \tag{24.2}$$

and

$$\frac{dI}{dz'} = -\alpha(I)I, \tag{24.3}$$

where z' is the propagation depth in the sample, I is the irradiance given by $I = n_0 c\varepsilon_0 |E|^2$, with n_0 the linear refractive index, ε_0 the vacuum permittivity, and c the speed of light in a vacuum. The value of $\alpha(I)$ in general includes linear and NLA terms, while Δn includes only nonlinear terms for the index change.

For third-order nonlinearities, where irradiance-induced changes in refraction and absorption are directly proportional to the irradiance, the nonlinear refractive index in the form given in Eq. (24.1) is usually used with the MKS units system. In the Gaussian or cgs units system, n_2 is usually defined as

$$\left\{\frac{n_2}{2}|E|^2\right\}_{esu} = \{n_2 I\}_{MKS}, \tag{24.4}$$

where n_2 is the nonlinear index of refraction, E is the peak electric field (cgs), and I denotes the irradiance (MKS) of the laser beam within the sample. The values of n_2(esu) and n_2(MKS) are related through the conversion formula, $n_2(\text{esu}) = (cn_0/40\pi)n_2(\text{MKS})$. We will use MKS units in this chapter and n_2 will refer to n_2 (MKS). While we are using n_2 here for *any* third-order nonlinearity, it may not be the best description for nonlinearities that have a response slower than the temporal changes in I [11, 12]. The nonlinear absorption may sometimes be written as,

$$\alpha(I) = \alpha_0 + \Delta\alpha = \alpha_0 + \beta I, \tag{24.5}$$

where α_0 is the linear absorption coefficient and β denotes the third-order nonlinear absorption coefficient, which for ultrafast NLA is equal to the two-photon absorption (2PA) coefficient. We qualify this with "sometimes" because often other nonlinear processes cannot be ignored. We will discuss this further in Section 24.5.

When the amplitude and the phase of the beam exiting the sample are known by integrating Eqs. (24.2) and (24.3), the field distribution at the plane of detection can be calculated using diffraction theory (Huygen's principle). The simplest distribution to use, and one that can be experimentally obtained, is a Gaussian beam. Assuming this, and that there is no nonlinear absorption, allows us to calculate the field at a position z' within the thin sample as:

$$E(z', r) = \sqrt{2I_0(z')/nc\varepsilon_0}\, e^{-\left(\frac{r}{w}\right)^2} \exp\left\{ik_0 n_2 I_0(z') e^{-2\left(\frac{r}{w}\right)^2}\right\}, \tag{24.6}$$

where the peak, on-axis input irradiance within the sample is I_0 with a spot size w (half-width at the $1/e^2$ maximum in the irradiance, $HW1/e^2M$). This shows a peak-induced phase distortion at the center of the beam, $\Delta\Phi_0$, of

$$\Delta\Phi_0 = \frac{2\pi}{\lambda} n_2 I_0(L) L_{\text{eff}}, \tag{24.7}$$

with, $L_{\text{eff}} = (l - e^{-aL})/\alpha$. Taking account of the possible temporal structure of the field gives an electric field at the exit of the sample as $E_e(Z,r,t)$, where Z is the position of the sample measured with respect to the input beam waist (in anticipation of Z-scan). In general for radially symmetric systems, a zeroth order Hankel transform of the input electric field will give the field distribution at a distance d from focus.

$$E(Z+d,r,t) = \frac{2\pi}{i\lambda d'} e^{\frac{i\pi r}{\lambda z}} \int_0^\infty E\left(Z, r', t - \frac{d'}{c}\right) e^{\frac{i\pi r'^2}{\lambda d'}} J_0\left(\frac{2\pi r r'}{\lambda d'}\right) r' dr', \tag{24.8}$$

where $d' = d - Z$ is the distance from the sample to the position where the field is monitored (again written in this way in anticipation of Z-scan). Here $J_0(x)$ is the zeroth-order Bessel function. There are other ways to calculate the far-field irradiance distribution for Gaussian input beams. See, for example, Ref. [13].

The effects of self-lensing can be easily seen by monitoring the distribution on a camera placed in the far field as shown in Fig. 24.1. Here the sample is placed at or near the beam waist of a Gaussian spatial distribution beam. The peaks are normalized to the same value, and the wings of the beam clearly show the spatial broadening in the far field at high irradiance. This is due to self-focusing in NaCl for which $n_2 > 0$ (known from other data) [14]. For this

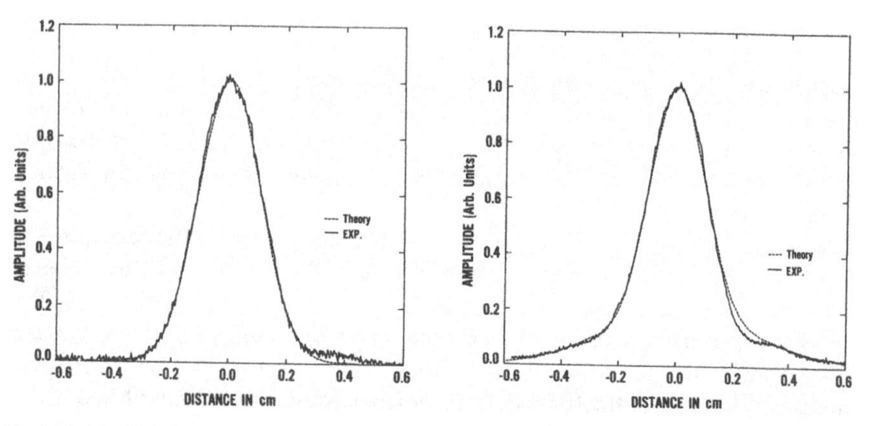

Fig.24.1 Far field fluence distribution after transmission through a 0.5-cm-thick NaCl sample placed at the beam waist of a $\lambda = 532$ nm, ~40 psec FWHM pulse (**a**) at I = 4.7 GW/cm^2 (*left*) and (**b**) I = 57 GW/cm^2 (*right*). Taken from Ref. [14].

geometry, the sign of the nonlinear refraction cannot be deduced from the far-field distribution. In Fig. 24.1(a) the induced-phase distortion is quite large ($\sim$0.5 λ), showing that the sensitivity using this method is not high. The invention of the Z-scan technique provided a simple and sensitive method to determine the sign of the nonlinear refraction which was previously difficult to determine. In order to deduce the sign of the nonlinear refraction the sample can be placed prior to and after the input beam waist. As described in Section 24.3, the beam distortion will then show opposite changes for different signs of Δn. This is key to the success of the Z-scan technique discussed next.

In addition, it is also difficult to separate the contributions of NLA and NLR with only beam distortion measurements. Even two-photon absorption alone leads to beam shape changes with propagation. For example, a Gaussian beam is spatially broadened after propagation through a 2PA material because the center portion of the Gaussian is preferentially absorbed and therefore the diffraction is reduced. This effect is hence similar to self-focusing.

24.3 Z-Scan

Since its invention in 1990, the Z-scan method has quickly gained acceptance as a rapid and sensitive technique for separately determining the nonlinear changes in index and changes in absorption [15–17]. This is primarily due to the simplicity of the technique. In most experiments the index change, Δn, and absorption change, $\Delta\alpha$, can be determined directly from the data without the need for computer fitting. However, the physical mechanisms for Δn and $\Delta\alpha$ cannot be unambiguously determined without other information.

The standard "closed aperture" Z-scan apparatus (i.e., aperture in place in the far field) for determining nonlinear refraction is shown in Fig. 24.2. The input beam is focused and the sample is moved through this focal position in the Z (propagation) direction while the transmittance is monitored in the far field

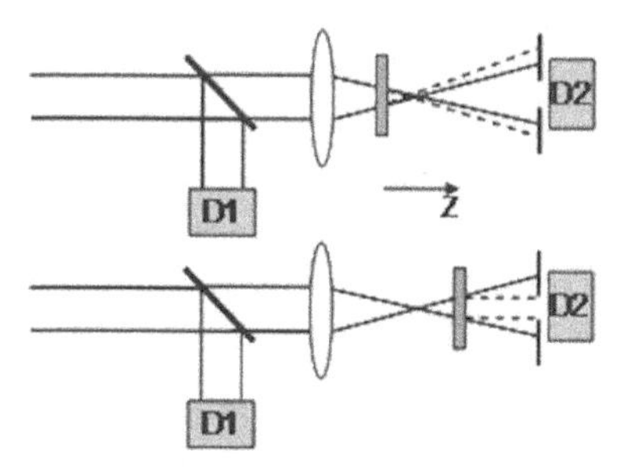

Fig. 24.2 "Closed aperture" Z-scan apparatus. The sample is scanned along "Z," monitoring the transmittance. The solid lines denote the linear focusing while the dotted lines depict the effects of the sample on this propagation assuming a self-focusing nonlinearity. The ratio of the outputs of detectors D2 and D1 is monitored.

through an aperture. The required scan range in an experiment depends on Z_0 and on the sample thickness L. If w_0 is the focal spot size (HW1/e^{2}M), Z_0 is defined as $\pi w_0^2/\lambda$ for a Gaussian beam. For thin samples, the scanning range in Z should be $\sim 10\ Z_0$ although all the information is theoretically contained within a scan range of $\pm Z_0$. This allows the full shape of the Z-scan to be observed and makes interpretation simpler.

A closed aperture Z-scan for a thin sample of BaF_2 exhibiting purely non-linear refraction at 532 nm, is shown in Fig. 24.3 (left) (solid line). For this material, the change in refractive index, $\Delta n > 0$, resulting in self-focusing which leads to a valley followed by a peak in the normalized transmittance as the sample, is moved away from the lens in Fig. 24.2 (increasing Z). The normalization is performed so that the transmittance is unity for the sample far from focus where the nonlinearity is negligible (i.e., for $|Z| >> Z_0$). The positive lensing in the sample placed before the focus moves the focal position closer to the sample resulting in a greater far-field divergence and a reduced aperture transmittance. On the other hand, with the sample placed after focus, the same positive lensing reduces the far-field divergence allowing for a larger aperture transmittance. The signal for the same magnitude of NLR but with the opposite sign (self-defocusing) is its mirror image, i.e., peak followed by valley (Fig. 24.3, right). Clearly, with the sample at focus the effect of NLR on the transmitted beam is minimized. This explains why the transmitted beam profile shown in Fig. 24.1 shows a very small change despite a quite large nonlinear phase shift.

The change in normalized transmittance for the Z-scan is linear in the induced phase distortion. This is best seen by looking at the change in

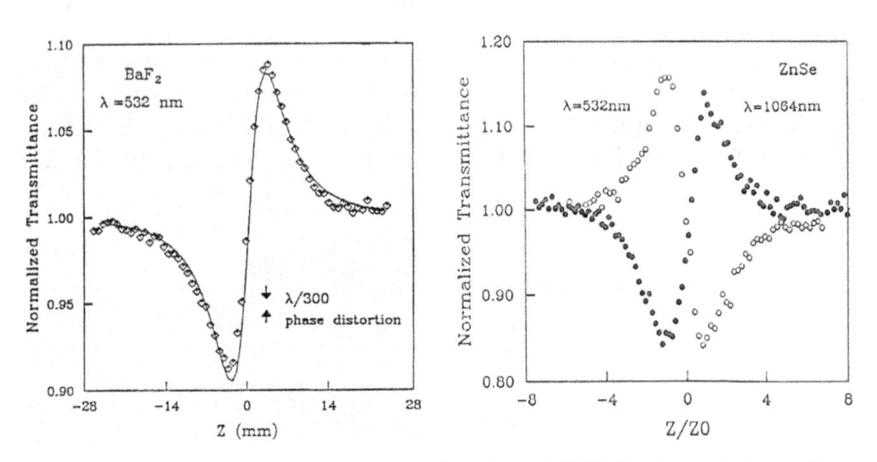

Fig. 24.3 (*left*) Z-scan of BaF_2 at $\lambda = 532$ nm with ~40 ps (FWHM) pulses with the irradiance turned down to show the signal-to-noise ratio allowing phase distortions of $\sim \lambda/300$ to be measured. The overall phase distortion in BaF_2 is $\sim\lambda/14$. Taken from Ref. [16]; (*right*) Closed aperture Z-scan transmittance curves for ZnSe at 1064 nm (*closed circles*) and 532 nm (*open circles*), clearly showing the dispersion of n_2 as it changes from positive at 1064 nm to negative at 532 nm. The second figure is from [18]

transmittance calculated by the difference between the normalized transmittance at the peak, T_p, and valley, T_v, i.e., $\Delta T_{pv} = T_p - T_v$. The relation between the induced phase distortion, $\Delta\Phi_0$, and ΔT_{pv} for a third-order nonlinear refractive process in the absence of NLA is empirically determined to be,

$$\Delta T_{pv} \cong 0.406(1 - S)^{0.27}|\Delta\Phi_0|, \tag{24.9}$$

where S is the transmittance of the aperture in the absence of a sample [16]. This relation is accurate to within ±3% for $\Delta T_{pv} < 1$. As an example, if the induced optical path length change due to the nonlinearity is $\lambda/250$, $\Delta T_{pv} \approx 1\%$ for an aperture transmittance of $S = 0.4$. Figure 24.3 shows an experimental sensitivity of better than $\lambda/300$.

This interferometric sensitivity is one of the most useful features of the Z-scan technique. At first sight this appears rather remarkable given that it is a single-beam method. However, Z-scan is based on propagation to the far field. Propagation results in diffraction, and diffraction is really an interference phenomenon, i.e., interference between different spatial portions of the beam, here the center of the beam interfering with the wings. Thus, Z-scan is effectively a single-beam interferometer and the sensitivity is a result of this interference. In addition, although the optics used are not as good as the sensitivity of the experiment, we must remember that we are looking at the *change* in phase and not the absolute phase.

The size of the aperture in a Z-scan experiment is specified by its transmittance, S, in the linear regime, i.e., when the sample has been placed far away from the focus. The sensitivity to the induced-phase distortion depends on S, going from its highest value for S very small to 0 for $S = 1$ (so-called "open aperture" Z-scan, which is only sensitive to nonlinear absorption, as discussed later). However, values of S from 0.1 to 0.4 work well for only a small loss in sensitivity as seen from Eq. (24.9). This allows a large signal on the detector and averaging over any local spatial beam inhomogeneities.

Equation (24.9) does not include the time averaging that occurs upon detection for short pulse inputs that are normally used in Z-scan experiments. The linear relationship between ΔT_{pv} and $\Delta\Phi_0$ allows us to use a simple multiplication factor, A_τ, which for pulses much shorter than the nonlinear response time, e.g., bound-electronic responses, is given by:

$$A_\tau = \frac{\int_{-\infty}^{+\infty} f^2(t)dt}{\int_{-\infty}^{+\infty} f(t)dt}, \tag{24.10}$$

where $f(t)$ is a function describing the irradiance pulse shape in time. For NLR with an instantaneous response, A_τ depends on the pulse shape. For example, for Gaussian pulses $A_\tau = 1/\sqrt{2}$. However for pulses much shorter than the response time of the material nonlinearity, $A_T = 1/2$ independent of the pulse shape. The temporal averaging must be reevaluated in cases involving higher-order nonlinearities.

The accuracy of the measurements of $\Delta\Phi_0$ and thus n_2 depend on how well the laser beam parameters are known, i.e., pulse energy (or power), and temporal and spatial properties [19]. In addition, the Z-scan signal is sensitive to all nonlinear optical mechanisms that give rise to a change of the refractive index (and absorption). Thus, it is not possible from single measurements to determine the origin of the nonlinearity (or nonlinearities). Other information such as the temporal dependence measured in pump-probe experiments is needed for this.

Another quite useful feature of the Z-scan signal is that the distance between peak and valley in Z, ΔZ_{pv} gives a direct measure of the diffraction length of the incident beam. Assuming a third-order nonlinear response using a Gaussian spatial profile beam,

$$\left|\Delta Z_{pv}\right| \approx 1.7 Z_0. \tag{24.11}$$

Given a known thin sample nonlinear response, this is a fast method for determining the spot size and helps to self-calibrate the irradiance for Z-scan.

The previous discussion assumed no nonlinear absorption. Another invaluable feature of Z-scan is its capability to separately and simultaneously measure NLR and NLA. This can be done even when both are present. We do not discuss this in detail here, but by performing two Z-scans, one "closed aperture" (S small) and one "open aperture," $S = 1$, the phase distortion can be extracted. This is easily accomplished in a single experiment using a beam splitter where one of the beams is sent to a detector with an aperture in place and the other beam goes to a detector set to collect all of the transmitted light. The open aperture Z-scan is insensitive to the induced-phase distortion for thin samples, and thus the NLA can be easily determined. The closed aperture Z-scan can then be fit with the known NLA and unknown NLR. However, a quick method to obtain the NLR is to divide the closed aperture Z-scan data (after normalization) by the open aperture Z-scan data (again after normalization). The resulting curve can be used to determine $\Delta\Phi_0$.

The ΔT_{pv} for this curve is essentially the same as the curve for a material with the same NLR but without NLA as long as the ratio of NLA to NLR does not get too large. The criteria for this are given in Ref. [16]. Typical curves for "open" and "closed" aperture Z-scans and their division are shown in Fig. 24.4 where $\Delta\Phi_0 = -0.5$. Other methods such as degenerate four-wave mixing (DFWM) present difficulties in separating the effects of NLR from NLA.

There have been a great number of publications discussing variations of the Z-scan technique. These include: using different beam profiles to increase the sensitivity [21]; eclipsing the Z-scan (EZ-scan) [22], which provides enhanced sensitivity by using a central obscuration disk instead of an aperture; the "2-color Z-scan" that collinearly focuses two beams of different wavelength to

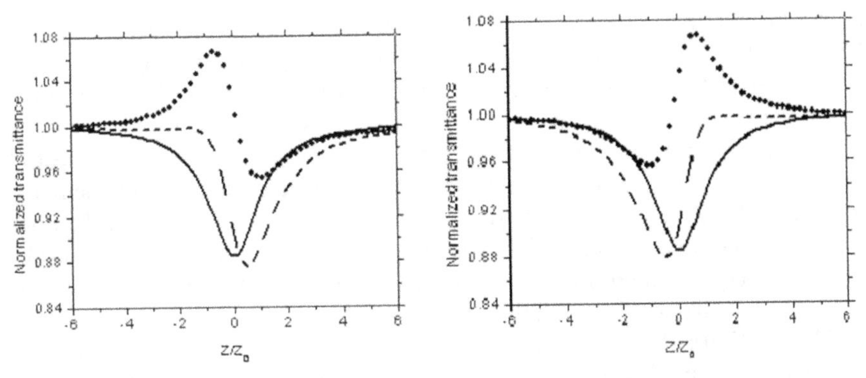

Fig. 24.4 Calculations of closed (*solid line*, $S = 0.5$) and open (*dashed line*) aperture Z-scan data along with their ratio (*dotted line*) for (*left*) self-defocusing and (*right*) self-focusing

measure nondegenerate nonlinearities by Z-scanning [23, 24]; and the time-resolved pump-probe Z-scans, where refractive changes induced by a strong pump pulse are measured [25, 26]; etc. Simple methods have also been developed for analyzing Z-scans where the sample is too thick for the external self-action approximation to hold [27–30].

24.4 Measuring Nonlinear Dispersion

In principle the dispersion of the nonlinear refractive index can be determined by performing Z-scans at many wavelengths, and excellent results have come from such measurements [31]. The same is true of determining nonlinear absorption spectra. Unfortunately, however, this can be difficult and time consuming. It is usually difficult to tune a laser or optical parametric source while keeping the same beam parameters. Hence, careful and time-consuming beam characterization is required at each wavelength to determine the irradiance. The Z-scan helps in determining the spot size (assuming prior knowledge of the order of the nonlinear response), but it gives no information about the temporal dependence. In addition, experience shows that these beam parameters can change from day to day as the laser/parametric source is tuned repeatedly.

Advances in white-light-continuum (WLC) generation [32–34] have resulted in an alternative source for Z-scans in the visible and near IR. Because of the sensitivity of the Z-scan, for many materials only a few nJ of energy are needed when using femtosecond pulses. The spectral energy density of the continuum turns out to be sufficient for measuring materials showing strong nonlinearities [35–37]. However, for materials with low nonlinearities, higher spectral irradiance is sometimes needed. Producing WLC in ~1-m-long cells filled with high-pressure noble gas–filled cells gives much higher spectral irradiance than in solids and liquids [38, 39]. For example, the WLC spectrum produced by focusing 0.7 mJ of 775 nm, ~140 fs pulses into a 1-m-long chamber filled with Krypton gas at 2.4 atm produces a useful continuum from 400 nm to > 800 nm [39, 40]. This continuum has sufficient spectral energy density over this spectral range for Z-scan. The calculated Rayleigh range within the sample is 8.5 cm [39]. In any ~10 nm spectral band there are many nJ of energy in ~100 fs pulses, which is sufficient for measuring NLR in most materials, e.g., organic dyes in solution [39]. In both methods of continuum generation it is important to maintain good spatial profiles. This requires careful control of the input beam parameters and energy to assure "single filament" operation.

The WLC are often produced by mechanisms that include self-focusing and, in particular, "small-scale" self-focusing can lead to multifilament operation which results in unusable spatial profile beams (see the chapter by Campillo in this book). A single filament operation results in high quality Gaussian-shaped beams. The spatial profile for the continua produced in Kr are Gaussian over the entire 400–800 nm spectral range; however, for longer wavelengths, while there is considerable energy, the spatial profile becomes doughnut-shaped, making Z-scans more difficult to analyze. We briefly describe this WLC Z-scan method below [39].

It is not possible to simply replace the source in the standard Z-scan with the WLC. The problem is that nondegenerate nonlinearities will accompany the degenerate response and, for example, the overall NLR will be considerably increased with no simple way to distinguish the relative contributions of degenerate and nondegenerate NLR. The simplest way around this is to simply spectrally filter the input using narrow band filters, NBF's (or "spike" filters). Other methodologies for spectrally dispersing in space or even in time using group velocity dispersion are possible [36], but are more complicated for measuring NLR given the importance of the spatial profiles in closed aperture Z-scans. The spectral selection can be done by simply introducing NBFs into the beam prior to the sample with care to ensure that they do not disturb the spatial or temporal profile. Automation of such filtering can be done using computer-controlled motorized filter wheels [39].

In addition, there are variable frequency filters available in certain frequency bands (e.g., the entire visible spectrum) where the tuning is continuous (so-called "linear variable filters"). These filters combine spatially varying high pass and low pass filters that are combined to give a spatially varying and wavelength-adjustable NBF. These can be moved laterally changing the transmitted narrow wavelength range continuously. A requirement for any of these NBFs is that the band pass is wide enough to allow short pulse transmittance. For the ~100 fs pulses of many sources, e.g., Ti:sapphire, the ~10 nm bandwidth works well.

With knowledge of the energy, beam size, and pulse duration for each spectral component of the WLC source, standard Z-scans can be performed. The pulsewidths for each spectral region can be determined by various standard techniques [41, 42]. The measured pulsewidths for the aforementioned WLC

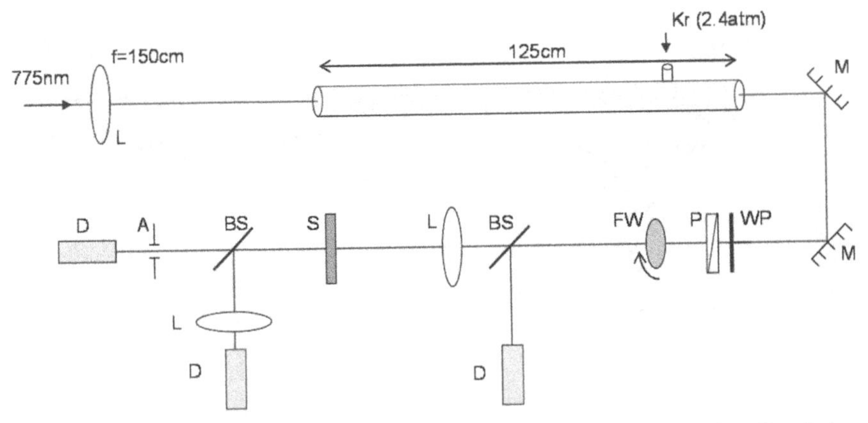

Fig. 24.5 WLC Z-scan experimental set-up: L, lens; M, mirror; WP, half-waveplate; P, polarizer; FW, filter wheel; BS, beamsplitter; D, detector; A, aperture; S, sample. Taken from Ref. [35]

produced in Kr averaged around 90–100 fs and had a time-bandwidth product of ~0.44 for the 10-nm bandpass filters. The advantage of this methodology is that in practice, these WLC are very reproducible from one day to the next so that once characterized, measurements on multiple samples can be rapidly taken over the entire spectral range of the WLC. A typical experimental arrangement is shown in Fig. 24.5.

An example of data taken with this method, displaying data for ZnSe, is shown in Fig. 24.6 (energy gap (E_g) = 2.7 eV with a thickness of 0.5 mm). Here the group velocity dispersion at the shortest wavelengths (<550 nm) is important to take into account as it affects the pulsewidth within the sample (at 480 nm a 28% change is calculated for the pulse between the front and back surfaces, this is the largest effect for ZnSe which linearly absorbs at shorter wavelengths).

The values of β and n_2 corresponding to ZnSe obtained from fits at different wavelengths are presented in Fig. 24.7 along with the theoretical predictions of Refs. [40, 44].

The important point to make concerning these data for n_2 is that the NLR can be measured in the presence of relatively strong NLA and with either positive or negative sign of n_2. In this case, n_2 changes from positive to negative as the ratio of $(\hbar\omega/E_g$ goes above $\sim 0.7\ E_g$. This feature of the Z-scan enabled the elucidation of nonlinear Kramers-Kronig relations for ultrafast nonlinearities to connect n_2 to two-photon absorption [45].

24.5 Physical Mechanisms Leading to Nonlinear Refraction

We have discussed two of the primary methods for measuring nonlinear refraction related to propagation of a beam with a near-field phase mask to the far field to redistribute the energy of the beam in space. There are many physical

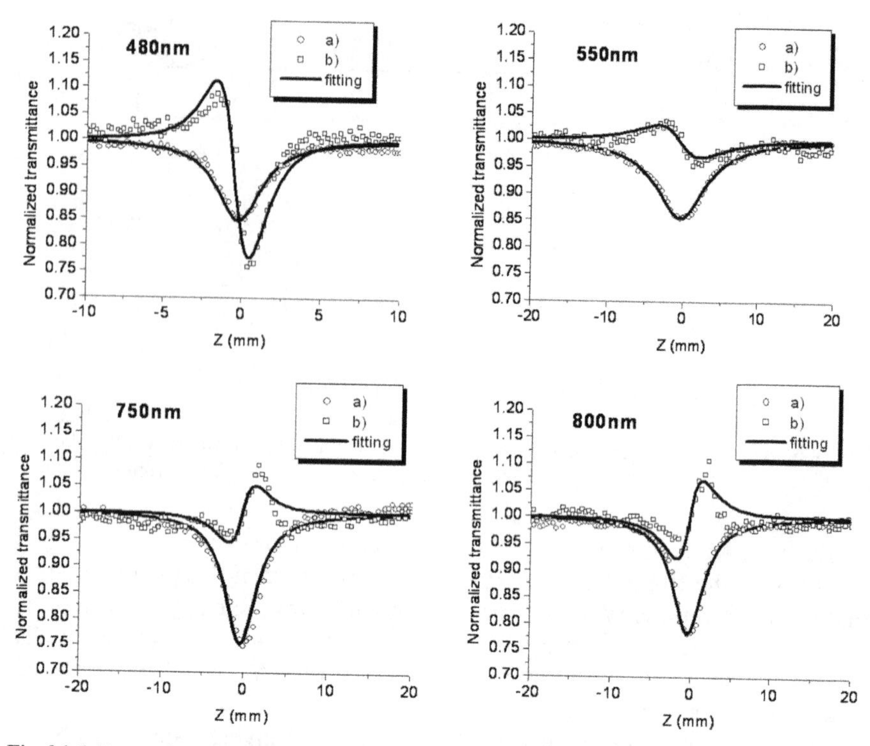

Fig. 24.6 Z-scan data at 480 nm, 550 nm, 750 nm, and 800 nm (**a**) open aperture and (**b**) closed aperture (the result of the division with open aperture). Taken from Ref. [39]

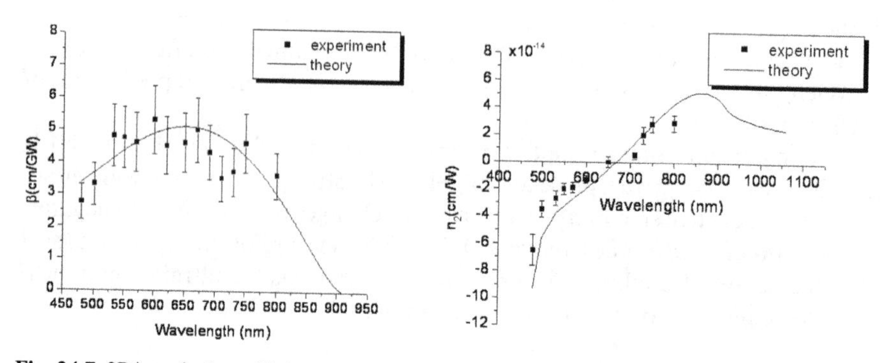

Fig. 24.7 2PA and n2 coefficients of ZnSe obtained from theory and from the experimental data fitting. Taken from Ref. [39]

processes that can lead to the initial phase mask. In this section we briefly discuss a few of these. We begin with the ultrafast, bound-electronic nonlinear refraction from the third-order nonlinear susceptibility, which is accurately described by the coefficient n_2 in Eq. (24.1). This n_2 is related to the nonlinear absorption of 2PA, etc., through causality and nonlinear Kramers–Kronig relations as described in Refs. [44, 45].

The Z-scan method was essential in allowing NLR to be measured in the presence of NLA in order to see the change in sign of n_2 at photon energies of about 0.7 of the bandgap energies in semiconductors. This sign change of n_2 was a crucial element leading to the full understanding of nonlinear Kramers– Kronig relations [45]. Measurements of these ultrafast responses can now be easily distinguished from slower nonlinearities with the use of femtosecond pulses; however, longer pulses often show other nonlinearities that can mask the faster responses. For semiconductors these include free-carrier nonlinearities from carriers produced via 2PA. Free-carrier refraction becomes substantial for picosecond or longer pulses, and is always negative. (An oscillator is produced with a zero resonance frequency so that one is always above resonance.) For 2PA created carriers, this is an effective fifth-order nonlinearity resulting from the third-order 2PA in combination with the first-order change in the refractive index [18]. In this case, the index change, is given by

$$\Delta n(t) = n_2 I(t) + \sigma_r N(t), \tag{24.12}$$

where σ_r is the free carrier refraction coefficient, and $N(t)$ is the photoexcited carrier density. For 2PA excitation, $N(t)$ is governed by

$$\frac{dN}{dt} = \frac{\beta I^2}{2\hbar\omega} - \frac{N}{\tau}. \tag{24.13}$$

For semiconductors, the lifetime, τ, is typically on the order of nanoseconds, so for picosecond pulses these carriers accumulate with time during the pulse. Because the carrier refraction is quadratic in I, above some input irradiance the carrier refraction will dominate the NLR. For long pulses, the free-carrier effects are almost always dominant unless very low irradiance can be used with a very sensitive technique [18].

In Fig. 24.8, we show streak camera measurements of the transmitted spatial beam profile of a 15 μJ, 30 ps, 532 nm pulse through a 2-mm-thick sample of ZnSe, which exhibits 2PA at this wavelength. We see from the beam profiles, measured every 9.3 ps, that the beam distortion due to nonlinear refraction is much stronger later in the pulse than at early times, as expected where free carrier refraction dominates the bound electronic n_2.

However, extracting values for the n_2 and free carrier refraction coefficient from the data in Fig. 24.8 is not particularly easy. It is preferable to perform a series of Z-scan experiments in order to extract these coefficients. This can be done by taking advantage of the quadratic and linear irradiance dependences of the respective carrier and bound electronic contributions to NLR. First, Z-scans can be taken at low irradiance where the bound electronic n_2 dominates.

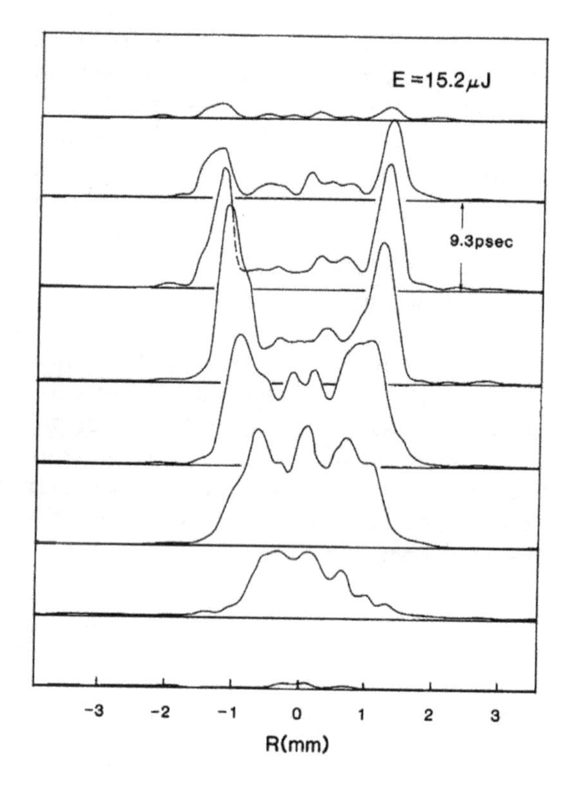

Fig. 24.8 Spatial energy distribution at 11 cm behind a thin (0.2 cm) ZnSe sample at 9.3-ps time intervals as detected by a streak-camera vidicon system for an input energy of 15.2 μJ. Taken from Ref. [46]

As always, we have to perform open and closed aperture Z-scans. In Fig. 24.9, we show open and closed aperture Z-scans on ZnSe with 30-ps pulses at 532 nm. The open aperture Z-scan [Fig. 24.9 (a)] gives us the 2PA coefficient. We find that absorption due to free carriers is negligible in this experiment. This value for the 2PA coefficient is used in fitting a value of n_2 to the closed aperture Z-scan shown in Fig. 24.9(b). [18]

At higher energies, free carrier refraction becomes significant, eventually dominating. The values obtained for β and n_2 at low energies may then be used to fit the final remaining parameter, which is σ_r, the free carrier refraction coefficient. Results of this fitting for higher energies are shown in Fig. 24.10. It should be noted that in these experiments, even at these higher energies, absorption due to the 2PA-excited carriers was insignificant. However, should free-carrier absorption be significant, its value can be found from high-energy open aperture Z-scans, and the rest of the procedure remains the same.

The nonlinearities seen here are also observed in organic dyes; however, the interpretation is nonlinear absorption and refraction from the 2PA generated excited states as opposed to free carriers. Here the sign of the NLR will depend on which side of resonance is the input photon energy [43].

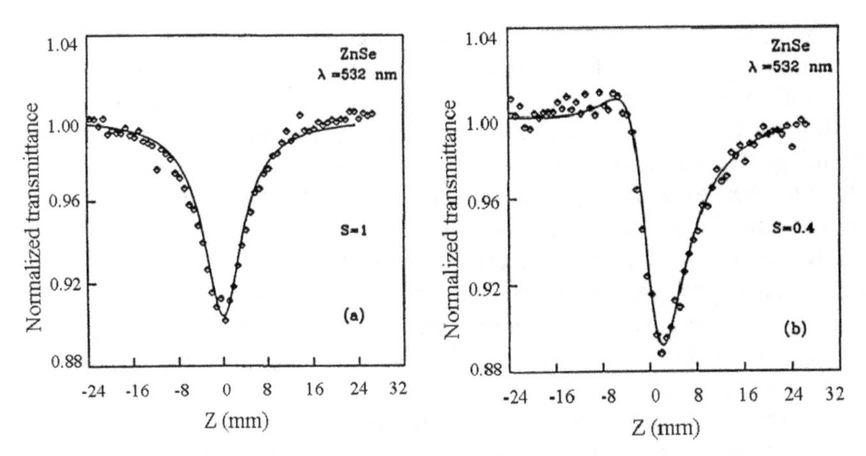

Fig. 24.9 Normalized Z-scan data of a 2.7-mm ZnSe sample measured with 27 ps (FWHM) pulses and $\lambda = 532$ nm at low irradiance ($I_0 = 0.21$ GW/cm^2). The solid curves are the theoretical fits. (**a**) Open-aperture data ($S = 1$) were fitted with $\beta = 5.8$ cm/GW. (**b**) 40%-aperture data were fitted with $\beta = 5.8$ cm/GW and $n_2 = -6.8 \times 10^{-14}$ cm^2/W. Taken from Ref. [18]

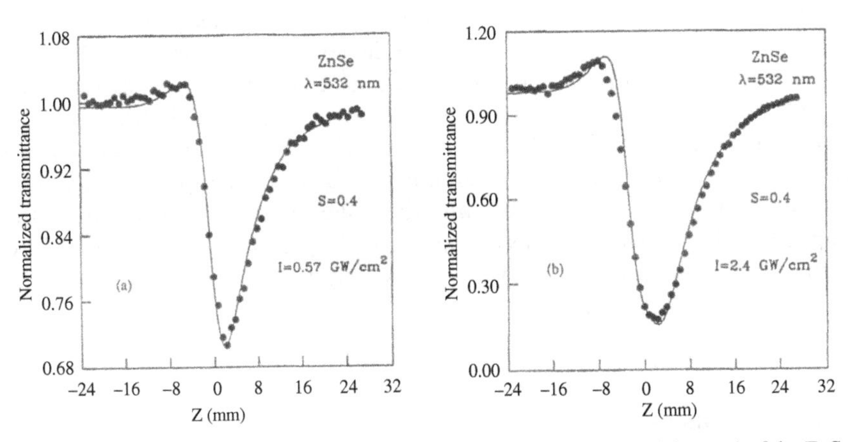

Fig. 24.10 Closed-aperture Z-scan data ($S = 0.4$) and theoretical fits (*solid curves*) of the ZnSe sample taken at high irradiance levels of (*left*) $I_0 = 0.57$ GW/cm^2 and (*right*) $I_0 = .4$ GW/cm^2, where free-carrier refraction is significant. The data were fit with $\beta = 5.8$ cm/GW, $n_2 = -6.2 \times 10^{-14}$ cm^2/W and $\sigma_r = -0.8 \times 10^{-21}$ cm^3. Taken from Ref. [18]

Irradiance dependence studies using beam propagation effects can help in determining the responses along with pulsewidth-dependent studies; however, complementary methods that provide direct information on the temporal response of the nonlinearity such as pump-probe techniques and four-wave mixing are also useful in determining the physical processes involved. [48, 49]

Examples of NLR processes that are easily confused with the bound electronic n_2 because they are effectively third-order, are excited-state refraction from linear absorption–created states (or free carriers), and the refraction associated with saturation. Both are related to population redistribution via causality, i.e., Kramers–Kronig relations [45]. In the case of excited-state refraction, excited states (or free carriers) are produced via linear or 1-photon absorption, and then the refractive index changes due to these excited states (or free carriers) or due to the reduction in the number of ground state absorbers [50]. Here again, the pulsewidths used are important, specifically in relation to the excited-state dynamics. If the pulsewidth is much longer than the decay times, the process will appear just like the ultrafast n_2, while if the pulses are shorter, the material nonlinearity grows within the pulse.

In addition, cascaded second-order nonlinearities [51] appear as third-order responses for low inputs. Here, using second-harmonic generation (SHG) as the example, the loss of two photons at the fundamental frequency to produce the SHG is analogous to 2PA, and the change in phase of the beam for nonphase matched operation is analogous to an index change below or above 2PA resonance (the sign of the effective n_2 from cascading changes from below phase match to above going through zero on phase match) [51].

Some other nonlinear refraction mechanisms including electrostriction and thermal nonlinear refraction are nonlocal requiring propagation in the transverse direction in the beam, again making the temporal response key [52].

The last nonlinearity to mention here is molecular reorientation, also known as the AC Kerr effect. In this nonlinearity, responsible for the $\sim$2 ps nonlinear response of CS_2, the input beam's electric field applies a torque to the molecules which causes a reorientation of the induced dipoles, which then increases the polarizability and thus the index along the direction of the applied field [53]. This nonlinear response of CS_2 is often used as a standard response to compare with; however, this only works for pulses long with respect to their response time.

24.6 Conclusion

Although it has been shown that relative measurements of n_2 can be performed [54] for absolute measurements, in order to give reliable values of the nonlinear refractive index, it is important to note the importance of accurately measuring the laser mode and pulse parameters because n_2 is irradiance dependent. Thus, given the pulse energy, we need to know both the beam area (i.e., spatial beam profile) and the temporal pulsewidth (i.e., temporal shape) in order to determine the irradiance. Any errors in the measurement of irradiance translate to errors in the determination of n_2 as well as any other nonlinear coefficients.

Using beam propagation to allow the induced phase mask from nonlinear refraction to propagate to give a redistribution of irradiance can greatly facilitate measurements of n_2. Indeed, viewing this propagation or diffraction as an interference between different portions of the beam is what leads to the interferometric sensitivity of the Z-scan. One of the most useful features of this method is its ability to separately measure NLR and NLA, even when both are present. We do not discuss this in detail here (details are given in Ref. [16]) but by performing two Z-scans, one "closed aperture" (S small) and one "open aperture," $S = 1$, the phase distortion can be extracted. Many other methods such as degenerate four-wave mixing (DFWM) have a difficult time in separating the effects of NLR from NLA. Determining the sign of n_2 is key to understanding nonlinear Kramers–Kronig relations.

A series of Z-scans at varying pulsewidths, frequencies, focal geometries, etc., along with a variety of other experiments, are often needed to unambiguously determine the relevant mechanisms. It is always advisable to use several complementary characterization techniques if possible to verify the nonlinear response, and the dispersion of the nonlinear refraction is key to understanding the physical mechanisms. In addition, the spectrum of nonlinear absorption can help determine the dispersion of nonlinear refraction as these quantities are related by causality. Simultaneous knowledge of both will further help in understanding these phenomena.

Acknowledgments We gratefully acknowledge the support of the National Science Foundation over many years, current grant ECS# 0524533. The work presented in this paper represents many years of effort involving colleagues and many former and current students as well as post-doctoral fellows. We thank all those involved and acknowledge their contributions through the various referenced publications. We explicitly thank Mansoor Sheik-Bahae for his many contributions, and Mihaela Balu for help in preparing this manuscript.

References

1. W. Kaiser, C.G.B. Garrett: Two-photon excitation in $CaF_2:Eu^{2+}$, *Phys. Rev. Lett.*, **7**, 229–231 (1961).
2. G.A. Askar'yan: Effects of the gradient of strong electromagnetic beam on electrons and atoms, *Soviet Phys JETP-USSR* **15**(6), 1088–1090 (1962).
3. R.Y. Chiao, E. Garmire, C.H. Townes: Self-trapping of optical beams, *Phys. Rev. Lett.*, **13**, 479–482 (1964).
4. V.I. Talanov: Self-focusing of waves in nonlinear media, *JETP Lett.*, **2**, 138 (1965).
5. Dmitriy I., Kovsh S. Yang et al.: Nonlinear optical beam propagation for optical limiting, *Appl Opt.*, **38**, 5168–5180 (1999).
6. H.P. Nolting, R. Marz: Results of benchmark tests for different numerical BPM algorithms, *IEEE J. Lightware Technol.* **13**, 216–224 (1995).
7. G.P. Agrawal: *Nonlinear Fiber Optics*, Academic Press, New York (1989).
8. M.D. Feit, J.A. Fleck, Jr.: Simple method for solving propagation problems in cylindrical geometry with fast Fourier transforms, *Opt. Lett.*, **14**, 662–664 (1989).
9. S. Hughes, J.M. Burzer, G. Spruce et al.: Fast Fourier transform techniques for efficient simulation of Z-scan measurements, *J. Opt. Soc. Am. B*, **12**, 1888–1893 (1995).
10. A.E. Kaplan: External self-focusing of light by a nonlinear layer, *Radiophys. Quant. Electron.*, **12**, 692–696 (1969).

11. R.W. Boyd: *Nonlinear Optics*, Academic Press, San Diego (2003).
12. R.W. Hellwarth: Third-order optical susceptibilities of liquids and solids, *Progr. Quant. Electron.*, **l.5**, 1–68 (1979)
13. D. Weaire,. B.S. Wherrett, D.A.B. Miller et al.: Effect of low-power nonlinear refraction on laser-beam propagation in InSb, *Opt. Lett.*, **4**, 331–333 (1979).
14. W.E. Williams, M.J. Soileau, E.W. Van Stryland: Optical switching and n_2 measurements in CS_2, *Opt. Commun.*, **50**, 256 (1984).
15. M. Sheik-bahae, A.A. Said, E.W. Van Stryland: High-sensitivity, single-beam n_2 measurements, *Opt. Lett.*, **14**, 955–957 (1989).
16. M. Sheik-Bahae, A.A. Said, T.H. Wei, D.J. Hagan, E.W. Van Stryland: Sensitive measurement of optical nonlinearities using a single beam, *IEEE J. Quant. Electron.*, **26**, 760 (1990).
17. P.B. Chapple, J. Staromlynska, J.A. Hermann et al.: Single-beam Z-scan: measurement technique and analysis, *J. Nonl. Opt. Phys. Mat.*, **6**, 251 (1997).
18. A.A. Said, M. Sheik-Bahae, D.J. Hagan et al.: Determination of bound and free-carrier nonlinearities in ZnSe, GaAs, CdTe, and ZnTe, *J. Opt. Soc. Am. B*, **9**, 405–414 (1992).
19. S. Hughes, J.M. Burzler: Theory of Z-scan measurements using Gaussian–Bessel beams, *Phys. Rev. A* **56**, R1103 (1997).
20. E.W. Van Stryland, M. Sheik-Bahae: Z-scan. In: *Characterization Techniques and Tabulations for Organic Nonlinear Optical Materials*, pp. 655–692, M. Kuzyk, C. Dirk (eds.), Marcel Decker, New York (1998).
21. W. Zhao, P. Palffy-Muhoray: Z-scan measurements of χ^3 using Top-Hat beams, *Appl. Phys. Lett.,* **65**, 673–675 (1994). See also, W. Zhao, J.H. Kim, P. Palffy-Muhoray: Z-scan measurements on liquid crystals using Top-Hat beams, *Appl. Phys. Lett.*, **65**, 673–675 (1994).
22. T. Xia, D.J. Hagan, M. Sheik-Bahae et al.: Eclipsing Z-scan measurement of $\lambda/10^4$ wavefront distortion, *Opt. Lett.*, **19**, 317–319 (1994).
23. H. Ma, A.S. Gomez, Cid B. de Araujo: Measurement of nondegenerate optical nonlinearity using a two-color single beam method, *Appl. Phys. Lett.*, **59**, 2666 (1991).
24. M. Sheik-Bahae, J. Wang, J.R. DeSalvo et al.: Measurement of nondegenerate nonlinearities using a 2-color Z-scan, *Opt. Lett.*, **17**, 258–260 (1992).
25. J. Wang, M. Sheik-Bahae, A.A. Said et al.: Time-resolved Z-scan measurements of optical nonlinearities, *J. Opt. Soc. Am. B*, **11**, 1009–1017 (1994).
26. V.P. Kozich, A. Marcano, F. Hernandez et al.: Dual-beam time-resolved Z-scan in liquids to study heating due to linear and nonlinear light absorption, *Appl Spectrosc*, **48**, 1506–1512 (1994). See also J. Castillo, V. Kozich, A. Marcano: Thermal lensing resulting from one- and two-photon absorption studied with a two-color time-resolved Z-scan, *Opt. Lett.*, **19**, 171–173 (1994).
27. M. Sheik-Bahae, A.A. Said, D.J. Hagan et al.: Nonlinear refraction and optical limiting in "thick" media, *Opt. Eng.*, **30**, 1228–1235 (1990).
28. J.A. Hermann, R.G. McDuff: Analysis of spatial scanning with thick optically nonlinear media, *J. Opt. Soc. Am. B*, **10**, 2056–2064 (1993).
29. J.-G. Tian, W.-P. Zang, C.-Z. Zhang et al.: Analysis of beam propagation in thick nonlinear media, *Appl. Opt.*, **34**, 4331–4336 (1995).
30. P.B. Chapple, J. Staromlynska, R.G. McDuff: Z-scan studies in the thin- and the thick-sample limits, *J. Opt. Soc. Am B*, **11**, 975–982 (1994).
31. B. Lawrence, W. Torruellas, M. Cha et al.: Identification and role of two-photon excited states in a π-conjugated polymer, *Phys. Rev. Lett.*, **73**, 597–600 (1994).
32. R.R. Alfano, S.L. Shapiro: Emission in the region 4000 and 7000 • via four-photon coupling in glass, *Phys. Rev. Lett.*, **24**, 584–587 (1970).
33. P.B. Corkum, C. Rolland: Supercontinuum generation in gases, *Phys. Rev. Lett.*, **57**, 2268–2271 (1986).

34. A. Brodeur, S.L. Chin: Ultrafast white-light continuum generation and self-focusing in transparent condensed media, *J. Opt. Soc. Am. B*, **16**, 637–650 (1999).
35. G.S. He, T.C. Lin, P.N. Prasad et al.: New technique for degenerate two-photon absorption spectral measurements using femtosecond continuum generation, *Opt. Express*, **10**, 566 (2002).
36. M. Balu, J. Hales, D.J. Hagan et al.: White-light continuum Z-scan technique for nonlinear material characterization, *Opt. Express*, **12**, 3820 (2004); see also, M. Balu, J. Hales, D.J. Hagan et al.: Dispersion of nonlinear refraction and two-photon absorption using a white-light continuum Z-scan, *Opt. Express*, **13**, 3594 (2005).
37. L. De Boni, A.A. Andrade, L. Misoguti et al.: Z-scan measurements using femtosecond continuum generation, *Opt. Express*, **12**, 3921 (2004).
38. H.R. Lange, G. Grillon, J.-F. Ripoche et al.: Anomalous long-range propagation of femtosecond laser pulses through air: moving focus or pulse self-guiding? *Opt. Lett.*, **23**, 120–122 (1998).
39. M. Balu, D.J. Hagan, E.W. Van Stryland: High spectral irradiance white-light continuum Z-scan, *Proceedings Ultrafast Phenomena Conference*, Monterey, CA, Springer-Verlag, New York (2006).
40. M. Kolesik, E.M. Wright, A. Becker et al.: Simulation of third-harmonic and supercontinuum generation for femtosecond pulses in air, *Appl. Phys. B*, **85**, 531–538 (2006).
41. R. Trebino, Frequency-Resolved Optical Gating: The Measurement of Ultrashort Laser Pulses, Kluwer Academic Publishers, New York (2000).
42. W. Rudolph, M. Sheik-Bahae, A. Bernstein, L.F. Lester: Femtosecond autocorrelation measurements based on two-photon photoconductivity in ZnSe, *Opt. Lett.*, **22**, 313–315 (1997).
43. E.W. Van Stryland, M.A. Woodall, H. Vanherzeele et al.: Energy band-gap dependence of two-photon absorption, *Opt. Lett.*, **10**, 490 (1985).
44. M.Sheik-Bahae, D.C. Hutchings, D.J. Hagan et al.: Dispersion of bound electronic nonlinear refraction in solids, *IEEE J. Quant. Electron.*, **27**, 1296 (1991).
45. D.C. Hutchings, M. Sheik-Bahae, D.J. Hagan et al.: Kramers–Kronig relations in nonlinear optics, *Opt. Quant. Electron.*, **24**, 1–30 (1992).
46. E.W. Van Stryland, Y.Y. Wu, D.J. Hagan et al.: Optical limiting with semiconductors, *J. Opt. Soc. Am. B*, **5**, 1980–1989, (1988).
47. A.A. Said, C. Wamsley, D.J. Hagan et al.: Third- and fifth-order optical nonlinearities in organic materials, *Chem. Phys. Lett.*, **228**, 646–650 (1994).
48. R.S. Lepkowicz, O.V. Przhonska, J.M. Hales et al.: Excited-state absorption dynamics in polymethine dyes detected by polarization-resolved pump-probe measurements, *Chem. Phys.*, **286**(2–3), 277–291 (2003).
49. E.J. Canto-Said, D.J. Hagan, J. Young et al.: Degenerate four-wave mixing measurements of high-order nonlinearities in semiconductors, *IEEE J. Quant. Electron.*, **27**, 2274–2280 (1991).
50. T.H. Wei, D.J. Hagan, M.J. Sence et al.: Direct measurements of nonlinear absorption and refraction in solutions of phthalocyanines, *Appl. Phys. B*, **54**, 46–51 (1992).
51. R. DeSalvo, D.J. Hagan, M. Sheik-Bahae et al.: Self-focusing and self-defocusing by cascaded second-order effects in KTP, *Opt. Lett.*, **17**, 28–30 (1992).
52. D.I. Kovsh, D.J. Hagan, E.W. Van Stryland: Numerical modeling of thermal refraction in liquids in the transient regime, *Opt. Express*, **4**, 315 (1999).
53. D. McMorrow, W.T. Lotshaw, G. Kenney-Wallace: Femtosecond optical Kerr studies on the origin of the nonlinear responses in simple liquids, *IEEE J. Quant. Electron.*, **24**, 443–454, (1988).
54. R.E. Bridges, G.L. Fischer, R.W. Boyd: Z-scan measurement technique for non-Gaussian beams and arbitrary sample thickness, *Opt. Lett.*, **20**, 1821 (1995).

Index

K

L

M

N

O

P

Q

R

CPSIA information can be obtained
at www.ICGtesting.com
Printed in the USA
LVHW051228220723
753115LV00005B/80

9 780387 321479